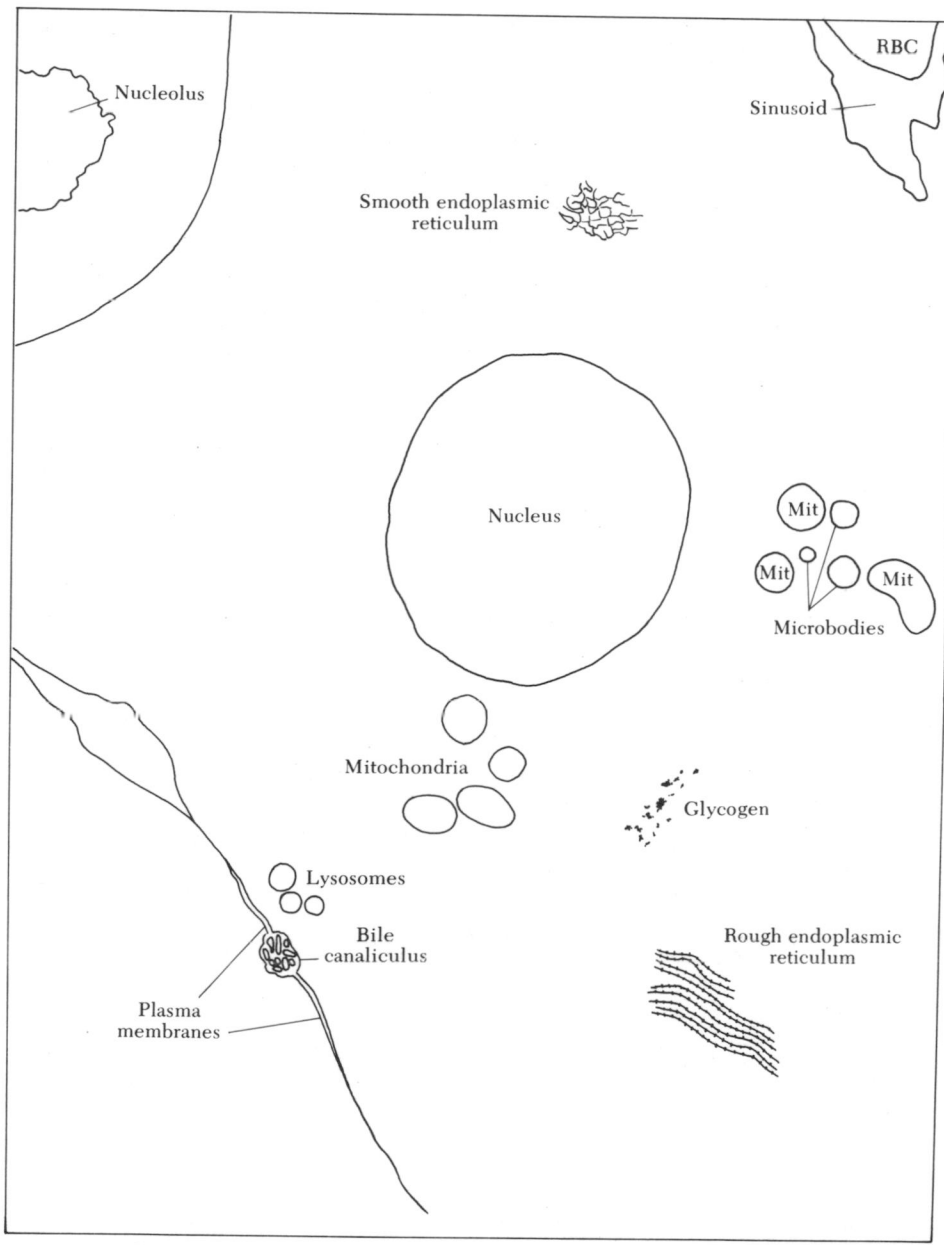

is contained within the rough endoplasmic reticulum, which appears as parallel arrays of membranes studded with dark dots that are ribosomes. The small dark clumps are glycogen. A few microbodies can be seen here and there. Much of the space between these structures is occupied by the intricate network of the smooth endoplasmic reticulum. A portion of sinusoid through which blood circulates can be seen in the upper right corner, and an erythrocyte, or red blood cell, is present in the sinusoid. Magnification 7000 ×. (Courtesy of Dr. Robert R. Cardell, Jr.)

A book is a mirror: if an ass looks in, no apostle will look out.

Georg Cristoph Lichtenberg

BIOCHEMISTRY

a functional approach

R. W. McGILVERY
Professor of Biochemistry
University of Virginia School of Medicine

W. B. SAUNDERS COMPANY *Philadelphia • London • Toronto*

W. B. Saunders Company: West Washington Square
Philadelphia, Pa. 19105

12 Dyott Street
London, WC1A 1DB

1835 Yonge Street
Toronto 7, Ontario

Biochemistry: a functional approach

SBN 0-7216-5911-X

Print No.: 9 8 7 6 5 4 3

PREFACE

Great advances have been made in recent years toward understanding biochemical processes and the molecular mechanisms involved in the processes. The growing complexity of the subject has made communication between specialists in these areas increasingly difficult. That is a pity, for each kind of knowledge illuminates the other. This book is an effort in part to show how many kinds of information can be used to understand biochemical function—noting the fronts on which current advances are being made without ignoring what was learned many decades ago.

The approach differs from that of the usual textbook in many ways. The customary initial catalogue of carbohydrates, lipids, and proteins presented in taxonomical array has been omitted; instead, I have given an introduction of the various compounds in a functional context where the biological significance of some of the properties can be appreciated. I have tried to make the book readable, agreeing with Ernest Hooton that you ought to treat science seriously, but you don't have to act as if you are in church. One of the tools used to this end is to make the facts a part of a general picture through the use of rationalization, even speculation, wherever possible, because nearly everyone has trouble in reasoning about new situations with isolated facts. My own experience suggests that a newcomer to the field not only retains the facts better through this approach, but also cultivates great delight in challenging old theory with new observations as they appear. In other words, he isn't led astray by being exposed to current thinking, even if it later proves to be wrong.

Hopefully, a complete grasp of the principles of the subject can be obtained by simply reading this book. However, I have great respect for the value of the individual teacher and his quite proper tendency to go his own way. This leads to another departure from tradition. Some principles are developed by close examination of a few examples rather than by attempting to mention all good examples. Much is said about hemoglobin, not because it is so much more important than other proteins nor because everything said about it here is something every biochemist ought to know, but so that the reader can see how to think for himself about the structure of globular proteins with precision.

This is not to say that the book is deliberately skimpy. It is intended to give sufficient basic biochemical knowledge for a variety of professions. The major emphasis is on mammalian biochemistry, especially human biochemistry. Not only is more known about biochemical variations in man than in any other species, but most people find themselves especially interesting, and anything creating interest helps the novice. Most biochemists would agree that if one thoroughly understands the biochemical economy of one organism, it then becomes relatively easy to understand the economy of any other organism.

Three colleagues were induced to read most of the manuscript. I rarely dared to defy their combined disfavor on any point, factual or syntactical. However, they often differed; this gave me the advantage of being able to do as I liked, but the disadvantage of being unable to assign blame to them, singly or collectively, for defects. Dr. H. J. Sallach, Department of Physiological Chemistry, University of Wisconsin, and Dr. Gerald Goldstein, Department of Internal Medicine, University of Virginia, read the entire manuscript. Dr. Adrian Gear, Department of Biochemistry, University of Virginia, read Parts I through IV.

Other colleagues at the University of Virginia gave unhesitating assistance. Dr. Julian Kitay of the Department of Internal Medicine read Chapter 23 from the endocrinologist's viewpoint, and Dr. James Ogilvie of the Department of Biochemistry read Chapters 28 and 29. Dr. Carlo Bruni of the Department of Pathology and Dr. Robert Cardell of the Department of Anatomy made their collections of electron micrographs freely available. Still others in various ways stimulated my desire to complete the book.

Laurel McGilvery bravely contracted to do much of the typing.

The staff of the W. B. Saunders Company lived up to their reputation for patience and sympathetic help. I am especially grateful for the encouragement of John Dusseau, Vice-president and Editor.

R. W. McGILVERY

Charlottesville, Virginia

CONTENTS

INTRODUCTION

CHAPTER 1

It is a social nicety to greet an acquaintance after several years of separation with the words, "You haven't changed a bit." We secretly note the coarsened skin, the deepened lines, the graying hair, but in a quantitative physical sense this conventional courtesy is frequently not far from the truth. During his absence, this one man, using the simplest of tools, could have diverted a stream, cut down hectares of forest, begot a hundred children, and in other ways changed his environment so drastically that even the most unobservant would know of his existence. Yet at the end of all this, we might well be hard put to measure more than a small change in the physical dimensions or chemical composition of the man himself.

All of this is well known, even trite, but contains in succinct form the two great problems that have provided much of the intellectual motivation for the development of modern biochemistry: *How* does the living organism develop the energy with which it can manipulate the environment to its advantage, and *how* does it manage to have this great flow of energy through it and still maintain its own form? These problems are enormously complex, but a current of excitement is running through biochemistry today from the realization that the general principles required for complete answers appear to be at hand.

It is our purpose in this book to develop these principles as far as we can, realizing that occasionally we shall be on shaky ground in applying them because of lack of sufficient factual information, but nevertheless feeling some security in our grasp of the subject, because a coherent understanding is a powerful and flexible tool for the assimilation of new information as it appears.

ENERGY AND KINETICS

Our flint-chipping ancestors probably realized that work requires food—although it is apparently still profitable to restate this truism in magazine advertisements. With the development of the concept of energy—a major turning point in human culture—it was possible to say that the energy we use to change our environment comes from the *difference* between the energy contents of the complex mixture of compounds in our diet and the equally complex mixture in our excreta. This now appears obvious, but few biologists would have seen any real application for this abstraction a century ago.

We know that a chemical reaction does not differ from any other physical process in the requirement that there must be release of energy if the process is to occur. Water flows down, not up, but which way is up for a chemical reaction? That is not so easy to say, and yet we know that every one of the hundreds of chemical reactions involved in transforming food into excreta must go in the direction that releases energy. We shall try as we go along to prevent intellectual vertigo by developing this fundamental dimension.

Knowing which chemical reactions are possible is not enough. Let us take a relatively simple example. Acetaldehyde occurs in living cells, and in our tissues there is an eventual consumption of oxygen accompanying the conversion of acetaldehyde to CO_2 and H_2O. But how does this happen? Even by restricting ourselves to a direct reaction with one atom of oxygen, there are many possibilities for the very first step:

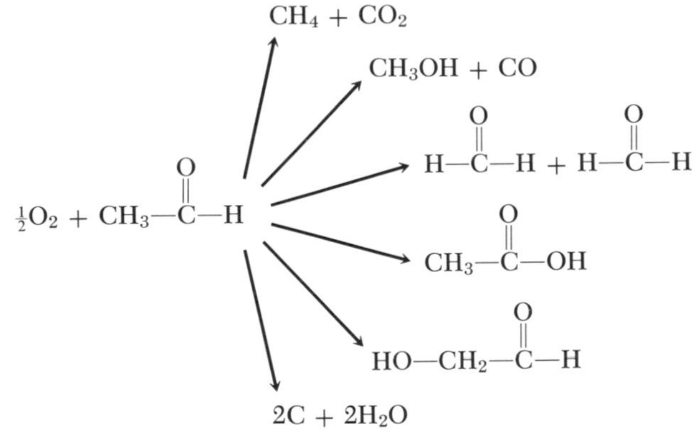

$$
\frac{1}{2}O_2 + CH_3\!-\!\overset{\displaystyle O}{\overset{\|}{C}}\!-\!H
\longrightarrow
\begin{cases}
CH_4 + CO_2 \\[4pt]
CH_3OH + CO \\[4pt]
H\!-\!\overset{\displaystyle O}{\overset{\|}{C}}\!-\!H + H\!-\!\overset{\displaystyle O}{\overset{\|}{C}}\!-\!H \\[4pt]
CH_3\!-\!\overset{\displaystyle O}{\overset{\|}{C}}\!-\!OH \\[4pt]
HO\!-\!CH_2\!-\!\overset{\displaystyle O}{\overset{\|}{C}}\!-\!H \\[4pt]
2C + 2H_2O
\end{cases}
$$

Each of these reactions would liberate energy. For each, we could write further possibilities, arriving at CO_2 and H_2O in every case, but with a bewildering variety of alternate routes. When we add to these all of the other conceivable reactions that acetaldehyde could undergo to arrive at a lower energy state, we should have an almost hopelessly complicated situation were it not for the fact that all of the possible reactions do not occur at the same rate, either in the test tube or in a living organism. Some of the reactions proceed many orders of magnitude more rapidly than others. The organic chemist marks this by saying, for example, that he can get a high yield of acetic acid from acetaldehyde. The biochemist notes the same circumstance by saying that a major metabolic pathway for the oxidation of acetaldehyde is *via* acetate.

Given all of the possible chemical reactions, as defined by the energy balance, the striking thing in the description of the living organism is how few general kinds of reactions have been selected, with the exquisite balance of their rates maintaining a relatively constant composition and at the same time channeling energy into the environment in such a way as to promote the survival of the species.

This precise control of the metabolic reactions is possible through the use of catalysts, the enzymes, to increase many-fold the rates of the selected few of the possible reactions. The use of enzymes not only enables preferential selection of a desired route, but also enables control of the amount of most of the compounds in the organism through adjustments of the catalyst itself. The quantity and the activity of an enzyme can be modified, and we shall see many examples.

In biochemistry, then, energetics defines the possible reactions, but kinetics defines which of the many are actually chosen, and the study of biochemical kinetics is a study of enzymes and their function. Since enzymes are proteins, the composition of the proteins in a cell is almost a definition of its function, and we shall begin the subject by considering the nature of proteins and their formation, and then use this information to consider the nature of the catalysis more closely. We shall then look at the development of the individual reactions into coherent sequences present in most cells, and finally we shall consider the chemistry of the more complex organization that exists when functions are distributed among several tissues.

PART I

The Proteins

THE GENERAL NATURE OF PROTEINS: HEMOGLOBIN

CHAPTER 2

ARGUMENT:

Proteins contain peptide chains made from amino acids by amide linkages. Some proteins contain only one peptide chain, but most contain several. For example, the most abundant hemoglobin in the blood of human adults has four peptide chains—two each of two kinds. Twenty different amino acids are commonly found in proteins, and the sequence in which these amino acids appear in the peptide—its primary structure—determines its shape and its association with other peptides, and therefore the properties of the protein. The sequence is fixed by the genetic information of the cell, and is the same in individuals of the same heredity. The structures provided by the 20 amino acids are not sufficient to provide all of the kinds of chemical reactivity required for biological function, and some peptides combine with other compounds—prosthetic groups—having additional kinds of structures. Such is the case with hemoglobin, which contains heme as a prosthetic group. Heme is an example of a metal chelate, with ferrous ions bound to a porphyrin and providing the site for attachment of oxygen. The amino acid sequence of hemoglobin peptides is such that the most likely three-dimensional folding forms stable complexes with heme. The sequence also favors association of four heme-peptide complexes into a complete molecule that can efficiently transport oxygen. This biological function can be served by molecules that differ in amino acid composition, and hemoglobin is actually a term for several kinds of molecules. There are three kinds in the usual human adult, but each of these kinds is identical from one individual to another. Some humans have still different hemoglobins, reflecting a different heredity. The hemoglobins of other species differ still more, to a degree dependent upon evolutionary divergence, but all of the proteins of this name have the general function of transporting oxygen.

THE PEPTIDE CHAIN

Proteins are large molecules, *macromolecules*, and their size is a biologically important attribute, because it prevents them from leaking out of cells through the plasma membrane. Formally, proteins are polymers of amino acids. The minimum size at which a polymer is called a protein is usually taken to be between the molecular weights of 5000 and 10,000.

More precisely, proteins are composed of *polypeptides*, and a peptide is a compound formed by head-to-tail amide bonds between *α-amino acids* (Fig. 2-1). We can see immediately an important chemical difference between amino acids and peptides. Every amino acid has at least one positive charge and one negative charge in a relatively small total volume. These obligatory charges disappear when the linear peptide is formed, except for a single positive charge on one end, the

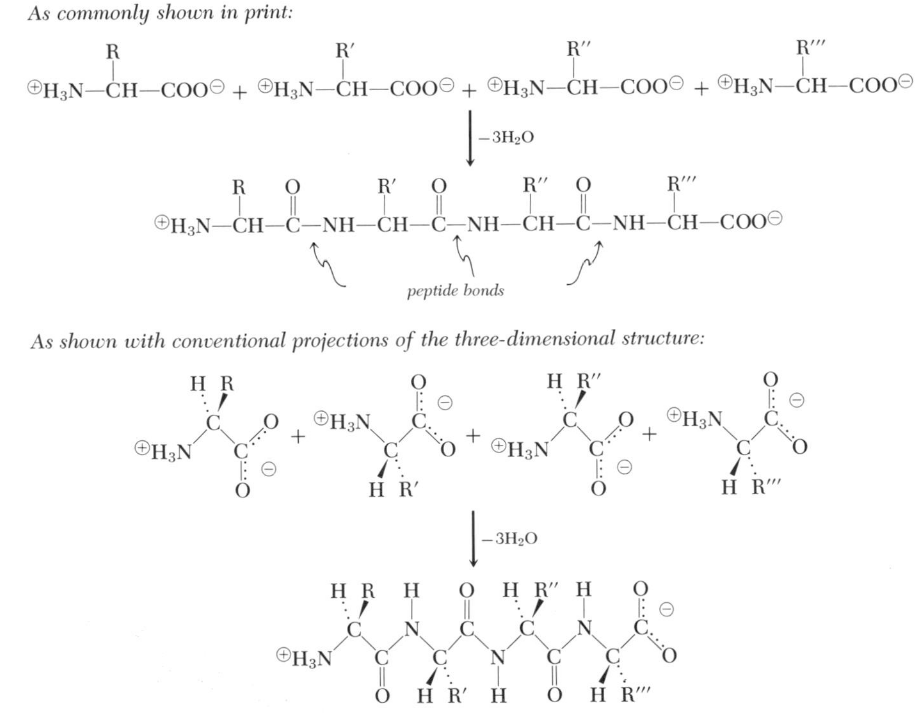

As commonly shown in print:

As shown with conventional projections of the three-dimensional structure:

Figure 2-1 Combination of L-α-amino acids into peptide chains by formation of amide bonds. Both of the illustrations use common formal conventions and neither is an accurate projection of the real geometry of the molecule.

amino end, and a single negative charge on the other, the carboxyl end. Now, the side chains of the amino acids, designated as R-groups in Figure 2-1, may carry positive or negative charges beyond those shown, but the number of these charges depends upon the particular kinds of side chains present—upon the kinds of amino acids that have been linked to form the peptide. This is especially important, because charge determines the degree of association with water; uncharged portions of the chain will tend to coalesce, excluding water, while charged portions will remain on the outside at the water interface. In other words, a peptide is not like free amino acids; various parts of it may, or may not, associate with water, and this property will be determined by the nature of its constituent amino acids.

A protein may be only a single, long, peptide chain, but most proteins are made of several peptide chains associated together, and these chains need not be identical. The most common type of hemoglobin in the blood of human adults is an example. This hemoglobin, known as hemoglobin A_1, is a combination of four peptide chains of two kinds, designated α and β, and the chain formula for hemoglobin A_1 (abbreviated Hb A_1) is $\alpha_2\beta_2$.

Hemoglobin illustrates many of the general principles of protein structure and function. Its function in transporting oxygen and the visible changes in color when oxygen is added or removed are well known, but there is also a large body of information on the way in which this function is created by the structure of the molecule. The formation of hemoglobin from the constituent amino acids has also been intensively studied, and several genetic alterations are known. We shall talk about the nature of hemoglobin in a general way in this chapter, and then discuss it more intensively in the following chapters to illustrate the way in which all proteins are made and how the nature of their constituent amino acids determines their properties.

Hemoglobin is a representative of the *globular* proteins—those with a compact structure that are usually quite soluble in the water phase of tissues. The molecule is roughly spheroidal, with dimensions of $6.4 \times 5.5 \times 5.0$ nanometers.° Its molecular weight is greater than 64,450, depending upon the degree of hydration, and it is easily retained within the erythrocytes of the blood.

All of the properties of hemoglobin are determined by the character of the amino acids in the two kinds of peptide chains found in the molecule. There are 20 different amino acids found in most proteins, and hemoglobin A_1 contains all but one of them. We shall consider these amino acids as we come to them; suffice it for now to note that they have side chains of differing chemical properties. The character of the peptide chains, and therefore of the hemoglobin molecule, is determined by the particular kind of amino acid residues and their spacing in the chain.

In other words, the α chains of the Hb A_1 in one individual are made with a particular order of amino acids, and the β chains have a different sequence; it is these sequences that determine the propensity of pairs of α and β chains to associate into a molecule capable of transporting oxygen.

Furthermore, the chains isolated from the hemoglobin of one human have exactly the same number and sequence of amino acids found in the same hemoglobin

---

° Nanometer $=$ nm $=$ millimicron $=$ mμ $=$ 10 Ångstrom units $=$ 10 Å.

from any other human. Sequence of amino acids determines function, and the sequence in a particular protein follows heredity. Therefore, function follows heredity. The genetic information transmitted from one cell to another specifies the combination of amino acids that will appear in each of the many different proteins of the cell, and the nature of these newly made proteins will only differ from those of the parent if the genetic information is altered.

We know there are thousands of different proteins in one human, probably millions of different proteins in all of the mammals, and perhaps trillions in all kinds of living organisms. Yet these differences must come from the particular way in which some 20 amino acids are combined. We shall somewhat slowly develop in the next chapters a full sense of how this can be, but an inkling can be obtained from considering part of the structure of the α chain of hemoglobin.

It is conventional to describe a peptide beginning with the N-terminal amino acid—the amino end. The α chain starts with a residue of *valine* (Fig. 2-2), followed by residues of *leucine* and *serine*, and continuing through a total of 141 amino acid residues to a terminus with residues of *lysine, tyrosine,* and *arginine.* Arginine is thus the C-terminal amino acid. We can give this peptide a trivial name: valyl leucyl seryl . . . lysyl tyrosyl arginine, but it is common to use abbreviations: Val-Leu-Ser . . . Lys-Tyr-Arg. A casual look at the structural formulas shown in the figure reveals the strikingly different chemical functions conveyed by the presence of these different amino acids. The only charge in the first three residues is on the terminal ammonium group, and the side chains have a compact nonpolar structure. In contrast, the C-terminal end is highly polar, not only having the negative charge of the carboxylate group, but also having positive charges on the ammonium group of the lysyl side chain and the guanidinium group of arginine, and these positive charges are carried on freely rotating chains capable of assuming a variety of positions relative to the peptide backbone. In addition, the flat phenyl ring of tyrosine has a potentially reactive phenolic hydroxyl group.

Even with this small sampling of six out of the 141 residues in the peptide, we can see that any substitution of one residue for another can easily change the reactivity of that segment.

THE PROSTHETIC GROUP

The 20 amino acids have the necessary variety of structures for making the different proteins, but still do not provide all of the kinds of chemical groups necessary for every biological function performed by the proteins. These additional groups are supplied by building appropriate peptides in such a way that they will strongly associate with other low-molecular weight compounds that do have the desired structure. The result of the association is a protein composed in part of peptide chains, the *apoprotein,* and in part of a non-peptide compound, the *prosthetic group* (from *prosthesis,* an additional part).

Hemoglobin is an example of such a protein. The name of the protein is an old contraction meaning blood globulin. Hemoglobin was found to contain a porphyrin bound with iron in addition to amino acids. It was a logical extension of the word to call the iron porphyrin, *heme,* and the peptide portion, *globin.*

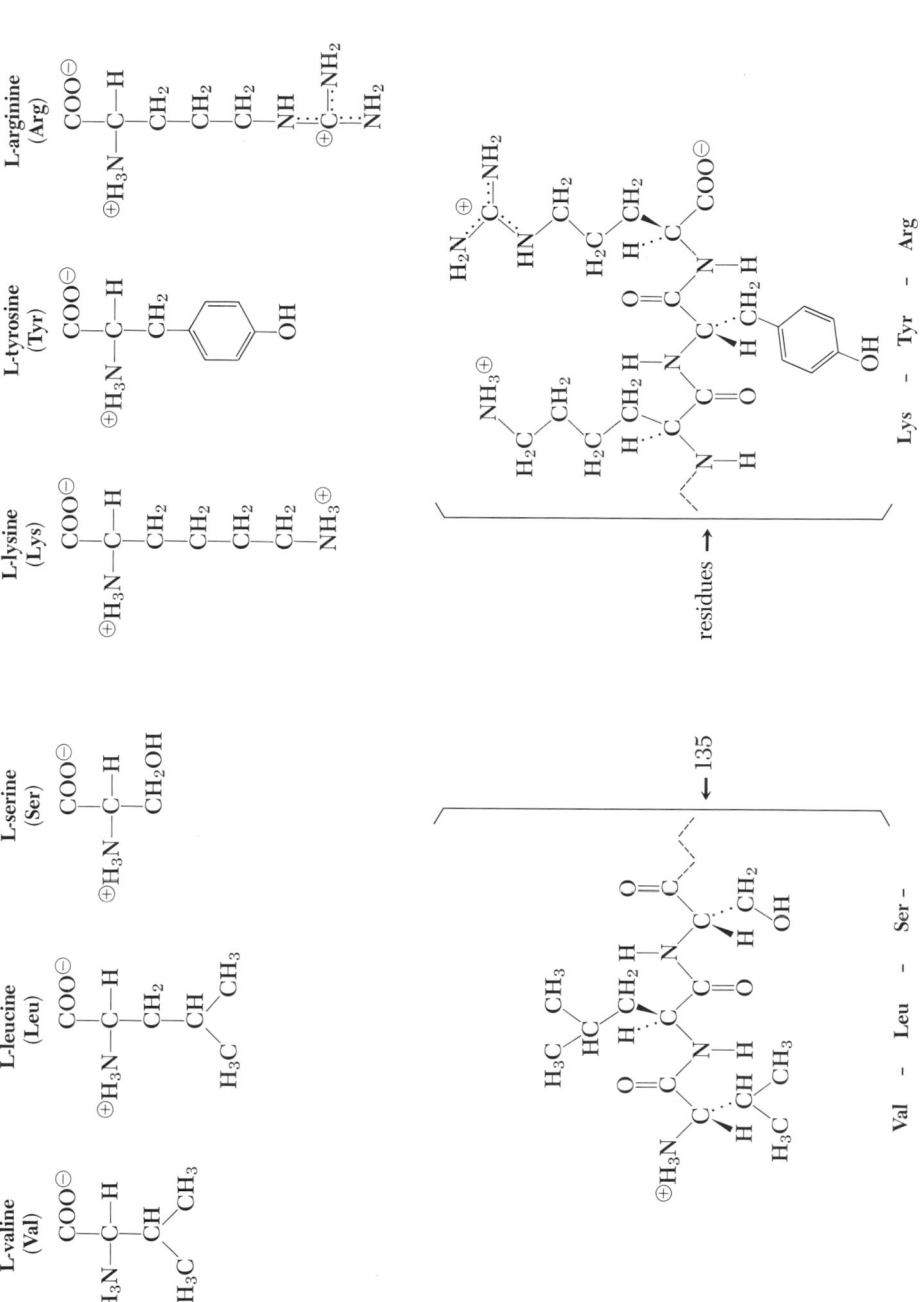

Figure 2-2 The two ends of an α-chain from human hemoglobin. The constituent L-amino acids are shown at the top with the conventional formulas ordinarily used to show stereoisomerism and amino acids. See Chapter 28 for a review of stereoisomerism and amino acids.

Globulin is a term for a general type of protein dating back to the time when not much was known about the structure and function of proteins, and they were classified on the basis of superficial physical properties, such as solubility in various salt solutions (globulins are precipitated at lower salt concentrations than are albumins). The classification as such is largely of historical interest, but many of the class names, including globulin and albumin, are in current use.

Heme is a combination of ferrous iron and *protoporphyrin IX* (Fig. 2-3). We shall say a great deal more about the porphyrins later, and concentrate for now on the generalities.

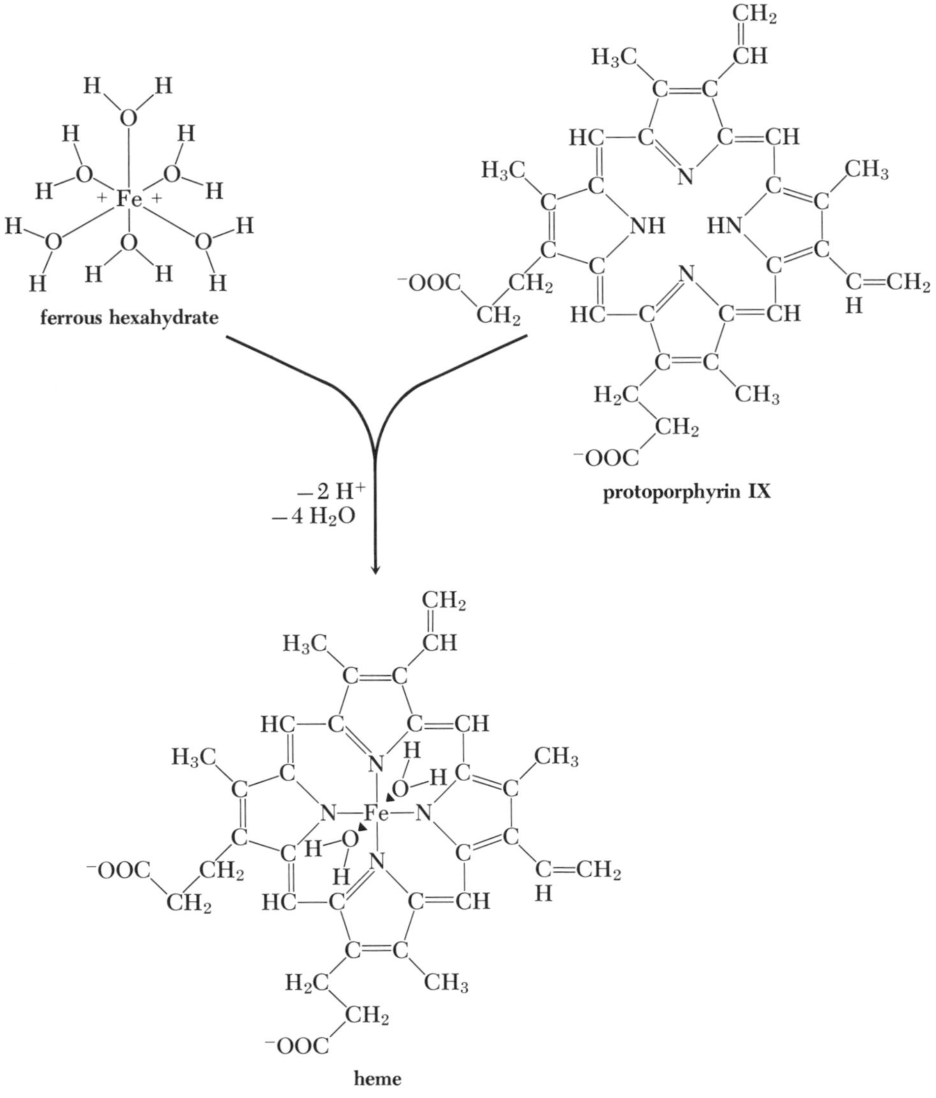

Figure 2-3   The formal stoichiometry of the combination of ferrous ion and protoporphyrin IX to form heme.

Heme is a classic example of a metal chelate—a compound formed between a metallic ion with unfilled orbitals and a ligand with unshared electron pairs. Metallic ions participate in many biological processes, mainly by forming complexes with proteins or other components of the cell. Some of these complexes, such as those of magnesium and calcium, are easily dissociated, so that a given atom of metal is rapidly exchanged from one complex to another. Other ions, such as cupric, cobaltous, zinc, and manganous, are bound more tightly, sometimes so tightly that the metal, for all practical purposes, is a fixed part of the structure as long as it exists. This is also the case with ferrous ion in heme.

The porphyrins contain a highly resonant planar ring system in which four nitrogen atoms are fixed in the center at a spacing that is ideal for bonding their unshared electrons with a metal ion. The porphyrins therefore act as tetradentate (four-toothed) ligands, and protoporphyrin IX has an especial affinity for iron.

Ferrous ions have a coordination number of six, which means that they can associate with six electron pairs per atom. In an aqueous solution of ferrous salts, the electron pairs are contributed by the oxygen of water molecules or of hydroxide ions. When ferrous ion forms a chelate with porphyrin, two of its coordination positions are still unfilled, so free heme in aqueous solution will also be hydrated as shown in the figure.

However, each peptide chain of globin has a sequence of amino acids that makes certain three-dimensional arrangements more likely than others, and the most likely of all is one that, in the presence of heme, causes it to fold around the heme. It is as if the globin formed a crevice into which the heme just fits. The important feature is that the folding brings a *histidyl* residue of the peptide chain (Fig. 2-1) into a position where a nitrogen atom of the residue can link to the fifth coordination position of the iron. The sixth position of the iron is left open and now has the ability to bind either oxygen or water.

The biological function of hemoglobin is therefore derived from both the heme and the peptides, which are combined in a particular structure with the desired reactivity. Each of the four peptide chains has its own heme, and we shall later see that the complete molecule has properties vital for oxygen transport under the conditions found in higher animals that are not attainable in a single heme-peptide.

We shall also later see that there are other proteins having metalloporphyrins as prosthetic groups serving quite different purposes from oxygen transport. In each case, the amino acid sequence of the peptide chain is such as to modify the chemical reactivity of the combined porphyrin in the desired way, so that even in the case of proteins with prosthetic groups, it is fundamentally the amino acid sequence that governs the behavior of the combination.

OTHER HEMOGLOBINS

After all of the emphasis on the fixed character of the amino acid sequence in a particular protein, it may seem like an anticlimax to discover that the usual human adult has at least two other hemoglobins in his blood, each of which can transport oxygen, that contain peptide chains differing from those of Hb A_1. The other two are Hb A_2 and Hb F (A = adult, F = fetal). One ml of adult blood

L-histidine
(His)

$$COO^{\ominus}$$
$$^{\oplus}H_3N—\overset{|}{\underset{|}{C}}—H$$
$$CH_2$$

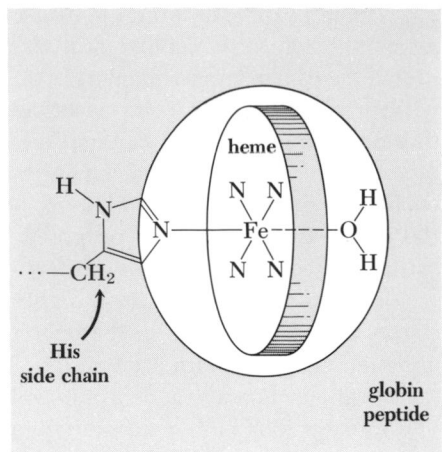

Figure 2-4 Schematic illustration of heme in a pocket formed by the globin peptide. A histidyl side chain from the globin acts as an additional ligand for the iron atom in heme. The free amino acid, histidine, is shown at the left.

contains about 0.15 g of Hb A_1, 0.004 g of Hb A_2, and only 0.0007 g of Hb F. The function of Hb A_2 is not clear. Hb F is a carry-over into adult life of a kind of hemoglobin that is the principal form during fetal development, and it has properties of oxygen transport more nearly suited to the conditions found in the uterus.

All of these hemoglobins have the same kind of α chains, but the other kind of chain differs in the kind of amino acid present at some positions, and it is designated a δ chain in Hb A_2, and a γ chain in Hb F. Therefore,

$$Hb\ A_1 = \alpha_2\beta_2$$
$$Hb\ A_2 = \alpha_2\delta_2$$
$$Hb\ F\ = \alpha_2\gamma_2$$

The existence of these different hemoglobins does not really shake our picture of constant amino acid sequence. It only adds the qualification that the same function can be carried out by different proteins, each with its own constant amino acid composition specified by separate genetic information for that protein. The different proteins may exist to function at different times during development, in different tissues of the body, or as a safeguard against changing environments of the cell.

However, we shall see that there are some humans who have still other hemoglobins, usually created by the substitution of a single amino acid residue in one kind of peptide chain. These people represent variants from the usual heredity, carrying a mutation that has affected the genetic information for hemoglobin.

The formation of a new species is the result of the accumulation of a number of persistent mutations. It is therefore common to find different amino acid compositions for the same kind of protein in different species. For example, no primate yet studied has a major adult hemoglobin identical to that of any other primate. The farther one gets from man in the evolutionary scale, the greater the differences in composition of particular proteins.

A Semantic Point

It is common in biochemical literature to use the name of a protein as a class word when talking about properties shared by similar proteins from all biological sources. Thus, the statement, "Hemoglobin is red," is automatically understood to apply to all related proteins for all species. Similarly, "human hemoglobin" without qualification usually means the collection of proteins found in most humans. It is important to note that such usage will frequently not be meant to include some of the less common varieties of hemoglobin found in humans. This kind of idiom is frequently used for many proteins, and the reader must supply the necessary qualifications and mental reservations for himself. It may seem woefully imprecise, but if every mention of proteins were studded with all of the phrases and clauses necessary to limit exactly what is meant, biochemical literature would be considerably more difficult to understand than it already is.

DESCRIPTION OF SIZE— SVEDBERG UNITS

We frequently think in terms of molecular weight in comparing the size of small molecules, or we use spatial dimensions of length when the information is available. Determination of molecular weight isn't easy for macromolecules, and the actual three-dimensional shape is known for only a few. It is therefore common to compare large molecules in terms of the velocity with which they move through a solution in a centrifugal field. This is a pragmatic practice, handy because it is common to use centrifuges in separating macromolecules for both analytical and preparative purposes. The measure that is used is the *sedimentation constant,* given in *Svedberg units* (S). (T. Svedberg was responsible for much of the theoretical and practical development of centrifugation as a tool in studying macromolecules.) Svedberg units describe the velocity attained per unit of applied force by a particle moving through a liquid medium.

Physics tells us that force accelerates, and force applied to a particle in a liquid will indeed accelerate it. However, there will be increasing frictional resistance to motion through the medium as the velocity increases, and acceleration continues only until a velocity is reached at which the frictional counter-force just balances the applied force. It is this limiting velocity with which we are concerned.

We know from experience that a large chunk of material falls through water faster than a fine powder of the same material, even though the total gravitational force is the same. The powder settles more slowly because of its larger surface in frictional contact with the medium. Generally speaking, larger particles with higher molecular weights will have higher sedimentation constants, if everything else is equal. Unfortunately, everything else rarely is equal. Different molecules have different shapes and different surface areas, even if the molecular weight is nearly the same. Long, thin molecules have lower sedimentation constants than spherical molecules of the same mass and volume.

We know from Archimedes' principle that the net force on an object in a fluid medium depends upon the difference between the mass of the object and the mass of the fluid displaced by it. Something suspended in water and spun in a centrifuge rotor will sink toward the periphery if it is more dense than water and float toward the axis if it is less dense. A given macromolecule may sink, float, or not move at all, depending upon the relative density of it and the medium. To avoid having a variety of sedimentation coefficients quoted for the same molecule, it is customary to correct the data so as to obtain the constant describing the sedimentation velocity in pure water.

The second factor determining the force is the centrifugal field, which depends upon the radius of the rotor and the square of the angular velocity. Commercial instruments used routinely today can accelerate samples up to 250,000 times gravity (250,000 $\times$ g), which is sufficient to enable practical study of molecules with molecular weights as low as 10,000.

In the succeeding chapters, we shall have occasion to refer to some cellular components by their S values for purposes of identification. To make sense out of the discussion, it is important to have in mind that S values, unlike molecular weights, are not additive when molecules combine. The union of two 5S particles does not create a 10S particle. (Two cannonballs glued together don't sink twice as fast as one alone.)

Although we can't directly compare sedimentation constants and molecular weights, it is helpful to have a rough idea of the corresponding ranges. A value of 2S will be obtained for nearly spherical proteins with molecular weights somewhat over 10,000; long, thin molecules having sedimentation constants of 2S may have molecular weights over 50,000. Similarly, 4S crudely corresponds to a molecular weight of 50,000 for spherical proteins, 8S to 160,000, and 16S to 400,000; the molecular weights of long, thin molecules with the same sedimentation constants are several-fold greater. (A more detailed description of the application of centrifugation and the necessary calculations is given by H. K. Schachman in Colowick, S. P., and N. O. Kaplan, eds.: *Methods in Enzymology*, Vol. 4, p. 32. Academic Press [1957].)

THE
SYNTHESIS
OF
PROTEINS

CHAPTER 3

ARGUMENT

The description of the order of amino acids necessary to form the peptide chains of proteins such as hemoglobin is contained within the nucleic acids, which constitute the genetic apparatus of the cell. The primary source is a deoxyribonucleic acid (DNA), which is duplicated through generations of dividing cells. DNA consists of a backbone of sugar phosphate esters with side chains of nitrogenous bases—the purines and pyrimidines. It is the order of arrangement of the side chains that constitutes the necessary information. Each amino acid is specified by combinations of three of these bases. This combination is a codon. The complete array of codons necessary to specify a peptide chain is a cistron. DNA occurs in the nucleus of most cells as a double strand, twisted together into a helix. One of the strands is the active source of information in the cell; the other serves as a matrix for the formation of the new active strand upon cell division. The active strand is used as a template to make another kind of nucleic acid, messenger ribonucleic acid (mRNA), which carries the coding of one or more cistrons from the nucleus to the site of protein synthesis in the cytoplasm. Within the cytoplasm are granules, the ribosomes, to which the messenger RNA is attached. The ribosomes also carry the enzymes necessary to catalyze synthesis of peptides according to the order specified by the messenger RNA.

The amino acids are identified for purposes of protein synthesis by a combination with transfer RNA (tRNA). These molecules, small in comparison with other nucleic acids, are each specific for a particular amino acid, and combine with the codon on messenger RNA designating that amino acid. After the tRNA joins with mRNA on a ribosome, a growing peptide chain is added onto the new amino acid carried by the tRNA, thereby making the chain one residue longer. The process of transferring the peptide chain onto new amino acids carried by different kinds of tRNA is repeated until the chain is completed.

Special codons signal the end of a peptide chain, and when one of these terminator codons occurs in the messenger RNA, the peptide chain is released from the ribosome. The released chains then spontaneously combine to form finished protein molecules, with any necessary prosthetic groups attached. Thus, the original DNA of erythroblasts bears specifications only for amino acid sequence, but thereby also specifies the complete formation of hemoglobin with four peptide chains, each with its associated heme.

GENERAL DESCRIPTION OF PROTEIN SYNTHESIS

Many details of protein synthesis deserve close attention, but it is imperative that the general picture not be obscured by the details. Figure 3-1 gives a schematic summary of the materials and events discussed more intensively later, so that the point of the discussion can be kept in mind.

The starting materials for protein synthesis are the free amino acids, present in solution inside the cell. Five problems have been solved in the evolution of the

---

Figure 3-1 Schematic summary of protein synthesis. *Top, step 1.* A molecule of DNA in the nucleus unfolds, and one of its strands is used as a template to direct the formation of messenger RNA from nucleoside triphosphates, which lose inorganic pyrophosphate (PP_i) as they attach to the growing RNA chain. The completed mRNA moves to the cytoplasm (*bottom*), where it binds ribosomes into a polysome, and acts as a template for protein synthesis.

The following steps are shown on separate ribosomes for clarity, but in fact they are repeated in sequence on each ribosome. The successive ribosomes grow longer and longer peptide chains as they move down the molecule of mRNA.

Step 2. Meanwhile, amino acids are combined with specific molecules of transfer RNA (tRNA) in the cytoplasm by a reaction that also involves the cleavage of adenosine triphosphate (ATP) into adenosine monophosphate (AMP) and PP_i.

Step 3. The tRNA molecules, carrying the amino acids in the form of aminoacyl groups, diffuse to the polysome, where the growing peptide chain is on another molecule of tRNA already attached. The incoming tRNA, which bears the next group required for the growing peptide (in this case a leucyl residue), has the proper configuration to complex with mRNA on the ribosome.

Step 4. When the proper tRNA is in place, the peptide chain is transferred onto the amino group of the new residue brought in by tRNA, so that the chain is now one residue longer.

Step 5. When the transfer of the previous step is completed, the previously bound tRNA no longer carries a peptide chain and is free to dissociate from the ribosome, returning to the mixed pool of tRNA in the soluble cytoplasm, where it is available for transport of another molecule of its specific amino acid. The ribosome now moves along the mRNA molecule to the position where the placement of the next amino acid will be directed.

Step 6. Steps 3, 4 and 5 are repeated. As each amino acid residue adds to the peptide chain, the ribosome moves down the mRNA molecule. When a ribosome has reached the end of the molecule, the peptide is completed and is detached into the soluble cytoplasm. The ribosome itself can then move free of the mRNA and be available for attachment to the beginning of yet another molecule of mRNA (not shown).

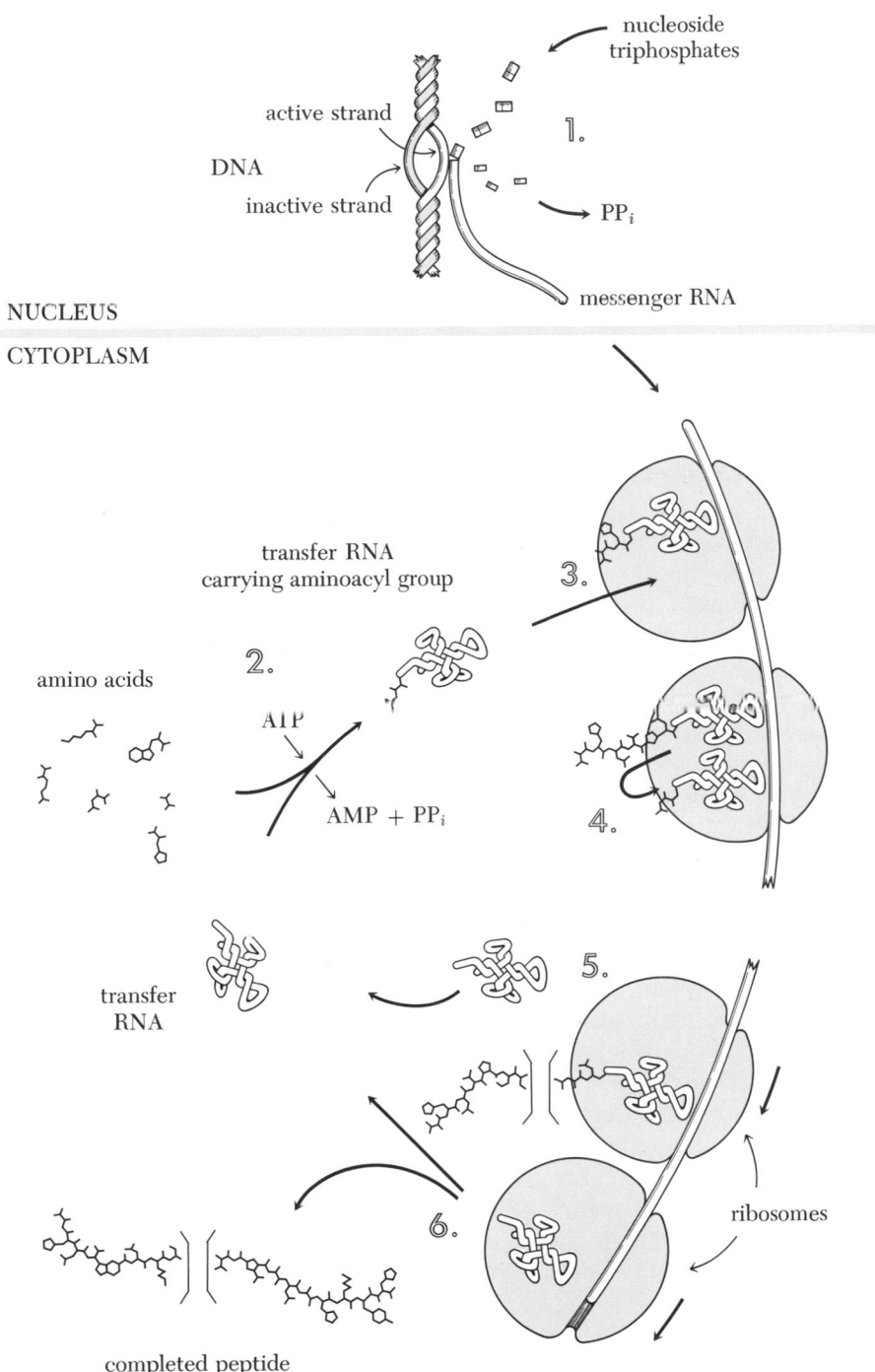

Figure 3-1   *Legend on opposite page.*

mechanism for making complex and specific peptides out of this mixture of amino acids:

1. The order of the amino acids in the peptides must be specified in a genetically stable fashion. Daughter cells must retain the ability of the parent to make particular proteins.

2. The source of the information for order of amino acids must be used again and again. This is true because information is carried through chemical structure, and the physical bulk of the information specifying a protein will therefore be comparable to that of the protein itself. There isn't enough space in the cell to store one informational molecule for each protein molecule to be made later during the life of the cell.

3. There must be a means of identifying the proper amino acid. A particular amino acid must be made to match at the right time and the right place during the building of a peptide chain, and it must be prevented from being put in at the wrong location.

4. The amino acid must be in a chemically active form so that formation of peptide bonds can proceed. This demands an input of chemical energy.

5. The peptides must combine with any prosthetic groups required, and associate in the proper way to make the final protein molecule. We have already stated, and shall present further evidence in the next chapter to support the statement, that this requirement is met by the particular combination of amino acids specified, so that the most stable configuration of the synthesized peptides is the desired protein.

The first three of these requirements are met by the use of a class of compounds, the nucleic acids. *Deoxyribonucleic acid (DNA)* is found in the nucleus, and is the primary carrier of information. A molecule of DNA remains for the life of the cell, and it may contain information for several peptides.

The information present on the DNA molecules is transcribed for use in protein synthesis, and the copy exists in the form of a ribonucleic acid, *messenger ribonucleic acid (mRNA)*. The formation of messenger RNA, which is the transcription of information, occurs in the nucleus. The completed messenger moves into the cytoplasm, and the original DNA molecule is then free to be used in the formation of additional messenger RNA carrying the same information.

Within the cytoplasm are granules—the *ribosomes*—built from proteins and other types of RNA. In some cells, but not all, the ribosomes may be studded on the *endoplasmic reticulum* (see the frontispiece). *Ribosomal RNA (rRNA)* apparently carries no information, but serves as part of a matrix to which messenger RNA and the necessary enzymes are attached. Most molecules of messenger RNA are so long that they bind a cluster of several ribosomes. This cluster, a *polysome*, is the anatomical site of protein synthesis.

The mechanism for the transfer of information involves the development of specific binding forces between nucleic acids, without participation of the amino acids as such. Still another nucleic acid is used to translate this information into a recognition of the proper amino acids. This is a smaller nucleic acid, known as *transfer RNA (tRNA)*, formerly called soluble RNA (sRNA).

There is at least one particular type of transfer RNA for each kind of amino acid, and a particular kind of transfer RNA will combine with the messenger RNA in place on a ribosome only when the information on the messenger RNA specifies the corresponding amino acid. Moreover, the transfer RNA will only combine tightly when it has an aminoacyl group attached, which is to say it must be carrying an amino acid residue. Such a compound is abbreviated as an *aminoacyl tRNA*, for example, valyl tRNA, which designates a valyl-specific tRNA with the valyl residue attached.

Once the first aminoacyl tRNA is in place on the ribosome and in contact with messenger RNA, the next also attaches, and formation of the peptide bonds begins.

The formation of the aminoacyl tRNA and the formation of the peptide bond both use as sources of energy a class of compounds known as nucleoside triphosphates. The particular ones used here are *adenosine triphosphate (ATP)* and *guanosine triphosphate (GTP)*, respectively. The energetics of these compounds is considered in Chapter 9, and we shall be concerned with their formation in most of Part II. They are mentioned now to note the way in which chemical energy is put into protein synthesis.

When formation of the peptides is complete, the participation of the nucleic acids is also finished, and all of the other biological functions of the cell come about as a result of the various chemical reactivities of the proteins themselves. The requirement for a stable source of information on the order of the amino acids, capable of being used repeatedly, has been met. We have seen that the remainder of the events in the formation of a protein molecule, which includes the combination of peptide chains and the association with prosthetic groups, do not require further genetic direction.

NUCLEIC ACIDS AND CODING

Deoxyribonucleic Acids

The formal structural relationship of the components to DNA is shown in Figure 3-2. The backbone of DNA is made from molecules of a sugar, 2-deoxy-D-ribose, which are linked as esters of phosphoric acid. We follow custom in referring to DNA as an acid, but in fact the physiological form is the ionized deoxyribonucleate illustrated. Attached on the sugar-phosphate backbone are *purine* and *pyrimidine* groups, which constitute the specific side chains. The repeating unit, both in a chemical and a biological sense, is a *nucleotide*—a combination of a nitrogenous heterocyclic compound with a sugar and phosphate. Therefore, DNA is an example of a *polynucleotide*, which is also a *macromolecule*.

The purines of DNA are *adenine* and *guanine*, and the pyrimidines are *cytosine* and *thymine*, as shown in the figure. We shall again follow custom in referring to these heterocyclic compounds as the "bases" of the nucleic acids, even though some behave as acids at pH 7.

Nomenclature of the Nucleotides

1. Nucleic acids and other nucleotides are described as nucleoside phosphates. A *nucleoside* is the combination of a heterocyclic base with a sugar, and the sugar is understood to be D-ribose unless otherwise specified. If the sugar is 2-deoxy-D-ribose, deoxy is

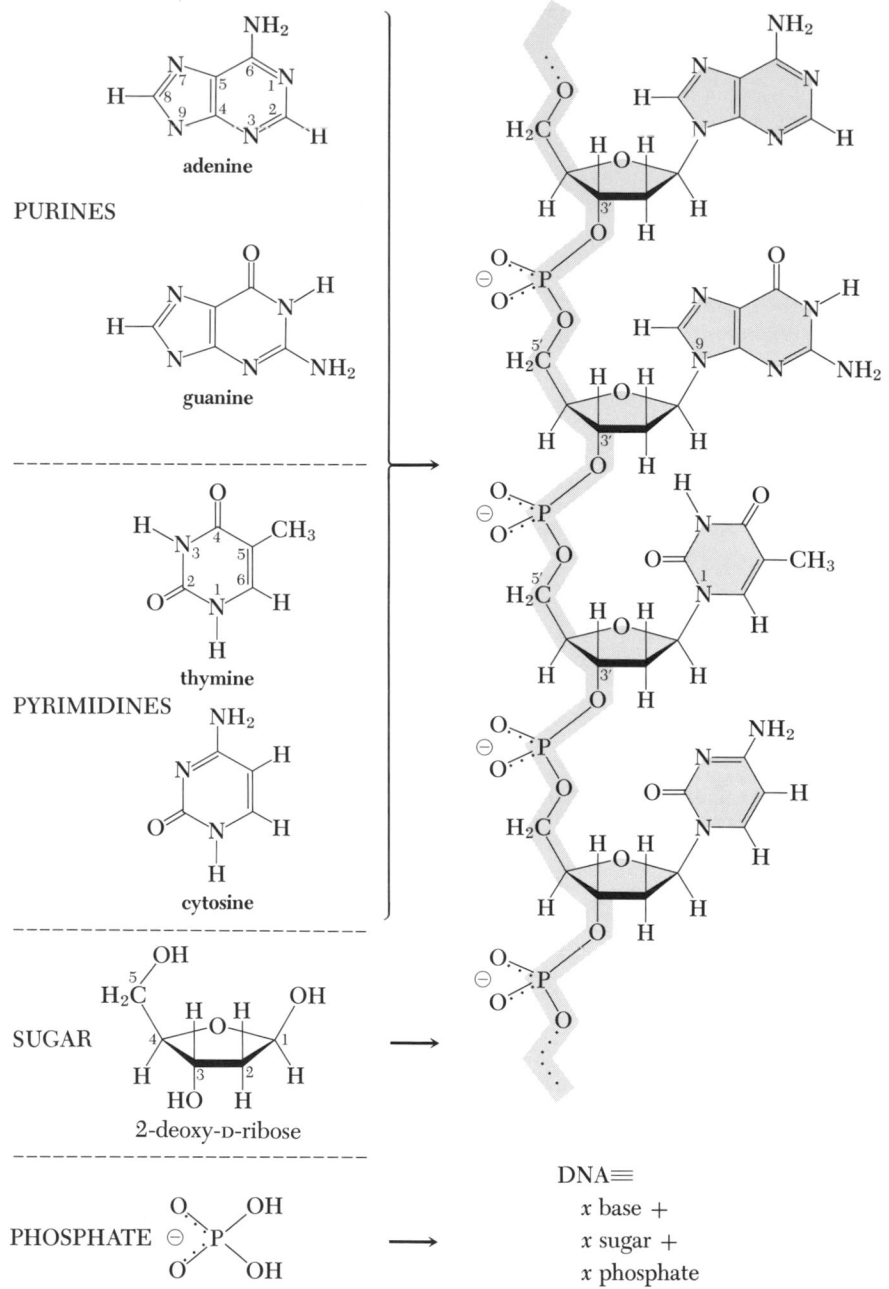

Figure 3-2   The formal stoichiometry for the combination of inorganic phosphate (P_i), 2-deoxy-D-ribose, and purines or pyrimidines into deoxyribonucleic acid (DNA). The conventional abbreviation for the polynucleotide fragment shown is d-ApGpTpCp. Adenine and guanine are substituted purines; thymine and cytosine are substituted pyrimidines.

The numbering of atoms in ribose, purines, and pyrimidines is shown. In the nucleotides, the atoms of the sugar are designated 1', 2', etc., to distinguish these positions from the similarly numbered positions on the purine or pyrimidine rings.

used as a prefix. Thus, the nucleosides of DNA are deoxyadenosine, deoxyguanosine, deoxy-cytidine, and deoxythymidine. (Caution: It used to be thought that thymine occured only in DNA, and thymidine without the deoxy- prefix is frequently used to describe the deoxy-nucleoside. Such semantic booby traps are among the perils of a growing science.)

2. The position of the phosphoryl ester bond on the sugar is specified by a numeral. Thus, 5'-deoxyadenosine monophosphate has the phosphate ester group on the fifth carbon of the deoxyribose portion of deoxyadenosine.

3. The nucleosides are abbreviated A, G, C, or T, with d- as a prefix for deoxynu-cleosides. When occurring in free form, MP is used as an abbreviation for monophosphate. Thus, dG is deoxyguanosine: 3'-dGMP is deoxyguanosine esterified with phosphate on the third carbon of the deoxyribose. We shall later see that the 5'-nucleotides are commonly used in metabolic reactions, and in such cases the number is frequently deleted, with the ab-breviation being understood to designate only the 5' isomer. Thus, ATP is understood as meaning adenosine-5'-triphosphate.

4. Polynucleotide chains, such as those occurring in DNA, are abbreviated with p as a symbol for phosphate. Used to the left of the nucleoside symbol, it indicates a 5' linkage, used to the right, a 3' linkage. Thus, d-pA is deoxyadenosine-5'-phosphate; dAp is deoxy-adenosine-3'-phosphate; and d-pApC is made by linking deoxyadenosine-5'-phosphate through its 3' hydroxyl to the phosphate of deoxycytidine-5'-phosphate. Another example is given in Figure 3-2.

The only structural variable along the polynucleotide chain of DNA is the location of the four bases. It has been recognized for some time that the coding for amino acids must involve the particular arrangement of at least three of these bases. The reason for this is that four bases taken two at a time could only yield 4^2 (or 16) different combinations, and there are 20 different amino acids to be coded. However, there are 64 possible combinations of four bases taken three at a time, and experimentation has shown that this is the kind of a combination actually used for protein syntheses. A particular arrangement of three nucleotide units in the DNA polynucleotide specifies a particular amino acid. This is the smallest unit of genetic information, and the triplet of nucleotides is known as a *codon* for a particular amino acid. For example, the sequence d-CpApA on the active strand of DNA specifies leucine, while the same bases arranged in the sequence d-ApApC specifies valine; d-ApCpA specifies still another amino acid, cysteine.

One can see that this kind of coding can only function if it is 'read' from a particular point in the nucleotide chain. For example, the sequence, d-CpApApCpApApCpApApCpCpApAp · · · , codes Leu-Leu-Leu-Leu read from the end, because d-CpApAp specifies leucine. If the reading begins with the second nucleotide, it codes Val-Val-Val- (d-ApApC), and with the third, Cys-Cys-Cys- (d-ApCpAp).

The minimum possible length of a DNA chain is three nucleotide units for each amino acid in the shortest peptide made under genetic control. In fact, the DNA molecules of mammalian cells, which appear to be smaller than those from most organisms, contain around 30,000 nucleotides in a typical molecule. There is some uncertainty, because molecules this size are easily sheared by ordinary mechanical forces during the manipulations required for their isolation from tissues. A chain of this length would represent the coding for 10,000 amino acid residues. Most peptide chains are much shorter than this—on the order of 100 to 1000 residues—so one DNA molecule must contain the coding for several peptide chains, or in genetic terms, several *cistrons*. The cistron is the unit of information for one chain, and therefore is a gene because it determines the amino acid sequence of a peptide. The sequence in turn fixes the properties, and therefore the function, of the peptide in the organism. A change in the nucleotide sequence at one position on a DNA molecule can alter the codon in a cistron, and therefore create a mutation in a gene.

Complementary Bases

DNA from the nuclei of most cells has two strands twisted into a *double helix*. This form exists because each base in one strand is matched by a base in the other strand that can form hydrogen bonds with it. However, all of the bases along the strands can only form these bonds when the strands have the particular orientation created by the helix. The formation of the bonds releases energy, and the double helix is therefore the most stable structure.

The matching bases are adenine with thymine, and guanine with cytosine (Fig. 3-3). This is a critical point in our discussion, because all of our reasoning on the mechanism of genetic transmission and on protein synthesis depends upon it. We are saying that *adenine and thymine are complementary* and *guanine and cytosine are complementary,* because they enable a stable fit of two strands of nucleic acid at regions where these pairs are matched.

The principle is so fundamental to the mechanics of reproduction and protein

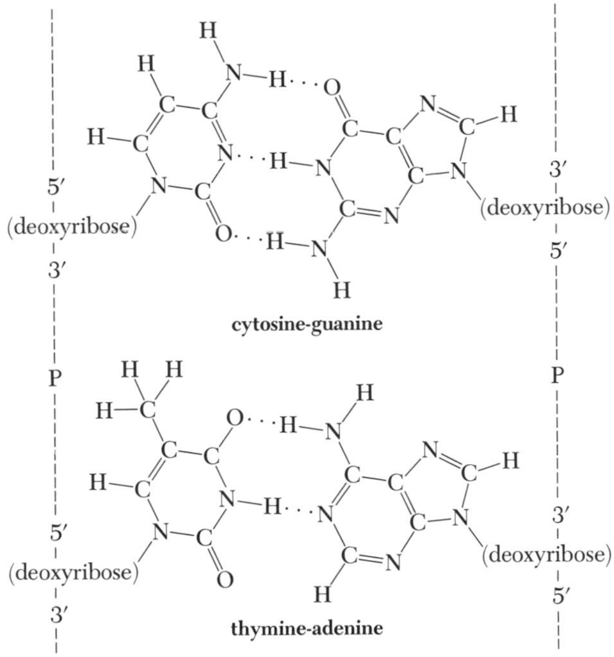

Figure 3-3 Hydrogen bonding of complementary bases between adjacent nucleic acid strands. The symbols are spaced to scale the actual positions of the atoms. Note the opposing direction of the phosphodiester bonds in the two strands being linked. Modified from Pauling and Corey, Arch. Biochem. Biophys., *65:*164 (1956).

synthesis that we appear safe in saying that these particular purines and pyrimidines are used because they have the structures necessary for the formation of the double helix and can also participate in the metabolic reactions that we shall consider later. Put another way, the double helix evidently conferred sufficient advantages for survival of the genetic line to cause preferential use of those bases enabling spontaneous formation of the structure.

The existence of the double helix in a sample of DNA can be shown in a simple way. Nucleotides absorb ultraviolet light with wavelengths near 260 nanometers, owing to the presence of the purine and pyrimidine rings. The light absorption of the rings is decreased when they are hydrogen-bonded, as in the double helix. If a solution of DNA is heated, the light absorption abruptly increases when the temperature at which thermal agitation is sufficient to disrupt the hydrogen bonds is reached. Other physical properties dependent upon the helix also disappear at this "melting point."

The two strands of the helix are actually *twisted in opposition,* so that if both strands are traced down the helix in the same direction, one will have phosphate linking the sugars 5', 3'—5', 3'—5', 3'—, whereas the other will be 3', 5'—3', 5'—3', 5'—. This makes for another problem in nomenclature, although the structural fact is simple enough. For example, if one strand has the sequence, d-pTpCpApG, within its structure, the other strand will have the complementary nucleosides dA, dG, dT, and dC in that order, looking down the complementary strand in the same direction as the first strand. However, convention demands that the strand be designated from the 5' end, so we have to reverse the order and designate the complementary segment as d-CpTpGpA. This is a case in which chemical nomenclature obscures the biological implications, because the chemical names of the two strands begin from opposite ends of the molecule.

A consequence of this pairing of bases is that double-stranded DNA ought to contain an amount of adenine equal to the amount of thymine, and an amount of guanine equal to the amount of cytosine, even with the amounts of the four bases distributed in a random fashion. Suppose that one strand has the following number of base residues:

Adenine—3800   *Thymine—2600*   Guanine—1700   *Cytosine—1900*

The number of residues of each base in the other strand will be the same as the number of its complement in the first strand:

Adenine—2600   Thymine—3800   *Guanine—1900*   Cytosine—1700

The total number in the whole molecule with both strands will be:

Adenine—6400   Thymine—6400   Guanine—3600   Cytosine—3600

This theoretical identity of composition of the complementary bases has been proven to be sufficiently close to the natural circumstances to enable it to be used as a test for the presence of double-stranded DNA from real organisms.

Replication of DNA

When a cell divides, both strands of DNA must be replicated to equip the nucleus of the additional cell. This is done by using both strands as templates on

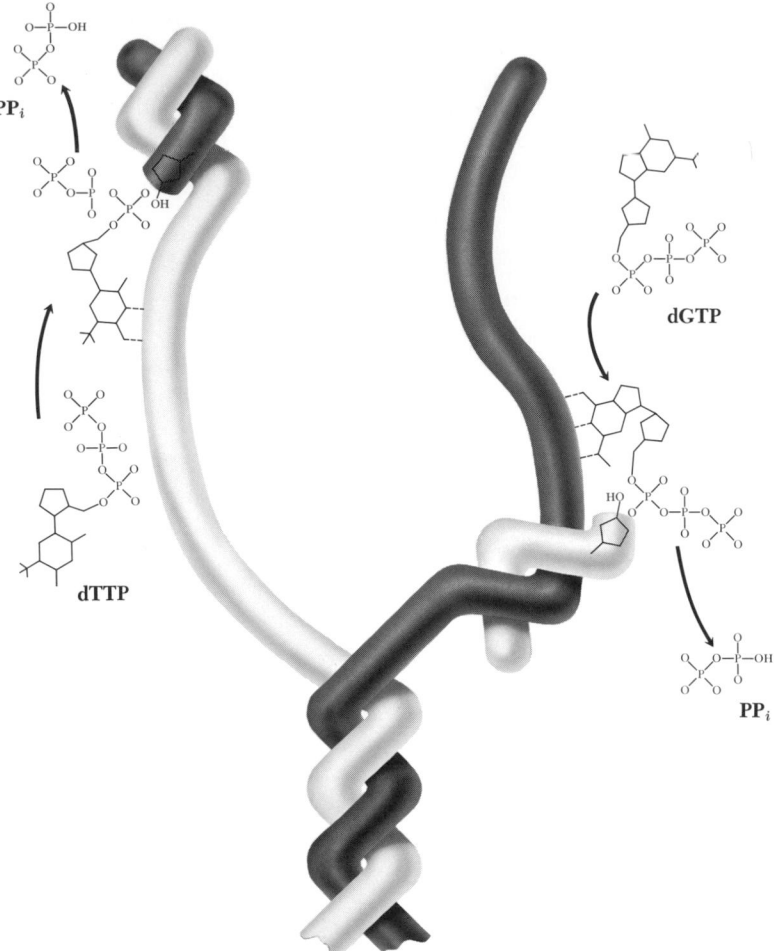

Figure 3-4 Highly schematic representation of the replication of DNA. As the strands of the parent molecule unwind, complementary nucleoside-5′-triphosphates are bound to the exposed bases, but at opposite ends of the separated strands. The triphosphates react with 3′-hydroxyl groups of the preceding nucleotide in the growing strands with the formation of a new 3′, 5′-phosphodiester linkage and the loss of inorganic pyrophosphate. As the new chains grow, new double helices are formed on each of the parent strands. The new polynucleotides are formed in segments, and are later joined in another reaction (not shown here) to make the very long finished strands. The unwinding was originally thought to occur only at the ends of the double helix, but is now believed to occur at several interior positions.

which complementary chains are formed. If we designate the original strands as A and B, and if there has been no mutation of either so that they remain exact complements, then the new strand formed on A will be exactly like B, creating a new molecule, AB′. Similarly, the complement of strand B, which is identical to strand A, will form to create A′B. The exact mechanism is unknown, but current evidence suggests something of the sort shown in Figure 3-4.

During division, the helix starts to unwind. As it does so, complementary nucleoside triphosphates from the surrounding solution are bound to each base on the strands. The nucleotides combine into new strands in a reaction catalyzed by the enzyme, *DNA polymerase,* with the release of *inorganic pyrophosphate* (PP_i). The new strands wrap into double helices with the parent strands.

The uncertain part of the mechanism has to do with the simultaneous replication of both strands. New nucleotides are added onto 3′-hydroxyl groups, but only one strand has a 3′ group at a given end of the parent molecule because the two strands are polymerized in opposing directions. Therefore, it is not possible for both strands to be reproduced in the same direction from one end by this mechanism. Figure 3-4 shows a mechanism in which one strand grows from the 3′ end, releasing the other strand until enough is free for this segment to be reproduced in the opposite direction, beginning at some point in the middle of the backbone and working toward the end. This explanation appeared unsatisfactory to some because it required one of the strands to be reproduced in unconnected segments that would have to be joined later to form a complete molecule.

This proved not to be as far-fetched as it first seemed. The DNA of some microorganisms, viruses, and of some animal organelles was shown to consist of a single-stranded ring, rather than a linear molecule. During reproduction of these molecules, the complementary nucleotides join around the ring, and a second enzyme, *DNA ligase,* is present to close the ends of the new strand (Fig. 3-5). There are similar chain-bonding enzymes in mammalian cells.

It now appears that both strands of double-stranded DNA (such as occurs in animals) are replicated in short segments, and that replication begins in the middle of the chain, not at the ends. It isn't known what triggers replication. Some believe that nicks are created in a strand by the action of the enzyme, *deoxyribonuclease* (*DNAase*), which catalyzes the hydrolysis of 3′-phosphate ester bonds. Such nicks may cause sufficient uncoiling of the double helix to expose complementary bases for the action of DNA polymerase.

In any event, the two molecules of DNA are separated between the daughter cells upon completion of division. Rearrangement of the strands may occur, but in most cases each daughter cell will now have identical molecules of doubly stranded DNA, the new molecules having one strand of parent DNA and one strand of new DNA. If there are no cell fatalities, the original two parent strands will persist separately through succeeding generations in two of the individual progeny cells, and the odds are fairly high that each of us still contains some of the original DNA strands contributed by his father and mother at conception.

The Goulian-Kornberg-Sinsheimer Experiment

M. Goulian, A. Kornberg, and R. L. Sinsheimer performed an elegant experiment to demonstrate that the processes of DNA replication we have described actually do suffice for the transmission of all of the necessary genetic information during biological reproduction (Proc. Natl. Acad. Sci. (U.S.), 58: 2321 [1967]. They used a virus, known as ΦX174 phage, that infects the microorganism *E. coli;* like all viruses, it reproduces inside the host cells. This phage contains a single-stranded molecule of DNA in the form of a closed ring, and therefore enables one to avoid the problems encountered with the linear double-stranded molecules found in most organisms when trying to separate the new complementary strands from the parent complementary strands.

The other critical feature of their experiment hinges upon the fact that phage DNA can retain its biological activity if the synthetic base, 5-bromouracil, is sub-

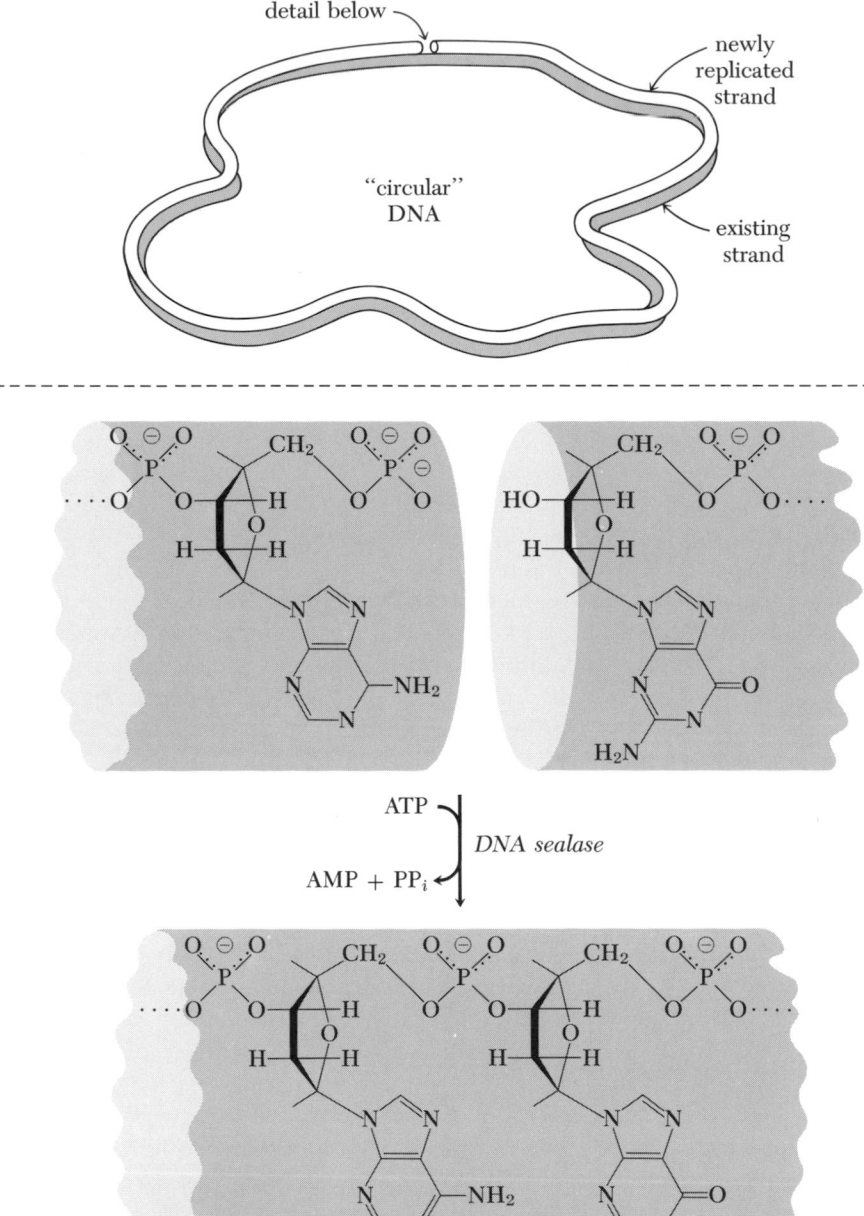

Figure 3-5 Replication of the single-stranded DNA molecules that occur in some microorganisms and in the mitochondria of animal cells. These single strands occur in closed "rings" without 3′ or 5′ terminal phosphates. A strand is replicated on the ring by a process similar to the replication of doubly stranded DNA in other organisms. However, this leaves a gap where the new strand begins at a 5′-phosphate and ends at a 3′-phosphate. The detail shows the closure of the gap in a reaction catalyzed by the enzyme, DNA sealase (also known as DNA ligase or joining enzyme). A separate molecule of ATP is also utilized as an energy source in this reaction, being cleaved to AMP and PP_i.

stituted for thymine, and DNA made with 5-bromouracil in place of thymine has a higher density because of the heavy bromine atoms.

5-bromouracil

Essentially what they did was to show that a complementary nucleotide made by DNA polymerase and closed into a ring by DNA ligase in the test tube could be separated. The new DNA was then used in the test tube to form *its* complement, which had the same biological infectivity as the DNA of the original phage harvested from natural cultures. They went about it in the following way:

The single strands of natural phage DNA, designated (+), were incubated with a mixture of the necessary deoxynucleotide triphosphates, except that 2-deoxy-5-bromouridine triphosphate was used in place of deoxythymidine triphosphate. Purified DNA polymerase and ligase enzymes were added to catalyze the reaction.

This resulted in the formation of a double ring containing a new complementary strand (−) in which bromouracil is incorporated (black circles). The double ring molecules were isolated and exposed to the enzyme, deoxyribonuclease, to catalyze the hydrolysis of 3′-phosphate ester bonds in the DNA backbone. The exposure was brief enough so that an average ring was nicked in only one position.

This treatment resulted in a mixture, with most of the molecules nicked on either the (+) or the (−) strands, or else not affected. The mixture was heated beyond the melting point of DNA and cooled. The nicked strands could not reassociate, so the mixture now contained linear strands from the nicked molecules as well as complete rings.

Mixture of nicked (+) and (−) strands, single (+) and (−) rings, and unaffected molecules. The single rings recombine to give double rings.

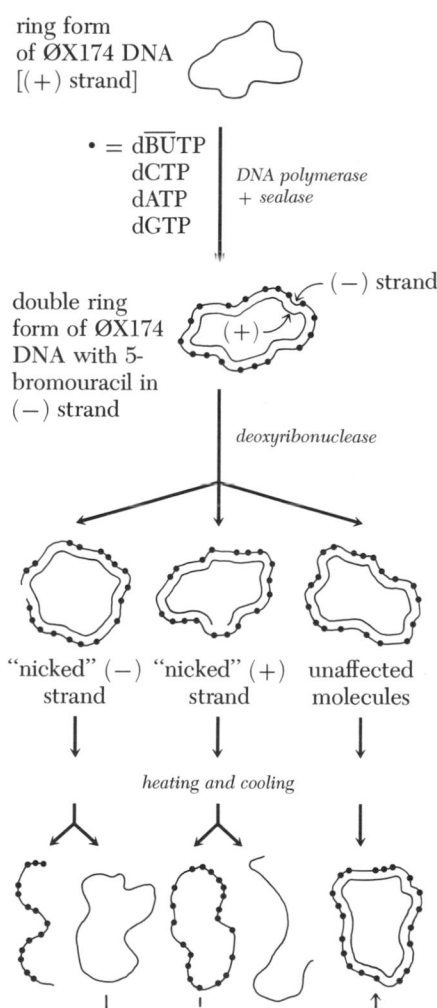

ring form of ØX174 DNA [(+) strand]

• = dB̄U̅TP
dCTP
dATP
dGTP

DNA polymerase + sealase

double ring form of ØX174 DNA with 5-bromouracil in (−) strand

(−) strand

(+)

deoxyribonuclease

"nicked" (−) strand   "nicked" (+) strand   unaffected molecules

heating and cooling

The mixture was then layered on top of a gradient of increasing salt concentration in a centrifuge tube. (Cesium chloride was used because it has high density and high solubility with few adverse effects.) Upon centrifugation in this density gradient, the single $(-)$ strands sedimented farther because their high percentage of bromine made them more dense. The single original $(+)$ strands, with no bromine, had the least density, and the double rings, with one bromine-containing strand, had intermediate density.

The bottom layer of $(-)$ strands was then separated, and this synthetic material was again incubated with a mixture of deoxynucleotide triphosphates, except that dTTP, rather than the bromouracil compound, was included, and the added CTP was labeled with radioactive tritium. A totally synthetic double ring formed, with the new $(+)$ ring incorporating radioactive cytosine (stars) to distinguish it from the bromine-labeled $(-)$ ring.

After a repetition of nicking of the rings by brief exposures to deoxyribonuclease, heat denaturation, and density gradient centrifugation, the synthetic $(+)$ strands were isolated—from the top of the gradient, since they had no heavy bromine.

Cultures of *E. coli* were inoculated with the isolated synthetic rings and were found to be infected in the same way as if the rings had been isolated from the original phage, and the infection resulted in the formation of complete virus particles.

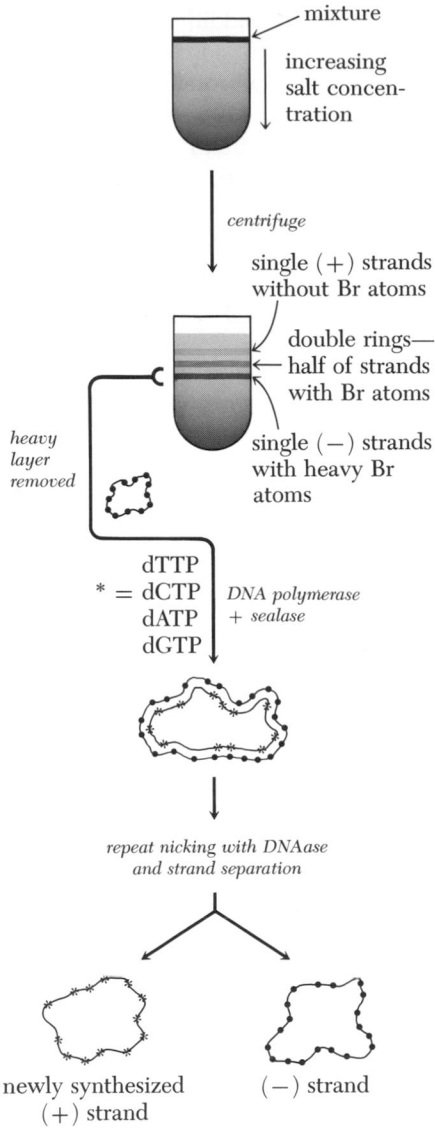

Messenger Ribonucleic Acids

When DNA is double-stranded, which is usually the case, only one of the strands is used as a source of information within the cell for protein synthesis. The inactive strand only serves as a template to make a new active strand for the daughter cell. (Of course, the active strand also is the template for making a new inactive strand, but it has the additional function of providing information during the interval between cell divisions.) The coding for amino acids is transferred from the active strand for use in protein synthesis by forming another nucleic acid, messenger RNA, complementary to the DNA. The mechanism of formation of messenger RNA is quite similar to that of DNA. The polynucleotide chain is formed from nucleoside triphosphates, with the loss of PP_i, in a reaction catalyzed by an enzyme, *RNA polymerase.*

The kinds of messenger RNA being transcribed at a particular moment determine the kinds of peptides being produced in the cell. The transcription therefore is influencing the physiological character of the cell. If all of the potential peptides were being made at once, all cells would be alike (if they could survive the resultant chaos). It turns out that only a small fraction of the total information in DNA is used by a cell at one time. Chromosomes contain DNA in combination with *histones,* which are small proteins rich in the amino acids, *lysine* and *arginine.* In the previous chapter we noted that the side chains of these amino acids carry positive charges, and the abundance of these charges on the histones readily enables them to combine with DNA, which has a negative charge on each phosphate group. It is a case of polycation combining with polyanion.

RNA polymerase will not catalyze the formation of messenger RNA when the necessary DNA is locked up in a compact complex with histones. It is quite clear that the formation of messenger RNA requires dissociation of histones to expose DNA to the polymerase, and that this dissociation is used as a primary means of regulating the types of peptides ultimately made by the cell. The DNA strands of chromosome segments visibly dissociate into puffs when the information in the particular segments is actively being transcribed. It is not clear what causes the histones to dissociate at particular points along the chromosome. We shall have more to say about the consequences of selective protein synthesis when we consider metabolic regulation in Chapter 23.

Messenger RNA differs from DNA in several important respects:

1. Some of the components are different. The sugar in the backbone is D-*ribose,* not 2-deoxy-D-ribose. *Uracil,* rather than thymine, is incorporated as the base complementary to adenine:

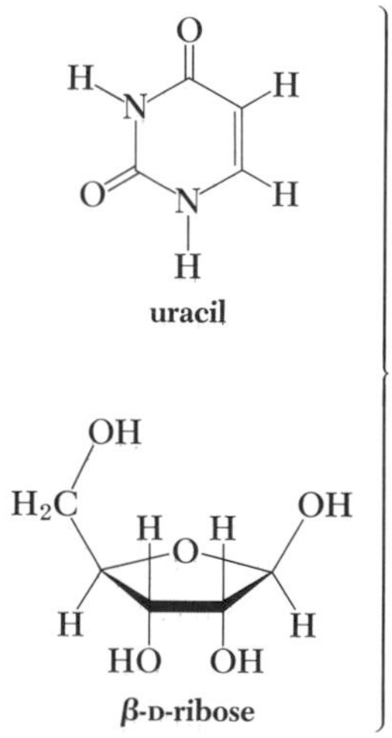

uracil

β-D-ribose

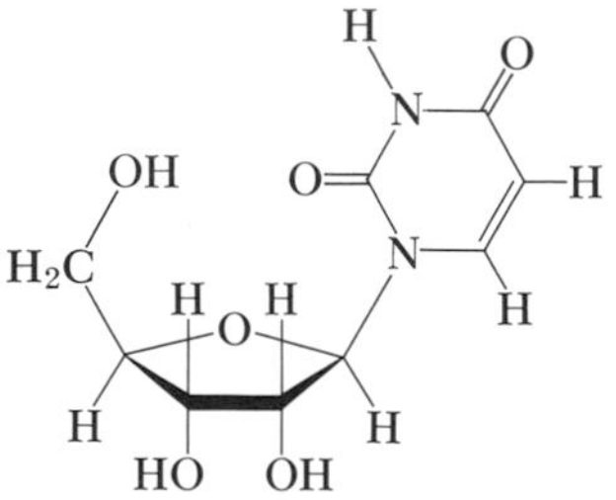

uridine

These differences probably have the evolutionary advantage of creating a chemical distinction between the stored coding transferred from generation to generation and the transient coding used in fluctuating amounts within a single cell for the immediate demands on protein synthesis.

2. The length of the polynucleotide chain is much shorter. A molecule of messenger RNA carries at most the coding for a few cistrons of related biological function. For example, the cistrons designating enzymes that catalyze successive metabolic reactions in a sequence may be on one mRNA molecule. Such a collection of several cistrons whose activity rises and falls together is called an *operon*. A molecule of DNA, on the other hand, can contain several operons. (Genetic evidence suggests that operons include some kind of coding that regulates the rate of production of messenger RNA, but chemical identification of such coding has not been made.)

3. Since messenger RNA is made only as the complement of one strand of DNA, the complement of a given kind of mRNA does not exist in the cell. Therefore, messenger RNA occurs as single-stranded molecules, not as double helices.

4. The life of a messenger RNA molecule is deliberately short. The mechanics of the turnover will be taken up in Chapter 20, but suffice it now to say that a typical messenger RNA molecule is hydrolyzed within a day. The lifetime depends upon the kind of messenger RNA and on the type of organism. The advantage of the short life and the resultant constant turnover of the various kinds of messenger RNA is that the cell is not committed indefinitely to the production of particular kinds of proteins. The turnover makes it possible to control the functions of the cell in response to environmental changes through alterations in the kinds of messenger RNA present, thereby altering the kinds of proteins being made. The adaptability conferred by this arrangement is still another reason for the existence of chemically distinct sources of information for cell division and for protein synthesis.

The codons in messenger RNA are the complements of the codons on the active strand of DNA, but read in the opposite direction and containing uridine (U) rather than thymidine (T). Thus, the codon d-CpApA specifying valine in the active DNA strand will appear as UpUpG in messenger RNA. (It is d-TpTpG in the inactive DNA strand.)

Unless otherwise specified, codon assignments for amino acids are given in the form present in messenger RNA. This convention has come about because it was convenient for the experimentalists, who have largely deduced the code from studies on changes in nucleotide composition of substitutes for messenger RNA. The complement of a messenger RNA codon, that is, the corresponding sequence found in the active DNA strand and in the transfer RNA molecules to be discussed shortly, is frequently referred to as the *anticodon*. (From a functional point of view, these designations seem reversed from what they ought to be, but it is too late to do anything about it.)

Many studies of the role of messenger RNA in protein synthesis have been made with red blood cells, which is one of the reasons we are using hemoglobin as a typical example of a protein. The advantages of the cells for such studies come from their unusual life cycle. The mature erythrocyte in mammals is a highly specialized cell, functioning largely as a bag of hemoglobin wandering through the circulation, and having low metabolic activity compared to other cells. This has permitted a little display of evolutionary thrift. Once the hemoglobin and small amounts of other proteins have been made, there is no further need for the apparatus of protein synthesis. Hence, we have the apparent paradox that the nucleus, including the DNA, starts to disappear from primitive red blood cells, the erythroblasts, when the synthesis of hemoglobin is reaching its peak. The secret lies in the relatively long life of messenger RNA in these cells. DNA can be removed, once it has served its purpose as a template, and hemoglobin formation will persist as long as the RNA is still present, which it is through the reticulocyte stage of development. The RNA also eventually disappears, and the mature cell survives for about four months without further protein synthesis because its limited metabolic activity causes little destruction of protein. The loss of the nucleus and many of the enzymes from the cell permits a large reduction in size, which may be an advantage to the function in oxygen transport, and also diminishes the demands of the cell on the oxygen supply. Erythrocytes carry oxygen, but consume little on the way, and this may be of especial benefit to large mammals with a high demand for oxygen.

In any case, the existence of reticulocytes has been a boon to students of protein synthesis because these cells contain RNA but lack DNA, so studies of protein synthesis are not complicated by changing rates of RNA formation. Furthermore, most of the messenger RNA in the cells contains coding for the peptides chains of hemoglobin, so there is not the complex mixture of RNA molecules found in most cells.

Nearly all of the messenger RNA in reticulocytes is about 150 nanometers long. One can calculate from this dimension that the molecules ought to have around 500 nucleotide residues. Now, the various peptides of hemoglobin contain 140 to 150 amino acid residues, so the theoretical number of nucleotide residues per cistron ought to be three times these numbers, or 420 to 450 residues. We can therefore deduce that the coding for one peptide chain of hemoglobin is carried on one molecule of messenger RNA.

This fact has genetic as well as structural implications, because there is genetic evidence that the cistrons for β and δ chains are part of a single operon in DNA. Complete operons evidently are not always coded on single messenger RNA molecules.

Ribosomal RNA and the Polysomes

Ribosomes are globular structures, about 23 nanometers in diameter in mammalian cells. Roughly half of their mass is ribonucleic acid, with the remainder composed of proteins. The order of nucleotides in rRNA carries no coding for amino acids, but the molecules are complementary to part of the DNA in the nucleus, meaning that they are made by using DNA as a template. It ought to be remembered that DNA must in some way direct the formation of *all* cellular components. This is done indirectly through the coding for enzymes catalyzing the formation of many components, but ribosomal RNA is directly specified by DNA. In short, not all of the genetic information in DNA represents coding for amino acids through messenger RNA. Some is used to create other, non-coding forms of RNA.

Ribosomal RNA is believed to exist only to bind messenger RNA and the particular enzymes necessary to catalyze peptide bond synthesis. Mammalian ribosomes are made in two parts, designated as the 40S and 60S components. (The commonly studied ribosomes of *E. coli* are slightly smaller, with 30S and 50S components.) Each part contains RNA (18S and 28S, respectively), along with several different proteins. Surprisingly, rRNA is made as a single chain (45S) in the nucleolus, and then hydrolyzed into at least three pieces. Two of these pieces spontaneously associate with the necessary proteins to form ribosomes.

The two pieces of ribosomes are separated at the beginning of peptide synthesis. The 40S component has an *initiator* site with the capacity for binding the initial codon on messenger RNA and the corresponding transfer RNA. The complex of the three substances—transfer RNA, messenger RNA, and 40S ribosomal component—then unites with the 60S component of the ribosomes. The 60S component provides two additional binding sites—an *aminoacyl tRNA site*, at which successive molecules bearing amino acid residues attach, and a *peptidyl site*, to which the growing peptide chain is moved.

The initial complete ribosome assembly physically is a compact mass made from the combined 40S and 60S components, with a long, thin thread of messenger RNA attached at one end, and with a tRNA molecule on its surface.

The protein components built into ribosomes have several functions. One of the proteins is necessary to attach incoming molecules of tRNA bearing successive amino acid residues to the aminoacyl tRNA site. Another is necessary to shift the ribosome along the messenger RNA as the peptide grows, thereby moving the tRNA associated with a particular codon from the aminoacyl site to the peptidyl site. This movement requires energy supplied by the hydrolysis of *guanosine triphosphate* (*GTP*). Still another protein is an enzyme, *peptide synthetase*, responsible for the actual formation of the peptide bond by moving the new aminoacyl groups from the tRNA carrying them onto the growing peptide chain. (The sequence of these steps will be more clear in a moment when we go over the process in detail.)

The ribosome moves down the messenger RNA molecule as the peptide chain grows. Once it has moved sufficiently, a second ribosome can also attach to the same molecule of messenger RNA, and begin the formation of a new peptide chain at the end while the first ribosome is still involved in decoding the middle of the messenger RNA. This can continue until the entire length of the messenger RNA is occupied by ribosomes, with separate molecules of peptide being made on each. The anatomical result is a cluster of ribosomes of a size determined by the length of the messenger RNA molecule. Such clusters, polysomes, are characteristic of cells actively synthesizing peptides.

Most of the polysomes in reticulocytes contain six or fewer ribosomes, because this is the maximum number easily accommodated on the messenger RNA coding for hemoglobin peptides. The polysomes in other kinds of cells occur in a wide range of sizes, and this would be expected since most tissues produce proteins of greatly varying chain lengths. When a cell in tissue culture is infected with polio virus, large polysomes containing as many as 60 ribosomes are formed on the very long RNA molecule found in this particular virus. (The viral RNA behaves like a messenger RNA in the host cell, dictating the formation of viral proteins in addition to its own replication.)

Transfer RNA

Knowledge of both the structure and function of transfer RNA molecules is rapidly accumulating, but some critical insights into the relation between structure and function are still lacking. Let us consider function first.

A given kind of transfer RNA carries one kind of amino acid residue, and it contains triplet coding so that it will combine with messenger RNA on the ribosome only at the location of the particular messenger codon specifying the amino acid; therefore, the bound amino acid matches the information in the messenger RNA.

However, the combination of the correct amino acid from the mixture in the *cytosol* (soluble cytoplasm of the cell) with a particular transfer RNA does not involve triplet coding. Instead, the specificity for attachment of the correct amino acid lies in enzymes catalyzing the combination. We shall discuss the general topic of enzyme specificity in Chapter 6, but let us say now that reaction of the correct amino acid to match the triplet coding depends upon specific bonding between the transfer RNA, the amino acid, and the enzyme catalyzing the reaction. The RNA-enzyme bonds depend upon the arrangement of bases in wide segments of the tRNA molecule and not on the triplet codes we have been talking about earlier.

This distinction is shown in another way. Triplet coding for genetic transfer of information is apparently constant from one organism to another over the complete range of phyla, so that the DNA of the bacterium, *E. coli*, has the same combination of three nucleotides to designate a particular amino acid as does the DNA of Marshall Nirenberg, who shared in the Nobel Prize for his contributions to elucidation of the code. (*E. coli* has proven to be a very convenient organism for studying coding. It is much easier to harvest microorganisms than Nobel Laureates, much as we might like to unravel the genetic makeup of the latter.) However, the transfer RNA of one organism cannot be substituted for the transfer RNA in all other organisms, because different arrangements of nucleotides in the regions away from the coding triplet will not bond effectively with the particular enzymes of another species.

It follows from this that the triplet coding in transfer RNA does not directly determine the attachment of the amino acid, and that the nature of the amino acid attached does not directly determine the recognition of triplets between transfer RNA and messenger RNA. The latter statement has further proof in an experiment from Fritz Lipmann's laboratory, often quoted because of its elegance, in which an amino acid residue on its proper transfer RNA was chemically altered and was found to be incorporated into peptides at the proper point for the original amino acid, rather than for the altered version (Fig. 3-6).

The starting material was cysteinyl-tRNA, normally used during protein synthesis to incorporate cysteine into peptides. When the compound was catalytically hydrogenated to remove the mercapto group, alanyl residues were created on transfer RNA molecules that ordinarily carried cysteine. Upon incubation with a system synthesizing peptides, it was found that these alanyl residues were incorporated at locations where cysteine would ordinarily appear, thereby showing that the messenger RNA-ribosome complex recognized the transfer RNA, not the amino acid attached to it.

We see that there must be at least one kind of transfer RNA for each different

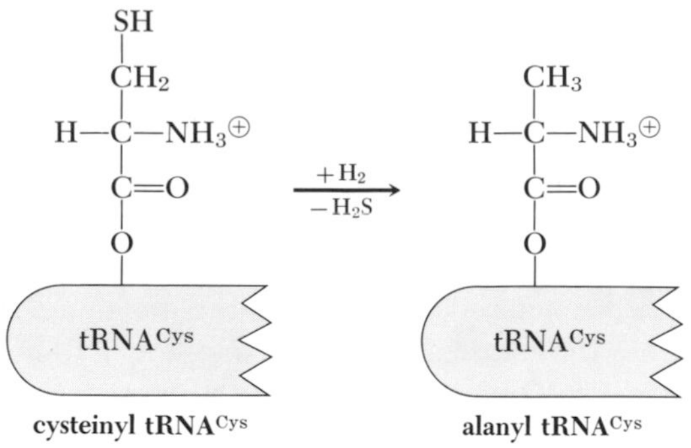

SH
|
CH_2
|
H—C—NH$_3^\oplus$
|
C=O
|
O
|
tRNA^{Cys}

$\xrightarrow[-H_2S]{+H_2}$

CH_3
|
H—C—NH$_3^\oplus$
|
C=O
|
O
|
tRNA^{Cys}

cysteinyl tRNA^{Cys} alanyl tRNA^{Cys}

Figure 3-6   Cysteinyl groups carried on transfer RNA can be converted to alanyl groups by hydrogenation in the presence of Raney nickel, but the RNA still carries coding for cysteine.

amino acid, and a specific enzyme for attaching the amino acid to the particular transfer RNA. In fact, there are several kinds of transfer RNA for most amino acids, and several specific enzymes. *The coding is redundant*—the same amino acid may be specified by more than one codon. This redundancy may be used in the regulation of the quantity of particular peptides specified by a single operon, because some kinds of transfer RNA react more slowly than do others carrying the same amino acid residue.

We have talked about the various transfer RNA's and their attachment to messenger RNA as if they reacted as single entities. In fact, there is still another small protein in the soluble cytoplasm that has the function of forming complexes with GTP and molecules of transfer RNA bearing aminoacyl residues. It is this complex that appears to unite with the aminoacyl site on ribosomes, giving up both the aminoacyl tRNA and the GTP later needed as a source of energy. For the sake of simplicity, let us continue to talk about tRNA and GTP separately, even though they may be carried together.

Turning now from function to structure, we find that the known varieties of transfer RNA are single polynucleotide chains with about 80 nucleotide units, which are also formed in the nucleolus as complements of part of DNA. *The chains always end as pCpCpA*, and this is the site for binding the aminoacyl group, which is attached as an ester with the 2′ or 3′ hydroxyl group of the terminal adenine nucleoside. (We say 2′ or 3′ because a molecule with an aminoacyl group at either position spontaneously attains an equilibrium between the two forms, with about 65 per cent being in the 3′ position at a given moment.) The attachment of the amino acid is catalyzed by an enzyme in a reaction using ATP as a source of energy (Fig. 3-7).

The portion of tRNA on which amino acids are carried is thus the same for every amino acid, but the remainder of the transfer RNA molecule must contain

---

Figure 3-7   The formation of an aminoacyl tRNA involves a reaction between the proper amino acid and adenosine triphosphate (top), releasing PP$_i$, and forming the aminoacyl adenosine monophosphate, which is a mixed anhydride of a carboxylic and phosphoric acids. The aminoacyl group is then transferred to the terminal ribosyl moiety of the corresponding tRNA (lower right). The group is shown on the 3′ oxygen (bottom center), but there is in fact an equilibration between the 2′ and 3′ positions. The AMP that is released (center left) is phosphorylated by the processes of oxidative metabolism, regenerating the original ATP.

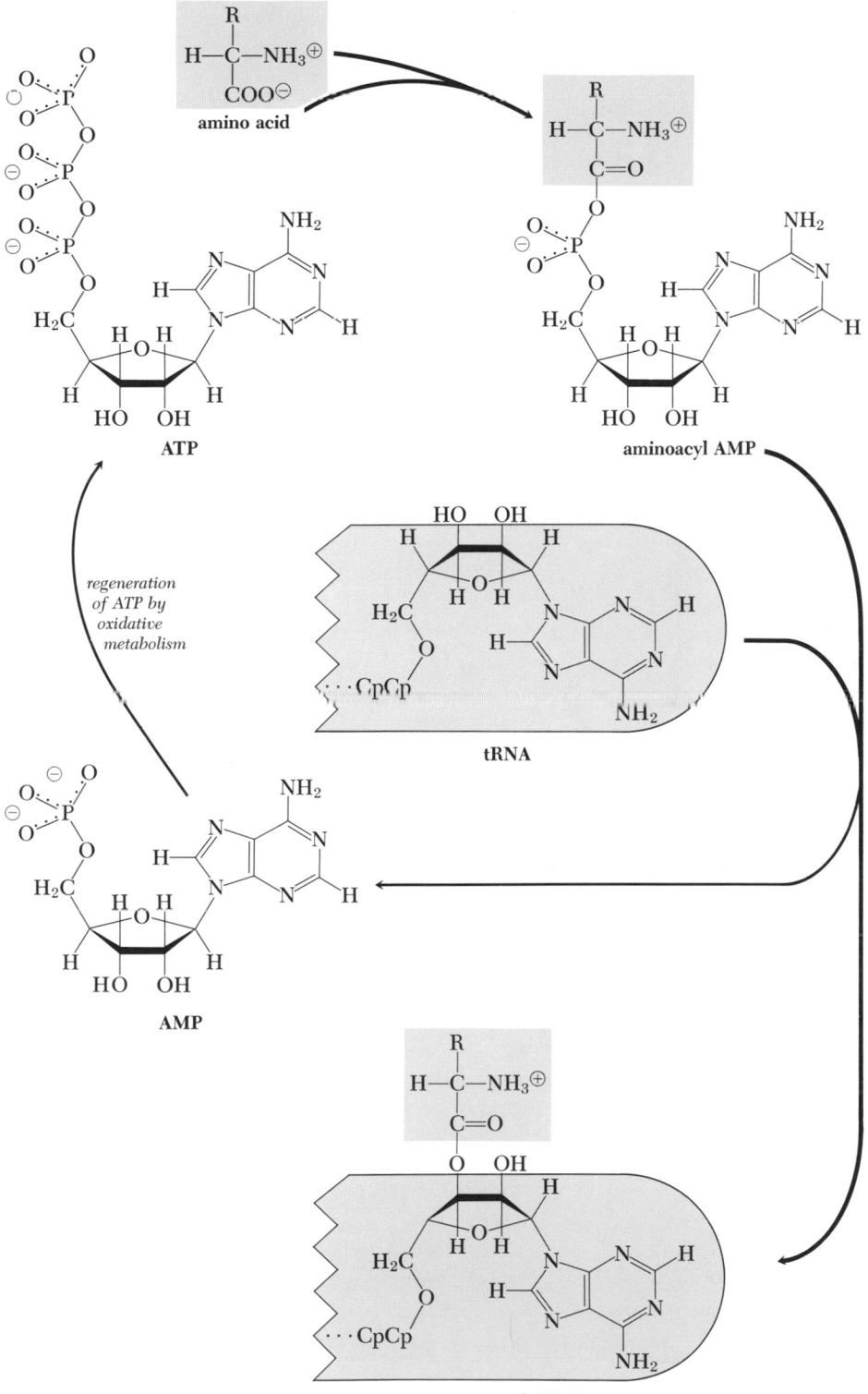

Figure 3-7 Legend on opposite page.

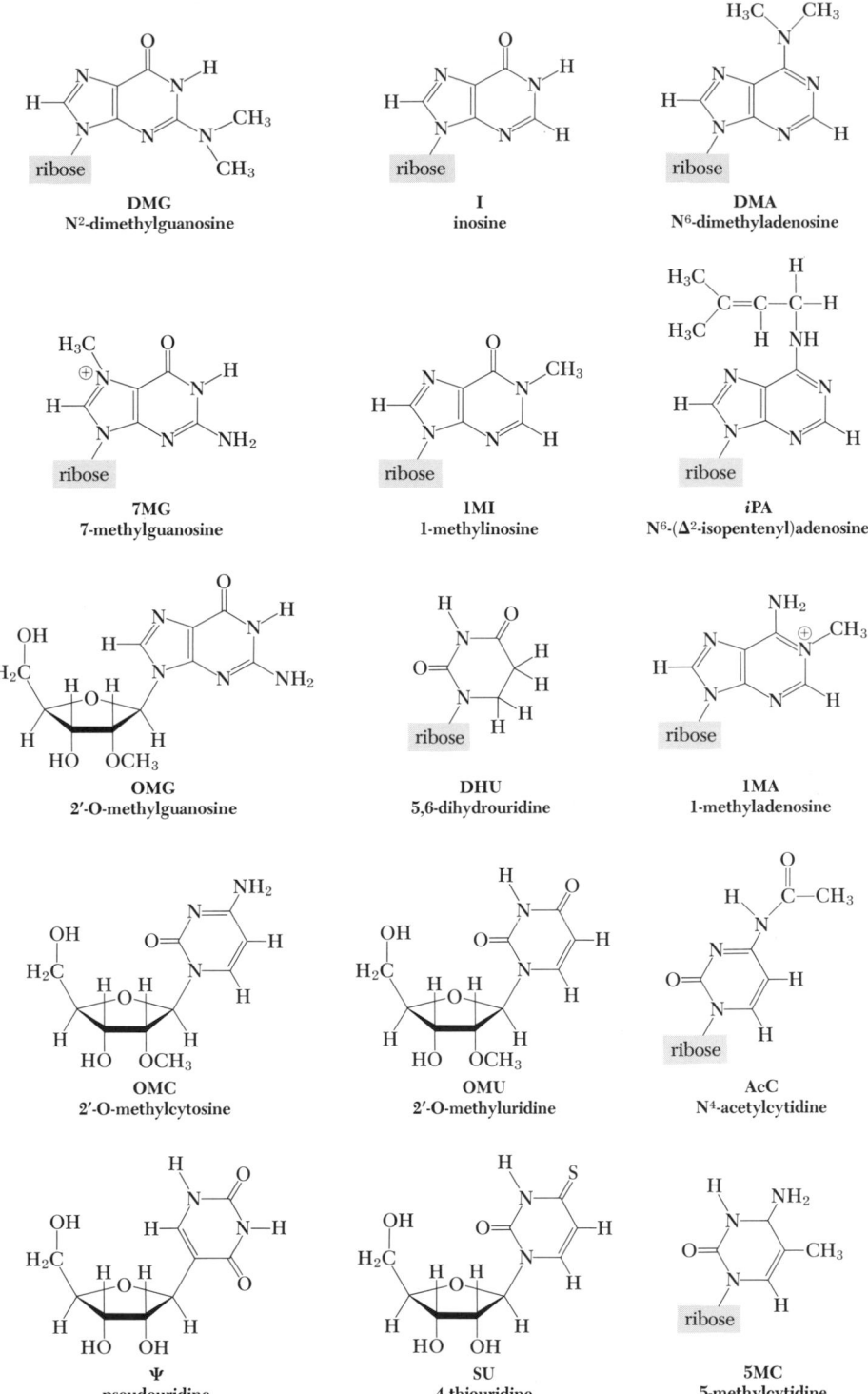

Figure 3-8   Some of the unusual bases discovered in tRNA. Not all of these occur in a single kind.

the coding triplet and the identifying structures to which the specific enzymes are bonded during attachment of the amino acid. The complete sequence of nucleotides has now been worked out for several kinds of transfer RNA. Unfortunately, most have been isolated from various kinds of yeasts, but the known properties of transfer RNA from other organisms are sufficiently like those of yeast to justify an assumption that the yeast RNA illustrates principles common to all.

Transfer RNA's contain a number of purines and pyrimidines differing from the ordinary constituents of other nucleic acids. The structures of some of these and their common abbreviations are shown in Figure 3-8. The formation of the unusual bases is discussed on p. 429. (Ribosomal RNA also contains methylated bases, but the sequences are not known.)

The transfer RNA's have a folded configuration created by internal bonding between complementary regions of bases within the molecule. The bonding frequently involves unusual bases, but our previous emphasis on the importance of complementary pairs among the four bases found in DNA ought not blind us to the realization that other purines and pyrimidines are capable of forming hydrogen bonds.

The configuration of one of the six kinds of transfer RNA for which nucleotide sequences are known is shown in Figure 3-9, and it illustrates the features shown by all:

1. All contain at least three loops of unpaired nucleotides, and the loop nearest the amino acid acceptor—the 3' terminal adenosine—invariably has the sequence TpΨpC in a seven-membered loop, with the T being the twenty-third nucleoside from the 3' end.

2. The triplet anticodon is in the center of the seven-membered middle loop, preceded by U and followed by a modified A. Dimethylguanosine is invariably the nucleoside in the eighth position before the triplet anticodon.

3. The loop closest to the 5' end has a variable number of unpaired nucleotides, but these bases are mainly dihydrouracil and various purines.

Some of these features may be peculiar to yeasts, but the process of bonding to ribosomes and messenger RNA is common to all cells, and some of the fixed structural elements may always be necessary for the bonding or for the recognition of the anticodon.

Anticodons in Transfer RNA

There is an additional complication to triplet coding in transfer RNA: unusual bases are also used in the coding. The table at the top of page 41 lists the amino acids carried by the tRNA's of known structure, the various messenger RNA triplets known to code for that amino acid, the exact complements of those triplets using the common bases, and the observed anticodons. (Remember that the exact complements and the observed anticodons are reversed from the order in mRNA codons because of the opposite direction of complementary chains.)

The striking observation is that none of the observed anticodons exactly matches the complement of any one of the messenger RNA codons known to cause the incorporation of the particular amino acid.

A little closer inspection shows that the terminal bases of the anticodons (corresponding to the initial bases of the codons) are exact complements, and the middle base is also an exact match, if one equates pseudouridine with uridine.

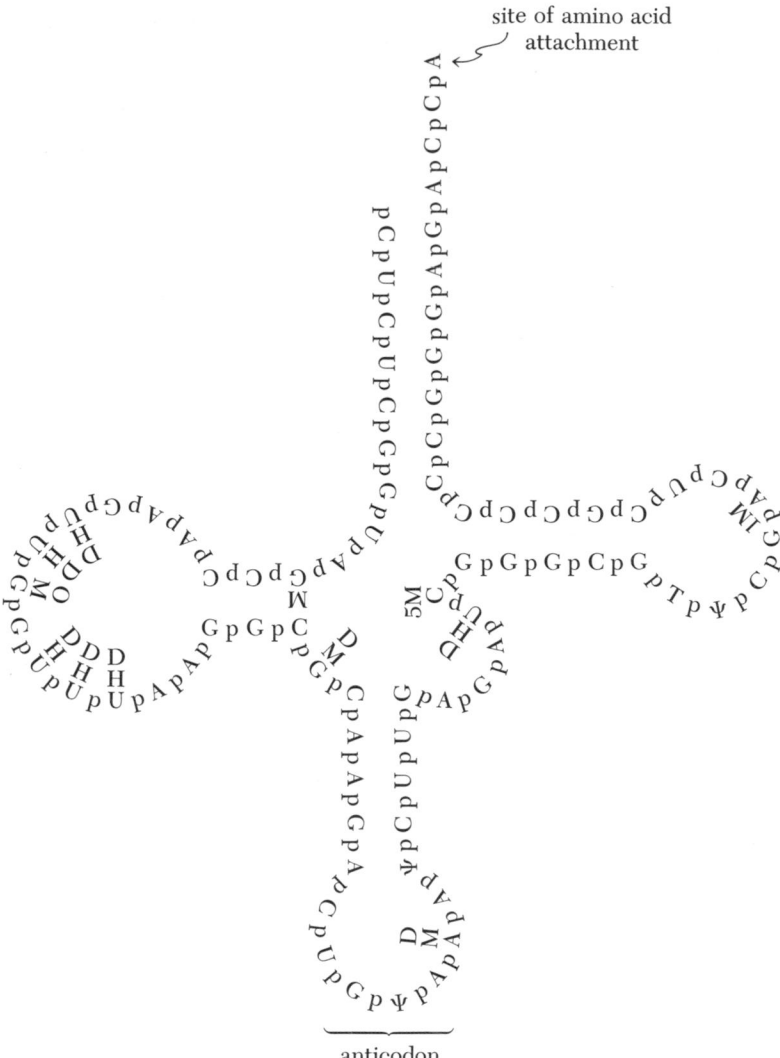

Figure 3-9   The probable base pairings in a tyrosine-specific tRNA from yeast. Modified from J. T. Madison, G. A. Everett, and H. K. Kung, in Cold Spring Harbor Symposia on Quant. Biol., *31:* 412 (1966). The structure is stretched out to illustrate the primary sequence; the actual three dimensional geometry is folded upon itself in more compact form. DHU = 5,6-dihydrouridine; DMA = N^6-dimethyladenosine; DMG = N^2-dimethylguanosine; 1MA = 1-methyladenosine; 5MC = 5-methylcytidine; MG = N^2-methylguanosine; OMG = 2′-0-methylguanosine; Ψ = pseudouridine (see Fig. 3-8 for structures of these bases).

However, the question remains as to which of the messenger RNA codons will bind with the transfer RNA codons having the unusual bases. Experimental tests showed that the alanine transfer RNA will bind with *all* of the mRNA triplets coding for alanine except GpCpG. This can only mean that the inosine of the anticodon bonds well with all of the coding nucleosides except guanosine. Tyrosyl transfer RNA will bond with either of the tyrosine mRNA codons, but will not bond with

Transfer RNA for:	mRNA Codons $\longrightarrow$	Complements $\longleftarrow$	Observed Anticodons $\longleftarrow$
Alanine	GpCpA GpCpG GpCpC GpCpU	UpGpC CpGpC GpGpC ApGpC	IpGpC
Phenylalanine	UpUpC UpUpU	GpApA ApApA	OMGpApA†
Serine°	ApGpC ApGpU UpCpC UpCpU UpCpA UpCpG	GpCpU ApCpU GpGpA ApGpA UpGpA CpGpA	IpGpA
Tyrosine	UpApC UpApU	GpUpA ApUpA	GpψpA†

° Two serine-specific transfer RNA's have been characterized, but both have the same anticodon.
† OMG is 2'-O-methylguanosine and ψ is pseudouridine (see Fig. 3-11).

UpApA or UpApG, indicating that the guanosine of the anticodon is also selective against adenosine or guanosine, but bonds well with either uridine or cytidine.

Considerations of this kind have led to the formulation of the *"wobble hypothesis,"* which was developed on the assumption that the first and second nucleosides of the messenger RNA codon are absolutely specific in the formation of hydrogen bonds with the corresponding complements in the anticodon. G is always paired with C, and U with A, and so on. Given these fixed positions, it was then possible to deduce what *other* bases in the third nucleoside of mRNA could also bond with particular bases of the anticodon, if the anticodon base could move through short arcs ("wobble"). The results of the study were as follows:

Anticodon Nucleosides	will pair with	Codon Nucleosides
A°		U
G		U or C
I		U or C or A
C		G
U		A or G

° Anticodons containing A in the first position are believed to be rare. The upshot is that only those mRNA codons ending in G can uniquely specify an amino acid (neglecting the possibility that A might sometimes occur in the first position of anticodons). Those terminating in U, C, or A can be replaced by some other codon that will bind the same anticodon. In other words, the 64 possible codons demand something less than 64 anticodons to bind the proper amino acids.

We shall consider the complete array of codons in the next chapter after becoming familiar with the structures of the remaining amino acids.

FORMATION OF THE PEPTIDE CHAIN

Having now considered the various nucleic acids and the nature of coding, let us turn to our present understanding of the actual events of peptide synthesis. More particularly, let us describe the steps by which the α chain of hemoglobin is formed, as an example of the general process followed in all sequences. The chain begins with the sequence, Val-Leu-Ser · · ·

The messenger RNA bearing the cistron for the chain must be present. Before going further, we have encountered an unsolved problem. How does the messenger RNA attach to the ribosome so as to begin at the beginning? Microorganisms solve the problem by having a special codon, ApUpG, at the beginning of each messenger RNA to signal the incorporation of the modified amino acid, N-formylmethionine. (The same codon causes the incorporation of ordinary methionine in the middle of the chain.)

$$
\begin{array}{cc}
O & COO^{\ominus} \\
\| & | \\
H-C-N-C-H \\
H CH_2 \\
CH_2 \\
S \\
CH_3
\end{array}
$$

N-formyl-L-methionine

Protein synthesis will only begin when N-formylmethionyl-tRNA is bound to this initial codon. After the entire peptide is made, the N-formylmethionyl residue is removed from it, so that the finished peptide then has the amino acid designated by the *second* codon in mRNA as the initial residue. The beginning ApUpG codon therefore has as its only apparent function the insurance that the coded message will be read from its beginning.

However, efforts to show the same type of mechanism involving N-formylmethionine in animals have failed. On the contrary, there is positive evidence that formation of hemoglobin peptides begins with the first residue of the finished peptides, the valyl residue. This doesn't mean there is no special initiating codon in animal mRNA. Current evidence suggests that the initiator codons bind particular kinds of tRNA that carry no amino acid residues, and are used only to get the ribosomes assembled at the beginning of the mRNA molecule. We shall assume this is indeed correct.

Step 1. An initiator tRNA binds to messenger RNA on a 40S ribosome unit at the position of the initiator codon. Attachment can only occur at the 5′ terminal, corresponding to the peptide N-terminal. This prevents peptide formation from beginning in the middle of the desired sequence.

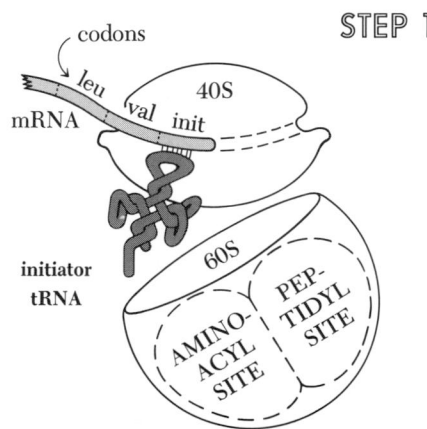

STEP 1

STEP 2

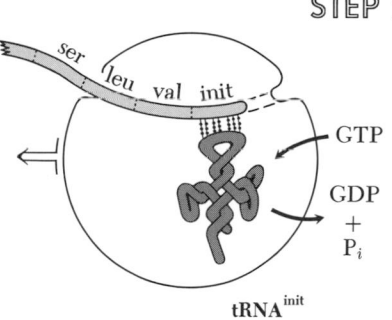

Step 2. The 60S ribosome unit attaches to the complex, and a transferase protein carried by the unit causes the ribosome to shift on mRNA by the length of one codon. The shift moves the initiator tRNA from the aminoacyl site to the peptidyl site, and requires energy supplied by hydrolysis of GTP.

STEP 3

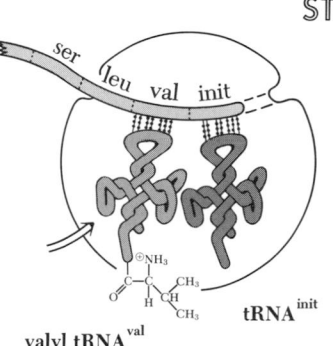

Step 3. A transfer RNA molecule carrying a valyl residue binds to the ribosome at the aminoacyl site, which is now adjacent to the valyl codon on messenger RNA. Attachment requires another transferase protein. The figure does not show that the incoming tRNA also is bound to a protein carrying GTP.

STEP 4

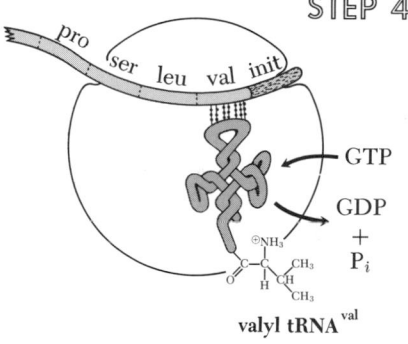

Step 4. The ribosome is again shifted by transferase action, causing the valyl tRNA to be moved from the aminoacyl to the peptidyl site. Initiator tRNA (not shown) is displaced from the ribosome by the shift, which also requires GTP.

STEP 5

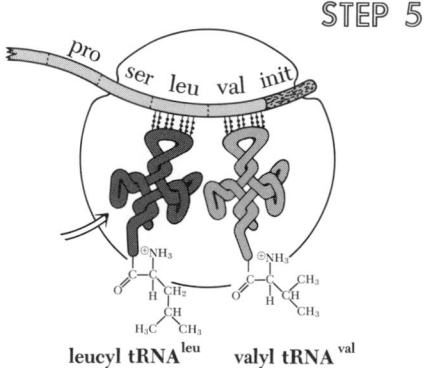

Step 5. A molecule of leucyl tRNA attaches to the now-vacant aminoacyl site, which has moved next to the leucyl codon on messenger RNA. The first two amino acid residues of the chain are in place, but each on its respective tRNA.

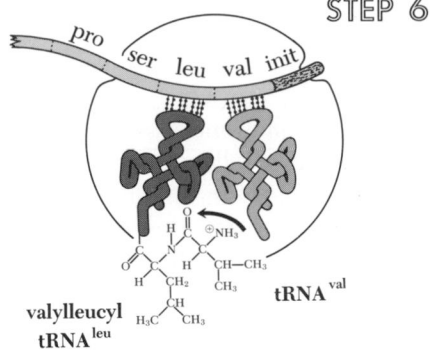

Step 6.   An enzyme, peptide synthetase, catalyzes the transfer of the valyl group from its transfer RNA onto the amino group of the leucyl residue, thereby forming the first peptide bond.

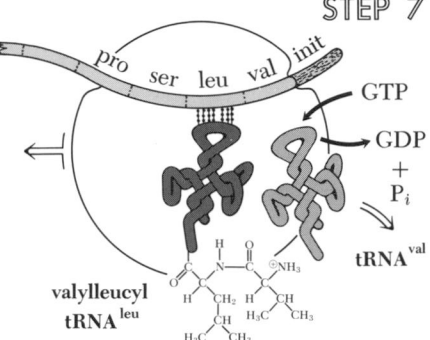

Step 7.   The valine-specific tRNA (tRNA^{val}) no longer has an aminoacyl residue attached, and is displaced by the valylleucyl tRNA when the ribosome shifts. Transfer RNA molecules without attached amino acids are only weakly bound, and the peptidyl site has an affinity for peptidyl tRNA, minimizing accidental displacements during ribosome shift.

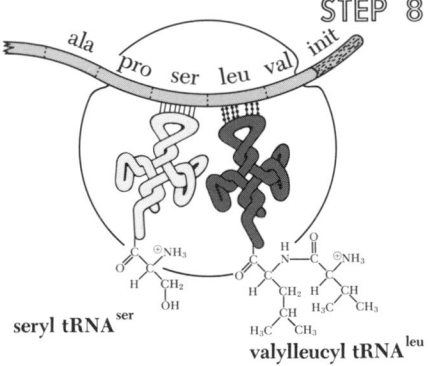

Step 8.   The aminoacyl site is again vacant, and a molecule of seryl tRNA now binds to the Ser codon present on messenger RNA next to the site.

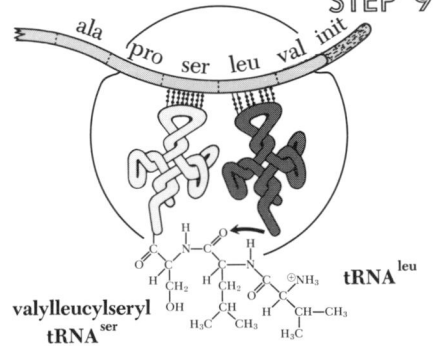

Step 9. Peptide synthetase catalyzes transfer of the valylleucyl group onto the newly-arrived seryl group, forming the second peptide bond.

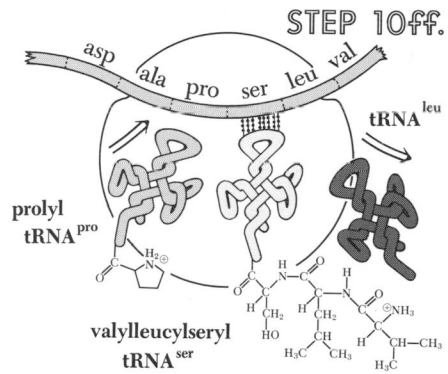

STEP 10ff.

Step 10 ff. The vacant tRNA^{leu} is displaced by movement of the ribosome, which brings valylleucylseryl tRNA^{ser} onto the peptidyl site. This also moves the aminoacyl site next to the fourth codon, and the entire process is repeated.

These steps will go on until the final . . . Lys-Tyr-Arg residues have been added. However, this does not complete the process. The residues for the α chain will be linked, but the peptides will still be attached to tRNA^{arg} on the peptidyl site, in contact with the Arg codon on messenger RNA.

Beyond the Arg codon is one of two codons, either UpApA or UpApG, which do not code for an amino acid, but signal the termination of a chain. When they appear at the aminoacyl site, a peptide-releasing enzyme is somehow activated so as to cleave the peptide-tRNA bonds, liberating the free peptide.

These punctuating codons are probably even more important within DNA and polycistronic mRNA molecules, keeping the coding for different peptides separated.

The liberation of the peptide is also accompanied by the release of the ribosome from its combination with messenger RNA. The release is accompanied by splitting of the ribosome into its 30S and 50S components in bacteria, which are then free to be used again in initiating peptide synthesis on another molecule of messenger RNA. Events in mammalian cells are not so clear. There are some free 80S ribosomes in these cells, and it has recently been suggested that the complete ribosome is released upon termination of the peptide chain, followed by an energy-requiring cleavage into the 40S and 60S components for chain initiation. If this is so, it introduces another potential process that can be regulated to control the rate of protein synthesis.

Current Problems

Our discussion makes it clear that there is much to be learned about the chemical structures or physical arrangements that define the beginning of cistrons and regulate their transcription. This is especially true of the replication of DNA during cell division and the formation of mRNA for peptide synthesis.

One fact we haven't mentioned may be a clue. In addition to the usual coding bases, thymine, cytosine, adenine, and guanine, DNA contains small amounts of 5-methylcytosine. This base, which bears the same relation to cytosine that thymine does to uracil, conceivably could be used as a marker to define divisions between cistrons.

There must be some kind of marker for cistrons, whatever its nature, and it seems reasonable to speculate that these markers are used to indicate the separate segments of DNA replicated during cell division, as well as the initial points for the formation of mRNA between divisions. Such special arrangements may also have the structures required for binding histones. We shall have to wait for the results of experimentation presently being conducted in a highly competitive atmosphere by a number of laboratories to see what the facts really are.

ANTIBIOTIC INHIBITORS OF PROTEIN SYNTHESIS

Selective control of protein synthesis is a valuable tool for a number of purposes. One is the control of malignant growth. Another is the control of invading microorganisms. Still another is as an experimental device.

A number of microorganisms produce *antibiotics*, substances that inhibit the life or growth of other organisms. In many cases, the inhibition is due to an interference with one or more of the steps in protein synthesis that we have outlined. The use of such antibiotics in the treatment of malignancies is under active study in a number of laboratories; some have proved to be very useful, and the results are provocative enough to cause renewed search for better agents. The problem is that malignant cells are derived from the normal cells of a patient, and have quite similar apparatus for protein synthesis. In such circumstances, it is difficult to find an agent that will be more lethal to the malignant cells than to the normal cells.

The success of antibiotics in treating infections is well known. The mechanism of protein synthesis is sufficiently different in microorganisms to make it easier to find agents affecting these organisms with little disturbance of protein synthesis in animal cells.

Let us consider a few of the antibiotics illustrating the various sites of action.

Inhibition of DNA Replication

A compound preventing the replication of DNA might be expected to prevent the division of cells without necessarily disturbing the formation of messenger RNA and later stages of protein synthesis. *Mitomycin* is an example of such a compound. It causes the formation of covalently bonded cross-links between DNA strands and prevents the separation necessary for replication. Mitomycin is therefore a prime candidate for trial in the control of the rapidly dividing cells in malignancies. Unfortunately, mitomycin affects other rapidly dividing tissues and in time is toxic to even the non-dividing cells, because it prevents the repair of the accidental breaks that are always occurring to DNA. Present clinical opinion is that the antibiotic is only of limited therapeutic value. However, it has limited application as an experimental tool for preventing DNA formation while permitting RNA formation. We shall consider more effective means of interfering with cell division in malignant tissues in Part IV.

Inhibition of Messenger RNA Formation

Actinomycin D is an example of a compound associating with the surface of DNA in such a way that the formation of messenger RNA cannot proceed. Antibiotics of this kind are generally toxic. They would be expected to have their greatest effects on cells dependent upon a rapid turnover of mRNA for maintenance of metabolic balance. Actinomycin has proven to be a valuable therapeutic agent in controlling a few special kinds of tumors, for example, Wilms' tumor. Wilms' tumor is a malignancy developing in the kidneys of children, being the most common malignancy of early childhood. It is an example of an *embryoma,* a tumor containing sufficiently undifferentiated cells that it may develop a variety of cell types. Other embryomas, such as occur in the testis, are also amenable to actinomycin therapy.

Actinomycin is also a potent experimental tool because it permits one to determine if changes in the content of particular proteins are results of changes in the rate of mRNA formation. Actinomycin prevents the formation of more mRNA, but commonly has little effect on the formation of new peptides using existing mRNA.

Inhibition of Ribosome Interactions

The mechanisms for chain initiation and the character of the ribosomes are different in bacteria and animals, and the most useful of the antibiotics—those effective against infections—frequently act by selectively combining with one or more of the ribosome components in microorganisms. Antibiotics effective by this mechanism include *tetracycline, streptomycin,* and *chloramphenicol.* Tetracycline, for example, prevents the combination of an aminoacyl tRNA with the initiator site on the 30S component. Streptomycin combines with the 50S component in a way that prevents normal combination with the 30S component during chain initiation.

Prevention of Peptide Chain Formation

As might be expected, compounds preventing normal formation of peptides are generally toxic and of limited clinical value. *Puromycin* is a compound that mimics the binding sites of tRNA and contains a free ammonium group to which a growing peptide chain can be transferred. However, the compound lacks the groups necessary to bind to the other enzymes of the ribosome complex, and it dissociates, carrying with it the aborted peptide chain.

Puromycin is also a valuable experimental tool because it enables one to prevent peptide synthesis at the final stages. If some physiological change is not sensitive to puromycin, then it does not depend upon the formation of new protein. If a change is sensitive to puromycin, but not to actinomycin, then it follows that the change is the result of an alteration of the rate of peptide formation using existing messenger RNA, not of the rate of messenger RNA formation.

Recapitulation of types of reactions

1. Nucleoside triphosphates may condense to form polynucleotides with the liberation of inorganic pyrophosphate:

$$n(\text{HO}-\text{R}-\text{O}-\text{P}_3\text{O}_9{}^{4-}) + \text{HO}-\text{R}'-\text{O}-\text{P}_3\text{O}_9{}^{4-} \longrightarrow$$
$$\text{H}-(\text{O}-\text{R}-\text{O}-\text{PO}_2-)_n{}^{-}\text{R}'-\text{O}-\text{P}_3\text{O}_9{}^{4-} + n\text{H}^+ + n\text{PP}_i^{3-}$$

Examples: The formation of DNA and of RNA.

2. Segments of polynucleotide may be joined with the simultaneous cleavage of an extra molecule of ATP:

$$—R—O—PO_3{}^{2-} + HO—R'— + ATP^{4-} \longrightarrow$$
$$—R—O—PO_2—O—R'— + AMP^{2-} + PP_i{}^{3-}$$

Example: The DNA ligase reaction.

3. Amino acids may react with ATP to form aminoacyl AMP, with the liberation of inorganic pyrophosphate:

$$
{}^+H_3N—R—COO^- + ATP^{4-} \longrightarrow {}^+H_3N—R—\overset{\overset{\displaystyle O}{\|}}{C}—O—AMP^- + PP_i{}^{3-} + H^+
$$

Example: The first step in the aminoacyl tRNA synthetase reaction.

4. Aminoacyl AMP may react with tRNA to form aminoacyl tRNA:

$$
{}^+H_3N—R—\overset{\overset{\displaystyle O}{\|}}{C}—O—AMP^- + (tRNA) \longrightarrow {}^+H_3N—R—\overset{\overset{\displaystyle O}{\|}}{C}—O—(tRNA) + AMP^{2-} + H^+
$$

Example: The second step in the aminoacyl tRNA synthetase reaction.

5. Aminoacyl tRNA molecules may react to form peptidyl tRNA, with a simultaneous hydrolysis of GTP:

$$
GTP^{4-} + {}^+H_3N—R—\overset{\overset{\displaystyle O}{\|}}{C}—O—(tRNA) + {}^+H_3N—R'—\overset{\overset{\displaystyle O}{\|}}{C}—O—(tRNA') \longrightarrow
$$
$$
{}^+H_3N—R—\overset{\overset{\displaystyle O}{\|}}{C}—NH—R'—\overset{\overset{\displaystyle O}{\|}}{C}—O—(tRNA') + tRNA + GDP^{3-} + P_i{}^{2-} + 2H^+
$$

Example: The peptide synthetase reaction.

Further reading

Frisch, L., ed.: *The Genetic Code.* Cold Spring Harbor Symp. Quant. Biol., *31* (1966). This volume contains a wealth of papers on work by many of the leading experimentalists in the entire field of protein synthesis.

Moldave, K.: *Nucleic Acids and Protein Biosynthesis.* Ann. Rev. Biochem., *34:* 419 (1965).

Singer, M. F., and P. Leder: *Messenger RNA: An Evaluation.* Ann. Rev. Biochem., *35:* 195 (1966).

Borek, E., and P. R. Srinivasan: *The Methylation of Nucleic Acids.* Ann. Rev. Biochem., *35:* 275 (1966).

Schweet, R., and R. Heintz: *Protein Synthesis.* Ann. Rev. Biochem., *35:* 723 (1966).

Novelli, G. D.: *Amino Acid Activation for Protein Synthesis.* Ann. Rev. Biochem., *36:* 419 (1967).

Osawa, S.: *Ribosome Formation and Structure.* Ann. Rev. Biochem., *37:* 109 (1968).

Stent, G.: *The Operon: On Its Third Anniversary.* Science, *144:* 816 (1964).

Crick, F. H. C.: *Codon-Anticodon Pairing: The Wobble Hypothesis.* J. Molec. Biol., *19:* 548 (1966).

Dayhoff, M. O., and R. V. Eck: *Atlas of Protein Structure and Sequence 1967–68.* Natl. Biomed. Res. Found. (This compendium also has base sequences for tRNA.)

Garen, A.: *Sense and Nonsense in the Genetic Code.* Science, *160:* 149 (1968).

Bonner, J., et al.: *The Biology of Isolated Chromatin.* Science, *159:* 47 (1968).

Kornberg, A.: *Active Center for DNA Polymerase.* Science, *163:* 1410 (1969).

Gottlieb, D., and P. D. Shaw, eds.: *Antibiotics.* Vols. 1 and 2. Springer-Verlag (1967).

The following papers provide leads to recent developments not covered in the preceding review articles. Those with limited time are advised to follow the Proceedings of the National Academy of Science as the best single source for current advances in this area. (This is not true of other areas.)

Colombo, B., C. Vesco, and C. Baglioni: *Role of Ribosomal Subunits in Protein Synthesis in Mammalian Cells.* Proc. Natl. Acad. Sci., *61:* 651 (1968).

Traub, P., and N. Nomura: *Structure and Function of* Escherichia coli *Ribosomes.* J. Molec. Biol., *40:* 391 (1969).

Skogerson, L., and K. Moldave: *Characterization of Aminoacyltransferase II with Ribosomes.* J. Biol. Chem., *243:* 5534 & 5361 (1968).

Parsons, J. T., and K. S. McCarty: *Rapidly Labeled Messenger Ribonucleic Acid-Protein Complex of Rat Liver Nuclei.* J. Biol. Chem., *243:* 5377 (1968).

Okazaki, R., et al.: *Mechanisms of DNA Chain Growth. I. Possible Discontinuity and Unusual Secondary Structure of Newly Synthesized Chains.* Proc. Natl. Acad. Sci., *59:* 598 (1968).

Luzzatto, L., D. Apirion, and D. Schlessinger: *Mechanism of Action of Streptomycin.* Proc. Natl. Acad. Sci., *60:* 873 (1968).

GLOBULAR PROTEINS: STRUCTURE IN HEMOGLOBIN AND ITS RELATION TO FUNCTION

CHAPTER 4

ARGUMENT

The four peptide chains of hemoglobin nest together into a compact molecule with the four heme groups toward the outside. Each of the four peptide chains is folded in such a way as to make this structure because it is the most stable of the possible configurations. The complex folding is created by a combination of relatively simple geometrical arrangements found in many proteins. Among these is the α-helix, in which the peptide chain is twisted into a cylindrical coil with the side chains directed toward the outside of the coil. The helix is created by hydrogen bonding between adjacent turns of the peptide chain. A helix may be prevented from forming in two ways. The residue of the amino acid, proline, contains a heterocyclic ring that cannot form the bond angles necessary in a continuous α-helix, and the presence of this residue necessitates an interruption of the helix. If some configuration of the peptide chain other than the helix favors bond formation with even greater energy loss, such a configuration will be more stable than the helix, and will be preferentially formed. Non-helical bonding may include hydrophobic interactions between non-polar side chains, hydrogen bonding with side chains, or van der Waal's interactions between closely packed segments of the chain.

Many of these interactions are especially well demonstrated in the heme pocket, a structure peculiar to hemoglobin, which is created by the particular folding of the peptide chain. The interior of the pocket is hydrophobic, and the insertion of the hydrophobic porphyrin ring excludes water, with a consequent loss of energy. Other features of the pocket geometry cause additional losses of energy upon combination with heme, so that the porphyrin-peptide complex is very stable. The pocket also includes a pair of histidyl residues, one that chelates with the heme iron and one that stabilizes a complex with oxygen.

Peptide subunits join to make the tetrapeptide because of interactions between side chains of residues exposed at the meshing surfaces; these interactions are of the same kind as those found in the interior of the peptides. Random association is in part prevented by a central core of polar side chains, which form a pocket filled with water.

The entire molecule is built to fulfill the biological requirements for oxygen transport; it can release oxygen at relatively high pressures so that the concentration of gas can be maintained at the elevated level necessary for rapid diffusion into tissues. This cannot be accomplished with a single peptide unit. Tetrapeptides can interact in such a way that the equilibrium between hemoglobin and oxygen will result in nearly complete formation of oxyhemoglobin in the lungs and nearly complete release of the oxygen to the tissues when the demand is high. The molecule is also constructed to release more oxygen in response to a drop in pH, such as that caused by products of an active metabolism, and this response, known as the Bohr effect, makes more oxygen available during peak demand.

It is necessary for hemoglobin to have a high solubility because it is concentrated within erythrocytes, and the high solubility is achieved by the compact globular shape with polar groups distributed on the outside to interact with water.

The origin of the structure in the amino acid composition of the peptides is confirmed by genetic alterations. The composition varies among species, but the critical bonding interactions are preserved by the substitution of related amino acids. For example, some positions in the peptide chains are always occupied by bulky non-polar amino acids. Among the many known human variants, only those substitutions in which the critical interactions are not affected result in fully functional molecules. The tendency to preserve structure upon mutation suggests a relationship between the genetic code and structure. The known human variants are all consistent with single alterations of triplet codons. The code is made in such a way that single alterations will frequently result in the substitution of an amino acid of a similar character with less likelihood of serious disruption of structure due to the mutation. It is possible that the code has evolved so that conservation of function is inherent in it.

THE SHAPE OF HEMOGLOBIN

The shape of the hemoglobin molecule is shown in an idealized sketch in Figure 4-1. We see that the four peptide chains are folded in a somewhat complex way and fit together into a compact structure roughly resembling a tetrahedron. The

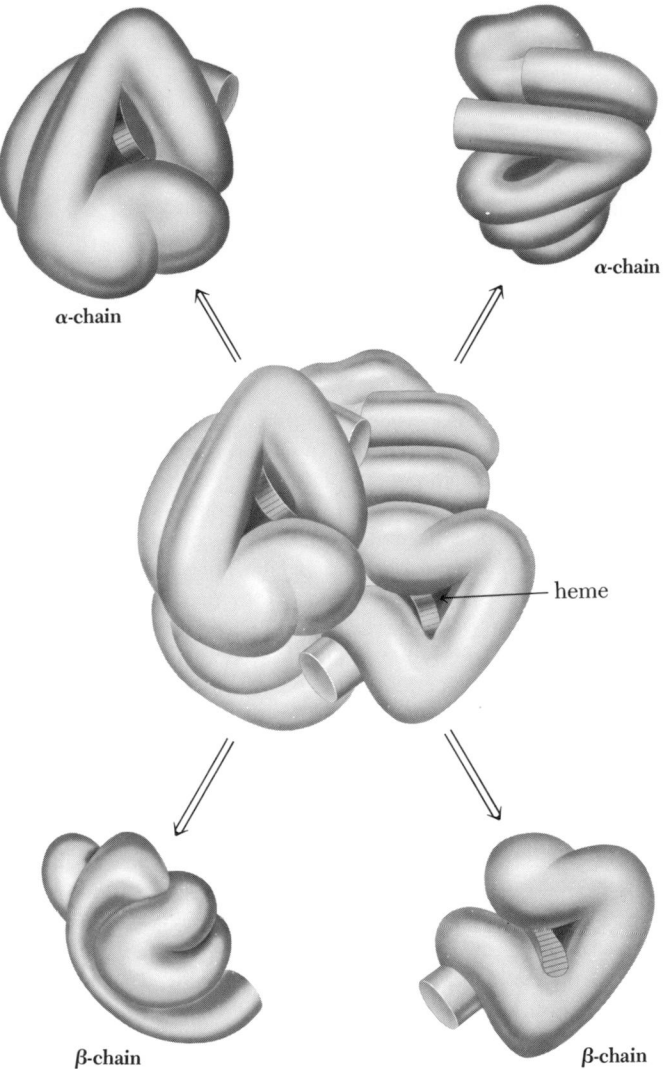

Figure 4-1   A schematic drawing of the most abundant adult hemoglobin. The molecule contains pairs of each of two kinds of peptide chains, termed α and β, which are also shown in exploded view at the approximate orientation they have in the complete molecule. Two of the four heme units are visible in pockets formed by folds of the peptide chains.

The drawing is of an idealized molecule—a sort of emaciated hemoglobin in which some of the external side chains have been plucked off to expose the underlying skeleton of the chains. In fact, it is difficult to distinguish chains at the surface of a complete model because the space between them is filled with protruding atoms.

The drawing is based on reports by Cullis, Muirhead, Perutz, and Rossman, Proc. Roy. Soc. (London), ser. A, 265:161 (1962); and by Perutz, J. Molec. Biol., 13:646 (1965), but the angle at which the molecule is viewed differs from those presented in the references.

four heme groups are near the outside surface. Even in the idealized representation, we can sense that the loops and turns must be derived from interactions between parts of the chain, with chemical groupings that can react strongly brought into proximity by the particular structure. As we have already emphasized, the necessary chemical reactivity is in the nature of the amino acid residues of the peptides.

We have noted that there are many kinds of hemoglobin, varying among species, and one individual in a given species frequently has more than one kind in his blood. It turns out that the various kinds of hemoglobin have an amazingly similar shape in their peptide chains. There is variation in the particular amino acids found at some locations, but the changes do not affect the ability of the peptides to fit together or to bind heme, either because of similar chemical nature of the amino acids or because of the position at which the change occurs. We can deduce that it could not be otherwise, because an animal that has lost the ability to specify peptides that fit together and bind heme could hardly survive the resultant disruption of oxygen transport.

The tetrapeptide, hemoglobin, has evidently evolved from a single peptide ancestor. Even the mammals contain a more direct descendant of the primitive precursor in the form of *myoglobin* (from myo, or muscle, and hemoglobin). This hemoglobin, found in muscles, consists of a single peptide chain. However, α and β chains of hemoglobin and the single chain of myoglobin have the same general shape (Fig. 4-2). The difference is that the chains of hemoglobin have exposed amino acid residues that make the chains combine into a tetrapeptide, and myoglobin does not.

GEOMETRY OF THE PEPTIDE UNITS

The contorted appearance of the peptide chains is obtained from relatively simple geometrical elements, which are a part of the structure of many proteins. It is the particular combination of the elements that creates the individual shapes

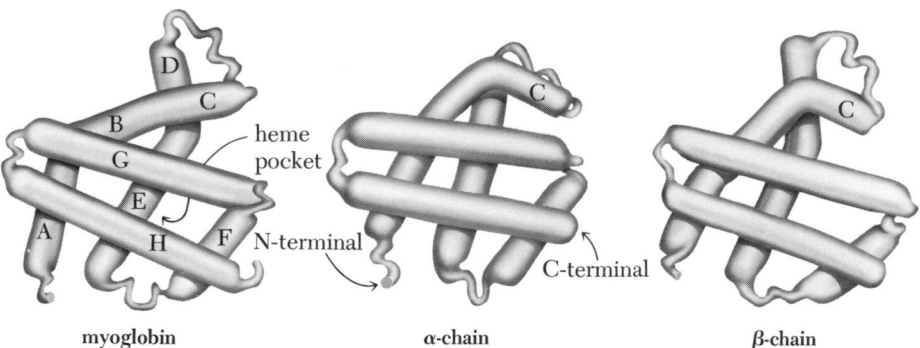

myoglobin α-chain β-chain

Figure 4-2 The arrangement of the backbone of the peptide chains in myoglobin, and the α and β chains of hemoglobin. The drawings are highly schematic renditions of models made by Perutz and associates (Proc. Roy. Soc. [London], ser. A, 265:161 [1962]). The cylindrical segments designated by letters are coils of peptide chain (α-helices). The space occupied by side chains of residues is not indicated.

of various protein molecules, and the principles governing the combinations are especially well defined with hemoglobin.

The α-Helix

The parts of the molecule indicated as straight cylinders in Figure 4-2 are coils of peptide chain, built as if the peptide backbone were wound in a spiral around nearly straight tubes. The particular coil is the α-helix, and this is the structure formed when a *single* peptide chain, not two or three together, reacts within itself to create hydrogen bonds between the —NH— and —C=O groups of the peptide backbone. It permits a regular alternation of hydrogen bonds to neighboring turns of the coil, with every available group bonded (Fig. 4-3).

The α-helix is of such a diameter that each amino acid residue occupies 1.7 radians when viewed down the tube, so there are roughly 3.6 residues for each turn.

The α-helix is a part of the *secondary structure* of the protein, as opposed to the *primary structure* represented by the sequence of amino acids. The secondary structure is the kind of regular configuration of the peptide backbone. (The tertiary structure is the complete combination of elements making the three-dimensional shape of the entire peptide, and the quaternary structure is the combination of peptides into a complete protein molecule. This kind of chemical taxonomy is somewhat misleading in implying that the various kinds of structure have independent existences arising from different causes, whereas in fact the presence of all depends upon the amino acid sequence.)

The Bends in the Chain

The straight helical segments of the peptide chain are linked at various angles, created at some points by sharp bends and at others by long loops. Why do these bends and loops exist—why isn't the entire peptide a single long helix?

One important reason is the presence of residues of the amino acid, *proline,* in the peptide chain:

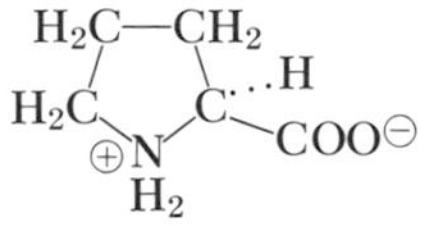

$$H_2C—CH_2$$
$$H_2C \qquad C \cdots H$$
$$_{\oplus}N \qquad C—COO^{\ominus}$$
$$H_2$$

L-proline

The nitrogen atom of proline is in a heterocyclic ring, and the shape of the ring prevents the peptide backbone from assuming the angle necessary to make an α-helix. A prolyl residue may be in the first turn of a helix, that is, in the first, second, or third position, but a helix can't form around it at any other position, and the presence of a prolyl residue therefore represents an obligatory interruption of a helix and a new direction of the peptide chain compared to the preceding segment.

A prolyl residue is found at each of two positions in every kind of hemoglobin

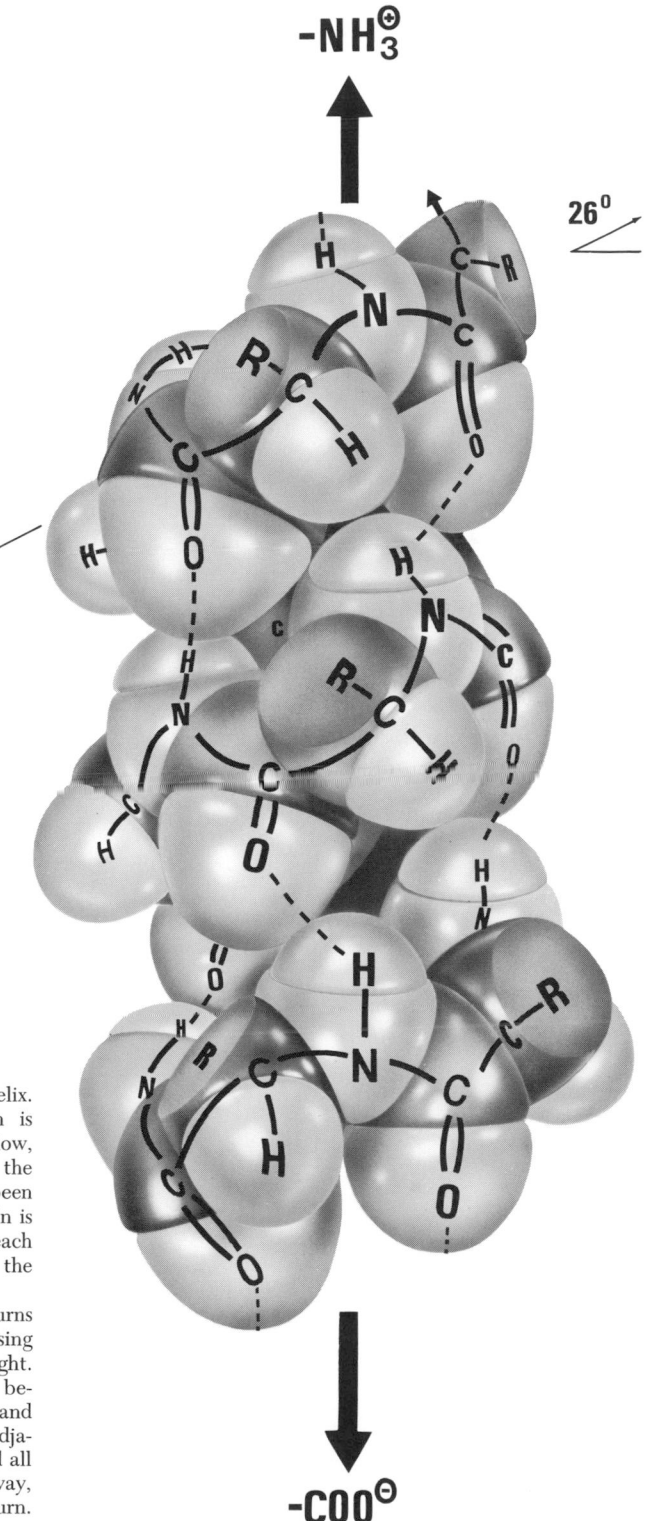

Figure 4-3 A segment of α-helix. The N-terminus of the chain is above, and the C-terminus below, the segment shown. For clarity, the side chains of the residues have been removed. Their link to the chain is shown by the symbol R, and in each case is directed outward from the core of the helix.

The drawing shows three turns on one side of the spiral, each rising at an angle of 26° from left to right.

Note the hydrogen bonds between carbonyl oxygen atoms and the amide hydrogen atoms of adjacent turns of the spiral. To bind all of the possible atoms in this way, there must be 3.6 residues per turn.

chain examined, and the chains have sharp bends immediately before the location of these residues:

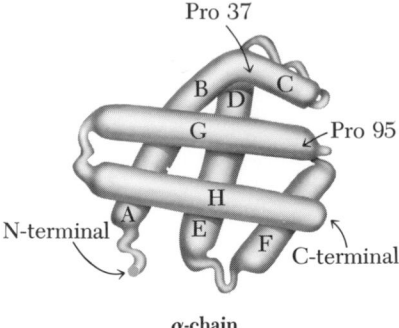

α-chain

The other bends and loops of the molecule contain prolyl residues in some kinds of hemoglobin, but not all, which tells us that bends can be formed even though a helical structure is formally possible. The formation of such bends reflects the creation of the configuration of lowest energy content. The conversion of a random configuration to a helix will release energy, but there may be other possibilities that release still more energy. For example, an arrangement in which *non-polar side chains are kept in the interior and polar side chains are in contact with water at the surface* may be more stable than a continuous helix in which all side chains are indiscriminately exposed.

Further, the bending of some segments into proximity may permit the formation of hydrogen bonds between them, in addition to the hydrogen bonds found in helical segments, thereby resulting in additional energy loss. A conspicuous example of this kind of stabilization involves a *tyrosyl* residue, which is invariably present next to the C-terminal residue of all hemoglobin peptides and forms a bond with a carbonyl group in the peptide backbone of a neighboring chain segment (Fig. 4-4).

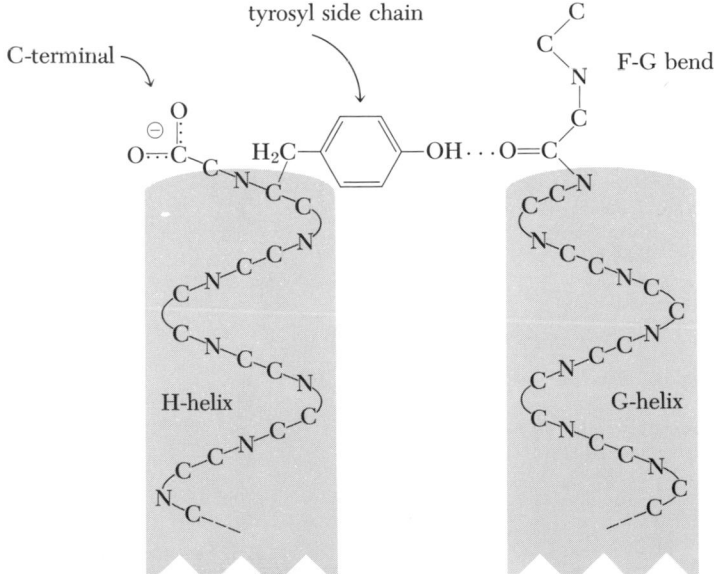

Figure 4-4   The hydroxyl group on tyrosyl side chains may form hydrogen bonds to fix structure. In the example shown, such a hydrogen bond ties the C-terminus of the H-helix in hemoglobin to an adjacent peptide chain at a point immediately before the G-helix.

In crude terms, this keeps the end of the chain from flapping around loose in the solution. Put in a more sophisticated way, the formation of the hydrogen bond contributes to the stabilization of that structure in which the chains are bent so as to bring the two regions together. The importance of the bond is attested by the retention of the tyrosyl residue throughout the long evolution of vertebrates.

There are other ways in which bends can be stabilized. The second and fifth segments of helix cross each other. Both of these segments have *glycyl* groups at the surface of contact:

$$^{\oplus}H_3N—CH_2—COO^{\ominus}$$

glycine

This is the simplest amino acid, without side chains that would prevent close fit at the crossover. (What would happen if the glycyl residues were shifted one turn on the helices? Would the crossover positions be shifted one turn by this simple change in amino acid sequence? We don't know.)

In summary, given the two obligatory bends created by the bond angles of prolyl residues, the remainder of the bends appearing in the molecule can be accounted for by the bonding they make possible throughout the molecule, not necessarily close to the bend itself. The sum of the energy liberated by forming these bonds is greater than the energy that could be liberated if the α-helix continued without bends.

THE HEME POCKET

The crevice into which heme fits is a structural feature of hemoglobin peptides that is not found in most proteins. However, a general principle is illustrated—*proteins can be constructed to interact specifically with some other compound.* The crevice is created by the way the peptide chain is folded, but the mere existence of a slot with room enough to contain heme won't cause the formation of a stable complex. Stability of the complex comes from several kinds of interactions between the porphyrin ring and the particular amino acid residues lining the pocket.

Exclusion of Water

The pocket has no charged groups surrounding the porphyrin ring. Most of the residues lining the pocket have bulky hydrocarbon side chains, such as those of *phenylalanine, valine, leucine,* and *isoleucine:*

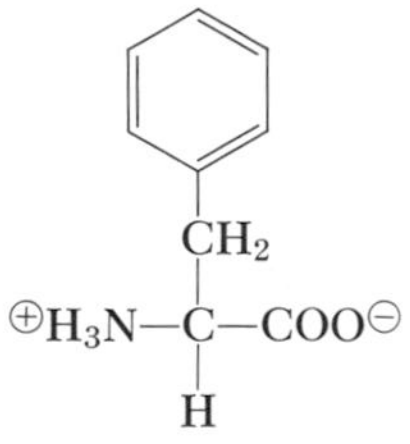

L-phenylalanine 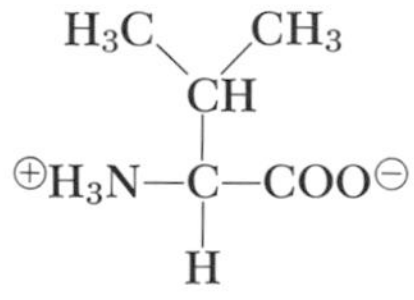   L-valine 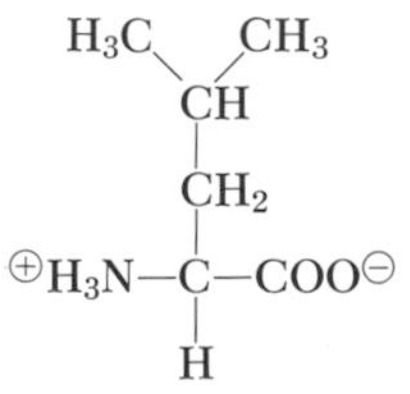   L-leucine 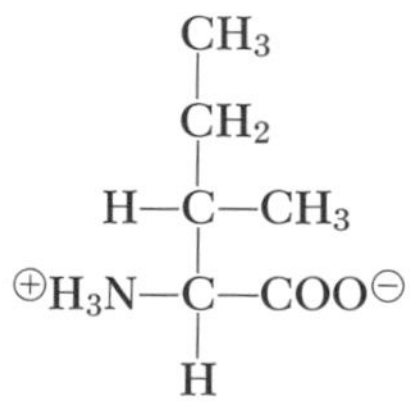   L-isoleucine

There are residues of *serine* and *threonine* nearby, but even the weakly polar hydroxyl groups of these residues are hydrogen-bonded to adjacent segments of peptide backbone, so that they have little hydrophilic character:

$$
\begin{array}{cc}
\text{OH} & \text{OH} \\
| & | \\
\text{CH}_2 & \text{H—C—CH}_3 \\
| & | \\
{}^{\oplus}\text{H}_3\text{N—C—COO}^{\ominus} & {}^{\oplus}\text{H}_3\text{N—C—COO}^{\ominus} \\
| & | \\
\text{H} & \text{H} \\
\text{L-serine} & \text{L-threonine}
\end{array}
$$

The lining of the pocket is therefore strongly hydrophobic, with little affinity for water.

Now, the porphyrin ring is itself a resonant, non-polar structure. Insertion of this hydrophobic ring into the hydrophobic pocket excludes water, and the result is a release of energy. This comes about because *water assumes an ordered structure, like ice, near hydrophobic groups,* and this structure has a higher energy content than the more random orientation of water molecules near polar groups with which they can interact. The effect of introducing the porphyrin ring into the pocket is the elimination of ice-like structures previously present around the unbound porphyrin and in the unfilled pocket. The energy of the ice structures is liberated when water is excluded from the hydrophobic regions. Internal hydrophobic interactions of this sort are among the strongest forces stabilizing the structure of globular proteins, and we see that most of the energy of the hydrophobic bond really results from exclusion of water. (Hydrocarbons interact very weakly in non-polar solvents.)

Van der Waal's Forces

The residues of the pocket are arranged so that many atoms of the peptide and the enclosed porphyrin are in close contact. Close contact results in interactions lumped together as van der Waal's forces. The release of energy from proximity of one pair of atoms is small, but so many are involved in this case that the total is large enough to contribute a significant part of the bonding energy. The increased release of energy upon making compactly nestled structures is probably of importance in determining the stable configuration of many proteins.

π-Bonding

The nature of the amino acid residues used to make a non-polar environment varies somewhat among different hemoglobins, but all hemoglobin peptides have a phenylalanyl residue at a particular location, arranged so that the aromatic ring is parallel to the porphyrin ring. The result may be the formation of π-bonds by resonant interaction between the two rings, releasing energy and fixing the angle taken by the porphyrin ring in the pocket. The extent of use of this kind of bonding in proteins is uncertain, because coplanar aromatic rings also develop forces through water exclusion and van der Waal's interactions.

Electrostatic Bonding

The two negatively charged propionate side chains of the porphyrin are exposed at the surface of the molecule, where they are in contact with positively charged side chains of the peptide, forming salt linkages. Here again, the amino acid sequence is such that folding of the molecule brings the positive charges into the proper location to interact with the porphyrin groups, in this case at the surface of the molecule in contact with water, not in the hydrophobic interior. There are three "basic" amino acids that can be positively charged—*lysine, arginine,* and *histidine:*

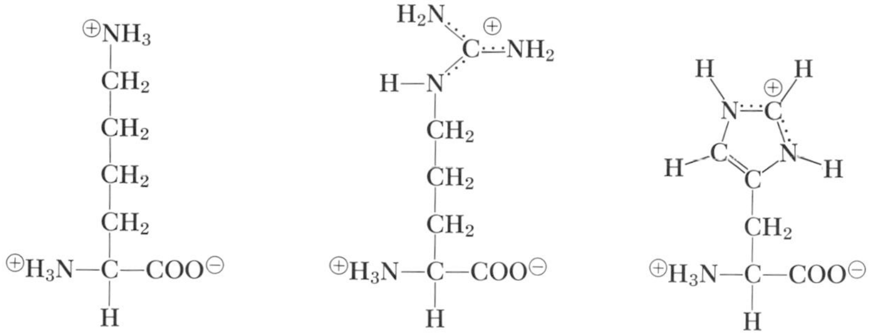

L-lysine   L-arginine   L-histidine
                                                 (*cationic form*)

A pair of lysyl residues supplies the charge to bond the propionate side chains in human hemoglobin, but various combinations of the three possibilities occur in hemoglobins from other species.

Salt linkages such as these can also occur between positively and negatively charged side chains, and contribute to the stability of structure in many proteins, but they are usually of lesser importance than hydrogen bonds and hydrophobic exclusion of water.

Interactions with Heme Iron

None of the interactions between peptide and porphyrin we have described involves the iron atom normally present in heme. Indeed, the atom is not necessary for the bonding, and this has been shown by mixing iron-free porphyrins with globin. Complexes formed that were nearly as stable as hemoglobin itself.

However, the bonding of the complete iron-porphyrin is the basis for the transport of oxygen, and we have already described the way in which the coordination number of iron is satisfied by a histidyl residue contributed by the peptide and by water (p. 14). In effect, the strong bonding of the surrounding porphyrin to the peptide holds the central iron atom in its functional position next to a histidyl residue.

In addition to the histidyl residue used for chelating iron, there is a second histidyl residue on the opposite side of the porphyrin ring, but not close enough to the metal for direct bonding. It appears likely, although not absolutely certain, that oxygen is transported because it forms a bridge between this histidyl residue and the iron atom, thereby displacing the water molecule previously present.

Although the exact nature of the bonding of molecular oxygen is not clear, it is known that the unpaired electrons found in both oxygen and hemoglobin are no longer present in oxyhemoglobin, so that the paramagnetism of these molecules disappears in the complex.

The pair of histidyl residues in the pocket occurs in all hemoglobins. We shall later see that pairs of histidyl residues are frequently involved in catalysis by proteins, and the function of the pair in hemoglobin may be an evolutionary modification of this more primitive and fundamental function.

COMBINATION OF SUBUNITS

When both α and β chains, all combined with heme, are present, the result is invariably a tetrapeptide with two of each kind of chain. The shapes of the chains are nearly alike, and it might appear that there is some chance for formation of molecules having three of one kind of chain and one of the other. This doesn't happen, and the reason is that there are sufficient differences in the structure of the two kinds of chains so that *unlike chains associate.*

The α and β chains have sections of surface composed of non-polar amino acid residues of complementary shape—they nest together. The same kinds of forces that hold the porphyrin in its pocket therefore operate to hold these unlike chains together at the matching surfaces, and the resultant $\alpha\beta$ dimer is the fundamental unit of the tetrapeptide, which is made from pairs of these dimers.

The bonds holding pairs of dimers to make tetrapeptides are more scattered, and include salt linkages between the terminal ammonium and carboxylate groups of β chains and hydrogen bonds involving other polar amino acid residues.

REQUIREMENTS FOR OXYGEN TRANSPORT

Equilibrium with Oxygen

The formation of a carrier for oxygen was a necessary development in the evolution of large animals with an active metabolism. The carrier must meet two specifications to be efficient. It must combine nearly completely with oxygen at the concentrations found in the environmental source. It must also combine relatively weakly with oxygen at the concentration to be maintained in the tissues, so that the carrier will release the oxygen upon demand.

The simplest kind of carrier would be one that combines reversibly with oxygen: $X + O_2 \rightleftharpoons XO_2$. The extent of combination would be determined by the concentrations according to the equilibrium expression:

$$\frac{[XO_2]}{[X][O_2]} = K_{eq}$$

The total concentration of the carrier—combined and uncombined—will be essentially constant during a cycle of transport, since the hemoglobin molecule is neither made nor destroyed very rapidly compared to the rate of circulation. We can therefore calculate the fraction of the total hemoglobin that will be oxygenated at varying oxygen concentrations from the above equation, and obtain a curve of the shape shown in Figure 4-5.

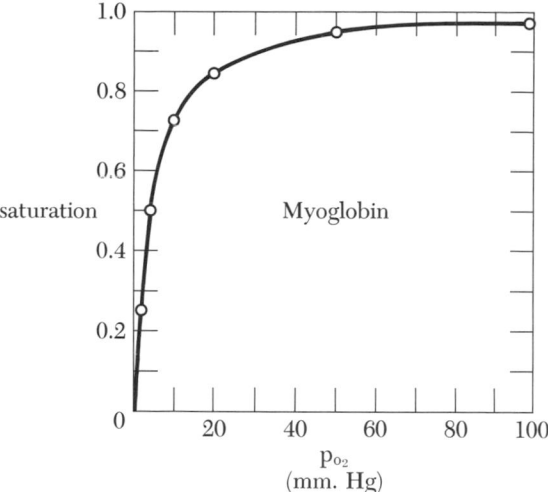

Figure 4-5 The saturation of a carrier with oxygen when the reaction between them has a simple equilibrium. Myoglobin, the hemoglobin of muscle, combines with oxygen in this way.

Such a carrier can only be efficient when the concentration of oxygen at the point of release is low compared to the concentration at the point of uptake. If this isn't so, the carrier will not be saturated with oxygen at the source, or will still retain most of the oxygen at the tissue to which it is to be delivered.

Despite the handicaps, this kind of a simple carrier will work under some circumstances. The primitive vertebrates have a hemoglobin with a single peptide chain and one heme, which combines with oxygen in the way illustrated in Figure 4-5. Even the mammals have such a hemoglobin, but it is the myoglobin of skeletal muscles, not the circulating carrier, and it is used as an auxiliary source of oxygen for these tissues during the many-fold increase in demand for exercise with the concomitant drop in concentration of the gas.

When animals grew larger and increased their oxidative metabolism, further modification of the hemoglobin was required. *There can only be a net diffusion of any compound, including oxygen, when there is a concentration gradient.* The greater the gradient, the faster the diffusion. To keep the physical bulk of the circulatory system within bounds, it is not possible to have every cell of a large animal in direct contact with the circulation, and yet the diffusion of oxygen to these cells must be fast enough to supply them during all activities necessary for survival of the species. This can be done by increasing the concentration gradient for oxygen between the interior of the cells and the blood. It would be self-defeating to accomplish this by lowering the concentration within the cell. The increase in the gradient has to come through a rise in the concentration of oxygen in the circulation, without preventing the release of oxygen from the carrier at this increased concentration. Put another way, the carrier-oxygen complex must dissociate so as to maintain a high concentration in the venous return of the circulation. The equilibrium shown in Figure 4-5 won't do, because a high concentration of oxygen and its nearly complete release are mutually exclusive in this system.

The evolution of hemoglobin as a tetrapeptide solved the problem. Each of the four hemes will bind oxygen, but binding to one affects the binding to the others. Within the tetrapeptide, each of the strongly bonded $\alpha\beta$ dimers tends to act as a unit in the combination with oxygen. The addition of oxygen to one peptide in the dimer increases the affinity of the second peptide for oxygen. The stepwise sequence of addition of oxygen can be represented as:

$$\alpha\beta-\alpha\beta + O_2 \rightleftharpoons \alpha O_2\beta-\alpha\beta + O_2 \rightleftharpoons \alpha O_2\beta O_2-\alpha\beta$$

The second association has a much greater equilibrium constant than the first. This means that oxygen tends to saturate the first $\alpha\beta$ dimer before it is bonded to the second, and that only a small amount of hemoglobin carries single molecules of oxygen and even less has two molecules distributed between dimers so as to form molecules such as $\alpha O_2\beta \cdot \alpha O_2\beta$. There may at particular oxygen concentrations be an *average* of one molecule of oxygen per molecule of hemoglobin, but this will represent mixtures of molecules having no oxygen bound and molecules with pairs of oxygen bound.

The combination with oxygen can therefore be represented as pairs of reactions in which two molecules of oxygen are attached:

$$\alpha\beta-\alpha\beta + 2O_2 \rightleftharpoons \alpha O_2\beta O_2-\alpha\beta + 2O_2 \rightleftharpoons \alpha O_2\beta O_2-\alpha O_2\beta O_2$$

Now, the combination of the first dimer with oxygen affects the affinity of the second dimer for oxygen, so that the equilibrium constant for the second reaction is about six times greater than that for the first reaction.

The result is that the equilibrium with oxygen assumes the complex sigmoid shape shown in Figure 4-6, which also indicates typical values for the concentration of oxygen in the circulation of human adults at rest and during exercise.

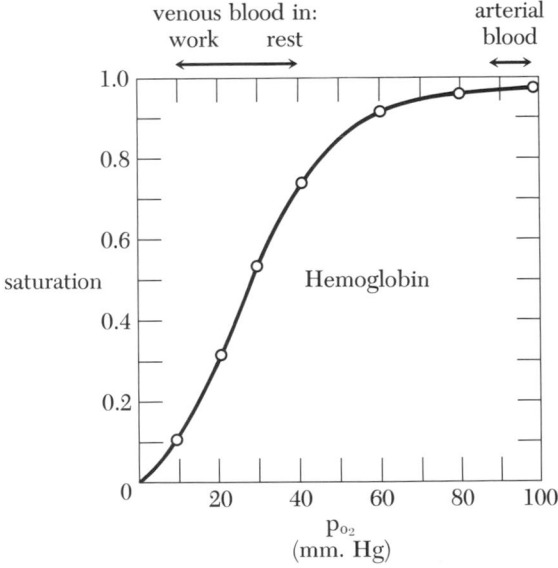

Figure 4-6   The saturation of hemoglobin A_1 exposed to varying partial pressures of oxygen. At the top are shown the ranges of oxygen tension found in venous blood of adult humans at work and at rest and in the arterial blood. The oxygen tension of arterial blood does not vary widely with exercise.

It is customary to give concentrations of oxygen in terms of the partial pressure of gas, expressed in mm of mercury, that would be in equilibrium with the solution. This is also known as the *oxygen tension.*

Given the equilibrium shown in Figure 4-6, oxyhemoglobin will dissociate even at relatively high oxygen tensions, and thereby build up the necessary concentration gradient between the circulation and the cells. The actual values found in the circulation at rest show that only part of the oxygen is released for the basal demands of the organism, but much of the remainder is readily available when increased oxygen consumption causes a drop in oxygen tension in the capillary bed.

The chemical nature of the changes in the complete molecule that generate the complex sigmoid character of the oxygen dissociation curve is not clear, but a physical change is evident. As oxygen is added, the β chains slide toward each other, and this somehow affects their reactivity. It is not a question of direct interaction between the heme groups that modifies their affinity for oxygen, because they are not that close to each other, but a modification of the surrounding peptides.

The advantages of the sigmoid dissociation curve are evidently great enough to have caused the evolution of the tetrapeptide oxygen carrier before the appearance of the bony fish. The frog retains the use of a single peptide in the tadpole stage, but shifts to production of a tetrapeptide upon metamorphosis. However, mammals use tetrapeptides for carrying oxygen even in the fetus.

The actual levels of oxygen tension maintained in the circulation vary among animals, and this variation comes from modification of the peptides so as to shift the position of equilibrium between hemoglobin and oxyhemoglobin.

The Bohr Effect

The changes in configuration accompanying oxygenation cause an increase in the acidic dissociation of the molecules, equivalent to the release of 0.7 H^+ for each O_2 added at pH 7.4. It is likely that increased dissociation of imidazolium ions (the charged form of the imidazole group) is the direct source of H^+, but the effect may be a diffuse one, involving small releases from several groups. This increase in acidity of hemoglobin upon complexing with oxygen is known as the Bohr effect.

The Bohr effect is an important physiological process, not a point only of technical interest. It adds an additional equilibrium to the oxygen-hemoglobin system, in which increased oxygen will displace H^+ from the protein and increased H^+ will displace oxygen from the protein. In other words, the dissociation curve of oxyhemoglobin (Fig. 4-6) is displaced toward higher oxygen tensions (to the right of the figure) as the pH falls. The tissues utilizing oxygen are also generating CO_2 and other acidic products, which lower the pH of blood. The drop in pH causes an increase in the release of oxygen beyond that which would have occurred at constant pH at a given oxygen tension, thereby making more oxygen available. The greater the output of CO_2, the greater the release of O_2. The effect therefore adjusts the amount of oxygen released to its demand, because a greater CO_2 output results from an increased utilization of O_2. Just the reverse occurs in the lungs. The CO_2 is blown off, and the resultant rise in pH increases the affinity of hemoglobin for oxygen.

The second useful result of the Bohr effect is the diminution of changes in the pH of the blood. It provides a buffering action because the drop in pH that would occur from the introduction of CO_2 is limited by the uptake of H^+ on hemoglobin. We shall consider the quantitative aspects of this buffering action in Chapter 25.

Solubility of Hemoglobin

We have seen that the peptide must bind heme, supply histidyl groups for interacting with oxygen, interact to generate the sigmoid dissociation curve at levels of oxygen tension appropriate for the organism, and react with hydrogen ions so as to modify oxygen binding. The combination of peptides must also contribute high solubility.

Evolution of erythrocytes as packages for the oxygen carrier gained several advantages over invertebrates that circulate their oxygen carriers in free solution. The carrier is protected against loss by excretion and against damage by cellular processes in ways we shall consider later. There is also the advantage that the erythrocyte must be squeezed through the capillaries, bringing the concentrated solution of carrier within the erythrocyte into close contact with the capillary wall. A simple solution of hemoglobin would tend to develop an annular flow, with a relatively stagnant layer held at the periphery of the capillary bore by friction.

If the carrier is to be confined within erythrocytes, it must have high solubility. The hemoglobin content of whole blood ranges from around 70 grams per liter in turtles to 150 grams per liter in humans. Confining this already high content of protein into erythrocytes results in concentrations ranging around 300 grams per liter within the cells.

To attain this high solubility, the protein must have a globular, as opposed to a linear, shape, so that there is little tendency for the molecules to associate in a stacked array and form large fibers. The protein must also have a number of charged groups on the outside surface that can interact with water.

We have seen that these requirements are met by an arrangement of non-polar residues that associate within peptides and between peptides so as to create the compact tetrahedron. We have said nothing about the polar residues on the surface, other than the positively charged groups near the propionate side chains of the porphyrin and the terminal residues of chains that interact at the ends of the molecule.

The polar residues, all of which are on the surface with the exception of the two histidyl residues in the heme pocket, include the three "basic" amino acids, *arginine, histidine,* and *lysine,* previously mentioned (p. 59). There are two "acidic" amino acids, ionized to have negative charge—*glutamate* and *aspartate:*

$$
\begin{array}{cc}
\overset{\displaystyle COO^{\ominus}}{\underset{}{|}} & \\
CH_2 & \overset{\displaystyle COO^{\ominus}}{\underset{}{|}} \\
| & CH_2 \\
CH_2 & | \\
^{\oplus}H_3N-C-COO^{\ominus} & ^{\oplus}H_3N-C-COO^{\ominus} \\
| & | \\
H & H \\
\text{L-glutamate} & \text{L-aspartate}
\end{array}
$$

In addition, residues of the corresponding amides, *glutamine* and *asparagine,* are present at the surface:

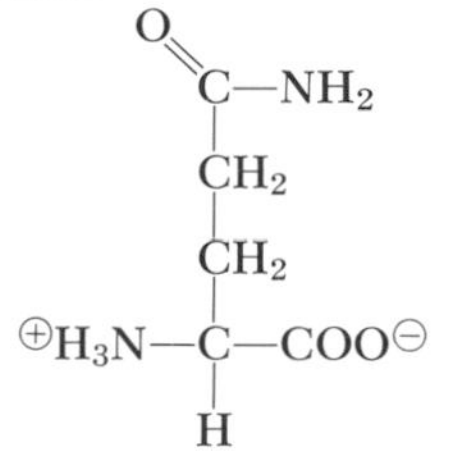

$$\begin{array}{cc}
\overset{\displaystyle O}{\underset{\displaystyle \|}{C}}-NH_2 & \\
CH_2 & \overset{\displaystyle O}{\underset{\displaystyle \|}{C}}-NH_2 \\
CH_2 & CH_2 \\
{}^{\oplus}H_3N-\underset{\displaystyle H}{C}-COO^{\ominus} & {}^{\oplus}H_3N-\underset{\displaystyle H}{C}-COO^{\ominus} \\
\text{L-glutamine} & \text{L-asparagine}
\end{array}$$

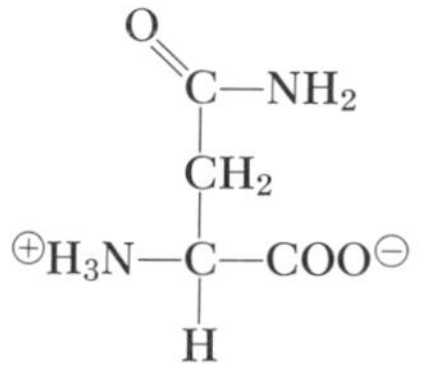

It ought to be fixed in mind that glutamine and asparagine, although structurally and metabolically related to glutamate and aspartate, are distinctive amino acids in their own right, coded by separate triplets during protein synthesis. There is a tendency to neglect this fact, because it is experimentally difficult to isolate amino acids from proteins without hydrolyzing the amide groups, and analyses of proteins therefore are frequently reported with the content of glutamate + glutamine and aspartate + asparagine lumped together without necessarily noting that this has been done.

The complete hemoglobin tetrapeptide has a cavity in the center of the molecule that contains water. The cavity, created from part of the faces of the peptide residues, also has polar amino acid residues. In other words, part of the interior surface of the individual peptides is polar so that these surfaces will not interact strongly, and part is non-polar where the peptide chains are in intimate contact. This may well be a lock and key arrangement, so that the four peptides must form the particular tetrahedron that brings the non-polar portions into contact and leaves the polar portions spaced apart to admit water.

EVIDENCE FOR STRUCTURE FROM GENETIC ALTERATIONS

We have developed the principles believed to govern the formation of functioning structure in globular proteins, such as hemoglobin, in the preceding sections. There are two general types of genetic evidence available to reinforce our understanding of these principles. One is from knowledge of the amino acid composition of various hemoglobin peptides from different species. We know that all of these hemoglobins work as oxygen carriers, and similarities in amino acid composition among them give positive testimony to the contribution to function made by these particular residues.

The second type of evidence comes from known genetic variants in the human population. Some mutants are fairly common, but most are quite rare. The disturbance or lack of disturbance in structure and function caused by these amino acid substitutions has in some cases confirmed already recognized principles, and in others has caused a re-examination of dogma.

We want to examine all of these variations in close detail because the knowledge thereby gained of the effect of each kind of amino acid will serve in reasoning on all other proteins.

Variations among Species

Most humans will synthesize five different peptides during their lifetime to make four hemoglobins. All of these hemoglobins contain the same two α chains, but differ in the second pair, which is β in Hb A_1, δ in Hb A_2, γ in Hb F, and ϵ in the embryonic hemoglobin. The complete sequence of amino acid residues is known for all except the ϵ chain.

The sequence is also known for chains from at least one hemoglobin in each of several species of mammals. Fragments are characterized from many other species on down the evolutionary scale to the carp and lamprey.

The complete sequence is known for the myoglobin from sperm whales and partially known for human myoglobin. (Sperm whale was chosen as a source because of the ease of isolation of myoglobin from its tissues, not for some profound phylogenetic purpose.)

The complete sequences for some of these hemoglobins are summarized in Table 4-1.

The sequences outlined are consistent with an evolution of all of them from a common precursor, and the apparent order in geological time of appearance of the chains as separate entities in myoglobin, α, γ, β—δ. The order is based on arguments such as these: There is a closer resemblance in sequence of the known fragments of human myoglobin (not shown in the table) to whale myoglobin than there is of either myoglobin to any of the hemoglobin peptides in the human; therefore, myoglobin and hemoglobin appeared before the divergence of whales and humans and before the divergence of the hemoglobin peptides. The α chain of the human resembles the α chain of other mammals more than it does the β chain of the human; therefore, α and β chains diverged before the divergence of the listed mammals. The α chains have fewer residues corresponding to β, γ, and δ chains than these chains have with each other: therefore the α chain diverged before the others. There are fewer differences between β and δ chains than between the others; therefore these diverged last of all.

Within a given type of chain, the number of changes from one species to another reflects the time since evolutionary divergence of the species. The β chain of the gorilla differs from that of a human in only one amino acid residue out of the 146, whereas the chain from the horse differs in 25 positions.

Constant Residues

There are few positions with the same amino acid residue in all of these proteins. It appears justifiable to assume that these positions are imperative for the function of the protein as an oxygen carrier without requiring rearrangement into a drastically different configuration. Such a rearrangement would require the simultaneous change of many amino acids, and therefore of many codons of DNA. The probability of such a simultaneous change is infinitesimal. The two *histidyl* residues on either side of the heme are always present, as might be expected from their role in binding iron and oxygen (α58 and α87). The *prolyl* residues creating the two critical bends in the neighborhood of the heme pocket are always present (α37 and α95). The *glycyl* residues at the crossover of helical segments (α25 and α59) are always present with one known exception—the lamprey has the next smallest

TABLE 4-1. THE AMINO ACID SEQUENCES IN HEMOGLOBIN CHAINS°

1. Each chain is numbered in columns from the N-terminus. Residues with corresponding positions in the molecular structure are in horizontal rows.
2. All of the kinds of residues known to occupy a position in various species are listed together under α or β-δ chains, respectively. The human chains are given in bold type. Where the human δ chain differs from the β, its differing residue is the second bold listing. The kinds of residues shown for the human may also occur in other species at the same location.
3. A zero (0) indicates the residue is absent in at least one species. Hyphenated residues, such as Asn-Asp at lamprey 83, indicate two residues in the space occupied by one in the other hemoglobins.

SINGLE PEPTIDES		TETRAPEPTIDE CHAINS			Helix	SIMILARITY OR COMMON FUNCTION OF RESIDUES
Lamprey	Whale Myoglobin	α	γ	β-δ		
1Pro						
Ile						
Val						
Asp						
5Ser						
Gly						
Ser						
Ala						
Pro						
10Val	1Val	1Val,Ser	1Gly	1Val,(0)		
			His	His,Gln,Met		
Leu	Leu	Leu	Phe	Leu		bulky non-polar
Ser	Ser	Ser	Thr	Thr,Ser	↑	H-bond (?)
Ala	Glu	Pro,Ala,Asp,Gly	5Glu	5Pro,Ala		
Ala	5Gly	5Ala,Glu,Lys	Glu	Glu		
15Glu	Glu	Asp	Asp	Glu,Asp		(−) charge
Lys	Trp	Lys	Lys	Lys		
Thr	Gln	Thr,Ser,Ala,Gly	Ala	Ser,Thr,Ala		
Lys	Leu	Asn,Ala	10Thr	10Ala,Glx,His		
Ile	10Val	10Val,Ile	Ile	Val	A	bulky non-polar
20Arg	Leu	Lys	Thr	Thr,Asn,Gly,Leu		
Ser	His	Ala,Ile,Thr	Ser	Ala,Gly,Ser		
Ala	Val	Ala	Leu	Leu,Phe		non-polar
Trp	Trp	Trp	15Trp	15Trp,Phe		π-bond
Ala	15Ala	15Gly,Ala,Glu,Ser	Gly	Gly,Ala,Asp,Ser		
25Pro	Lys	Lys	Lys	Lys		(+) charge (not lamprey)
Val	Val	Val,Ile	Val	Val	↓	bulky non-polar
Tyr	Glu	Gly,Ser				
Ser	Ala	Ala,Gly,Pro,Ser				
Asn	20Asp	20His,Lys	Asn	Asn,His,Leu,Lys	↑	polar (?)
30Tyr	Val	Ala,Gly	20Val	20Val,Glu		
Glu	Ala	Gly,Asp	Glu	Asp,Glu		
Thr	Gly	Glu,Asp	Asp	Glu,Ala		
Ser	His	Tyr,Gly,Ile	Ala	Val,Ala		
Gly	25Gly	25Gly	Gly	Gly		helical contact
35Val	Gln	Ala	25Gly	25Gly,Ala		
Asp	Asp	Glu	Glu	Glu		(−) charge
Ile	Ile	Ala	Thr	Ala	B	
Leu	Leu	Leu,Val	Leu	Leu		bulky non-polar
Val	30Ile	30Glu,Gly	Gly	Gly		

° This is a composite of information from various sources. Some of these make no distinction between Asp and Asn, or Glu and Gln; in such cases one of the forms is assigned when phylogenetic correspondence justifies it. Otherwise, the uncertainty is indicated by Asx or Glx. The following list of the more complete sources was used:

Braunitzer, G., Adv. Prot. Chem., 19: 1 (1964).
Poop, R. A., Federation Proc., 24: 1252 (1956).
Schroeder, W. A., Ann. Rev. Biochem., 32: 301 (1963).
Zuckerkandl, E., Sci. American, 212: no. 5, 110 (May, 1965)
Dayhoff, M. O., and R. V. Eck, Atlas of Protein Sequence and Structure, 1967–68, Natl. Biomed. Res. Found.

However, some disputed, or unconfirmed, or partial sequences have been omitted. Those used include the carp, horse, donkey, pig, rabbit, mouse, gorilla, and human α chains; and the horse, pig, rabbit, llama, cow, sheep, gorilla and human β chains.

TABLE 4-1 CONTINUED ON FOLLOWING PAGE

Table 4-1. *(continued)*

Single peptides		Tetrapeptide chains			Helix	Similarity or common function of residues
Lamprey	*Whale Myo-globin*	α	γ	β-δ		
^{40}Lys	Arg	**Arg**	^{30}Arg	^{30}Arg		(+) charge
Phe	Leu	**Met**	Leu	Leu		bulky non-polar
Phe	Phe	**Phe,Leu**	Leu	Leu		bulky non-polar
Thr	Lys	**Leu,Ala,Thr**	Val	Val		
Ser	^{35}Ser	^{35}**Ser,Gly,Val**	Val	Val		
^{45}Thr	His	**Phe,Tyr**	^{45}Tyr	^{45}Tyr		
Pro	Pro	**Pro**	Pro	Pro		bend at heme
Ala	Glu	**Thr,Gln**	Trp	Trp		
Ala	Thr	**Thr**	Thr	Thr	C	H-bond (not lamprey)
Gln	^{40}Leu	^{40}**Lys**	Gln	Gln,Arg		
^{50}Glu	Glu	**Thr**	^{40}Arg	^{40}Arg		polar
Phe	Lys	**Tyr**	Phe	Phe,Tyr		
Phe	Phe	**Phe**	Phe	Phe		π-bond
Pro	Asp	**Pro,Ala**	Asp	Glu,Asp,Ser		
Lys	^{45}Arg	^{45}**His**	Ser	Ser,His		
^{55}Phe	Phe	**Phe,Trp**	^{45}Phe	^{45}Phe		π-bond
Lys	Lys	**(0),Ala**	Gly	Gly		
Gly	His	**Asp**	Asn	Asp		
Met	Leu	**Leu,Phe,Val**	Leu	Leu		bulky non-polar
Thr	^{50}Lys	**Ser,Thr**	Ser	Ser		H-bond (?)
^{60}Ser	Thr		^{50}Ser	^{50}Thr,Ser,Asx		H-bond (?)
Ala	Glu		Ala	Pro,Ala		
Asp	Ala		Ser	Asp,Asn,Gly		
Glu	Glu		Ala	Ala		
Leu	^{55}Met		Ile	Val,Ile	D	bulky non-polar
^{65}Lys	Lys	^{50}**His,Pro**	^{55}Met	^{55}Met,Leu		
Lys	Ala	**Gly**	Gly	Gly,Asn		
Ser	Ser	**Ser**	Asn	Asn		
Ala	Glu	**Ala,Asx,Glu,Gly**	Pro	Pro,Ala		
Asp	^{60}Asp	**Gln,Glu,Pro**	Lys	Lys		
^{70}Val	Leu	^{55}**Val,Ile**	^{60}Val	^{60}Val		bulky non-polar
Arg	Lys	**Lys**	Lys	Lys		(+) charge
Trp	Lys	**Gly,Ala,(0)**	Ala	Ala		
His	His	**His**	His	His	E	O$_2$ complex
Ala	^{65}Gly	**Gly**	Gly	Gly		helical contact
^{75}Glu	Val	^{60}**Lys**	^{65}Lys	^{65}Lys,Ser		
Arg	Thr	**Lys**	Lys	Lys		
Ile	Val	**Val**	Val	Val		bulky non-polar
Ile	Leu	**Ala,Ser,Ile**	Leu	Leu		low polarity (?)
Asn	^{70}Thr	**Asn,Glu,Met-Gly**	Thr	Gly,Asp, Ala,His,Thr		
^{80}Ala	Ala	^{65}**Ala,Gly**	^{70}Ser	^{70}Ala,Ser		small size (?)
Val	Leu	**Leu,Val**	Leu	Phe		bulky non-polar
Asn-Asp	Gly	**Thr,Ala,Gly**	Gly	Ser,Cys,Gly		
Ala	Ala	**Asn,Asp,Leu,Lys**	Asp	Asp,Glu		
^{85}Val	^{75}Ile	**Ala**	Ala	Gly		non-polar
Ala	Leu	^{70}**Val,Gly**	^{75}Ile	^{75}Leu,Val,Ile,Met		non-polar
Ser	Lys	**Ala,Glu,Gly,Ser**	Lys	Ala,Gly,Glx,His,Ser,Lys		
Met	Lys	**His,Lys**	His	His,Gln		
Asp	Lys	**Val,Ile,Leu**	Leu	Leu,Cys		
^{90}Asp	^{80}Gly	**Asp**	Asp	Asp		
Thr	His	^{75}**Asp**	^{80}Asp	^{80}Asn,Asp,Lys		polar (?)
Glu	His	**Met,Leu**	Leu	Leu		
Lys	Glu	**Pro**	Lys	Lys		
Met	Ala	**Asn,Gly**	Gly	Gly		
^{95}Ser	^{85}Glu	**Ala,Gly**	Thr	Thr,Ala		
Met	Leu	^{80}**Leu**	^{85}Phe	^{85}Phe,Trp		bulky non-polar
Lys	Lys	**Ser,Ala**	Ala	Ala,Ser		
Asp	Pro	**Ala**	Gln	Thr,Gln,Ala,Lys,Ser		bulky non-polar
Leu	Leu	**Leu**	Leu	Leu		3-carbon
^{100}Ser	^{90}Ala	**Ser**	Ser	Ser	F	
Gly	Gln	^{85}**Asp,Asn,Glu**	^{90}Glu	^{90}Glu		
Lys	Ser	**Leu**	Leu	Leu		

TABLE 4-1. *(continued)*

SINGLE PEPTIDES		TETRAPEPTIDE CHAINS			Helix	SIMILARITY OR COMMON FUNCTION OF RESIDUES
Lamprey	*Whale Myo-globin*	α	γ	β-δ		
His	His	His	His	His	↓	Fe chelate
Ala	Ala	Ala	Cys	Cys		3-carbon
105Lys	95Thr	His,Ser	Asp	Asp		polar (?)
Ser	Lys	90Lys	95Lys	95Lys,Glu		polar (?)
Phe	His	Leu	Leu	Leu		
Gln	Lys	Arg	His	His		polar
Val	Ile	Val	Val	Val		bulky non-polar
110Asp	100Pro	Asp	Asp	Asp	↑	(−) charge (not whale)
Pro	Ile	95Pro	100Pro	100Pro		bend at heme
Gln	Lys	Val,Ala	Glu	Glu		
Tyr	Tyr	Asn	Asn	Asn		H-bond (?)
Phe	Leu	Phe	Phe	Phe		bulky non-polar
115Lys	105Glu	Lys	Lys	Arg,Leu,Lys		
?	Phe	100Leu,Ile	105Leu	105Leu		bulky non-polar
?	Ile	Leu	Leu	Leu	G	bulky non-polar
?	Ser	Ser,Ala	Gly	Gly		small size
?	Glu	His,Asn	Asn	Asn		polar
120?	110Ala	Cys,His	Val	Val		
?	Ile	105Leu,Ile	110Leu	110Leu,Ile		bulky non-polar
?	Ile	Leu,Val	Val	Val,Ala,Ile		non-polar
?	His	Val	Thr	Cys,Ile,Leu,Val		
?	Val	Thr,Gly,Ser	Val	Val		
125?	115Leu	Leu,Ile	Leu	Leu,Val		bulky non-polar
?	His	110Ala,Met	115Ala	115Ala,Gly,Ser		
?	Ser	Ala,Asp,Phe,Ser,Val	Ile	His,Arg	↓	
?	Arg	His,Tyr	His	His,Asn,Arg	↑	polar
?	His	Leu,His,Val	Phe	Phe,His,Met		
130?	120Pro	Pro	Gly	Gly		
?	Gly	115Ala,Asn,Gly,Ser	120Lys	120Lys,Asn,Ser		
?	Asn	Glu,Asp	Glu	Glu,Asp		polar
?	Phe	Phe	Phe	Phe		π-bond (?)
?	Gly	Thr,Pro	Thr	Thr,Asx,Ser		
135?	125Ala	Pro	Pro	Pro		non-polar
?	Asp	120Ala,Glu	125Glu	125Pro,Gln,Ala,Leu,Val		
?	Ala	Val	Val	Val,Met,Ala,Leu		non-polar
?	Gln	His	Gln	Gln		polar
?	Gly	Ala,Met	Ala	Ala		non-polar
140?	130Ala	Ser	Ser	Ala,Asp,Ser,Glx	H	
?	Met	125Leu,Val	130Trp	130Tyr,His,Phe		bulky
?	Asn	Asp	Gln	Gln		polar
Lys	Lys	Lys	Lys	Lys		(+) charge
Ala	Ala	Phe	Met	Val		non-polar
145Leu	135Leu	Leu,Phe	Val	Val		bulky non-polar
Ala	Glu	130Ala,Gln,Ser	135Thr	135Ala,Thr		
Gly	Leu	Ser,Asn,Asp	Gly	Gly		
Val	Phe	Val,Leu	Val	Val		bulky non-polar
Ala	Arg	Ser,Ala	Ala	Ala		
150Ile	140Lys	Thr,Leu	Ser	Asn,Asp,Ala		
Val	Asp	135Val,Ala	140Ala	140Ala		
Leu	Ile	Leu	Leu	Leu		bulky non-polar
Arg	Ala	Thr,Ser	Ser	Ala		
Ser	Ala	Ser,Glu	Ser	His		
155Ala	145Lys	Lys	Arg	Lys,Arg		(+) charge (not lamprey)
Tyr	Tyr	140Tyr	145Tyr	145Tyr		H-bond
	Lys	Arg	His	His	↓	(+) charge
	Glu					
	Leu					
	150Gly					
	Tyr					
	Gln					
	Gly					

amino acid, *alanine*, on one of these positions. Evidently, the lamprey can tolerate this slightly larger bulk at the crossover.

The *tyrosyl* residue used to tie down the C-terminal end of the chains by hydrogen bonding is always present (α140), as is the *phenylalanyl* residue in contact with the porphyrin ring (α43).

There are other constant, or nearly constant, residues. One is a *leucyl* residue (α83) in the heme pocket, which is one of the residues in van der Waal's contact with the porphyrin ring, and it may be that no other residue would fit exactly. A *lysyl* residue, with its positive charge, is always present at the position corresponding to α127, but it is not known why, because there is no obvious interaction of the charged group with other parts of the molecule.

There is a residue of *tryptophan* (α14) in myoglobin and in all of the tetrapeptides, except that a hemoglobin from llamas and another from sheep have *phenylalanine* at this position. The function is obscure, but tryptophan has an aromatic ring and may be orienting the molecule in some unsuspected way:

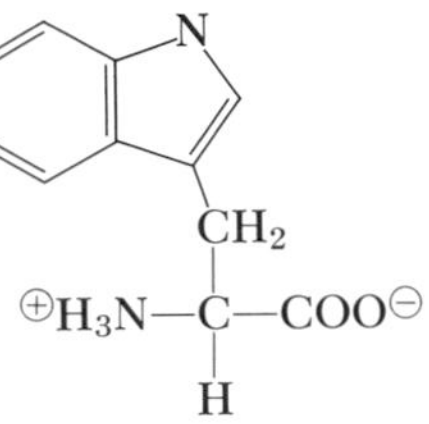

$$\oplus H_3N-\underset{\underset{H}{|}}{\overset{\overset{CH_2}{|}}{C}}-COO^{\ominus}$$

L-tryptophan

Similarly, there is a second *phenylalanyl* residue (α46) in the heme pocket of myoglobin and all of the tetrapeptides with the exception of a carp hemoglobin, which has a *tryptophanyl* residue, suggesting that tryptophan and phenylalanine may sometimes be interchangeable for π-bonding.

There is a *threonyl* residue near one of the sharp bends (α39) in all except the primitive lamprey hemoglobin, and it may be that its side chain hydroxyl group is used to form a critical hydrogen bond stabilizing the position of the bend, as shown in Figure 4-7.

Similar Residues

The constant residues may seem like very few out of the 140 odd residues in each peptide, but there are many more cases in which the same kind of function is always displayed by *similar* amino acids in every known chain. According to our

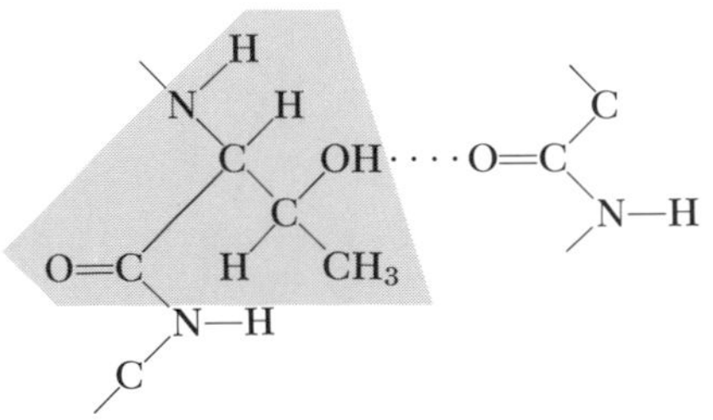

Figure 4-7 The hydroxyl group of threonyl residues may form hydrogen bonds with neighboring parts of a peptide chain, thereby stabilizing the structure.

analysis of the structure, the positioning of the non-polar side chains that are folded into the interior is critical. Now, consider the residues of the amino acids, *valine, leucine, isoleucine, methionine,* and *phenylalanine:*

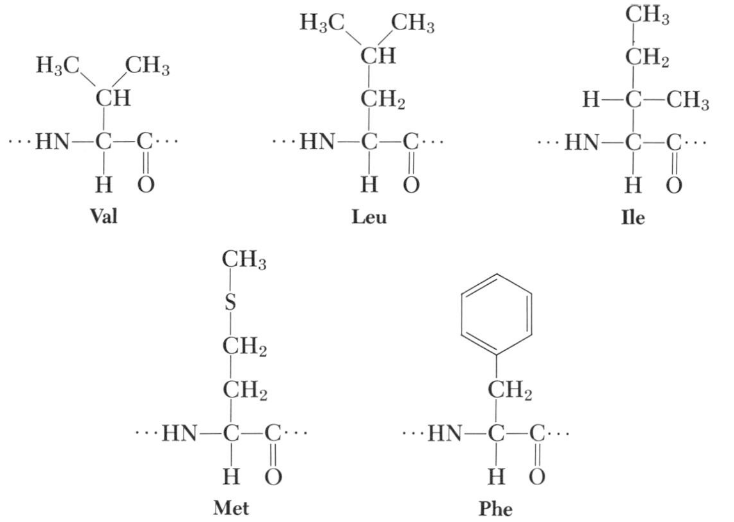

Val   Leu   Ile

Met   Phe

All of these residues have what might be described as *bulky non-polar* side chains. A substitution of one for the other will not change the non-polar character of the position in the peptide. The only effect of the substitution will be to modify the fit of adjacent segments of the molecule because of the different geometry of the space occupied by the side chain and because of changes in van der Waal's interactions. (Modifications of this type may well contribute to the differences in affinity for oxygen of hemoglobins from different species.)

All told, there are 22 positions that are invariably occupied by these bulky, non-polar residues. The list is particularly impressive in the helical segments, where these highly non-polar positions tend to occur at intervals of three or four residues, which is just what one would expect in a helix of approximately 3.6 residues per turn with one side of the helix creating a non-polar region.

Six positions always have one of the basic amino acids that may be positively charged, but only two positions always carry the negatively charged *glutamyl* or *aspartyl* residues. This is not to say that the surface can be low in total negative charges. The net charge of hemoglobin is slightly negative at the physiological pH, but the particular location of most of the negative charges does not appear to be used for structural integrity. In some positions, molecules from related species have the amide residues of *glutamine* or *asparagine* in place of glutamate or aspartate. These have no net negative charge, but their shape and volume is close to that of the free carboxylates, which may be sufficient to maintain shape and have the necessary interaction with water at the surface.

All in all, there are about 70 positions in which residues of similar function occur in each of the known chains. This gives us encouragement to believe that our interpretation of the factors maintaining structure is correct. There have been many opportunities to test the elimination of these particular residues during the several hundred million years of evolution of the chains.

The evolution of another heme protein, *cytochrome c,* has been studied in detail. The function of this protein will be discussed in Chapter 10. Suffice it now to note that the protein is involved in the transport of electrons to oxygen. This is a more ancient and critical function than the transport of molecular oxygen by hemoglobin. Cytochrome c from human hearts contains a single chain with 104 amino acid residues. When this protein is compared with cytochrome c from other species down the phylogenetic scale to yeast, one finds 35 positions that are occupied by identical residues in all plant and animal sources studied. There are many more positions that are consistently occupied by residues of similar reactivity toward water or of similar charge.

Human Variants

Human blood is probably more readily accessible than any other experimental tissue, and samples are drawn each year from millions of individuals. Human hemoglobin therefore represents a unique opportunity for studying genetic variation of a single protein in a single species, each variation representing an alteration in those DNA molecules affecting the synthesis of the protein.

Human chromosomes are paired, so that each nucleated cell will contain two molecules of each type of DNA. A mutation involving the substitution of one nucleotide for another will be transmitted if it has occurred in the molecule retained in the germ cell used for conception. The offspring will usually have the original type of molecule from the other parent, and will be *heterozygous* in genetic language—each of the child's cells will have one molecule of DNA that carries the change and one that does not. In clinical language, the child would have a trait for whatever organic change the molecular substitution has caused.

If the mutation is not deleterious, it may persist in later generations. If the particular individuals carrying the mutation survive by chance, there is an increasing probability of the mating of two such individuals with the offspring having the mutation in both molecules of the particular type of DNA. These offspring are now *homozygous* to the mutation.

Should the mutation convey some reproductive advantage to the particular individuals who first are affected and there are sufficient fertile matings among them to get the change established, the occurrence of the new molecule will gradually increase among the population.

Most mutations are deleterious, and will gradually be eliminated, the rate depending upon the severity of the reproductive damage resulting. It must not be overlooked that advantage or disadvantage in these cases refers only to the rate of increase of the number of descendants. (A mutation absolutely limiting human life span to 50 years, if such were possible, might be an advantage for offspring in some cultures and a disadvantage for others.)

Thalassemias

We shall mainly be concerned with mutations affecting coding, but mutations may affect any function of DNA, including the coding for ribosomal or transfer

RNA and the regulator genes that determine the rate of synthesis of particular kinds of messenger RNA. The latter kind of mutation occurs in humans at sites affecting hemoglobin synthesis, and the results are informative.

Mutations causing a diminished rate of formation of messenger RNA bearing the coding for one of the hemoglobin chains result in conditions known as *thalassemia*. Homozygous thalassemia is obviously a serious condition, and especially so if the defect is a failure to stimulate α chain production, because all of the usual human hemoglobins contain α chains. Production of β, γ, and δ chains will continue. These chains can associate to make tetrapeptides, such as β_4, γ_4, and δ_4, which are composed of only one type of chain. However, the association of peptides is weaker than the dimer associations in normal hemoglobin. The molecules have the myoglobin type of oxygen dissociation curve and lack the Bohr effect, and therefore cannot effectively replace the normal molecule. The discovery of the nature of the abnormality reinforces our picture of the normal hemoglobin molecule as one in which the peptides are specifically aligned to create the sigmoid dissociation curve and the Bohr effect.

The thalassemias in which β chain production is repressed are interesting because they result in an increased production of Hb F in the adult. In geneticists' jargon, the operator gene for fetal hemoglobin is derepressed by the failure to make normal amounts of Hb A_1. Our picture of the mechanism of protein synthesis requires that the capability for producing messenger RNA with coding for γ chains must still be present in the adult, because he ought to be carrying his original kinds of DNA. This is confirmed by the restoration of γ chain production when β chain production is impeded.

Codon Changes

Now let us turn to those variants in whom codons have been changed so that an amino acid at one position in one kind of peptide chain is replaced by another. The substitution may substantially change the properties of the hemoglobin molecule, but even if it does not, it sometimes interferes with the normal rate of production of messenger RNA, so that the cells of a heterozygote will contain less of the new hemoglobin than of Hb A_1, and will have increased amounts of Hb F to compensate for the deficiency. The homozygote may be definitely anemic because of the slower hemoglobin synthesis.

Humans with long ancestry in a region of the world ranging from west Africa across the Mediterranean to southeast Asia commonly have four variant hemoglobins—*Hb S, Hb C, Hb D Punjab*, and *Hb E*—as well as thalassemia. All of these variants are so common that they cannot in any sense be regarded as abnormal. As many as 30 per cent of the individuals in some areas will have a trait for one or the other variant, that is, will be heterozygous to one of the hemoglobin genes.

Variants resulting from amino acid substitutions are designated by letters or geographical locations, or both. The variants were originally discovered because some of them involve a change in the total charge on the molecule, causing the molecules to migrate at a different rate in an electric field (Fig. 4-8) so that they will separate from unmodified hemoglobin. The letters are intended to designate types of hemoglobins with similar *electrophoretic mobilities* (rate of movement in an electric field). Later, each hemoglobin was named after the location at which

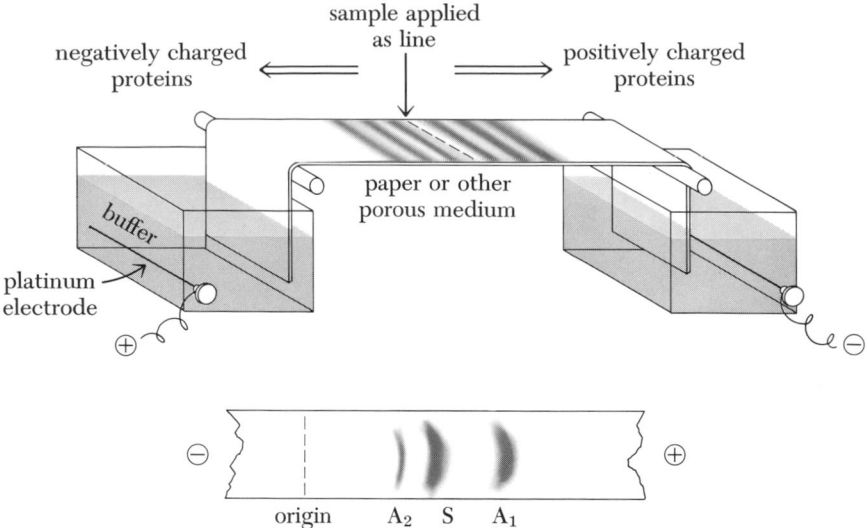

Figure 4-8 *Top.* Zone electrophoresis. A strip of porous material (paper, cellulose acetate) is saturated with a buffer solution and suspended between two containers of the buffer that are fitted with inert electrodes. A mixture of proteins is applied as a thin stripe. Those proteins having different net charges will separate when a voltage is applied. For example, the protein with the greatest excess of negative over positive charges, *i.e.*, the greatest net negative charge, will experience the greatest force propelling it toward the positive electrode and it will move as a separate band ahead of all the others. Likewise, proteins with more positive than negative charges will separate toward the negative electrode. The drawing is highly schematic. Practical apparatus accommodates several strips at once, and they are enclosed to minimize evaporation.

Bottom. The actual result of electrophoresis of the contents of erythrocytes from a human with sickle cell trait. The buffer used was at pH 8.6—a frequent choice at which nearly all proteins will have a net negative charge and be separated toward the positive electrode from the point of application. The proteins were stained with a dye after separation. This individual has two major and one minor hemoglobin components, identified as hemoglobins A_1, S, and A_2. New variants of hemoglobin are frequently discovered by this simple procedure.

an individual carrying it was first discovered. The nomenclature is clumsy, although sometimes picturesque. For example, hemoglobins M Saskatoon, M Emory, M Kurume, and M Chicago have a splendid cosmopolitan flavor, but they all turned out to be identical mutations.

It is frequently more helpful to use a nomenclature specifying the amino acid substitution. For example, the first variant discovered is Hb S (S = sickle), which is a result of the substitution of a valyl residue for a glutamyl residue at position 6 of the β chain. The substitution is represented as $\beta^{6\text{Glu}\rightarrow\text{Val}}$ (read as Glu *becoming* Val), and the hemoglobin as $\alpha_2\beta_2{}^{6\text{Glu}\rightarrow\text{Val}}$ or simply as $\alpha_2\beta_2{}^{6\text{Val}}$.

The other common variants we mentioned (Hbs C, D, and E) also have substitutions in the β chain, which will only affect the structure of Hb A_1, and not of Hb A_2 or of Hb F. These and the less common known variants, which include some in the α, γ, or δ chains, are listed in Table 4-2.

The high frequency of occurrence of Hb S was not realized for some time, because the mutation is relatively innocuous to heterozygotes, although the less common homozygotes develop an anemia. Replacement of a glutamyl residue by

TABLE 4-2. VARIANTS OF HUMAN HEMOGLOBIN WITH KNOWN PRIMARY STRUCTURES[°]

α-CHAIN VARIANTS		β-CHAIN VARIANTS		δ-CHAIN VARIANTS		γ-CHAIN VARIANTS	
Substitut⁻	Name	Substitution	Name	Substitution	Name	Substitution	Name
5 Ala → Asp	J Toronto	6 Glu → Lys	C	2 His → Arg	Sphakiá	6 Glu → Lys	F Texas
12 Ala → Asp	J Paris	6 Glu → Val	S	16 Gly → Arg	A′₂		
15 Gly → Asp	J Oxford	6 Glu → Val	C Harlem	22 Ala → Glu	Flatbush		
16 Lys → Glu	I	+73 Asp → Asn					
22 Gly → Asp	J Medellin	7 Glu → Gly	G San Jose				
23 Glu → Gln	Memphis	7 Glu → Lys	Siriraj				
30 Glu → Gln	G-Honolulu	16 Gly → Asp	J Baltimore				
47 Asp → Gly	L-Ferrara	22 Glu → Ala	G Coushatta				
51 Gly → Arg	Rus	22 Glu → Lys	E Saskatoon				
54 Gln → Glu	Mexico	22 Glu → Ala	G Saskatoon				
54 Gln → Arg	Shimonoseki	26 Glu → Lys	E				
57 Gly → Asp	Norfolk	28 Leu → Pro	Genova				
58 His → Tyr	M Osaka	46 Gly → Glu	K Ibadan				
68 Asn → Lys	G Philadelphia	47 Asp → Asn	G Copenhagen				
68 Asn → Asp	Ube-2	56 Gly → Asp	J Meinung				
87 His → Tyr	M Iwate	61 Lys → Asn	Hikari				
92 Arg → Leu	Chesapeake	63 His → Tyr	M Saskatoon				
92 Arg → Gln	J Capetown	63 His → Arg	Zurich				
115 Ala → Asp	J Tongariki	67 Val → Ala	Sydney				
116 Glu → Lys	O Indonesia	67 Val → Glu	M Milwaukee				
		69 Gly → Asp	J Cambridge				
		70 Ala → Glu	Seattle				
		79 Asp → Asn	G Accra				
		87 Thr → Lys	D Ibadan				
		90 Glu → Lys	Agenogi				
		95 Lys → Glu	N Baltimore				
		98 Val → Met	Köln				
		99 Asp → His	Yakima				
		102 Asn → Thr	Kansas				
		113 Val → Glu	New York				
		120 Lys → Glu	Hijiyama				
		121 Glu → Gln	D Punjab				
		121 Glu → Lys	O Arab				
		126 Val → Glu	Hofu				
		132 Lys → Gln	K Woolwich				
		136 Gly → Asp	Hope				
		143 His → Asp	Kenwood				
		145 Tyr → His	Rainier				

[°] The change in amino acid residues is indicated by listing the usual residue first and the variant residue second, so Ala → Asp is to be read as Ala becoming Asp. The number indicates the position of the residue in the particular chain.

the non-polar valyl residue makes it possible for the N-terminal portion of the chain to coil up on itself. This has little effect on oxygen transport, but it lowers the solubility of the hemoglobin by making it easier for molecules to stack up on each other in an ordered array. The Hb S concentration exceeds the solubility in erythrocytes from homozygotes. The molecules form long strands, associated into tubules. The molar volume of the molecule is thereby diminished, and the usual discoid erythrocyte puckers into a sickle shape. The erythrocyte becomes more fragile, and the increased rate of destruction causes an anemia—*sickle-cell anemia.* Oxygenated Hb S is more soluble than is Hb S itself, so the sickling phenomenon occurs in more cells at the lower oxygen tension of the venous circulation. Even the cells of heterozygotes will sickle if the oxygen tension is low enough. Heterozygotes are said to have *sickle-cell trait,* and they sometimes have difficulty at high altitudes.

Once the prevalence of sickle-cell trait was realized, a dilemma was presented. The homozygous condition is usually lethal in early life, and this would cause a constant loss of the mutation from the hereditary pool. Since it has persisted, its presence must have given some compensatory advantage. The advantage, it turned out, was a protection against a particularly dangerous species of malarial parasite, the tropical *Plasmodium falciparum,* which spends part of its life cycle within erythrocytes. The incidence of Hb S in Africa parallels the incidence of falciparum malaria, and it has been clearly shown that the chances of a child's surviving malaria are greatly enhanced if he has Hb S in his blood. Those who move from the malarial belt lose this precarious advantage over Hb A_1 homozygotes, and there is evidence that the incidence of Hb S has declined in their descendants, because only the deleterious aspect of the mutation can then exhibit itself.

It has been concluded that the other variants common to the malarial belt convey the same type of protection, but the evidence in terms of quantitative data is skimpy. The conclusion rests largely on such findings as the observation that the incidence of thalassemia in Italian villages in which there has been a minimum of breeding with outsiders also parallels the severity of the malarial infestation in the villages.

RARE CODON VARIANTS. Most of the known mutations listed on p. 75 have been found in only a few individuals, and have little importance as medical problems, but are important for the information they give on the function of hemoglobin. What can be learned from these mutants? Let us first consider those variants designated as *hemoglobin M.* These are variants in which the iron of hemoglobin is spontaneously oxidized to the ferric state. The resultant ferriporphyringlobin is called *methemoglobin,* hence the use of M as a designation. Since the variant molecules differ from Hb A_1 in only two of the four chains, the molecule will have two peptides with the oxidized hemin (ferriporphyrin IX) and two with heme (ferroporphyrin IX).

Hemin peptides cannot complex with oxygen, and therefore will not function as a carrier. The unaffected heme peptides in the variant molecules can complex, and in a nearly normal way if they are α chains, but this still leaves only half of the normal capacity for oxygen binding. Persons homozygous to the Hb M mutation might not survive, but the heterozygotes are amazingly normal in all but cosmetic respects. Methemoglobin has a brown color, and a person with a high concentration has the cyanotic appearance usually associated with oxygen deprivation. (The color is less red, and therefore is interpreted as more blue even though it is not really blue.) Despite the abnormality, there is little real sign of oxygen insufficiency in the individuals, and their erythrocytes have nearly normal lifetimes. The known variants are as follows:

HEMOGLOBIN M VARIANTS

Substitution	Name	Location of Affected Residue
$\alpha^{58\mathrm{His}\rightarrow\mathrm{Tyr}}$	M Osaka	heme pocket—complexes with oxygen
$\alpha^{87\mathrm{His}\rightarrow\mathrm{Tyr}}$	M Iwate	heme pocket—binds iron
$\beta^{63\mathrm{His}\rightarrow\mathrm{Tyr}}$	M Saskatoon	heme pocket—complexes with oxygen
$\beta^{67\mathrm{Val}\rightarrow\mathrm{Glu}}$	M Milwaukee	heme pocket—in contact with porphyrin

Three of the substitutions replace histidyl residues in the heme pocket, and the resultant susceptibility of the iron to spontaneous oxidation serves as a reminder of the importance of the entire peptide-iron-porphyrin chelate. When the chelate is intact, oxygen forms an unreactive complex. When it is disrupted, oxygen spontaneously reacts with the ferrous porphyrin, as is commonly the case with ferrous compounds, oxidizing it to the ferric porphyrin.

M Milwaukee is an interesting case in which a valyl residue is replaced by a polar residue in the hydrophobic lining of the heme pocket. The position is four residues removed from the oxygen-binding histidyl residue in the E helix, and just enough to direct the residue back into the pocket, alongside the histidyl residue. The introduction of a charged glutamyl residue at this position in place of the non-polar valyl residue disrupts the electronic configuration of the oxygen complex and permits reaction between heme iron and oxygen, forming the ferric porphyrin, hemin.

Some substitutions even closer to the histidyl residue, such as $\beta^{61Lys \rightarrow Asn}$ (Hb Hikari) and $\alpha^{57Gly \rightarrow Asp}$ (Hb Norfolk), do not cause methemoglobin formation and have little effect on function because the helix turns these residues away from the heme pocket.

However, we are reminded that our understanding is not complete by the existence of Hb Zurich ($\beta^{63His \rightarrow Arg}$) in which there is a direct substitution of an arginyl residue for the histidyl residue, with little effect on function (compare M Saskatoon). The arginyl side chain does carry a positive charge and has a resonant structure, but it is sufficiently different from the imidazole group of histidine to make one cautious about positive statements on the actual mechanism of oxygen binding. There is some consolation in knowing that the substitution is not fully equivalent to the original molecule. Hb Zurich is rapidly oxidized to methemoglobin if the mutant individual is given sulfonamides, and this is how it was discovered.

Let us turn now to the few variants involving substitutions for non-polar residues:

Substitution	Name	Location of Affected Residue
	VARIANTS WITH SUBSTITUTIONS FOR NON-POLAR RESIDUES	
$\alpha^{5Ala \rightarrow Asp}$	J Toronto	surface
$\alpha^{12Ala \rightarrow Asp}$	J Paris	surface
$\alpha^{115Ala \rightarrow Asp}$	J Tongariki	surface
$\beta^{28Leu \rightarrow Pro}$	Genova	interior near crossover of B and E helices
$\beta^{67Val \rightarrow Ala}$	Sydney	heme pocket—porphyrin contact
$\beta^{70Ala \rightarrow Glu}$	Seattle	heme pocket—porphyrin contact
$\beta^{98Val \rightarrow Met}$	Köln	heme pocket—porphyrin contact
$\beta^{113Val \rightarrow Glu}$	New York	surface
$\beta^{126Val \rightarrow Glu}$	Hofu	crevice on surface

Five of these mutations involve residues at the surface of the molecule, even though they are non-polar, and the substitution of charged carboxylates for these residues has little effect on function. This is consistent with the deduction made from the large observed variation in these residues among species—the exact placement of charged groups on the surface is often not critical for maintenance of structure.

The remaining four mutations are definitely deleterious. They cause the hemoglobin molecule to be unstable, which results in an increased destruction of erythrocytes and hemolytic anemia. The leucyl residue at $\beta 28$, which is replaced by a prolyl residue in Hb Genova, is directed toward the interior from the middle of helix E, and the only known substitution in peptides from all species is by the similar isoleucyl residue, except for this one human variant. Since the prolyl residue must interrupt the helix, according to our view, the observed instability of the hemoglobin confirms the expectation.

Three substitutions are in the heme pocket. Hb Sydney is of interest because it is a substitution for Val at the same position changed in Hb M Milwaukee, but with an alanyl rather than the charged glutamyl residue. M Milwaukee is readily oxidized but stable; Sydney binds oxygen without oxidation, but it is unstable. The glutamyl side chain of M Milwaukee is more nearly the size of the valyl side chain that has been replaced than is the single methyl group of the alanyl side chain. This shows that size and shape, as well as lack of charge, are necessary in the heme pocket to preserve the geometry of the interior surface. This would also explain the instability of Hb Köln, in which the bulky non-polar side chain of valine is replaced by the considerably more bulky non-polar side chain of methionine.

Hb Seattle, like Hb M Milwaukee, is created by a substitution of a charged glutamyl residue for a non-polar residue within the heme pocket. However, the substitution in this case is on the opposite side of the porphyrin ring from the oxygen-complexing histidyl residue, and more toward the periphery of the ring, so the result is the creation of instability in the peptide without disrupting the oxygen complex.

There are undoubtedly many more substitutions for non-polar residues that have not been detected because they involve the substitution by other non-polar groups. In such cases, the hydrophobic interior would remain hydrophobic, and the disruption of function might well be too small to attract clinical attention. Unless there were symptoms, such a substitution might escape discovery indefinitely because there would be little, if any, difference in electrical charge of the molecule, and it would not separate from ordinary Hb A_1 in routine screening of blood samples by electrophoresis.

Three other hemoglobin variants are of interest because they produce a marked change in the oxygen affinity of the molecule. There is a varying degree of loss of heme-heme interaction and increased oxygen affinity, so that the oxygen dissociation curve approaches that obtained with myoglobin. These are:

Substitution	Name	Location of Variant Residue
α^{92}Arg→Leu	Chesapeake	surface, at contact between dimers
β^{99}Asp→His	Yakima	surface, next to sharp bend
β^{145}Tyr→His	Rainier	next to C-terminal—hydrogen bonding

It is easy to rationalize the change in oxygen affinity caused by the first and last of these. As we have seen, there is ample evidence for a belief that the character of the oxygen dissociation curve depends upon the interaction between the four peptide chains. The substitution of a residue so as to change the charge at a point of contact between the two pairs of dimers, or the loss of the hydrogen-bonding tyrosyl residue at the C-terminal portion of the chain—an end in contact with a second chain—might well be expected to alter the nature of the interactions. Hb

Yakima would be less easy to predict with present knowledge. The substitution is adjacent to the ubiquitous prolyl group at one of the critical bends, and may change the angle of the bend somewhat, but it is on the surface away from the peptide-peptide contacts.

GENETIC CODING AND STRUCTURE

The Relationship Between Code and Structure

We have gone over evidence that peptide chains evolved so as to retain amino acids of related function in particular portions of the chain, and have looked at known mutations in humans. These data can also be used to test the validity of our knowledge of the genetic code. A simple mutation involves a change in one base of DNA, and the existence of evolution requires favorable mutations. Thus, there are two reasonable expectations in a genetic code. First, only one base change ought to be required to produce the known human variants. Secondly, changes of single nucleotides ought to be likely to substitute amino acids of similar structure, so that the mutant peptide is able to retain biological function. Let us now see how well these expectations are fulfilled. The codons for each amino acid according to current knowledge are listed in Table 4-3.

Coding for Human Variants

When we match the known variants (p. 75) with the proposed codons, it turns out that every one can be accounted for by a single substitution in a coding triplet. This is such a striking confirmation of the codon dictionary that the tables have been turned, and the codon dictionary has been used to detect an error in the determination of structure of a variant hemoglobin. Hemoglobin I was reported to be a Lys → Asp substitution at a time before the codon dictionary was established. Such a variant would require a double mutation according to the dictionary—a change of two nucleotides in a codon. While this is theoretically possible, the probability is quite low. For example, if a single change within a codon occurs once in 10^8 individuals, a double change ought to occur only once in 10^{16} individuals. When the improbability of the substitution was realized, the amino acid composition of Hb I was reinvestigated, and the original report was shown to be in error, with the actual substitution being Lys → Glu, which involves only a single mutation.

It ought to be pointed out that there are some apparently double mutations that may be discovered. Hemoglobin C Harlem is an example representing a second mutation of the relatively common Hb S, except that the second change occurs at another position in the chain. There is a reasonable chance that a mutation of one or the other of the common variants will be discovered that represents an additional single mutation of the variant residue, but which will appear to be a double mutation of Hb A_1.

Even a casual inspection of the previous table of human mutants shows that certain kinds of amino acid substitutions, for example, Glu → Lys, occur at several different locations. This particular substitution could occur by GpApA → ApApA, or by GpApG → ApApG, either of which involves a substitution of adenine for guanine in the triplet.

TABLE 4-3.   MESSENGER RNA CODONS

Listed by Codon

ApApA—Lys	CpApA—Gln	GpApA—Glu	UpApA—Term°
ApApG—Lys	CpApG—Gln	GpApG—Glu	UpApG—Term°
ApApC—Asn	CpApC—His	GpApC—Asp	UpApC—Tyr
ApApU—Asn	CpApU—His	GpApU—Asp	UpApU—Tyr
ApCpA—Thr	CpCpA—Pro	GpCpA—Ala	UpCpA—Ser
ApCpG—Thr	CpCpG—Pro	GpCpG—Ala	UpCpG—Ser
ApCpC—Thr	CpCpC—Pro	GpCpC—Ala	UpCpC—Ser
ApCpU—Thr	CpCpU—Pro	GpCpU—Ala	UpCpU—Ser
ApGpA—Arg	CpGpA—Arg	GpGpA—Gly	UpGpA—Cys (?)
ApGpG—Arg	CpGpG—Arg	GpGpG—Gly	UpGpG—Trp
ApGpC—Ser	CpGpC—Arg	GpGpC—Gly	UpGpC—Cys
ApGpU—Ser	CpGpU—Arg	GpGpU—Gly	UpGpU—Cys
ApUpA—Ile (?)	CpUpA—Leu	GpUpA—Val	UpUpA—Leu
ApUpG—Met	CpUpG—Leu	GpUpG—Val	UpUpG—Leu
ApUpC—Ile	CpUpC—Leu	GpUpC—Val	UpUpC—Phe
ApUpU—Ile	CpUpU—Leu	GpUpU—Val	UpUpU—Phe

Listed by Amino Acid

Ala—GpCpA	Gly—GpGpA	Leu—CpUpA	Thr—ApCpA
GpCpG	GpGpG	CpUpG	ApCpG
GpCpC	GpGpC	CpUpC	ApCpC
GpCpU	GpGpU	CpUpU	ApCpU
		UpUpA	
Arg—ApGpA	Gln—CpApA	UpUpG	Trp—UpGpG
ApGpG	CpApG		
CpGpA		Met—ApUpG	Tyr—UpApC
CpGpG	Glu—GpApA		UpApU
CpGpC	GpApG	Phe—UpUpC	
CpGpU		UpUpU	Val—GpUpA
	His—CpApC		GpUpG
Asp—GpApC	CpApU	Pro—CpCpA	GpUpC
GpApU		CpCpG	GpUpU
	Ile—ApUpC	CpCpC	
Asn—ApApC	ApUpU	CpCpU	
ApApU	ApUpA (?)		
		Ser—ApGpC	
Cys—UpGpA (?)	Lys—ApApA	ApGpU	
UpGpC	ApApG	UpCpA	
UpGpU		UpCpG	
		UpCpC	
		UpCpU	

° Term are terminator codons. The codons ApUpA and UpGpA may code for Ile and Cys, respectively, but this is not certain. Note that amino acids within a block of codons, or in the corresponding position in the same row or column of blocks, are interchangeable by substitution of one nucleotide.

For some reason, the adenine-guanine exchange appears to be a particularly prevalent mutation. There are six possible interchanges of the four kinds of bases, but 30 out of the 60 listed human variants represent only the adenine-guanine interchange. The reason for this is not clear.

Evolutionary Changes and Coding

The analysis we made of the evolution of the hemoglobin chains showed that there is an apparent conservation of function, with non-polar amino acids substituted for other non-polar amino acids, small side chains substituted for other small side chains, and so on. One would expect this conservation to be reflected in the coding, and the actual agreement is exciting.

For example, phenylalanyl, isoleucyl and some leucyl and valyl residues can be interchanged by single substitutions of nucleotides. The remaining leucyl and valyl residues are interchangeable with each other and with methionyl residues by a single mutation. Seryl residues may be converted by single substitutions to the three-carbon alanyl or cysteinyl residues, or to the other alcohol, the threonyl residue. The smallest residues, glycyl and alanyl, can be exchanged in one step, as can glutamyl and glutaminyl, aspartyl and asparaginyl, or aspartyl and glutamyl.

All in all, the success in matching the coding with the perpetuation of particular properties in the proteins is amazingly good. It reinforces the validity of our analysis of the function of both nucleic acids and proteins, and more importantly, raises the possibility that the genetic code itself is an evolutionary development to minimize the number of deleterious mutations.

Recapitulation of types of reactions

1. Heme proteins may complex with oxygen without oxidation of the iron:

$$[Fe^{2+}\text{-porphyrin-peptide}] + O_2 \rightleftharpoons O_2{=}[Fe^{2+}\text{-porphyrin-peptide}]$$

Example: The binding of oxygen to various hemoglobins.

2. Heme proteins may react with oxygen so as to oxidize the iron:

$$4[Fe^{2+}\text{-porphyrin-peptide}] + O_2 + 4H^+ \rightleftharpoons 4[Fe^{3+}\text{-porphyrin-peptide}] + 2H_2O.$$

Example: The oxidation of hemoglobins to methemoglobins.

Further reading

Neurath, H., ed.: *The Proteins.* Academic Press (1963). A multi-volume authoritative, but advanced, reference work.

Schroeder, W. A.: *The Hemoglobins.* Ann. Rev. Biochem., *32:* 301 (1963).

Cullis, A. F., H. Muirhead, M. F. Perutz, and M. G. Rossmann: *The Structure of Hemoglobin IX.* Proc. Roy. Soc. London A, *265:* 161 (1962).

Perutz, M. F., J. C. Kendrew, and H. C. Watson: *Structure and Function of Hemoglobin I and II.* J. Molec. Biol., *13:* 646 and 669 (1965).

Guidotti, G.: *Studies on the Chemistry of Hemoglobin I, II, III.* J. Biol. Chem., *242:* 3673, 3685, and 3694 (1967). These papers develop the importance of the $\alpha\beta$ dimer in determining the equilibrium with O_2.

Bryson, V., and H. J. Vogel: *Evolving Genes and Proteins.* Academic Press (1965).

Stevens, F. C., A. N. Glazer, and E. L. Smith: *The Amino Acid Sequence of Wheat Germ Cytochrome c.* J. Biol. Chem., *242:* 2764 (1967).

Park, S. G., and E. M. Scott: *The Hereditary Methemoglobinemias.* Also H. Lehmann, R. G. Huntsman, and J. A. M. Ager: *The Hemoglobinopathies and Thalassemia.* In J. B. Stanbury, J. B. Wyngaarden, and D. S. Frederickson, eds.: *The Metabolic Basis of Inherited Disease* (p. 1090 and p. 1100). McGraw-Hill (1966).

Schroeder, W. A., et al.: *Evidence for Multiple Structural Genes for the* γ *Chain of Human Fetal Hemoglobin.* Proc. Natl. Acad. Sci., *60:* 537 (1968).

Winterhalter, K. H., G. Amiconi, and E. Antonini: *Functional Properties of a Hemoglobin Carrying Heme Only on* α *Chains.* Biochemistry, *7:* 2228 (1968).

The Genetic Code. Cold Spring Harbor Symposia on Quantitative Biology, Vol. 31 (1966).

FIBROUS PROTEINS

CHAPTER 5

ARGUMENT

The properties of the proteins used in the formation of structural elements are quite different from those of the highly soluble, globular proteins. The characteristics of low solubility and high mechanical strength that are necessary for the maintenance of structural integrity are attained by twisting long peptide chains into rope-like coils and by forming covalent bonds between peptide subunits.

Collagen is an example; it is a constituent of most connective tissues and the most abundant protein in vertebrates. Three peptide chains, each in a tight helix, are counter-twisted into a tight triple helix to form tropocollagen units. Covalent linkages are formed between peptide chains. Tropocollagen aggregates and polymerizes to form collagen—a stiff rod that can be stacked into arrays having great strength. The polymerization of tropocollagen involves the formation of covalent bonds by the condensation of aldehyde groups, which are derived from lysyl residues. Conditions are known in which the normal formation of collagen is prevented, either by blocking the aldehyde groups, as in lathyrism, or by preventing the oxidation of prolyl residues to the hydroxyprolyl residues peculiar to this protein. The oxidation is a process that requires ascorbic acid. Scurvy is a deficiency of this vitamin, and many of its symptoms are a result of diminished collagen formation.

Keratin is another structural protein and is found in skin, hair, and nails. It has fibers made from single peptide chains coiled into long α-helices. Bundles of three of these coils are counter-twisted. The resultant fibers are embedded in the matrix of a different protein, which is polymerized by the formation of disulfide bonds between molecules; the combination has strength and flexibility. Disulfide bonds, which are made by the oxidation of adjacent cysteinyl residues, are also used to make intrapeptide bonds in a variety of other proteins, and provide another way in which particular three-dimensional configurations can be achieved.

Our previous discussion has been concerned with proteins of globular structure, easily soluble in water, and this is not always a desirable property. Some proteins are used to provide structural elements, and maintenance of structural integrity requires low solubility in water. This is as true for an organelle within a micro-organism as it is for the integument of an elephant. Many structural elements must also have mechanical strength. It is desirable for the organism to resist the slings and arrows of outrageous fortune, and imperative that it not thaw and resolve itself into a dew.

Vertebrate skin contains striking examples of molecules formed to meet these requirements. Two populations of cells are in skin, separated sharply by a thin layer made of a network of fibers. To the outside of the layer, the cells produce an insoluble protein aggregate, *keratin*. These cells and their products constitute the *epidermis*.

Inside the layer is the *dermis*, the deep layer of the skin. Here the cells are producing fibers made of the protein, *collagen*, and the fibers are criss-crossed in a felt-like mat.

The division between dermis and epidermis is as sharply marked by the separate location of collagen and keratin as it is by the visible differences between the cells of the two layers. Both of these proteins are very insoluble. Both have high tensile strength. Both contain fibers. But here the resemblance ends, and it is interesting to see how their structures vary to satisfy the separate requirements of the tissues in which they occur.

COLLAGEN

Occurrence of Collagen

Collagen is the predominant protein of vertebrates. It is not limited to the skin, but occurs in all organs, with its fibers serving to stiffen the ground substance in which tissue cells are embedded, and also to form more discrete connective tissue. Collagen makes up about a third of the total solid matter in bone. Its fibers are arranged at an angle to each other so as to resist mechanical shear from any direction. Cartilage is about half collagen on a dry weight basis, and acquires its toughness from the arrangement of the protein.

The peptide chains comprising collagen are formed in fibroblasts, or in the related osteoblasts of bone and chondroblasts of cartilage. The chains combine into a complete protein molecule, still fairly soluble, called *tropocollagen*. Tropocollagen is secreted into the extracellular space, where it condenses into the rigid collagen structure.

Fibroblasts can move about through tissues. For example, they migrate into the area of a wound and lay down networks of collagen to strengthen the tissue and avoid further damage during healing.

Structure of Collagen

Collagen has a unique kind of amino acid composition. Almost a third of the residues are of the smallest amino acid, glycine. More than one in 10 are prolyl residues. Another one in 10 are prolyl residues with a hydroxyl group at position 4, or less commonly at position 3, of the ring:

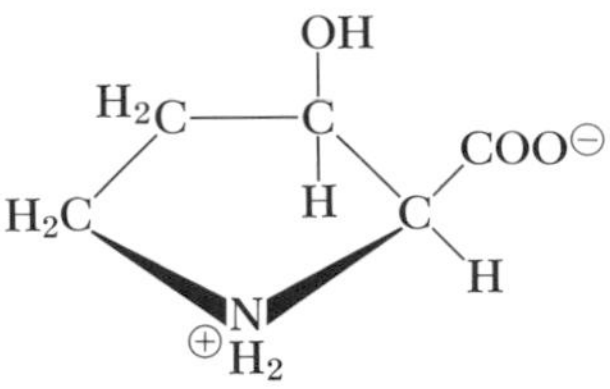

3-hydroxy-L-proline 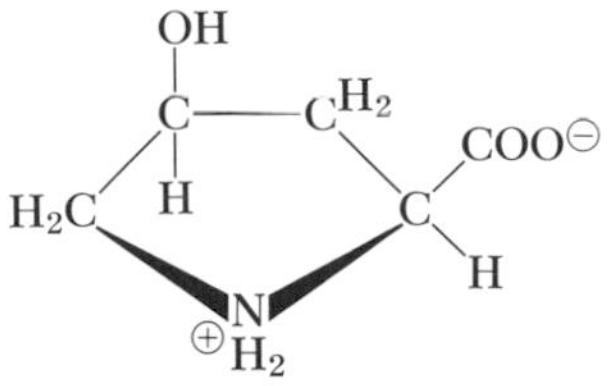   4-hydroxy-L-proline

These *hydroxyprolines*, abbreviated Hyp, have been found only in collagen, and analysis of their content in a tissue is used as a measure of the amount of collagen present. A residue of another unusual amino acid, *5-hydroxylysine*, occurs in collagen to the extent of about 1 per cent of the total.

Hydroxyproline was not listed in our codon dictionary for the good reason that no codon exists for it. Instead, it is formed from unsubstituted prolyl residues after they have been incorporated into the growing collagen peptides on the ribosomes of fibroblasts. The conversion involves an oxidation catalyzed by an uncharacterized enzyme (Fig. 5-1 and p. 93). It hardly seems likely that the oxidation occurs at random, and this raises a question. How can any sort of specific peptide chain be made when prolyl residues are being transformed into hydroxyprolyl residues without the direction of messenger RNA? We don't know, but a possible clue lies in the frequent occurrence of the sequence, —Gly-Pro-Hyp—, and it may be that only the second prolyl residue of a pair can be attacked by oxidation, so that the direction by messenger RNA of the position of pairs of prolyl residues in the chain would thereby also direct the position of the later oxidation.

In any event, we have chains in which every third residue is glycyl, repeated regularly at this spacing throughout most of the length of the peptide. Over one in five of the residues are either prolyl or hydroxyprolyl, although not at regular spacings. The ring structure of these residues prevents α-helix formation, and their abundance eliminates the possibility that this structure is in the collagen molecule.

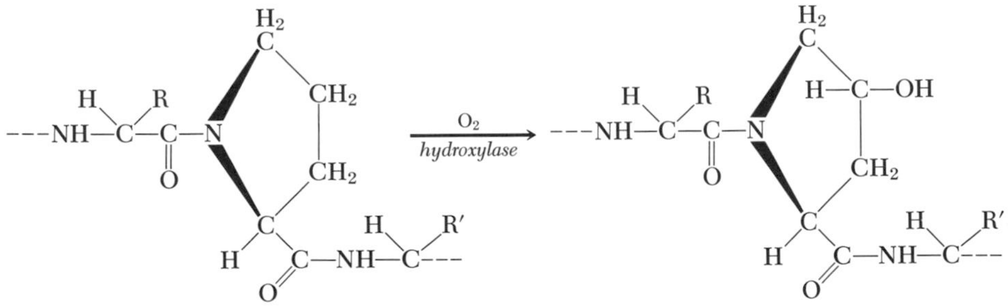

Figure 5-1   The oxidation of prolyl residues in collagen peptides utilizes molecular oxygen and forms hydroxyprolyl residues. The enzyme catalyzing the reaction is in the class of hydroxylases, which will be discussed in more detail in Part IV.

However, the heterocyclic ring does not exclude *all* helical structures, and the particular bond angles it imposes on the chain, together with the opportunity for close packing given by the glycyl residues, makes the most stable structure one in which three peptide chains are twisted together as in a rope. The result is a helix, but it is not to be confused with the α-helix found in many other proteins.

The complete primary structure of the peptide chains is not known; two of the strands in the triple helix appear to be quite similar, perhaps identical, but the third is different, so collagen is like hemoglobin in being made of at least two different peptide chains.

The amino acid composition causes each strand to be twisted to the left one turn for each three residues. This brings all of the glycyl residues to the same side of each strand, and the resultant absence of side chains on this side permits three strands to be tightly packed at this surface. The bundle of three strands has a twist to the right—in the opposite direction of the twist within the strands (Fig. 5-2). The opposing twists give strength against lengthwise deformation, because an unwinding of one tightens the other.

Figure 5-2   The strands of collagen are in a triple helix. Each of three peptide chains is individually coiled in a left-handed helix, and the three together are twisted in a right-handed helix. There are 10 turns in the individual chains per turn of the triple helix.

To give further stability, the strands are linked lengthwise by covalent bonds, which are not seen in hemoglobin, and the current picture of the bonds is this (Fig. 5-3): A single peptide chain constitutes a subunit. Four of these subunits are laid out lengthwise with overlapping ends. The overlap is held together by pairs of *ester-like bonds,* which are formed between carboxyl groups on aspartyl side chains contributed by one of the subunits and groups of unknown character contributed by the next subunit in line. The effect is that of end-to-end splices in the individual strands of triple helices.

The ends of the peptide chains, and perhaps some interior segments, do not have the characteristic coiled form of the remainder of the length, and can be spaced

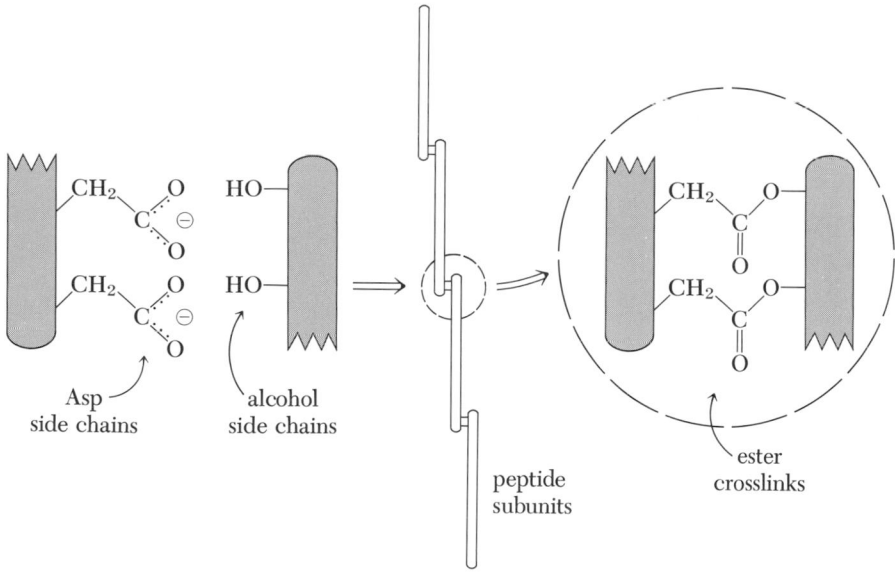

Figure 5-3   The end-to-end linkage of four peptide subunits in a single collagen strand may involve the ester-like bonds known to be present. Each subunit is a part of a triple helix rather than the isolated strand shown here for simplicity. A single strand is made of four subunits; the tropocollagen molecule is three such strands joined in a triple helix. Recent evidence suggests the bonds are aspartyl amides rather than the esters shown here.

more loosely. These portions contain *aldehyde* groups, which are capable of condensing to form additional covalent cross-links. Some of these aldehyde groups are formed by simple reduction of the C-terminal carboxylate groups. Some are formed by oxidation of lysyl residues (Fig. 5-4), and the figure also shows the kinds of structures actually found in collagen that arise by condensation of lysinal and lysyl residues. (The C-terminal aldehydes are more abundant, but the nature of the bridges formed from them has not yet been defined; they probably are similar to those illustrated for lysyl residues.)

The use of the aldehyde side chain is perhaps common in structural proteins. Another protein of connective tissue, *elastin,* has resisted efforts to define its structure, but it is known to contain unusual amino acids, *desmosine* and *isodesmosine* (Fig. 5-5), which are made by linking lysyl groups and the corresponding aldehydes from *four* peptide chains.

The C-terminal aldehydes probably supplement the aspartyl ester-like bridges and the triple helix in locking together the three strands, each composed of at least four subunits linked end-to-end. The result is a tropocollagen molecule with a width of only 1.4 nanometers and a length of 300 nanometers. Its outline is that of a narrow line. There are many opportunities for bond formation on the large surfaces of linear molecules, and they frequently aggregate side-by-side. Tropocollagen does aggregate, except that the staggering of four subunits in each strand makes the most compact structure one in which the molecules overlap by the length of one subunit, or a quarter of their length. Current belief is that the aggregate is stabilized by the formation of additional covalent bonds between the triple-stranded tropocolla-

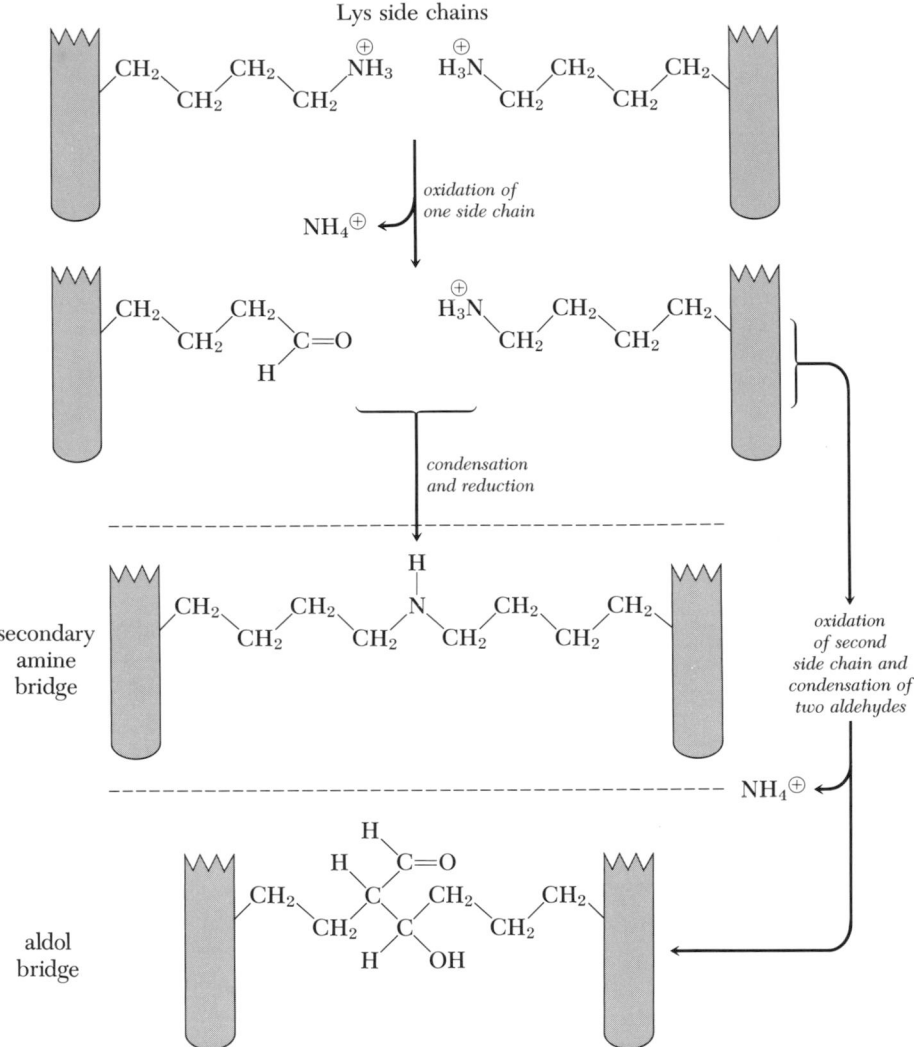

Figure 5-4   The use of lysyl side chains in forming bonds between strands after converting them to aldehydes. The "lysinal" may react with a lysyl ammonium group, followed by reduction of the resulting Schiff's base, or it may react with another aldehyde by an aldol condensation. Both types of linkage have been found in collagen.

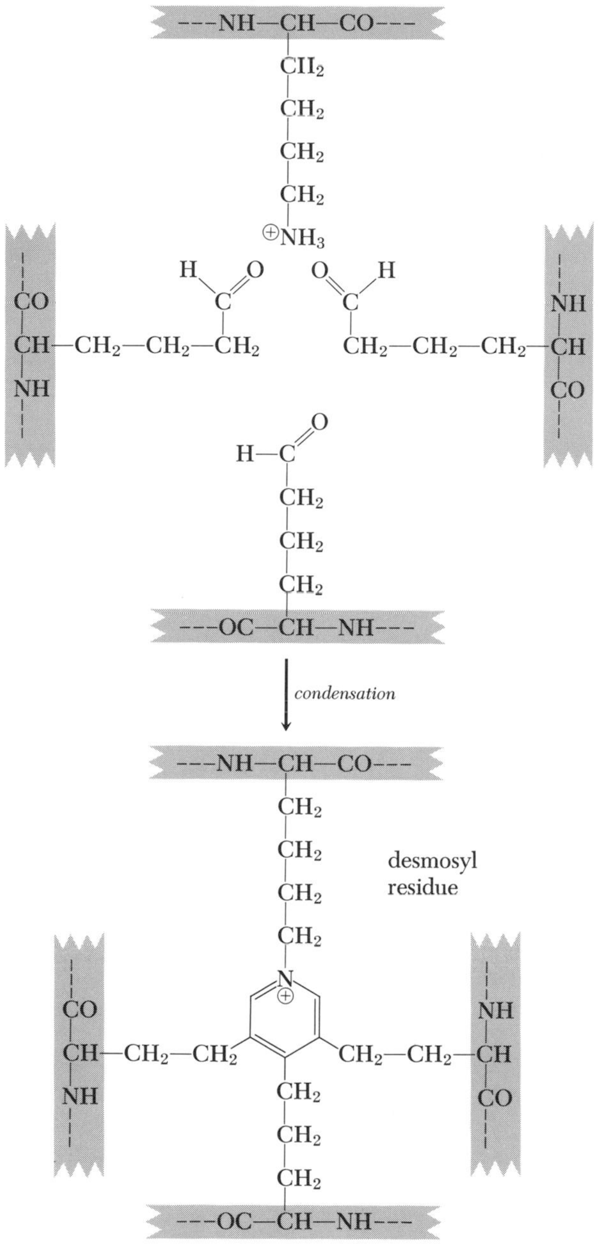

Figure 5-5 Aldehydes formed by oxidation of lysyl residues ("lysinal") are also involved in cross-linking peptide chains of elastin. In this case, one lysyl residue and three lysinal residues condense to form a pyridine ring, which ties together four segments of peptide chain. The result is a residue of the amino acid desmosine, which has four each of ammonium and carboxylate groups, and is known only in elastin. The condensation may also result in an isodesmosyl residue, which differs only in having one of the —CH_2—CH_2— peptide groups at position 2 rather than position 3 of the pyridine ring.

gen molecules. It is possible that molecules are locked together by a shift of some of the intramolecular bonds so as to link strands from adjacent molecules (Fig. 5-6). It is also possible that some of the linkages involving aldehyde groups are not formed until this stage, so that condensation occurs *between* tropocollagen molecules, rather than within them. Whatever their origin, the various kinds of bonds make the bundle rigid by hindering slippage of tropocollagen molecules within it.

The result is collagen. Collagen is not definable as a compound with molecules of definite size. It is composed of collections of subunits, frequently so large that the word molecule loses its meaning. (A crystal is technically a molecule, but it does not help communication to call it that.) We shall see when we discuss all of the structural components of tissues in more detail (Chapter 24) that the situation

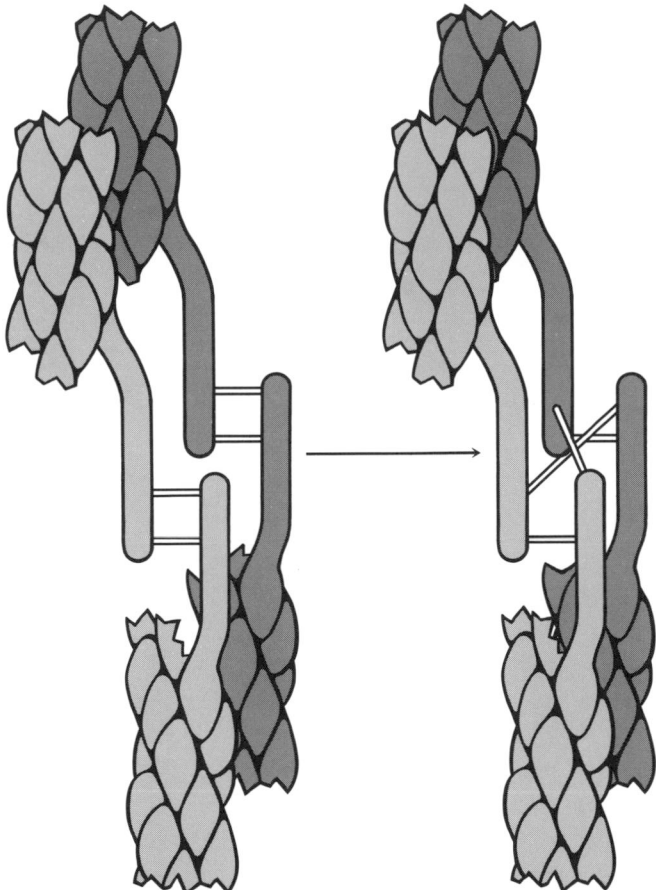

Figure 5-6 Adjacent molecules of tropocollagen may be covalently linked by a shift of bonds between subunits. The diagram indicates only one of the three peptide strands from each of two adjacent tropocollagen molecules. The diagram illustrates the principle, but the actual nature and position of such bonds is unknown. The bonds may include both ester-like and aldehyde-condensation linkages. Some or all of the intermolecular linkages, as shown on the right, may arise by direct reaction rather than by the kind of exchange shown. Either route seems feasible with present information.

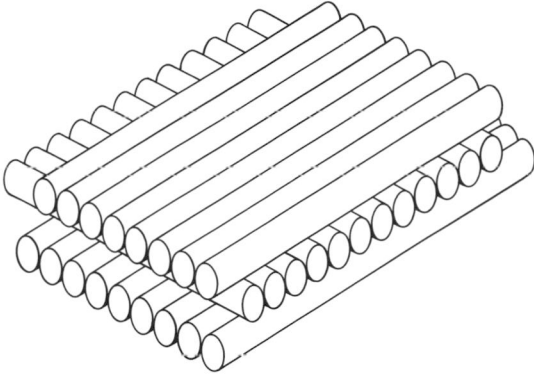

Figure 5-7   Collagen fibers are stacked crosswise to give rigidity to such tissues as the cornea of the eye.

is further complicated by the attachment of carbohydrates to collagen, but the arrangement of peptide chains described here provides mechanical strength to the entire assembly.

Let us emphasize again that the paramount property of collagen, mechanical strength, depends upon the formation of stiff rods that are resistant to stretching, and this property is created through an arrangement of amino acids making an aggregation into tightly twisted chains, the most stable configuration, and by reactions causing the formation of cross-links between chains. Given these rods as structural materials, they may be arranged at various angles in a loose array, as in the dermis, giving the tissue cohesion with retention of flexibility in all directions; or they may be arranged in stacks with successive layers at right angles to make a rather stiff structure, as in the cornea of the eye (Fig. 5-7).

Disturbances of Collagen Formation

A new investigation of an old problem has given support to the concept of utilization of aldehyde groups in cross-linking collagen chains. The ancients knew, and Hippocrates reported, that humans eating peas from some species of the genus *Lathyrus* developed neurological symptoms, burning or tingling sensations (paresthesias) and muscular weakness and spasms. Later, at least one Teutonic duke decreed a maximum permissible adulteration of bread flour with these peas, so as to limit the ill effects.

Within the last two decades it was found that farm animals that were fed the seeds of the ornamental sweet pea, *Lathyrus odoratus,* failed to form collagen and elastin in proper amounts. The result might be deformities of the skeleton, or it might be a failure to form normal arterial walls so that blood could force its way between the layers of the aorta, thereby causing a dissecting aneurysm and consequent fatal rupture of the vessel.

The human and animal syndromes seemed at first to be separate entities; this idea was reinforced by the discovery that β-aminopropionitrile is the constituent of sweet peas that causes their toxic effect, but this compound does not occur in the species causing neurological disturbances in humans. However, a re-examination of Greeks who ate seeds of *Lathyrus* species other than sweet peas during the famine of World War II, and who were known to have had the neurological symptoms, showed that they also had skeletal malformations, indicating that the two syndromes may be variants of a single type of lesion. (It is common for neurological symptoms of biochemical disorders to be emphasized over other abnormalities—partly because of the discomfort and fear they provoke in the patient, and partly because of the immediately dangerous consequences.)

It now appears that *lathyrism* is caused by compounds that inhibit the oxidation of lysyl residues to the aldehyde form. The exact mechanism is not known, but there is reason to believe that the oxidation may require the presence of a prosthetic group, *pyridoxal phosphate*, about which we shall have a great deal to say in the next chapter. There is also reason to suspect that the lathyrogenic agents will react with pyridoxal phosphate. If so, this may well account for the effects on normal connective tissue formation.

But what about the neural form of the disease? The nervous system is particularly susceptible to any circumstances causing an interference in the function of pyridoxal phosphate. This is due to the necessity for certain metabolic reactions requiring pyridoxal phosphate that have nothing to do with connective tissue formation. To pile speculation on speculation, it may well be that particular lathyrogenic agents may have greater access to, or greater affinity for, some cellular components in the brain, while others are more reactive in connective tissue.

The parent toxic compound is evidently β-cyano-L-alanine which remains as such in some species of *Vicia*, a genus related to *Lathyrus*. In some species, β-cyano-L-alanine is converted to α,γ-diaminobutyrate, while in the sweet pea it is decarboxylated to form β-aminopropionitrile. All of the differences of response

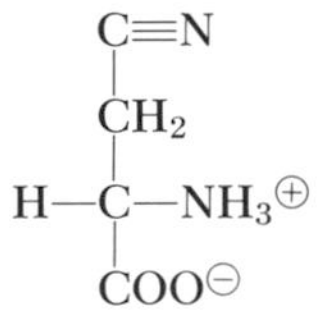

$$C\equiv N$$
$$|$$
$$CH_2$$
$$|$$
$$H-C-NH_3^{\oplus}$$
$$|$$
$$COO^{\ominus}$$

β-cyano-L-alanine

$$C\equiv N \qquad\qquad CH_2-NH_3^{\oplus}$$
$$| \qquad\qquad\qquad |$$
$$CH_2 \qquad\qquad CH_2$$
$$| \qquad\qquad\qquad |$$
$$CH_2-NH_3^{\oplus} \qquad H-C-NH_3^{\oplus}$$
$$\qquad\qquad\qquad\qquad |$$
$$\qquad\qquad\qquad\qquad COO^{\ominus}$$

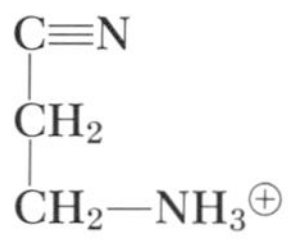

β-aminopropionitrile   L-α,γ-diaminobutyrate

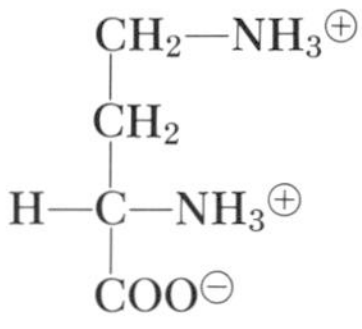

would be explained if the first two compounds affect pyridoxal phosphate in the nervous system more than they do in fibroblasts. The β-aminopropionitrile has been shown to prevent normal cross-linking of tropocollagen, and the product of the fibroblasts accumulates as this soluble, incomplete molecule rather than as collagen fibers. The effects of the other derivatives of β-cyano-L-alanine have not been extensively studied.

Scurvy

Scurvy is another disruption of normal collagen synthesis known to the ancients, and it is caused by a deficiency of *ascorbic acid* (vitamin C). The exact function of the vitamin is unknown, despite much work by a number of people. We shall say more about the compound later (p. 678).

However, it is known that ascorbic acid is required for the oxidation of prolyl residues to hydroxyprolyl residues. This is a necessary reaction for the formation of the peptide chains comprising collagen, but not in the formation of elastin, and one would expect scurvy to cause some, but not all, of the disturbances of connective tissue function seen in lathyrism. There are indeed some features in common. Blood vessels are weakened in scurvy, but the effect is seen in small vessels, and there are small hemorrhages instead of the massive failure of the great vessels due to the failure of elastin formation in lathyrism. Collagen deposition in the bones is definitely impaired, so that formation of new bone may be prevented and old bone becomes brittle.

KERATIN

Structure and Function of Keratin

The keratins are largely confined to exterior surfaces of the vertebrate body— the epidermis, hair, nails, horns, and hoofs—although some may be formed on epithelial surfaces. The epidermis is the typical location.

The epidermis must be flexible, yet tightly compact so as to prevent leakage of the body constituents or entry of contaminants from the environment. A structure meeting these requirements cannot be formed from stacks of fibers such as those composing collagen. Keratin does contain fibers; they are made from peptide chains in the right-handed α-helical configuration, but with a much longer helical segment than those we described in hemoglobin. Three of these helical chains are twisted together in a left-handed direction, but this rope differs from collagen in both twist and dimensions. It is not known how the units are linked together to give the great length seen in keratin.

The required strength and desired degree of imperviousness are gained in keratin by embedding the fibrils in a matrix of protein. The arrangement in epidermis is unknown, but in hairs, the fibrils appear from electron micrographs to be arranged in rings of nine around cores of two (Fig. 5-8). This kind of arrangement of fibers in a matrix may be fairly common, because it is seen with different proteins (for example, in the flagella of some microorganisms).

Fibers embedded in a matrix can give strength and flexibility to a structure, as humans have discovered. The principle is used in such diverse applications as concrete reinforced with steel rods and plastics filled with glass fibers. The matrix distributes stresses among the fibers, and the fibers prevent the matrix from cracking, so that the combination of the two is stronger than either alone.

Cells of the epidermis near the basement membrane, which is the layer next to the dermis, are capable of dividing. The increase in the number of cells pushes

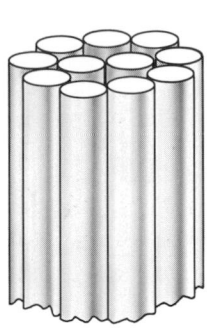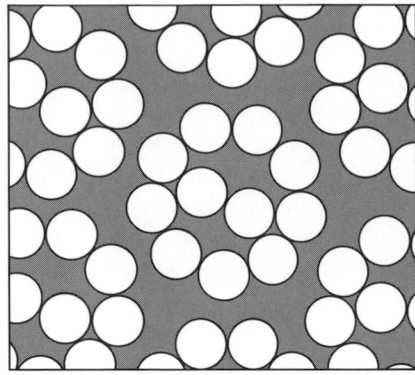

Figure 5-8 *Left*. Hair keratin is made of bundles of microfibrils embedded in a matrix. The bundle is a ring of nine around a core of two fibrils. It is not shown, but each fibril is a rope of three peptide chains, each an α-helix with the three twisted together in a left-handed helix.

Right. Cross section of hair keratin. The bundles of microfibrils are in hexagonal arrays buried in a matrix (shaded) of disulfide-linked protein.

some of them toward the surface of the skin. As they mature, fibrils form within them, running from wall to wall across the cell, with the ends gathered at thickenings of the cell wall (*desmosomes*) that are spaced around the periphery. The cells develop interlacing spiny processes. Adjacent cells fuse together at the desmosomes, so that the fibrils and desmosomes constitute a cabled network tying all of the cells together.

At the same time, the cells accumulate a protein containing a high number of *cysteine* residues, with their side-chain mercapto- or sulfhydryl groups. This protein is the precursor of the matrix. As the cells move toward the surface, the sulfhydryl groups are oxidized to form *disulfide bridges* by an uncharacterized enzyme:

$$
\begin{array}{ccc}
\text{--NH--CH--CO--} & & \text{--NH--CH--CO--} \\
| & & | \\
CH_2 & & CH_2 \\
| & & | \\
SH & \xrightarrow{\ +\frac{1}{2}O_2\ } & S \\
| & & | \\
SH & & S \\
| & & | \\
CH_2 & & CH_2 \\
| & & | \\
\text{--NH--CH--CO--} & & \text{--NH--CH--CO--}
\end{array}
$$

The parent cysteinyl residues need not be on the same peptide chain, or even in the same molecule. The resultant disulfide bonds cement the smaller molecules into a solid mass around the fibrils. As this proceeds, the cell structures lose their form, the original cell fills with keratin, and it becomes a part of a horny layer, only a fraction of a millimeter thick in human skin, but resisting tear and effectively confining the fluid medium within the body.

The disulfide bridges are the characteristic chemical feature of keratin, and oxidized cysteine residues comprise nearly one-fifth of the molecule in some keratins.

Keratin is even less describable in molecular terms than collagen because it comprises a composite of fiber and matrix, and its composition differs from site to site within the same organism.

The oxided form of free cysteine is *cystine:*

$$H_3\overset{\oplus}{N}-\underset{\underset{CH_2-SH}{|}}{\overset{\overset{COO^\ominus}{|}}{C}}-H \quad + \quad H-\underset{\underset{HS-CH_2}{|}}{\overset{\overset{\overset{\oplus}{NH_3}}{|}}{C}}-COO^\ominus + \tfrac{1}{2}O_2 \quad \longrightarrow \quad H_3\overset{\oplus}{N}-\underset{\underset{CH_2-S-}{|}}{\overset{\overset{COO^\ominus}{|}}{C}}-H \quad H-\underset{\underset{S-CH_2}{}}{\overset{\overset{\overset{\oplus}{NH_3}}{|}}{C}}-COO^\ominus + H_2O$$

2 L-cysteine   L-cystine

Therefore, keratin has a low content of cysteine, but a high content of cystine. In conversation, it is well to avoid confusion by emphasizing the extra syllable in cysteine: sis-TEH-een.

DISULFIDE BONDS IN OTHER PROTEINS

The disulfide bond represents an effective means for fixing parts of peptide chains in close proximity. Keratin is an extreme example of its use; the bond also appears in soluble proteins, but in smaller amounts.

In the case of hemoglobin, the final configuration is achieved without the use of covalent linkages between any of the amino acid side chains; hydrogen bonds, non-polar bonding, and electrostatic attraction are sufficient. These types of bonding are weak, and the area of the peptide chain containing matching residues must

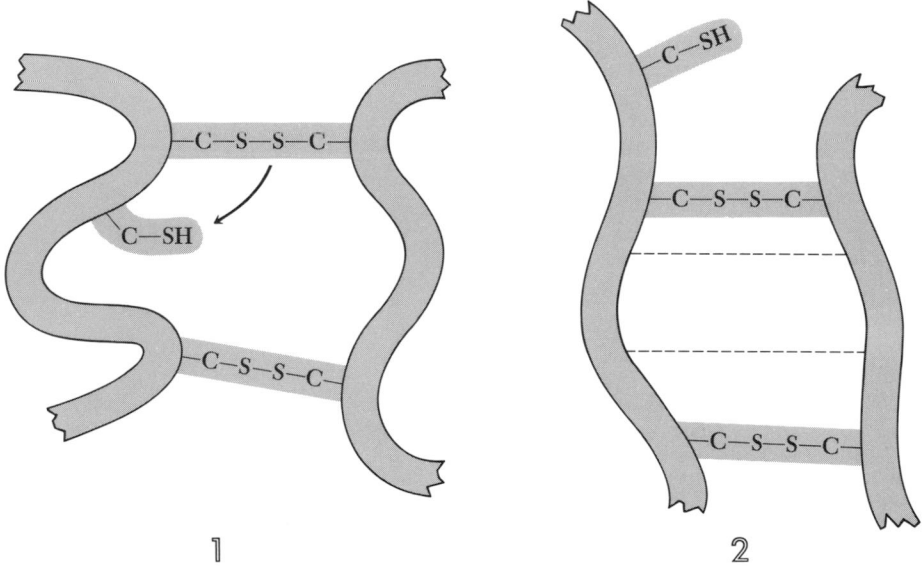

Figure 5-9 Exchange between disulfide and sulfhydryl groups enables formation of a more compact and stable structure. The equilibrium between structures 1 and 2 will favor the latter because the two chains are closer in it and additional forces, such as hydrogen bonds, can develop between them.

be large if the structure is stable. It is reasonable to expect that the covalent disulfide bonds give greater latitude to the possible arrangement of residues in a protein, making it possible to hold structures together in ways for which the weaker forces would not be sufficient. In any event, many proteins—probably most—do contain disulfide bridges, and frequently the structure unfolds if these are broken by reduction to sulfhydryl groups.

As in the case of hydroxyproline, there is no genetic coding for cystine. It is difficult to see how there could be, since the disulfide bridges are not used to link adjacent cysteinyl residues on a chain, but to link different parts of a peptide.

The details of the process have not been worked out, but the effect is as follows: Only cysteine is coded into a peptide chain. Once the chain has been formed, it comes in contact with a pair of enzymes on the endoplasmic reticulum. The first enzyme catalyzes a quite random oxidation of pairs of cysteinyl residues to a cystinyl residue. Which residues are linked will depend upon which happen to be close to each other upon contact with the enzyme. The second enzyme serves to correct those mistakes by catalyzing an exchange between these disulfide bridges and the remaining free —SH groups (Fig. 5-9). This exchange will continue until the most stable form is produced—the form in which the structure is the most compact and free from strain. This most stable form is the finished molecule.

Recapitulation of types of reactions

Note: Most of the reactions by which the special bonds of collagen and elastin are formed have not been studied in enough detail to be used as type examples, and are not listed.

1. Pairs of mercaptans may be oxidized to disulfides:

$$R{-}SH + HS{-}R' \longrightarrow R{-}S{-}S{-}R' + 2H^+ + 2e^-$$

Example: The oxidation of cysteinyl residue to a cystinyl residue.

2. Disulfides may exchange with another sulfhydryl group:

$$R{-}S{-}S{-}R' + R''{-}SH \longrightarrow R{-}SH + R''{-}S{-}S{-}R'$$

Example: The rearrangement of disulfide bridges in proteins.

Further reading

Schubert, M., and D. Hamerman: *A Primer on Connective Tissue Biochemistry.* Lea and Febiger (1968). Not as simple as implied.

New York Heart Association: *Connective Tissue: Intercellular Molecules.* Little, Brown and Co. (1964).

Montagna, W., and W. C. Lobitz, Jr., eds.: *The Epidermis.* Academic Press (1964).

Ramachandran, B. A., and B. S. Gould, eds.: *Treatise on Collagen,* Vols. 1 and 2. Academic Press (1968).

Piez, K. A.: *Cross-Linking of Collagen and Elastin.* Ann. Rev. Biochem., 37: 547 (1968).

Udenfriend, S.: *Formation of Hydroxyproline in Collagen.* Science, 152: 1335 (1966).

Manning, J. M., and A. Meister: *Conversion of Proline to Collagen Hydroxyproline.* Biochemistry, 5: 1154 (1966).

Harding, J. J.: *The Unusual Links and Crosslinks of Collagen.* Adv. Prot. Chem., 20: 109 (1965).

Schneider, A., et al.: *The Presence of Lysinal (2,6-Diaminohexanal) in Tropocollagen.* Biochem. Biophy. Res. Comm., *26:* 43 (1967).

Gallop, P. M., et al.: *Isolation and Identification of α-Aminoaldehydes in Collagen.* Biochemistry, *7:* 2409 (1968).

Page, R. C., and E. P. Benditt: *Collagen Has a Discrete Family of Reactive Hydroxylysyl and Lysyl Side-Chain Amino Groups.* Science, *163:* 578 (1969).

Tanzer, M. L.: *Experimental Lathyrism.* Intl. Rev. Conn. Tissue Res., *3:* 91 (1965).

Anwar, R. A., and G. Oda: *The Biosynthesis of Desmosine and Isodesmosine.* J. Biol. Chem., *241:* 4638 (1966).

Crosther, W. G., et al.: *The Chemistry of the Keratins.* Adv. Prot. Chem., *20:* 1919 (1965).

Seifter, S., and P. M. Gallop: *The Structure Proteins.* In Neurath, H., ed.: *The Proteins,* 2nd ed., Vol. 4 (p. 153). Academic Press (1967).

ENZYMES

CHAPTER 6

ARGUMENT

The enzymes are those proteins that catalyze the chemical reactions of living organisms. They contain the same amino acids and the same types of chemical bonds as other proteins. Their catalytic activity is created by the particular conformation of their peptide chains. The use of proteins as catalysts has made it possible to evolve a finely detailed architecture, so that a typical enzyme will catalyze the reaction of only a few compounds—it has a high specificity. The compounds that react by enzymatic catalysis are called substrates for the enzyme, and the specificity of the substrate-enzyme interaction is a reflection of the formation of multiple bonds between various groups on the substrate and on the enzyme.

The binding of substrate to enzyme may be studied by modifying the substrate molecule and noting what changes diminish binding; or it may be studied by using competitive inhibitors—compounds that will react with the binding sites, thereby displacing substrates. Inhibitors of enzymes may be of great practical importance; some are powerful drugs and others are used to kill unwanted organisms.

The catalytic mechanism frequently involves the imidazole group of histidyl residues, acting as acid-base catalysts. The nature of the mechanism is such that it is most effective near pH 7, and this may account for the evolution of histidyl residues and the nearly neutral internal environment of most cells.

Enzymatic catalysis sometimes involves the formation of intermediates covalently bonded to the protein, and the hydroxyl group of seryl residues is frequently used for this purpose during hydrolytic reactions. The groups available on amino acid residues are not sufficient to enable catalysis of all biochemical reactions, and additional groups are supplied in some enzymes by binding other compounds containing the necessary structures in the form of prosthetic groups, or coenzymes. Coenzymes frequently contain structures quite different from those found in the many transient intermediates being consumed by metabolism. The unusual structure assures a higher lifetime for the molecule and enables a highly specific combination between the peptide chains and the appropriate coenzyme. Many animals have lost the ability to synthesize the highly specialized structures of coenzymes, and precursors must be supplied in the diet. Such precursors are the vitamins.

Life depends upon the existence of an ordered array of chemical reactions. Order can be created because nearly all of the reactions are catalyzed by proteins created for the purpose—the enzymes—and the properties of the enzymes can be adjusted to control rates of reaction. (We are speaking here of reactions involving the formation or cleavage of relatively stable covalent bonds, and are excluding acid-base dissociations and reactions involving only hydrogen bonds or electrostatic interactions.)

The enzymes are proteins containing the same amino acids found in other proteins. As with other proteins, enzymes have a three-dimensional structure that represents the conformation of least energy content. All of our present information supports the view that enzymatic catalysis involves no novel bonds; catalysis proceeds by the same mechanisms seen in ordinary organic reactions, using structures supplied by amino acid residues and prosthetic groups. The binding of the prosthetic groups is in itself a result of the conformation of the amino acid residues.

The study of the nature of enzymes is therefore a specialized branch of protein chemistry, dealing with those particular proteins that have conformations that catalyze particular chemical reactions, and extending the principles of protein structure we have already developed.

Enzymes differ from simpler catalysts in important quantitative ways that make possible a fine control of metabolic processes. For one thing, effective catalysis by an enzyme may be limited to reactions of a few closely related compounds. To use the technical nomenclature, an enzyme may have a high *specificity* for its *substrates*.

Substrate is an old term, firmly embedded in the biochemical literature, for a compound whose reaction is catalyzed by an enzyme: substrate $\xrightarrow{enzyme}$ product. We can rephrase the concept of enzyme specificity: An enzyme may catalyze the reaction of only a few different substrates. Put still another way, only a few compounds may be capable of acting as substrates for an enzyme.

In any catalytic process, the reacting compound must combine with the catalyst—random collisions between reactant and catalyst must result in interactions of the kind we call bond formation before anything else can happen. It is very easy to see that an enzyme *might* have an arrangement of amino acid residues such that only a very few substrates have the necessary shape and chemical groupings to form bonds with several of these residues simultaneously.

This appears to be an eminently logical basis for the specificity of enzymes, which makes it even more embarrassing to admit that the exact nature of the binding groups has yet to be unequivocally established for a single one of the hundreds of known enzymes. Why is this so? Because a sufficiently complete definition of structure is only now being completed with a few enzymes, and rigorous analysis of their mechanism is yet to come.

This does not in any way mean that our knowledge of substrate-enzyme interaction reduces to a collection of vague speculations. An enzyme catalyzes a reaction, and a study of the reaction itself can tell much about the catalyst, even if it has never been isolated in pure form. This is sometimes difficult to accept with confidence. We feel more comfortable with a compound we can hold in our hand, but the catalysis of a reaction is in itself a definitive property like molecular weight or amino acid composition. This property, unlike the others, can be measured with quite impure samples. An example may be reassuring.

ACETYLCHOLINESTERASE AND THE BINDING OF SUBSTRATES

How does one go about developing a picture of enzyme-substrate binding? Let us analyze the procedure in detail for a single enzyme. The enzyme is an important one, but even so, our object is not to fix in mind the experimental details themselves, but to see how they are *used*, so as to be able to apply the same principles in reasoning about other enzymes.

The Function of Acetylcholinesterase

A nerve transmits a stimulus from one point in the body to another through a traveling wave of changes in concentration of ions. In many nerves the stimulus is carried beyond the terminus of the neuron in the form of *acetylcholine*. The arrival of the ionic wavefront at the terminus causes the release of acetylcholine, which diffuses into the intimately associated receptor cell, such as a muscle fiber or another neuron, and causes another traveling wave of excitation to begin in the receptor. Those receptors sensitive to acetylcholine are said to be *cholinergic*, and include all of the motor endplates in skeletal muscles, which activate contraction of the fibers.

Once the acetylcholine has provoked the desired response, it must be removed to permit recovery of the receptor for a future stimulus, or to prevent repeated and uncontrolled responses after a single stimulus. The removal is accomplished by hydrolysis of the compound, and is catalyzed by an enzyme, *acetylcholinesterase:*

$$H_3C - \overset{\overset{\displaystyle CH_3}{|}}{\underset{\underset{\displaystyle CH_3}{|}}{\overset{\oplus}{N}}} - CH_2 - CH_2 - O - \overset{\overset{\displaystyle O}{\|}}{C} - CH_3 + H_2O \longrightarrow$$

acetylcholine

$$H_3C - \overset{\overset{\displaystyle CH_3}{|}}{\underset{\underset{\displaystyle CH_3}{|}}{\overset{\oplus}{N}}} - CH_2 - CH_2 - OH + {}^{\ominus}OOC - CH_3 + H^+$$

choline   acetate

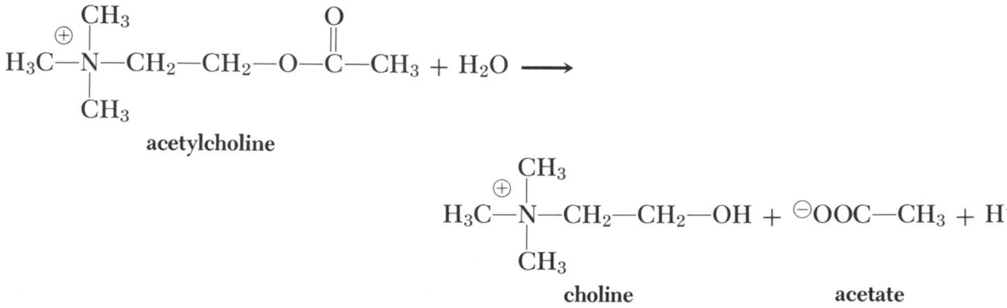

Enzymes have been given names in the past with the suffix -*ase* added to the name of the substrate. Sometimes the kind of reaction was also indicated (oxidase, decarboxylase), but not when hydration or hydrolysis was involved. In the latter cases, the names often included the kind of bond hydrolyzed, as in *-esterase* or *-amidase*. Attempts have been made in recent years to construct a systematic nomenclature, which is summarized in Chapter 29. The final test, usage, appears likely to arrive at a mixture of the new nomenclature with old trivial names, as is presently the case with accepted terminology for organic compounds. We shall use the names believed to be most common in current use by biochemists.

The quantity of acetylcholine released and the activity of the enzyme must be nicely balanced. The receptor must be triggered before all of the acetylcholine is destroyed, but any additional delay merely prolongs the time of recovery to the initial state with the receptor ready for renewed stimulation.

An animal possessing cholinergic nerves therefore depends upon acetylcholinesterase for survival. Modification of the activity of the enzyme directly affects the responses triggered by cholinergic nerves, and compounds causing such a modification can be employed for extermination (pesticides for invertebrates, war gases for people). They also have a more gentle use as drugs to alter the activity of the nervous system. There are other drugs that affect cholinergic nerves by enhancing or inhibiting the action of acetylcholine on the receptor itself, but these are not a part of our present concern.

Acetylcholinesterase has been intensively studied for many years because of the desire to understand the mechanism of neural transmission in which this enzyme has such a critical function and because of the large potential for practical use implied in any modification of its activity. As a result, we know a great deal about the binding of its substrate through studies with crude preparations of the enzyme—more than we do about some enzymes available as crystalline preparations in relatively large amounts. (Acetylcholinesterase has only recently been crystallized.)

Given an enzyme, how does one proceed to define the mechanism of binding of the substrate? One alters the chemical structure of the substrate, and then sees what happens. If binding involves a particular group on the substrate, then removal of that group ought to weaken the binding.

How does one know the binding of a substrate is weaker? If it is, then the reaction catalyzed by the enzyme will not proceed as rapidly at the same concentration of modified substrate. But there is a flaw, in that a slower rate also may be a result of some effect on the mechanism of catalysis after binding, so caution is required in interpretation. There is another approach that is helpful for a solution of the problem, which we shall consider next.

The Use of Competitive Inhibitors

A compound that is bound to the catalytic site on the surface of the protein will hinder any other compound from occupying the site. The compound may not undergo the reaction catalyzed by the enzyme, but its binding can be recognized by an inhibition of the reaction with true substrates. The inhibitor (I) will be in equilibrium with the enzyme (E) and the complex between them (EI):

1. $\quad$ I + E $\longleftrightarrow$ EI.

The removal of the free enzyme resulting from this equilibrium will slow the binding reaction of a true substrate, which is the first reaction in the catalytic process:

2. $S + E \longleftrightarrow ES$.

The effect is a competition between the inhibitor and the substrate for the enzyme. With a given concentration of inhibitor, the concentration of substrate can be raised to a high enough value that the free enzyme is nearly completely removed to form the enzyme-substrate complex. This will shift the equilibrium of reaction (1) to the left; little of the enzyme will be in combination with the inhibitor, and the reaction will be going as fast as if no inhibitor were present.

This, then, is the hallmark of *competitive inhibition*. The inhibition can be overcome by increasing the substrate concentration, and the reversal of inhibition can be described in terms of simple equilibrium. The quantitative analysis of the effect will be considered in the next chapter, but now we are concerned with the qualitative deductions made possible by the discovery of competitive inhibitors for a particular enzyme.

If a compound is a competitive inhibitor, then it must be bound to the enzyme in a position blocking the binding groups for an authentic substrate. We can't leap to the conclusion that the inhibitor is bound in exactly the same way, because there may be a simple competition for space on the surface of the enzyme even though the two compounds are fixed to the surface in different ways. Failure to recognize this point sometimes leads to an unprofitable investment of intellectual effort in attempts at molding a picture of an active site that will fit all known competitive inhibitors. Even so, by comparing enough inhibitors and looking at their structural relationship to the actual substrate, it is possible to make sound conclusions about the physiologically important sites on the protein.

The Functional Groups of Acetylcholine

Let us look once more at the structure of the substrate, analyze the possibilities for binding, and compare the possibilities with the actual data.

1. The compound is an ester:

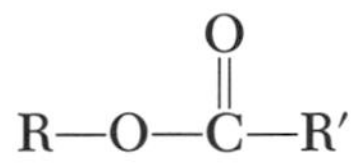

$$
\begin{array}{c}
O \\
\parallel \\
R-O-C-R'
\end{array}
$$

This group must contain atoms bound to the enzyme, because it is the group that is attacked in the catalytic process.

2. The compound has a quaternary nitrogen with an obligatory positive charge:

$$R_3N^+{-}R'$$

This looks like a good possibility for binding because the protein could supply negative charges on glutamyl or aspartyl side chains to form an electrostatic bond.

3. The compound contains three methyl groups and a methylene group around the tetrahedral nitrogen atom:

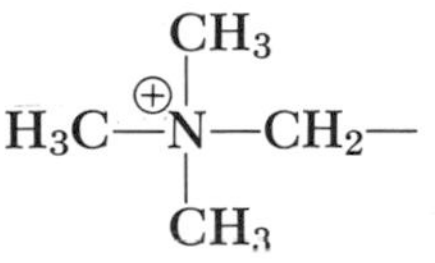

$$\underset{\overset{|}{\text{CH}_3}}{\overset{\overset{\text{CH}_3}{|}}{\text{H}_3\text{C}\overset{\oplus}{-}\text{N}-\text{CH}_2-}}$$

It is conceivable that the enzyme surface has non-polar groups arranged in such a way that they will bond specifically with these methyl groups. There are many possibilities for such residues in the complement of side chains, as we have seen in the case of hemoglobin.

4. Similarly, the methyl group of the acetyl portion of the substrate:

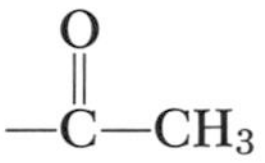

$$\overset{\overset{\text{O}}{\overset{||}{}}}{-\text{C}-\text{CH}_3}$$

may be bonded specifically by neighboring non-polar residues.

These exhaust the obvious possibilities for binding; now let us turn to the facts.

Experimental Observations with Acetylcholinesterase

First, the enzyme will catalyze the hydrolysis of a wide variety of acetyl esters, even ethyl acetate. However, the maximum velocity for hydrolysis of ethyl acetate is only 0.12 of the rate attainable with acetylcholine. Furthermore, the concentration of ethyl acetate required to obtain one-half of the maximum rate of hydrolysis is 500-fold larger than the concentration of acetylcholine required for one-half of its maximum rate. (This comparison of concentrations required for one-half of maximum velocity, $\frac{1}{2} V_{max}$, is useful for reasons that will be developed in the next chapter.) Comparing these substrates, we see that they differ only in the presence or absence of the trimethylammonium group, so we tentatively assume that this group forms important bonds with the enzyme.

The positive charge has been shown to be important by many lines of experimentation. For example, if a salt, such as NaCl, is added to a solution of the enzyme, more acetylcholine is required to maintain the same rate of hydrolysis. Raising the concentration of the total salts in a solution diminishes the effective concentration—the *activity*—of each ion, and it seems reasonable to conclude that the ionic activity of acetylcholine in part determines its association with the enzyme.

For quantitative assessment of this kind of an effect, it is convenient to express the salt concentration in terms of the *ionic strength*, μ, which is one-half the sum of the products of each ion's concentration and the square of its charge. This expression is a measure of the square of the charge density in a solution, and ionic activity is a function of the charge density and therefore of $\sqrt{\mu}$.

Example 1. 0.1 M KCl. $[\text{K}^+] = 0.1$, charge $= +1$; $[\text{Cl}^-] = 0.1$, charge $= -1$. $\mu = \frac{1}{2}(0.1 \times 1^2 + 0.1 \times (-1)^2) = 0.1$.
Example 2. 0.05 M MgSO$_4$. $[\text{Mg}^{+2}] = 0.05$, charge $= +2$; $[\text{SO}_4{}^{2-}] = 0.05$, charge $= -2$. $\mu = \frac{1}{2}(0.05 \times 2^2 + 0.05(-2)^2) = 0.2$.

The lowering of the affinity between acetylcholine and the enzyme by increased salt concentrations therefore supports the idea of an electrostatic bond between the substrate and enzyme. It does not positively rule out an indirect effect of salt through changes in the spatial relations of the peptide chain of the enzyme, which may also contain important electrostatic bonds.

Better yet, it has been shown that almost any positively charged ammonium ion is a competitive inhibitor for acetylcholine on the enzyme, and the inhibition is diminished as the $[H^+]$ is lowered across the K' value for the ammonium ion, which thereby loses a charged proton:

$$R-NH_3^+ \longrightarrow R-NH_2 + H^+; \qquad K' = [R-NH_2]\,[H^+]/[R-NH_3^+].$$

This indicates that the charged ammonium ion, but not the neutral amine, is being bound in the same way as is the substrate.

Best of all, a direct comparison has been made between acetylcholine and its carbon analogue:

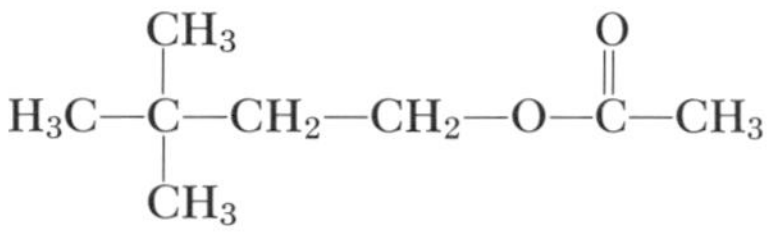

$$H_3C-\underset{\underset{CH_3}{|}}{\overset{\overset{CH_3}{|}}{C}}-CH_2-CH_2-O-\overset{\overset{O}{\|}}{C}-CH_3$$

3,3-dimethylbutyl acetate

This compound has a shape almost identical to that of acetylcholine, but it has no positive charge. It can be hydrolyzed just as rapidly as is acetylcholine in the presence of acetylcholinesterase, but its concentration must be eight-fold greater to attain $\frac{1}{2} V_{max}$. It seems very likely that the loss of positive charge is the cause of this weaker binding.

We have hedged on each one, but now have three independent observations. The alternatives for each observation, themselves independent, now become much less likely, and the common explanation becomes the probable one. We therefore offer it as a conclusion: There is electrostatic bonding between the charged quaternary nitrogen of the substrate and some negative charge on the enzyme.

The fact that 3,3-dimethylbutyl acetate is such a good substrate makes one suspect that the branched methyl groups themselves contribute greatly to binding with the enzyme. Data are also available for the series of methylammonium ions, which act as competitive inhibitors, and the dissociation constants for the enzyme-inhibitor complex, K_I, have been calculated from measurements of their effectiveness as inhibitors:

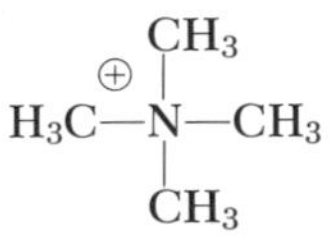

$H_3C-\overset{\overset{CH_3}{|}}{\underset{\underset{CH_3}{|}}{\overset{\oplus}{N}}}-CH_3$	$H_3C-\overset{\overset{CH_3}{|}}{\underset{\underset{H}{|}}{\overset{\oplus}{N}}}-CH_3$	$H_3C-\overset{\overset{H}{|}}{\underset{\underset{H}{|}}{\overset{\oplus}{N}}}-CH_3$	$H-\overset{\overset{H}{|}}{\underset{\underset{H}{|}}{\overset{\oplus}{N}}}-CH_3$
tetramethyl-ammonium	trimethyl-ammonium	dimethyl-ammonium	methyl-ammonium
$K_I = 1.6 \times 10^{-3}$ M	$K_I = 1.4 \times 10^{-3}$ M	$K_I = 11 \times 10^{-3}$ M	$K_I = 63 \times 10^{-3}$ M

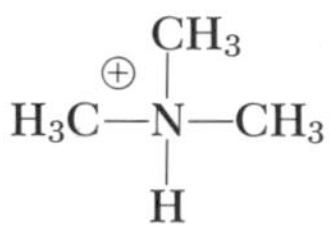

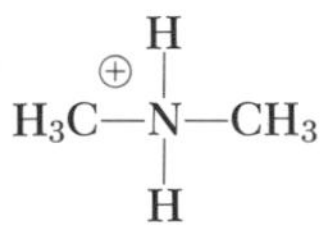

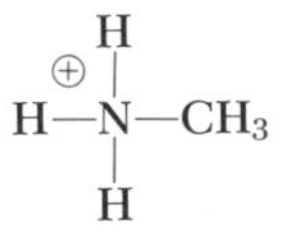

As more and more methyl groups are added, the ion becomes more tightly bound to the enzyme, except that addition of the last methyl group to form the tetramethylammonium ion makes no difference. The exception isn't disturbing because a methyl group projecting from the surface of the enzyme couldn't form bonds anyway, and one of the four bonds on the nitrogen atom must be projecting outward from the surface. On the whole, these data strongly support the assumption that non-polar bonding of the aliphatic carbons is an important factor for substrate attachment.

Looking at the other end of the substrate molecule, we find that acetyl esters, whether of choline or of simpler alcohols, are hydrolyzed more rapidly than are propionyl esters (propionyl $= -CO-CH_2-CH_3$). The next homologue, butyryl choline, is not hydrolyzed at a measurable rate, and is a competitive inhibitor for acetylcholine (butyryl $= -CO-CH_2-CH_2-CH_3$). The most likely explanation is that the size of the butyryl group prevents the ester bond from being brought into contact with the catalytic site, although the choline portion of the molecule can be bound. These data support the idea that the enzyme is constructed so as to fit the $-CH_3$ portion of the acetyl group.

Conclusion and Applications

From the information presented and a large body of confirmatory data of like kind, we can conclude with some confidence that acetylcholine is bonded to its esterase by electrostatic attraction, by non-polar bonding with three out of the four aliphatic groups on the nitrogen atom, and by the carbonyl group. There is a strong likelihood that the methyl group of the acetyl moiety is also bonded by dispersion forces.

The kind of reasoning employed here has general application to enzymes. An enzyme catalyzing a specialized reaction in particular tissues ought to be expected to bond with many of the groups on its substrate, so as to define the compound undergoing the reaction more closely. Most of the intracellular reactions require such a high specificity. In those other cases in which it is economical to have a single enzyme catalyze a reaction by a wide variety of substrates, the binding will be limited to a few groups on the substrate. There are examples of such reactions, particularly with the enzymes of the digestive tract that hydrolyze the components of the diet. In all of these cases, testing of related compounds can do much to define the origin of the specificity and to suggest the nature of the groups on the enzyme necessary for substrate binding.

Beyond the understanding of enzymatic processes gained by the sort of study we have outlined, a further benefit is obtained from a rational approach to the design of compounds deliberately employed to change the activity of enzymes in a living organism. The design of a drug still contains much of an empirical approach, because all of the parameters operating in an animal cannot be defined with accuracy, but our growing knowledge of enzymatic function already permits experimentation with greater chance of success than a purely empirical approach, besides offering rational correlation of information on a large collection of drugs.

However, more than a little imagination is sometimes required. Take, for example, this ion:

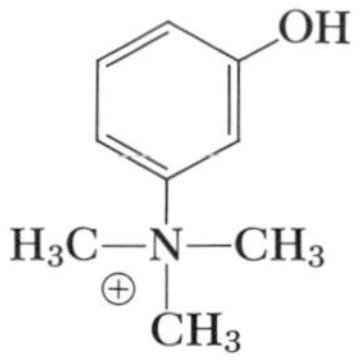

(3-hydroxyphenyl)trimethylammonium

Superficially, it looks like just another trimethylammonium compound, but the spacing between that group and the oxygen atom is almost exactly right for a fit on the active site of acetylcholinesterase, and it is bound very tightly to the enzyme ($K_I = 10^{-7}$M). If one administers it to an animal, the pupils of its eyes will constrict sharply, and its skeletal muscles will start to twitch from the effects of the accumulation of acetylcholine. Derivatives of this compound are used therapeutically to counteract the effects of curare. Curare is used by South American Indians and by anesthesiologists of various ancestries to relax the muscles, which it does by inhibiting the effect of acetylcholine on the motor endplates. Much of the subject of pharmacology is occupied with effects and counter-effects of drugs on the nervous system.

RIBONUCLEASE AND THE CATALYTIC MECHANISM

A simple combination between substrate and enzyme does not result in catalysis. An increase in the rate of reaction requires that bonds be formed in such a way that they are created more easily, *i.e.*, have a greater possibility of formation, than are the bonds involved in the same reaction of the substrate without a catalyst. The enzyme must have side chains capable of forming such bonds.

The details of the mechanism of enzymatic catalysis are as uncertain as those of substrate binding, and for the same reason: lack of precise definition of the architecture of the enzymes. However, there is general agreement on some important principles, and these are illustrated by the enzyme, pancreatic ribonuclease A.

The Reaction Catalyzed by Pancreatic Ribonuclease A

Ribonuclease A is one of the several enzymes found in the mammalian pancreas that catalyze the hydrolysis of ribonucleates. It is a relatively small protein, containing only 124 amino acid residues in a known sequence, and hence has been studied intensively in attempts to formulate a specific mechanism for at least this one enzyme.

The enzyme is secreted from the pancreas into the small bowel to catalyze the hydrolysis of ribonucleates contained in the diet. (The diet of animals consists mainly of tissues from other organisms, and will contain DNA and RNA.) The catalysis is specific for the 5′-phosphate ester bond at positions where this phosphate is also attached to the 3′-carbon of a pyrimidine nucleoside, such as cytidine or uridine. The polynucleotide chain will be hydrolyzed into fragments, and each fragment will be a nucleotide, or polynucleotide, bearing a free —OH group on the 5′-carbon at one end and an ester phosphate group on the 3′-carbon at the other end. The latter end must be a pyrimidine nucleotide, and the former end will be a purine nucleotide if hydrolysis is complete.

Thus, the hydrolysis of a portion of an RNA chain might proceed like this:

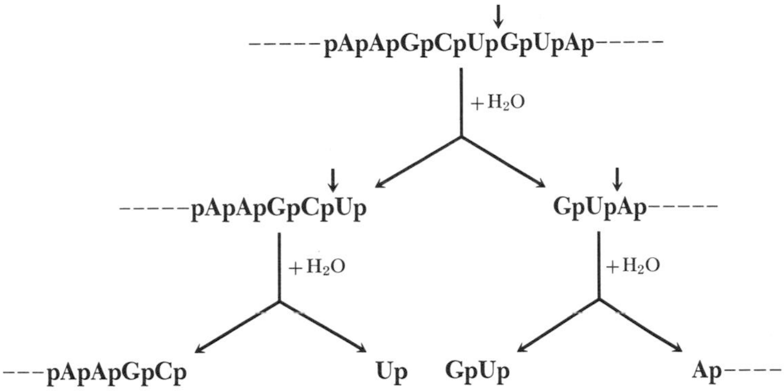
-----pApApGpCpUpGpUpAp-----
+H₂O
-----pApApGpCpUp
GpUpAp-----
+H₂O
+H₂O
---pApApGpCp
Up
GpUp
Ap----

Let us discuss a possible mechanism for catalysis by this ribonuclease, with an important disclaimer. No mechanism for the enzyme is generally accepted, although several have been proposed, but all recent proposals have several elements in common that could be illustrated by any one of them.

A. The substrate is bound to the enzyme in such a way as to bring the phosphate ester group to be hydrolyzed near the imidazole side chains of two histidyl residues in the enzyme. These residues are known to be at positions 12 and 119 in the peptide chain. The R-groups in the figure represent the long polynucleotide on either side of the bond to be broken. Their bulk presents no problem, because they can extend from the active site on either side of the cross-section illustrated.

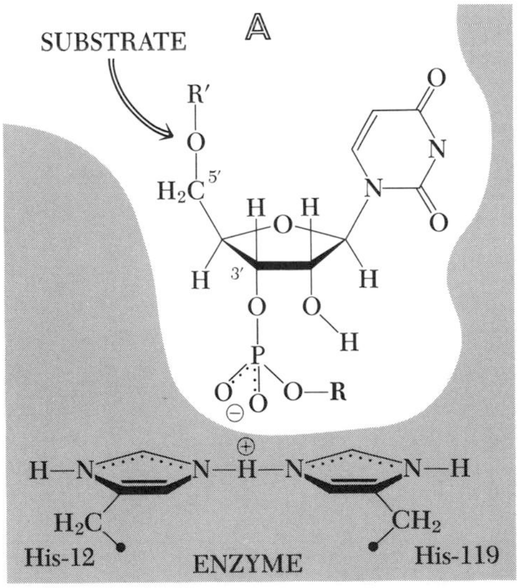
A
SUBSTRATE
R′
O
N
H₂C
5′
H
H
O
N
O
H
3′
H
O
O
H
P
O
O
O—R
H—N
N—H—N
N—H
H₂C
CH₂
His-12
ENZYME
His-119

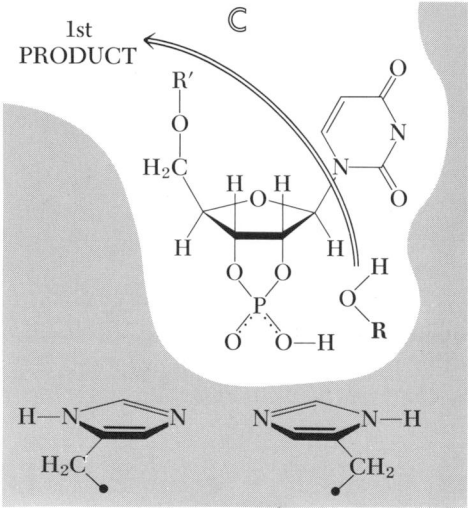

B. Catalysis begins by transfer of a proton from the imidazolium ion to the phosphate ion. A simultaneous shift of electrons breaks the P—O bond to the 5'-carbon of the adjacent nucleotide and establishes a new bond.

C. The results are the formation of a 2',3'-cyclic phosphate ester of the pyrimidine nucleoside and the release of a polynucleotide fragment (R—OH) that now has a free 5'-hydroxyl group. This polynucleotide can diffuse out into solution as one of the products of the hydrolysis.

D. Water now occupies the position freed by release of the product (HOH replacing ROH).

E

R′
O
H₂C
H O H
H H
O O
P
O O O H
H
H—N N N N—H
H₂C CH₂

E. Electrons again shift, transferring the P—O bond from the 2′ position onto the water molecule, with an accompanying release of a proton to the imidazole group.

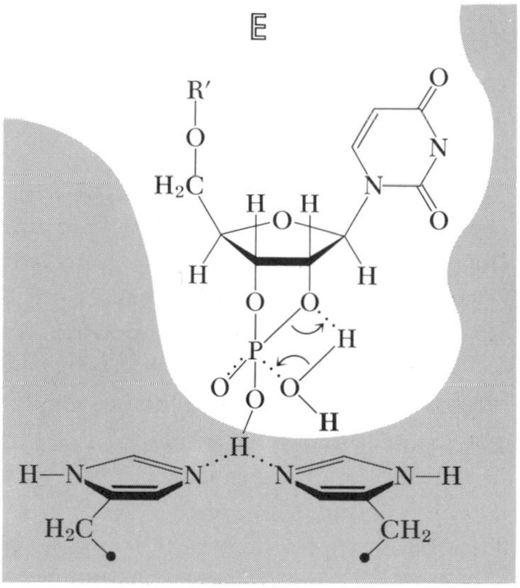

2nd
PRODUCT
F

R′
O
H₂C
H O H
H H
O O
H
P
O O O—H
⊖
⊕
H—N N—H—N N—H
H₂C CH₂

F. The resultant monophosphate ester on the remaining polynucleotide is the other product of the reaction. When it diffuses out into free solution, the enzyme is left in its original state, ready to repeat the whole sequence with a new substrate molecule.

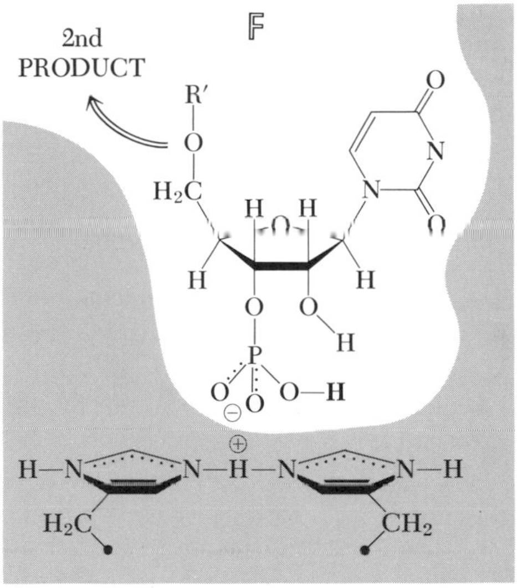

Generalities Illustrated by the Ribonuclease Mechanism

The mechanism we illustrated is of a type also found in simpler organic reactions—*acid-base catalysis*. Belief is solidifying that this is the common, if not ubiquitous, kind of catalysis for enzymatic reactions. Partly, this belief is a result of analysis of rates of reactions with changes in pH, and of model reactions catalyzed by simpler compounds showing behavior similar to enzymatic reactions. More importantly, it comes from the discovery of a *histidyl* residue, either paired with another or associated with a *lysyl* residue, at the active site of many enzymes.

It is generally accepted that the pair of histidyl residues participates in the ribonuclease reaction. Chemical modification of the residues destroys the catalysis. There is disagreement on whether the pair act in concert, as shown in the illustration, or separately, with one side chain accepting a proton while the other releases a proton. The scheme illustrated has a certain elegance. The two imidazole rings are mechanistically equivalent.

The scheme also stands a particular test that must be met by any proposed mechanism, and that is microscopic reversibility. Microscopic reversibility means that the forward and back reactions must traverse the same route. Each step must be freely reversible, and the reaction in one direction must proceed by an exact reversal of the steps in the opposite direction.

Beyond this, there ought to be an overall symmetry. The first step's mechanism ought to be the reverse of the last step's, the second's the reverse of the next-to-the last's, and so on. Why is this so? Enzymes do not determine the position of equilibrium of a reaction, they only change the rate. Any reaction can proceed to some extent in either direction, given the proper concentration of starting materials. Therefore, what we are calling products could act as substrates. If they are to act as substrates, they must react with the enzyme in the same sequence of steps by which they could be formed. Therefore, the kind of step by which they are formed must also be the kind of step by which they can react. In the case at hand, R—O—P—O—R must be mechanistically equivalent to R—O—P—O—H, and R—O—H mechanistically equivalent to H—O—H, no matter which way one looks at the reaction.

Some Evolutionary Considerations

Generalizations on the mechanism of enzyme action face the same tests as are required for any biochemical hypothesis: They must be consistent with the total knowledge of the particular organism being considered, including its evolution. Fascinating as the reactions may be in themselves, we ought not forget that we are dealing with *biological* reactions. It is the happy agreement with this test that makes the concept of acid-base catalysis by histidyl residues so appealing, even without the more direct evidence supporting the same position.

The question at hand is whether or not there is a particular advantage in the development of proteins containing histidine, with the use of this residue for catalysis. Now, a reaction proceeding by acid-base catalysis will go faster if the proton donor and acceptor have equal dissociation constants, so that the fraction of each with or without protons attached will be equal at a given pH. Otherwise, one or the other of the transfer reactions will be slowed by the lower concentration of the donor form.

Further, even if acid catalysis and base catalysis are used in separate reactions, one catalyst can serve both functions only if its acidic form and its basic form are both present. Added to this is the fact that catalysis will involve reactions analogous to these:

1. $catalyst—H^+ \longrightarrow catalyst + H^+$
2. $catalyst + H_2O \longrightarrow catalyst—H^+ + OH^-$

and this pair of reactions will only go at equal rates when $[H^+] = [OH^-]$. The reason for this comes from the fact that dissociation (1) and solvolysis (2) frequently have about equal rates, but are much slower than the respective back-reactions, which are limited only by diffusion, and therefore by the concentrations of H^+ and OH^-. Unless $[H^+] = [OH^-]$, the regeneration of one of the forms of the catalyst will be favored over the other, which will be depleted, and the rates of the forward reactions will thereby be unbalanced. A more complete discussion can be found in the reviews by Eigen and Hammes cited in the bibliography.

These requirements are enough, but there is the additional fact that a wide variety of enzymatic reactions involve phosphate esters, as we shall see. The final dissociation constant for most phosphate monoesters is near 10^{-7} (pK = 7). If the acidic and basic forms of these are to be approximately equal in concentration, the pH must be near 7.

In sum, we have the requirements:

$$[\text{catalyst}] = [\text{catalyst}{-}H^+],$$
$$[H^+] = [OH^-],$$
$$[R{-}OPO_3{}^{2-}] = [R{-}OPO_3H^-].$$

The last requirements fix the pH for most effective catalysis at 7, and what is the pH of the cells in most living organisms? Near 7!

The first requirement then becomes one of finding a catalyst with a dissociation constant near 10^{-7}. Superimposed on this are requirements that the group employed must associate easily with water (demonstrated by a high solubility of compounds containing the group) and yet be stable in water. It must be resistant to oxidation or reduction, and must differ sufficiently from other kinds of groups involved in metabolic reactions, so that it itself is not destroyed by enzymatic catalysis. In short, it must be chemically inert under the conditions prevailing in cells; otherwise it would not last long in its prescribed function.

Compounds meeting all of these requirements are hard to come by, and *imidazole* itself can hardly be improved upon. It is very soluble in water, highly stable under chemical attack, and has a pK near 7. The histidyl residue may represent a device for introducing the imidazole ring into proteins as a catalyst for enzymatic reactions, with less satisfactory groups dropped from this function long ago.

The histidyl residue is also relatively isolated in terms of biological reactivity. It is the only example of a singly substituted imidazole involved in general metabolic processes.

Therefore, we do have a possible explanation for the fact that biological systems usually operate near pH 7 and for the evolution of the amino acid, histidine, with its use in enzymatic catalysis.

An interesting offshoot from this analysis is the question of the origin of the histidyl residues used on either side of the heme group in hemoglobin. It is not likely that acid-base catalysis is involved in oxygen binding, but it may well be that hemoglobin represents a modification of a heme enzyme, with the histidyl residues' original function modified to fit the new situation.

Other Catalytic Residues

The emphasis given to concerted acid-base mechanisms does not preclude other mechanisms involving protonation and deprotonation, such as general acid or general base catalysis. The histidyl residue, with its dissociation constant near 7, would not always be the best catalyst in such mechanisms. Some enzymes may have mechanisms that utilize the carboxylate groups of *glutamyl* or *aspartyl* residues; others use the *lysyl* ammonium group, and still others the carboxylate or ammonium groups at the ends of the peptide chains.

In the past, hypothetical mechanisms for some enzymes overemphasized these groups, and histidyl residues were later found to be actually present at the active sites, but the rebound from this fad ought not obscure their true participation in some reactions. The principles involved are much the same as in the concerted mechanism we have illustrated. The important additional consideration is that, while it is true that the optimum pH for a mixture of enzymes using both acid and base catalysis will be near 7, an individual enzyme can be made to operate most effectively at a pH removed from this value—especially toward the acidic range at which carboxylates become good catalysts. Organisms sometimes have evolved enzymes to this specification. Microorganisms in particular may have reactions proceeding better in an acidic solution. A mammal has strongly acidic conditions in its stomach, and enzymes secreted in that organ are most active at low pH.

PROVISION OF REACTIVE GROUPS BY THE PEPTIDE: HYDROLASES

The peptide side chains of an enzyme, in addition to providing groups for substrate binding and for catalysis, may also provide reactants in a more conventional chemical sense, that is, groups forming covalent bonds with part of the substrate in the course of the reaction. Let us discuss two of the many examples now. More will be seen in later chapters.

The Formation of Acylated Hydrolases

Enzymes catalyzing hydrolyses are hydrolases. The existence of covalently linked substrate-enzyme compounds has been proven for several, notably among the *esterases* and *amidases:*

$$H_2O + R-\overset{\overset{\displaystyle O}{\|}}{C}-O-R' \xrightarrow{\textit{esterase}} R-COO^- + HOR + H^+$$

$$H_2O + R-\overset{\overset{\displaystyle O}{\|}}{C}-NH-R' \xrightarrow{\textit{amidase}} R-COO^- + {}^+H_3N-R'$$

Ribonuclease is an esterase, and if we look again at its mechanism we see that it involves the formation of cyclic phosphodiester intermediate, and the intermediate is later cleaved by the addition of water. Formally, the intermediate is created by transferring the phosphoryl group from the oxygen atom of the remaining polynucleotide chain onto an adjacent oxygen atom of the same ribose unit:

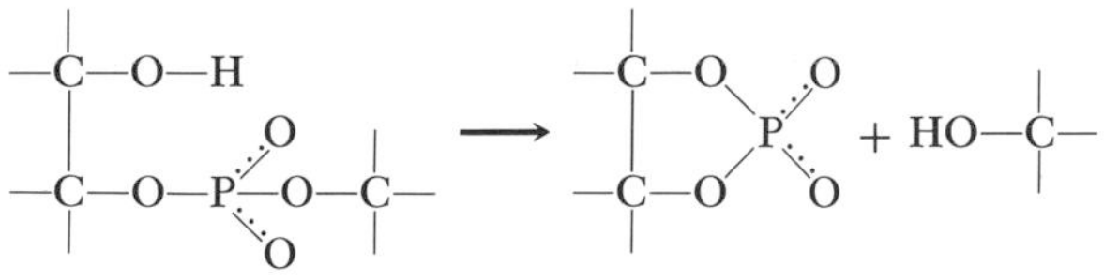

Not all ester and amide substrates have a hydroxyl group conveniently at hand to form an intermediate. In order to employ this kind of mechanism, the hydroxyl group must be provided by the enzyme, which can be done through the use of *seryl, threonyl,* or *tyrosyl* residues:

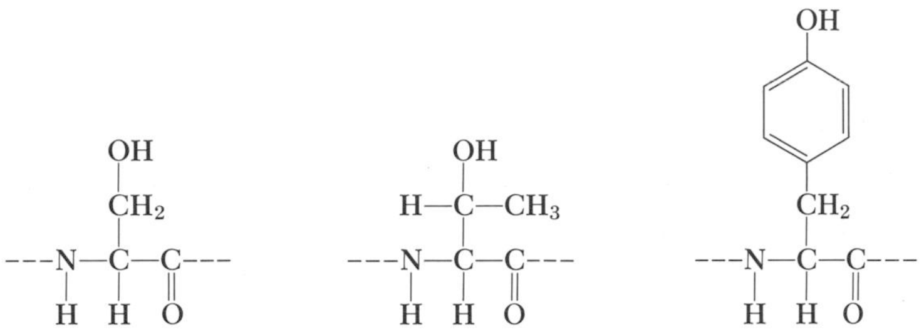

Among these possibilities, the primary alcohol represented by the seryl residue appears to be most commonly employed.

Consider the enzymes *trypsin* and *chymotrypsin*. These enzymes are found in the intestinal lumen. We shall later discuss their formation; now we are concerned only with their function, which is to catalyze the hydrolysis of the peptide chains in dietary proteins. They are in a special class of amidases, the *endopeptidases*. That is, they catalyze the hydrolysis of peptide bonds in the middle of a peptide chain, as opposed to the *exopeptidases*, which attack terminal amino acid residues.

The two enzymes differ in specificity, with trypsin attacking only lysyl or arginyl peptides, while chymotrypsin is less specific, but with some preference for peptides of the aromatic amino acids, tyrosine, phenylalanine, or tryptophan. Despite these differences in specificity, the active site is very similar in the two enzymes, which might be expected from the identity of the kind of bond affected by the two enzymes. Differing specificity in the two cases will be created by substrate binding to differing residues on the peptide, whereas the catalytic mechanism involves other groups, which may be the same in many enzymes of the same type. In both enzymes, the catalytic site has two histidyl residues, expected from the kind of mechanism we have discussed. The site also has a seryl residue contributed by another portion of the peptide.

This seryl group is esterified during the course of the reaction, and the intermediate acylated enzyme can be isolated, if simple substrates are used. Therefore, the reaction goes by this route:

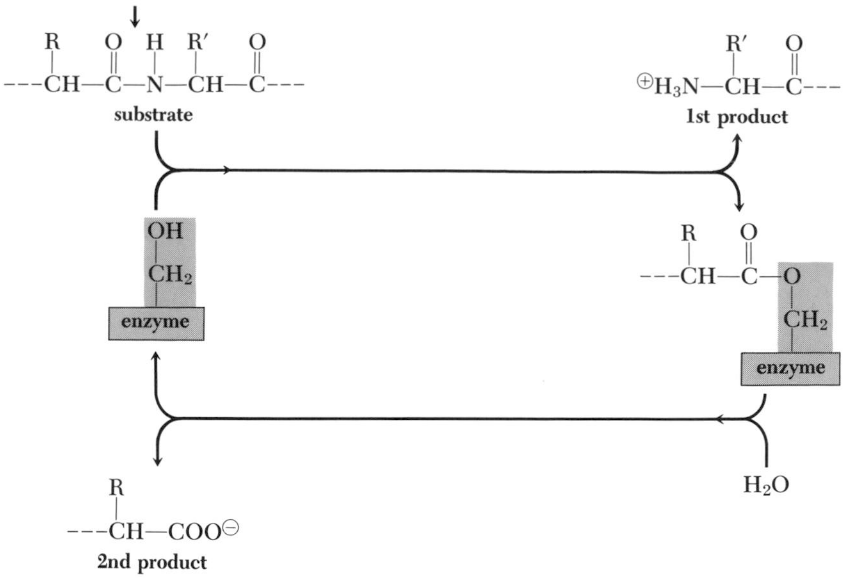

Although the mechanism closely resembles that of ribonuclease, it is the enzyme that provides the —OH group used in forming the intermediate ester. The participation of seryl groups in this way is a part of the mechanism of several kinds of enzymes.

Blocking Agents for the Seryl Group: Biochemical Reagents and Lethal Weapons

Alcohols react with acid anhydrides to form esters, and this is true of the seryl groups on proteins. In principle, one ought to be able to mix a protein with an acid anhydride, and thereby esterify the seryl groups in the protein. Uncharged anhydrides of phosphoric acid do react well in this way, and uncharged mixed anhydrides of phosphoric and hydrofluoric acid react even better. Uncharged anhydrides can be made by blocking any remaining acidic groups, for example, by converting them to esters. The ester anhydrides react with proteins to yield a phosphate triester, with one of the ester bonds formed with the protein:

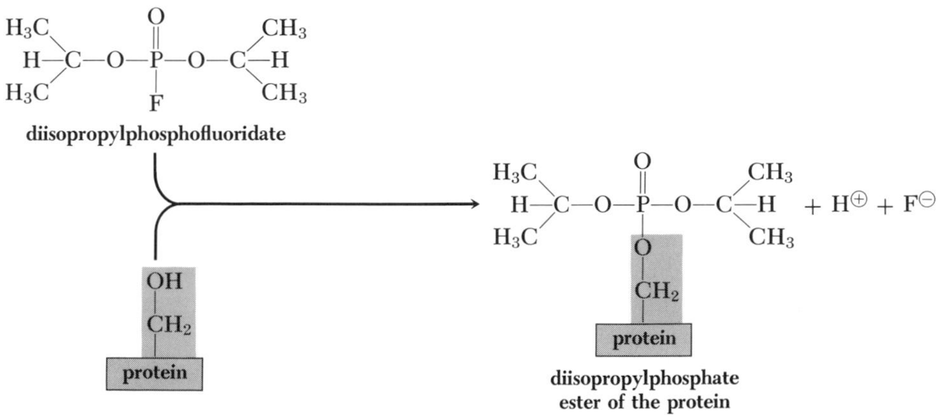

The particular agent shown, also called *diisopropylfluorophosphate (DFP)*, is an effective tool for demonstrating the participation of seryl groups in an enzymatic reaction because the formation of the triester on these groups prevents the completion of the hydrolytic mechanisms we have illustrated. The enzyme is inhibited, and the inhibition is non-competitive, because there is no form of the enzyme in equilibrium with both substrate and inhibitor. The enzyme will be inactive as a catalyst, even with high concentrations of substrate. If an enzyme is not inhibited in this way by DFP, seryl residues are not involved in its mechanism.

The effectiveness of the esters of phosphorofluoridic acid was discovered by accident. When they were being prepared for the first time in the laboratory of Dr. Willie Lang in Germany, the chemists developed the mental confusion, sense of constriction of the larynx, and painful loss of accommodation of the eye now known to be characteristic of an acetylcholinesterase inhibitor. The original workers concluded, on empirical grounds, that the compounds might have value as insecticides. This suggestion was ignored, but with the approach of the Second World War, it was realized they could provide very potent war gases. A variety of volatile derivatives, many still shrouded in secrecy, were prepared in several of the belligerent countries during the war, and their development continues.

These observations in turn stimulated research into the mechanism of action, which disclosed the combination with seryl groups and established the participation of this group in the acetylcholinesterase reaction. The compounds will react with enzymes other than acetylcholinesterase, but attention was directed to phosphate esters with configurations preferential for that particular enzyme, so that the nervous system would be selectively disabled by small concentrations

After World War II, the original suggestion that the compounds ought to be good insecticides was resurrected, and proved to be correct. The problem with insecticides is one of lowering the volatility and yet retaining the small aliphatic groups, thereby diminishing hazards to humans without sparing the bugs. The better compounds are still dangerous, but *malathion*, a sulfur analogue that is converted to the oxy-compound in tissues, is thought to be safe enough to permit general sale. Its formula is:

$$H_3C-O \diagdown \quad S$$

$$\qquad\qquad P \diagup\diagdown$$

$$H_3C-O \diagup \quad S-CH-\underset{\underset{O}{\|}}{C}-O-CH_2-CH_3$$

$$CH_2-\underset{\underset{}{\|}}{\overset{\overset{O}{\|}}{C}}-O-CH_2-CH_3$$

malathion

PROVISION OF REACTIVE GROUPS BY COENZYMES: TRANSAMINASES

Many enzymes differ from those we have previously discussed in that effective catalysis requires the presence of chemical groups that do not occur in the side chains of the common amino acids. These chemical groups must be added in the form of a compound other than an amino acid, in other words, as a prosthetic group or *coenzyme*. In such cases the enzyme will consist of the coenzyme and the associated peptide chains, which are called the *apoenzyme*. Apoenzyme + coenzyme = enzyme.

The peptide chains of such enzymes have a third function. In addition to binding the substrate and providing groups required for catalysis, they also must be capable of binding the coenzyme. Let us discuss an example of an important group of enzymes, the *transaminases*, to illustrate these functions.

Transamination: The Reaction and Coenzyme

A critical reaction in the metabolism of all organisms is the transfer of an amino group from one carbon skeleton to another by the process of transamination. Frequently, the most active exchange is between glutamate and aspartate:

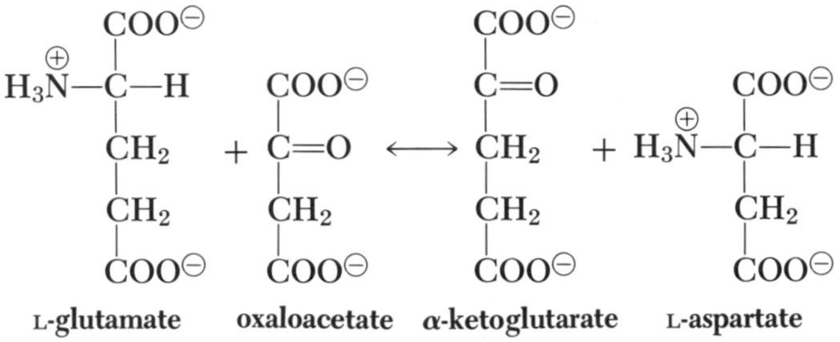

$$
\begin{array}{ccccc}
\overset{\oplus}{H_3N}-\overset{\displaystyle COO^{\ominus}}{\underset{\displaystyle CH_2}{\overset{|}{\underset{|}{C}}}-H} & & \overset{\displaystyle COO^{\ominus}}{\underset{\displaystyle CH_2}{\overset{|}{\underset{|}{C}}=O}} & \overset{\displaystyle COO^{\ominus}}{\underset{\displaystyle CH_2}{\overset{|}{\underset{|}{C}}=O}} & \overset{\displaystyle COO^{\ominus}}{\underset{\displaystyle CH_2}{\overset{\oplus}{H_3N}-\overset{|}{\underset{|}{C}}-H}}
\end{array}
$$

L-glutamate   oxaloacetate   α-ketoglutarate   L-aspartate

This reaction is catalyzed by the enzyme *glutamate-aspartate transaminase* (also known as glutamate-oxaloacetate transaminase).

The coenzyme for the transaminases is *pyridoxal phosphate:*

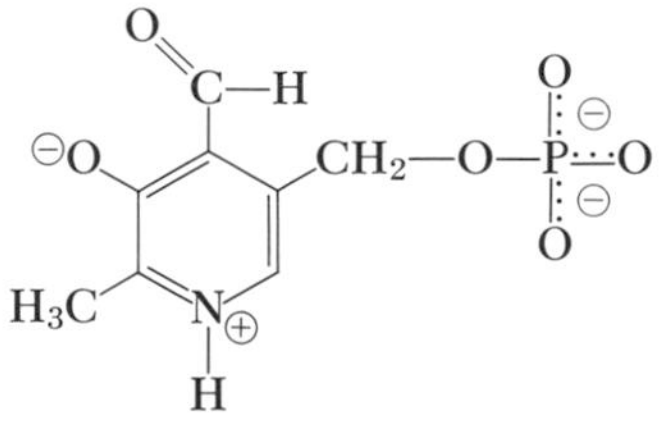

pyridoxal phosphate

Pyridoxal phosphate is fixed by the peptide chains of the transaminases in such a way that its aldehyde group is brought into proximity with a lysyl side chain. The lysyl residue supplies an ammonium group on a tail of carbon atoms waving out some distance from the remainder of the peptide. With any organic compound, neighboring aldehyde and amino groups will react spontaneously to form an *aldimine*, and the transaminases are no exception. The resultant aldimine is the reactive structure in the enzyme (Fig. 6-1). In the case of the glutamate-aspartate transaminase from mammalian hearts, there are two active sites on each molecule of the enzyme, and therefore two molecules of pyridoxal phosphate are bound. *Many enzymes have more than one active site,* but act by the same mechanism at each site.

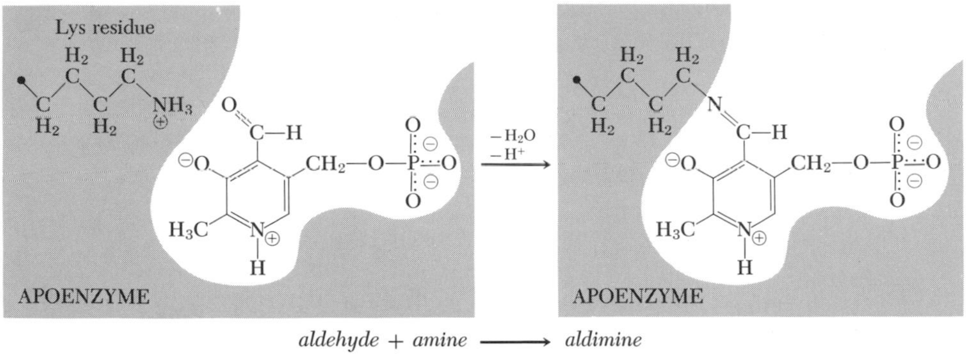

aldehyde + amine ⟶ aldimine

Figure 6-1 The combination of the peptides of a transaminase with the coenzyme, pyridoxal phosphate. A lysyl residue reacts with the aldehyde group of the coenzyme to form an aldimine, a form of Schiff's base.

Faith in the economy of chemical evolution would lead to the conclusion that the remainder of the pyridoxal phosphate molecule—the ionized phenolic group, the methyl group, the nitrogen of the pyridine ring, and the charged phosphate ester group—serve some function, and this function may well include the binding of the coenzyme at the correct location on the apoenzyme. This kind of an argument ought not be dismissed, but we have more concrete experimental evidence to justify the conclusion. If analogues are synthesized that lack one of these groups, they frequently are less effective coenzymes, and also are not bound to the apoenzyme so tightly. These two effects are quite independent. An analogue could be bound loosely, but react better when it is bound. Contrariwise, it might be bound very tightly, but be a poor reactant, thus being a good competitive inhibitor for the natural coenzyme. This is the same kind of situation found with substrates. Indeed, when we think on it, a coenzyme *is* a substrate, but of a special class, in that it is used repeatedly and dissociates much less readily than ordinary substrates.

Pyridoxal phosphate illustrates a general property of coenzymes, which is to have a small part of the structure, perhaps a single group, actively participating in the catalytic reaction. The active structure is a part of a more complex molecule, with the remainder of the structure serving to accentuate the reactivity of the effective group and to provide a framework for binding the coenzyme to the apoenzyme.

The Steps in Transamination

We shall detail the individual steps in the transamination reaction (Fig. 6-2) as an example of one of the more lengthy sequences. Catalysis hinges upon the fact, known from study of ordinary organic compounds, that an aldimine or ketimine will form more readily by exchange with a pre-existing aldimine than it will by direct combination of an amine and a carbonyl compound. The enzyme does have the pre-existing aldimine between pyridoxal phosphate and a lysyl residue, so the reaction proceeds by straightforward steps.

A. A molecule of L-glutamate collides with the enzyme and is bound, probably by its carboxylate groups, so that its ammonium group is near the aldimine structure on the enzyme.

B. The double bond shifts from the lysyl group to the glutamate, releasing the lysyl ammonium group and forming the aldimine of glutamate.

C. The double bond shifts, forming a ketimine.

D. The ketimine is hydrolyzed to form α-ketoglutarate and pyridoxamine phosphate. The α-ketoglutarate is free to dissociate, but the pyridoxamine phosphate is still held to the enzyme by the other groups on the molecule. (However, the loss of the covalent binding makes it experimentally easier to dissociate coenzyme and apoenzyme at this stage, and this has aided the discovery of the mechanism of the reaction.)

E. A molecule of oxaloacetate combines with the protein, also by attachment of carboxylate groups.

F. A ketimine is formed between the oxaloacetate and the pyridoxamine phosphate.

G. The double bond again shifts, now forming the aldimine of L-aspartate and pyridoxal phosphate.

H. The aldimine double bond shifts from the nitrogen of L-aspartate to the lysyl residue on the protein, leaving L-aspartate free to dissociate and restoring the original condition of the enzyme so that it can begin the whole process over again with a new molecule of L-glutamate.

The mechanism can operate in the exact reverse of the steps given, so that one can start with L-aspartate + α-ketoglutarate and arrive at L-glutamate + oxaloacetate.

The sequence details only the actual organic reactions, *i.e.*, those in which there is a change in covalent bonding, and does not describe the electron shifts responsible for catalysis. This has not been established with certainty, but there is some evidence that at least one histidyl residue is involved, and that the catalysis is of an acid-base type.

COENZYMES AND VITAMINS

In addition to being the first example of a coenzyme we have discussed, pyridoxal phosphate is also the first example of a compound for which a precursor must be supplied in the human diet because the entire compound cannot be synthesized by human tissues.

Coenzymes frequently contain structures not found in ordinary metabolites and behave as catalysts in the sense that they are eventually recovered in their original

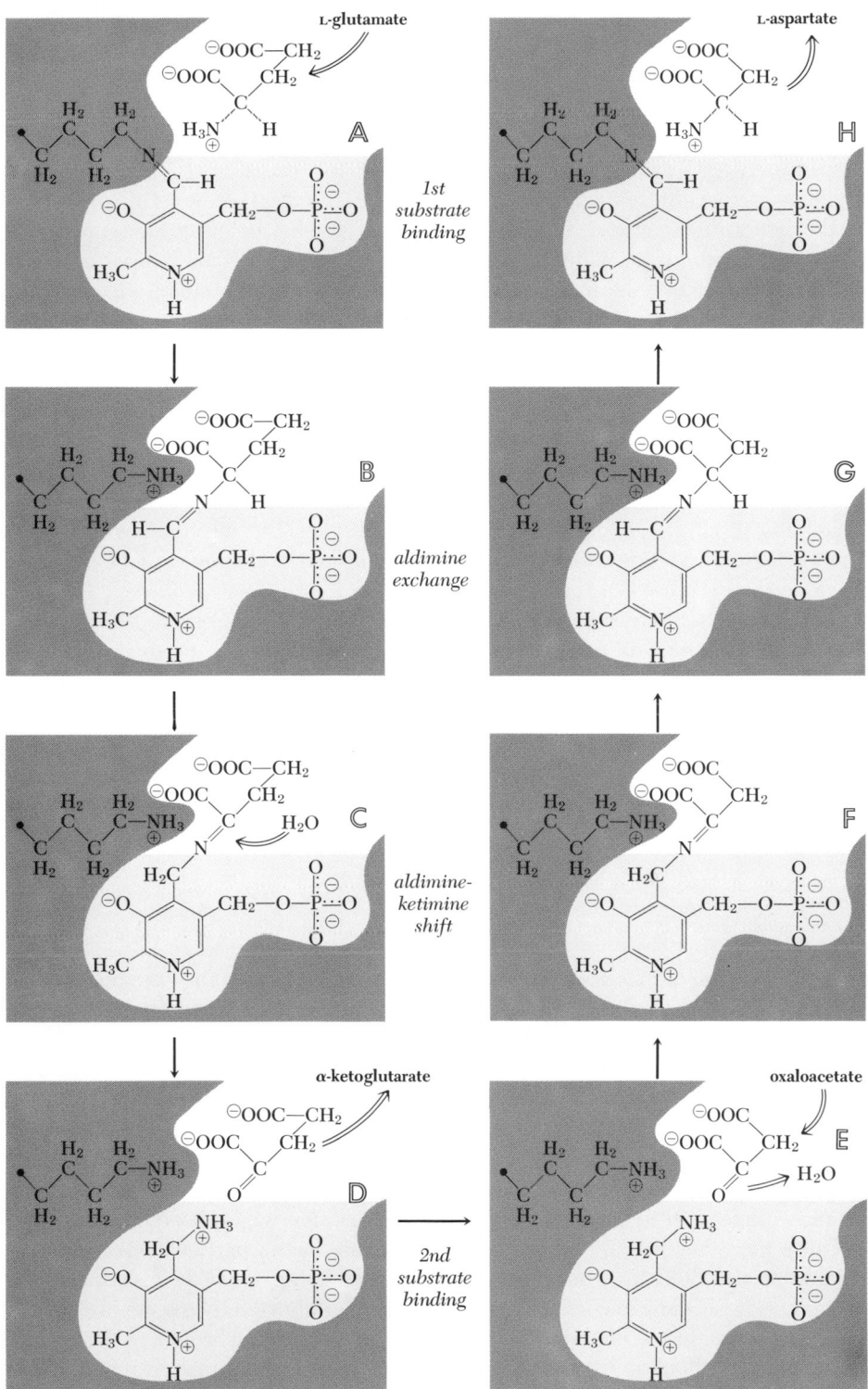

Figure 6-2   The steps in the glutamate-aspartate transaminase reaction.

form. Therefore, they need not be present in high concentrations in the tissues, any more than the apoenzyme need be. We have noted the widespread occurrence of transamination reactions among plants and animals. Most other kinds of enzymatic reactions, with the necessary coenzymes, also have a wide occurrence.

Putting these facts together, it is easy to rationalize the evolutionary truth—organisms whose principal nutrition is supplied by tissues from other organisms frequently have lost their ability to synthesize the structural elements of coenzymes. If the structure is unusual, it need not be made for the quantitatively important metabolic reactions, in which moles of compounds may be handled per day, and its sole function may be in the coenzyme. But only a small amount of coenzyme is required, and if the structure is constantly appearing in dietary compounds, its synthesis becomes unnecessary. Therefore, mutations deleting some of the reactions peculiar to the formation of coenzymes will not cause the death of the animal, and may even give it some advantage by making room in the cell for increased amounts of the enzymes that are absolutely necessary to form other compounds.

When such a deletion has occurred, and small amounts of the necessary organic structure must be supplied in the diet, the dietary compound is a *vitamin*. The formation of all coenzymes does not require vitamins, but all vitamins for which the biological function has been exactly established are known to be required because they are used to form coenzymes.

The substituted pyridine ring of pyridoxal phosphate is an example of a structure that cannot be synthesized by vertebrates, and its dietary precursors are lumped together as *vitamin B_6*, which includes *pyridoxal, pyridoxamine,* and the corresponding alcohol, *pyridoxine:*

$$\ominus O \quad \overset{CH_2OH}{\underset{H_3C \quad \overset{|}{\underset{N}{\oplus}} \quad}{\diagup\diagdown}} CH_2OH$$

$$H$$

pyridoxine

We can also rationalize the complexity of coenzymes from the example at hand. As shown, the binding of pyridoxal phosphate and the basis for the transamination reaction depend upon a simple reaction between an aldehyde and an amine. There are many aldehydes involved in metabolism. Suppose that one of these many aldehydes came in contact with the lysyl group at the active site after the formation of the peptide chains making up the apoenzyme of transaminase. The formation of an aldimine might well occur. However, it would eventually dissociate, and even though similar accidents might occur, eventually a molecule of pyridoxal phosphate would collide with it. Once this happened, dissociation would become infrequent, because the other binding groups on the molecule fit the particular peptide configuration and hold the coenzyme in place. Two things have been gained. The combination of coenzyme and apoenzyme has been stabilized by the use of multiple binding groups. Just as importantly, a variety of aldehydes can be used in the general metabolism without any significant disruption of the transamination reaction, because only the one very specific aldehyde structure will fit the apoenzyme.

The same kind of an advantage applies to the lysyl residue. We shall see that these residues are employed in other enzymes to bind completely different coenzymes, participating in reactions bearing little resemblance to transamination. Each of these kinds of coenzymes has a unique configuration, not capable of being confused with any of the others, and each apoenzyme can't bind the wrong coenzyme to its lysyl groups because it is built to conform to the complex structure of the proper coenzyme.

This represents a beginning on an important concept in metabolism: *The kinds of structures used in biological compounds and the types of reactions they undergo are relatively few.* The close regulation of the complex assembly of reactions that we lump together as metabolism doesn't depend upon each compound's having a unique kind of structure. It depends upon the compound's having a particular *combination* of structures, and the matching of the combination by a configuration built into only a few, perhaps one, of the hundreds of enzymes made by the cell.

Recapitulation of types of reactions

1. Acyl esters may be hydrolyzed:

$$R-\overset{\overset{\displaystyle O}{\|}}{C}-O-R' + H_2O \longrightarrow R-COO^- + HO-R' + H^+$$

Example: The reactions catalyzed by acetylcholinesterase.

2. Phosphate diesters may be hydrolyzed:

$$R-O-PO_2^- -O-R' + H_2O \longrightarrow R-O-PO_3^{2-} + HO-R' + H^+$$

Example: The reactions catalyzed by ribonuclease.

3. Peptides may be hydrolyzed:

$$R-\overset{\overset{\displaystyle O}{\|}}{C}-NH-R' + H_2O \longrightarrow R-COO^- + {}^+H_3N-R'$$

Examples: The reactions catalyzed by trypsin or chymotrypsin.

4. Transamination:

$$R-\overset{\overset{\displaystyle R'}{|}}{C}=O + R''-\overset{\overset{\displaystyle R'''}{|}}{C}H-NH_3^+ \longrightarrow R-\overset{\overset{\displaystyle R'}{|}}{C}H-NH_3^+ + R''-\overset{\overset{\displaystyle R'''}{|}}{C}=O$$

Example: The reaction catalyzed by glutamate-aspartate transaminase.

5. Lysyl residues may form Schiff's bases with carbonyl compounds:

$$R-(CH_2)_4-NH_3^+ + O=C-R'' \longrightarrow R-(CH_2)_4-N=C-R'' + H^+ + H_2O$$

Example: The binding of pyridoxal phosphate to transaminases.

Further reading

Gutfreund, H.: *An Introduction to the Study of Enzymes*. John Wiley (1965).

Koelle, G. B., ed.: *Cholinesterase and Anticholinesterase Agents*. Handbuch der Exptl. Pharmakologie, suppl. 15, Springer Verlag (1963). This is a compendium of reviews; some are in English.

Wilson, I. B.: *Acetylcholinesterase*. In Boyer, P. D., H. L. Lardy, and K. Myrbäck, eds.: *The Enzymes*, 2nd ed., Vol. 4 (p. 501). Academic Press (1960).

Witzel, H.: *The Function of the Pyrimidine Base in the Ribonuclease Reaction*. Progress in Nucleic Acid Res., *2:* 221 (1964).

Hummel, J. P., and G. Kalnitsky: *Mechanisms of Certain Phosphotransferase Reactions: Correlation of Structure and Catalysis in Some Selected Enzymes*. Ann. Rev. Biochem., *33:* 15 (1964).

Kartha, G., J. Bello, and D. Harker: *Tertiary Structure of Ribonuclease*. Nature, *213:* 862 (1967).

Wyckoff, H. W., et al.: *The Structure of Ribonuclease S at 3.5 A Resolution*. J. Biol. Chem., *242:* 2984 (1967).

Webb, J. L.: *Enzymes and Metabolic Inhibitors*. Academic Press (1963).

Eigen, M., and G. G. Hammes: *Elementary Steps in Enzyme Reactions*. Adv. Enzymology, *25:* 1 (1963).

Westheimer, F. H.: *Mechanisms Related to Enzyme Catalysis*. Adv. Enzymology, *24:* 441 (1962).

Snell, E. E.: *Chemical Structure in Relation to Biological Activities of Vitamin B_6*. Vitamins and Hormones, *16:* 78 (1958).

Neurath, H.: *Evolution of the Structure and Function of Proteases*. Science, *158:* 1638 (1967).

Wang, J. H.: *Facilitated Proton Transfer in Enzymatic Catalysis*. Science, *161:* 328 (1968).

Leuzinger, W., A. L. Baker, and E. Cauvin: *Acetylcholinesterase II. Crystallization, Absorption Spectra, Isoionic Point*. Proc. Natl. Acad. Sci., *59:* 620 (1968).

Harte, R. A., and J. A. Rupley: *Three-dimensional Pictures of Molecular Models—Lysozyme*. J. Biol. Chem., *243:* 1663 (1968). The pictures of the complex of lysozyme with its substrate ought to be seen for their beauty, even though the reaction is too complex for introduction at this point.

THE
RATES OF
ENZYMATIC
REACTIONS

CHAPTER 7

ARGUMENT

Quantitative measurements of the activity of enzymes are necessary for understanding their biological function and for determining the amount present in biological samples. A complete quantitative description is mathematically complex because of the many intermediate steps in an enzymatic reaction, but useful simplifications can be made. One of the most effective is to assume that only substrates, but no products, are present, so that the reaction is proceeding in only one direction. Using this assumption, estimates can be made of the maximum velocity (V_{max}) attainable by given amounts of enzyme, and the substrate concentration necessary to achieve half of that maximum velocity. This concentration is known as the *Michaelis constant* (K_M) for the enzyme.

Analysis of these kinetic parameters sometimes enables recognition of competitive inhibitors, whose effects can be relieved by increasing the substrate concentration, and non-competitive inhibitors, which are not affected by substrate concentration and have the effect of removing part of the enzyme.

Modification of the rate of enzymatic reactions is an important part of normal regulation of the chemical economy in organisms. Those reactions that are easily reversible at substrate concentrations existing within cells are ordinarily not regulated in this way, because the necessary enzyme must be present in high concentrations to insure useable rates of reactions. Regulation ordinarily occurs in reactions proceeding only in one direction under normal physiological circumstances. In some cases, regulation is achieved by developing enzymes with relatively weak affinity for the substrates, so that the reaction will accelerate when the substrate concentration rises, thereby increasing its utilization when the supply is high. In other cases, regulation is accomplished through the development of enzymes whose activity is changed by combining with other compounds. For example, in a sequence of metabolic reactions, an end product of the sequence may inhibit the enzyme catalyzing the first reaction in the series, even though the end product is not directly involved in that reaction. Any accumulation of the final product will therefore shut off its own production at the first step. Enzymes regulated in this way often combine with their substrates so as to give sigmoidal relationships between rate and concentration, like the kind of curve seen for the combination of oxygen and hemoglobin. The development of such kinetic characteristics depends upon the presence of multiple reactive sites, usually contributed by more than one peptide chain, in the enzyme molecule.

We dissected enzymatic reactions into individual steps and made qualitative analyses of the mechanisms in the previous chapter because it is important to understand *how* enzymes work. It is even more important to understand *what* is accomplished by the action of enzymes. One aspect is a description of the nature of the reactions that are catalyzed, but it is a mistake to assume that memorization of sequences of reactions is enough for understanding the metabolic economy. To be sure, the metabolic economy is composed of chemical reactions, many of which we must learn, but an inseparable part of the economy is the maintenance of quantitative balance. In short, we cannot begin to understand the physiological function of enzymes until we have some idea of the factors influencing reaction rates, and this idea requires numbers as well as qualitative facts. (It is also true that some with a facility in mathematics try to substitute symbolic manipulations for learning the facts, but we need not fear making this error in our later discussions.)

Secondly, there is a practical question involved. We want to be able to measure the concentration of enzymes in biological samples, and it is easiest to do this by measuring the action of each enzyme, determining how much the sample will accelerate the particular reaction for which the enzyme is specific. This is conveniently done by measuring the amount of substrate that disappears or the amount of product that is formed over a given period of time, and calculating the activity of the enzyme from these values. When we say that a *unit* of enzyme is the amount that will catalyze the formation of *one micromole* of product *per minute* under specified conditions, we are defining a quantity as surely as if we weighed the enzyme (which wouldn't be feasible anyway). This particular unit is the common expression of quantity of an enzyme.

THE OVERALL RATE FOR ENZYMATIC REACTIONS

We saw in the previous chapter that even an enzymatic reaction involving a single substrate and a single product will proceed in at least three steps: The enzyme and substrate combine to form an intermediate complex, X; X is transformed by internal catalysis to another complex, Y; and Y decomposes into the enzyme and product. Summarizing these reactions, and the reverse reactions that can also occur, we have:

$$E + S \rightleftharpoons X \rightleftharpoons Y \rightleftharpoons E + P$$

There are six different reactions in this example, which is much simpler than most enzymatic reactions actually are, and each of these reactions will involve a separate mathematical description of its rate. Rather than have the mathematics obscure the main ideas, let us begin with an even simpler case.

Description of the Simplest Model

The major complication in analyzing the rate of an enzymatic reaction involving a single substrate and a single product is the reverse reaction. The mathematics can be greatly simplified if measurements are made under conditions in which the reverse reaction is negligible compared to the forward reaction. Under these conditions, the overall formation of product is dependent only on the forward reaction, rather than on the net balance of forward and reverse reactions.

How can these simplifying circumstances be achieved? If an enzyme and its substrate are mixed together, it will only require a fraction of a second for the concentration of the first intermediate complex (X in the sequence above) to rise to a point where its rate of removal just balances the rate of its formation from enzyme and substrate. The interconversion of the intermediate complexes is the slow step in the process, and Y will rapidly decompose to give the product and enzyme. We will have achieved, at least for a time, a *steady state* in which the concentrations of E, X, and Y remain nearly constant. In most practical circumstances, this steady state is reached in the time it takes to thoroughly mix an enzyme with its substrate.

Now, if we have available a precise and sensitive method for measuring the formation of product, we can allow the reaction to proceed for a brief interval of time and measure the quantity of product formed before the concentration of product has increased to the point where the back reactions will be quantitatively important. The product obtained per unit time will therefore be an estimate of the *initial velocity* of the reaction. Hopefully, the initial velocity will be the rate of the enzymatic reaction after the steady state is reached, but before the overall reaction is slowed down because the substrate concentration is falling or the product concentration is rising.

Surprisingly, this approach frequently gives useful information. (There are many bad examples in the literature in which the limitations have been conveniently overlooked, but some rough guides to the behavior of enzymes can even be obtained from many of these.) If a series of determinations of initial velocity are made at varying substrate concentrations, a graph of the sort shown in Figure 7-1 will commonly be obtained. We can sense from the graph that the enzymatic reaction will only be proceeding at its *maximum velocity* (V_{max}) when the substrate concentration is approaching infinity. This would be the point at which the enzyme was

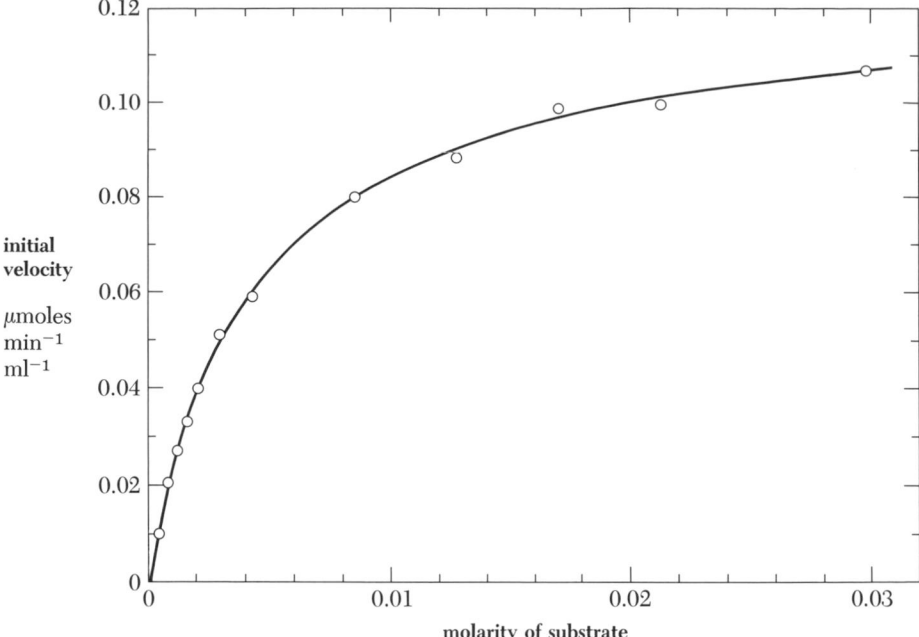

Figure 7-1   Measured values of the initial velocity of an enzymatic reaction with varying concentrations of substrate. The scatter apparent at the higher substrate concentrations amounts to 2 per cent of the highest velocity, which is quite good for measurements of most enzymes with standard techniques.

saturated with substrate, so that a molecule of enzyme released at the time of product formation would immediately collide with a molecule of substrate, and the concentration of free enzyme at any moment would be infinitesimal.

Unfortunately, we can't begin to approach infinite concentration of substrate. For one thing, extremely high concentrations of nearly all solutes will cause some disturbance of structure, which will mean loss of catalytic activity with most enzymes. However, we can estimate the maximum velocity mathematically. (It must not be forgotten that catalytic activity is a fundamental parameter of enzymes, and a measure of maximum velocity is as much a measure of the quantity of enzyme present as is the determination of any of its other properties.)

The mathematical form of the relationship of initial velocity to substrate concentration was recognized long before much was known about the mechanisms of enzymatic reactions, and the equation we use today was reported in 1913 by Michaelis and Menten:

1. $$v = \frac{V_{max}S}{K_M + S}$$

in which v = initial velocity at a given substrate concentration, V_{max} = maximum velocity at saturating substrate concentration, K_M = a constant, now known as the Michaelis constant, and S = concentration of substrate.

Let us now consider some of the useful aspects of the Michaelis constant.

Relationship Between Initial Velocity and Michaelis Constant

When the initial concentration of the substrate is made equal to the Michaelis constant, the Michaelis-Menten equation becomes:

2. $$v = \frac{V_{max}S}{K_M + S} = \frac{V_{max}S}{S + S} = \tfrac{1}{2} V_{max}$$

This tells us two things. First, it defines the Michaelis constant as the concentration of substrate at which the initial velocity will be half the maximum velocity. (The constant was originally defined in terms of the equilibrium constant for dissociation of the enzyme-substrate complex, but this definition has been replaced for most purposes by the kinetic definition just given, although Michaelis' name is retained.) Secondly, it tells us that any practical value for a saturating constant of substrate must lie someplace between a value equal to K_M and the impossible infinite concentration.

What is the practical degree of saturation attainable for laboratory measurements with an enzyme? It depends upon the precision desired in estimating V_{max}, which is the measure of the quantity of enzyme present for assay purposes. The fraction of V_{max} obtained when S is various multiples of K_M is shown in the following:

when $\dfrac{S}{K_M} =$

1	2	3	4	5	7	10	20	50	100

$\dfrac{v}{V_{max}} =$

0.50	0.67	0.75	0.80	0.83	0.88	0.91	0.95	0.98	0.99

Most people will settle for a substrate concentration 10- to 20-fold greater than the Michaelis constant when assaying enzymes, call that a saturating concentration, even though it isn't, and accept the 5 to 10 per cent error in estimating V_{max}.

This actually works fairly well for comparing the amount of enzyme in unknown samples, so long as the basic presumption remains correct that the product concentration is negligible. Suppose we have a solution of enzyme in the presence of $10 \times K_M$ concentration of substrate and allow the reaction to proceed until 10 per cent of the substrate has been converted to product. If the product concentration is truly negligible, the rate will only drop by 1 per cent during assay (S drops from $10K_M$ to $9K_M$ and v from $0.91V_{max}$ to $0.90V_{max}$). We can determine the amount of product formed and divide the value by the elapsed time to determine an average velocity over the period that will theoretically be within 0.99 of the true initial velocity. Now, if we have a second sample that has twice as much enzyme, so that V_{max} is doubled, and incubate it for the same period of time, the substrate concentration will drop by almost 20 per cent (from $10K_M$ to $8K_M$), but this will only cause the initial velocity to fall by about 2 per cent (from $0.91V_{max}$ to $0.89V_{max}$). Therefore, double the amount of enzyme will come within 1 per cent of giving double the amount of product formation in a fixed period of time under these conditions, and we can use the procedure for a fairly accurate *comparison* of the amount of enzyme in different samples, even though the system is operating at only $0.9V_{max}$. (This analysis presumes that there is no error in the measurement of the product formed. There always is a problem, and it is a nice problem—a part of the art—to balance analytical error from measuring too little product against theoretical error from too much reaction.)

Estimation of the Michaelis Constant

We see that knowledge of the Michaelis constant is useful for determining how much substrate to add when one wishes to measure an enzyme. More importantly, it is a guide for determining how effectively a particular enzyme is functioning in the tissues of a living organism. It is therefore one of the parameters frequently determined for enzymes, and the way in which this is accomplished is described in the following.

The Michaelis-Menten equation can be inverted and rearranged:

3. $$\frac{1}{v} = \frac{K_M + S}{V_{max}S} = \frac{K_M}{V_{max}} \times \frac{1}{S} + \frac{1}{V_{max}}$$

If we now take $1/v$ and $1/S$ as our variables, we have a simple linear equation of the type $y = ax + b$, which can either be solved by a linear regression analysis or by graphical analysis.

Most people seem to prefer pictures to algebra, and therefore use the graphical solution. The entire procedure is quite simple:

1. A fixed amount of the enzyme is incubated with varying initial concentrations of the substrate. The time of incubation is kept as small as is consistent with accurate analysis of the amount of product formed, so as to minimize changes in substrate concentration.

2. The amount of product formed is divided by the time of incubation to obtain the "initial" velocity. (If the analytical procedure utilizes recording instruments, the initial velocity can be estimated directly from the plot.) The reciprocal of each initial velocity is calculated.

3. The mean substrate concentration is calculated for the period of time over which the velocity was estimated in each case. This will be the initial concentration less half the concentration of product formed in the total time. The reciprocal of this mean substrate concentration is calculated for each measurement.

4. The reciprocals of initial velocities are plotted as functions of the reciprocals of substrate concentrations. If Michaelis-Menten kinetics describe the reaction, the result will be a straight line, such as is shown in Figure 7-2, which was derived from the same data shown in Figure 7-1 in a direct plot of velocity against concentration rather than the respective reciprocals. The reciprocal plots are known as Lineweaver-Burk plots, after the men best known for developing this type of analysis.

5. As is indicated on the figure, the intercept on the ordinate is the reciprocal of V_{max} and the intercept on the abscissa is the reciprocal of $-K_M$, so both of these constants can be estimated from the intercepts.

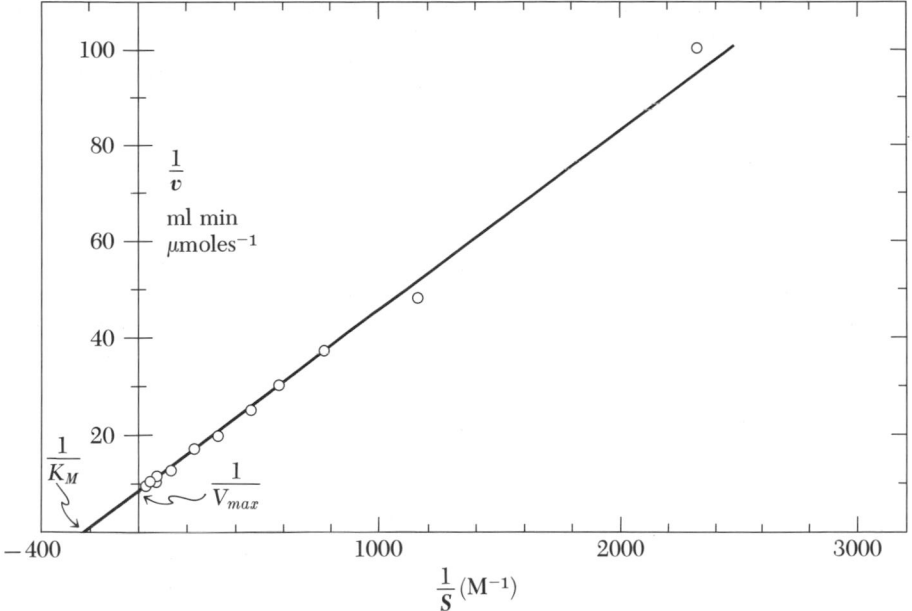

Figure 7-2 A Lineweaver-Burk plot of the data from Figure 7-1. Note the gross scatter evident in values obtained at low substrate concentrations (high values of $1/S$), which were not evident in the steeply rising portion of the regular plot given in Figure 7-1. The intercepts yield the following values:

$$\text{abscissa} = -\frac{1}{K_M} = -230 \text{ M}^{-1}; \quad K_M = 0.0043 \text{ M}$$

$$\text{ordinate} = \frac{1}{V_{max}} = 8.3 \text{ ml min } \mu\text{moles}^{-1}; \quad V_{max} = 0.120 \ \mu\text{moles min}^{-1} \text{ ml}^{-1}$$

Reactions with Two or More Substrates

Many enzymatic reactions involve more than one substrate: $A + B \xrightarrow{E} P$, or $A + B \xrightarrow{E} P + Q$, and so on. The initial velocity of these reactions will depend upon the concentration of each of the different substrates, and the actual form of the equation describing the initial velocity depends upon the mechanism of the reaction.

For example, two substrates may react in a *sequential* fashion, with both becoming attached to the enzyme before any products are formed: $A + E \rightleftharpoons AE$; $AE + B \rightleftharpoons ABE \rightleftharpoons PQE \rightleftharpoons Q + PE$; $PE \rightleftharpoons P + E$. On the other hand, the substrates may react in a *ping-pong* mechanism, in which one substrate reacts with the enzyme and dissociates into one product, leaving some functional group from the substrate attached to the enzyme. The modified enzyme then reacts with the second substrate so as to transfer the group to it, forming the second product and releasing the original form of the enzyme: $A + E \rightleftharpoons AE \rightleftharpoons PE' \rightleftharpoons P + E'$; $E' + B \rightleftharpoons BE' \rightleftharpoons QE \rightleftharpoons Q + E$. The overall result seems similar, but the two mechanisms have quite different kinetic properties.

However, if a large excess of all but one of the substrates is added, variation of the concentration of the remaining substrate will affect the initial velocity in the way we have described for single-substrate reactions, and the K_M for this substrate can be determined from simple Lineweaver-Burk plots. The process can be repeated in turn for each of the substrates. (This kind of analysis is only valid if the high concentration of one substrate does not change the catalytic properties of the enzyme in some way.)

The Effect of the Reverse Reaction

All of the simple analyses we have given depend upon the assumption that the reverse reaction can be neglected. Enzymes do catalyze the attainment of equilibrium, and, as we pointed out in the preceding chapter, what we call product in one case may act as a substrate in another. It follows from this that there will be kinetics of similar mathematical form for "product" reacting to yield "substrate," complete with Michaelis constants for the products and a maximum velocity in the reverse direction.

When the back reaction has to be taken into consideration, the situation becomes mathematically awkward if more than one substrate or product is involved, but we can develop the main points from a consideration of the simple one-substrate, one-product reaction: $S \rightleftharpoons P$. There is in effect a competition for the enzyme between the substrate and the product. Suppose that an enzyme has substantially lower affinity for the product than it does for the substrate. The Michaelis constant for the product will be high. It will require correspondingly high concentrations of product for the *reverse* reaction to reach half of its maximum velocity compared to the concentration of substrate required for the *forward* reaction to reach half of its maximum velocity. In these circumstances, the amount of enzyme tied up in combination with the product will remain low until a considerable amount of substrate has been converted to product.

If the contrary is true, and the affinity of the enzyme for the product is high, so that only a small concentration of product is necessary to accelerate the back reaction to half its maximum, then the formation of even a little product will substantially lessen the amount of free enzyme available to react with substrate, and the rate of formation of product will rapidly fall.

The effect of these possible variations is shown in Figure 7-3, in which the formation of product is plotted as a function of time for four hypothetical enzymes, all with identical Michaelis constants and maximum velocities for the forward reaction (K_F and V_F), but differing in the Michaelis constants and maximum velocities for the back reaction (K_B and V_B). These parameters are not completely independent, but are functions of the equilibrium constant for the overall reaction according to an expression:

$$4. \qquad K_{eq} = \frac{V_F K_B}{V_B K_F}$$

This is known as the Haldane relationship. Figure 7-3 shows two reactions with moderately high K_{eq} of 10^3, and two with $K_{eq} = 1$.

One of the examples in Figure 7-3, Curve A, illustrates the type of reaction beloved by those who assay enzymes, with a high equilibrium constant, low affinity for the product, and nearly linear kinetics for a large part of the reaction so that the amount of product formed will be directly proportional to the amount of enzyme added over a wide range.

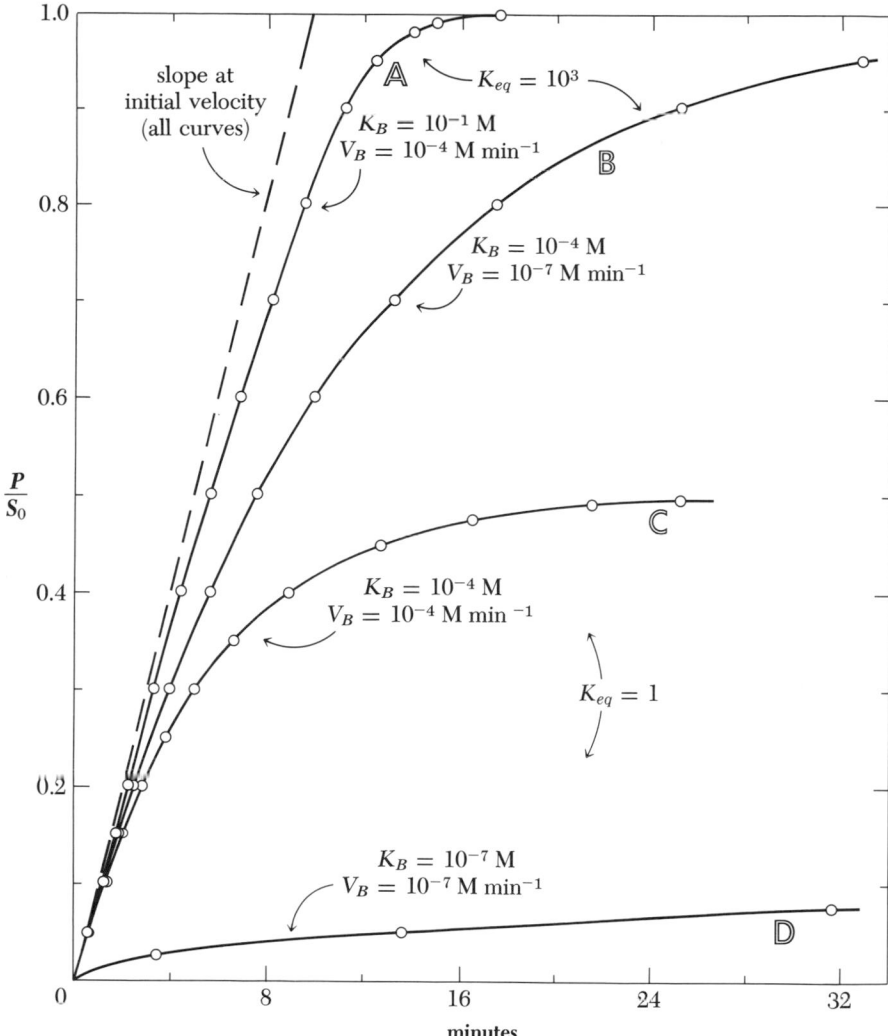

Figure 7-3 The effect of the back reaction on the time course of enzymatic reactions. The fraction of the initial substrate that has been converted to product (P/S_0) is plotted as a function of time for reactions catalyzed by four different enzymes. Each enzyme has identical maximum velocities and Michaelis constants for the forward reaction ($V_F = 10^{-4}$ M min^{-1} and $K_F = 10^{-4}$ M), with an initial substrate concentration of 10^{-3} M. The kinetic parameters for the back reactions are listed by the respective curves. Note that *all* of the reactions would follow the dashed line if substrate was continually added to maintain its concentration and if the product was removed as fast as it formed.

Another of the examples, Curve D, shows the kind of reaction that is almost impossible to tackle in the ordinary way. The affinity of the enzyme is so much higher for the product than it is for the substrate that the overall rate of product formation drops immediately after a little product is formed. Note that this is true even though the maximum velocity in the forward direction is 1000-fold greater than that in the reverse direction. Trying to measure the reaction in the reverse direction by using what we are calling "product" as the starting substrate wouldn't help much because the enzyme is 1000-fold less effective as a catalyst in that direction.

Fortunately, this kind of an enzyme is exceedingly rare, if it ever occurs. We can see that it would be most inefficient as a biological catalyst. Real enzymes are more like the other examples shown. If an enzyme is catalyzing a part of a sequence of reactions that serves its biological function by proceeding only in one direction, it will tend to have a high V_{max} in that direction, and a relatively low affinity for the product, as evidenced by a low K_F/K_B ratio. If the enzyme catalyzes a reaction that is biologically reversible—going sometimes in one direction and sometimes in the other direction, depending upon transient changes in concentrations of metabolites within the cell—the enzyme will usually have approximately equal maximum velocities in both directions. According to the Haldane relationship, this means that the Michaelis constants for the forward and back reactions will also be approximately equal if the reaction is easily reversible ($K_{eq} \sim 1$).

The Effect of Competitive Inhibitors on Michaelis-Menten Kinetics

We noted in the preceding chapter the ways in which inhibitors can be used to make deductions about the nature of substrate binding and the mechanism of enzyme action. These deductions also depend upon quantitative analysis of enzyme kinetics under conditions in which the rate of the back reaction is not significant.

Competitive inhibitors are especially useful because of the information they supply about interaction of substrate and enzyme. Simple inhibitions of this sort arise from the combination of an inhibitor, I, to yield an inactive enzyme-inhibitor complex: $E + I \rightleftharpoons EI$, and the combination is described by a simple equilibrium constant: $K_I = \dfrac{[EI]}{[E][I]}$. If the inhibition is competitive, combination of the inhibitor with the enzyme will prevent the combination of the substrate, and vice versa, because both combinations involve the same part of the enzyme surface. The initial velocity of the reaction in the presence of a competitive inhibitor is described by the equation:

$$5. \qquad v = \frac{V_{max}S}{S + K_M(1 + I/K_I)}$$

The equation can be rearranged in terms of reciprocals of initial velocity and substrate concentration, as we did with the simple Michaelis-Menten equation, but the essential features are apparent as written. Mathematically, the presence of inhibitor has the effect of increasing K_M in the equation by the factor $\left(1 + \dfrac{I}{K_I}\right)$, but it *does not change the value of* V_{max}. This is what would be expected, because an infinite concentration of substrate should effectively displace all of a competitive inhibitor.

The equation also shows us that K_I is equal to the concentration of inhibitor that doubles the *apparent* value of K_M. Now, if the inhibitor doesn't change the maximum velocity but does alter the apparent value of the Michaelis constant, differing concentrations of inhibitor ought to give a series of straight lines intersecting at $1/V_{max}$ on the ordinate, when Lineweaver-Burk plots are made of the effect of substrate concentration on rate of reaction (Fig. 7-4).

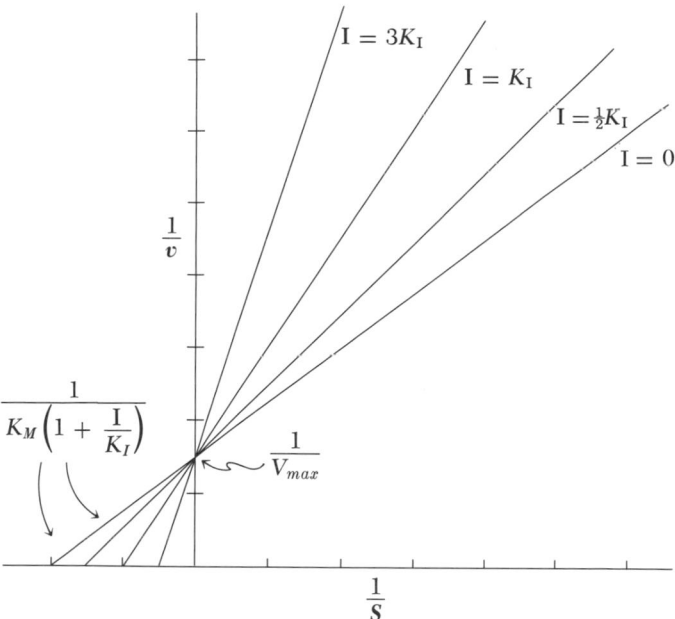

Figure 7-4   Lineweaver-Burk plots obtained with varying concentrations, I, of a purely competitive inhibitor. There is a single intercept on the ordinate because V_{max} is the same in all cases, but the intercept on the abscissa changes.

The value of K_I can be estimated from the value of the apparent K_M in the presence of inhibitor, and this value is obtained from the negative reciprocal of the intercept on the abscissa. Figure 7-4 shows an example.

Non-Competitive Inhibitors and Michaelis-Menten Kinetics

Compounds may inhibit an enzymatic reaction by combining with an enzyme in such a way that they are not displaced by the substrate, but prevent the enzymatic reaction from occurring. For example, the reaction of the phosphofluoridate esters with acetylcholinesterase is essentially irreversible, and the esterified enzyme will not catalyze the hydrolysis of acetylcholine. The effect of these inhibitors is to remove enzyme from the solution, thereby lowering V_{max}, but the remaining enzyme displays the same kinetics as does a more dilute solution of the enzyme with no inhibitor added. In other words, K_M will not be affected, and Lineweaver-Burk plots at varying inhibitor concentrations will give a series of lines that intersect on the abscissa (Fig. 7-5).

We now have two straightforward cases that can be recognized from Lineweaver-Burk plots of the kinetics at different inhibitor concentrations: straight lines intersecting on the ordinate indicate classical competitive inhibition, and straight lines intersecting on the abscissa indicate purely non-competitive inhibition.

However, these cases do not by any means exhaust the possibilities for inhibition. An inhibitor may react more readily with the intermediate enzyme-substrate

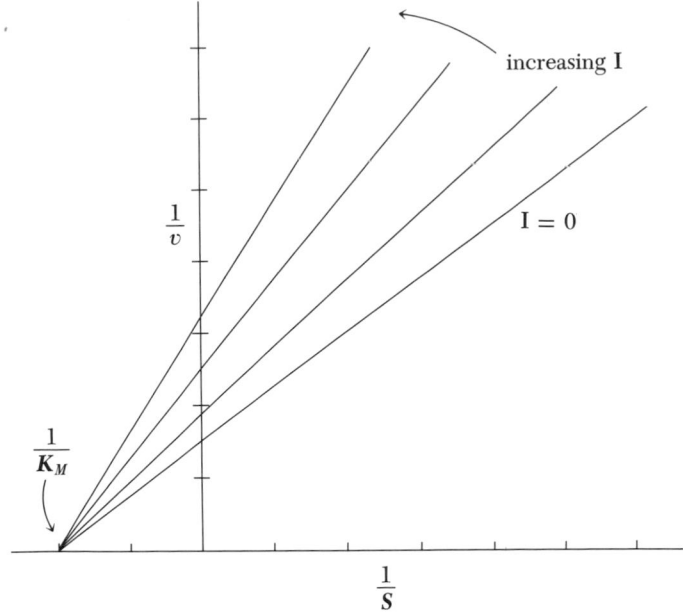

Figure 7-5   Lineweaver-Burk plots obtained with varying concentrations of a purely non-competitive inhibitor. K_M is constant in all cases, so there is a single intercept on the abscissa. The inhibitor has the same effect as removing enzyme, so V_{max}, and the intercept on the ordinate, change.

complexes than it does with the free enzyme. The inhibitor-enzyme complex may be catalytically active, but with altered values of K_M or V_{max}. The result of the various possibilities may be that varying inhibitor concentrations will produce Lineweaver-Burk plots that are straight lines, but which do not all intersect at one point, so that the inhibition is "mixed" rather than being competitive or non-competitive. In other cases, Lineweaver-Burk plots produce curved, rather than straight, lines. Permutations of these various kinds of effects can be analyzed in as much detail as patience permits, but such extensive analysis is not required for our purposes. More complete treatment can be found in the references cited at the end of the chapter, especially in Webb's treatise.

PHYSIOLOGICAL CONTROL OF KINETICS

The Concept of the Rate-Limiting Step

The important principle in understanding metabolic regulation is the idea that one reaction in a sequence may be governing the rate of the entire sequence. Such a reaction is referred to as the rate-limiting step.

Consider the sequence:

$$A \xrightarrow{E_1} B \xrightarrow{E_2} C \xrightarrow{E_3} D \xrightarrow{E_4} E \xrightarrow{E_5} F$$

Let us presume the following: (1) compound A is being supplied at a constant concentration, (2) compound F, the product of the sequence, is being removed in such a way as to hold its concentration constant, (3) the equilibrium constant for each of the individual reactions is identical, (4) all of the enzymes involved have identical Michaelis constants, and (5) the maximum velocity of all of the enzymes is identical, except that E_3, the enzyme catalyzing C $\longrightarrow$ D, has a substantially lower V_{max} than the others.

Now let us set up a hypothetical case in which a cell begins with concentrations of all of the intermediates equal to the concentration of the starting material:

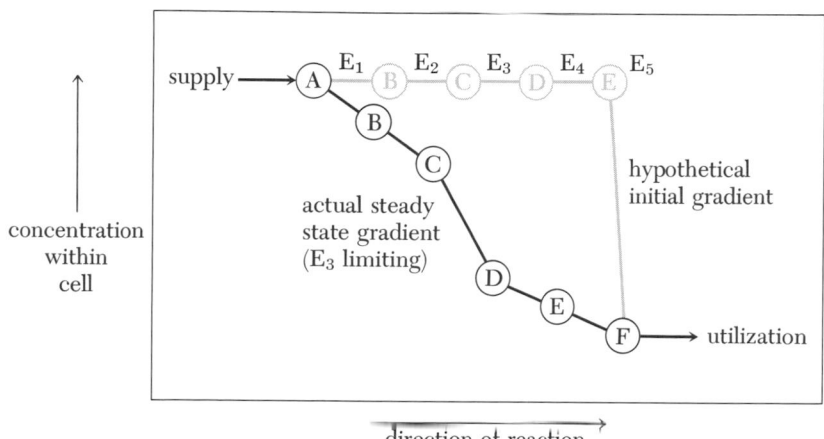

What will happen? There is a large difference in concentrations between E and F, so the last reaction will rapidly proceed. This will drop the concentration of E, so the conversion of D to E will accelerate, thus lowering the concentration of D. This will tend to accelerate the conversion of C to D, *but there is less of this enzyme, so the conversion will be accelerated less by the same fall in product concentration.* The result will be that the drop in concentrations of D and E will continue, and eventually cause the conversion of D $\longrightarrow$ E $\longrightarrow$ F to slow down. The reaction C $\longrightarrow$ D does not cause as sharp a drop in the concentrations of B and C, because the reactions A $\longrightarrow$ B $\longrightarrow$ C are being catalyzed by more active enzymes, and it takes *less difference in concentration to make these reactions go fast enough to replace the C being removed.*

The end result is the attainment of concentrations such that *all* of the reactions are proceeding at the same rate, and in this steady state the difference in concentrations of substrate and product for the example we are using will be greatest for the reaction with the least active enzyme.

If all are now proceeding at the same rate, why are we justified in calling E_3 rate-limiting? The answer is that if we now increase the activity of this enzyme in any way, there will be a corresponding increase in the rate of conversion of C $\longrightarrow$ D, less difference in concentration between these two compounds, and a corresponding increase in rate of all of the other reactions in the sequence. On the other hand, if we increased the activity of one of the other enzymes, little would happen, because the difference in concentrations is already small between their substrates and their products.

(It ought not be assumed from this example that the substrate concentration is always much larger and the product concentration is always much smaller for rate-limiting reactions than for neighboring reactions. The equilibrium constants and kinetic parameters for neighboring reactions in a sequence are rarely identical, as we have assumed. It is the degree of displacement from the equilibrium values that is greater for the rate-limiting step.)

The idea of a rate-limiting step therefore is that the rate of an entire sequence is fixed by the activity of the single enzyme that has the lowest V_{max}.

Physiologically Reversible Reactions

There are some reactions in tissues that are biologically reversible, with normal daily variation of the concentrations of the reacting components causing displacement from the equilibrium concentrations sometimes in one direction and at other times in the opposite direction. Such enzymes must be built to have comparable kinetic parameters (K_M and V_{max}) in either direction.

Enzymatic reactions, like other chemical reactions, slow down as the concentrations approach equilibrium values, and speed up as the concentrations shift from equilibrium. In that sense, reversible reactions are "controlled" in that they tend to minimize deviations from equilibrium. However, they are not control points in the sense of being reactions at which entire sequences are regulated. The problem usually met during evolution of the chemical economy has been the converse— adjusting conditions so that the rate of easily reversible reactions does not become limiting.

Consider the reaction catalyzed by glutamate-aspartate transaminase:

$$\text{glutamate} + \text{oxaloacetate} \rightleftharpoons \alpha\text{-ketoglutarate} + \text{aspartate}$$

All four of the compounds are essential metabolites, with each being used for a variety of purposes, and they are constantly present within the cell so that the partial reactions shown above are simultaneously proceeding in both directions. Yet, it is important that the net reaction represented by the *difference* between the two partial reactions be rapid if the concentrations are displaced from equilibrium in either direction. This can only be accomplished by having a large concentration of the enzyme, as compared to the amount of those enzymes catalyzing essentially irreversible reactions in which the rate of the back reaction is inconsequential. The actual velocity of the net transaminase reaction may only be a small fraction of the maximum velocity theoretically obtainable if no product were present, but there is so much of the enzyme that this small fraction is still comparable to the rate of other enzymatic reactions in the tissue.

This principle generally applies. Enzymes catalyzing reactions that are reversible under biological conditions are usually present in high concentrations, so that even small displacements from equilibrium will be sufficient to achieve a rapid rate of reaction. However, this circumstance sacrifices any possibility of infallible regulation of the rate of the reaction by substrate concentration because it cannot be the primary rate-limiting step in a sequence. A regulated step must have the concentrations displaced from equilibrium because the products are being removed and the substrates accumulating.

Inherent Control by Substrate Concentration

If a reaction normally proceeds in one direction in a tissue, its rate may be controlled through changes in the substrate concentration relative to the K_M. This is an obvious possibility for reactions with a high equilibrium constant that are practically irreversible. Such controls may also be used for reactions that are readily reversible in the test tube, but which are used in only one direction within cells because of the constant removal of the products of the reaction.

A prime example of effective physiological use of substrate concentration as a controlling mechanism is found in the enzymes catalyzing the transfer of phosphate from ATP to glucose:

$$\text{ATP}^{4-} + \text{glucose} \longrightarrow \text{ADP}^{3-} + \text{glucose-6-phosphate}^{2-} + \text{H}^+$$

This is the first reaction in the utilization of glucose by cells. The reaction in liver is catalyzed by an enzyme called *glucokinase* and in the brain by another enzyme given the general name *hexokinase* because it is not very specific for the kind of hexose. The suffix, *kinase*, was invented to suggest that the reaction results in the activation of glucose for metabolism. As is implied by the single arrow, the reaction is essentially irreversible under the conditions prevalent in tissues.

The measured Michaelis constants for these two enzymes are as follows:

	K_M (ATP)	K_M (glucose)
brain hexokinase	4×10^{-4} M	5×10^{-5} M
liver glucokinase	1×10^{-4} M	2×10^{-2} M

The concentration of ATP in these tissues is held near 0.001 M or greater. This is substantially greater than K_M for either enzyme, and this substrate is therefore not limiting in either case.

Now, the concentration of glucose in the blood supply to the brain and other peripheral tissues usually ranges around 0.005 M, although it may increase to around 0.008 M after consumption of a large amount of carbohydrate and may drop to around 0.003 M during fasting or heavy muscular work. The liver receives blood from the intestine through the portal circulation as well as from the arterial circulation. The portal blood may have substantially higher concentrations of glucose during absorption of a carbohydrate-containing meal.

All of the values for glucose concentration we have mentioned are nearly two orders of magnitude higher than the K_M for glucose with the brain hexokinase. This provides a considerable margin of safety for the brain, which is almost completely dependent upon the glucose supply for its major fuel, even allowing for a large concentration gradient between the concentration of glucose in the blood and in the interior of the neurons. The observed facts are consistent with this, because the concentration of glucose in blood must drop to nearly 0.001 M before disturbances of the central nervous system are noticed in most humans, and even lower to produce unconsciousness.

On the other hand, the K_M of the liver glucokinase is greater than the concentration of glucose usually found in arterial blood. Therefore, the uptake of glucose catalyzed by this enzyme will be very sensitive to glucose concentration, and this is consistent with the physiological role of the liver. The liver ordinarily produces glucose, rather than consuming it, and it only removes more glucose from the blood than it produces during temporary overloads after eating. The balance between output and uptake occurs with blood concentrations near 0.006 M.

A rise in glucose concentration will accelerate the liver glucokinase reaction because the enzyme is operating far below its saturating substrate concentration, whereas the brain hexokinase is always nearly saturated, and is relatively insensitive to fluctuations in glucose concentration.

Control of Enzyme Activity
(Allosteric and Related Effects)

Any compound affecting the structure of the peptide chains in an enzyme may change the catalytic activity; the result may be a stimulation or an inhibition. We have already seen an analogous mechanism in the case of hemoglobin (p. 62), in which the binding of oxygen molecules to one pair of subunits affects the binding of the other pair in the molecule so as to give a sigmoid association curve, and this effect is created through alterations in the shape of the peptides caused by the attachment of oxygen.

Similarly, the attachment of substrates to an enzyme that has more than one catalytic site may affect the affinity of the remaining sites for additional substrate; it commonly increases their affinity so that a plot of initial velocities against substrate concentration has a sigmoid shape, like the oxygen saturation curve of hemoglobin, rather than the familiar hyperbola obtained when simple Michaelis-Menten kinetics are obeyed (Fig. 7-6).

Such an enzyme will only begin to catalyze its reaction effectively when the substrate concentration approaches the effective K_M value. Put another way, an

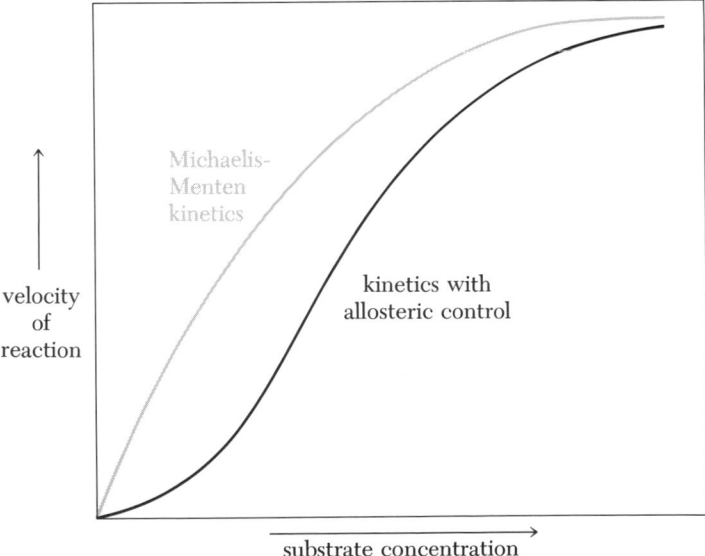

velocity of reaction

Michaelis-Menten kinetics

kinetics with allosteric control

substrate concentration

Figure 7-6   The effect of substrate concentration on the initial rate of an enzyme subject to allosteric regulation. Michaelis-Menten kinetics are shown for comparison.

enzyme with simple kinetics will catalyze the reaction at 0.33 V_{max} when $S = 0.5$ K_M, but an enzyme with sigmoid kinetics may only operate at 0.1 V_{max} or less at the same relative concentration of substrate. The effect will be to utilize the substrate from the cellular supply only when its concentration has risen above some limiting level.

This kind of regulation through alteration of the enzyme has even further potent possibilities for control, because the compound altering the structure of the enzyme need not be a substrate for the enzyme, and the site at which it binds need not be the site at which enzymatic catalysis occurs. It has only been realized in recent years that effects of this sort are major mechanisms for the control of metabolic pathways. Attention was focused more sharply on the idea by the invention of the term *allosteric effect* by Monod and Jacob to describe the control of enzymatic catalysis through the ability of the enzyme to exist in two conformations. (They shared a Nobel prize for which this was part of their contribution.) Although there are other known mechanisms, the term is frequently used today for any kind of control. To avoid confusion, we shall refer to *modulation* of enzyme activity by *modifiers* as a general term that includes allosteric effects, and reserve the specific term for the kind of regulation originally intended.

Efficient regulation simply means avoiding the presence of too much or too little of any particular compound within any part of an organism. The evolution of enzymes with the property of being modulated by modifiers was probably an essential step in achieving efficiency, especially the evolution of such responses to compounds other than the substrates or products of the immediate reaction. We shall see that there are extended sequences of reactions existing only to produce a single ultimate compound. In such cases, it is a scarcity or superfluous abundance of the final product that is potentially inefficient and therefore biologically damaging. The concentration of the end product can be held in a narrow range by using it as an inhibiting modifier of the very first reaction in the long sequence by which it is produced. This way, a rise in its concentration will prevent any wasteful accumulation, not only of itself, but of all the intermediates peculiar to its formation. A fall in its concentration will enable the initial enzyme to accelerate toward its full capacity, thereby replacing the deficit.

The classic example is an enzyme, *aspartate transcarbamylase*, catalyzing the first step in the formation of the pyrimidines used for making nucleotides. (The formation of pyrimidines is discussed in Chapter 20.) In many microorganisms, this enzyme is inhibited by cytidine triphosphate.

The primary use of the pyrimidine nucleotides is as precursors for nucleic acids. Nucleic acid formation is, of course, critical for cell division, and therefore for the survival of the microorganisms. It is a logical deduction that the species would be at a disadvantage if cell division were impaired for want of the pyrimidine precursors, and yet an overproduction of these precursors would be wasteful because they could only take up space within the cell or diffuse out into the surrounding medium and be lost.

The inhibition of aspartate transcarbamylase by CTP solved the problem. The cell can have an ample supply of enzymes making pyrimidines, capable of meeting the demands for nucleic acids when the environment permits growth of more cells, and yet the pyrimidines will not be accumulating because a rising concentration of CTP will shut off the first reaction in the process by which pyrimidines are formed.

The term *negative feedback* was borrowed from electronics by Dr. Arthur Pardee to describe this elegant effect, which he discovered. (Perhaps if Pardee had thought of some term more obviously derived from the ancient languages for his pioneering contributions, he might also have received a Nobel prize.)

The control is even more exquisite than was first realized. It is now known that the control arises from the presence of two kinds of peptide subunits within the enzyme molecule. One kind, the *catalytic subunit,* has the sites at which the substrates are bound and caused to react. The other kind, the *regulatory subunit,* has no catalytic activity, but binds CTP. For example, the aspartate transcarbamylase of the microorganism, *Escherichia coli,* has two catalytic subunits, each with two catalytic sites, making a total of four such sites in the complete enzyme. The enzyme also has four regulatory subunits, each capable of binding one molecule of CTP.

When the catalytic subunits are separated from the regulatory subunits in the laboratory, they show no tendency to associate, and display ordinary Michaelis-Menten kinetics. However, the complete enzyme has a sigmoidal response of initial velocity to added aspartate, *even in the absence of CTP.* This means that the catalytic subunits are interacting with each other through the regulatory subunits in the complete molecule, and doing so in such a way that addition of substrate to one subunit is affecting the architecture of the other subunit.

Now, the addition of CTP to this complete molecule changes the kinetics, and it does so *by shifting the sigmoidal substrate concentration curves to higher concentrations.* In other words, CTP acts to increase the effective K_M of the enzyme for its substrate, aspartate, and thereby slows the reaction at a given substrate concentra-

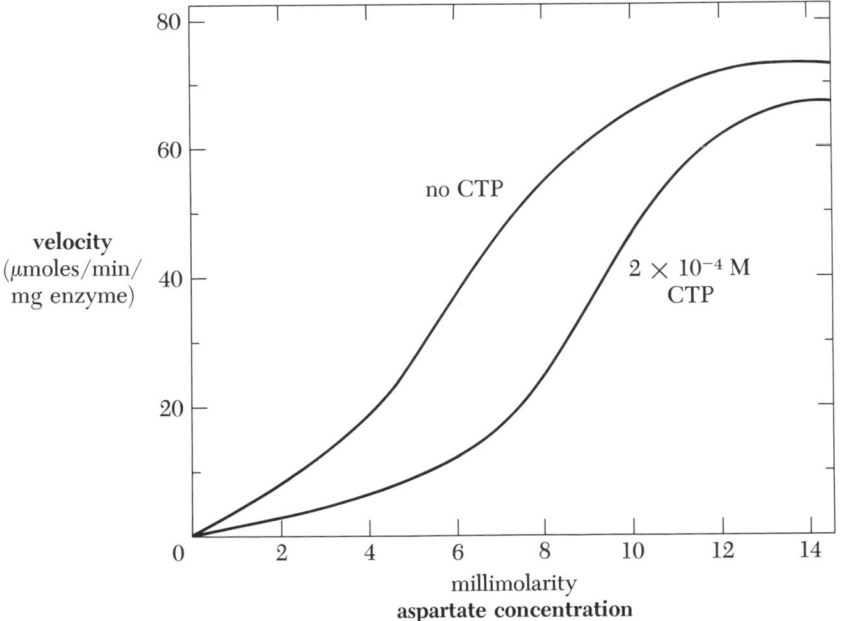

Figure 7-7   Effect of CTP on the kinetics of aspartate transcarbamylase. This modifier has the effect of increasing the K_M of the enzyme toward the substrate, aspartate, thereby decreasing the rate at a given substrate concentration. Modified from J. C. Gerhart and A. B. Pardee, J. Biol. Chem., *237:* 891 (1962).

tion (Fig. 7-7). The combination of the regulatory subunits with CTP somehow affects the ability of the catalytic subunits to interact; there is a modification of the forces transmitted between catalytic subunits through the regulatory subunits. Put more accurately, the presence of CTP may stabilize the conformation of the enzyme that has the least affinity for aspartate.

This enzyme also illustrates another important kind of regulation—modulators may have competitive inhibitors in the same way that substrates have competitive inhibitors, because they are reacting with proteins in similar ways. Such inhibitors prevent the modification of enzyme activity. In the case of aspartate transcarbamylase, ATP will compete with CTP for combination on the regulatory subunits, and binding of ATP does not change the enzyme's affinity for aspartate. What good does this do? It tends to balance the concentration of purine and pyrimidine nucleotides in the cell. If ATP (and therefore GTP, since the purine nucleotides have common origins) is in excess of CTP, the inhibition of the aspartate transcarbamylase by CTP will be relieved, and more pyrimidine made to balance the purine supply. On the other hand, if the cell is running short of purine nucleotides, it can't make complete nucleic acids, and this situation will be met by the full inhibitory effect of CTP on its own synthesis because the inhibition is not being prevented by ATP.

As we proceed later with our discussion of metabolism, we shall meet many examples of modulation of enzyme activity, some of which are simple inhibitions or activations. There are others that are like aspartate transcarbamylase from microorganisms in having modulators and inhibitors of modulation. In such cases it is well to remember that *inhibition of an activation is equivalent to an inhibition,* but *inhibition of an inhibition is equivalent to an activation.*

Recapitulation of types of reactions

1. ATP may react with alcohols to form phosphate esters:

$$R-OH + ATP^{4-} \longrightarrow R-OPO_3{}^{2-} + ADP^{3-} + H^+$$

Example: The phosphorylation of glucose in the glucokinase or hexokinase reactions (considered in more detail in Chapter 13).

Further reading

Gutfreund, H.: *An Introduction to the Study of Enzymes.* John Wiley (1965).

Webb, J. L.: *Enzymes and Metabolic Inhibitors.* Academic Press (1963).

Reiner, J. M.: *Behavior of Enzyme Systems.* Burgess (1963).

Christensen, H. N., and G. A. Palmer: *Enzyme Kinetics.* Saunders (1967). This is written in learning program format.

Stadtman, E. R.: *Allosteric Regulation of Enzyme Activity.* Adv. Enzymology, 28: 41 (1966).

Atkinson, D. E.: *Regulation of Enzyme Activity.* Ann. Rev. Biochem., 35: 85 (1966).

Monod, J., J. P. Changeux, and F. Jacob: *Allosteric Proteins and Cellular Control Mechanisms.* J. Molec. Biol., 6: 306 (1963).

Changeux, J. P., J. C. Gerhart, H. K. Schachman, and M. M. Rubin: *Allosteric Interactions in Aspartate Transcarbamylase.* Biochemistry, 7: 531, 538, and 553 (1968).

Weber, K.: *Aspartate Transcarbamylase from* Escherichia coli. J. Biol. Chem., 243: 543 (1968). A different view of the exact architecture of the enzyme.

Bergmeyer, H. U.: *Methods of Enzymatic Analysis.* Verlag Chemie (1965). A most valuable practical guide.

DISRUPTION OF PROTEIN STRUCTURE

CHAPTER 8

ARGUMENT

We have seen that the stability of a protein structure comes from the summation of forces arising from particular arrangements of amino acid residues. These forces are expressed through the formation of hydrogen bonds, van der Waal's interactions, electrostatic interactions, and water exclusion. They may result in the formation of regular arrangements of peptide chains, such as a helix, followed by further folding of segments of chains into particular overlapping arrangements, and the entire structure may be further stabilized through the association of separate peptide chains into a single protein molecule. Any alteration of the forces creating the assembly can alter one or more elements of the structure, and an alteration affecting only a small fraction of the total bonds may be sufficient to cause extensive disruption of structure.

Any agent or condition causing a general change in bonding, such as thermal vibration, a change in hydrogen bonding, or an alteration of the interaction with water, will cause a disruption of structure in most proteins. Proteins are generally sensitive to elevated temperatures, to extremes of pH, to non-polar solvents, and to high concentrations of compounds such as urea.

Since the fundamental source of the structure is the arrangement of amino acids, chemical alteration of the residues frequently causes disruption, just as genetic substitution of one residue for another may cause changes in structure. For example, the position of the disulfide bridges fixes the structure. The structure is frequently drastically altered if new bridges are formed by oxidation of sulfhydryl groups, or if existing bridges are destroyed by reduction. Simple removal of amino acid residues by hydrolysis or cleaving internal peptide bonds by hydrolysis often cause disruption of structure.

Proteins contain many side chain groups that are effective ligands for metallic ions, and chelation of a protein with these ions, especially with the heavier metals that form very stable chelates, may disrupt the structure.

Those handling proteins that are removed from their biological sources usually want to preserve the original structure, and therefore take pains to avoid exposing the proteins to conditions favoring disruption. They therefore usually work at low temperatures, avoid contamination by metallic ions, do not expose the protein to solutions at extreme pH, and avoid the creation of foams, with their large surface area.

TYPES OF DISRUPTION

Since the architecture of a protein is complex, alterations in it can range from a simple ionization of an acidic side chain, producing only slight changes in the biological properties of the molecule, to cleavage into fragments by rupture of the covalent bonds.

As an aid to thinking, it is customary to classify alterations of structure into types, but it is important to remember that an actual change in a real protein may simultaneously involve several of the types. Preoccupation with semantics in this field may cloud, rather than aid, thought.

1. Covalent bonds may be broken. Such changes are said to affect the primary structure of the protein. The bonds may be in bridges between peptide chains, in the side chains of the amino acid residues, or in the peptide chain itself. The cleavage of covalent bonds frequently alters the character of a protein so drastically that it completely loses biological function. We have seen that function depends upon the presence of particular side chains in a defined spatial arrangement. Changing these side chains obviously destroys function, but rupture of covalent bonds in a part of the molecule completely removed from the active site may also alter the necessary spatial arrangement. This is the chemical analogue of the effects of genetic substitution of amino acid residues, which we discussed in connection with hemoglobin. For example, ribonuclease loses its enzymatic activity if the disulfide bonds are reduced to pairs of sulfhydryl groups because these covalent bonds are necessary in holding the peptide chains together.

However, rupture of covalent bonds does not always cause immediate loss of function. The bacterial endopeptidase, *subtilisin*, catalyzes the hydrolysis of the peptide chain of ribonuclease between residues 20 and 21, but this does not immediately destroy the catalytic activity of ribonuclease, even though there are no other covalent bonds attaching the first 20 residues to the remainder of the molecule. Hydrogen bonds and non-polar interactions are sufficient to hold the molecule together.

2. Helical structures may be disrupted so that the peptide chain has a more open, and random, configuration. This is a change in the secondary structure of a protein, and usually causes a loss of biological activity. The change can be measured as a diminution in the rotation of a beam of plane-polarized light by the solution because a regular helix causes a high rotation at wavelengths in the ultraviolet.

3. The three-dimensional arrangement, or tertiary structure, of proteins may be altered so that the peptide chains are no longer folded into a compact form. The molecules then tend to associate in linear array, and the protein is less soluble. The decreased solubility is particularly noticeable when the pH of the solution is at the isoelectric point of the protein, because the absence of net charge eliminates the net electrostatic repulsion between like charges that would hinder association into an insoluble particle. In addition, molecules without net charge have fewer small ions attracted to the surface. These ionic clouds interact with water and raise the solubility, because it requires more energy to remove the "bound" water so that the protein molecules can approach each other and precipitate.

Increases in the viscosity of a solution can be used as an index of the degree of unfolding of a protein, because the open molecule has a greater surface in contact with the solvent and greater friction upon movement.

4. Finally, the association of peptide chains, or quaternary structures, may be altered. Even a molecule normally composed of a single chain may associate into dimers or higher polymers. Molecules composed of several peptide chains can dissociate as well as aggregate. Some will behave as dimers, dissociating into identical halves, each containing nearly the same secondary and tertiary structure present in the original molecule. Changes of this kind usually have a quantitative effect on biological function, but do not necessarily cause a complete loss of function. We shall see examples of enzymes whose activity is controlled in intact organisms by dissociation and aggregation.

Denaturation

Denaturation implies a change of a protein from its "native" or functional state. Fairly obvious criteria were used before much was known about the structure of proteins. For example, a protein was denatured if it coagulated into a visible precipitate or irreversibly lost some biological activity, such as enzymatic function. The techniques used in those days for handling proteins were rather crude by present standards, and only irreversible changes were easily observed.

Now there is a dichotomy in the use of the term. Specialists in the structure of proteins would like to abolish the word, or confine it to those alterations of secondary and tertiary structure producing an open and more random configuration of the protein. Others still use it as a general term for any change resulting in a loss of function or in some marked alteration in solubility. One may be regarded as a structural definition, the other as functional. We shall use the term in the latter sense, despite occasional ambiguity, because of the lack of an adequate substitute for changes of unknown mechanism, whereas structural changes of known character are amenable to short and accurate description.

MEANS OF DISRUPTION

Change in Temperature

Changes in the bonds of proteins are chemical reactions and are accelerated by an increase in temperature, as are all chemical reactions. However, there is an important difference in the behavior of proteins, with its manifestations so obvious that few give it thought as a chemical phenomenon.

The internal temperature of a human ranges around 37°C, and the proteins of human tissues retain their function for weeks at this temperature. Let a pot of boiling water be spilled over a person's arm, and the soluble proteins of his skin coagulate in a few seconds. This can only mean that the rates of the reactions of disruption are accelerated over a million-fold by this rise of about 60°, whereas most chemical reactions would only be accelerated approximately 100-fold.

It turns out that most proteins have a sharp transition temperature. At 5° below this temperature they will be stable for hours, whereas they will be disrupted in a fraction of a minute at 5° above this temperature. The increased temperature causes rupture of weak bonds maintaining the secondary and tertiary structure. The cleavage of one bond increases the probability of cleavage of a second, which in turn increases the probability of cleavage of another, and so on until all are broken in a rush. The sharpness of the transition with rising temperature depends upon the extent of ordered structure present, for example, upon the number of turns in a helix. The transition temperature is in effect a melting point for the ordered structure.

A rise in temperature will also increase the rate of any other reaction affecting protein structure, such as oxidation or hydrolysis, but the increases are more like those in other chemical reactions, with the rate increasing 2- to 3-fold for a 10° rise.

Change in pH

Changes in pH can have complex effects on the structure of a protein, owing to the number of acidic and basic groups contained within the molecule. We have explored the use of histidyl and lysyl residues in the mechanism of many enzymatic reactions, with other charged groups probably being used for substrate binding in many cases. In the case of hemoglobin, we saw the use of the phenolic hydroxyl group of a tyrosyl residue for binding adjacent segments of a peptide chain, and the use of aspartyl residues for the same purpose.

Changes in the hydrogen ion concentration that affect the ionization of these or similar groups will also affect the structure of the protein, and it is sometimes difficult to make the essentially artificial distinction between a denatured and a native protein.

For example, the degree of ionization of a group used in an enzyme for acid-base catalysis changes with pH. This is a change in structure affecting biological function. However, it is frequently easy to reverse this immediately by restoring the pH to the proper value. None would regard this change in dissociation of a group at the active site as a denaturation.

As the pH is shifted further, more and more groups will be affected, and the extent of the change in the structure of the molecule will become greater, but what is the tolerable limit of change at which the molecule can be regarded as native? This is not a very profitable distinction to attempt in some cases. We saw that a shift in structure of hemoglobin with changes in pH has a very definite biological value in regulating the transport of oxygen. A similar shift in structure may cause complete loss of function of another protein, but still be readily reversible.

However, the change in structure can become great enough to satisfy anyone's definition of a denatured protein. A shift in pH that causes a change in the ionization of groups so that they no longer are a part of hydrogen bonds makes it easier for further disruptions to occur, and the transition temperature for the protein is thereby lowered. The greater the change in pH, the greater the number of affected groups, and the lower the transition temperature. When the transition temperature is lowered across the actual temperature of the solution, there will be a sudden unfolding of the structure and a sharp loss of biological function.

Few proteins will withstand exposure to pH 2 or pH 12, even at 0°C. Some won't tolerate pH 5 or pH 9 without extensive disruption.

It is sometimes possible to detect major changes by following biological activity with variations in pH. For example, an enzyme using both acidic and basic groups in its mechanism will have a pH-activity curve of the shape illustrated in Figure 8-1 A, provided that changes in the dissociation of these groups is the only thing affecting the activity. However, if dissociation of groups involved in structural bonding also occurs in the same pH range, with a resulting unfolding of the structure, there will be an abrupt loss of activity over a small range in pH, as shown in Figure 8-1 B.

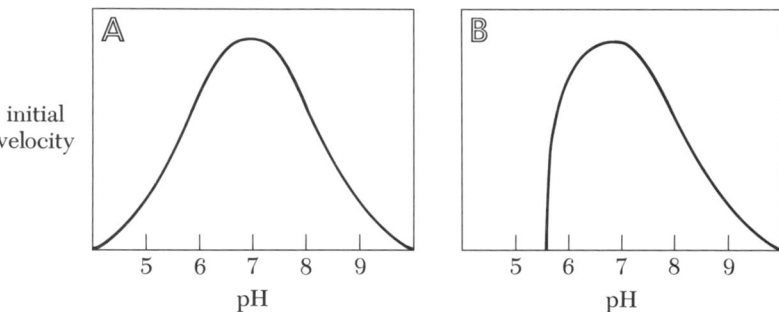

Figure 8-1   The effect of pH on the rate of enzymatic reactions. *A.* The enzyme is stable over the pH range tested, and the mechanism of catalysis requires both acidic and basic groups. *B.* The mechanism is similar to that of the above enzyme, but the transition temperature for disruption of the enzyme's structure falls at pH 5.7 below the temperature used for assay.

Denaturation by exposure to extreme pH has a physiological function. Ingested proteins are hydrolyzed to amino acids for absorption through the combined action of several peptidases in the digestive tract. Compact globular proteins frequently resist the action of peptidases, probably because the atoms of the peptide bond are also linked in hydrogen bonds and because many of the bonds capable of attack by particular enzymes are buried in the interior.

This is no longer true after denaturation, and denaturation of most ingested proteins is accomplished by exposure to a low pH in the stomach. The parietal cells of the gastric mucosa secrete 0.16 M hydrochloric acid. Even after dilution by a meal, the pH will usually be 2 or lower, and most proteins will be disrupted with their peptide bonds exposed to hydrolysis.

Organic Solvents

The distribution of charges on a protein molecule will be changed if a less polar solvent is substituted for water, thereby lowering the *dielectric constant* of the medium. The charges on the molecule will interact more with each other, owing to the decrease in the dipolar character of the solvent. This may lead only to an association of the molecules causing decreased solubility if the temperature is kept low, but few proteins will withstand precipitation by ethanol or acetone at room temperature without changes in secondary and tertiary structure.

Urea and Related Compounds

Urea, and related compounds such as the guanidinium ion, will cause a disruption of bonding in proteins through unknown mechanisms:

$H_2N-\overset{\overset{O}{\|}}{C}-NH_2$ $H_2N\cdots\overset{\overset{NH_2}{\vdots}}{\underset{\oplus}{C}}\cdots NH_2$

urea   guanidinium ion

The effect is as if these compounds formed hydrogen bonds on the peptide chain, thereby competing with the normal chain interactions, but this mechanism has been disputed.

Whatever may be the real mechanism for their effect, they are useful tools in the study of proteins. At concentrations up to a few moles per liter, urea will cause many proteins to dissociate into separate chains. At concentrations between 5 and 10 moles per liter, the secondary structure unfolds. The advantage of these reagents comes from fewer side reactions, compared to other means of experimental disruption discussed above.

Metal Chelation

Atoms with unshared electrons, such as oxygen, nitrogen, and sulfur, can act as ligands for metallic ions. These atoms are abundant in proteins, which therefore are effective chelating agents for many metallic ions.

An art museum owes its existence to exploitation of this property of proteins. A proprietary silver proteinate, Argyrol, has a low enough dissociation constant to avoid extensive tissue damage while maintaining an antiseptic equilibrium concentration of the free ion, and once was widely employed as an antiseptic agent for surface application. The income from it enabled the proprietor to assemble the extensive Barnes art collection.

In the present discussion we are treating random chelation, not the specific binding of metallic ions necessary for biological function of proteins such as hemoglobin. Those proteins contain arrangements of ligating groups that are especially effective for a particular metal, and the association constant is orders of magnitude greater than it is for non-specific interactions.

The behavior of inorganic compounds provides a qualitative guide to the strength of the various kinds of bonds. For example, mercuric sulfide is extremely insoluble, and is not dissolved by ammonia. Mercuric oxide has the low solubility of 52 micrograms per ml of water at $25°C$, but this is over 1000-fold greater than the solubility of the sulfide, and the oxide is soluble in concentrated ammonia solutions. The solid sulfides and oxides are lattices in which the multiple pairs of unshared electrons on sulfur or oxygen can combine with at least two of the metallic ions, and the strength of these bonds determines the difficulty of dissociating separate ions into free solution, and therefore the insolubility. Ammonia, on the other hand, forms soluble complexes because it has only one unshared pair of electrons, and formation of a similar lattice is not possible.

The behavior of the mercuric compounds indicates that the Hg—S bond is stronger than the Hg—N bond, which is stronger than the Hg—O bond, and this is confirmed by experiments in which the dissociation constants of organic compounds containing these groups are measured.

Therefore, one would expect mercuric ion to be bound by sulfhydryl groups on proteins in preference to binding by amine or carboxylate groups, and this indeed occurs. The binding is so strong that univalent organic mercurials can be used to titrate sulfhydryl groups on proteins. One such mercurial is p-chloromercuribenzoate:

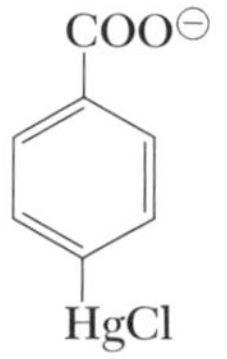

$$COO^{\ominus}$$

HgCl

p-chloromercuribenzoate

Repeated small doses of mercuric ion will accumulate to a toxic level in an intact organism because the retention is so high.

On the other hand, ferric or ferrous ions have such a high affinity for oxygen that they can be kept in solution only by lowering the pH to reduce the concentration of hydroxide ion in water. The alkaline earth ions (Ca^{2+}, Mg^{2+}, etc.) also have a higher affinity for oxygen, although the binding is much weaker than it is with the heavier metallic ions. It is sufficiently strong to keep much of the calcium circulating in the blood in the form of complexes with protein. Ions with a high affinity for nitrogen include Cu^{2+}, Co^{2+}, and Mn^{2+}.

Binding of a metallic ion often involves a competition with H^+ for the ligand group:

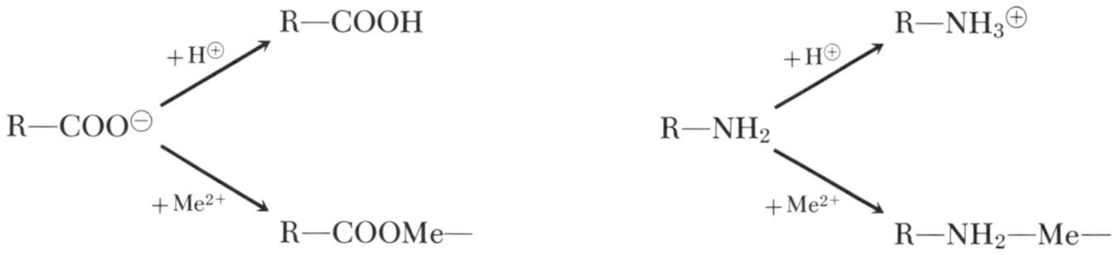

$$R-COO^{\ominus} \underset{+Me^{2+}}{\overset{+H^{\oplus}}{\rightleftharpoons}} \begin{matrix} R-COOH \\ R-COOMe- \end{matrix} \qquad R-NH_2 \underset{+Me^{2+}}{\overset{+H^{\oplus}}{\rightleftharpoons}} \begin{matrix} R-NH_3^{\oplus} \\ R-NH_2-Me- \end{matrix}$$

Similar competition exists with the imidazole group of histidyl side chains and the sulfhydryl group of cysteinyl side chains. Therefore, metallic ions will be bound more tightly as the pH is raised. Conversely, binding of metallic ions displaces protons.

Addition of trace amounts of strongly chelated metallic ions will frequently affect biological activity by combining with surface groups on a protein, but without necessarily disrupting the remainder of the structure. When large quantities are present, particularly at higher pH values, the metallic ion can compete effectively for groups used to bond the structure together, and disruption occurs.

In general chemical usage, a ligand is a molecule that donates the necessary electrons to form coordinate covalent bonds with metallic ions. Unfortunately, there is a tendency for some protein chemists to refer to any small molecule or ion that reacts with proteins as a ligand for the protein. In some cases, this is the same as redefining black as white. For example, the protein is as clearly the ligand in interactions with manganous ion in the language of most chemists today as it was in the past, and it is most confusing to this majority to have the metallic ion called a ligand. The practice perhaps arises from a misunderstanding of references to oxygen as a ligand for hemoglobin; this is technically correct, since the oxygen is being bound to the central iron atom of the complex, and does not represent a redefinition of the term.

Oxidizing and Reducing Agents

Strong oxidizing or reducing agents will attack proteins as they do other organic compounds containing similar groups. We are more interested in the milder reactions affecting sulfhydryl and disulfide bonds.

The addition of a mild oxidizing agent, even some disulfide of low molecular weight, may cause the formation of new disulfide bridges that change the conformation of the molecule. The likelihood of such a change depends upon the number of accessible sulfhydryl groups and their relative location.

Somewhat stronger oxidizing agents can oxidize the sulfur to still higher states—the sulfenic, sulfinic, or sulfonic acids ($R—SOH$, $R—SO_2H$, or $R—SO_3H$). Even if this causes no further disruption in structure, the loss of sulfhydryl groups in this way usually destroys function.

Molecular oxygen will react with sulfhydryl groups, but at a very slow rate. Many of these groups are not even accessible, being either bonded firmly or buried in the interior. All in all, proteins are relatively safe against oxidation in the intact organism. When they are handled in solution, it is a different matter, because they may become contaminated with trace metals, such as Cu^{2+} or Mn^{2+}, that catalyze the reaction with O_2.

The catalysis involves an oxidation and reduction of the metallic chelate. This can be demonstrated nicely by adding a small amount of a manganous salt to a solution of cysteine at pH 9, forming the green manganous cysteinate chelate. The solution turns violet when it is shaken in air because the supply of oxygen is rapid enough to keep most of the metal in the form of the oxidized manganic cysteinate. It again turns green upon standing, because the rate of reduction of the metal then exceeds the rate of diffusion of oxygen into the solution. The violet color can be restored by reshaking, and the sequence can be repeated until all of the cysteine is oxidized. The metal is successively oxidized by oxygen and reduced by cysteine, the cysteine becoming cystine. Cystine only weakly binds the ion, so remaining cysteine will remove it, and the process is repeated. The end result is the formation of a precipitate of cystine, which is only very slightly soluble.

Oxidation is also accelerated by elevated temperatures and by any disruption of the structure exposing hitherto buried sulfhydryl groups. Hence, many of the disruptive agents mentioned earlier result in damage to the protein beyond their immediate effect if the solution contains oxygen. This will be especially true at higher pH values because of augmented chelation of trace metals, and because the ionized sulfhydryl group is more sensitive to oxidation than the protonated form.

Surface Effects

A protein molecule at a water-gas interface is subjected to the surface tension of the solvent, and this is often sufficient force to disrupt the structure. French cuisine leans heavily on this phenomenon. The principal protein of egg whites, *ovalbumin,* is especially susceptible to denaturation in this way, and vigorous beating of egg whites forms a foam that is stiffened by the insoluble denatured protein, thereby creating a light meringue. (This is not to be taken as a biochemical endorsement or condemnation of any national cookery.)

Mechanical Shear

Simple stirring of a solution is usually not considered to be a chemically effective force, but it can be. Stirring involves acceleration of part of the liquid to higher velocities than the remainder. Large molecules may have dimensions great enough that the difference in velocity of liquid passing opposite sides of the molecule isn't negligible. Two factors are involved. One is surface area, because area determines the total force generated by obstruction of the faster-moving liquid. The other is length of the molecule, because it determines the effective distance between the part of the molecule being accelerated by higher-velocity liquid and the part being restrained by slower-moving liquid. If the molecule is large enough, the resultant shearing moment arising from a combination of greater force acting over greater distance may become sufficient to break a covalent bond. Not much can be done to a protein with a glass rod in a beaker, but a high-speed impeller can destroy a protein in solution. The speed required to do this becomes less as the molecular weight and the longest dimension increase.

The intense local force created by hand grinding of a dried protein with a mortar and pestle also may destroy many molecules. We already noted that grinding and stirring are particularly destructive to deoxyribonucleic acids, which have longer strands and higher surface areas than most proteins.

REVERSIBILITY

In discussing the biological synthesis of proteins, it was emphasized that the sequence of amino acids dictates the final configuration of a protein, and this configuration is assumed spontaneously, or as we later saw, with the aid of an enzyme forming the necessary disulfide bridges.

From this it follows that any disruption of protein structure that does not involve the primary structure ought to be self-healing once the cause of the disruption is removed. Even the formation or reduction of disulfide bridges ought to be reversible.

This has been shown to be true in many cases. Some proteins, notably trypsin, are denatured in hot water, but return to their active form upon cooling. The small single chain of ribonuclease uncoils when the disulfide bridges are reduced, but the enzymatic activity can be restored by careful oxidation, showing that the natural configuration is the preferred configuration.

Even insulin, which consists of two peptide chains held together by disulfide bridges, can be partially restored by reoxidation from its dissociated reduced state, even though the chances for error in putting two chains together are much greater than they are with the single-chain form in which insulin is originally made in the pancreas (see p. 557).

Removal of an inhibiting metal from a protein frequently leads to full recovery of activity.

Proteins dissociated by urea can often be recovered by removing the urea. This kind of treatment seems to be especially easy to reverse, and this is another reason that urea is a favored reagent among those studying protein structure.

This is all very well, but in far more cases a boiled protein is a spongy precipitate that shows no signs of ever again being a nicely soluble globular protein, no matter how long it may be chilled. A solution of a protein lowered to pH 4 will probably develop a precipitate. Even if the precipitate will go back into solution at pH 7, it often shows none of its original biological activity. These are the seemingly irreversible effects for which the term "denaturation" was originally invented.

How is this behavior reconcilable with our concept of the spontaneous formation of the proper structure? There is no totally satisfactory answer, but a partial resolution of the apparent contradiction can be made.

In many cases, the irreversible change is due to accompanying changes in the covalent structure. There may be oxidations, hydrolyses of ester bonds, or even hydrolyses of the amide groups on glutamine side chains, so that the necessary primary structure is not being maintained.

In more cases, the disulfide bonds are rearranged into what was a more stable configuration under the denaturing conditions, and the reversion to the original form is difficult because the necessary conditions are not, or cannot, be maintained for a long enough time.

In still more cases, the denatured proteins are formed in a mixture of several different proteins, which interact with each other so as to prevent any one of them from recombining properly, particularly if they have formed an insoluble precipitate. It is likely that the concentration of unfinished protein molecules in a living cell never reaches the concentrations usually employed in experimental work, so that there are fewer possibilities for undesirable interactions.

Finally there is the possibility, not at all absurd in light of past experience, that future work will show how to reverse many of the disruptions now regarded as irreparable.

PRACTICAL CONSIDERATIONS

Removal of Proteins

Since the proteins are usually the major constituent of tissues, after water, they make the analysis for lesser constituents difficult, and their removal is the first step in many analytical methods. Denaturation into less soluble forms that can be removed by filtration or centrifugation is the classic method for this purpose. Almost all of the agents of denaturation we have described have been used. The most common ones currently employed for deproteinization are strong acids that have anions that will bind to the denatured protein and make it insoluble below its isoelectric point. Examples are *trichloroacetic acid* and *perchloric acid.* These reagents at 0.5 molar concentration will precipitate most peptides having molecular weights over 1000.

Organic solvents, such as *alcohol* or *acetone,* are also used, particularly in analyses of compounds that are unstable in acidic solutions.

Denaturation is also used to selectively destroy contaminants during purification of a protein. Because of the sharp transition temperature for proteins, it is possible to heat a solution to a temperature within a few degrees of the point at which the desired protein will be disrupted, and have it remain stable while all proteins that have a lower transition temperature are denatured.

Similarly, contaminating proteins that are more unstable at extreme pH values can be removed by exposure to these conditions.

Preventing Denaturation

Most of the means for increasing the stability of proteins are obvious from the previous discussion, and are constantly used by the knowledgeable when handling protein solutions, particularly when accurate assays of enzymes are desired:

1. The temperature ought to be kept low. There are a few alarming exceptions of proteins that are more stable at $30°C$ than they are at $0°C$, but the contrary is so often true that use of low temperatures is routine in handling proteins. The object is to diminish the effects of oxidation, of surface tension, and of reaction with the ions inevitably added during manipulation of a protein outside of its original host.

2. The pH ought to be near 7 unless the protein is known to be more stable at some other value.

3. Contamination by metallic ions ought to be rigorously avoided. This requires the use of water distilled from glass apparatus. Even then, it is frequently desirable to add a chelating agent as a safety factor to compete with the protein for any contaminants. Ethylenedinitrilotetraacetate (EDTA, also called ethylenediaminetetraacetate) is often used for this purpose:

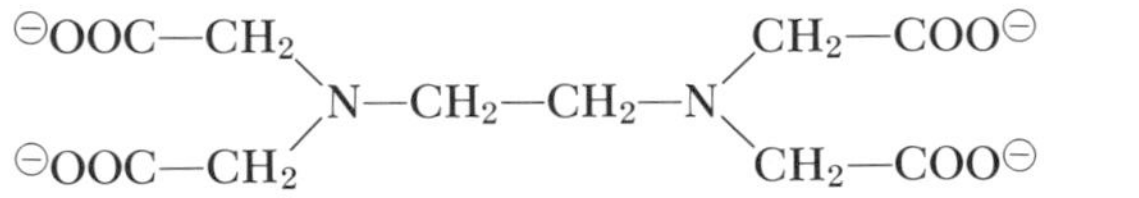

$$^{\ominus}OOC-CH_2 \diagdown \atop ^{\ominus}OOC-CH_2 \diagup N-CH_2-CH_2-N \diagup CH_2-COO^{\ominus} \atop \diagdown CH_2-COO^{\ominus}$$

ethylenedinitrilotetraacetate (EDTA)

4. Stirring air into the solution is avoided so as to minimize surface denaturation.

5. The protein concentration is kept as high as possible, so as to spread out contaminants among as many molecules as possible, thereby diminishing the possibility of rupture of enough bonds on any molecule to cause it to unfold. There also appears to be some other stabilizing effect from the presence of a high concentration of protein.

Further reading

Putnam, F. W.: *Protein Denaturation*. In Neurath, H., and K. Bailey, eds.: *The Proteins*, 1st ed., Vol. 1, part B (p. 807). Academic Press (1953). This is a more descriptive treatment than in the following references.

Steinhardt, J., and S. Beychok: *Interaction of Proteins with Hydrogen Ions and Other Small Ions and Molecules*, Vol. 2 (p. 139). Also H. A. Scheraga: *Intramolecular Bonds in Proteins. Noncovalent Bonds*, Vol. 1 (p. 477). Both in Neurath, H., ed.: *The Proteins*, 2nd ed. Academic Press (1963–4). These are heavy going, but represent important summaries of the forces holding proteins together.

Hermans, J., Jr., D. Puett, and G. Acampora: *On the Conformation of Denatured Proteins*. Biochemistry, 8: 22 (1969).

ENERGY BALANCE OF HYDROLYSIS AND SYNTHESIS

CHAPTER 9

ARGUMENT

A reaction, whether catalyzed by an enzyme or not, can only proceed toward equilibrium. A reaction must release energy to be able to occur, and the magnitude of the release is dependent upon the displacement of concentrations from the equilibrium condition. It is convenient to express the release in terms of the change in free energy upon reaction. A negative value represents a release of energy, so all real reactions have negative free energy changes. The free energy consists of two components, one of which is a measure of the actual energy of the molecules, usually expressed in terms of *enthalpy*, or heat content. The other is a measure of the number of energy states in which the molecules can exist—their randomness—and this is called *entropy*.

The change in free energy is a logarithmic measure of the relationship of actual concentrations to the equilibrium concentrations. Since identical concentrations may represent varying distances from equilibrium for different reactions, it is common to compare the potential energy available from various reactions by their standard free energies, which are defined as the energy released per mole reacting, when all components are at unit concentration. Standard free energies are really logarithmic comparisons of equilibrium constants. Free energies, actual or standard, are additive; the overall free energy for a sequence of reactions is the sum of the free energies for the individual reactions, and this makes it especially useful in quantitative analysis of metabolic sequences.

Although low standard free energies represent reactions with equilibrium

154

constants near one, this does not always mean that the reaction is easily reversed under physiological condition. For example, the hydrolysis of peptides, like many hydrolyses, has a low standard free energy, but hydrolysis of peptides is irreversible under real conditions, and cannot be used for protein syntheses. The synthesis, like the formation of many cellular components, must make use of the energy released in processes such as the oxidation of glucose, which has a large negative standard free energy.

In order to transfer energy from one reaction to another, both reactions must be modified so as to involve the same compounds. The adenine nucleotides are frequently used for this purpose. For example, ATP is formed by the oxidation of glucose and is used in the synthesis of proteins, so that it couples these two processes. They are no longer independent reactions, because the supply of adenine nucleotides is limited.

ATP has two pyrophosphate bonds, called high-energy phosphate bonds because their cleavage by hydrolysis has a highly negative standard free energy at neutral or higher pH values. The ester phosphate bond found in AMP releases considerably less energy upon hydrolysis and is classed as a low-energy phosphate bond.

The formation of peptide chains involves the cleavage of the pyrophosphate bonds in ATP and GTP so that the overall process has a highly negative standard free energy. However, this is not wasteful because it insures that the equilibrium lies far in the direction of synthesis, so it is possible to form proteins from low concentrations of amino acids.

The work done by contracting skeletal muscles is derived from the free energy of hydrolysis of ATP. The mechanism by which chemical energy is transduced to mechanical energy is not understood. The process involves two fibrous proteins, actin and myosin. It is believed that the hydrolysis of ATP, catalyzed by myosin, causes filaments of actin to slide toward each other over the myosin molecules. The filaments of actin are attached to strong protein bands lying across the muscle fibrils. Contraction represents the pulling of these bands toward each other by the sliding actin filaments.

Throughout our discussion to this point, we have concentrated on the kinetics and mechanisms of reactions, but no matter how effective an enzyme may be, it cannot make a reaction go in a direction that will consume energy—in a direction such that the concentrations of reactants and products are going away from their equilibrium values. This really was inherent in our discussion of kinetics and the effect of product concentration, but it is frequently more convenient to discuss the direction of a reaction in terms of the energy released rather than in terms of relative rates of opposing reactions.

The reactions of protein synthesis and hydrolysis provide ready illustrations of principles we shall frequently invoke in discussing other metabolic routes. We saw that the hydrolysis of proteins to amino acids requires only water and a catalyst, whereas the synthesis of proteins from amino acids involves the simultaneous hydrolysis of ATP and GTP. We shall use energetics to show that *both* of these processes are essentially irreversible, and that the synthesis of proteins could not proceed by a simple reversal of hydrolysis. There are many similar examples among metabolic reactions, in which a compound is formed by one route and degraded to the same starting materials by another, but with other compounds participating in one of the conversions so as to provide the necessary energy.

THE CONCEPT OF FREE ENERGY

Relation to the Equilibrium Constant

When a system, A $\rightleftharpoons$ B, is at equilibrium, there is no energy exchange with the environment from the opposing chemical reactions. Whatever energy change is involved from a rearrangement of molecules of A to make molecules of B will be counterbalanced by the opposite change from the conversion of B to A, which must be going at an equal rate if the system is at equilibrium.

If we now raise the concentration of A, the formation of B will be proceeding faster than the reverse reaction. The overall effect, the net reaction, will be a conversion of A to B, and energy will be liberated, but how much? This will now be considered.

J. Willard Gibbs, who single-handedly created a large fraction of physical chemistry, introduced the term *free energy* to describe the energy available for useful work from changes occurring at constant pressure. The symbol G is now used to designate the Gibbs free energy, although it is designated F in older literature. The Gibbs free energy is a useful parameter for biochemical systems because they do operate at essentially constant pressure.

It can be shown that the free energy of a compound in solution is proportional to the logarithm of the concentration of the compound. The change in free energy, ΔG, which is created by transferring a mole of a compound from a solution at concentration c_1 to a solution at a concentration c_2, is given by the equation:

$$1. \qquad \Delta G = -RT \ln \frac{c_1}{c_2} = -2.303RT \log_{10} \frac{c_1}{c_2}$$

$$R = 1.987 \text{ calories per degree per mole}$$
$$T = \text{absolute temperature in kelvins.}$$

The free energy change will be negative if the first concentration is greater than the final concentration, which means that free energy is being released from the system.

We can use this equation to calculate the free energy liberated during a net reaction of A to B. We know that the conversion of a mole of A to a mole of B at equilibrium has no free energy change, and the problem is essentially to find the differences in energy between the concentrations [A], [B] and $[A_{eq}]$, $[B_{eq}]$. Let us go at in this way (Fig. 9-1).

A.   We have two solutions, the first with concentrations at [A], [B], and the second with concentrations at $[A_{eq}]$, $[B_{eq}]$.

B.   We remove a mole of A from the first vessel at concentration [A] and add it to the second at concentration $[A_{eq}]$.

C.   We add a mole of B to the first vessel at [B], obtaining it from the second at $[B_{eq}]$.

D.   We are in effect converting one mole of A to B at the concentrations in the first solution and converting one mole of B to A at the equilibrium concentrations in the second solution, at which $\Delta G = 0$.

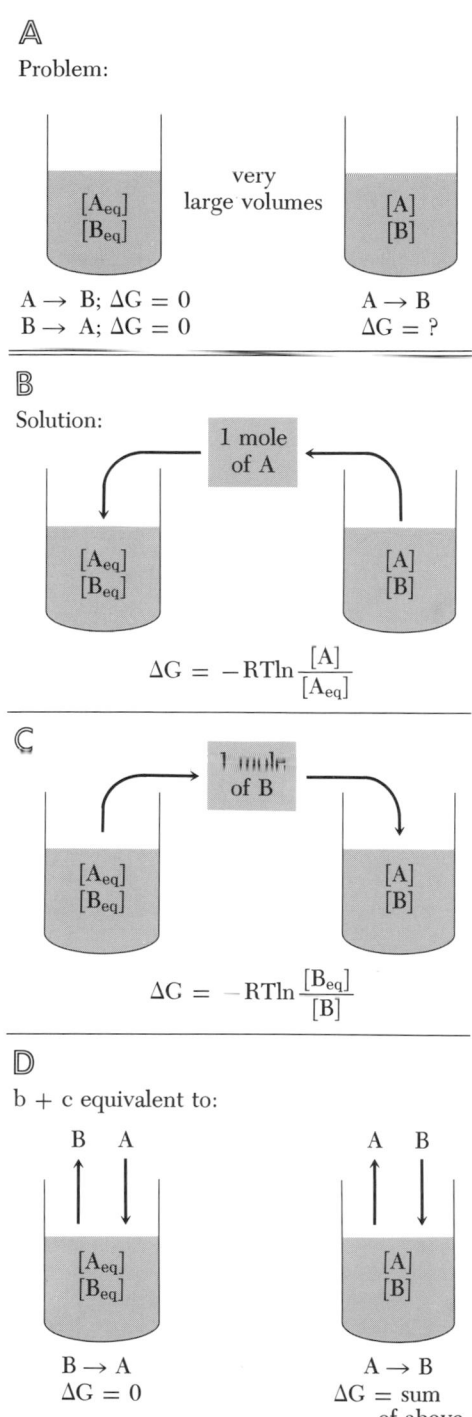

Figure 9-1   Derivation of free energy of reaction from free energy of dilution. See text on facing page for description of steps.

The total free energy change from the transfers is therefore identical to the free energy change of reaction of one mole of A to form B at the concentrations in the first vessel. Making these calculations, we arrive at this equation:

$$2. \qquad \Delta G = -RT \ln \left(\frac{[A]}{[A_{eq}]} \right) - RT \ln \left(\frac{[B_{eq}]}{[B]} \right) = -RT \ln \left(\frac{[A][B_{eq}]}{[B][A_{eq}]} \right)$$

We see right away that the combined form contains the equilibrium constant, $\frac{[B_{eq}]}{[A_{eq}]}$, so the equation can be rewritten:

$$3. \qquad \Delta G = -RT \ln \left(\frac{[A]}{[B]} \right) - RT \ln K_{eq}.$$

Using this equation, the free energy liberated per mole of A converted to B can be calculated if the equilibrium constant and the concentrations of reactant and product are known. We see that the free energy liberated is proportional to the logarithm of the ratio of concentrations of the reactant and product, plus some fixed value determined by the logarithm of the equilibrium constant.

This fixed value, given by the second term of the equation, $-RT \ln K_{eq}$, is called the standard free energy change for the reaction. The free energy change is equal to the *standard free energy* change when the concentration of each reactant and product is fixed at 1 M, because the logarithm of 1 is zero:

$$\text{standard free energy change} = \Delta G^0 = -RT \ln K_{eq} = -2.303 RT \log K_{eq}$$

The standard free energy change is therefore a logarithmic scale for comparing equilibrium constants of reactions.

Numerical Values

It is handy to have in mind a rough idea of the relationship between numerical values for standard free energy changes and equilibrium constants, because we shall later be using these numbers in appraising the position of equilibrium for metabolic sequences. Figure 9-2 plots the relationship at $0°$, $25°$, and $38°C$ ($273°$, $298°$, and $311°K$). Looking at the figure, we see that a reaction with $\Delta G^0 = -1$ calorie per mole is likely to be freely reversible, whereas a value of $\Delta G^0 = -10,000$ calories per mole means that the equilibrium lies so far in one direction that it would be difficult to measure the reverse reaction. But what about intermediate values? Here we have to be careful. Let us compare two important physiological reactions with quite similar standard free energy changes:

	ΔG^0 cal/mole	K_{eq}
cis-aconitate^{3-} + $H_2O \longrightarrow$ citrate^{3-}	-2040	31.4
glucose-6-phosphate^{2-} + $H_2O \longrightarrow$ glucose + P_i^{2-}	-2200	41.0 M

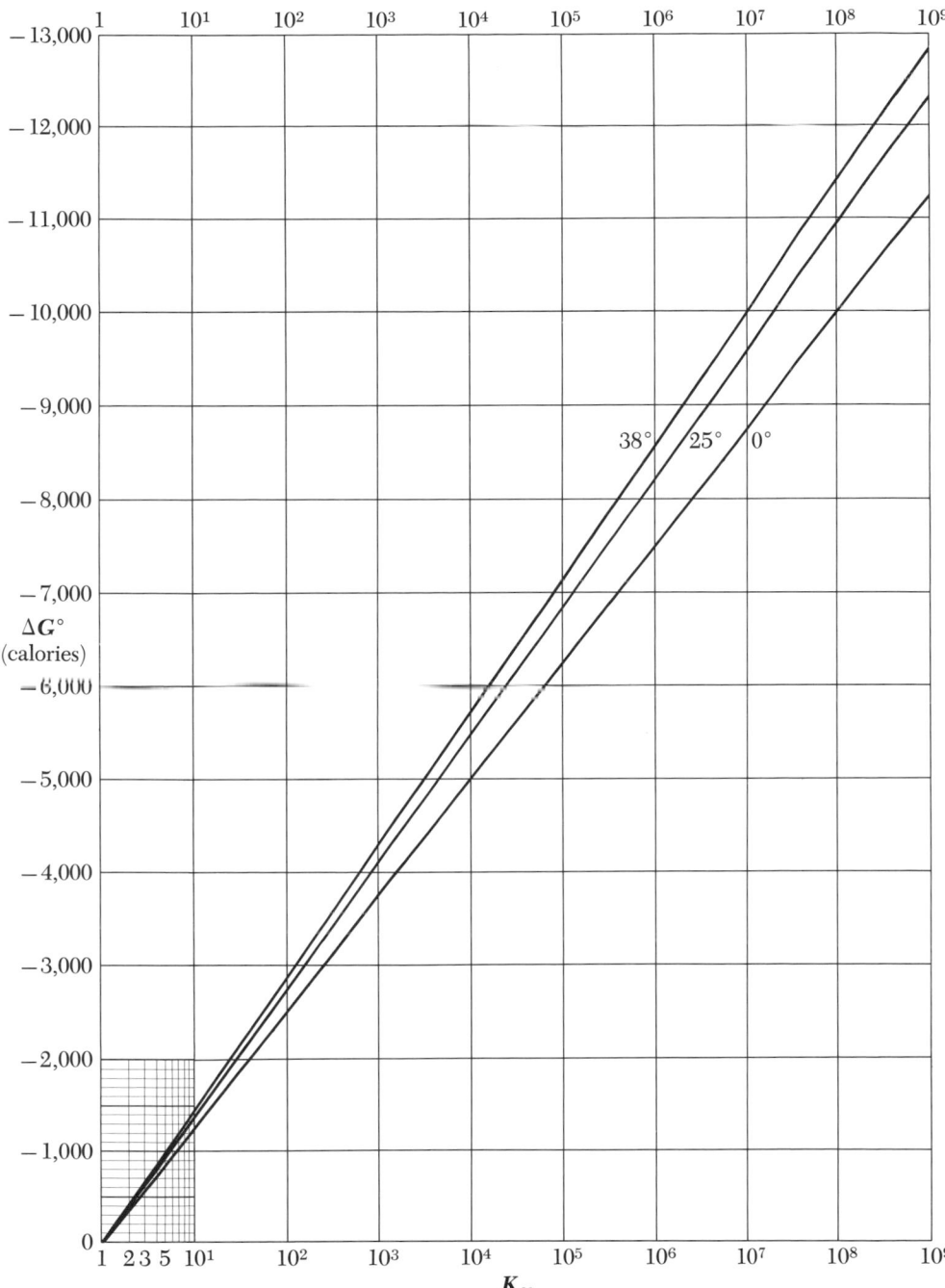

Figure 9-2 The relation of standard free energy change and equilibrium constant at three temperatures. The ΔG^0 scale is linear; the K_{eq} scale is logarithmic. The same plots apply to positive free energies if the exponents of the K_{eq} scale are made negative and its intermediate digits are reversed. (That is, read 9, 8, 7, 6 · · · from left to right, rather than 2, 3, 4, 5 · · ·)

The values for free energy change and equilibrium constants used here are based on a standard convention in which the concentration of water is taken as unity, since it remains substantially constant during reactions in aqueous solution.

The two reactions superficially appear quite similar in their ease of reversibility, with equilibrium constants of the same range, but there is a catch. The equilibrium constant for the second reaction has a molarity dimension, whereas the first constant is dimensionless. Calculation of the position of equilibrium will show the consequences of this difference in dimension.

Assume that we begin with 0.01 M reactant in both cases and wait for the reactions to come to equilibrium. The amount of reactant remaining, x, can be calculated from the equilibrium expression:

$$\text{for } cis\text{-aconitate: } \frac{0.01 - x}{x} = 31.4 \qquad\qquad x = 3.09 \times 10^{-4}$$

$$\text{for glucose-6-phosphate: } \frac{(0.01 - x)(0.01 - x)}{x} = 41.0 \text{ M} \qquad x = 2.4 \times 10^{-6}$$

The concentration of cis-aconitate remaining at equilibrium will be over 100-fold greater than that of glucose-6-phosphate—over 3 per cent of the starting material against less than 0.03 per cent. The relative ease of reversibility of the two reactions is easily checked by calculating what will happen in the two cases if the concentration of the products is raised by 0.001 M. When this is done, the result shows that 3.1 per cent of the added citrate will be converted back to cis-aconitate, but only 0.05 per cent of the added glucose and inorganic phosphate will be converted back to glucose-6-phosphate.

There is a general principle here. A dimensionless equilibrium constant means that a given number of reactants yield the same number of products; for example, $A + B \rightarrow C + D$ (but not $A \rightarrow B + C$, or $A + B \rightarrow C$). In such cases, the equilibrium constant is a direct statement of the ratio of concentrations of reactants and products at equilibrium, and the reverse reaction may be biologically significant if $-\Delta G^0$ for the forward reaction is no greater than 5000 calories per mole. In the example chosen, it turns out that the somewhat unfavorable conversion of citrate to cis-aconitate actually is the physiological direction, due to the constant removal of the latter compound by another reaction (p. 212).

If the number of products exceeds the number of reactants, the equilibrium constant will have some molarity dimension and the position of equilibrium will depend upon the actual concentrations, with the equilibrium shifting in the forward direction as the concentrations are lowered. With the range at which most compounds occur in tissues (10^{-5} to 10^{-2} M), only those reactions of this kind that have $-\Delta G^0$ values of less than 2000 calories per mole are likely to be significantly reversible.

In particular, hydrolyses create more products than there are reactants, since molecules are being cleaved, and most hydrolyses of esters, amides, etc. have standard free energy changes more negative than -2000 calories per mole. This means that a simple reversal of hydrolysis is rarely a useful reaction in biological systems. The example given, the hydrolysis of glucose-6-phosphate, is used for the formation of glucose (p. 279), and not for creating glucose-6-phosphate.

Energy, Enthalpy, and Entropy

We want to pause a little and mention some of the other thermodynamic quantities, not so much because we are going to use them intensively in our thinking, but because they are necessary to those wishing to pursue some specialized aspects of the subject. Even those without a need for intensive physical theory are likely to encounter the expressions, and ought to be prepared to distinguish their proper use from misuse. (The words have had a persistent vogue in the construction of rather vague philosophies about the mysteries of life.)

Classical thermodynamics arose from a consideration of heat engines, and the principles were later extended to chemical reactions. It is convenient to regard the free energy change in chemical reactions as arising from two components:

$$\Delta G = \Delta H - T\Delta S$$

H designates the heat content, or *enthalpy*, of the compounds. A change in heat content can be measured in a calorimeter.

The second term includes changes in the *entropy*, S, which was originally developed in terms of heat transfer to make the answer come out right, and it is this derivation that is commonly seen in textbooks of physical chemistry. It makes it much simpler to admit this at the beginning, and our forebears had nothing to be ashamed of in proceeding in this empirical way.

It startles some to learn that heat and temperature, like entropy, are human concepts rather than inherent natural entities. They are very useful concepts, simplifying treatment of changes in motion of large groups of particles for many purposes, but the particles and their motion are the reality, and there is no obligation to think in terms of heat and entropy if more profitable concepts are at hand. The physical sensation of heat depends upon additional factors, and the confusion of this neural stimulus with the thermodynamic abstractions sometimes creates a barrier to understanding.

A later, and intellectually much more satisfying, derivation was made from statistical mechanics, showing that the second term in the free energy equation can be regarded as an expression of the probability of existence of energy states. This is much less formidable than it sounds, and we can arrive for ourselves at a recognition of the existence of this second component without making a quantitative derivation.

Let us assume once more a compound A in equilibrium with a compound B, but with the new requirement that A and B have identical heat contents—the same inherent energy. Now let us suppose that B, unbeknownst to us, actually exists in a thousand different configurations, all in rapid equilibrium and with identical heat contents, but with only one of the configurations being in equilibrium with A. If we now set out to measure the reaction of A to B, we shall find that the change in heat content during the reaction is zero, but in our analytical method we shall not be able to distinguish the thousand different kinds of B that are present. Since they have the same inherent energies, each of the thousand different kinds will be present in equal amounts, and the amount of A at equilibrium will also be equal to the amount of each one of the thousand forms of B. We shall conclude that the concentration of B is 1000 times greater than the concentration of A, and that the equilibrium constant is 1000, and estimate the free energy change to be $-RT \ln 1000$. The energy can't come from a change in heat content, which is zero, and it must arise solely from the existence of B in a thousand different forms.

Real molecules do not exist in a thousand identical energy states, and an exact derivation of free energy change is made by computing, when possible, the distribution of the real energy states, but the idea is the same. Here is the basis for the common statement that the entropy is a measure of the randomness of a system. There is nothing exotic about it. Even a redheaded girl is more difficult to find in a crowd.

Entropy is a useful concept for attempting to define the bases for the structure of proteins, but has found little fruitful application in a discussion of metabolic processes, which are more easily handled by relating energy changes to equilibria, and it is the latter approach that will occupy most of our attention.

Changes in Free Energy Are Additive

The overall change in free energy for any series of reactions is the sum of the free energy changes in the individual reactions. This makes free energy an especially useful concept for reasoning about the complex series of reactions that constitute metabolism.

The additive nature of free energy changes can be demonstrated by assuming three consecutive reactions, $A \to B \to C \to D$:

1. $A \longrightarrow B$ $K_{eq_1} = \dfrac{[B_{eq}]}{[A_{eq}]}$ $-\Delta G_1^0 = RT \ln K_{eq_1}$

2. $B \longrightarrow C$ $K_{eq_2} = \dfrac{[C_{eq}]}{[B_{eq}]}$ $-\Delta G_2^0 = RT \ln K_{eq_2}$

3. $C \longrightarrow D$ $K_{eq_3} = \dfrac{[D_{eq}]}{[C_{eq}]}$ $-\Delta G_3^0 = RT \ln K_{eq_3}$

4. (sum) $A \longrightarrow D$ $K_{eq_\omega} = \dfrac{[D_{eq}]}{[A_{eq}]}$ $-\Delta G_\omega^0 = RT \ln K_{eq_\omega}$

If we solve the first equilibrium expression for $[B_{eq}]$, substitute this value in the equation for the second reaction, solve this for $[C_{eq}]$, and substitute it in the third, we arrive at this expression:

$$K_{eq_1} K_{eq_2} K_{eq_3} = \frac{[D_{eq}]}{[A_{eq}]}$$

This ratio of concentrations also gives the equilibrium constant for the overall reaction. Therefore, the product of the individual equilibrium constants is the overall equilibrium constant:

$$K_{eq_1} K_{eq_2} K_{eq_3} = K_{eq_\omega}$$

Taking the logarithm of the equation, and multiplying by $-RT$, we have:

$$-RT \ln (K_{eq_1} K_{eq_2} K_{eq_3}) = -RT \ln K_{eq_\omega}$$

Since the logarithm of a product is the sum of the individual logarithms, this is equivalent to saying:

$$\Delta G_1^0 + \Delta G_2^0 + \Delta G_3^0 = \Delta G_\omega^0$$

and this is what we set out to show.

Not only is the additive nature of free energy changes useful in analyzing metabolic processes, but it can also be used to calculate the free energy change for a reaction by simply adding the values for two completely independent reactions whose theoretical sum happens to be the reaction of interest.

For example, it is important to have some idea of the free energy change for the reaction:

1. $\quad$ $ATP^{4-} + H_2O \longrightarrow ADP^{3-} + P_i^{2-} + H^+$

The equilibrium for this reaction lies so far to the right that it is technically difficult to measure the remaining concentrations of reactants with sufficient accuracy for reasonable estimates of the equilibrium constant. However, values can be obtained from two other enzymatic reactions having lower equilibrium constants, thereby enabling reasonable measurements of equilibrium concentrations and calculation of their standard free energies:

2. $\quad$ $ATP^{4-} + glucose \longrightarrow ADP^{3-} + glucose\text{-}6\text{-}phosphate^{2-} + H^+$
$$\Delta G^{0'} = -4500 \text{ cal mole}^{-1}$$

3. $\quad$ $glucose\text{-}6\text{-}phosphate^{2-} + H_2O \longrightarrow glucose + P_i^{2-}$
$$\Delta G^{0'} = -3100 \text{ cal mole}^{-1}$$

If we add reactions (2) and (3), the overall result is reaction (1), and we can find its free energy change:

$$\Delta G_2^{0'} + \Delta G_3^{0'} = \Delta G_1^{0'} = -7600 \text{ cal mole}^{-1} \text{ (at pH 7 and 30°C)}$$

The prime symbols are used to designate free energy changes at *constant pH.* When H^+ is a reactant or a product, changes in its concentration would obviously change the free energy and the position of equilibrium for a reaction. Its fixed concentration is already taken into consideration when the prime symbol is a part of the standard free energy designation. This is a convenience for biochemical purposes, because the pH of tissues does remain relatively constant in most cases, and a repetitive calculation is thereby eliminated.

HYDROLYSIS OF PEPTIDES
AND FREE ENERGY

The standard free energy change for the hydrolysis of one peptide bond in the interior of a protein molecule is estimated to be -450 cal/mole at 37°C and pH 7. (This assumes an open chain rather than a helix, and no charged residues near the bond.) The corresponding equilibrium constant is approximately 2 M, and a value so close to unity might make one think it would be easy to make the peptide bond by simply reversing hydrolysis. This would be quite wrong, and the example deserves close examination to show the necessity of caution in making judgments about hydrolyses having low standard free energy changes. It also shows what a powerful tool actual calculations of equilibria can be in reasoning about biological systems.

The hydrolysis of one peptide bond cleaves a chain into two fragments:

$$\text{peptide} + H_2O \longrightarrow \text{fragment}_a + \text{fragment}_b;$$

$$K_{eq} = \frac{[\text{fragment}_a][\text{fragment}_b]}{[\text{peptide}]} = 2 \text{ M}$$

We can estimate the extent of possible hydrolysis, and conversely, the extent of possible reversal to form a peptide, with this equation. Let us estimate a peptide concentration biased toward a high value, which will exaggerate the possible reversal. Assume:

1. A tissue with 250 grams of protein per kilogram.
2. There is a particular peptide chain, with a molecular weight of 25,000, that represents 0.01 of the total protein.
3. All of this chain is in solution, and the entire tissue can be treated as a solution with a density of 1. The concentration of the chain is therefore 1×10^{-4} M.

Now, if the hydrolysis of one bond in the peptide is brought to equilibrium, there will be x concentration of each of the fragments formed, and $(10^{-4} - x)$ concentration of the original peptide remaining. Using the equilibrium equation, we find:

$$\frac{x \times x}{(10^{-4} - x)} = 2 \text{ M}; \qquad x = 9.9995 \times 10^{-5}\text{M}; \qquad (10^{-4} - x) = 5 \times 10^{-9}\text{M}$$

We have arrived at the conclusion that only one molecule of the peptide can exist per 20,000 molecules of each of the peptide fragments at equilibrium, and combination of such fragments to make peptides by a reversal of hydrolysis is not a practical reaction.

Further, we have dealt here with the hydrolysis of a single peptide bond. If the peptide had 200 amino acid residues, the complete hydrolysis would be:

$$\text{peptide} + 199\ H_2O \longrightarrow 200 \text{ amino acids.}$$

The standard free energy change for this reaction would be greater than $199 \times (-450)$ calories per mole, corresponding to an equilibrium constant of 10^{63}! The hydrolytic breakdown of a protein is definitely not a biologically reversible reaction.

THE HIGH-ENERGY PHOSPHATES

We know in fact that proteins are made from amino acids by series of reactions involving the hydrolysis of pyrophosphate bonds on ATP and GTP (Chapter 3). *This is an example of one of the truly fundamental principles of metabolism.* The nucleoside polyphosphates act as intermediates in many different sets of reactions, and the effect is as if these compounds transported energy so as to drive processes with inherently unfavorable equilibria. In fact, the common participation of these compounds makes the entire complex of reactions behave as an energetic unit.

For example, humans are like many organisms in that they consume large amounts of glucose and oxygen, with correspondingly large excretion of carbon dioxide and water. The complete oxidation of glucose in itself is highly *exergonic;* that is, it releases a large amount of free energy:

$$\text{glucose} + 6\,O_2 \longrightarrow 6\,CO_2 + 6\,H_2O \qquad \Delta G^{0'} = -675{,}000 \text{ cal mole}^{-1}\ (25°C)$$

This is roughly 1600-fold greater than the standard free energy changes of $+450$ calories per mole for the formation of a peptide bond by the reversal of hydrolysis. Yet, if we mix together glucose, oxygen, and amino acids in water, and bring the two reactions to equilibrium, the amino acids will sit there essentially unchanged, even though oxygen is being consumed, glucose is disappearing and carbon dioxide is bubbling out of the solution. There isn't any way of transferring energy between completely unrelated reactions at constant temperature.

The problem is solved in organisms by changing *both* reactions so that adenine nucleotides become reactants. Neglecting stoichiometry for the time being, the processes become:

$$\text{glucose} + O_2 + (ADP + P_i) \longrightarrow CO_2 + H_2O + ATP$$
$$\text{amino acids} + ATP \longrightarrow \text{peptides} + ADP + P_i)$$

ATP is being produced by one process and consumed by the other. The adenine nucleotides are in limited supply, and if these two reactions were the only ones in the body, glucose could not be oxidized unless peptides were being formed from amino acids. In a sense, the oxidation of glucose and the synthesis of peptides now become a single reaction, and the overall free energy change is the sum of the changes for the two separate processes.

We shall return to this point again and again. The formation and utilization of ATP cannot be divorced from each other. One is controlled by the other, and this is a major factor in the regulation of the entire chemical economy of the organism.

Free Energy of Hydrolysis of ATP

ATP can be hydrolyzed through several routes involving the cleavage of P—O bonds, some of which are summarized in Figure 9-3. For example, each phosphate group may be removed in turn, finally producing adenosine. However, these hydrolyses involve the cleavage of different structures. In two cases, an acid anhydride—a *pyrophosphate*—is being hydrolyzed:

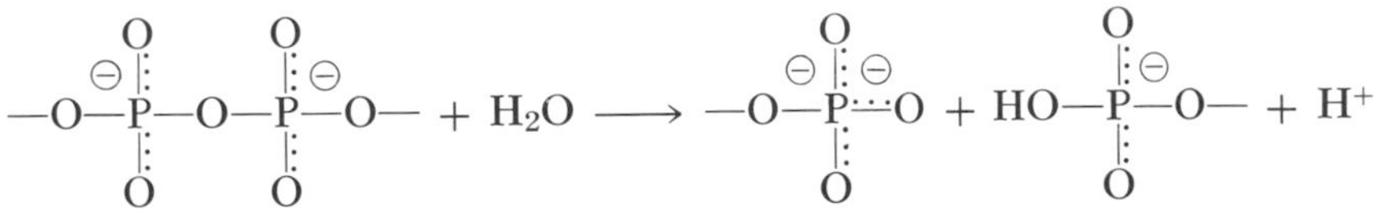

Formally, pyrophosphoric acid represents the combination of two molecules of phosphoric acid with the loss of water, and more molecules of phosphoric acid

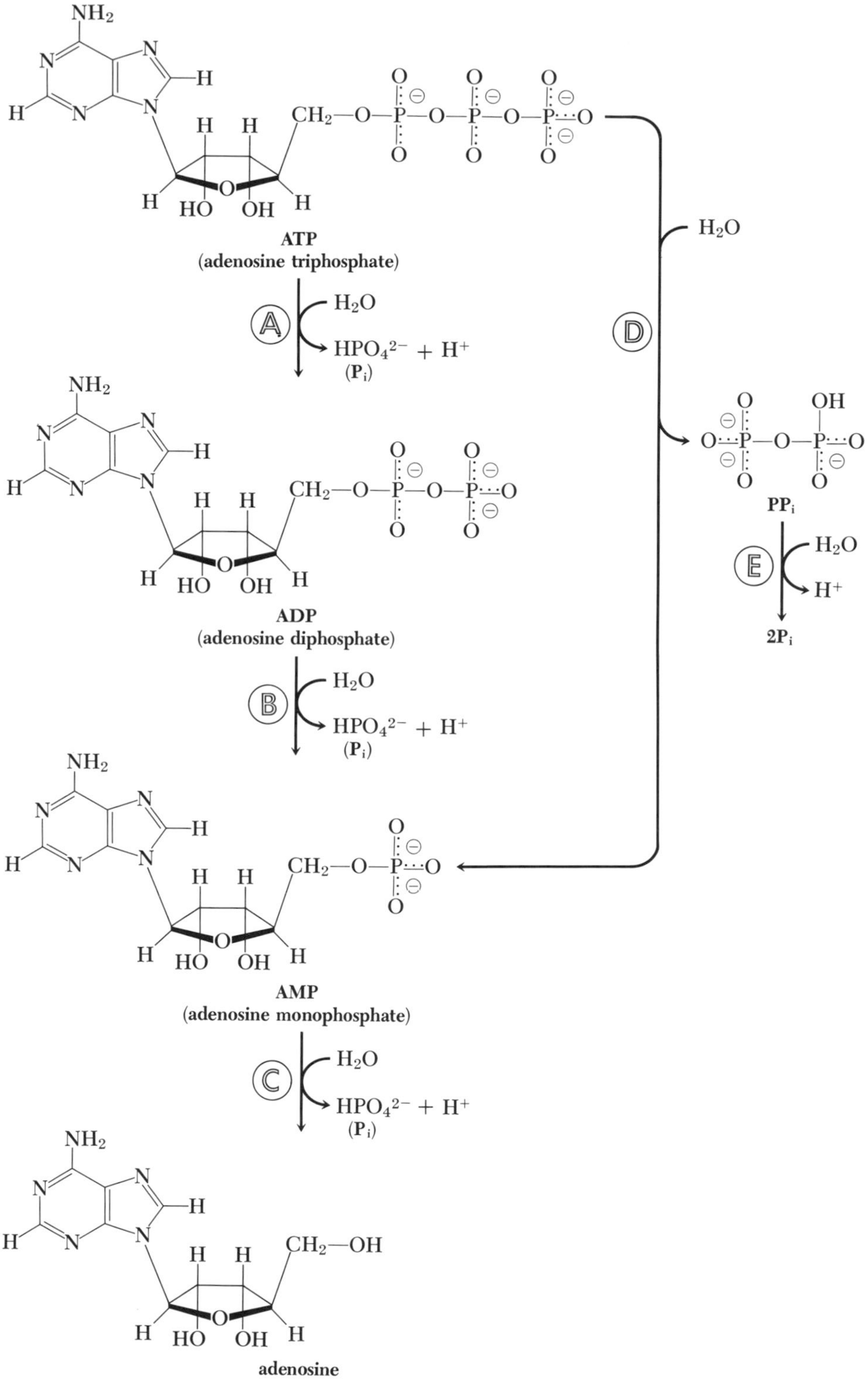

Figure 9-3 Possible hydrolyses of phosphate bonds in ATP.

may be added with the loss of water to make extended chain polyphosphoric acids, which are polyanhydrides:

$$
\underset{\text{OH}}{\overset{\text{O}}{\text{HO}-\overset{\parallel}{\underset{|}{\text{P}}}-\text{OH}}} + \underset{\text{OH}}{\overset{\text{O}}{\text{HO}-\overset{\parallel}{\underset{|}{\text{P}}}-\text{OH}}} \xrightarrow{-\text{H}_2\text{O}}
$$

$$\text{P}_i + \text{P}_i$$

$$
\underset{\text{OH} \quad \text{OH}}{\overset{\text{O} \qquad \text{O}}{\text{HO}-\overset{\parallel}{\underset{|}{\text{P}}}-\text{O}-\overset{\parallel}{\underset{|}{\text{P}}}-\text{OH}}} \xrightarrow[-n\text{H}_2\text{O}]{+n\text{H}_3\text{PO}_4}
$$

$$\text{PP}_i + n\text{P}_i$$

$$
\underset{\text{OH} \quad \text{OH} \quad \text{OH} \quad \text{OH}}{\overset{\text{O} \qquad \text{O} \qquad \text{O} \qquad \text{O}}{\text{HO}-\overset{\parallel}{\underset{|}{\text{P}}}-\text{O}-\overset{\parallel}{\underset{|}{\text{P}}}-\text{O}-\overset{\parallel}{\underset{|}{\text{P}}}-\text{O}-\overset{\parallel}{\underset{|}{\text{P}}}}} \cdots\cdots\cdots -\text{O}-\underset{\text{OH}}{\overset{\text{O}}{\overset{\parallel}{\underset{|}{\text{P}}}}}-\text{OH}
$$

$$\text{PPP}\cdots\cdots\text{P}_i$$

ATP is a polyanhydride with two pyrophosphate bonds. When they are both cleaved, AMP remains. However, AMP is an ester, $R \quad O—P—O—$, not an anhydride, and hydrolyses of esters are different reactions than hydrolyses of anhydrides. This is readily seen from the standard free energies:

Reaction	$\Delta G^{0'}$ (37°C, pH 7.0)[*]
$\text{ATP}^{4-} + \text{H}_2\text{O} \longrightarrow \text{ADP}^{3-} + \text{P}_i^{2-} + \text{H}^+$	-8600
$\text{ADP}^{3-} + \text{H}_2\text{O} \longrightarrow \text{AMP}^{2-} + \text{P}_i^{2-} + \text{H}^+$	-7800
$\text{ATP}^{4-} + \text{H}_2\text{O} \longrightarrow \text{AMP}^{2-} + \text{PP}_i^{3-} + \text{H}^+$	-8100
$\text{PP}_i^{3-} + \text{H}_2\text{O} \longrightarrow 2\ \text{P}_i^{2-} + \text{H}^+$	-8300
$\text{AMP}^{2-} + \text{H}_2\text{O} \longrightarrow \text{A} + \text{P}_i^{2-}$	~ -3000

Hydrolysis of a pyrophosphate bond has a standard free energy near $-8,000$ calories per mole, whether the bond be in ATP, ADP, or PP_i; but the hydrolysis of the phosphate ester bond in AMP has a substantially smaller negative standard free energy—so much so that the equilibrium constant for its hydrolysis is less than 0.001 that for pyrophosphate hydrolyses (see Fig. 9-2). Because of these differences in standard free energy of hydrolysis, the pyrophosphates are said to be high-energy compounds, compared to phosphate esters, and the pyrophosphate bond is said to be a *high-energy* or *energy-rich* bond. This does not imply that the free energy

---

[*] There is some disagreement on the precise values for these standard free energies. The values given are from a tabulation by K. Burton in *The Biochemist's Handbook*, p. 94, D. Van Nostrand (1961). A more recent estimate for the first reaction has been made by R. A. Alberty (*J. Biol. Chem.*, 243: 1337 [1968]).

liberated upon hydrolysis is concentrated in the atoms constituting the bond. It is only a convenient way of designating bonds that are hydrolyzed to yield products having considerably less free energy than the parent compound.

The distinction between high-energy and low-energy bonds is empirical, and the classification ought not be applied too rigidly. If we set the dividing line at -5000 calories per mole for the standard free energy of hydrolysis, then we should be calling a compound releasing 5001 calories per mole "high energy," and one releasing 4999 calories per mole "low energy." This would be perniciously wasteful nit-picking.

Despite this disclaimer, we shall see that it is useful to speak of "high-energy phosphates," and biochemists talk of "moles of high-energy phosphate," understanding that this inherently means moles of substituted pyrophosphates with negative standard free energies of hydrolysis substantially greater than 5000 calories per mole.

In this language, the moles of high-energy phosphate per mole of compound are two for ATP, one each for ADP and PP_i, and none for AMP. Other phosphate esters, such as glucose-6-phosphate (p. 243), have none. The symbol $\sim P$ is frequently used to designate high-energy phosphate bonds, so that ATP may be designated as $A—P\sim P\sim P$.

Much of the difference in standard free energy of hydrolysis of the anhydride and ester phosphates comes from the difference in energy of ionization of the products. We can see from the previous tabulation of the stoichiometry that hydrolysis of the high-energy phosphates at pH 7 involves the formation of H^+, but the hydrolysis of an ester phosphate does not. Esters of carboxylic acids have substantially higher free energies of hydrolysis because the acids do ionize at pH 7:

$$EtOAc + H_2O \longrightarrow EtOH + OAc^- + H^+ \qquad \Delta G^{0\prime} = -5100 \text{ cal mole}^{-1}$$

FREE ENERGY OF PEPTIDE SYNTHESIS

One mole of ATP and at least one mole of GTP are used for each mole of amino acid incorporated into a peptide. The overall reaction is like this:

$$x(\text{amino acids}) + x\text{ATP}^{4-} + x\text{GTP}^{4-} + (x + 1)H_2O \longrightarrow$$
$$\text{peptide} + x\text{AMP}^{2-} + x\text{PP}_i^{3-} + x\text{GDP}^{3-} + x\text{P}_i^{2-} + 2x\text{H}^+$$

We already have the standard free energy values necessary to estimate the position of equilibrium for the reaction:

$$\Delta G^{0\prime}$$
calories per mole

	$\Delta G^{0\prime}$ *calories per mole*
amino acid + peptide $\longrightarrow$ elongated peptide + H_2O	$+400$
$\text{ATP}^{4-} + H_2O \longrightarrow \text{AMP}^{2-} + \text{PP}_i^{3-} + \text{H}^+$	-8100
$\text{GTP}^{4-} + H_2O \longrightarrow \text{GDP}^{3-} + \text{P}_i^{2-} + \text{H}^+$	$\underline{-8600}$
	Sum: $\overline{-16{,}300}$

The equilibrium constant calculated from this standard free energy change for the addition of a single amino acid residue is approximately 2×10^{11}, which assures that the formation of peptides by the above reactions will never be restrained by an approach to equilibrium concentrations so long as there is a continual supply of amino acids and nucleoside triphosphates.

Although the equilibrium constant might already seem high, there is an additional cleavage of high-energy phosphate, beyond that which we have listed, that makes the position of equilibrium even further toward synthesis. We shall repeatedly see that evolution does not incorporate wasteful processes into the metabolic economy, and when there is an apparent exception it pays to examine the presumptions more closely.

We must first recognize that the overall reaction lumps together reactions that are physically separate in cells. Aminoacyl-tRNA compounds are formed in the soluble cytoplasm, whereas the combination into peptides occurs on the endoplasmic reticulum. If diffusion is not to be limiting, or the many kinds of peptides being formed are not to compete among themselves for precursors, each separate process must have a high equilibrium constant.

The Formation of Aminoacyl-tRNA

Aminoacyl-tRNA is produced in the soluble cytoplasm by this enzymatic reaction:

$$\text{amino acid} + \text{ATP}^{4-} + \text{tRNA} \longrightarrow \text{aminoacyl-tRNA}^+ + \text{AMP}^{2-} + \text{PP}_i^{3-}$$

The equilibrium constant is near 0.34, corresponding to a standard free energy change of $+650$ calories per mole! The acylated RNA cannot be efficiently produced by this system as it stands, and therein lies the reason for invoking still another cleavage of a pyrophosphate bond.

The soluble cytoplasm contains an enzyme, inorganic pyrophosphatase, catalyzing the hydrolysis:

$$\text{PP}_i^{3-} + H_2O \longrightarrow 2 \text{ P}_i^{2-} + H^+$$

and the overall reaction is now the combined reactions of this enzyme and of the aminoacyl-tRNA synthetase:

$$\text{amino acid} + \text{ATP}^{4-} + \text{tRNA} + H_2O \longrightarrow$$
$$\text{aminoacyl-tRNA}^+ + \text{AMP}^{2-} + 2 \text{ P}_i^{2-}$$

and $\Delta G^{0\prime} = -7650$ calories per mole.

Correspondingly, $K_{eq} = 2.4 \times 10^5$, which is more than enough to insure efficient utilization of amino acids.

In general, *reactions producing inorganic pyrophosphate as one of the products have a favorable position of equilibrium, owing to the constant removal of inorganic pyrophosphate*. The effective result is the cleavage of an additional high-energy phosphate bond in these reactions. The activation of amino acids is only one of many reactions that depend upon this effect.

The Formation of the Peptide Bonds

The assessment of the energetics for the closure of the peptide bond is considerably more difficult. The standard free energy for the reaction:

$$\text{aminoacyl-tRNA}^+ + \text{GTP}^{4-} + \text{peptide} + H_2O \longrightarrow$$
$$\text{elongated peptide} + \text{GDP}^{3-} + \text{P}_i^{2-} + \text{tRNA} + 2 H^+$$

is $-16,950$ calories per mole, which would indicate that the hydrolysis of GTP is an unnecessary part of the reaction from the standpoint of energetics. However, this reaction differs from the others we have been considering in that it occurs on large particles, and peptide synthesis involves an actual physical movement of ribosomes relative to mRNA, which also requires energy. The actual energetics of this process are unexplored.

The important point of this discussion is that the cell invests sufficient energy in the critical process of protein formation to insure two important conditions. Synthesis can proceed with low concentrations of the starting amino acids. The rate of synthesis is controlled by factors affecting activity of the enzymes, rather than by the relationship of substrate concentration to the equilibrium position of the reaction. Life depends upon the maintenance of specific proteins in proper proportions, and it would not do to have their balance upset through prevention of formation by transient fluctuations in amino acid concentration.

Many cellular syntheses proceed by mechanisms resembling that of protein synthesis, with the process being driven by the cleavage of pyrophosphate bonds. In all of these cases, the nucleoside pyrophosphates are simply the agents by which the syntheses are coupled to the energy-yielding oxidative metabolism of the cell. We separate the oxidative metabolism and the syntheses in our discussion for ease of thought, but they are not separate events. The fact that the adenine nucleotides are products of one series of reactions and substrates for another links the two series into a common process in which one segment cannot proceed without the other.

ATP AND MUSCULAR CONTRACTION

The skeletal muscles have a greater capacity for consuming ATP than any other kind of tissue in the body. A considerable portion of our discussion of metabolism will be devoted to the generation of ATP for muscular contraction. Despite the obvious importance of muscles, we still do not have a picture of the molecular mechanisms by which chemical energy is transformed into mechanical work. We can point out, and emphasize strongly, that the overall process of *muscular con-traction involves the hydrolysis of ATP to ADP and P_i*. This hydrolysis is the direct source of energy for contraction. In order to utilize the energy obtained by oxidation of foods for muscular work, the oxidation of foods must cause the formation of ATP from ADP and P_i.

The structure and function of the contractile proteins in skeletal muscle are currently under intensive investigation, and it is possible that a detailed picture will be available within a few years. A broad outline of the major features will suffice for our purposes now—what is known of the details suggests that the proteins are specialized combinations of the kinds of structure we have already examined intensively.

The mechanics of muscular contraction can be described in terms of three structural elements (Fig. 9-4). A muscle fibril is divided crosswise at regular intervals by rigid bands of protein, the *Z-bands*. Rigid rods protrude from the Z-bands on both sides. These are the thin filaments. In a relaxed muscle, the ends of the thin filaments from adjacent Z-bands are a considerable distance apart.

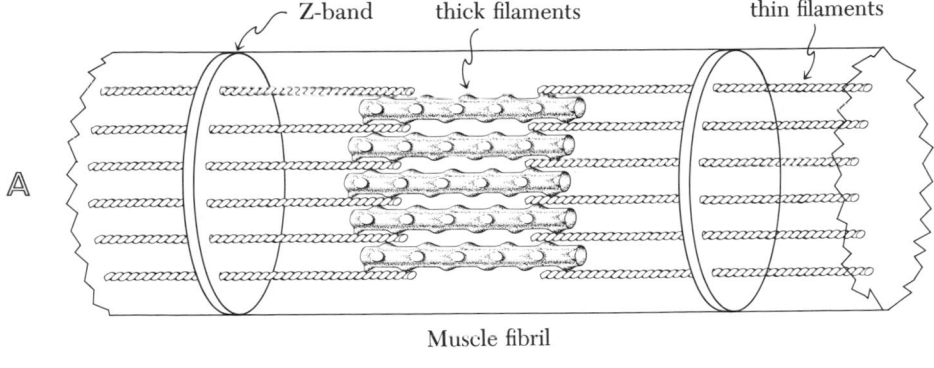

Muscle fibril

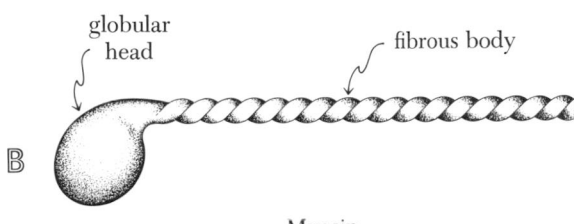

Myosin

Figure 9-4   *A.* Muscle fibers are made of bundles of fibrils. A fibril is a bundle of filaments. Thin filaments are rigid rods containing the protein, actin. One end of each thin filament is firmly attached to the Z band, which is a transverse disk of protein. The ends of thin filaments are overlapped by thick filaments, which have protruding knobs containing sites for binding the actin in thin filaments. The thick filaments are made of bundles of the protein, myosin. Contraction involves the pulling of the ends of thin filaments toward each other through the lattice of thick filaments, thereby also pulling the attached Z-band proteins closer together and shortening the muscle fibril. (The sketch is not to scale; the diameter of the filaments is exaggerated roughly $10\times$ compared to the diameter of the fibril. A real fibril is a dense mass of filaments.)

   B. The myosin molecule has a long body made of twisted helices and a globular head. When myosin molecules associate to form thick filaments, the body portions stack together and the globular heads protrude from the surface of the filament. The head contains the site for binding actin and for hydrolyzing ATP, which causes the thin filament to move to the next head toward the center.

The space between the ends of the thin filaments is occupied by thick filaments, which resemble logs with large protruding knots spaced over the surface. The thick filaments overlap the ends of the thin filaments. Muscular contraction is simply a sliding of the thin filaments toward each other between the thick filaments, pulling the rigid Z-bands closer together. The fibril becomes shorter and thicker. Bundles of fibrils constitute a muscle fiber, and bundles of fibers constitute a muscle, so the entire muscle also becomes shorter and thicker. (It staggers the imagination to think of the number of filaments sliding at Muscle Beach and similar display grounds.)

   The movement of the thin filaments relative to the thick filaments involves the interaction of proteins in the filaments. The thin filaments contain *actin,* along with other proteins of less certain function. Actin has peptide chains, mainly in an α-helix configuration and twisted into a rope.

   The thick filaments are composed of *myosin.* Myosin contains five peptide chains. Two of these are huge, with a molecular weight of 210,000. (During the synthesis of this extraordinarily long peptide, polysomes with as many as 50 to 60 ribosomes appear.) Most of the length of the large peptides is coiled as an α-helix and twisted again to make a two-stranded rope. One end of each peptide is used to make a globular head on the molecule, which also contains three additional small peptide chains (M. W. of 20,000).

The fibrous parts of the myosin stack together to make the body of the thick filaments. The stacking occurs in such a way that the globular heads are exposed on the surface. The heads have two important functions: They bind actin. They hydrolyze ATP to ADP and P_i.

It appears that the hydrolysis of ATP causes some kind of structural change in the neighboring actin, or in the globular head (or both) so that the site of binding shifts to the next globular head down the thick filament, where the process can be repeated to cause further contraction. We are in dark ignorance of the molecular mechanism of this critical step.

Muscles are caused to contract by neural stimuli arriving at specialized terminals—the motor end-plates. We know the stimulus causes a release of acetylcholine. We also have a strong suspicion that the movement of the filaments is precipitated by the presence of calcium ion, which activates the ATPase site of myosin. (ATPase = adenosinetriphosphatase; the term is used for any enzyme causing a hydrolysis of ATP). We are in a considerable fog as to what happens between these two events, except that it undoubtedly involves the traveling wave of changing ion concentrations that can be demonstrated in stimulated tissue.

The activating Ca^{2+} is removed through a system of tubules permeating the spaces between filaments. These *sarcoplasmic tubules* are specialized forms of endoplasmic reticulum terminating in vesicles near the Z-bands. It is probably the vesicles that have the capacity to concentrate Ca^{2+}, removing the ion from the filaments so the fiber can relax.

Before going on to Part II, in which we shall consider the formation of ATP for muscular contraction and other purposes, we might note that current calculations indicate that a skeletal muscle is a very efficient machine. A muscle is capable of using 55 per cent of the internal energy released by ATP hydrolysis for mechanical work. Mechanical work utilizes nearly all of the change in free energy derived from the hydrolysis.

Recapitulation of types of reactions

1. Inorganic pyrophosphate may be hydrolyzed:

$$HOP_2O_6^{3-} + H_2O \longrightarrow 2\,HOPO_3^{2-} + H^+$$

Example: The reaction catalyzed by inorganic pyrophosphatase.

Note: Other new reactions mentioned, *e.g.*, the aconitase reaction, are outlined with structural formulas in later chapters.

Further reading

Florkin, M., and H. S. Mason, eds.: *Comparative Biochemistry*. Academic Press (1960). These articles are especially good:

Huennekens, F. M., and H. R. Whiteley: *Phosphoric Acid Anhydrides and other Energy-Rich Compounds*, Vol. 1 (p. 107).

Eyring, H., R. P. Boyce, and J. D. Spikes: *Thermodynamics of Living Systems*, Vol. 1 (p. 16).

Atkinson, M. R., and R. K. Morton: *Free Energy and the Biosynthesis of Phosphates*, Vol. 2 (p. 1).

Wilkie, D. R.: *Muscle*. St. Martin's Press (1968). A readable simple survey.

Huxley, H. E.: *Electron Microscope Studies on the Structure and Synthetic Protein Filaments from Striated Muscle*. J. Molec. Biol., 7: 281 (1963).

Davis, R. E., M. J. Kushmerick, and R. E. Larson: *ATP Activation and the Heat of Shortening of Muscle*. Nature, *214:* 148 (1967).

Trotta, P. P., P. Dreizen, and A. Stracher: *Studies on Subfragment I, A Biologically Active Fragment of Myosin*. Proc. Natl. Acad. Sci., *61:* 659 (1968).

Gibbons, I. R.: *The Biochemistry of Motility*. Ann. Rev. Biochem., *37:* 521 (1968).

The following interesting pair of articles discuss the logical basis of thermodynamics; each asserting the other's approach is wrong:

Dixon, J. R., and A. H. Emery, Jr.: *Semantics, Operationalism, and the Molecular-Statistical Model in Thermodynamics*. Amer. Scientist, *53:* 428 (1965).

Tribus, M.: *Micro- and Macro-Thermodynamics*. Amer. Scientist, *54:* 201 (1966).

The following illustrate methods of determining free-energy changes, as well as containing important values:

Carpenter, F. H.: *The Free Energy Change in Hydrolytic Reactions: The Non-Ionized Compound Convention*. J. Amer. Chem. Soc., *82:* 1111 (1960).

Jencks, W. P., S. Cordes, and J. Carriuolo: *The Free Energy of Thiol Ester Hydrolysis*. J. Biol. Chem., *235:* 3608 (1960).

PART II

The Generation of ATP

OXIDATIVE PHOSPHORYLATION

CHAPTER 10

ARGUMENT

Reaction of substrates with molecular oxygen supplies most of the energy for cells. The way in which the energy becomes available is through production of ATP by phosphorylation of ADP. In order to obtain a maximum yield of ATP, the transfer of electrons from substrates to oxygen is broken into several steps so that at least two (and more often three) distinct phosphorylations can occur for each pair of electrons transferred. These steps involve a succession of electron carriers, compounds that can be reduced by accepting electrons from the preceding carrier and can again become oxidized by transferring electrons to the next carrier.

The mechanism of the phosphorylation is not known, but three hypotheses have been advanced. Past reasoning, still respected by many, invoked the formation of chemical intermediates during electron transfer that are eventually decomposed by P_i with the formation of high-energy phosphate. The newer Mitchell hypothesis argued that electron transfer acts as a proton pump, creating a concentration gradient across a membrane that is discharged to produce high-energy phosphate. The newest hypothesis, from David Green's laboratory, asserts that electron transfer causes mitochondrial proteins to assume configurations of high energy content, which is discharged by the formation of high-energy phosphate.

The enzymes of oxidative phosphorylation are a part of the inner membrane of mitochondria. Electrons are transferred through increasingly stronger oxidizing agents from substrates to oxygen. That is, successive carriers have increasingly more positive electrode potentials. (Electrode potential is related to free energy change during the oxidation-reduction reaction.) The initial carriers are derivatives of the vitamins, nicotinate and riboflavin, in the form of nicotinamide adenine dinucleotide (NAD) and riboflavin-5′-phosphate. Later carriers are hemin proteins, the cytochromes, with structures modified so as to make the hemin group an increasingly more powerful oxidant.

When electron transfer and phosphorylation are coupled, neither process can occur without the other. Oxygen is not consumed unless ADP is phosphorylated. As a result, if there is not a great demand for high-energy phosphate, the oxidation of substrates will automatically slow because ADP is not available—the adenine nucleotides will be in the form of ATP.

Oxidative phosphorylation can also be impeded artificially through the use of inhibitors, some of which have practical applications beyond their use as experimental tools. Most of the inhibitors are highly toxic because of the absolute need for oxidative phosphorylation. Some of the inhibitors affect electron transfer. Others disturb the phosphorylation reactions—either by inhibiting phosphate transfer, which also inhibits the coupled electron transfer, or by causing hydrolysis of the high-energy phosphate intermediate. Uncoupling in the latter way is toxic because of the resultant loss of ATP, but it accelerates oxygen consumption, which is no longer limited by needing ADP to accept the high-energy phosphate.

We have developed in the previous discussion some idea of the general character of proteins and of the special nature of the enzymes. The rest of the book will be concerned mainly with what these proteins do—the functions for which they exist. Metabolism is complex, so much so that the available information has not been completely assimilated and coordinated by any one person. We have to accept the fact that there is a limit to the number of bits of information that can be handled simultaneously by the human brain (some believe seven bits is the usual maximum). We must divide the metabolic processes into segments that can be comprehended and then attempt to rebuild the entire metabolic economy out of the segments, even though sets of reactions do not proceed as isolated entities in real organisms. We begin, then, with a consideration of those segments that lead to a production of high-energy phosphate.

Oxygen probably appeared in the earth's atmosphere as a possible nutrient only after the evolution of photosynthetic organisms that produce the gas. Reactions between oxygen and organic compounds liberate more energy than most of the others that would be possible in an aqueous solution with substances available on the earth's surface. Hence, reactions were evolved in which the generation of high-energy phosphate, principally as ATP, is coupled to utilization of oxygen. This is the predominant process now existing in most animals for production of high-energy phosphate, and is even utilized in modern green plants to make part of their supply. In short, the formation of high-energy phosphates associated with oxidations by molecular oxygen has become a vital process.

Most of the things we consciously associate with life—growth, motion, and thought—are processes utilizing ATP, but we have no direct awareness of the equally important events producing ATP except in our hunger for oxygen. Deprivation of oxygen is almost immediately lethal to many cells because of their constant requirement for the ATP produced as a result of transfer of electrons from organic compounds to oxygen.

The arrangement for the transfer of electrons is an example of the beautiful economy of reactions frequently seen in living organisms. There are many substrates to contribute electrons, but the electrons are channeled into a common route so that only a few more enzymatic reactions are necessary for the further transfer of electrons to oxygen, and the high-energy phosphate is produced during these transfers regardless of the original source of the electrons.

Transfer of electrons with an accompanying phosphorylation to produce high-energy phosphate is referred to as *oxidative phosphorylation*. Oxidative phosphorylation includes some of the most critical of biological processes, and it is a key concept in our present methods of reasoning about metabolism.

MITOCHONDRIAL STRUCTURE

Oxidative phosphorylation occurs in the mitochondria of cells, and the generation of ATP by this process appears to be the principal function of these organelles. The mitochondria of various tissues differ considerably in size, shape, and location within the cell. They may be spheres or may be elongated into cylinders. Typical dimensions range from 300 to 10,000 nanometers. For example, the mitochondria of skeletal muscles are neatly arranged in fixed locations among the muscle fibers, whereas in more globular cells, such as those of liver, the mitochondria are not only scattered through the cytoplasm, but can be observed to move about and change shape—probably in response to changing nutrient concentrations at various locations in the cytoplasm.

Mitochondria are visible with the light microscope, but the important internal anatomy could not be described before the electron microscope became available, and it is still a subject for active investigation accompanied by heated assertions.

The outside of a mitochondrion is a membrane in the form of a relatively smooth closed container. There is a second membrane within this container, also in the form of a closed bag, but it is usually invaginated to form ridges or tubes protruding into the center of the mitochondrion. These dimples, called *cristae*, are analogous to the forms created by poking one's fingers, either end-on or sideways, into an inflated balloon. The cristae may be so numerous, or of such complex shape, that it is difficult to recognize them as part of the continuous membrane (Fig. 10-2).

The presence of cristae increases the total area of the inner membrane. The number of these folds determines the maximum rate at which mitochondria can utilize oxygen, because the enzymes of electron transport are located on the inner membrane. The greater the membrane surface, the larger the amount of enzymes that can be accommodated.

The outer and inner membranes of a mitochondrion have different functions, but they have a similar kind of structure, which appears to be quite common among biological membranes—a non-polar core with polar faces, so that each membrane appears to be triple-layered in electron micrographs. About half of a mitochondrial membrane is composed of structural proteins, which may have large areas of non-polar surface.

An important part of the membrane core is made from another class of compounds, the *phospholipids*. *Lipids* are the fatty and oily compounds of organisms, which we shall consider in detail later. They comprise a variety of compounds, but they are chemically similar in having a large fraction of the molecule composed of hydrocarbon chains or rings. Phospholipids, as the name implies, are phosphate

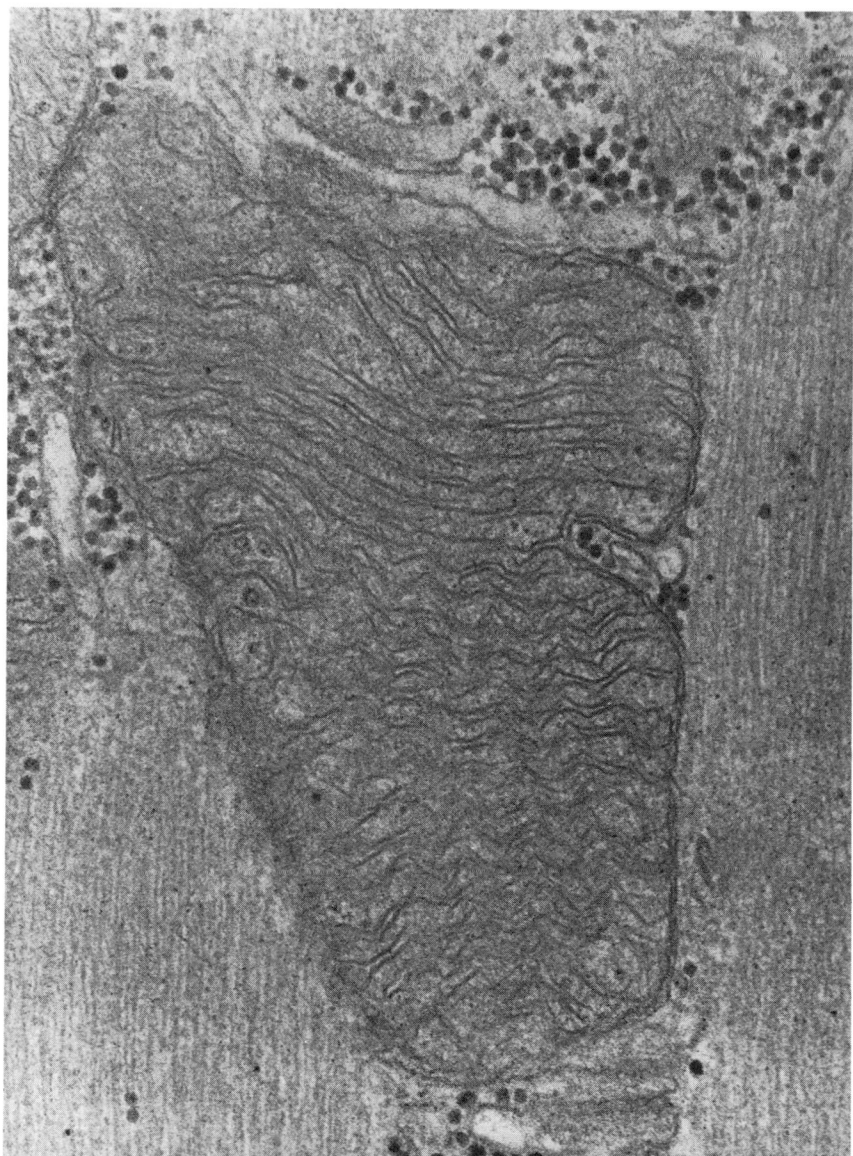

Figure 10-1   A mitochondrion in rat thigh muscle. The mitochondrion is enclosed within a membrane. The interior is filled with processes of a second membrane, and these cristae have the wavy shape characteristic of mitochondria producing ATP. The mitochondrion is intimately associated with muscle fibers to either side. More mitochondria lie in a band above and below the one seen here. Magnification 80,000×. (Courtesy of Dr. Carlo Bruni.)

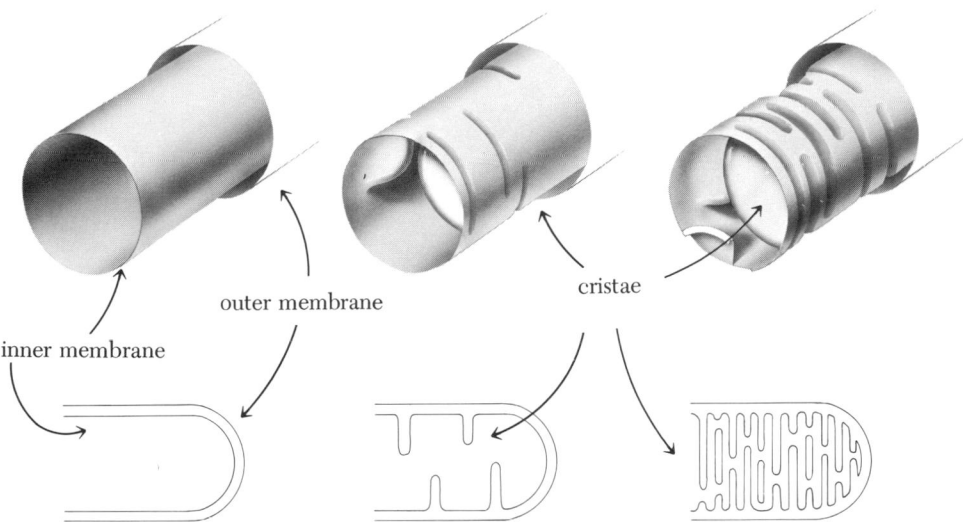

Figure 10-2 Idealized gross structure of mitochondria, shown in cut-away view at the top and in longitudinal section at the bottom. The structure is made from two concentric bags. The membrane comprising the inner bag may be indented into ridges, called cristae. As shown, there may be only a few cristae, or there may be so many that they nearly fill the interior, with cross-sections appearing to have nearly even laminations.

esters, which makes them different than many other lipids in an important respect: They have charged groups in addition to a bulky non-polar structure. They can bridge two environments, with the charged part of the molecule associated with water and the remainder associated with other non-polar compounds.

The *phosphatidyl cholines*, or *lecithins*, are typical examples, differing among themselves only in the nature of the long-chain acyl groups that are a part of the molecule. A particular phosphatidyl choline is shown in Figure 10-3. (Those who read the fine print on food packages will see the lecithins mentioned frequently. They are widely used because of the emulsifying property resulting from their ability to associate with both aqueous and oily phases.)

It appears that the phospholipids are aligned in membranes with their hydrocarbon tails in the center and their charges toward the outside aqueous phases. The hydrocarbon tails alone won't completely fill the space in a flat sheet; the membranes can be curved so that they do, or the remaining space may be filled by neutral lipids. Some believe that the changes of shape seen in mitochondria are due to the migration of these neutral lipids from one part of the membrane to another, thereby causing a change in its curvature. The principle of the general arrangement is shown in Figure 10-4, but the detailed position of the constituent molecules is not known.

The interior surface of the inner membrane appears in high-resolution electron micrographs to be studded with spherical particles on stalks. These particles are believed to contain the enzymes of oxidative phosphorylation. It is reasonable to expect these enzymes to be on the surface in contact with the surrounding aqueous medium, which contains the ADP, P_i, and other polar compounds used as substrates.

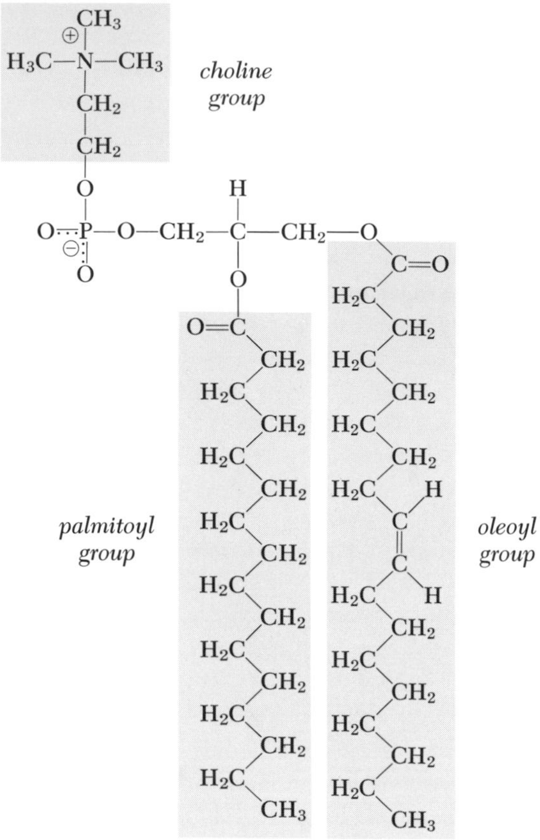

Figure 10-3 A phosphatidyl choline, or lecithin. These compounds are esters of glycerol with two molecules of long-chain aliphatic acids and one molecule of phosphoric acid. The phosphoric acid is also esterified with the alcohol group of choline.

Figure 10-4 A schematic illustration of various proposals for the structure of mitochondrial inner membranes. Many believe that biological membranes in general are composed of a layer of lipids sandwiched between two layers of protein (*A*). The lipid layer is in itself a double layer containing phospholipids with hydrocarbon tails pointing inward and the charged groups in the outside (*B*).

Some suggest that the protein layers are composed of structural units with a shape defined by the nature of the constituent proteins and made so that the units fit together in a mosaic (drawn here as hexagons, but with the shape actually not known).

The particles on stalks seen studding the membrane in high-resolution electron micrographs are believed to contain the enzymes responsible for phosphorylation of ADP to form ATP. Some propose that the base-pieces of the stalks contain the enzymes of electron transport.

Nearly all of these proposals are in dispute, except that there is little doubt of the non-polar nature of the interior. The major disagreements are: (1) The lipid interior may not be continuous. There are indications of protein bridges between the two surfaces. (2) The enzymes contained in the membrane may be scattered widely throughout it, especially those involved in electron transport. (3) The particles protruding on stalks may be artifacts of the procedures used in preparing mitochondria for the electron microscope. (In fairness, it ought to be noted that the Golgi apparatus of cells was once "proven" to be an artifact, and later proved to be quite real. The existence of phosphorylating particles of corresponding dimensions has been shown quite clearly in the laboratory of Dr. Efraim Racker, but it may be that they are normally buried in the membrane and only protrude during the fixing procedures.

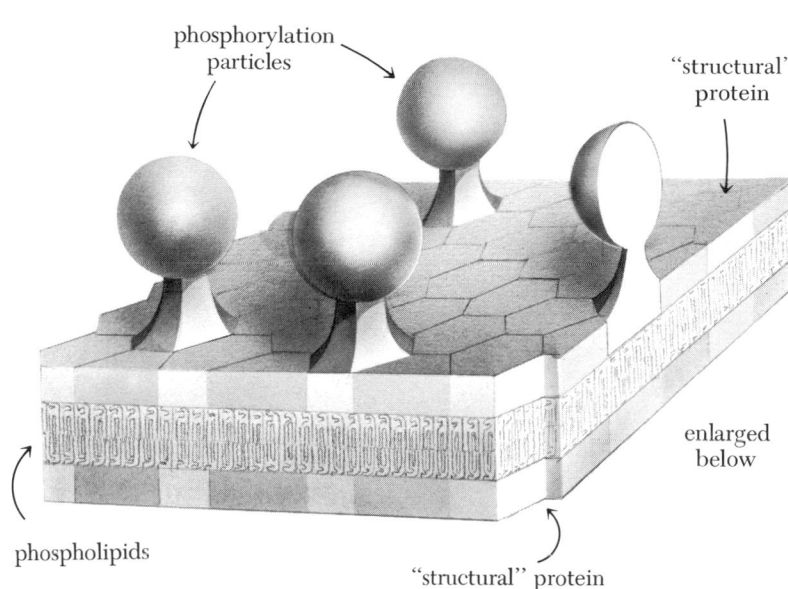

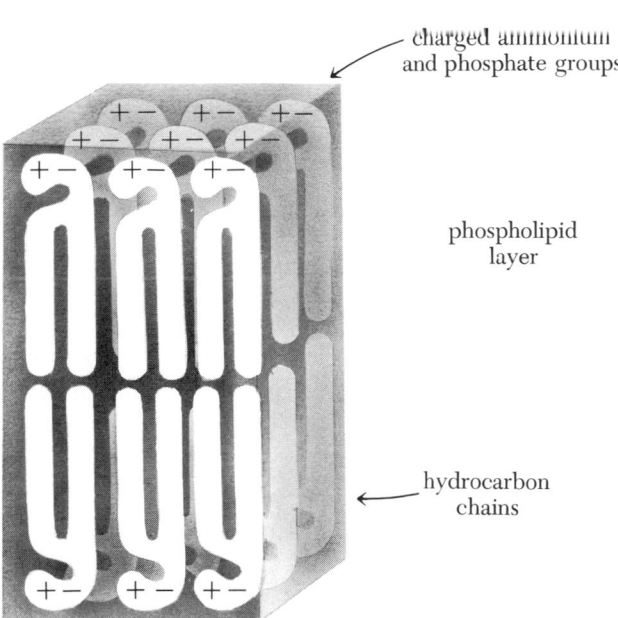

Figure 10-4   See legend on opposite page.

ELECTRON CARRIERS IN MITOCHONDRIA

General Description

Electrons are transferred in mitochondria from the substrates to oxygen through a series of intermediate carriers (Fig. 10-5). A carrier is reduced by the addition of the electrons, and then reoxidized by transfer of the electrons to the next in line, which now becomes reduced. It in turn is reoxidized by the following carrier, until the electrons finally arrive at oxygen, which is the final electron acceptor. Let us look at the process in general, and then consider the individual carriers in more detail.

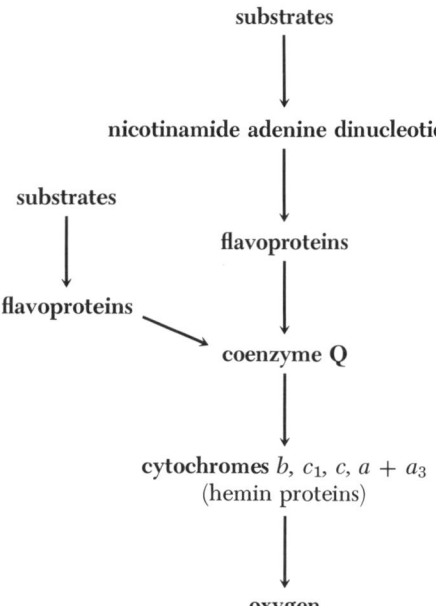

Figure 10-5   The flow of electrons in mitochondria. The oxidation of one compound is accompanied by the reduction of the next along the line of arrows. The eventual products are oxidized substrates and reduced oxygen (water). Some substrates are oxidized by nicotinamide nucleotides, others directly by flavoproteins.

The initial donors of electrons are the various substrates found in mitochondria, with the ingested food being their original source. As soon as electrons are lost from a substrate, it becomes oxidized. For every oxidation there must be a corresponding reduction. When many substrates are oxidized, the corresponding reduction requires *nicotinamide adenine dinucleotide (NAD)*, which we shall discuss in a moment, as an electron acceptor. Before going further, let us carefully review the language employed with oxidation-reduction reactions so that there is no later confusion. The reaction at hand is:

$$\text{substrate} + \text{NAD} \longrightarrow \text{oxidized substrate} + \text{reduced NAD}$$

In this reaction:

1. The substrate is being oxidized because it is losing electrons. The substrate is an electron donor.

2. The nicotinamide nucleotide, NAD, is being reduced because it is gaining electrons. NAD is an electron acceptor.

3. NAD is an oxidizing agent for the substrate because electrons are removed from the substrate when they react together.

4. Similarly, the substrate is a reducing agent for NAD, because electrons are added to NAD when they react together.

The language is confusing, but is in constant use for communication, and it is easier to understand it than to change people. Once more, a reducing agent becomes oxidized while it is reducing other compounds. An oxidizing agent becomes reduced while it is oxidizing other compounds.

In the series of compounds shown in Figure 10-5, each successive compound acts as an electron donor for the one following after it has gained electrons from the one preceding. Taking the longest series as an example, some substrates transfer electrons to NAD. This produces oxidized substrate, which no longer participates in the process, and reduced NAD.

A *flavoprotein* enzyme, *NADH dehydrogenase,* oxidizes the reduced NAD. NAD is now in its original condition, ready to react with more substrate, but the flavoprotein is reduced.

The reduced flavoprotein reacts with *coenzyme Q,* which restores the original oxidized form of the flavoprotein, but produces reduced coenzyme Q.

These successive oxidations and reductions of each carrier proceed through a series of hemoproteins, the *cytochromes,* terminating with molecular oxygen. Everything is returned to its original state except the substrate and oxygen, which have been converted to oxidized substrate and water, respectively.

Figure 10-5 also shows that some substrates do not donate electrons to NAD, but are directly oxidized by additional specific flavoproteins, which in turn transfer electrons to coenzyme Q in the same way as does NADH dehydrogenase.

Some of the electron transfers in the figure involve a production of high-energy phosphate. This is one of the reasons for the evolution of the complicated series of transfers in place of direct reaction between a substrate and molecular oxygen. A direct reaction could not produce three high-energy phosphates per pair of electrons transferred because it is nearly impossible to have a simultaneous collision of the necessary four reactants (one molecule of substrate and three molecules of either ADP or P_i) on the enzyme. The stepwise transfer of electrons, with individual phosphorylations requiring only binary collisions, makes a maximum yield of high-energy phosphate possible.

Nicotinamide Adenine Dinucleotide (NAD)

Nicotinate and its amide occur in all organisms:

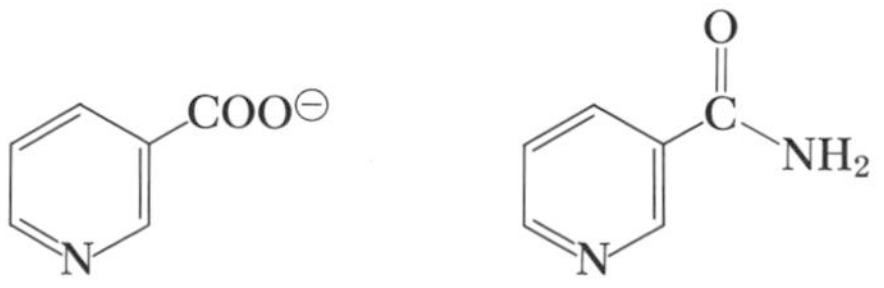

nicotinate   nicotinamide

It is a vitamin, frequently listed as the euphemism, *niacin,* which was coined to avoid any connotation of association with nicotine in the lay mind (honi soit qui mal y pense). The sole function of the compound is to act as a precursor for nucleotides that are used in some oxidation-reduction reactions.

The name, nicotinamide adenine dinucleotide, implies its structure, in which nucleotides containing nicotinamide and adenine are linked through a pyrophosphate bond:

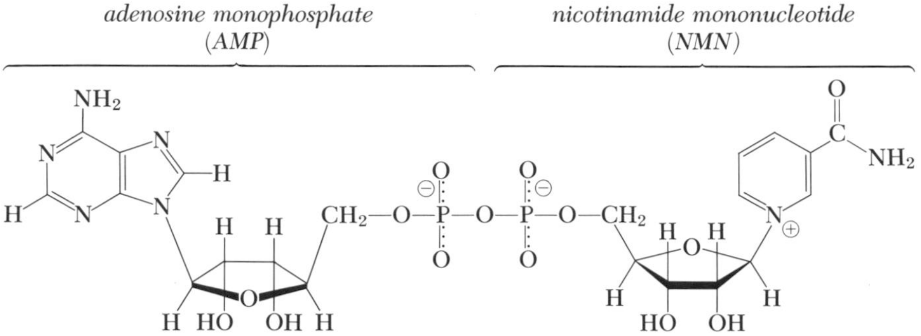

adenosine monophosphate
(AMP)

nicotinamide mononucleotide
(NMN)

nicotinamide adenine dinucleotide
(NAD)

The pyridine ring of NAD is the part of the molecule that changes during oxidation and reduction. The combination of nicotinamide with ribose creates a quaternary nitrogen with a positive charge in the ring, and the nucleotide is a stronger oxidizing agent than free nicotinamide itself.

The bulky remainder of the coenzyme serves to affix the compound to the proper enzymes. The principle here is the one we noted earlier in connection with pyridoxal phosphate—the additional structures of the molecule beyond the reactive pyridine ring serve to bind it in a distinctive way to specific enzymes.

The reduction of NAD requires the addition of a hydride ion—a proton with two electrons—and the nucleotide loses its positive charge in the process (Fig. 10-6). The hydride ion is contributed by the substrate being oxidized.

Many of the substrates that are oxidized by NAD are primary or secondary alcohols. A typical, and important, example is the oxidation of L-malate to oxaloacetate (Fig. 10-7). The oxidation of malate actually involves removal of hydrogen as a hydride ion, not an addition of oxygen, and it has long been recognized that biological oxidations frequently, although not always, are dehydrogenations. Therefore, the enzyme catalyzing this reaction is a malate dehydrogenase, or to be more specific, NAD-malate dehydrogenase, to indicate the nature of the oxidizing agent. (See p. 723 for a discussion of a more systematic nomenclature.) NAD is said to be a coenzyme for this particular malate dehydrogenase because the same molecules of NAD can be used repeatedly to oxidize many molecules of malate, provided there is some additional means of reconverting NADH to the oxidized NAD.

The semantics here are a little cloudy. A molecule of NAD is not so tightly bound to a molecule of malate dehydrogenase that it "belongs" to it. It may, and

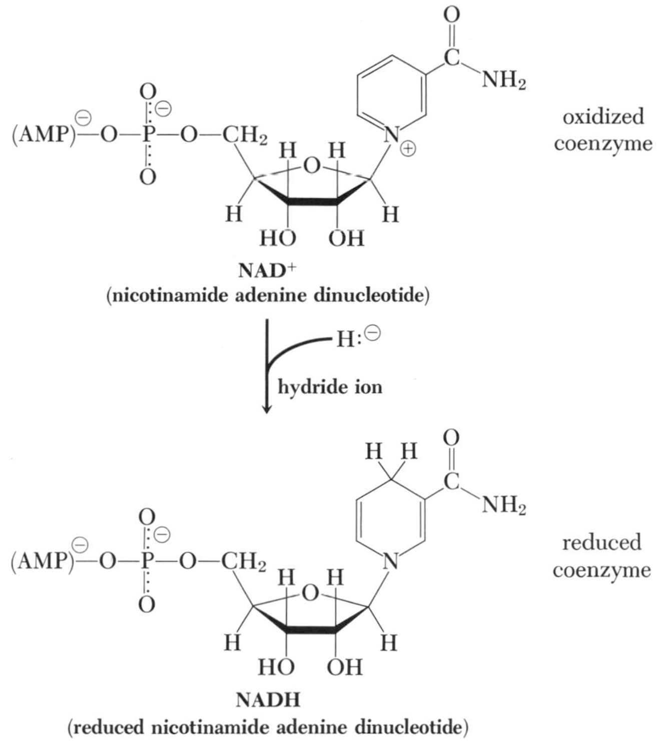

Figure 10-6 The coenzyme, nicotinamide adenine dinucleotide, and its reduced form. The abbreviations, NAD⁺ and NADH, show only the *change* in charge resulting from reduction of the pyridine ring, not the total charge.

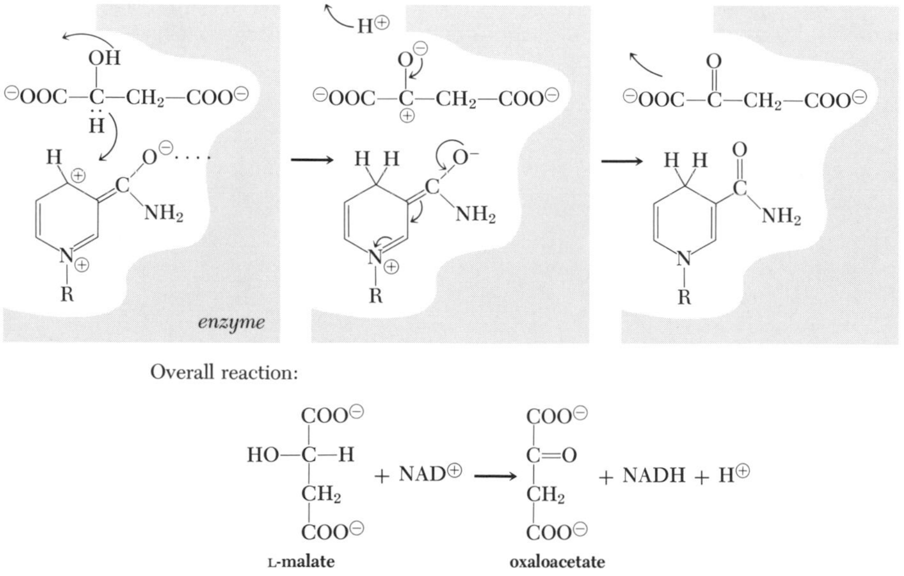

Figure 10-7 Postulated mechanism for enzymatic oxidation of an alcohol by NAD⁺. NAD⁺ is modified by attachment to the enzyme so as to form a cationic site attracting a hydride ion from the substrate. Note that the H⁺ released and the additional hydrogen on the nucleotide are both derived from the substrate. Oxidation of alcohols to carbonyl compounds is among the most common types of biochemical reactions.

187

does, move from one enzyme to another. In fact, NAD is a second substrate for malate dehydrogenase, and the only justification for making a distinction between it and other substrates, such as malate, is that a given molecule of NAD has a long lifetime in the cell, being used over and over, whereas substrates like malate are rapidly converted to products that migrate out of the cell. Other compounds called coenzymes, such as pyridoxal phosphate in the transaminases, frequently are recovered in their original form during the single enzymatic reaction in which they are involved without the intervention of any additional enzymes, and are fixed to given molecules of enzyme in the way implied by the term coenzyme.

NAD has an earlier name, *diphosphopyridine nucleotide (DPN)*, which is still in common use. In still older literature, it was referred to as coenzyme I.

Flavoproteins

Most of the NADH produced in mitochondria is reoxidized by a single enzyme, NADH dehydrogenase, located in the inner membrane. The NADH may arise from the oxidation of a number of substrates catalyzed by many different specific dehydrogenases located in the various parts of the mitochondrion, but the NADH produced from all of them can diffuse to the cristae for re-oxidation by NADH dehydrogenase.

NADH dehydrogenase contains riboflavin-5′-phosphate as a coenzyme, or prosthetic group, and is therefore a flavoprotein. Riboflavin is another example of a vitamin whose sole function is to serve as a constituent of coenzymes used in oxidation-reduction reactions. It has an intense yellow color (*flavus* means yellow), which is characteristic of the flavoproteins. The monophosphate is often referred to as flavin mononucleotide (FMN). This is unfortunate nomenclature, because riboflavin formally consists of the heterocylic compound, 6, 7-dimethyl-isoalloxazine, combined with ribitol. Ribitol is the alcohol corresponding to ribose rather than the sugar itself, so riboflavin is not a nucleoside:

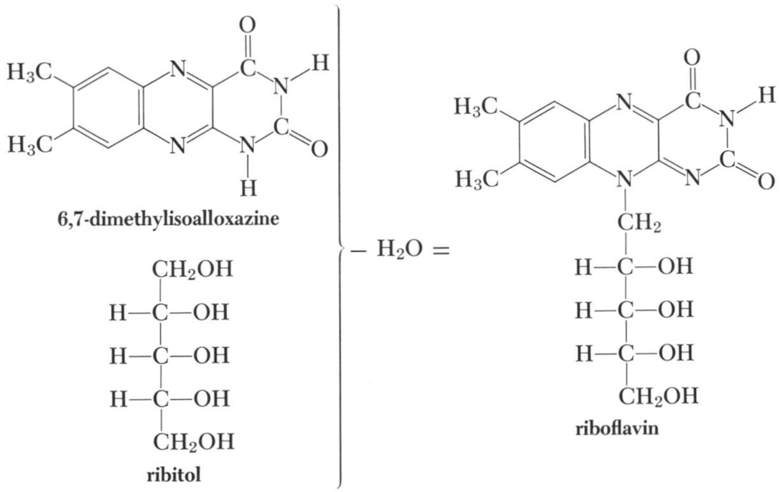

6,7-dimethylisoalloxazine

ribitol

$- H_2O =$

riboflavin

The ribityl group is a part of the flavin, not something extra. Despite this bemoaning of the name, anyone wishing to find the compound in an index is well advised to check flavin mononucleotide as well as riboflavin monophosphate.

The flavoproteins are a versatile group of enzymes. Some of the flavoproteins form quite stable free radicals, with the flavin reduced by only one electron, so that a single electron may be carried by alternating the flavin between its oxidized state and the free radical or by alternating the free radical with the fully reduced form; there are examples known of each of these mechanisms (Fig. 10-8). Other flavoproteins never form the free radical, and use two electrons to change the fully oxidized to the fully reduced form.

Many flavoproteins also contain metallic ions that undergo oxidation and re-

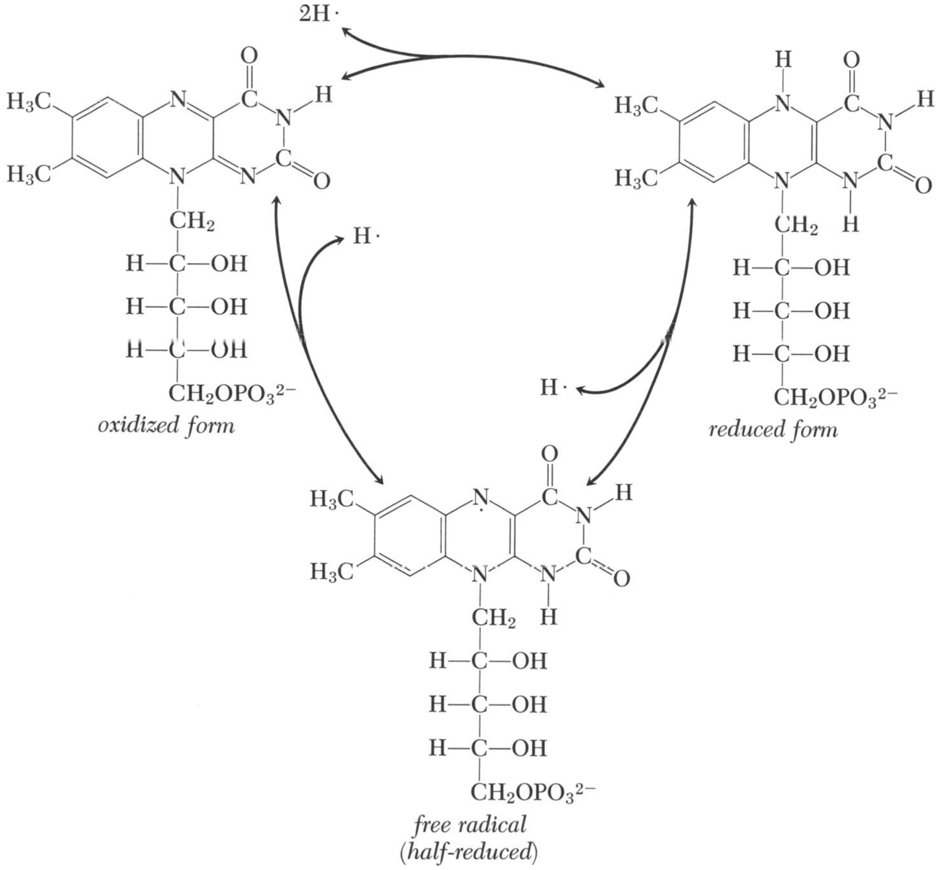

riboflavin-5'-phosphate
(flavin mononucleotide, FMN)

Figure 10-8 Alternate routes of electron transfer with FMN. Some enzymes with flavin prosthetic groups catalyze a complete reduction in one step (*top*) by transfer of two hydrogen atoms with the accompanying electrons. Others utilize successive single electron transfers (*bottom*), involving the formation of an intermediate flavin free radical.

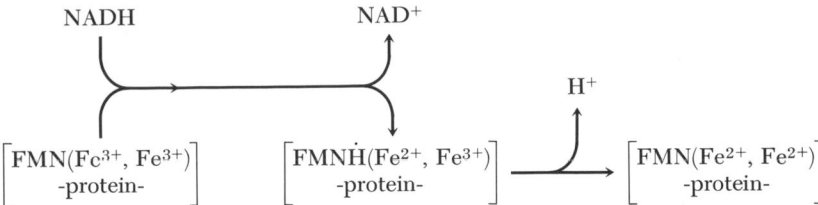

Figure 10-9   NADH dehydrogenase is an example of an iron-containing flavoprotein. The metal is also oxidized and reduced during electron transfer by enzymes of this type. A possible mechanism involves the initial transfer of a pair of electrons to the enzyme, one being used to form a flavin-free radical and the other to reduce one atom of ferric iron. An intramolecular oxidation-reduction occurs, with the flavin-free radical being oxidized back to its initial state by a second atom of iron.

duction. Our present important example, NADH dehydrogenase, contains iron. The iron is bound to the dehydrogenase by direct chelation with the protein (the specific ligands are not known), and these iron-containing proteins therefore differ from the hemoproteins, in which a porphyrin is used to ligate iron.

The presence of iron adds a complication to the mechanism, and there isn't enough information to be certain of the number of electrons transferred at one time. Figure 10-9 sketches a likely possibility, but it is easy to imagine other permutations of the possible sequences of reduction that will involve both the iron and the flavin on the protein.

Whatever may be the mechanism, the important end result is the conversion of NADH to NAD, available for the further oxidation of substrates, and the formation of a reduced flavoprotein, which must transfer its pair of electrons to another acceptor if it is to be available to react with more NADH. Flavoproteins in general have a strong association between the prosthetic group and the peptides, so it is not possible to exchange a reduced flavin molecule for another in the oxidized state on the same enzyme molecule.

This is an important distinction between NADH dehydrogenase, along with the other flavoproteins and hemoproteins we shall be mentioning later, and the NAD-coupled substrate dehydrogenases, such as malate dehydrogenase. There are many molecules of NAD available to a single molecule of malate dehydrogenase, so a given molecule of the enzyme can continue to catalyze the oxidation of malate even if the NADH it has previously produced is not completely removed. However, NADH dehydrogenase can only transfer electrons to the molecules of riboflavin-5'-phosphate that are actually bound to its peptides, and the reduced riboflavin must be immediately reoxidized if the enzyme is to be a catalyst and react with more molecules of NADH.

Coenzyme Q (Ubiquinone)

The second substrate of NADH dehydrogenase, the electron acceptor, is coenzyme Q, also known as ubiquinone. Coenzyme Q (CoQ) is actually a group of compounds, all containing the same quinone structure, but substituted with a side chain composed of varying numbers of isoprene groups linked head-to-tail:

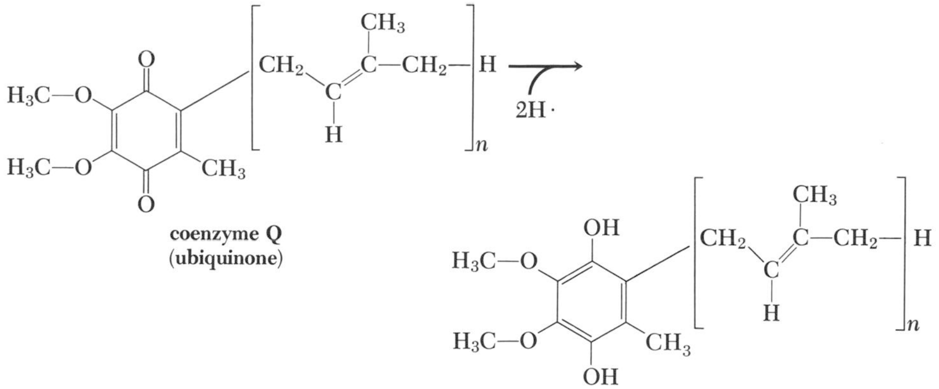

coenzyme Q
(ubiquinone)

dihydrocoenzyme Q
(ubihydroquinone)

These compounds can be synthesized by animals, and have no vitamin precursors. The most common compound in mammalian tissues has 10 isoprene groups, and is designated CoQ_{10}. Coenzyme Q can receive two electrons, and one mole is reduced for each mole of substrate originally oxidized.

The long hydrocarbon tail of coenzyme Q gives it a high solubility in the lipid matrix of the mitochondrial membrane, and a low solubility in water. This means that NADH dehydrogenase must be in contact with both the polar and the non-polar phases of the mitochondrial inner membrane, since it is receiving electrons from NADH in aqueous solution and transferring them to CoQ in lipid solution.

This organization of the electron transport chain on an insoluble membrane makes it difficult to isolate the constituent enzymes in their original association. The membrane must be disrupted, and it is difficult to assess the degree of alteration in the components caused by the disruption, which inevitably exposes some constituents to a medium of the wrong water or lipid content. As a result, there has been much controversy over the routes of electron transport; it has been especially sharp over the nature of the primary acceptor for the electrons transferred by mitochondrial flavoproteins such as NADH dehydrogenase. In other words, the second substrate for this enzyme was in dispute. Some doubted the importance of CoQ in electron transport long after its discovery. It is important to know this background, because a person reading reports from different laboratories is sometimes hard put to recognize that the authors were talking about the same process, or that a paper represented a change in position over earlier publications. The participation of CoQ is now generally recognized, but even today there are some who believe it transfers electrons one step beyond the point at which we are placing it. The weight of opinion and of the evidence is presently against this view.

The Cytochromes

The remainder of the electron carriers after coenzyme Q are all cytochromes— proteins combined with iron porphyrins, but differing from hemoglobin in that the iron is alternately reduced and oxidized during their physiological function. The prosthetic group changes from hemin to heme and back again as a part of the catalytic mechanism.

The different cytochromes are classified according to characteristic light absorption bands in the visible region and are designated by letters; the order in which electrons are transferred is cytochromes b, c_1, c, $(a + a_3)$. Subscript numbers are added to designate cytochromes with similar spectra discovered later. (The number is a historical, not functional, designation.) The nature of the spectrum is determined by the side chains on the porphyrin rings.

The first cytochrome in the sequence, *cytochrome b*, reacts with reduced coenzyme Q, oxidizing it to its original quinone form. Cytochrome b contains protohemin IX, the same ferric porphyrin found in methemoglobin. Electrons transferred from reduced coenzyme Q reduce two of these groups to the ferrous porphyrin, protoheme IX:

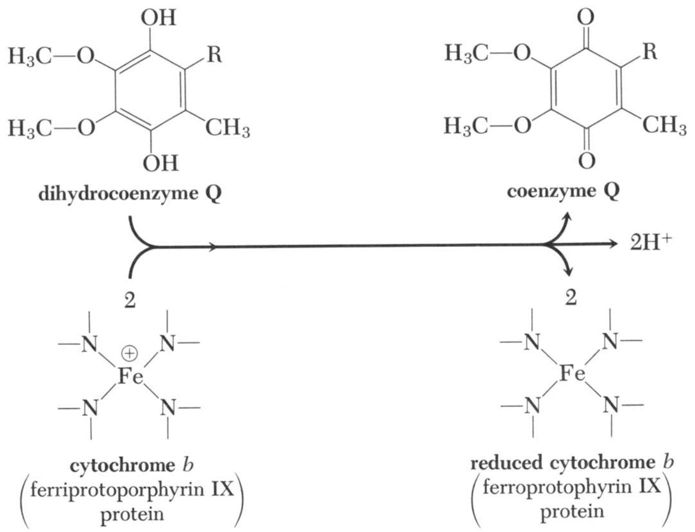

The reduced cytochrome b then reacts with *cytochrome c_1*, probably with the mediation of still another protein containing non-heme iron, which is omitted from the scheme for the sake of brevity.

Cytochrome c and c_1 contain a porphyrin that is covalently bound to cysteinyl residues of the respective peptides. The effect is as if a sulfhydryl group of the peptide added across the double bond of each vinyl group on protoporphyrin IX (p. 12), creating a pair of thioether bonds:

$$\text{porphyrin}-CH_2-CH_2-S-CH_2-\text{peptide}$$

The differences in the structures of the cytochromes serve to make them increasingly stronger oxidizing agents, but the reaction between cytochromes b and c_1 illustrates the general kind of reaction with all, in which a ferroporphyrin is oxidized by transfer of its electron to the ferriporphyrin of the succeeding cyto-chrome:

ferrocytochrome b + ferricytochrome c_1 $\longrightarrow$

$\qquad$ ferricytochrome b + ferrocytochrome c_1

In the same way, the electrons now on ferrocytochrome c_1 are transferred to ferri*cytochrome c*. Cytochromes b and c_1 behave as enzymes, so no further catalyst is necessary for the sequence we have considered for the transfer of electrons from reduced coenzyme Q to cytochrome c. Cytochrome c is a much smaller protein than the other cytochromes of the sequence, and also differs from them in being found in the cytosol as well as in the mitochondria.

The final reactions of electron transfer involve *cytochrome $(a + a_3)$*. This odd designation is used because there is still some argument about the existence as separate entities of proteins having the two different absorption spectra that lead to separate designations of a and a_3. The complex as a whole is termed *cytochrome oxidase*, because it reacts directly with molecular oxygen to catalyze the removal of electrons from ferrocytochrome c (Fig. 10-10).

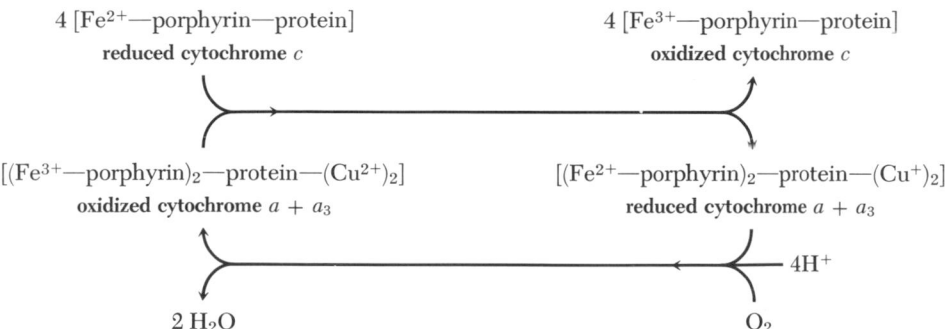

Figure 10-10   The transfer of electrons from reduced cytochrome c to oxygen by cytochrome $a + a_3$. Cytochrome $a + a_3$ contains both hemin and cupric ions that are reduced during the transfer. Four electrons are required for the complete reduction of a molecule of O_2, and the intermediate steps by which they are accumulated from the single electron available on reduced cytochrome c are not clear. Current thought is that the electrons pass through heme a to Cu^{2+} and then to heme a_3.

The porphyrin a of cytochrome oxidase differs from that of the other cyto-chromes and has not been completely characterized; a distinguishing feature is the presence of a formyl group, R—CH=O, as a side chain. The oxidase also contains *copper*, and the $Cu^{2+} \rightleftharpoons Cu^+$ oxidoreduction is a part of the mechanism of elec-tron transport.

The presence of copper in this critical protein would in itself necessitate a supply of the element in the diet, and we shall see that copper is also a part of other enzyme systems. Even though the requirement is absolute, the total amount necessary to form all of these enzymes is small, and any losses are easily replaced by the traces of copper found in nearly every food. A human deficiency has been demonstrated only in four starved infants. Potential insufficiency of copper is of more concern to livestock ranchers, because herbivores grazing over deficient soils may have an inadequate intake.

The Complete Electron Transfer Chain

Figure 10-11 summarizes the interrelationships of the electron carriers we have discussed, and shows the steps at which phosphorylations occur. Not only are the individual reactions connected in the way shown in the figure, but they are also linked by physical proximity of the constituent proteins on the mitochondrial cristae. This can be shown experimentally. Mild disruption of the cristae will liberate intact particles that catalyze the entire sequence. More vigorous treatment of these particles will produce fragments, which can be separated, but the fragments contain those proteins catalyzing neighboring reactions. For example, one does not obtain a fragment containing cytochrome oxidase and cytochrome b, but lacking the intermediate cytochrome c and cytochrome c_1. The various cytochromes occur in integer molar ratios, which also supports the concept of organized particles.

Except for NAD and possibly cytochrome c, the electron carriers are closely associated with lipid in the mitochondrial membrane. It may be that electron transport occurs in a hydrophobic environment, diminishing the chances for accidental oxidation of the wrong compound by the successively stronger oxidizing agents in the chain. Furthermore, oxygen is more soluble in non-polar media than it is in water. We still have much to learn about oxidation-reduction reactions in non-aqueous environments.

The surface of the phosphorylating particles must be polar, because the enzymes must have access to phosphates dissolved in the aqueous phase.

Little is certain about the fine detail of mechanism at any stage of electron transport. There is some evidence that the NAD-coupled substrate dehydrogenases utilize a *tryptophan* residue in the oxidation-reduction mechanism. (One would like to believe that this is so to have some rational explanation for the presence of this enigmatic amino acid; there frequently are only a few residues in a protein.)

The small protein, cytochrome c, has been highly purified, and the amino acid sequence is known for the protein from a variety of sources. (Known kinds of mammalian cytochrome c have 104 residues; the number varies in other classes of animals.) Yet, the mechanism of oxidation and reduction is not clear for even this simple case, even though it is known that a *histidyl* residue and a *methionyl* residue are used for binding the iron from opposite sides of the porphyrin ring. It is especially difficult to understand how electrons get from the surface of the molecule into the hemin crevice.

COUPLED PHOSPHORYLATION

The Known Steps

The study of the phosphorylations accompanying electron transport is also plagued by experimental difficulties in separating enzymes from their membrane

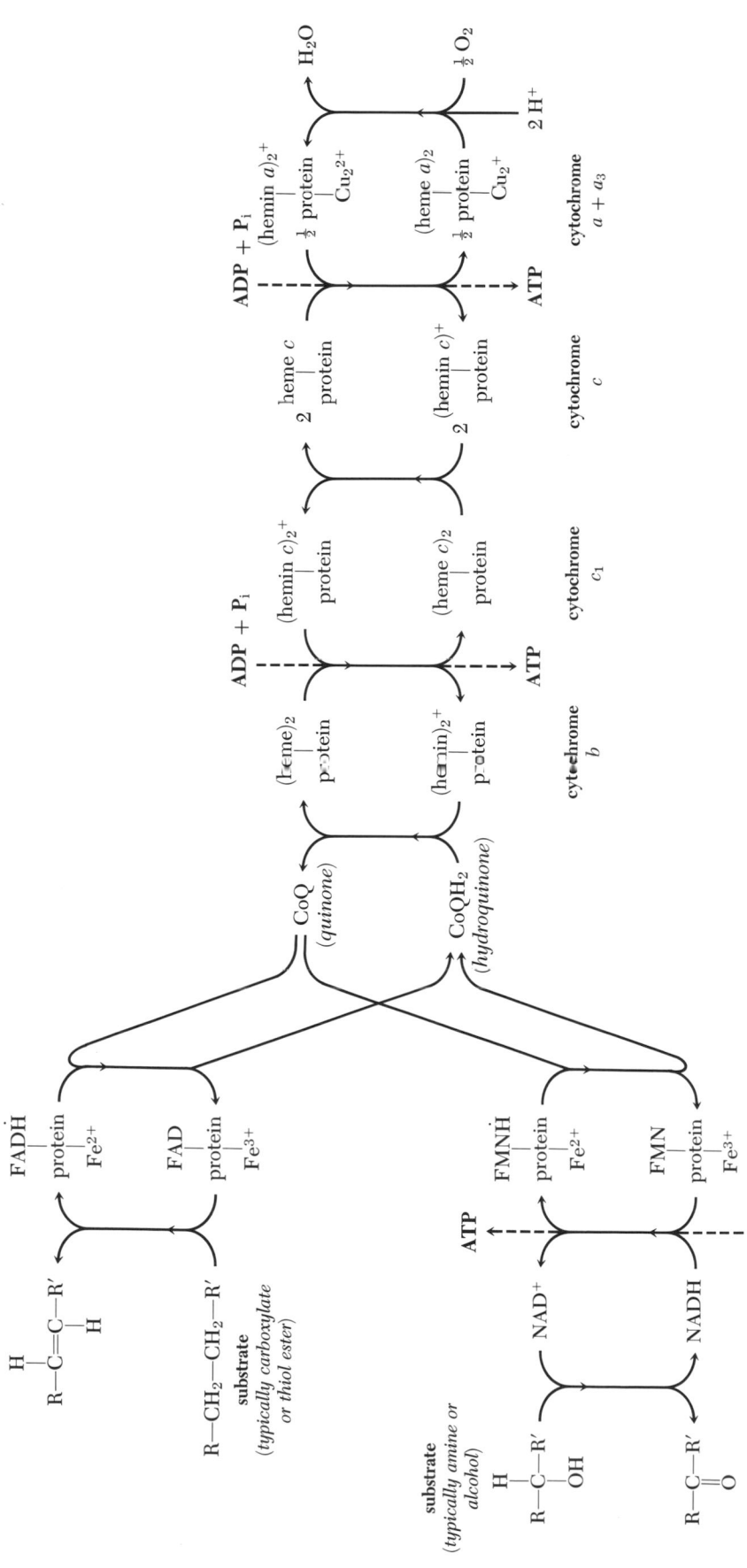

Figure 10-11 The electron transport chain of mitochondria. Alcohol and amine groups are oxidized by NAD^+; hydrocarbon chains are oxidized by flavoproteins. Electrons from either kind of oxidation arrive at coenzyme Q and are then transported *via* the cytochromes to oxygen. The reactions associated with phosphorylation are also indicated.

195

matrix. It is important that we separate what appear to be definite facts from optimistic assumptions based on inconclusive experiments. Here is what we must have firmly in mind:

1. ATP is generated from ADP and P_i by the transfer of electrons from substrates to molecular oxygen in mitochondria. This is the principal means of linking the oxidation of nutrients to the synthesis of body constituents or to the contraction of muscles.

2. The ATP generated in mitochondria is available to the remainder of the cell. ATP, ADP, and P_i can cross the mitochondrial membranes. (The transport appears to involve a one-for-one exchange of ATP inside for ADP outside, so the total concentration of adenine nucleotides doesn't shift.)

3. ATP is generated at three sites on the electron transport chain, associated with the following electron transfers: (a) from NADH to coenzyme Q by the flavoprotein, NADH dehydrogenase; (b) from coenzyme Q to cytochrome c_1 through cytochrome b; and (c) from cytochrome c to O_2 through cytochrome $(a + a_3)$.

4. Therefore, three moles of ATP may be generated by the passage of electrons from a mole of substrate through NAD to molecular oxygen, but only two moles may be generated if electrons are transferred directly from the substrate to coenzyme Q through a flavoprotein, because this transfer bypasses the first phosphorylation site.

5. NADH generated in the cytosol will not pass into mitochondria, and cannot be used directly for oxidative phosphorylation. Therefore, mitochondrial electron transfer must begin with substrates present within the mitochondria.

6. The final two steps of the phosphorylation process involve: (a) the formation of a high-energy phosphate of unknown character from P_i and another reactant, perhaps a protein; and (b) the decomposition of the high-energy phosphate intermediate by ADP, with the formation of ATP.

When we look over this collection of near certainties, we see that we really know a great deal about oxidative phosphorylation in terms of what must be fed into the system and what comes out; it is this kind of information that is most valuable in reasoning about events affecting other reactions.

Even so, it is frustrating that we are still puzzled about the actual mechanism of these key metabolic processes. There have been many speculations. All perforce begin with the assumption that electron transfer does something that can be translated into the formation of a high-energy bond. Most speculation has hinged on the hypothetical existence of an unknown reactant, X. If we let A: represent the reduced coenzyme and B the oxidized second coenzyme for each reaction in which ATP is generated, the minimum sort of scheme consistent with available data is something like this:

$$
\begin{aligned}
&1. && \text{A:} + \text{X} \longrightarrow \text{A:X} \\
&2. && \text{A:X} + \text{B} \longrightarrow \text{A}{\sim}\text{X} + \text{B:} \\
&3. && \text{A}{\sim}\text{X} + \text{P}_i \longrightarrow \text{A} \;+\; \text{X}{\sim}\text{P} \\
&4. && \text{X}{\sim}\text{P} \;+\; \text{ADP} \longrightarrow \text{X} + \text{ATP}
\end{aligned}
$$

The essential feature of this and of similar schemes is the combination of the hypothetical reactant with a coenzyme and the conversion of this combination to a high-energy form by the transfer of electrons. The high-energy combination is later split by inorganic phosphate to produce a high-energy phosphate, which is transferred to ADP, producing ATP. The same general kind of process is believed to occur at the three sites of phosphorylation.

Several variants on this scheme, usually invoking additional hypothetical intermediates, have been made. This is not to say that the schemes were the product of random imagination; they were devised to account for experimental observations of the effects of a variety of conditions and substances on oxidative phosphorylation. However, they all suffered until recently from the inability to identify any of the hypothetical intermediates.

Efraim Racker's laboratory succeeded in separating an enzyme complex involved in the last steps of the sequence, which are concerned with the phosphorylation of ADP. This *coupling factor* complex was dissociated from the enzymes of electron transfer, which were still capable of oxidizing substrates, but without accompanying phosphorylation. The capacity for oxidative phosphorylation was restored by adding back the coupling factor.

In isolated form, coupling factor transfers the intermediate high-energy phosphate to water, so that the final reactions run in the reverse direction and cause the hydrolysis of added ATP:

$$ATP^{4-} + H_2O \longrightarrow ADP^{3-} + P_i^{2-} + H^+$$

The complex therefore behaves like an *adenosinetriphosphatase (ATPase)* when removed from its natural environment.

Racker has also recently suggested that the *non-heme iron*, that is, iron not associated with porphyrins, long known to be present in mitochondrial membranes, may also be associated with the phosphorylating complex rather than with the electron transfer system. We shall have to wait and see if this proves to be part of the missing link for generating high-energy intermediates.

The Mitchell Hypothesis

Electron transfer may also drive a movement of ions across the mitochondrial membrane in addition to the formation of ATP. For example, K^+ or Ca^{2+} accumulates in exchange for H^+. The same ion movements can be driven in the absence of electron transfer by adding ATP, which is hydrolyzed to ADP and P_i as the ions are transported ("ion-stimulated ATPase"). These events, which share some features with the general ability to concentrate ions across biological membranes at the expense of high-energy phosphate, are not directly accounted for by the sort of scheme we outlined earlier.

Peter Mitchell proposed that the hypothetical intermediates of oxidative phosphorylation could not be discovered because they did not exist, and that high-energy phosphate was generated during electron transfer as a result of the establishment of a difference in ion concentrations across the mitochondrial inner membrane, particularly of H^+, which was relieved by the formation of the high-energy phosphate. He therefore offered a theory which attempted in one package to correlate ion movement with electron transfer and to account for the experimental failure to establish the nature of any high-energy intermediates.

Since a number of strong, even arrogant, personalities quickly became either proponents or opponents of the Mitchell hypothesis, the intensity of the response to such a drastic and speculative revision of current theory obstructed reason even longer than usual. In time, Mitchell was forced to re-introduce high-energy intermediates, their character still unknown, into his speculations to reconcile them with experimental observations. Despite this important retraction, his hypothesis served the very useful function of disrupting what had essentially become a private contest among a few laboratories with little interest in the general biological context of the problem. He also focussed attention on the necessity of explaining the ion translocations, which are not speculations but observable facts.

The essential features of the present modification of the Mitchell hypothesis are shown in Figure 10-12.

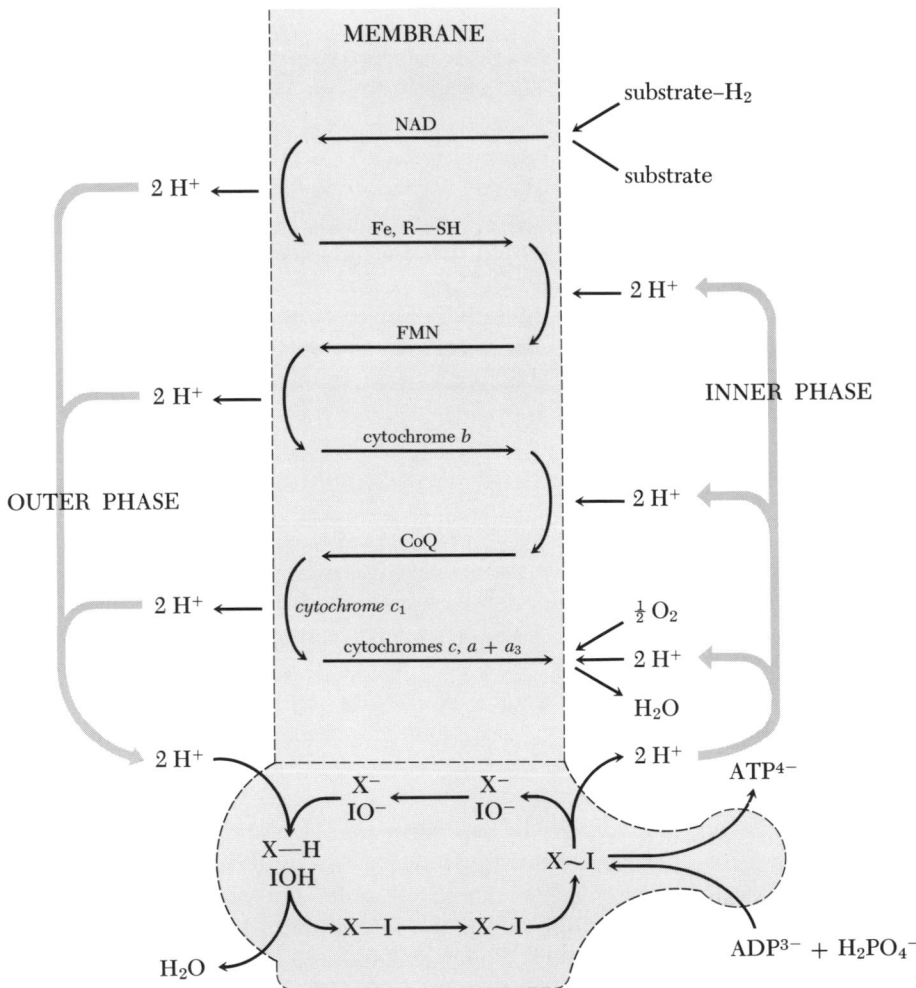

Figure 10-12   The Mitchell hypothesis of oxidative phosphorylation. See the text for details.

1. Protons and electrons are added at one side of the mitochondrial membrane to a carrier within the membrane. They may be obtained as a hydride ion from substrates, or separately, as in the case of the cytochrome-mediated electrons transfers.

2. The carrier is reoxidized on the other side of the membrane by the next component of the electron transfer chain. Two protons are released from the membrane.

3. A proton gradient across the membrane has been created. The basic forms of the components of high-energy intermediates within the membrane will combine with more protons on the side of the membrane at which their concentration is highest. These components are now in their acidic, non-ionized, forms.

4. The acid components combine, with the liberation of water, to form the hypothetical high-energy intermediate, which diffuses across the membrane to the proton-poor side.

5. The high-energy intermediate is decomposed by P_i, and ATP is formed from ADP. At the same time, the components of the intermediate are released. Since they are on the proton-poor side of the membrane, they ionize and thereby replace the two protons previously removed by electron transfer in the first step. The overall process of oxidation plus phosphorylation therefore does not involve a net transfer of protons from one side of the membrane to the other.

The current Mitchell hypothesis includes many of the important features of older speculations based on chemical reactions, including the production of three moles of high-energy phosphate per pair of electrons carried through the complete chain and the formation of unknown intermediates within the membrane. Continued convergent evolution of ideas seems likely.

The Mechanical Coupling Hypothesis

It has long been known that mitochondria change shape when they are deprived of substrates, of oxygen, or of ADP. David Green and co-workers now propose that these changes are extreme manifestations of a change in membrane conformation normally a part of oxidative phosphorylation. According to this view, the membrane exists in three states: a relaxed state of sheet-like appearance, a vesicular state, and a twisted state. The latter states are energized physical modifications, the vesicular form resulting from electron transfer and the twisted form from combination with phosphate. Exposure of the twisted form to ADP causes it to be discharged to the relaxed state with formation of ATP.

The changes in shape are real. There is no theoretical reason that mechanical energy could not be transduced into chemical energy by a sort of reversal of the events of muscular contraction. There also is no evidence that such a transduction ever occurs.

REGULATION OF OXYGEN
CONSUMPTION BY ADP

When electron transfer is coupled to phosphorylation, *oxygen cannot be consumed unless ADP and P_i are available.* Of these two compounds, it turns out that ADP has the lower concentration in most tissues, and is frequently the limiting factor in oxidative phosphorylation.

We shall develop the theme repeatedly and in greater detail, but it is helpful to understand it early. Consider the skeletal muscles as an example. The actin-myosin complex doesn't hydrolyze ATP while the muscle stays relaxed—at least, it doesn't anywhere near the rate seen during contraction. Under these conditions, most of the adenine nucleotides remain in the form of ATP and there is little ADP available for oxidative phosphorylation. Therefore, few electrons can be transferred to oxygen; few substrates are oxidized and little oxygen is consumed.

Here we have an important self-regulatory effect in metabolism: *The rate of oxidation of substrates tends to adjust automatically to the rate of energy demand.*

THE EFFICIENCY OF OXIDATIVE
PHOSPHORYLATION

Assessing the biological economy requires some judgment about the efficiency of high-energy phosphate formation. Has evolution been prodigal? Could there be the formation of more than three moles of high-energy phosphate during the oxidation of one mole of NADH? There have been various suggestions that this is indeed the case. We can make an estimate of the efficiency from the known free energies for the overall reactions, but first ought to discuss free energy in oxidation-reduction reactions.

Oxidation-Reduction Potential and Free Energy

The free energies for oxidation-reduction reactions are frequently expressed in terms of the electrical potential that can be generated in an electrical cell. We shall not use these expressions directly in our thinking, but the cell potential is the basis for a number of laboratory measurements, and it is helpful to have the principle in mind for interpreting these numbers when they are encountered.

An electrical cell consists of two electrodes, with an oxidation occurring at one and a reduction at the other. It is formally convenient to separate the reactions occurring at each electrode and express their potentials individually, although this cannot be done in an absolute way. For purposes of comparison, the potential of an electrode at which electrons are being stripped from hydrogen gas to form H^+, or are being added to H^+ to form H_2, is taken as zero. Therefore, when one sees a value of 0.320 volts for the reaction:

$$NADH \longrightarrow NAD^+ + H^+ + 2e$$

it is really the potential for a cell in which the overall reaction is:

$$NADH + H^+ \longrightarrow NAD^+ + H_2$$

and the other electrode reaction is:

$$2H^+ + 2e \longrightarrow H_2$$

Cell potentials can be added, so the potential for the reaction between NADH and oxygen can be obtained from the individual electrode potentials:

$$
\begin{array}{ll}
NADH \longrightarrow NAD^+ + H^+ + 2e & E^{0'} = 0.320 \text{ volts} \\
\tfrac{1}{2}O_2 + 2H^+ + 2e \longrightarrow H_2O & E^{0'} = 0.816 \text{ volts} \\
\hline
NADH + \tfrac{1}{2}O_2 + H^+ \longrightarrow NAD^+ + H_2O & E^{0'} = 1.136 \text{ volts}
\end{array}
$$

The potentials we are listing are standard potentials, in which the pH is fixed at a specified value and all other reactants are at 1 molar concentration. (The reverse reactions have identical negative standard potentials.) The sum of the potentials is related to the change in standard free energy by the equation:

$$\Delta G^{0'} = -nFE^{0'}$$

in which n = number of electrons transferred in the reaction; and F = Faraday constant = 23,060 calories per mole per volt.

Therefore, the standard free energy change for the oxidation of NADH by O_2 at pH 7 and 25°C is:

$$-2 \times 23,060 \times 1.136 = 52,400 \text{ cal mole}^{-1}$$

The Equilibrium of Oxidative Phosphorylation

The oxidation of alcohols is a typical reaction for the generation of high-energy phosphate. Since NAD is used as the primary oxidant, 3 moles of ATP are formed per mole of alcohol consumed, owing to the later passage of electrons from NADH to oxygen through the complete electron transport chain. The malate dehydrogenase reaction was mentioned earlier as a typical example, and when this reaction is coupled to the electron transport scheme in mitochondria, the overall oxidation-reduction becomes:

$$malate^{2-} + \tfrac{1}{2}O_2 \longrightarrow oxaloacetate^{2-} + H_2O$$
$$\Delta G^{0'} \sim -46,000 \text{ cal mole}^{-1} \text{ (pH 7, 37°C)}$$

Since the standard free energy of formation of ATP from ADP and P_i is only 8600 calories per mole, it might appear that at least 5 moles of high-energy phosphate could be made per mole of malate oxidized, but we have already learned that standard free energies cannot be used for this kind of reasoning when there are molarity dimensions in equilibrium constants, and there must be a calculation of the actual equilibrium conditions.

The calculation is laborious, but the result is important, and let us rapidly skim through the method. Essentially, the problem is one of finding the possible P : O ratio, that is, how many moles of high-energy phosphate can be formed per atom of oxygen consumed. Let us call this ratio x, and the stoichiometry of oxidative phosphorylation for the oxidation of malate will then be:

$$malate^{2-} + xADP^{3-} + xP_i^{2-} + \tfrac{1}{2}O_2 + xH^+ \longrightarrow$$
$$oxaloacetate^{2-} + xATP^{4-} + (x+1)H_2O; \Delta G^{0'} = 8600x - 46,000 \text{ cal mole}^{-1}$$

The corresponding equilibrium constant is:

$$K_{eq} = \left(\frac{[oxaloacetate]}{[malate]}\right) \times \left(\frac{[ATP]}{[ADP]}\right)^x \times \left(\frac{1}{P_i}\right)^x \times \left(\frac{1}{[O_2]}\right)$$

How to tackle this unwieldy equation? An easy way is to assume that x is 3, 4, or 5, and compute the resultant ratio of oxaloacetate to malate concentrations, using known concentrations of the phosphate compounds in tissues. The concentration of oxygen in mitochondria is unknown, but is probably near 1/400 atmosphere partial pressure in a moderately working skeletal muscle. Going through all of this we arrive at these results:

(basis: $[ATP]/[ADP] = 2$; $[P_i] = 0.01$ M; $[O_2] = 0.0025$ atmospheres)

x	$\Delta G^{0'}$ $cal\ mole^{-1}$	K_{eq}	$\dfrac{[oxaloacetate]}{[malate]}$
3	$-20,200$	1.7×10^{14}	1×10^6
4	$-11,600$	1.5×10^8	5×10^{-3}
5	$-3,000$	1.3×10^2	2×10^{-11}

What an astounding result! Malate can be converted to oxaloacetate almost quantitatively and generate 3 moles of high-energy phosphate per mole of malate oxidized, but only 0.005 of the malate could be oxidized if a single additional mole of high-energy phosphate were formed; the generation of 5 moles would be quite impossible.

We gain from this the idea that evolution has not been at all prodigal, that the P : O ratio of 3 for malate oxidation is the maximum that could be attained at the actual range of concentration of intermediates within the cell. The reaction is efficient, and there is a margin for variation of concentrations. The reaction would still proceed if the $[ATP]/[ADP]$ ratio built up to 10 or higher, rather than the value of 2 actually observed. (This ratio has a critical effect on the position of equilibrium because it is raised to the power of the number of moles of ATP created.)

Anyone who studies the literature carefully will soon realize that the values for standard free energy changes are more shaky than we have been implying. As an example of the uncertainty, the phosphate compounds can form chelates of different free energy content with Mg^{2+}, and excess magnesium ion causes a drop of nearly 1000 calories per mole in the standard free energy of ATP formation. The actual degree of chelation within tissues is difficult to assess because there are so many other ligands competing for the metal, but it is proper that some skepticism ought to be created about the argument given above. However, all is well, because it turns out that an error of 2000 calories per mole in the standard free energy change does not change the picture we developed. Calculation on that basis would show a P : O ratio of 4 to be barely possible, but the reaction would be so close to equilibrium at tissue concentrations of the intermediates that even a slight rise in ATP concentration would bring metabolism to a fatal halt.

INHIBITORS OF OXIDATIVE PHOSPHORYLATION

What is the effect of any compound that interferes with oxidative phosphorylation? It is surely lethal if the interference is severe, since ATP is required for so many vital functions and oxidative phosphorylation supplies most of the ATP. We might expect such compounds to have their major application outside of the laboratory as pesticides for killing animals. We might also expect them to be toxic to all animals, if they can enter the tissues. This is indeed the case. However, some of the compounds have important uses inside the laboratory as experimental tools. Let us consider some examples.

Inhibitors of Electron Transfer

CYANIDE.   The cyanides are among the poisons better known by the general public. They are not extraordinarily potent (the minimum lethal dose for humans is estimated to be of the order of 1 to 3 millimoles), but the small HCN molecule rapidly enters tissues, so that a sufficient quantity may be lethal within a few minutes. It is this quick effect, forestalling effective countermeasures in the absence of advance preparation, that has gained the cyanides so much respect, and has led to their use as pesticides against such diverse enemies as rats and insects in ships, moles in lawns, and murderers in some states.

His lawyer probably doesn't realize it, but the felon executed in a gas chamber is having his mitochondrial electron transport blocked at its terminal reaction. The cyanide ion, CN^-, combines with cytochrome oxidase, perhaps through bonds to the cuprous ion associated with the partially reduced enzyme. (Cyanide also combines with iron porphyrins, especially in the ferric state, but with much less affinity than it has for cytochrome oxidase.) The consequence is the cessation of transfer of electrons to oxygen. The previous electron carriers in the chain accumulate in their reduced state, and the generation of high-energy phosphate ceases.

The effect of cyanide is as fundamental as deprivation of oxygen, and like the latter, causes rapid damage to the brain. Indeed, there have been clinical trials of sub-lethal doses of cyanide as a means of causing corrective disturbances in the brains of psychotics, much as with electric or insulin shocks used for the same purpose. No advantage appeared to counterbalance the disadvantage of the dangerously small margin of safety in dosage.

Hydrogen sulfide will also combine with cytochrome oxidase, and few realize that it is as toxic as HCN. Its bad odor gives more warning, but there have been fatalities from only a few inspirations at high concentrations of the gas. (In the test tube, 0.1 mM sulfide inhibits cytochrome oxidase more than does 0.3 mM cyanide—96 per cent against 90 per cent.)

ANTIMYCINS.   The antimycins are antibiotics produced by species of *Streptomyces*. They strongly associate with mitochondria and block the passage of electrons from cytochrome b to cytochrome c_1; 0.07 micromole per gram of mitochondrial protein is effective. As laboratory tools, they permit experimental distinction between events in the earlier and later parts of the electron transfer chain. Since they kill the host as well as invaders, they are not therapeutic agents.

ROTENONE.   Rotenone is a compound extracted from the roots of tropical plants (*Derris elliptica, Lonchocarpus nicou*) that complexes avidly with NADH dehydrogenase; only 30 nanomoles per gram of mitochondrial protein are effective. It is a valuable tool for distinguishing routes of electron flow beginning with NAD-coupled dehydrogenases from those beginning with flavoproteins, since it does not affect the latter. The results of its effect on oxidative metabolism have long been exploited by primitive people. Rotenone is relatively non-toxic to mammals because it is absorbed poorly, although exposure of the lungs to the dust is a little more dangerous. However, the compound readily passes into the gills of fish and the breathing tubes of insects, and is intensely toxic to these animals. Fish-eaters apply preparations of the appropriate plant roots to ponds, collect the floating fish whose mitochondrial NAD remains reduced, and eat them with impunity. Even today, rotenone is in favor as a relatively safe insecticide, with low toxicity to land vertebrates and a short lifetime.

Inhibitors of Oxidative Phosphorylation

OLIGOMYCINS.   These antibiotics from various *Streptomyces* inhibit the transfer of high-energy phosphate to ADP. Therefore, they also inhibit electron transfers coupled to phosphorylation, but have no effect on oxidation-reduction reactions that are not coupled. They are widely used as experimental tools for discriminating between the two kinds of reactions.

ATRACTYLOSIDE.   This compound from *Atractylis gummifera*, a plant native to Italy, attracted renewed attention when several children were poisoned by eating rhizomes of the plant, with three deaths. Investigation showed that atractyloside blocks oxidative phosphorylation by preventing the passage of ADP into mitochondria, so that the extra-mitochondrial store of ATP cannot be renewed. It is used as a tool for separating intra-and extramitochondrial changes.

2,4-DINITROPHENOL.   Dinitrophenol has the effect of *uncoupling* oxidation from phosphorylation, with an effective concentration of 10 micromolar. It causes the spontaneous breakdown of the high-energy phosphate intermediate, thereby creating the effect of an ATPase. Since ADP is no longer required for the removal of the intermediate, the rate of electron transfer is not regulated by the concentration of ADP. The rate of oxygen consumption with and without dinitrophenol is an index of the degree of respiratory control, that is, the extent to which oxygen consumption is limited by the supply of ADP.

The compound has the same effect on the whole body, causing an increase in the *metabolic rate,* which is a term for the rate of oxygen consumption at rest. It once had a vogue as a means of losing weight through its uncoupling effect. Since oxidations produce less ATP in the presence of dinitrophenol, more substrates are consumed to maintain normal ATP concentrations. However, its effects on tissue metabolism were shortly shown to have dangerous consequences and it was abandoned as a proper agent for remission of obesity.

Further reading

Lehninger, A. L.: *The Mitochondrion.* W. A. Benjamin (1964). A readable survey.

Tager, J. M., S. Papa, E. Quagliariello, and E. C. Slater, eds.: *Regulation of Metabolic Processes in Mitochondria.* Elsevier (1966). This contains valuable technical discussions of many aspects of oxidative phosphorylation.

Mitchell, P.: *Translocations through Natural Membranes.* Adv. Enzymology, *29:* 33 (1967).

Estabrook, R. W., and M. E. Pullman, eds.: *Oxidation and Phosphorylation.* In Colwick, S. P., and N. O. Kaplan, eds.: *Methods in Enzymology,* Vol. 10. Academic Press (1967).

Pullman, M. E., and G. Schatz: *Mitochondrial Oxidations and Energy Coupling.* Ann. Rev. Biochem., *36:* 539 (1967).

Sanadi, D.: *Energy-Linked Reactions in Mitochondria.* Ann. Rev. Biochem., *34:* 21 (1965).

Boyer, P. D., H. Lardy, and K. Myrbäck: *The Enzymes,* 2nd ed. Academic Press (1960). This multivolume work includes detailed articles on most of the coenzymes and enzymes involved in oxidative phosphorylation.

Green, D. E., and D. H. MacLennan: *The Mitochondrial System of Enzymes.* In Greenberg, D., ed.: *Metabolic Pathways,* 3rd ed., Vol. 1 (p. 48). Academic Press (1967).

The following papers from the voluminous literature on the subject illustrate current lines of investigation:

Yamashita, S., and E. Racker: *Reconstitution of the Mitochondrial Oxidation Chain from Individual Components, I, II.* J. Biol. Chem., *243:* 2446 (1968) and *244:* 1220 (1969).

Wharton, D. C., and Q. H. Gibson: *Studies of the Oxygenated Compound of Cytochrome Oxidase.* J. Biol. Chem., *243:* 702 (1968).

Chan, T., and K. A. Schellenberg: *Studies on the Presence and Role of Tryptophan in Pig Heart Mitochondrial Malate Dehydrogenase.* J. Biol. Chem., *243:* 6284 (1968).

Sanadi, D. R., K. W. Lam, and C. K. R. Kurup: *The Role of Factor B in the Energy Transfer Reaction of Oxidative Phosphorylation.* Proc. Natl. Acad. Sci., *61:* 277 (1968). Includes discussion of many proposed factors.

Bulos, B., and E. Racker: *Partial Resolution of the Enzymes Catalyzing Oxidative Phosphorylation, XVII, XVIII.* J. Biol. Chem., *243:* 3891 and 3901 (1968).

Harris, A., et al.: *The Conformational Basis of Energy Conservation in Membrane Systems, I, II.* Proc. Natl. Acad. Sci., *59:* 624 and 830 (1968).

Lowenstein, J. M., and B. Chance: *The Effect of Hydrogen Ions on the Control of Mitochondrial Respiration.* J. Biol. Chem., *243:* 3940 (1968). Evidence against Mitchell hypothesis.

Eisenhardt, R. H., and O. Rosenthal: *Studies on Energy Transfer in Oxidative Phosphorylation, III.* Biochemistry, *7:* 1327 (1968). Evidence for high-energy intermediate.

Winkler, H. H., F. L. Bygrave, and A. L. Lehninger: *Characterization of the Atractyloside-sensitive Adenine Nucleotide Transport System in Rat Liver Mitochondria.* J. Biol. Chem., *243:* 20 (1968).

THE
OXIDATION
OF
ACETYL
COENZYME A

CHAPTER 11

ARGUMENT

Most of the carbon atoms of ingested food are eventually excreted as CO_2. Some two-thirds of the atoms appear in acetyl groups during intermediate stages of the conversion. The oxidation of acetyl groups is therefore one of the major metabolic processes, and more than two-thirds of the ATP used in the body is generated as a result of the transfer of electrons from these groups to O_2 in mitochondria. During metabolism, the acetyl groups are linked as thiol esters to coenzyme A, which is a mercaptan used for transfer of acyl groups in general.

The oxidation of acetyl coenzyme A occurs in a series of reactions known as the citric acid cycle, which begins by a condensation of the acetyl group with oxaloacetate to form citrate and ends with the regeneration of oxaloacetate. Oxaloacetate is effectively acting as a carrier for the two carbons of the acetyl group in the cyclical process, and there is no net production or consumption of the compound as a result. The citric acid cycle includes the production of two molecules of CO_2 and the transfer of four pairs of electrons. Three pairs of electrons are removed by forming NADH and the remaining pair is transferred directly to coenzyme Q by a flavoprotein. Each of the four pairs then moves through the mitochondrial electron transport chain to an atom of oxygen. Three moles of ATP are generated for each mole of NADH and two moles for each mole of reduced coenzyme Q, making a total of 11 moles of ATP produced by oxidative phosphorylation per mole of acetyl group oxidized. There is an additional mole of high-energy phosphate generated in the form of GTP by the reaction of an energy-rich thiol ester involved in the cycle, that is, by a substrate-level phosphorylation rather than oxidative phosphorylation.

The citric acid cycle is an arrangement of types of reactions frequently used in other metabolic processes. These include the oxidation of a hydrocarbon chain by a flavoprotein, the hydration of the resultant unsaturated compound to form an alcohol, and the oxidation of the alcohol to a carbonyl compound. Also involved are the oxidative decarboxylation of a 2-ketocarboxylate, which produces a coenzyme A ester, and the oxidative decarboxylation of a 3-hydroxycarboxylate, which produces a ketone.

Approximately two-thirds of the carbon atoms in the compounds eaten by an animal are transformed into acetyl groups at some point in the processes by which the atoms are eventually converted to CO_2. The acetyl group is attached to a nucleotide, coenzyme A, which has the general function of transporting acyl groups within cells. The simple two-carbon acetyl group, in the form of acetyl coenzyme A, is critically important in the chemical economy of cells. (Fig. 11-1). It is an intermediate common to the metabolism of nearly all kinds of biological compounds. Mitochondria generate more than two-thirds of the total high-energy phosphate used in animals by oxidizing the acetyl group to CO_2 and H_2O. Furthermore, the group is the source of nearly all of the carbon atoms found in fats and related lipids.

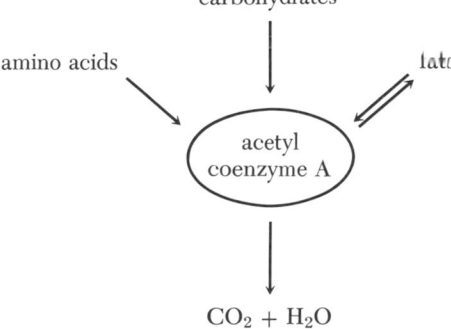

Figure 11-1 Acetyl coenzyme A is formed from carbohydrates, fats, and amino acids. Whatever the source, a large fraction of it is oxidized to CO_2 and H_2O, but any excess may be used to form fats for storage.

ACETYL COENZYME A: THE COMPOUND

The Structure of Coenzyme A

Coenzyme A contains one sulfhydryl group that is the reactive part of the coenzyme, combining with acyl groups to form thiol esters. It is therefore commonly abbreviated as CoA-SH. Using this shorthand description, acetyl coenzyme A is:

$$CH_3-\overset{\overset{\displaystyle O}{\|}}{C}-S-CoA$$

and its simple hydrolysis in a test tube is written as:

1. $$CH_3-\overset{\overset{\displaystyle O}{\|}}{C}-S-CoA + H_2O \longrightarrow CH_3-COO^- + H^+ + CoA-SH.$$

This handy way of abbreviating the structure properly emphasizes the acyl group being transported and its linkage as a thiol ester, but coenzyme A is actually quite complex. The reactive sulfhydryl group is at one end of a molecule comprising an adenine nucleotide and the three other compounds named in Figure 11-2. We have already noted that coenzymes frequently contain unusual structures—unusual in the sense that animals have lost the ability to form them, and they must be supplied in the diet. Coenzyme A has such a structure, contained in the vitamin, D-*pantothenic acid.*

Surprisingly, the vitamin does not in this case contain the reactive group of the coenzyme. The unusual branched chain structure of pantothenate may act as a marker primarily for those enzymes fabricating the complete coenzyme from its components rather than for the enzymes with which it acts as a coenzyme. Some of the acyl transferring enzymes can use synthetic β-alanylaminoethylthiol peptides lacking the pantothenate structure. Be this as it may, animals lacking pantothenate die, and it is no consolation to know some of the enzymes could have functioned with another thiol if it had been at hand. Fortunately, a natural deprivation of pantothenate is very rare because coenzyme A is present in all living cells, and therefore in the natural foods of all animals.

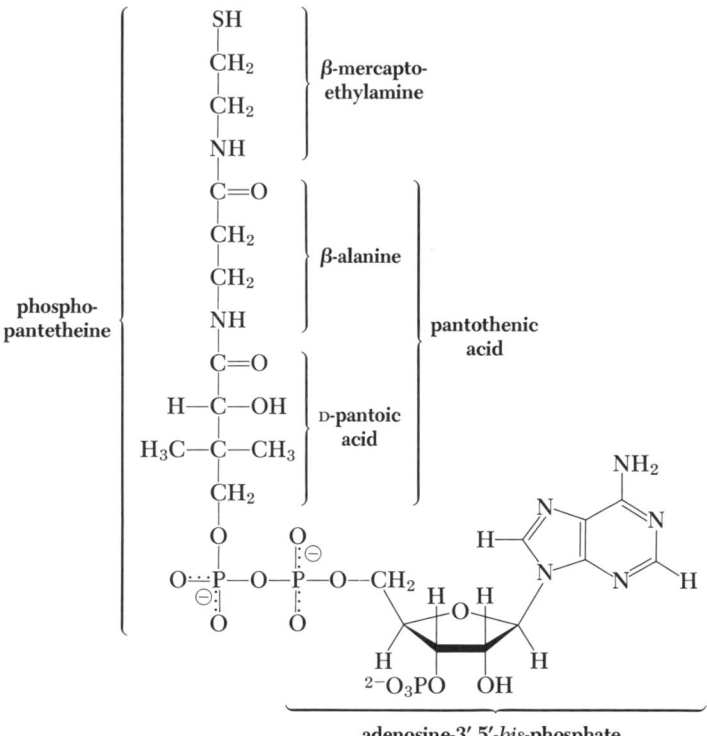

adenosine-3′,5′-*bis*-phosphate

Figure 11-2   The structure of coenzyme A and the names of some of its constituents.

Energetics of Thiol Esters

Thiol esters such as acetyl coenzyme A are high-energy compounds, with the standard free energy for hydrolysis according to equation (1) being near -7700 calories per mole at pH 7 and 39°C. This corresponds to a K'_{eq} of 2.5×10^5, so acetyl coenzyme A won't form simply because acetate and coenzyme A are present in the same solution.

In other words, acetyl coenzyme A must be made by the same sort of general process used for the formation of peptides—coupling the formation to some reaction that would by itself have a highly negative standard free energy, thereby counterbalancing the highly positive standard free energy for the combination of acetate and coenzyme A. An example will show the principle.

Acetic acid occurs in the diet; vinegar is a familiar source. Ingested acetic acid is brought into the mainstream of intracellular metabolism by converting it to acetyl coenzyme A. The conversion is accomplished by an enzyme that catalyzes the simultaneous conversion of ATP to AMP and PP_i. The enzyme is called *acetate thiokinase* because it "activates" acetate in a metabolic sense.

The inorganic pyrophosphate produced by the reaction is in turn hydrolyzed by the ubiquitous inorganic pyrophosphatase with an additional liberation of free energy, so the formation of acetyl coenzyme A is another process in which there is an effective cleavage of two high-energy phosphate bonds during the synthesis of a compound because of the cooperative action of the primary enzyme catalyzing the synthesis and the pyrophosphatase acting on PP_i. The overall reaction is:

2. $CH_3COO^- + CoA—SH + ATP^{4-} + H_2O \longrightarrow$

$$CH_3—\overset{\overset{\displaystyle O}{\|}}{C}—S—CoA + AMP^{2-} + 2P_i^{2-} + H^+$$

with $\Delta G^{0\prime} = -8700$ calories per mole.

The combined reaction is energetically feasible, and it is convenient to think about it as a combination of the energetically favorable hydrolysis of ATP to AMP and 2 P_i with the energetically unfavorable combination of acetate and coenzyme A. The common intermediate between the two processes is acetyl AMP on the enzyme, analogous to the aminoacyl AMP formed in amino acid activation (p. 37).

We shall later encounter several more important thiokinases that catalyze similar reactions, but with different carboxylate substrates. An important thing to bear in mind with all of them is that the thiol ester has an energy-rich bond, and the coenzyme A derivatives are formed to participate in reactions that would not be possible without the favorable equilibrium created when the thiol ester is later cleaved.

THE ROUTE OF OXIDATION OF ACETYL COENZYME A

General Description of the Citric Acid Cycle

Acetyl coenzyme A is oxidized in a sequence of reactions known variously as the citric acid cycle, the tricarboxylic acid cycle, or the Krebs cycle (after Sir Hans A. Krebs, its discoverer). The cycle begins with a condensation between acetyl coenzyme A and oxaloacetate to form the six-carbon tricarboxylate, *citrate*. Citrate undergoes a sequence of transformations and oxidations in which two moles of CO_2 are produced, with the remaining four carbons left in the form of oxaloacetate, ready to condense again with acetyl coenzyme A.

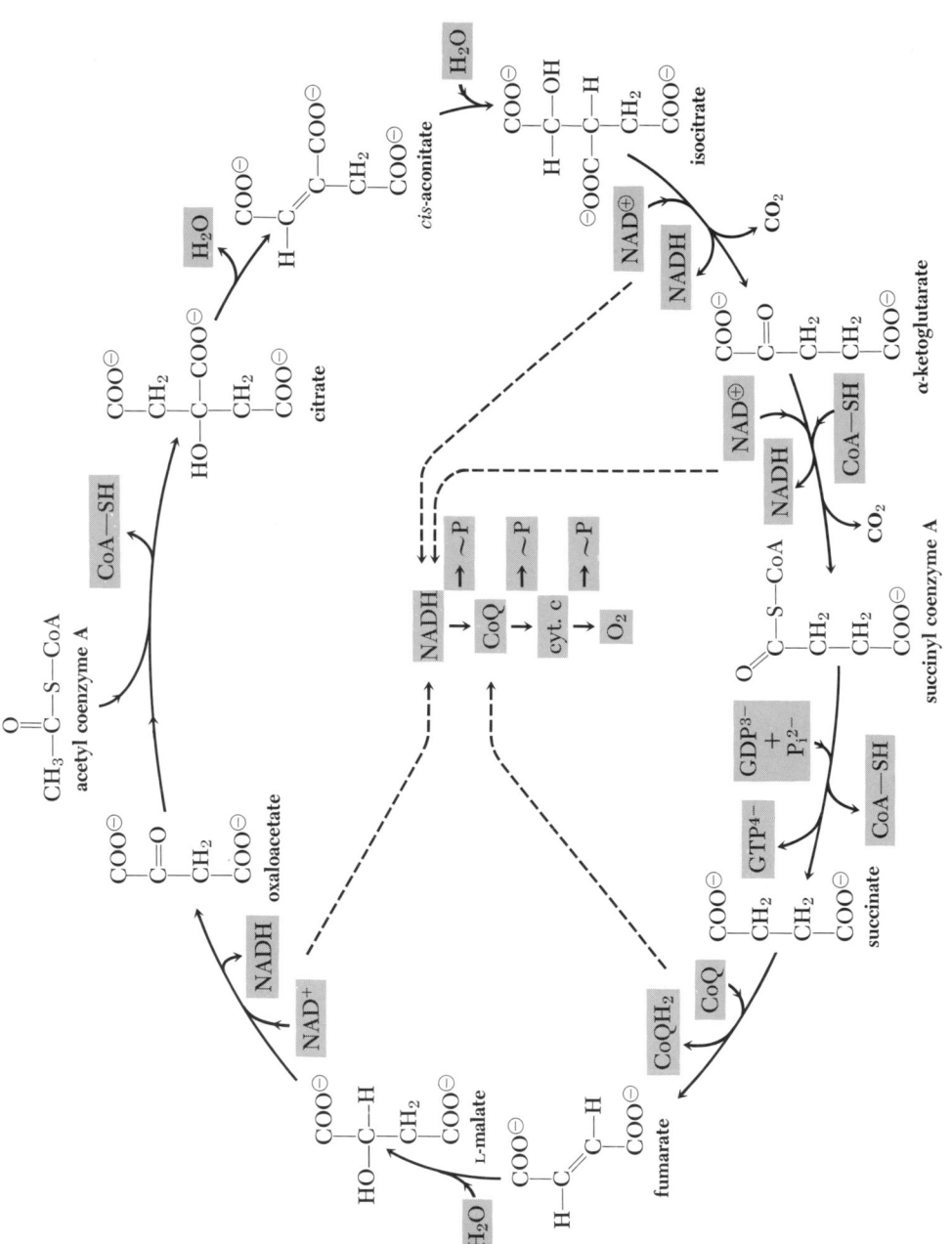

Figure 11-3   The citric acid cycle. See the text for details.

The net effect of the citric acid cycle is the complete oxidation of the acetyl group, with the four-carbon skeleton of oxaloacetate acting as a carrier. Considering only this sequence of reactions, oxaloacetate is behaving as a coenzyme because only a small amount of it can be used repeatedly in carrying out the oxidation of large quantities of acetyl groups. (Oxaloacetate is not classed among the coenzymes because it is also a substrate for other reactions in which it is consumed, and also because the same carbon skeleton is not used repeatedly, even in the citric acid cycle.)

What is the purpose of the citric acid cycle? Why have the enzymes catalyzing these reactions evolved in mitochondria rather than others catalyzing a direct attack on acetate itself? We don't know exactly, but we can make some likely guesses:

1. The enzymes catalyzing the reactions of the citric acid cycle have sharp specificity for the particular substrates found in the cycle, but the reactions are of *types* commonly used throughout other metabolic sequences, and the cycle may represent a specialized evolution from more general enzymes.

2. The reactions of the cycle permit the use of the ordinary oxidants, NAD and flavoproteins, as initial electron carriers, which can be reoxidized through the mitochondrial electron transport scheme discussed in the previous chapter, thereby using a pathway common to many metabolic processes for the generation of ATP and putting the citric acid cycle under the same metabolic control.

3. The intermediates of the cycle are also intermediates in several other metabolic processes, as we shall see later. The result is an integration of the metabolism of the acetyl group with these processes.

The Formation of Citrate

The organic chemist frequently makes use of reactions involving a condensation between a carbonyl group and an activated CH group, of which the classic example is an aldol condensation:

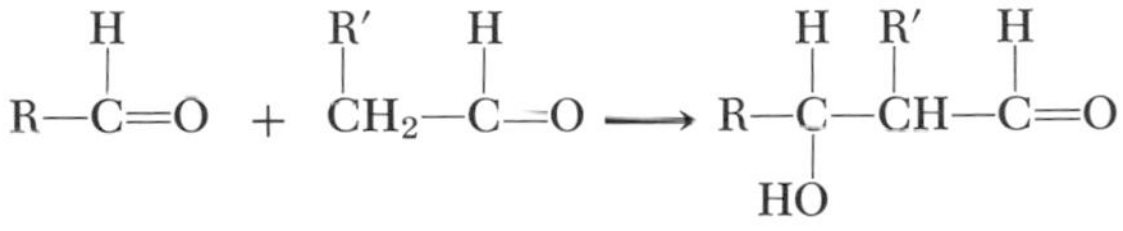

aldehydes   aldol

Such condensations are also common in biological systems, and the formation of citrate from acetyl coenzyme A and oxaloacetate is an example:

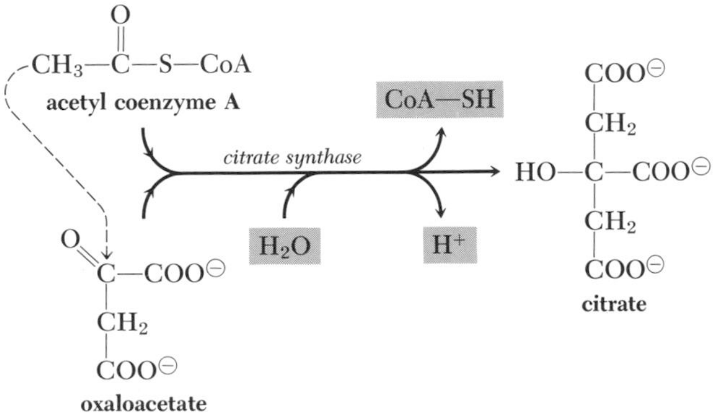

However, the formation of citrate also involves the cleavage of a thiol ester bond, and the resulting additional liberation of free energy makes the condensation essentially irreversible under intracellular conditions.

The Conversion of Citrate to α-Ketoglutarate

The initial pair of reactions in the citric acid cycle are a means of converting citrate into an alcohol that can be oxidized. The hydroxyl group of citrate is on a tertiary carbon, and could not be attacked without breaking a carbon bond, but rearrangement into the isomer, *isocitrate*, permits oxidation of the resultant secondary alcohol:

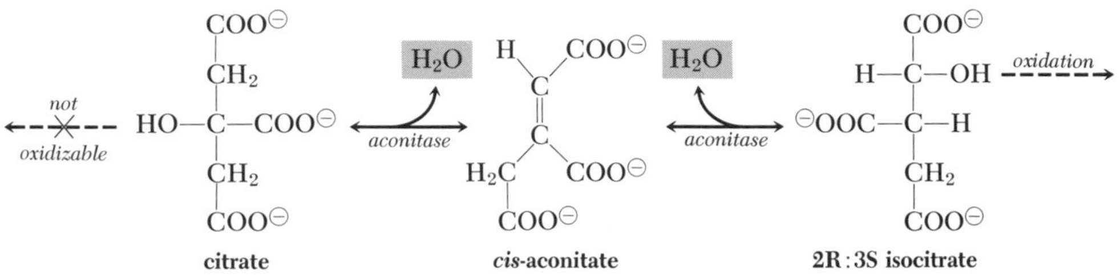

(2R:3S designates the particular isomer of isocitrate, which has two asymmetric carbon atoms. Stereoisomers and their nomenclature are reviewed in the Appendix, p. 697.)

The equilibrium between these three compounds is catalyzed by a single enzyme, *aconitase*. The hydroxyl group may be placed on either carbon when water is added across the double bond of aconitate. The enzyme will likewise catalyze either reverse reaction—the removal of water from citrate or isocitrate to form aconitate. The interconversion of alcohols and unsaturated compounds by addition or removal of water is a common type of biochemical reaction, and the standard free energy change is usually small enough so that the reaction can go in either direction to a biologically significant extent.

In this case, the equilibrium mixture contains about 91 per cent citrate, 3 per cent aconitate, and 6 per cent isocitrate, but the intramitochondrial concentration of citrate is greater than 1 mM, so that the equilibrium concentration of the other two compounds is of the order of 0.1 mM, which is a reasonable level for metabolic intermediates. There is nothing in aconitase itself directing the reaction through the sequence: citrate ⟶ aconitate ⟶ isocitrate. The enzyme simply accelerates the attainment of equilibrium between the three compounds. The fact that the sequence within mitochondria does go in the direction indicated is a result of the continual formation of citrate from the condensation of acetyl coenzyme A and oxaloacetate and the continual removal of isocitrate in the next reaction we shall consider.

Isocitrate is oxidized by NAD in a reaction catalyzed by a specific *isocitrate dehydrogenase*. If this was a simple alcohol dehydrogenase, the products of the reaction would be NADH and the corresponding ketone, oxalosuccinate:

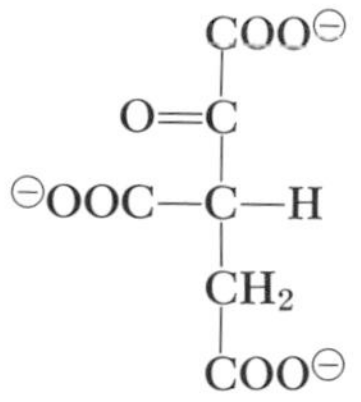

$$\begin{array}{c} COO^{\ominus} \\ | \\ O{=}C \\ | \\ {}^{\ominus}OOC{-}C{-}H \\ | \\ CH_2 \\ | \\ COO^{\ominus} \end{array}$$

oxalosuccinate

However, oxalosuccinate is never released from the enzyme. Notice that its carbonyl group occupies the α position relative to the top carboxylate group as drawn in the figure, but the carbonyl group is a carbon removed, or in the β position, relative to the middle carboxylate group. 3-Keto acids are known to decarboxylate readily in the test tube. That is, they lose CO_2 to form a ketone, and oxalosuccinate reacts in this way. The middle carboxylate group, which is the one conjugated with the carbonyl group, is lost, and the product is the five-carbon dicarboxylate, α-*ketoglutarate*.

Catalysis by the NAD-isocitrate dehydrogenase of mitochondria couples the oxidation and decarboxylation in a single reaction:

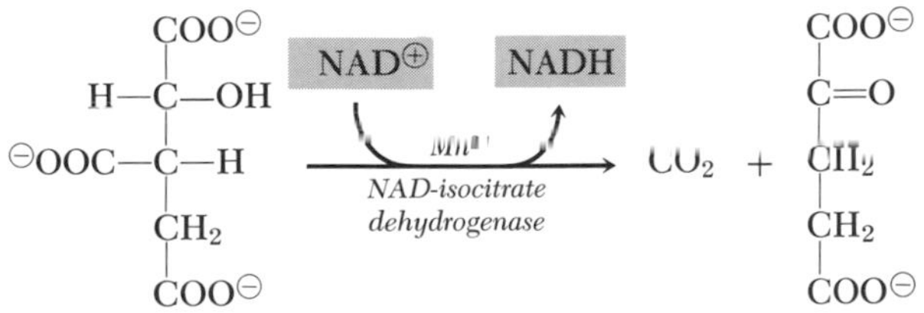

$$\begin{array}{c} COO^{\ominus} \\ | \\ H{-}C{-}OH \\ | \\ {}^{\ominus}OOC{-}C{-}H \\ | \\ CH_2 \\ | \\ COO^{\ominus} \end{array} \quad \xrightarrow[\substack{NAD\text{-}isocitrate \\ dehydrogenase}]{NAD^{\oplus} \qquad NADH} \quad CO_2 \;+\; \begin{array}{c} COO^{\ominus} \\ | \\ C{=}O \\ | \\ CH_2 \\ | \\ CH_2 \\ | \\ COO^{\ominus} \end{array}$$

2R:3S isocitrate α-ketoglutarate

Notice that *manganous ion* is required for the activity of this enzyme. This is not ordinarily true for alcohol dehydrogenases, and the metal evidently is involved in the decarboxylation process. (Manganese is another metal that must be present in trace amounts in the diet to maintain activity of enzymes for which it is a cofactor.)

At this point in the citric acid cycle, one mole each of acetyl coenzyme A, oxaloacetate, and NAD have been consumed with the production of one mole each of CO_2, α-ketoglutarate, and NADH. The CO_2 represents half of the total produced by the stoichiometric oxidation of the acetyl group, and the NADH represents one quarter of the necessary electron transfers in this oxidation.

The Oxidative Decarboxylation of α-Ketoglutarate

The oxidation of α-ketoglutarate involves three separate proteins aggregated into a complex within the inner membrane of mitochondria, and the entire complex constitutes α-*ketoglutarate dehydrogenase*. Complexes such as this are the next step in organization beyond a single protein. One protein may contain several peptide chains aggregated spontaneously into a stable molecule. Similarly, different proteins can be constructed so that they will spontaneously associate in fixed proportions to form complexes. Each of the three kinds of protein in the α-ketoglutarate dehydrogenase catalyzes a distinct step in the overall reaction, but the association of what might be distinct enzymes into a single complex makes it possible to lock the three reactions together, because the intermediate forms of the substrate are not released into solution.

A convenient formalism is to regard the oxidation of 2-ketocarboxylates as the sum of two organic reactions; one being a simple decarboxylation to produce CO_2 and an aldehyde:

$$R-\overset{\overset{\displaystyle O}{\|}}{C}-COO^- + H^+ \longrightarrow R-\overset{\overset{\displaystyle O}{\|}}{C}-H + CO_2 \qquad \Delta G^{0'} \sim -5{,}000 \text{ cal mole}^{-1}$$

and the other being the oxidation of the aldehyde:

$$R-\overset{\overset{\displaystyle O}{\|}}{C}-H + NAD^+ + H_2O \longrightarrow$$

$$R-COO^- + NADH + 2H^+ \qquad \Delta G^{0'} \sim -12{,}000 \text{ cal mole}^{-1}$$

These steps could, in theory, proceed independently, and do so in some microorganisms, but the two steps combined have a standard free energy release ample for the creation of a high-energy bond, and the use of an enzyme complex permits the efficient utilization of this chemical potential. The high-energy bond actually formed is a thiol ester with coenzyme A, and the overall reaction of α-ketoglutarate dehydrogenase becomes:

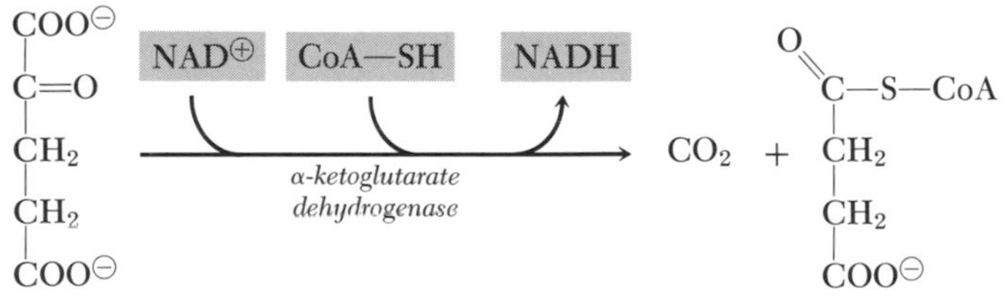

α-ketoglutarate · · · · · · · · · · succinyl coenzyme A

Such high-energy bond formation is quite independent of the high-energy phosphate generated by mitochondrial electron transport; it serves to tap the extra chemical potential of the oxidative decarboxylation of α-ketocarboxylates.

The immediately preceding isocitrate dehydrogenase reaction is also an oxidative decarboxylation, but of a 3-hydroxycarboxylate. The standard free energy change for that reaction is only $-3{,}500$ calories per mole, which is too little to support the formation of an extra high-energy bond.

Conversion of chemical potential in the reaction first involves decarboxylation of α-ketoglutarate without releasing the aldehyde (succinsemialdehydate), which is accomplished through the use of the coenzyme, *thiamine pyrophosphate*, in the *decarboxylase* component of the enzyme complex (Fig. 11-4). Thiamine is another of the compounds with an unusual structure whose synthesis has been deleted in the evolution of mammals, and is therefore a vitamin.

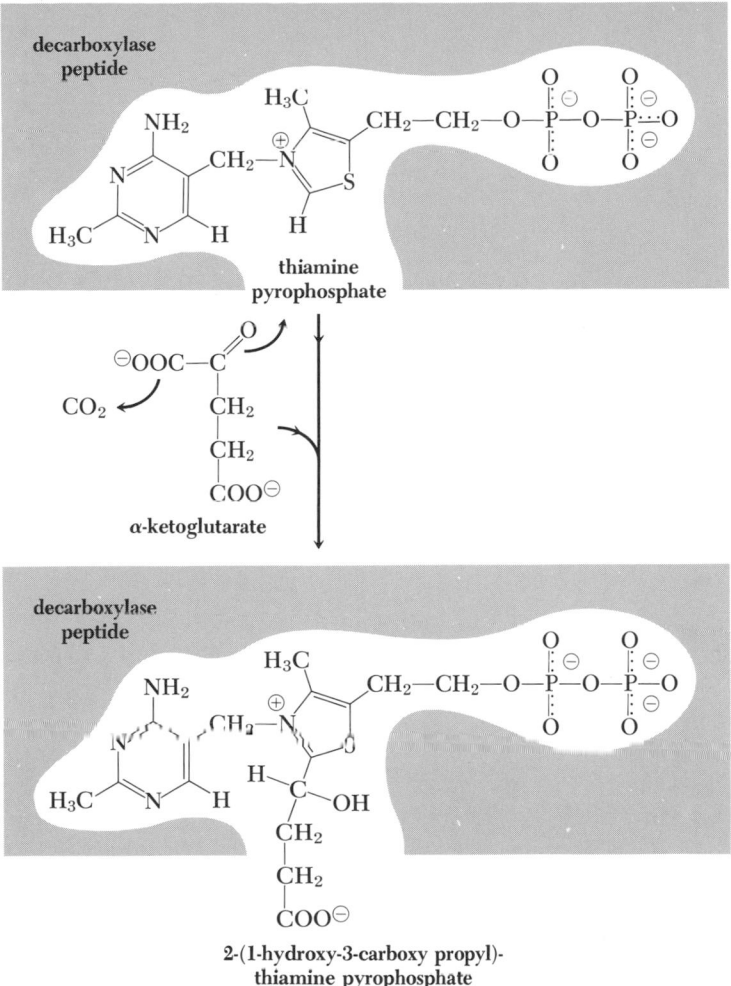

Figure 11-4   The decarboxylation of α-ketoglutarate by reaction with thiamine pyrophosphate, which is attached to catalytic peptides.

The decarboxylase component also catalyzes the transfer of the succinaldehydate group to the second protein of the complex, *transsuccinylase,* with a resultant oxidation to form a succinyl group (Fig. 11-5 A). (Some name the decarboxylase component as a dehydrogenase for this reason, but we reserve the latter name for the entire complex.) The oxidizing agent is a coenzyme containing a disulfide bond, *lipoate,* which is attached to a lysyl residue in the peptide chain of transsuccinylase. In effect, the disulfide oxidizes the aldehyde group with the simultaneous creation of a thiol ester bond. The transsuccinylase peptide then catalyzes transfer of the succinyl group from a thiol group of the dihydrocoenzyme to the thiol group of coenzyme A, thereby creating succinyl coenzyme A (Fig. 11-5 B).

The reaction is now complete except for the regeneration of the original disulfide bond in the lipoyl group. *Dihydrolipoate dehydrogenase,* which is a flavo-

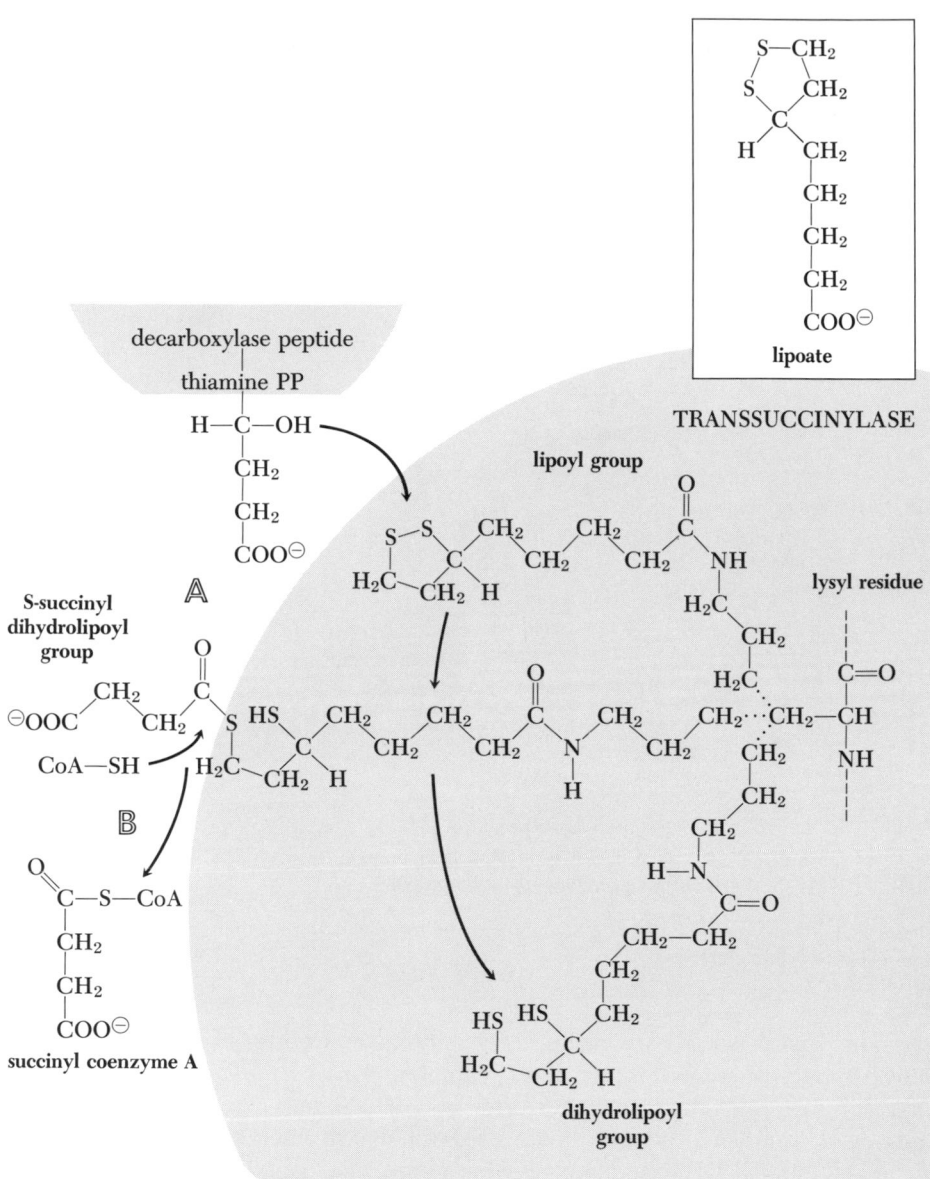

Figure 11-5 When a succinate semialdehyde group is transferred from thiamine pyrophosphate to a lipoyl group on transsuccinylase, it is oxidized to a succinyl group. The lipoyl group itself is the oxidizing agent. The succinyl group is then transferred to coenzyme A.

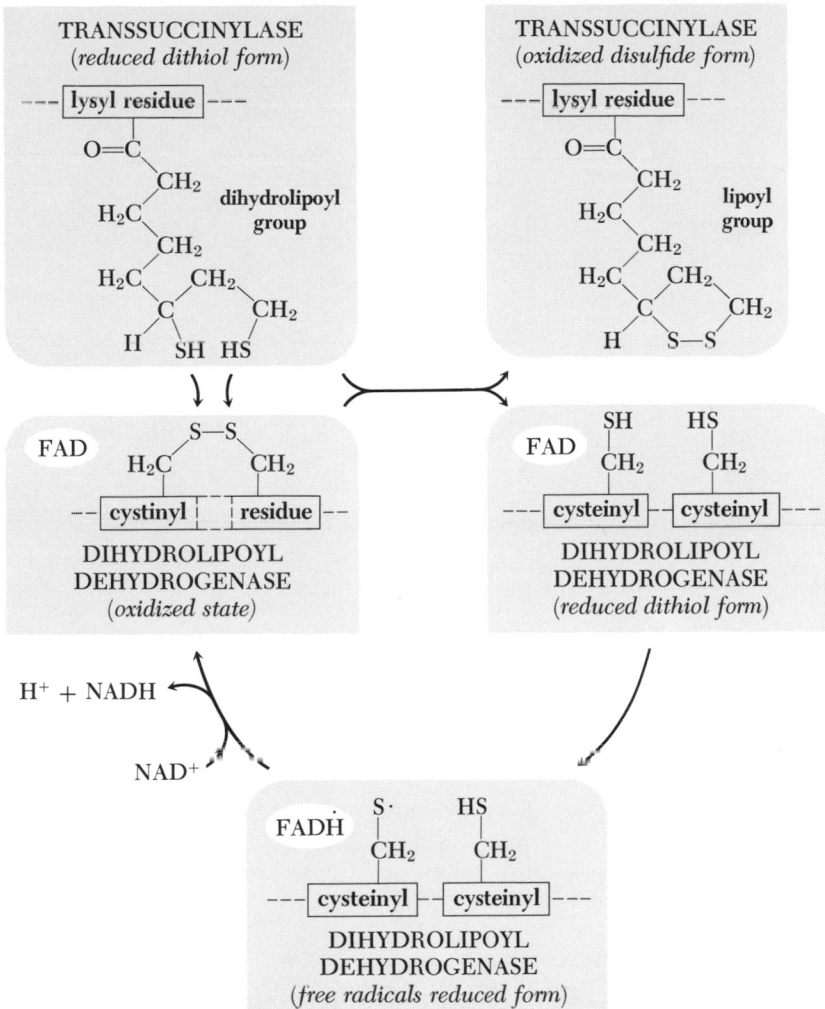

Figure 11-6 The final steps of the α-ketoglutarate dehydrogenase reaction involves the reconversion of the reduced lipoyl group to its oxidized form. The electrons are carried through both an FAD molecule and a disulfide group on the dihydrolipoyl dehydrogenase component of the enzyme complex. The final electron acceptor is NAD.

protein and the third enzyme of the complex, accomplishes this final step by transferring the electrons of the dihydrolipoyl group to NAD (Fig. 11-6).

The key structural feature making the enzyme complex possible evidently is the presence of the long side chain represented by the lipoyllysyl group, which can swing about on the surface of the transsuccinylase protein and successively come in contact with the decarboxylase, coenzyme A, and the dihydrolipoate dehydrogenase (Fig. 11-7). The complex has a definite architecture, with eight molecules of transsuccinylase occupying the corners of a cube, but the location of the molecules of decarboxylase and dehydrogenase in the cube has not been determined.

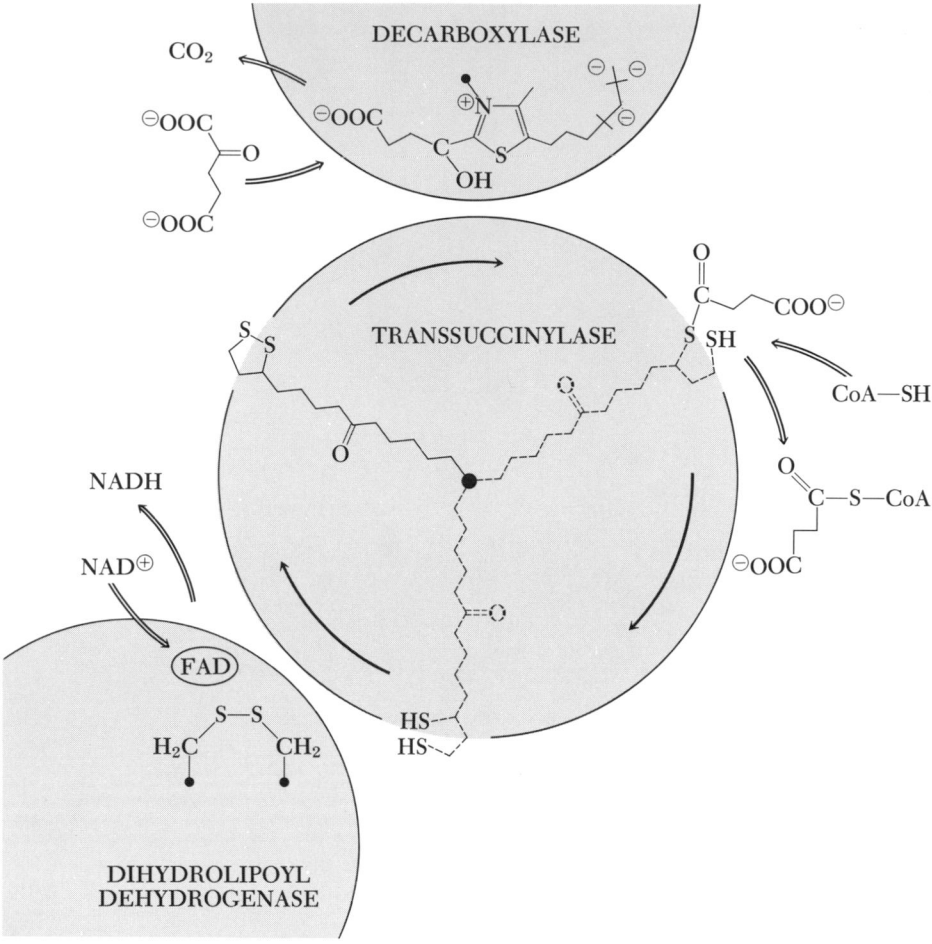

Figure 11-7   The α-ketoglutarate dehydrogenase reaction is believed to depend upon the wide area that can be covered by the long lipoyl-lysyl group (*center*) on the transsuccinylase peptides.

Succinyl Coenzyme A Synthetase

The oxidation of α-ketoglutarate produces the second, and final, mole of CO_2 released in the citric acid cycle. The remainder of the steps include two more oxidations, but first the potential energy of succinyl coenzyme A is recovered as a high-energy phosphate (Fig. 11-8). This reaction involves the cleavage of succinyl coenzyme A on the enzyme surface by P_i, forming an intermediate phosphorylated enzyme and liberating coenzyme A. The high-energy phosphate is transferred onto GDP, forming GTP. (The reaction may proceed *via* succinyl phosphate, as shown in the figure.)

Since the standard free energies of hydrolysis of succinyl coenzyme A and of GTP are nearly equal, the reaction is freely reversible, and the enzyme is called a *synthetase* (an enzyme catalyzing the formation of a compound at the expense

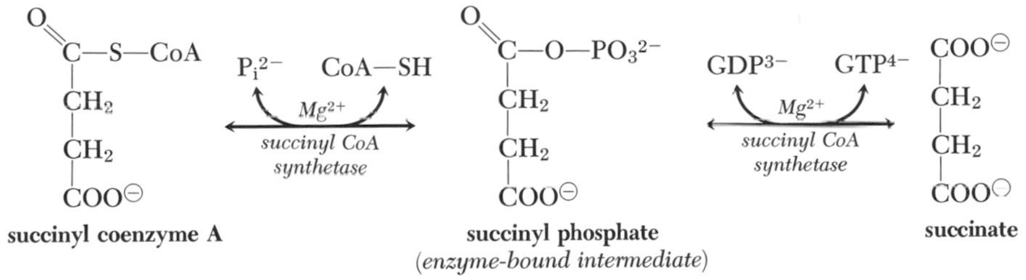

Figure 11-8 The reversible cleavage of succinyl coenzyme A with an accompanying phosphorylation of GDP takes place in at least two stages. The initial step may involve a phosphorolysis of the substrate to form succinyl phosphate, as shown. There also is the formation of a high-energy phosphorylated enzyme, not shown, which is the intermediate reacting with GDP.

of high-energy phosphate) to be consistent with other enzymes of the same type, even though the physiological direction of this particular example is toward succinate in the citric acid cycle.

Succinyl coenzyme A synthetase requires Mg^{2+} for activity. Most enzymes catalyzing the transfer of phosphate groups require this ion, presumably to bind the substrate to the active site of the enzyme.

The Oxidation of Succinate

Succinate is oxidized to *fumarate* by *succinate dehydrogenase,* which is a flavoprotein intimately associated with the electron transport system in the mitochondrial inner membrane. In discussing electron transport, we mentioned a direct transport of electrons from some substrates to coenzyme Q by flavoprotein dehydrogenases. Succinate dehydrogenase is an example of such a flavoprotein:

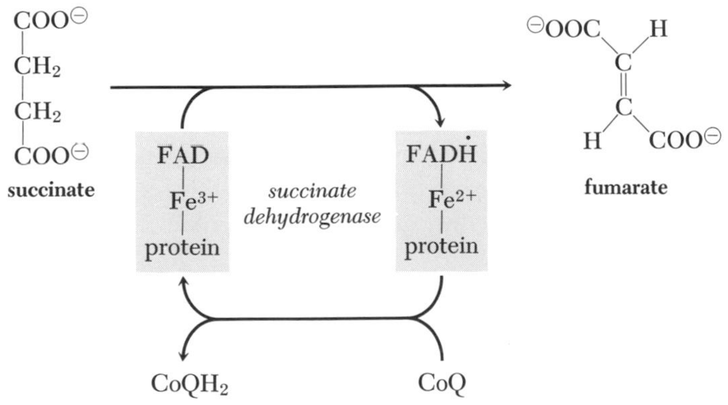

Since the introduction of electrons into the transport scheme at the level of coenzyme Q bypasses the first site of oxidative phosphorylation, the oxidation of succinate and the resultant passage of electrons to oxygen can only result in the formation of two high-energy phosphate bonds. Why doesn't succinate dehydrogenase utilize NAD, as do the isocitrate and α-ketoglutarate dehydrogenases, so as to generate three high-energy phosphate bonds in the electron transport scheme? The answer is that the standard free energy change for the oxidation of succinate is some 10,000 calories less per mole than it is for the oxidation of most alcohols.

The reaction could not support the generation of more than two high-energy phosphate bonds, and to make it work electrons must be fed into the common electron transport chain at the coenzyme Q level, rather than through NADH dehydrogenase. It is true that succinate dehydrogenase and NADH dehydrogenase are both flavoproteins, but NADH dehydrogenase is coupled to phosphorylation, and succinate dehydrogenase is not, so there is an inherent difference of -8600 calories per mole in the $-\Delta G^{0\prime}$ for the two routes of electron flow.

The Regeneration of Oxaloacetate

Fumarate is hydrated to form the hydroxycarboxylate, L-*malate*, which is then oxidized by NAD to form oxaloacetate, thereby replacing the oxaloacetate originally used at the beginning of the citric acid cycle:

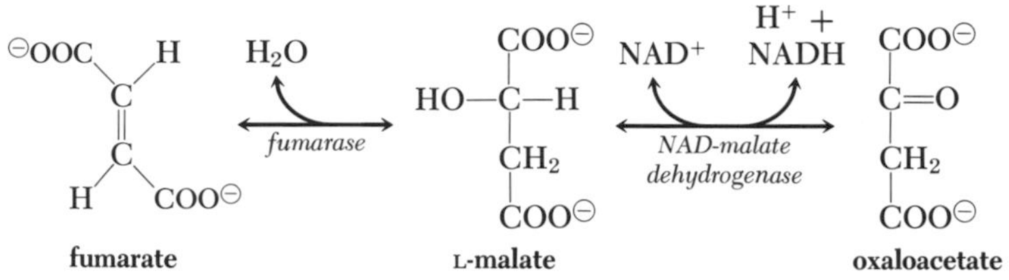

These reactions are not especially complex, and the important physiological fact is the transformation of the substrates as shown. However, let us digress to discuss some of the general structural implications illustrated by these reactions. The hydration catalyzed by fumarase is quite similar to that of aconitase, except that only one product, L-malate, can be produced by hydration of fumarate; fumarate is symmetrical with reference to the double bond, and adding water either way gives the same result.

There is a somewhat more tricky point in this hydration. Fumarate is a *trans* compound (maleate is the corresponding *cis* isomer), and hydrates to yield the L-enantiomorph. Aconitase acts on the *cis* isomer of its substrate, and the resultant 2R:3S isocitrate has its hydroxyl group in the D-configuration relative to the adjacent carboxylate group. Enzymatic addition of water to double bonds conjugated with a carbonyl or carboxylate group frequently, but not always, yields the D-alcohol relative to the closest carbonyl if the unsaturated compound is *cis* and the L-alcohol if it is *trans:*

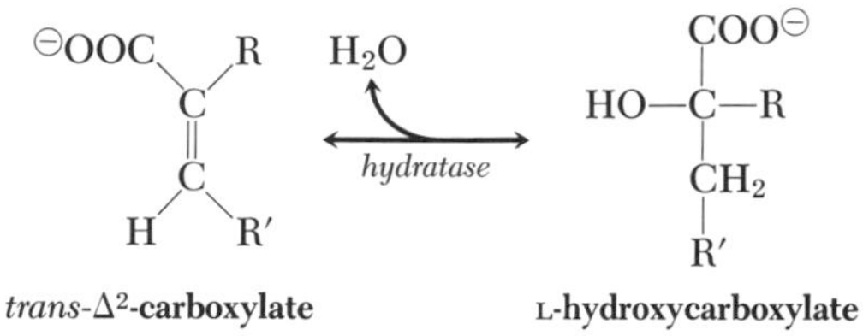

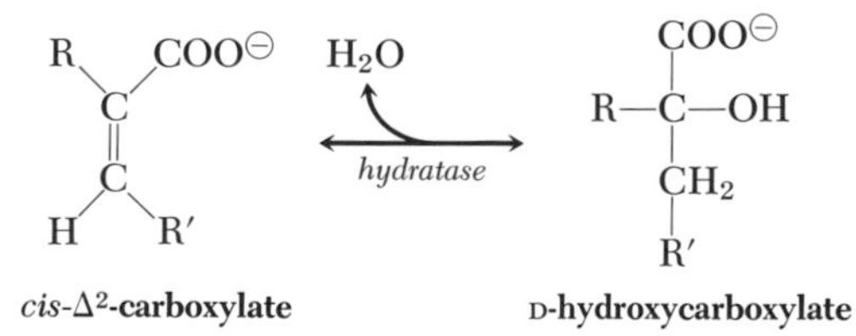

The NAD-malate dehydrogenase and isocitrate dehydrogenase act on quite similar substrates and comparison of the reactions shows an important point. Malate and isocitrate both have hydroxyl groups on a carbon that is α to one carboxylate group and β to another, We noted that β-keto carbonyl compounds are readily susceptible to decarboxylation, which indeed occurs with isocitrate dehydrogenase. The reaction catalyzed by NAD-malate dehydrogenase is an exception, even though its product, oxaloacetate, can be readily decarboxylated in the test tube. We shall see in later chapters that there are indeed other reactions in which oxaloacetate is decarboxylated, but it is imperative for the operation of the citric acid cycle that this does not always occur, and the separation of the oxidation of malate from the decarboxylation of oxaloacetate is the basis for important regulatory purposes that we shall explore later. No such purpose would be served by making oxalosuccinate a discrete intermediate in the oxidation of isocitrate, and the oxidation and decarboxylation are economically combined in the function of isocitrate dehydrogenase.

STOICHIOMETRY OF THE CITRIC ACID CYCLE

CO$_2$ Production and O$_2$ Consumption

If the citric acid cycle is really oxidizing the acetyl group of acetyl coenzyme A to CO_2 and H_2O, the sum of all of the equations for the individual reactions must add up to the formal stoichiometry:

$$CH_3{-}\overset{\overset{\displaystyle O}{\|}}{C}{-}S{-}CoA + 2O_2 \longrightarrow 2CO_2 + CoA{-}SH + H_2O$$

after any high-energy phosphate produced has been utilized and converted back to the starting materials. Is this the case? If we add all of the reactions in the cycle, we arrive at the sum:

$$CH_3{-}\overset{\overset{\displaystyle O}{\|}}{C}{-}S{-}CoA + 3NAD^+ + CoQ + GDP^{3-} + P_i^{2-} + 2H_2O \longrightarrow$$

$$2CO_2 + CoA{-}SH + 3NADH + CoQH_2 + 2H^+ + GTP^{4-}$$

Neglecting ATP production, electrons will be transferred eventually to oxygen according to the equation:

$$3\ NADH + 3H^+ + CoQH_2 + 2\ O_2 \longrightarrow 3\ NAD^+ + CoQ + 4\ H_2O$$

When GTP is utilized in other reactions, the effect will be:

$$GTP^{4-} + H_2O \longrightarrow GDP^{3-} + P_i^2 + H^+$$

If we add these two equations to the sum for the cycle, we indeed have the formal stoichiometry for the oxidation of acetyl groups.

Actually, we could arrive at the same result in a simpler way, noting that oxaloacetate and acetyl coenzyme A go into the cycle, but oxaloacetate and two molecules of CO_2 come back. There are four dehydrogenases, and the electrons from each of these reactions will eventually be transferred to one atom of oxygen.

Either way we approach it, we arrive at another important conclusion: *The stoichiometry of the citric acid cycle does not involve the production or consumption of oxaloacetate or any other constituent of the cycle. The reactions of the citric acid cycle do not provide a route for making oxaloacetate from acetyl groups.* Failure to appreciate this point has led intelligent men into serious error in the past, and it must be kept firmly in mind to avoid making the same mistakes in the future.

High-Energy Phosphate Yield

The transport of electrons from the three moles of NADH and the one mole of reduced coenzyme Q produced in the cycle to molecular oxygen results in the formation of ATP. Mitochondria having tightly coupled oxidative phosphorylation will produce three moles of ATP with the oxidation of each mole of NADH, for a total of nine moles, and a further two moles of ATP with the oxidation of $CoQH_2$.

In sum, 11 moles of ATP are formed by oxidative phosphorylation during the electron flow generated by one mole of acetyl coenzyme A going through the citric acid cycle. There is an additional mole of GTP formed during the cleavage of succinyl coenzyme A, making a grand total of 12 moles of $\sim P$ formed per acetyl group completely oxidized.

The P : O ratio, the moles of high-energy phosphate formed per atom of oxygen consumed, is therefore 3.00 for the citric acid cycle.

Recapitulation of types of reactions

1. Carboxylates with saturated chains may be oxidized by flavoproteins to form 2,3-unsaturated carboxylates:

$$
\begin{array}{ccc}
\text{COO}^{\ominus} & & \text{COO}^{\ominus} \\
| & & | \\
\text{CH}_2 & \xleftarrow{\;(FAD\text{-}protein)\;} & \text{CH} \\
| & & \| \\
\text{CH}_2 & & \text{CH} \\
| & & |
\end{array}
$$

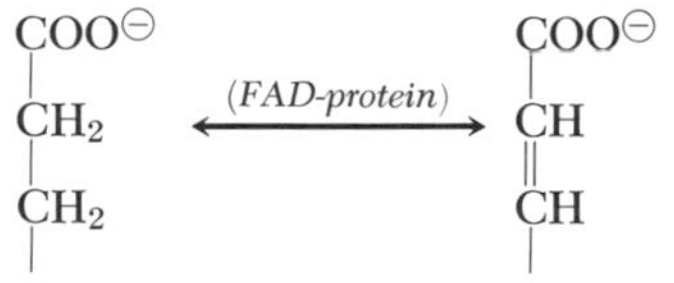

Example: The oxidation of succinate to fumarate.

2. 2,3-Unsaturated carboxylates may be hydrated to form 3-hydroxycarboxylates:

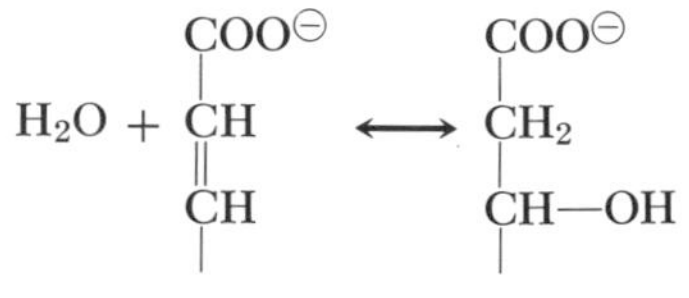

$$
\begin{array}{ccc}
& \text{COO}^{\ominus} & \text{COO}^{\ominus} \\
& | & | \\
\text{H}_2\text{O} + & \text{CH} & \text{CH}_2 \\
& \| & \longleftrightarrow \quad | \\
& \text{CH} & \text{CH}-\text{OH} \\
& | & |
\end{array}
$$

Examples: The hydration of aconitate to yield citrate or isocitrate; the hydration of fumarate to yield malate.

3. 3-Hydroxycarboxylates may be oxidized by NAD-coupled dehydrogenases. The oxidation may be a simple one, producing the corresponding ketocarboxylate:

$$\text{NAD}^+ + \begin{array}{c} \text{COO}^\ominus \\ | \\ \text{CH}_2 \\ | \\ \text{CH}-\text{OH} \\ | \end{array} \longleftrightarrow \begin{array}{c} \text{COO}^\ominus \\ | \\ \text{CH}_2 \\ | \\ \text{C}=\text{O} \\ | \end{array} + \text{NADH} + \text{H}^+$$

Example: The oxidation of malate to oxaloacetate.
Or, the oxidation may be accompanied by a decarboxylation:

$$\text{NAD}^+ + \begin{array}{c} \text{COO}^\ominus \\ | \\ \text{CH}_2 \\ | \\ \text{CH}-\text{OH} \\ | \end{array} \longrightarrow \begin{array}{c} \text{CH}_3 \\ | \\ \text{C}=\text{O} \\ | \end{array} + \text{CO}_2 + \text{NADH}$$

Example: The oxidation of isocitrate to α-ketoglutarate and CO_2.

4. 2-Ketocarboxylates may be oxidized by an enzyme complex that includes a thiamin pyrophosphate-linked decarboxylase, a lipoate-linked transacylase, and an NAD-coupled dihydrolipoate dehydrogenase that is a flavoprotein, so that the overall reaction produces an acyl coenzyme A, CO_2, and NADH:

$$\text{NAD}^+ + \text{CoA}-\text{SH} + \begin{array}{c} \text{COO}^\ominus \\ | \\ \text{C}=\text{O} \\ | \end{array} \xrightarrow[\substack{(\textit{lipoyl protein}) \\ (\textit{FAD-protein})}]{(\textit{ThPP-protein})} \begin{array}{c} \text{O} \\ \diagdown \\ \text{C}-\text{S}-\text{CoA} \\ | \end{array} + \text{CO}_2 + \text{NADH}$$

Example: The oxidation of α-ketoglutarate to succinyl CoA and CO_2.

5. Acyl esters of coenzyme A may be formed by cleaving a high-energy phosphate bond. The cleavage may be a simple removal of a single phosphate from a nucleoside triphosphate, in which case the reaction is freely reversible, and may also be used to form the high-energy phosphate from the thiol ester:

$$\begin{array}{c} \text{O} \\ \diagdown \\ \text{C}-\text{S}-\text{CoA} \\ | \end{array} + \begin{array}{c} \text{GDP}^{3-} \\ (\text{or ADP}^{3-}) \end{array} + \text{HOPO}_3{}^{2-} \xrightleftharpoons[\text{synthetase}]{\textit{acyl CoA}} \begin{array}{c} \text{COO}^\ominus \\ | \end{array} + \begin{array}{c} \text{GTP}^{4-} \\ (\text{or ATP}^{4-}) \end{array} + \text{CoA}-\text{SH}$$

Example: The formation of GTP by cleavage of succinyl coenzyme A to form succinate.
Or, the nucleoside triphosphate may be cleaved to release inorganic pyrophosphate, which is further hydrolyzed by inorganic pyrophosphatase. The removal of pyrophosphate makes the process essentially irreversible, so that this kind of reaction is used to form thiol esters from the free carboxylates:

$$\begin{array}{c} \text{COO}^\ominus \\ | \end{array} + \text{CoA}-\text{SH} + \text{ATP}^{4-} + \text{H}_2\text{O} \xrightarrow[\substack{+ \\ \textit{inorganic} \\ \textit{pyrophosphatase}}]{\substack{\textit{acylate} \\ \textit{thiokinase}}} \begin{array}{c} \text{O} \\ \diagdown \\ \text{C}-\text{S}-\text{CoA} \\ | \end{array} + \begin{array}{c} \text{AMP}^{2-} \\ + 2\,\text{HOPO}_3{}^{2-} + \text{H}^+ \end{array}$$

Example: There is no example in the citric acid cycle itself, but we discussed the formation of acetyl coenzyme A from acetate by this mechanism.

6. The free energy released by cleavage of a thiol ester bond may be used to drive an aldol-type condensation between a methylene carbon and a ketone:

$$H_2O + \overset{|}{\underset{|}{C}}=O + \overset{|}{CH_2}-\overset{O}{\overset{\|}{C}}-S-CoA \longrightarrow HO-\overset{|}{\underset{|}{C}}-\overset{|}{CH}-COO^{\ominus} + CoA-SH + H^+$$

Example: The condensation of acetyl coenzyme A and oxaloacetate to form citrate.

Further reading

Jaenicke, L., and F. Lynen: *Coenzyme A.* Also Massey, V.: *Lipoyl Dehydrogenase.* In Boyer, P. D., H. Lardy, and K. Myrback, eds.: *The Enzymes,* 2nd ed., Vol. 3 (p. 3) and Vol. 7 (p. 325), respectively. Academic Press (1960).
Lowenstein, J. M.: *The Tricarboxylic Acid Cycle.* In D. E. Greenberg, ed.: *Metabolic Pathways,* 3rd ed., Vol. 1 (p. 147). Academic Press (1967).
Reed, L. J., and D. Cox: *Macromolecular Organization of Enzyme Systems.* Ann. Rev. Biochem., 35: 57 (1966).

THE FORMATION OF ACETYL COENZYME A FROM FATTY ACIDS

CHAPTER 12

ARGUMENT

The neutral fats are triglycerides—esters of glycerol with three molecules of fatty acids—but the fatty acid residues make up the bulk of these fuels. When fat is oxidized for energy production, most of the energy comes from oxidation of the fatty acid residues. The common natural fatty acids are carboxylic acids with long hydrocarbon chains. They are oxidized in mitochondria by a sequence of reactions causing two carbons to be released as acetyl coenzyme A. The sequence is repeated until the entire chain has been cleaved into these two-carbon pieces. The initial oxidation reactions result in the formation of ATP from oxidative phosphorylation. However, twice again as much ATP results from the oxidation of the acetyl coenzyme A in the citric acid cycle.

Over half of the fatty acid residues in mammalian tissues contain one or more double bonds in the hydrocarbon chain, spaced at 3-carbon intervals in *cis* configurations. Additional enzymes exist to catalyze the shift of these double bonds and the racemization of D-hydroxy derivatives formed by their hydration so that they conform in structure to the intermediates of saturated chain oxidation. This makes it possible to use the same set of enzymes for catalyzing most of the reactions during oxidation of all types of fatty acids.

The oxidation of fatty acids has a P : O ratio near 2.8. This ratio is a useful guide in assessing the amount of oxygen that must be consumed to achieve energy balance with different fuels.

FATTY ACIDS

Fatty acids are major fuels for animals, both as a primary supply from the diet and as a secondary supply created from other dietary components. They are stored and later released through the blood stream to meet the demands of many tissues, especially muscles. The importance of the oxidation of fatty acids is not limited to the obese or devotees of greasy foods; it is a critical part of the metabolic economy in the lean as well as the lardy.

Fatty acids are oxidized in mitochondria to form acetyl coenzyme A. The resultant acetyl groups may be oxidized in the citric acid cycle to generate ATP, but the oxidation of the fatty acids themselves is also coupled to phosphorylation in mitochondria to produce even more ATP.

The Nature of Fatty Acids

Fatty acids are simple *aliphatic monocarboxylic acids*, R—COOH, existing in solution at physiological pH as the ionized *alkyl carboxylates*, R—COO⁻. The natural fatty acids vary in size, but the common ones contain an even number (from 14 to 24) of carbon atoms in a straight chain. This hydrocarbon chain may be completely saturated, or it may contain one or more double bonds.

Many of the naturally occurring fatty acids have trivial names in common use; here are some examples for the anionic form of saturated acids:

Formula	Trivial Name	Systematic Name	Type Designation
$CH_3(CH_2)_{12}$—COO⁻	myristate	tetradecanoate	14:0
$CH_3(CH_2)_{14}$—COO⁻	palmitate	hexadecanoate	16:0
$CH_3(CH_2)_{16}$—COO⁻	stearate	octadecanoate	18:0
$CH_3(CH_2)_{18}$—COO⁻	arachidate	eicosanoate	20:0

The type designations shown are abbreviations for the number of carbon atoms and double bonds. *Palmitate,* the ionic form of the most abundant fatty acid in humans, has 16 carbon atoms and no double bonds, and is 16:0.

As is the case with nucleic acids, it is common to speak of fatty acids, even though the corresponding anion is the predominant form in aqueous solutions at physiological pH values. We shall sometimes follow custom and other times refer to the carboxylate form, particularly in dealing with balanced equations.

As the name implies, fatty acids are derived from fats, which are esters of glycerol with three molecules, usually different, of fatty acids. These *triglycerides,* or *neutral fats,* are named from top to bottom when the formula is drawn so as to have a conventional L-configuration. Thus, myristoyl palmitoyl stearoyl glycerol is:

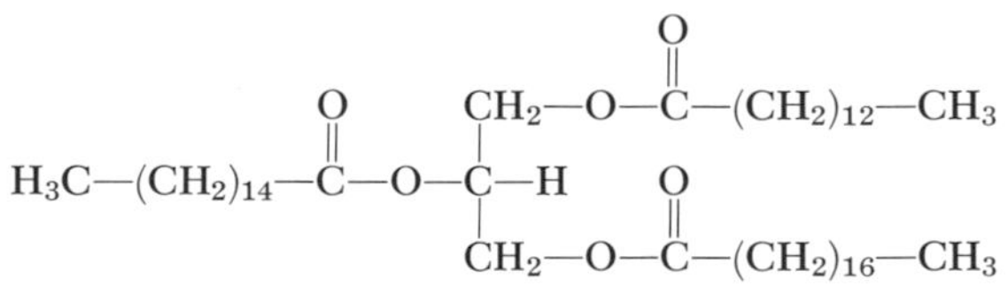

$$H_3C-(CH_2)_{14}-\overset{\overset{\textstyle O}{\|}}{C}-O-\overset{\overset{\textstyle CH_2-O-\overset{\overset{\textstyle O}{\|}}{C}-(CH_2)_{12}-CH_3}{|}}{\underset{\underset{\textstyle CH_2-O-\overset{\overset{\textstyle O}{\|}}{C}-(CH_2)_{16}-CH_3}{|}}{C}}-H$$

Triglycerides are hydrolyzed within cells by a class of esterases, the *lipases*, to release the fatty acids that will be used as a source of acetyl coenzyme A:

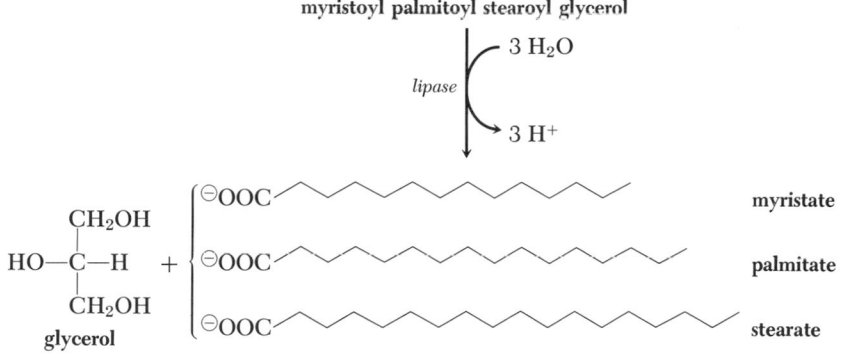

It only takes a glance at this equation to see that the fatty acids constitute by far the major part of the mass of the products; the metabolism of fats is mainly the metabolism of fatty acids in a quantitative sense.

We shall first consider the metabolism of the saturated fatty acids, and then go on to consider the unsaturated compounds.

The Conversion of Fatty Acids to Their Coenzyme A Esters

The first step in the metabolism of fatty acids is the formation of a thiol ester with coenzyme A. How will this be accomplished? It will certainly be at the expense of a high-energy phosphate bond, but will the route utilize a synthetase similar to succinyl CoA synthetase, or will it utilize a thiokinase like acetate thiokinase, with ATP being cleaved to AMP and PP_i followed by hydrolysis of the PP_i?

The answer is that both routes are used. Mitochondria contain an *acyl coenzyme A synthetase*, forming the thiol ester in a reversible reaction utilizing GTP:

$$R-COO^- + CoA-SH + GTP^{4-} \longrightarrow R-\overset{\overset{\displaystyle O}{\displaystyle \|}}{C}-S-CoA + GDP^{3-} + P_i^{2-}$$

This intramitochondrial reaction, which uses GTP rather than the more abundant ATP, has only recently been discovered. It is believed that it may tie a part of the uptake of fatty acids for oxidation to the rate of generation of GTP by the citric acid cycle, but this is speculative.

The most active route for the formation of acyl coenzyme A is through the action of *acyl thiokinase:*

$$R-COO^- + CoA-SH + ATP^{4-} \longrightarrow R-\overset{\overset{\displaystyle O}{\displaystyle \|}}{C}-S-CoA + AMP^{2-} + PP_i^{3-}$$

The cytosol is especially rich in acyl thiokinase, but mitochondria also have the enzyme. Since PP_i is attacked by inorganic pyrophosphatase, the formation of acyl coenzyme A by this process becomes irreversible, and the rate of uptake of the fatty acid in the cytosol is not limited by an approach to equilibrium.

In other words, fatty acids may be metabolized even though their concentrations are low. Furthermore, their uptake does not depend upon the maintenance of high concentrations of ATP, as it might with a freely reversible synthetase. The purpose of the oxidation of fatty acids is to generate high-energy phosphate, and it would not do to have the initial step dependent upon a high concentration of the very compound one is trying to produce.

Intracellular Transport of Acyl Groups

The oxidation of fatty acids occurs in mitochondria, but the compounds are presented to the cell outside of the mitochondria, either as free fatty acids from the blood or as triglycerides that are hydrolyzed in the cytosol. The fatty acids can be, and are, converted to the coenzyme A esters in the cytosol, but these esters cannot diffuse to the inner membrane of mitochondria. (This impermeability of mitochondria to most nucleotides other than ATP, ADP, and AMP serves to maintain internal control of metabolism within the organelle.)

It appears that acyl groups are moved in and out of mitochondria by transferring them from coenzyme A to another compound, *carnitine*. The acyl carnitine can penetrate mitochondria, and since the *carnitine-CoA acyl transferase* will function on either side of the membrane, the result is an equilibration of intramitochondrial and extramitochondrial acyl CoA compounds (Fig. 12-1).

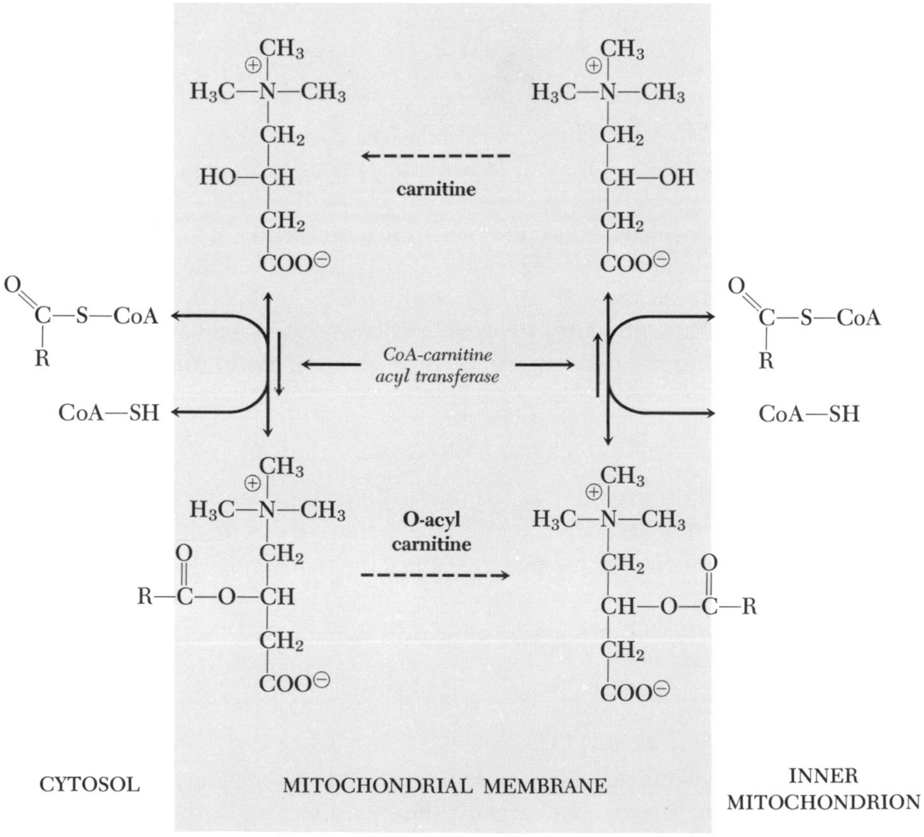

Figure 12-1 Long chain fatty acid residues appear as coenzyme A esters both within and without mitochondria, but the residues evidently are transferred to carnitine for movement from one side of the mitochondrial membranes to the other.

The details of this transfer, including the exact anatomical location and its regulation, if any, are still in dispute.

The Oxidation of Acyl Coenzyme A

Acyl coenzyme A compounds are oxidized in a straightforward way; for example, the initial sequence with palmitoyl coenzyme A is:

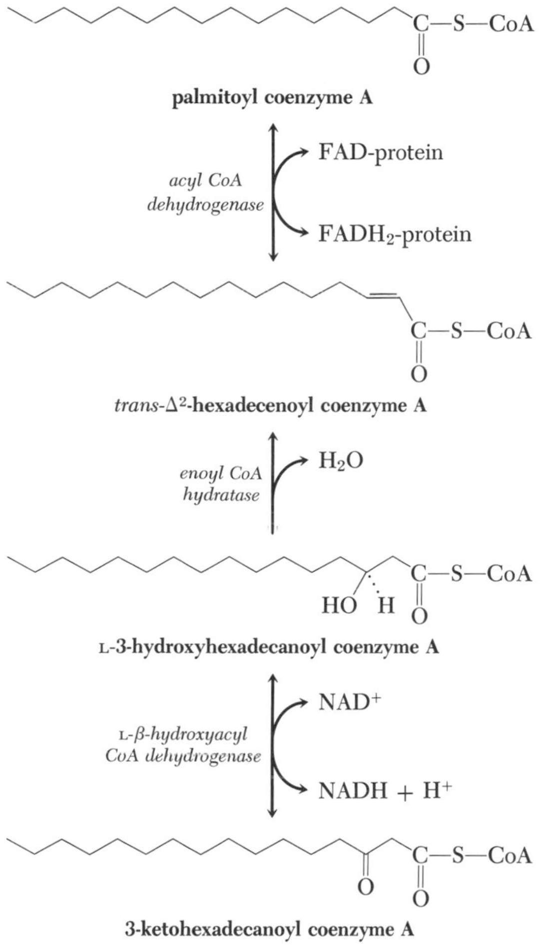

palmitoyl coenzyme A

acyl CoA dehydrogenase — FAD-protein → FADH$_2$-protein

trans-Δ^2-hexadecenoyl coenzyme A

enoyl CoA hydratase — H$_2$O

L-3-hydroxyhexadecanoyl coenzyme A

L-β-hydroxyacyl CoA dehydrogenase — NAD$^+$ → NADH + H$^+$

3-ketohexadecanoyl coenzyme A

All of these reactions are of familiar types, resembling the reaction sequence beginning with succinate in the citric acid cycle. The methylene carbons are oxidized to an unsaturated *trans* configuration, then hydrated to form an L-hydroxycarbonyl compound, followed by oxidation of the alcohol to a ketone.

Of course, the reactions of the acyl coenzyme A compounds are catalyzed by specific enzymes, not by the enzymes of the citric acid cycle, but the principles are the same, and the enzymes are also located in mitochondria.

An acyl coenzyme A is oxidized by a flavoprotein because this reaction, like the oxidation of succinate, requires a stronger oxidizing agent than NAD. Several of these flavoproteins with varying specificity occur in a mitochondrion, and the number differs from one tissue to another. Some react more rapidly with the longer chain derivatives, even larger than palmitoyl coenzyme A. Some react better with intermediate compounds, and others with short-chain derivatives, such as hexanoyl coenzyme A. (Caproyl coenzyme A is the trivial name for this six-carbon acyl ester.)

The mechanism of electron transfer by acyl coenzyme A dehydrogenase is different from that of succinate dehydrogenase. These flavoproteins contain no iron, and transfer electrons within mitochondria to still another flavoprotein, the *electron-transfer flavoprotein* (ETF). It is the electron-transfer flavoprotein that transfers electrons to non-heme iron on the way to coenzyme Q:

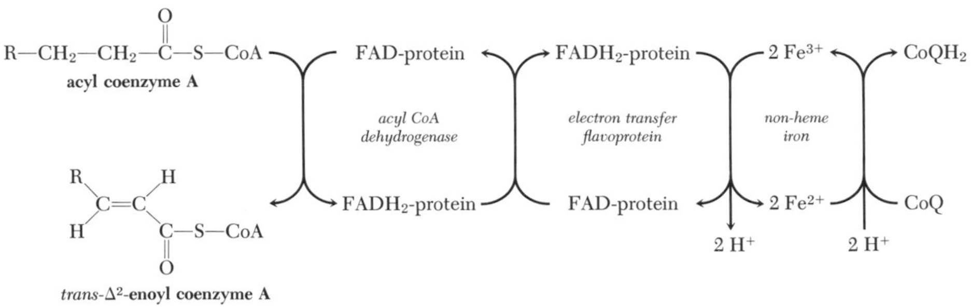

Why this extra flavoprotein is utilized for fatty acid oxidation is not clear, since the end result is similar to that in succinate oxidation—electrons are introduced into the mitochondrial electron transport chain at the level of coenzyme Q, with two moles of high-energy phosphate generated per mole of acyl coenzyme A oxidized. A likely possibility is that the distinctive pathway provides independent regulation of electron flow from the first step in the general process of fatty acid oxidation, but no direct experimental test of this speculation has been devised.

The hydration of unsaturated acyl coenzyme A compounds to form the corresponding L-3-hydroxyacyl derivatives is straightforward, quite similar to the hydration of aconitate or fumarate in the citric acid cycle, but catalyzed by still a third enzyme.

The oxidation of the alcohol group by NAD has no unusual features, except that there is no possibility of an accompanying decarboxylation such as we saw in the oxidation of isocitrate, because there is no free carboxylate group—it is fixed as a coenzyme A ester.

Thiolysis of 3-Ketoacyl Coenzyme A

Esters of 3-keto acids are subject to C-C scission both in the test tube and in biological systems through attack by a nucleophilic reagent. (The familiar synthesis of ethyl acetoacetate by a Claisen condensation is the reverse of this reaction.) The biological nucleophilic reagent frequently is coenzyme A, and the cleavage therefore is a *thiolysis*—a splitting by a thiol.

Enzymes exist to catalyze thiolysis of the 3-ketoacyl coenzyme A esters formed by fatty acid oxidation. For example, the 3-ketohexadecanoyl coenzyme A formed by the oxidation of palmitoyl coenzyme A is split in this manner:

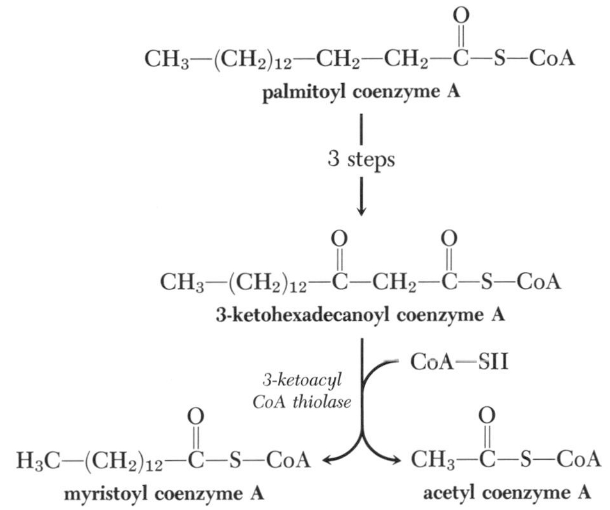

$$CH_3-(CH_2)_{12}-CH_2-CH_2-\overset{\overset{\displaystyle O}{\|}}{C}-S-CoA$$

palmitoyl coenzyme A

|

3 steps

↓

$$CH_3-(CH_2)_{12}-\overset{\overset{\displaystyle O}{\|}}{C}-CH_2-\overset{\overset{\displaystyle O}{\|}}{C}-S-CoA$$

3-ketohexadecanoyl coenzyme A

3-ketoacyl
CoA thiolase

CoA—SII

$$H_3C-(CH_2)_{12}-\overset{\overset{\displaystyle O}{\|}}{C}-S-CoA \longleftarrow \qquad \longrightarrow CH_3-\overset{\overset{\displaystyle O}{\|}}{C}-S-CoA$$

myristoyl coenzyme A   **acetyl coenzyme A**

What does this achieve? A molecule of acetyl coenzyme A is formed from the original palmitate molecule in a reaction that is nearly irreversible at intracellular concentrations, and the remaining 14 carbons are now in the form of the coenzyme A ester of myristic acid. Myristoyl coenzyme A is also a substrate for an acyl coenzyme A dehydrogenase, and undergoes the same sequence of reactions as that shown for palmitoyl coenzyme A. It is oxidized, hydrated, oxidized again, and then it, too, is cleaved by coenzyme A, releasing still another molecule of acetyl coenzyme A with the remaining 12 carbon atoms as dodecanoyl coenzyme A.

In short, oxidation of the coenzyme A esters of fatty acids followed by thiolysis removes two carbons as acetyl coenzyme A, but the remaining carbons are still present as coenzyme A esters of fatty acids, which will go through the whole process again. The sequence will be repeated until the final four-carbon chain (butyryl coenzyme A) has been oxidized and cleaved into two molecules of acetyl coenzyme A, at which point all of the carbons in the original palmitate will have been converted to acetyl coenzyme A (Fig. 12-2).

What an elegant economy of mechanism! Mitochondria use the same sequence of reactions, and chop a random mixture of fatty acids chains into units of acetyl coenzyme A, which is the standard fuel for the citric acid cycle.

What happens if a fatty acid with an odd number of carbon atoms is introduced into mitochondria? There are such fatty acids in nature, although they are much less common than those with an even number of carbon atoms. The sequence of reactions proceeds just as before. We expect this, because it is asking too much of enzyme specificity for substrates with chains of, say, 14, 16, or 18 carbon atoms to be bound to an enzyme without the enzyme also binding those of 15 or 17 carbon atoms. Acetyl coenzyme A units are successively cleaved from these odd-numbered chains until the very end of the sequence, when the three-carbon remainder is in the form of *propionyl coenzyme A:*

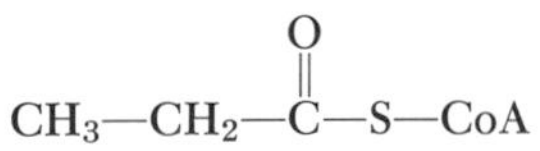

$$CH_3-CH_2-\overset{\overset{\displaystyle O}{\|}}{C}-S-CoA$$

propionyl coenzyme A

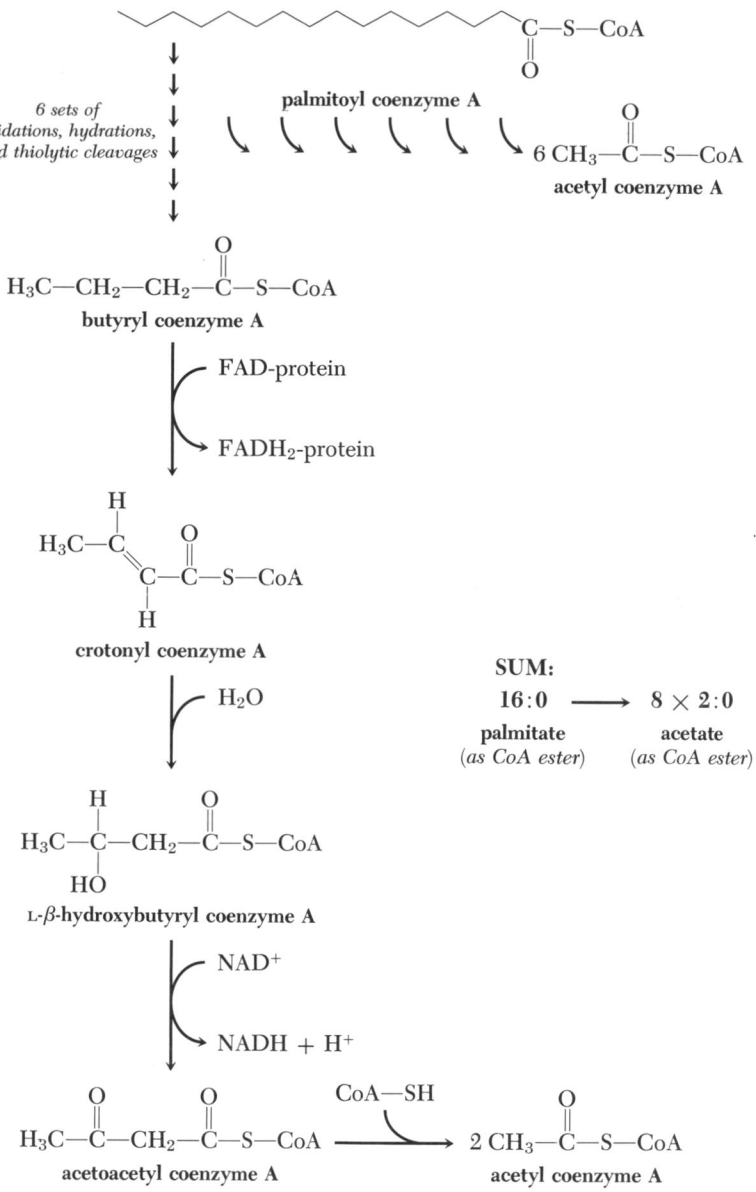

Figure 12-2 Total cleavage of palmitoyl coenzyme A to acetyl coenzyme A.

This compound is not a substrate for citrate synthase, and therefore does not enter the citric acid cycle. It undergoes a completely different metabolism, which is important in the breakdown of amino acids and will be considered in connection with them (p. 366). The odd-chain fatty acids are only a small fraction of the total, and even with these, only the terminal three carbons appear as propionyl coenzyme A. The metabolism of propionyl coenzyme A is therefore not of great quantitative significance in fatty acid oxidation.

UNSATURATED FATTY ACIDS

More than half of the total acyl residues in the tissues of a typical mammal contain one or more double bonds. *Oleic acid* is the most abundant of the unsaturated fatty acids. It has a single double bond in the middle of an 18-carbon chain with a *cis* configuration, and is therefore systematically named *cis-*Δ^9*-octadecenoate*:

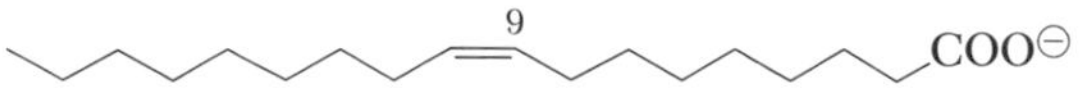

oleate
(18:1)

Most of the natural unsaturated fatty acids have all of the double bonds in *cis* configuration; some of the more common in addition to oleic acid are:

16:1

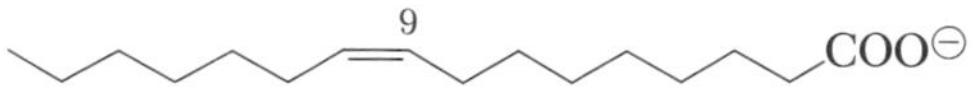

palmitoleate
(*cis-*Δ^9-hexadecenoate)

18:2

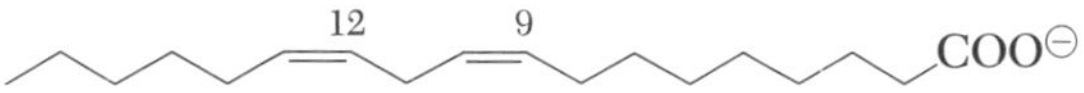

linoleate
(*cis,cis-*$\Delta^{9,12}$-octadecadienoate)

18:3

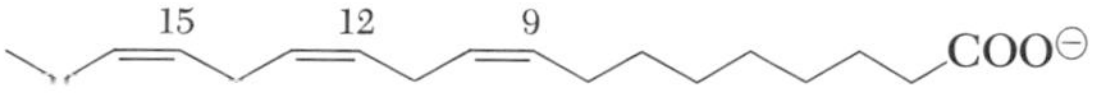

linolenate
(*all cis-*$\Delta^{9,12,15}$-octadecatrienoate)

20:4

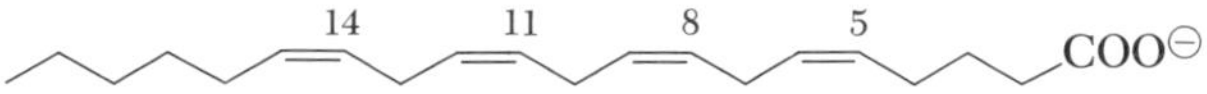

arachidonate
(*all cis-*$\Delta^{5,8,11,14}$-eicosatetraenoate)

There are dozens of known examples beyond those shown, but most share the same peculiarity—the double bonds are spaced at 3-carbon intervals. The function served by this spacing is not known; it may be structural, since the greatest variety of polyunsaturated residues is found in membrane phospholipids.

Higher animals can make oleoyl residues from stearoyl residues, as we shall discuss in connection with fatty acid synthesis (Chapter 16), but they have lost the necessary enzymes for introduction of double bonds beyond the ninth carbon atom in a chain. *Polyunsaturated* acids such as *linoleic* or *linolenic* acids must be supplied in the diet and are called *essential fatty acids*.

Oxidation of the Unsaturated Fatty Acids

Since over half of the fatty acid residues in body lipids are unsaturated, a large part of the high-energy phosphate generated in many tissues must come from the oxidation of this type of compound. The general route includes all of the enzymes involved in the oxidation of saturated fatty acids with the addition of a few more reactions that convert the unsaturated structures into intermediates of the regular pathway.

As an example, let us consider the metabolism of linolenate—all *cis-*$\Delta^{9,12,15}$-

octadecatrienoate. The double bonds are well removed from the carboxylate end, so it will begin to be metabolized as if it were an ordinary saturated fatty acid:

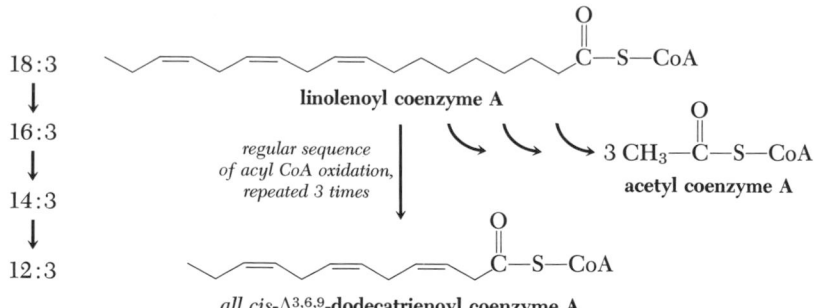

1. Acyl thiokinase catalyzes the formation of linolenoyl coenzyme A.

2. Linolenoyl coenzyme A is oxidized by an acyl coenzyme A dehydrogenase, forming the corresponding *trans*-Δ^2-enoyl coenzyme A, which, of course, will also have the original three double bonds at carbons 9, 12, and 15, in addition to the newly created one at carbon 2.

3. The new double bond is hydrated to form L-3-hydroxylinolenoyl coenzyme A, which is then oxidized by NAD to form 3-ketolinolenoyl coenzyme A.

4. 3-Ketolinolenoyl coenzyme A is cleaved with coenzyme A, liberating acetyl coenzyme A, and the whole process is repeated.

The sequence can proceed until three molecules of acetyl coenzyme A have been cleaved, but no farther, because the chain is then shortened so that the double bond that was originally on the ninth carbon of linolenate is now on the third carbon of the remaining chain, thereby preventing oxidation by the flavoprotein of acyl coenzyme A dehydrogenase, which attacks carbons 2 and 3.

One of the additional enzymes now comes into play by catalyzing a migration of the double bond from the third to the second carbon atom, and at the same time changing the configuration from *cis* to *trans*:

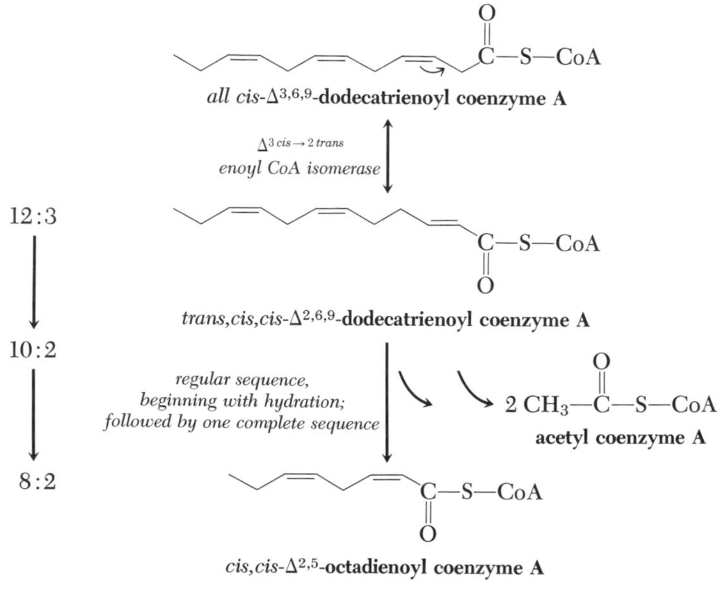

all cis-$\Delta^{3,6,9}$-**dodecatrienoyl coenzyme A**

$\Delta^{3\,cis}\rightarrow 2\,trans$
enoyl CoA isomerase

12:3

trans,cis,cis-$\Delta^{2,6,9}$-**dodecatrienoyl coenzyme A**

10:2

regular sequence, beginning with hydration; followed by one complete sequence

O
‖
2 CH$_3$—C—S—CoA
acetyl coenzyme A

8:2

cis,cis-$\Delta^{2,5}$-**octadienoyl coenzyme A**

Equilibrium for this reaction is reached with about $7/8$ in the *trans* form, so no energy donors or receptors are involved. After migration, the double bond is in the same position and has the same configuration as the regular intermediates of fatty acid oxidation, so the compound will once more enter the general pathway, beginning with a hydration. Two more molecules of acetyl coenzyme A will be cleaved from the chain before the second of the original double bonds in linolenate, still in its original *cis* configuration, reaches carbon 2 from the carbonyl end. This is in the same position as the double bond created during the regular sequence of oxidation, but of opposite configuration. It turns out that the responsible enzyme, enoyl coenzyme A hydrase, is not specific for configuration, and it will catalyze the hydration of either the *cis* or the *trans* form. However, it may be recalled from our discussion of the citric acid cycle that hydration of *cis* compounds often produces D-hydroxy derivatives, and the result of the reaction in this case is a D-3-hydroxyacyl coenzyme A, which is not a substrate for the L-3-hydroxyacyl coenzyme A dehydrogenase used in the regular route.

An additional enzyme, a racemase, exists to catalyze the conversion of these D-3-hydroxyacyl derivatives to the L-isomers, which can be attacked by the dehydrogenase (Fig. 12-3). The reaction sequence then proceeds with the cleavage of another molecule of acetyl coenzyme A, leaving *cis*-Δ^3-hexenoyl coenzyme A. This is a substrate for the isomerase mentioned above, and will be converted to *trans*-Δ^2-hexenoyl coenzyme A, which is in the mainstream of fatty acid oxidation and readily converted to three molecules of acetyl coenzyme A.

Here is another example of the beautiful economy of organization of metabolism. The introduction of two additional types of enzymes, an *enoyl coenzyme A isomerase* and a *3-hydroxyacyl coenzyme A racemase,* makes it possible to handle any combination of double bonds found in an unsaturated chain through the same route used for the saturated fatty acids. (It is an interesting exercise in mental gymnastics to test this on a random assortment of double bonds in a long chain, except that the allene structure, $-C{=}C{=}C-$, is forbidden. Allenes are rare among natural compounds, only being known as products of the metabolism of a few kinds of microorganisms.)

YIELD OF HIGH-ENERGY PHOSPHATE FROM FATTY ACID OXIDATION

As we saw, each cleavage of acetyl coenzyme A from a saturated acyl chain is preceded by a pair of oxidations, one catalyzed by a flavoprotein and the other by an NAD-coupled alcohol dehydrogenase. Electrons from the flavoprotein are inserted into the mitochondrial electron transport system at the coenzyme Q level, and therefore yield two high-energy phosphate bonds per molecule of substrate oxidized. Electrons from NADH go through the complete electron transport mechanism and yield three high-energy phosphate bonds per molecule of substrate. The sum of the two reactions is therefore five high-energy phosphate bonds generated per acetyl coenzyme A unit cleaved. How many such cleavages occur with a given

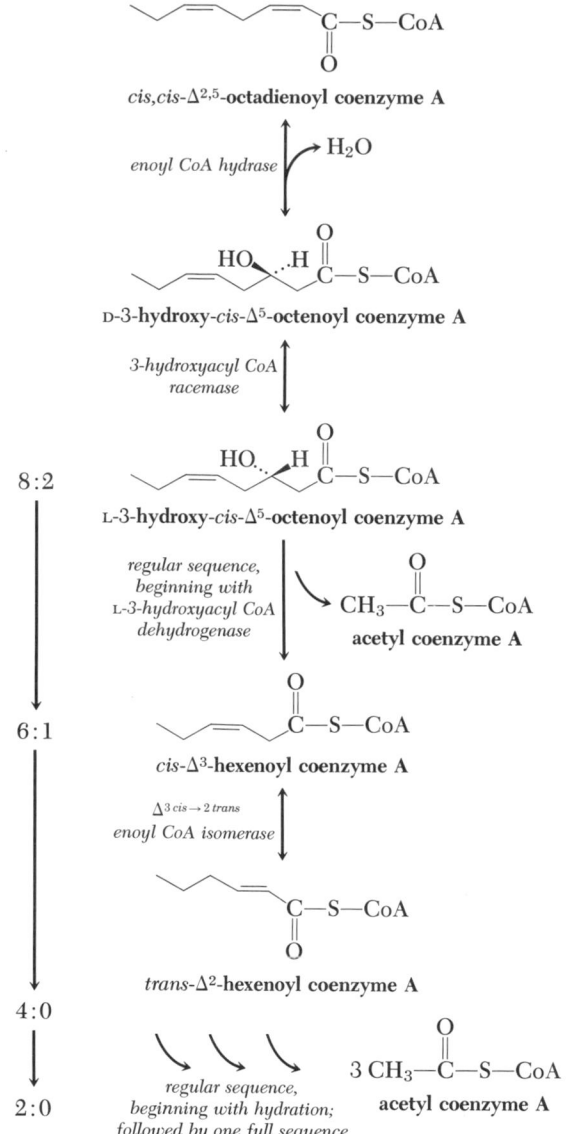

Figure 12-3   Complete metabolism of polyunsaturated fatty acids requires racemization of 3-hydroxyl groups and isomerization of double bonds.

molecule? There is one less than the number of pairs of carbon atoms in the chain, because the cleavage of two carbons from the last four carbons of the chain also leaves the terminal two carbons in the form of acetyl coenzyme A. (It only takes $n-1$ cuts to divide a string into n pieces.)

We must take into account the two high-energy phosphate bonds expended in forming the original coenzyme A ester and the 12 moles of high-energy phosphate formed as a result of the oxidation of each mole of acetyl coenzyme A in the citric acid cycle. We also must allow for the fact that each double bond already present in the original molecule eliminates one of the acyl coenzyme A dehydrogenase reactions necessary with the saturated compounds, and therefore diminishes the high-energy phosphate yield by two.

We can cast a general balance sheet for the mitochondrial oxidation of fatty acids containing n carbon atoms and x double bonds:

1. Subtract two high-energy phosphates for the formation of the coenzyme A ester:
$$-2$$

2. There are $\frac{n}{2}$ acetyl coenzyme A units formed, and therefore $\left(\frac{n}{2} - 1\right)$ pairs of oxidations, each resulting in the formation of 5 high-energy phosphate bonds:
$$5\left(\frac{n}{2} - 1\right)$$

3. The oxidation of each acetyl coenzyme A unit results in the formation of 12 high-energy phosphate bonds:
$$12\left(\frac{n}{2}\right)$$

4. Except that 2 high-energy phosphate bonds must be subtracted for each double bond already present in the original substrate:
$$-2x$$

These terms can be combined to give the general formulation: $\sim P = 8.5n - 2x - 7$

It is a simple matter then to show that the complete oxidation of *palmitate* (the most abundant saturated fatty acid in man) is accompanied by the formation of a theoretical maximum of 129 high-energy phosphate bonds, and the equation for this oxidation becomes:

$$C_{16}H_{31}O_2^- + 23\ O_2 + 129\ ADP^{3-} + 129\ P_i^{2-} + 130\ H^+ \longrightarrow$$
$$16\ CO_2 + 145\ H_2O + 129\ ATP^{4-}$$

The P : O ratio for this reaction is 2.80.

Similarly, the reaction for the complete oxidation of *oleate*, the most abundant of all of the acyl residues in man, is:

$$C_{18}H_{33}O_2^- + 25\tfrac{1}{2}\ O_2 + 144\ ADP^{3-} + 144\ P_i^{2-} + 145\ H^+ \longrightarrow$$
$$18\ CO_2 + 161\ H_2O + 144\ ATP^{4-}$$

and the P : O ratio is 2.82.

We have gone at some length into the method of arriving at the high-energy phosphate yield because we shall later need these, and similar values, for assessing the place of the fatty acids in the economy of the organism.

Recapitulation of types of reactions

Fatty acid oxidation involves three types of reactions not discussed in earlier chapters:

1. 3-Ketoacyl thiol esters may undergo C—C cleavage by thiolysis:

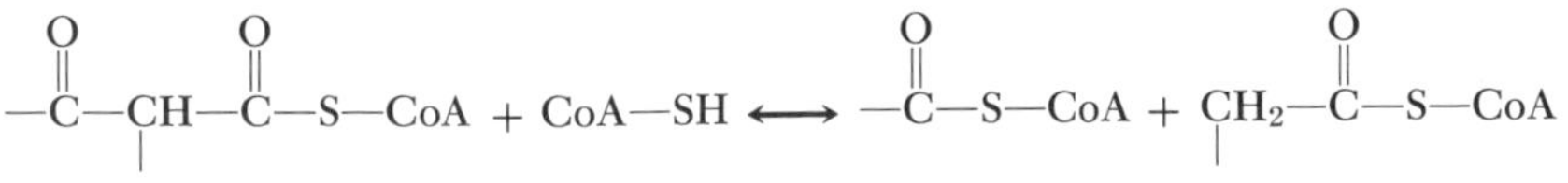

$$-\overset{O}{\overset{\|}{C}}-\overset{}{\underset{|}{CH}}-\overset{O}{\overset{\|}{C}}-S-CoA + CoA-SH \longleftrightarrow -\overset{O}{\overset{\|}{C}}-S-CoA + \overset{}{\underset{|}{CH_2}}-\overset{O}{\overset{\|}{C}}-S-CoA$$

Example: The reactions catalyzed by 3-ketoacyl CoA thiolase.

2. Double bonds may be shifted:

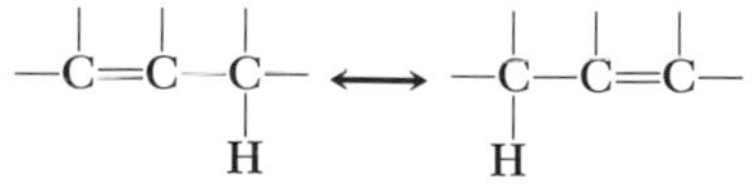

$$-\overset{|}{C}=\overset{|}{C}-\underset{H}{\overset{|}{C}}- \longleftrightarrow -\underset{H}{\overset{|}{C}}-\overset{|}{C}=\overset{|}{C}-$$

Example: The reactions catalyzed by *cis-3→trans-2* enoyl CoA isomerase.

3. Stereoisomers may be interconverted:

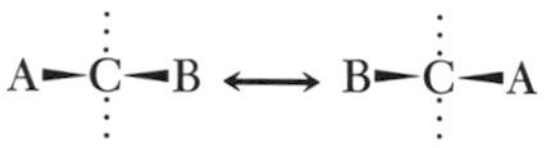

$$A\!-\!\overset{\vdots}{\underset{\vdots}{C}}\!-\!B \longleftrightarrow B\!-\!\overset{\vdots}{\underset{\vdots}{C}}\!-\!A$$

Example: The reactions catalyzed by 3-hydroxyacyl CoA racemase.

Further reading

Green, D. E., and D. W. Allmann: *Fatty Acid Oxidation.* In D. E. Greenberg, ed.: *Metabolic Pathways,* 3rd ed., Vol. 2 (p. 1). Academic Press (1968). A general description.

Wolf, G., ed.: *Recent Research on Carnitine.* M.I.T. Press (1965).

Norum, K. R., and J. Bremer: *The Localization of Acyl Coenzyme A-Carnitine Acyl Transferases in Rat Liver Cells.* J. Biol. Chem., *242:* 407 (1967).

Stoffel, W., and H. Caesar: *Zur β-Oxidation der Mono- und Polyenfettsauren.* Z. physiol. Chem., *341:* 76 and 85 (1965). In German, with summaries in English and clear figures.

The following are general sources of information on the character and metabolism of fats:

Kinsell, L. W., ed.: *Adipose Tissue as an Organ.* C. C Thomas (1962).

Rodahl, K., and B. Issekutz, eds.: *Fat as a Tissue.* McGraw-Hill (1964).

THE FORMATION OF ACETYL COENZYME A FROM GLUCOSE

CHAPTER 13

ARGUMENT

The hexose, glucose, is a prominent source of acetyl coenzyme A in many tissues. Glucose is converted by a series of reactions in the cytosol to two equivalents of pyruvate, which has three carbon atoms. These reactions are known as the Embden-Meyerhof pathway. Pyruvate is oxidatively decarboxylated in mitochondria to form acetyl coenzyme A.

The transport of glucose from the extracellular fluid into cells of tissues such as skeletal muscle is regulated, perhaps through the use of some carrier in the plasma membrane, and transport requires the presence of the hormone, insulin, in that tissue. Transport in muscle can only proceed from a higher to a lower glucose concentration; however, the intestinal mucosa is able to concentrate glucose. This osmotic work requires a coupling of glucose transport to some process with a negative standard free energy change, which appears to be a concomitant transport of Na^+ associated with oxidative phosphorylation.

The metabolism of glucose begins with a phosphorylation by ATP, followed by an isomerization and another phosphorylation to produce the keto sugar ester, fructose-1,6-diphosphate. The two phosphorylations are regulated so as to adjust the metabolism of glucose to the capacity for utilizing the products. A mole of fructose-1,6-diphosphate is converted to two moles of triose phosphates—esters of 3-carbon sugars. Beginning with an oxidation by NAD and followed by rearrangements to produce pyruvate, two moles of ATP are generated in the cytosol for each mole of triose phosphate metabolized. The NADH that is also produced in the cytosol cannot enter mitochondria to be oxidized directly. However, one of the triose phosphates, dihydroxyacetone phosphate, has the additional function of acting as a carrier of electrons in skeletal muscle and brain. It is reduced by NADH in the cytosol and reoxidized by a flavoprotein in mitochondria with passage of the electrons to coenzyme A and thence to oxygen. Dihydroxyacetone phosphate is not consumed in this process. The net effect is the oxidation of triose phosphates in the cytosol by molecular oxygen in mitochondria with a concomitant generation of two additional moles of ATP per triose unit.

Pyruvate is oxidatively decarboxylated in mitochondria by a pyruvate dehydrogenase complex resembling the α-ketoglutarate dehydrogenase used in the citric acid cycle; it involves thiamine pyrophosphate, lipoate, and FAD. A thiamine deficiency leads to a depletion of the coenzyme for this reaction and an accumulation of pyruvate. The resultant disturbance of carbohydrate metabolism especially affects the nervous system, producing some of the symptoms of beri-beri or of Wernicke's encephalopathy, which are different forms of thiamine deficiency in humans. Some of the same effects are mimicked in arsenite poisoning, which is a result of chelation with dihydro-lipolyl residues on these same enzyme complexes.

The complete combustion of one mole of glucose by way of the Embden-Meyerhof pathway, pyruvate dehydrogenase, and the citric acid cycle produces 36 moles of ATP, with an overall P : O ratio of 3.0.

The metabolism of glucose provides a major source of acetyl coenzyme A for most tissues, and is especially important for the brain. The supply of glucose in the blood may be obtained by absorption from the intestine, where it occurs as a product of the hydrolysis of dietary *starch*. Starch accounts for over half the fuel in most human diets. Glucose may also be produced from other metabolites in the liver, and to a lesser extent, in the kidney.

Glucose is a *hexose* (see p. 709 for a review of the chemistry of carbohydrates), and it exists in aqueous solution as a mixture of three forms (Fig. 13-1). These forms equilibrate spontaneously, but tissues also contain a *mutarotase* to accelerate the reaction. At 37°C and pH 7.0, an equilibrium mixture contains 0.635 of the total as β-glucopyranose, 0.365 as α-glucopyranose, and only 0.0008 as the open chain compound. The β-anomer is the substrate for some enzymes affecting glucose, but many are not specific for a particular anomer.

The metabolism of glucose is considerably more complex than the metabolism of fatty acids, and it is convenient to consider it in parts (Fig. 13-2): (1) the absorption of glucose by cells, followed by the formation of *glucose-6-phosphate;* (2) the transformation of glucose-6-phosphate into a form that can be split in half to yield two *triose phosphates* (trioses are 3-carbon sugars); (3) the conversion of the triose phosphates into the 2-ketocarboxylate, *pyruvate;* and (4) the oxidation of pyruvate to *acetyl coenzyme A*, which is then oxidized by the citric acid cycle.

The extramitochondrial part of this sequence, the conversion of glucose to pyruvate, is frequently called the *Embden-Meyerhof pathway,* after its discoverers.

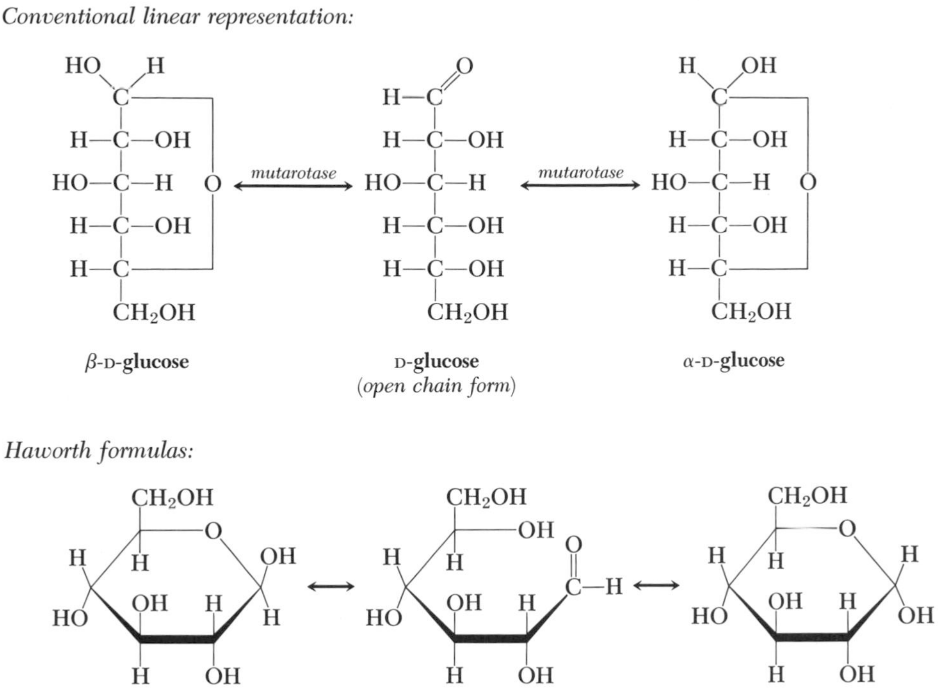

Conventional linear representation:

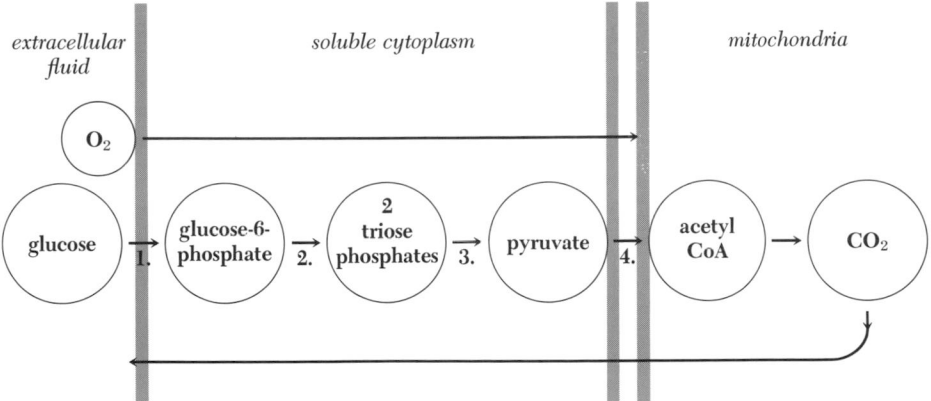

β-D-**glucose** D-**glucose** (*open chain form*) α-D-**glucose**

Haworth formulas:

Figure 13-1 The mutarotation of glucose, shown in both conventional linear formulas and the cyclical Haworth formulas. The central aldehyde form may condense with its own C-5 hydroxyl group to create 6-membered pyranose rings. Condensation may occur so that the C-1 oxygen is below or above the plane of the resultant ring, yielding α or β glucopyranose, respectively. Such hemiacetal isomers of sugars are called anomers. Their interconversion occurs spontaneously, but is also catalyzed by mutarotase; mutarotation describes the change in rotation of polarized light that accompanies equilibration of the anomeric forms.

Figure 13-2 General outline of the oxidation of glucose. The numbered segments of the scheme are discussed in the text.

CELLULAR UPTAKE AND PHOSPHORYLATION OF GLUCOSE

Glucose Transport

The transport of glucose into cells has been studied intensively because of its obvious importance in regulating the metabolic economy of the organism. There are many unsolved problems, but we do have answers to some important questions.

The plasma membrane of cells separates the interior of the cell from the extracellular fluid, which has a different chemical composition, so that there are concentration gradients for most compounds across the membrane. If the membrane is freely porous to these constituents, diffusion will occur until the effective concentrations, the activities, are equal on the two sides. The first question in considering the fate of a compound such as glucose that is brought to the cell through the extracellular fluid is: Does the compound simply diffuse through the plasma membrane as if the membrane were an inert porous sheet? If so, control over its passage into the cell can only be exerted by changing the relative concentrations of the compound outside and inside the cell.

It turns out that glucose does not pass into most, perhaps all, tissues by this simple, unregulated diffusion, and a second question must be asked: Does the compound always move across the membrane from the side of higher concentration to the side of lower concentration, but at a rate that can be regulated independently of the concentration gradient? The driving force for this kind of transport is the same as it is in simple diffusion—the difference in concentrations outside and inside the cell. However, the ability to slow or accelerate the attainment of equilibrium concentrations implies that the substance reacts with something in the membrane during its passage, and the reaction with this *carrier* can be controlled. Glucose evidently enters most of the tissues in the body by this sort of mechanism, and the control is especially pronounced in the case of the heart, skeletal muscles, and adipose tissue. Of course, the concentration of glucose in tissues using this mechanism cannot exceed the concentration in blood, which ranges from 0.004 M to 0.009 M in the usual human, the higher figure only being attained during the first hour after ingestion of meals containing mostly carbohydrate.

Regulated diffusion has its control through changes in the chemical nature of the plasma membrane. One of the striking examples comes from the effects of the hormone, *insulin*, which is a small protein secreted by the islets of Langerhans in the pancreas in response to increasing concentrations of blood glucose. It affects plasma membranes, especially in skeletal muscle and adipose tissue, so as to accelerate the movement of glucose through the membrane. The primary function of the action is to promote the storage of glucose as triglycerides or glycogen (Chapters 15 and 16) during times of plentiful supply. The mechanism by which insulin causes a more rapid transport of glucose is unknown, but the consequences of its failure are known in detail because about one person in 20 has a genetic tendency to develop *diabetes mellitus*. This disease is caused by failure to elaborate insulin, or to use it effectively. We shall consider it in more detail in Chapter 23. Suffice it for now to note that a failure of glucose uptake by peripheral tissues is a major factor in this biochemical abnormality.

Some tissues couple the absorption of glucose to a reaction that, by itself, proceeds with a large free energy loss. When this is the case, it is possible to raise the concentration of glucose within the cell to a higher level than the concentration outside, because the necessary work is being performed by the coupled reaction. This is the way glucose is transported by the mucosa of the small intestine, which will take up the sugar even after the concentration in the lumen falls below the concentration in the blood stream. The evolutionary advantage of this arrangement seems obvious—it permits the extraction of more fuel from the ingested food. Unfortunately, the mechanism by which energy is provided is not so obvious. The transport of glucose is in some way linked to the transport of sodium ions. It has long been known that cells can "actively" transport some ions; that is, they can be moved against the direction of diffusion so as to maintain a concentration difference across membranes. This is not much help, because the mechanism of ionic transport isn't known either, as we noted in the discussion of oxidative phosphorylation. However, it is evident that active transport is linked to the cleavage of high-energy phosphate, and it is likely that one of the uses for ATP is to concentrate ions in cells, and glucose is carried along by this process in the intestinal mucosa.

The association between intestinal absorption of glucose and sodium led to the recent development of an interesting technique for the management of cholera. This infection causes severe diarrhea, with a resultant depletion of water and salts that may be lethal. However, a constant infusion of glucose into the gut by means of a tube introduced through the mouth results in a dramatic alleviation of the diarrhea, evidently due to an associated increase in salt and water reabsorption.

After glucose crosses the plasma membrane, the major route of its metabolism in most tissues begins with its phosphorylation:

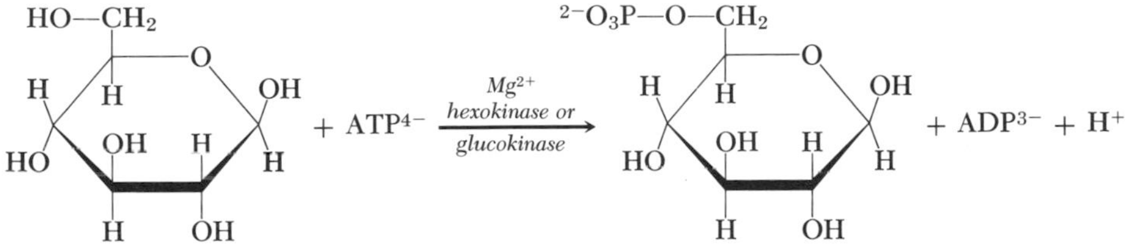

β-D-**glucose** β-D-**glucose-6-phosphate**

We used this reaction in Chapter 7 as an example, making these points:

1. The reaction is essentially irreversible at physiological concentrations of reactants and products.

2. There are two kinds of enzymes catalyzing the reaction—hexokinases, which also will phosphorylate other hexoses, such as fructose or mannose, and a specific glucokinase, reacting only with glucose.

3. The hexokinase of brain has a low K_M for glucose, enabling it to phosphorylate the compound even at low concentrations, whereas the glucokinase of liver has a high K_M, so that liver only introduces glucose into its metabolic routes when the blood concentration is relatively high.

In addition to the gross differences between the glucokinase of liver and hexokinases, there are differences between the hexokinases of different tissues. In other words, an organism makes several kinds of enzymes—*isozymes*—to catalyze the same reaction. Nothing has yet been discovered to contradict the rational explanation that these differences have evolved to permit regulation of enzymatic activity under the differing intracellular conditions required in the various tissues for their differentiated functions.

For example, three types of hexokinases have recently been discovered in rat tissues. The Michaelis constants for the substrates are as follows:

Type	K_M glucose	K_M ATP
I	5×10^{-5} M	4×10^{-4} M
II	3×10^{-5} M	8×10^{-4} M
III	7×10^{-6} M	1×10^{-3} M

The exact purpose of these variations is unknown. Liver contains the three types, and it is tempting to speculate that type III exists to phosphorylate more glucose when the ATP concentration is especially high, while the other types function only when glucose-6-phosphate is depleted compared to the available glucose.

Brain and kidney contain primarily the type I enzyme, whereas skeletal muscle and adipose tissue—the tissues especially sensitive to insulin—contain the type II enzyme. Heart, which is not as sensitive to insulin deprivation, contains both types I and II in about equal amounts. This rough correlation between the presence of type II hexokinase and insulin sensitivity may or may not be significant. It is too early to know.

The activity of the hexokinases can be regulated by the concentration of their own end-product. Glucose-6-phosphate is an effective inhibitor of the mixed type with all of the hexokinases. That is, the inhibition resembles purely competitive inhibition in that it is diminished by increasing the concentration of glucose, but it resembles purely non-competitive inhibition because no concentration of glucose can completely overcome the inhibition and restore full maximum velocity.

With the enzyme from brain, 0.23 millimolar glucose-6-phosphate inhibits about 50 per cent in the presence of 1 millimolar glucose and excess ATP (4 millimolar): The inhibition is even more effective at lower concentrations of ATP. We are justified in believing that the inhibition regulates the rate of glucose uptake to match the rate at which glucose-6-phosphate can be utilized in the tissue.

THE FORMATION OF TRIOSE PHOSPHATE

Conversion of Glucose-6-Phosphate to Fructose-6-Phosphate

The sequence of reactions forming acetyl coenzyme A begins with an isomerization of glucose-6-phosphate to the ketose ester, fructose-6-phosphate, catalyzed by the enzyme, *phosphoglucoisomerase:*

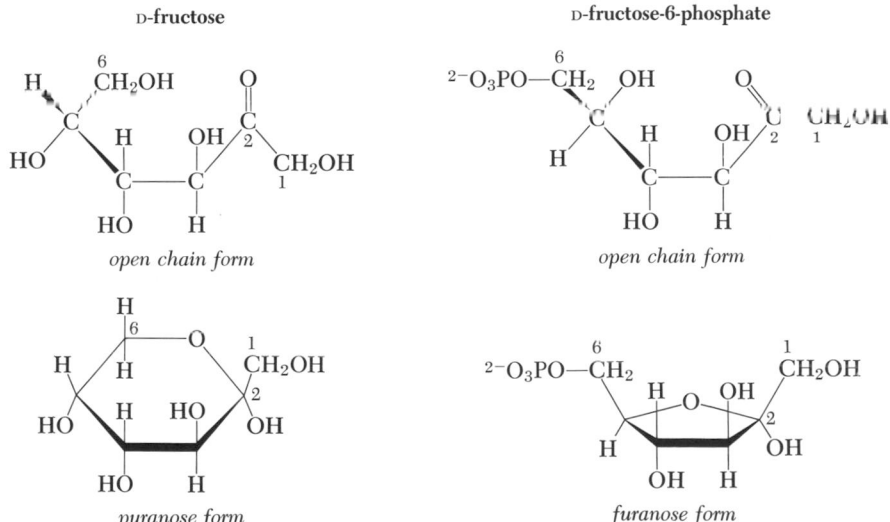

This is a prominent example of a general class of enzymes catalyzing the interconversion of aldoses and ketoses, and it interconverts the open-chain forms of the hexose phosphates, probably by way of an intermediate enediol on the enzyme, as the figure suggests.

A larger fraction of the phosphate esters exists in the open-chain form than is the case with the free sugars. This is especially true of fructose-6-phosphate, because the favored six-membered ring of the free sugar involves C-6, which is blocked by the phosphate in the ester. The only possible closure is the less favored five-membered furanose ring:

The phosphoglucoisomerase reaction is readily reversible because the aldose and ketose configurations have nearly equal free-energy contents, the equilibrium concentration of fructose-6-phosphate being about one-third of the total concentration at equilibrium. There is no known allosteric control for the reaction. This is an example of a reversible reaction in which the enzyme cannot catalyze the reaction near its maximum velocity because both substrate and product are present in significant amounts. A tissue capable of a rapid net reaction in either direction must have an unusually high concentration of the responsible enzyme. This is indeed the case in skeletal muscle, which can metabolize glucose-6-phosphate rapidly at times of heavy exertion. A large amount of the enzyme makes a rapid rate of conversion of glucose-6-phosphate to fructose-6-phosphate possible even though the ratio of concentrations of these compounds is not far from the equilibrium value.

Phosphorylation of Fructose-6-Phosphate

Fructose-6-phosphate is phosphorylated by ATP to form fructose-1,6-diphosphate in a reaction catalyzed by a specific phosphofructokinase:

$$
\begin{array}{ccc}
\begin{array}{c}
CH_2OH \\
| \\
C{=}O \\
| \\
HO{-}C{-}H \\
| \\
H{-}C{-}OH \\
| \\
H{-}C{-}OH \\
| \\
CH_2OPO_3{}^{2-}
\end{array}
&
+\ ATP^{4-}\ \xrightarrow[\substack{phosphofructo-\\kinase}]{Mg^{2+}}
&
\begin{array}{c}
CH_2OPO_3{}^{2-} \\
| \\
C{=}O \\
| \\
HO{-}C{-}H \\
| \\
H{-}C{-}OH \\
| \\
H{-}C{-}OH \\
| \\
CH_2OPO_3{}^{2-}
\end{array}
&
+\ ADP^{3-}\ +\ H^{+}
\end{array}
$$

D-fructose-6-phosphate   D-fructose-1,6-diphosphate

This phosphorylation is also essentially irreversible and is an important control point in metabolism. The responsible kinase requires the presence of magnesium ions for activity.

One of the controls over the rate of this reaction comes from an inhibition of phosphofructokinase by ATP, one of its own substrates, and from a relief of this inhibition by AMP (Fig. 13-3). An enzymatic reaction can't proceed unless substrate

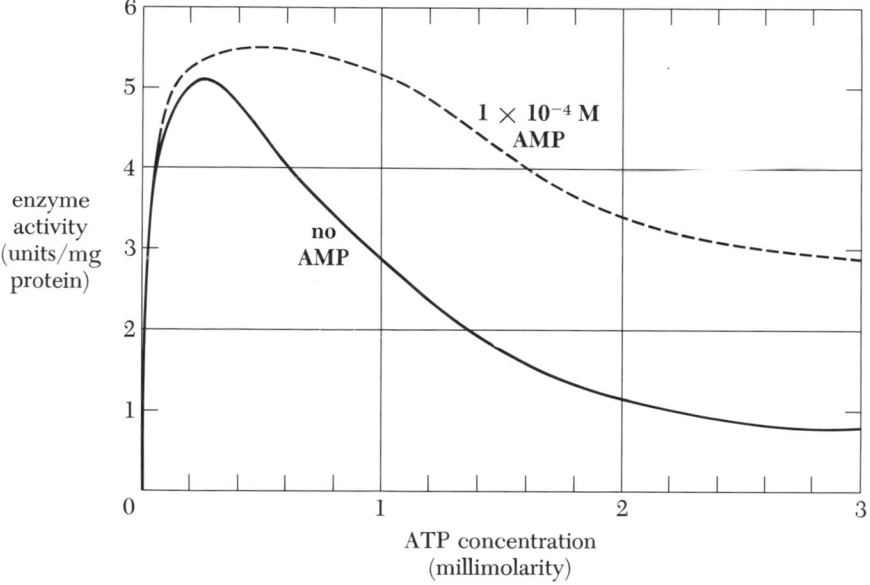

Figure 13-3   The regulation of phosphofructokinase by the concentrations of ATP and AMP. ATP is a substrate for the enzyme; increased concentrations cause increased rate of reaction up to 0.3 mM. Beyond that level, increased concentrations of ATP inhibit the enzyme due to combination with allosteric sites. However, AMP competes with ATP for the allosteric site and has no inhibitory action. It requires much higher concentrations of ATP to cause equivalent inhibition in the presence of AMP (*dashed line*).

is present to react, and phosphofructokinase would not be an enzyme if the reaction was not accelerated by raising the concentration of ATP from a zero level. However, there evidently is a second site on the protein, allosteric rather than catalytic, at which ATP also binds and causes a modification of the protein structure to an inactive form, and this effect becomes predominant and causes a decreased rate if the ATP concentration becomes high enough.

However, AMP will also bind to the allosteric site, thereby competing with ATP, and AMP does not cause the inhibiting modification of structure. Sufficient AMP will therefore prevent the inhibition of the enzyme by ATP, as is shown by the data in the figure.

Phosphofructokinase is also inhibited by citrate, which we shall see is eight steps removed from fructose-1,6-diphosphate in the metabolic scheme. This inhibition is also believed to be an important factor in metabolic control.

Why does the enzyme have these properties? What are their advantages? We can construct a reasonable interpretation. It is a truism to say that there wouldn't be any animal if its tissues didn't contain sufficient enzyme to handle metabolism at a rate consistent with survival, meaning consistent in the broadest sense—the enzymes must be capable of sustaining the functions of the tissue during environmental challenges of all kinds in the animal's particular ecological niche. However, the amount of an enzyme necessary to meet temporary, although serious, metabolic challenges is frequently greater than the amount necessary most of the time in the life of the animal. Such is the case with phosphofructokinase in muscles.

Glucose need not be oxidized as quickly during sleep or rest as it must be during work or combat. Measurements show that indeed it is not, and metabolism evidently adjusts to needs. However, this teleological statement must reflect concrete physical changes—enzymes don't respond to wishful thinking. The phosphofructokinase reaction is essentially irreversible and would catalyze the formation of fructose-diphosphate until all of the available glucose had been converted to this compound, whether needed or not, unless there was some external regulation of the enzyme. (Fructosediphosphate is prepared commercially in exactly this way, by damaging yeast cells so that they lose their internal control of carbohydrate metabolism.)

Control of the reaction by ATP means that the flow of carbons from glucose into the Embden-Meyerhof pathway will be accelerated when ATP is being utilized (causing a drop in its concentration), and slowed when a supply of ATP accumulates.

Inhibition by ATP can be countermanded by an accumulation of AMP. Now, AMP results from the adenylate kinase reaction, $\text{ADP} + \text{ADP} \rightleftharpoons \text{ATP} + \text{AMP}$, as well as from the activation of fatty acids and amino acids, so it acts as a signal, indicating that an increasing *fraction* of the total adenine nucleotides is in the form of ADP and there is a requirement for more substrates as electron donors in oxidative phosphorylation if the ADP is to be converted to ATP. The signal is very sensitive because the adenylate kinase reaction causes the AMP concentration to be proportional to the square of the ADP concentration.

Likewise, if citrate is not being removed as fast as it is formed in the citric acid cycle, its increased concentration will inhibit phosphofructokinase and prevent augmentation of the surfeit by continued formation of excess acetyl coenzyme A from the hexose phosphates.

An amateur critic of evolution might ask why phosphofructokinase isn't built to be inhibited by its own end product, fructose-1,6-diphosphate, in the same way that hexokinases are inhibited by glucose-6-phosphate. We shall see that enzymatic steps between fructosediphosphate and acetyl coenzyme A can generate ATP by substrate-level phosphorylations, and this mechanism provides an emergency supply of energy for the muscles, although limited in quantity, and it is probably advantageous to have the intermediates in these reversible reactions mobilized rapidly without control by anything within them other than the ATP supply.

Cleavage of Fructosediphosphate

The enzyme, *phosphofructoaldolase,* catalyzes the cleavage of fructose-1,6-diphosphate into two isomeric triose phosphates—the aldose, D-glyceraldehyde-3-phosphate, and the ketose, dihydroxyacetone phosphate:

$$
\begin{array}{ccc}
& & CH_2OPO_3{}^{2-} \\
& & | \\
& & C=O \\
& & | \\
CH_2OPO_3{}^{2-} & & CH_2OH \\
| & & \textbf{dihydroxyacetone phosphate} \\
C=O & & \\
| & & \\
HO-C-H & \xleftarrow{\textit{phosphofructoaldolase}} \quad \updownarrow \textit{triose phosphate isomerase} \\
| & & \\
H-C-OH & & \\
| & & \\
H-C-OH & & H-C{\nearrow}^{O} \\
| & & | \\
CH_2OPO_3{}^{2-} & & H-C-OH \\
\textbf{D-fructose-1,6-diphosphate} & & | \\
& & CH_2OPO_3{}^{2-} \\
& & \textbf{D-glyceraldehyde-3-phosphate}
\end{array}
$$

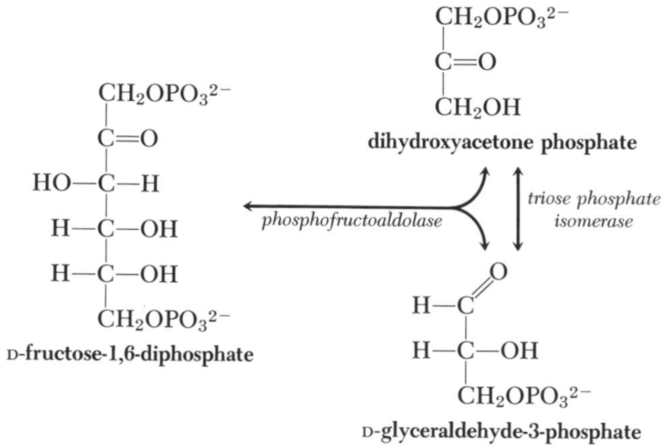

The reaction is freely reversible, and the condensation of the triose phosphates is a typical aldol condensation, which, as we already noted, is a common kind of biological reaction. Aldols are 3-hydroxy carbonyl compounds, and most sugars have this configuration as part of their structure. The standard free energy change of their formation by condensation is usually moderately low, so that they may be cleaved or formed by this type of reaction. Aldolases do exist to catalyze this reaction for many sugars, or their derivatives, but not all of them. The example at hand was the first discovered, and was named "aldolase" without designation of specificity; this practice is still followed by some.

Phosphofructoaldolase is another example of an enzyme in which a lysyl residue reacts with an aldehyde. The ketose phosphates, either fructose-1,6-diphosphate or dihydroxyacetone phosphate, form a Schiff's base with the lysyl residue in the same way that pyridoxal phosphate combines with apotransaminase (p. 117).

The figure also shows that the triosephosphates are interconverted by another enzyme, *triose phosphate isomerase.* This is a typical isomerase, analogous to phosphoglucoisomerase, and it occurs in high concentrations for the same reason: the assurance of rapid reaction with near-equilibrium concentrations of substrate and product. The existence of triose phosphate isomerase makes it possible to replace either of the triose phosphates at the expense of the other, so that all of the carbons of glucose can be funnelled through one of them. For example, the sequence of reactions considered in the next section constantly removes glyceraldehyde-3-phosphate, which is formed from carbons 4, 5, and 6 of fructosediphosphate by the phosphofructoaldolase reaction. When the utilization of glyceraldehyde-3-phosphate lowers its concentration to less than 0.044 of the concentration of dihydroxyacetone phosphate, which is the equilibrium level, the isomerase reaction will cause the formation of glyceraldehyde phosphate from dihydroxyacetone phosphate, which came from carbons 1, 2, and 3, of fructosediphosphate. All six of the carbons are thereby being utilized, and two moles of glyceraldehyde-3-phosphate can be removed for each mole of hexose phosphate originally supplied.

THE FORMATION OF PYRUVATE FROM TRIOSE PHOSPHATES

The formal stoichiometry for the conversion of a triose to pyruvate involves an oxidation:

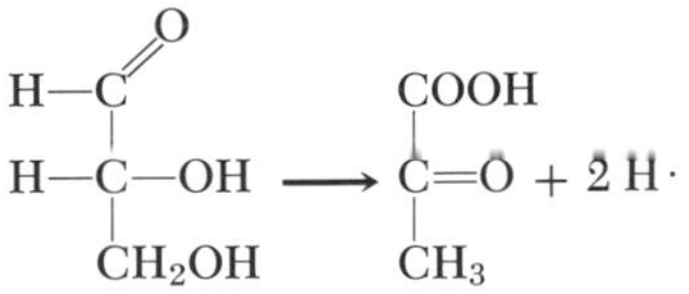

$C_3H_6O_3$ $C_3H_4O_3$

The conversion is actually accomplished in the soluble cytoplasm of cells by a sequence of reactions designed to capture free energy in the form of ATP.

The Oxidation of Glyceraldehyde-3-Phosphate

The sequence begins with the oxidation of the aldehyde group on glyceraldehyde-3-phosphate by NAD. Such oxidations have a standard free energy change sufficiently negative to support the formation of a high-energy bond, as we saw in the oxidation of α-ketoglutarate to succinyl coenzyme A (p. 214). The energy is captured in the oxidation of the triose phosphate by forming a high-energy phosphate anhydride, 1,3-diphospho-D-glycerate:

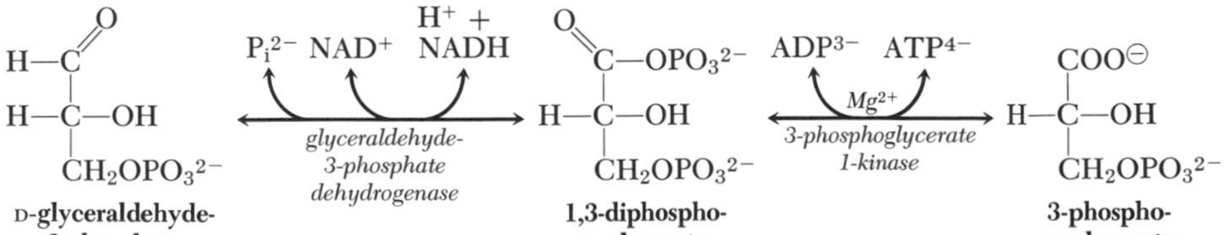

D-glyceraldehyde-3-phosphate **1,3-diphospho-D-glycerate** **3-phospho-D-glycerate**

This is a mixed anhydride of a carboxylic and phosphoric acids, analogous to the aminoacyl AMP compounds used in protein synthesis, with a standard free energy of hydrolysis of $-13,600$ calories per mole at pH 7 and 25°C.

As the figure shows, there is a specific kinase catalyzing the transfer of the high-energy phosphate to ADP to form ATP and release 3-phospho-D-glycerate.

The mechanism of *glyceraldehyde-3-phosphate dehydrogenase* shows some kinship to that of α-ketoglutarate dehydrogenase in creating the high-energy bond as a thiol ester, but they are quite different reactions. Glyceraldehyde-3-phosphate initially combines with a cysteinyl residue on the dehydrogenase as a thiohemiacetal, which is converted to a thiol ester by oxidation with NAD. Phosphorolysis of the ester creates the high-energy phosphate bond and liberates the sulfhydryl group (Fig. 13-4).

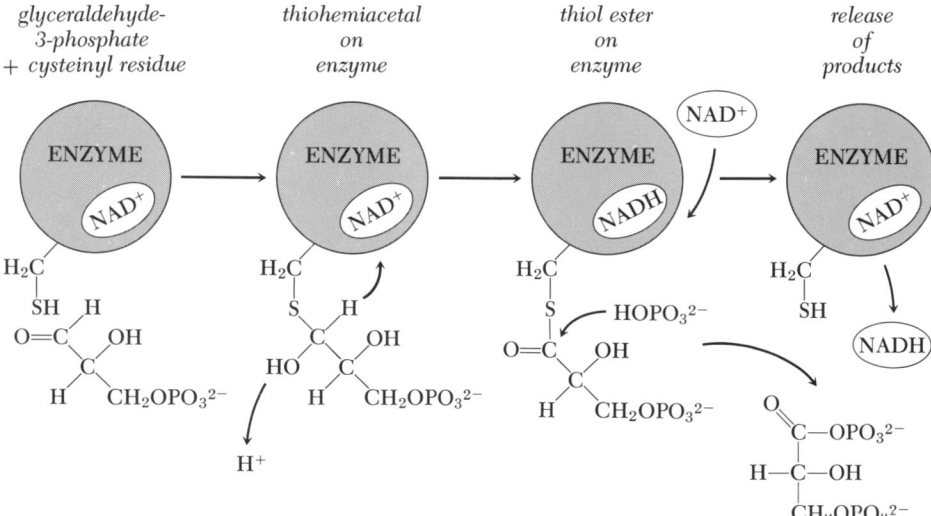

Figure 13-4   The oxidation of glyceraldehyde-3-phosphate involves its combination with a sulfhydryl group on the dehydrogenase peptide, forming a thio-hemiacetal. This structure is oxidized by NAD, also on the enzyme, to form a thiol ester, which is cleaved by inorganic phosphate to produce 1,3-diphosphoglycerate. The NADH produced on the enzyme is less tightly bound, and is displaced by a molecule of NAD from the solution.

We see that the cysteinyl residue doesn't participate in electron transfer in this reaction in the way that a lipoyl residue does in the oxidation of α-keto acids.

However, glyceraldehyde-3-phosphate dehydrogenase is susceptible to inhibition by reagents reacting with thiols. This provided a powerful tool for early investigations in carbohydrate metabolism because treatment of an intact tissue with *iodoacetate:*

stopped metabolism of the triose phosphates at this point, and was the means by which it was shown that high-energy phosphate is depleted during muscular work.

The Transfer of Electrons from Cytoplasmic NADH to Mitochondria

All of the reactions we have been discussing in this chapter occur in the soluble cytoplasm of cells, and NADH cannot diffuse from this compartment into mitochondria. Therefore, the NADH generated by the oxidation of glyceraldehyde-3-phosphate cannot be directly oxidized by the mitochondrial electron transport chain. Since the nicotinamide nucleotides are in limited supply, NADH must in some way be reoxidized within the soluble cytoplasm if carbohydrate metabolism is to continue. This is especially important to the brain and skeletal muscles, which have rapid conversion of glucose-6-phosphate to pyruvate without extensive utilization of NADH for other purposes, such as reductive synthesis.

The principal means of accomplishing this is to use NADH for the reduction of a substrate—in other words, to make a dehydrogenase reaction go in the opposite direction to the substrate oxidations we emphasized in discussing mitochondrial reactions. This isn't at all difficult, because most of the alcohol dehydrogenase reactions of the type:

$$H-\overset{|}{\underset{|}{C}}-OH + NAD^+ \longrightarrow \overset{|}{\underset{|}{C}}=O + NADH + H^+$$

have an equilibrium in favor of the alcohol. For example, the malate dehydrogenase reaction of the citric acid cycle favors the formation of malate, rather than oxaloacetate, and the citric acid cycle only works because NADH is constantly being removed by the electron transport system and oxaloacetate is constantly being removed to form citrate.

However, the NADH accumulating in the cytoplasm from the conversion of glucose to pyruvate tends to make all of the alcohol dehydrogenase reactions in that cellular compartment proceed in the direction of formation of the alcohol at the expense of ketone. Among such dehydrogenases is one catalyzing the interconversion of dihydroxyacetone phosphate and L-glycerol-3-phosphate, and the position of equilibrium for this reaction lies far in the direction of reduction of the ketone to the alcohol. This being so, the NADH generated by the oxidation of glyceraldehyde-3-phosphate can be used to reduce dihydroxyacetone phosphate (Fig. 13-5).

This combination of reactions results in an accumulation of glycerol-3-phosphate, along with 3-phosphoglycerate in the cytoplasm. However, glycerol-3-phosphate can diffuse into mitochondria. The mitochondria of brain and skeletal muscle contain a completely different glycerophosphate dehydrogenase, which is an iron-containing flavoprotein resembling succinate dehydrogenase in that it is membrane-bound and transfers its electrons to coenzyme Q in the regular electron transport chain. Therefore, the direction of equilibrium in mitochondria greatly favors the oxidation of glycerol-3-phosphate back to dihydroxyacetone

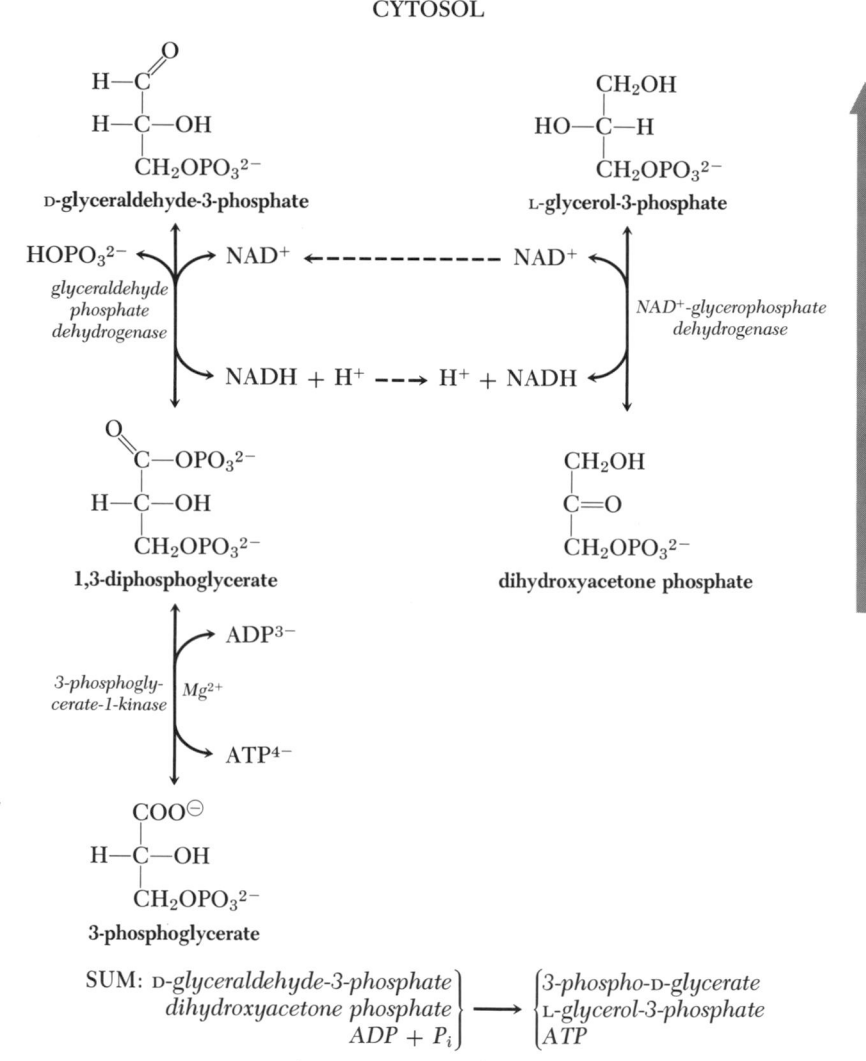

Figure 13-5   The oxidation of glyceraldehyde-3-phosphate results in the formation of 1,3-diphospho-glycerate and NADH. The high-energy phosphate is transferred to ADP, forming ATP and 3-phospho-glycerate. NADH is used to reduce dihydroxyacetone phosphate to glycerol-3-phosphate, thereby regenerating NAD. The latter reaction is a part of the glycerol phosphate shuttle for transferring electrons to mitochondria, shown in the next figure.

phosphate, and the accumulating dihydroxyacetone phosphate will diffuse back into the soluble cytoplasm:

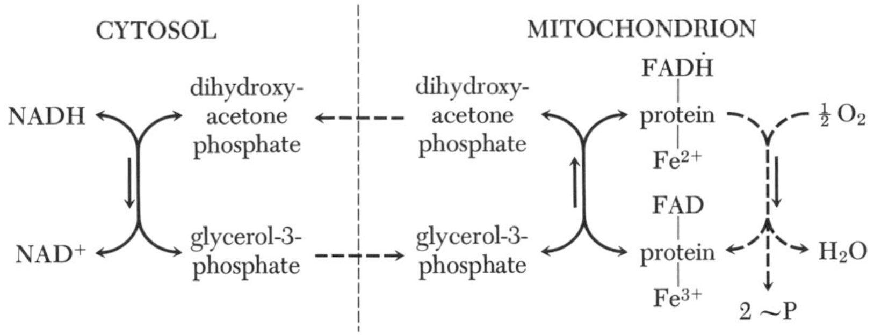

The triose phosphate is serving as a carrier for the electrons from the soluble cytoplasm to the mitochondria. There is no net consumption of dihydroxyacetone phosphate because the amount consumed in the cytoplasm is regenerated in the mitochondria, once the concentration gradients have been established between the compartments. Extra dihydroxyacetone phosphate produced from the aldolase reaction will form more glyceraldehyde-3-phosphate, as we discussed earlier, and not extra glycerol phosphate.

When this *glycerol phosphate shuttle* for electrons is added to the equations for the formation of 3-phosphoglycerate, the overall stoichiometry becomes:

$$\text{(glyceraldehyde-3-P)}^{2-} + \tfrac{1}{2}O_2 + 3 \text{ ADP}^{3-} + 3 \text{ P}_i^{2-} + 2 \text{ H}^+ \longrightarrow$$
$$\text{(3-phosphoglycerate)}^{3-} + 3 \text{ ATP}^{4-} + 3 \text{ H}_2O$$
$$\Delta G^{0\prime} = -38{,}700 \text{ cal mole}^{-1} \text{ (pH 7.0, 25°C)}$$

Remember again that this highly negative standard free energy does not necessarily indicate "waste." The equilibrium constant has dimensions, in this case of $M^{-3.5}$, and, as in oxidative phosphorylation, the production of 3 moles of high-energy phosphate is all that can be sustained at the actual concentrations within a cell.

The Conversion of 3-Phosphoglycerate to Pyruvate

The remaining reactions in the formation of pyruvate rearrange the three carbon compounds in such a way as to capture the free energy difference between 3-phosphoglycerate and pyruvate in the form of ATP (Fig. 13-6).

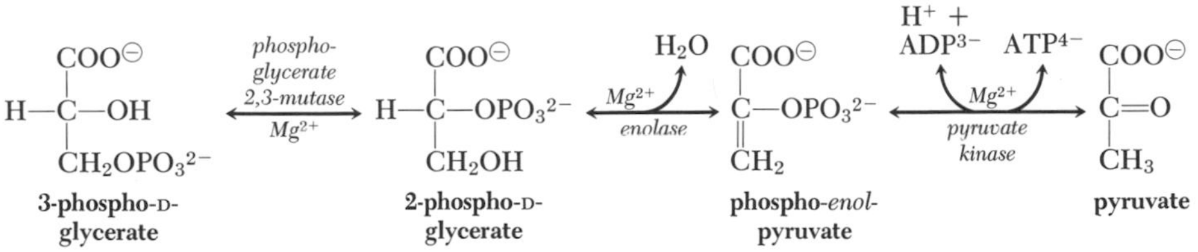

Figure 13-6 The final steps of the Embden-Meyerhof pathway involve the rearrangement of glycerate to pyruvate in such a way that the energy that would otherwise be released is captured as ATP. The phosphate group on 3-phosphoglycerate is transferred to carbon-2 so that removal of the elements of water creates the phosphate ester of the enol form of pyruvate. The phosphorylated enol is a high-energy phosphate compound, which can readily react with ADP to form ATP.

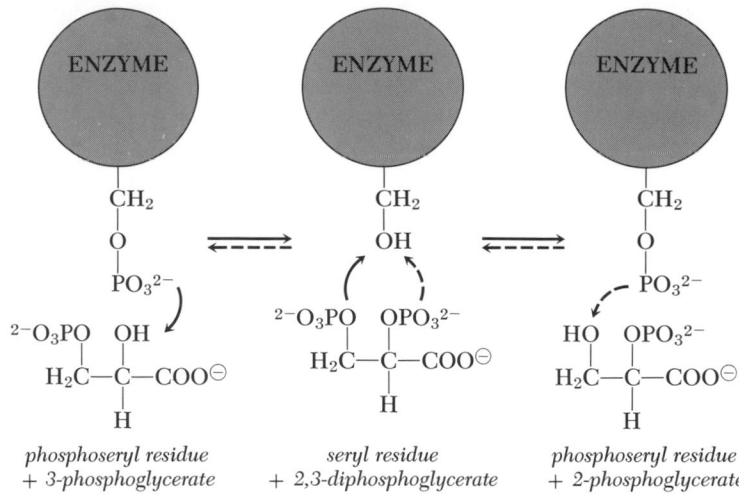

phosphoseryl residue
+ 3-phosphoglycerate

seryl residue
+ 2,3-diphosphoglycerate

phosphoseryl residue
+ 2-phosphoglycerate

Figure 13-7   Mutases, which catalyze intramolecular transfer of phosphate, frequently have mechanisms involving phosphorylation of a seryl group on the enzyme peptide by a doubly phosphorylated form of the substrate. As is shown in the center, either of the phosphate groups of 2,3-diphosphoglycerate may be transferred to a seryl group. Therefore, either 3-phosphoglycerate or 2-phosphoglycerate may be formed. Since the reactions are reversible, the enzyme will catalyze the interconversion of the phosphoglycerates. A small amount of 2,3-diphosphoglycerate is required initially to prime the enzyme with phosphate groups.

The sequence begins with a transfer of phosphate from the third to the second carbon of glycerate. Enzymes catalyzing these intramolecular transfers of phosphate are called *mutases,* and usually proceed by the kind of mechanism shown in Figure 13-7 for phosphoglycerate 2,3-mutase. The active enzyme contains a phosphorylated seryl residue, and the phosphate is first transferred to either 3- or 2-phosphoglycerate. This forms the transient intermediate, 2,3-diphosphoglycerate, on the enzyme surface, which in turn reacts to transfer one of its phosphate groups back onto the scryl residue. However, either of the phosphate groups may be transferred, and the remaining phosphate on the glycerate may be on either the second or the third carbon, so the effect of a continuation of the process is to bring 2-phosphoglycerate and 3-phosphoglycerate to equilibrium.

Small concentrations of the intermediate 2,3-diphosphoglycerate are necessary in the solution to prime the reaction. To insure that it is present, there is a correspondingly small amount of a diphosphoglycerate mutase in the cytoplasm, which catalyzes a transfer of the phosphate from the first to the second carbon of 1,3-diphosphoglycerate:

$$\underset{\textbf{1,3-diphospho-D-glycerate}}{\overset{\displaystyle O}{\underset{\displaystyle CH_2OPO_3{}^{2-}}{\overset{\displaystyle \|}{\underset{\displaystyle |}{\overset{\displaystyle C-OPO_3{}^{2-}}{\underset{\displaystyle H-C-OH}{|}}}}}} \quad \xrightarrow[\substack{\textit{diphosphoglycerate}\\ \textit{mutase}}]{Mg^{2+}} \quad \underset{\textbf{2,3-diphospho-D-glycerate}}{\overset{\displaystyle COO^{\ominus}}{\underset{\displaystyle CH_2OPO_3{}^{2-}}{\overset{\displaystyle |}{\underset{\displaystyle |}{\overset{\displaystyle H-C-OPO_3{}^{2-}}{|}}}}}$$

This priming reaction ought not be confused with the major pathway we are discussing, in which most of the 1,3-diphosphoglycerate is used to form ATP by the phosphoglycerate kinase reaction discussed above.

Continuing on the major pathway, 2-phospho-D-glycerate is dehydrated by the enzyme, *enolase,* to form phospho-*enol*-pyruvate. This is a straightforward transformation of an alcohol into an unsaturated compound, except that the product has a phosphate ester group on one of the unsaturated carbons, and is therefore an ester of the enol form of a keto compound, pyruvate. This is a high-energy phosphate with a standard free energy for hydrolysis of $-13,300$ calories per mole at pH 7.0 and 25°C.

Why is the free energy for the hydrolysis of an enol phosphate ester so much greater than that of an ester of an ordinary alcohol? The reason is that as soon as the enol is released, it will spontaneously come to equilibrium with the keto form:

$$\underset{\displaystyle -C=C-}{\overset{\displaystyle OH}{|}} \longrightarrow \underset{\displaystyle -C-C-}{\overset{\displaystyle O\ H}{\|\ |}}$$

and the free energy released by this equilibration will be added to the free energy of ester hydrolysis. In the case of pyruvate, the position of equilibrium is far in the direction of the ketone—so far that good quantitative measurements of remaining enol have not been obtained—and a correspondingly large amount of free energy is released.

As might be expected, there is another enzyme present to catalyze the capture of the potential energy in phospho-*enol*-pyruvate by transfer of the phosphate to ADP, thereby forming ATP, and releasing free pyruvate. This *pyruvate kinase* reaction completes the sequence by which a molecule of glucose can be converted to two molecules of pyruvate.

The Complete Sequence

With the complete sequence at hand, the individual steps can be seen as part of a rational progression. Glucose is phosphorylated to capture it within the cell by an irreversible process that can be regulated in rate. The aldose phosphate is transformed into a ketose phosphate, freeing the first carbon so that it can be phosphorylated, again by a regulated, irreversible reaction. The second phosphorylation makes it possible to split the molecule in the middle and get two molecules of triose phosphates, interconvertible because the phosphate is on the corresponding carbon of each. Now an energy-yielding oxidation of an aldehyde occurs, producing phosphoglycerate. The phosphate group is shifted from the third to the second carbon. The shift sets the stage for a dehydration to an *enol* configuration, thereby capturing energy as a high-energy phosphate bond—an *enol* configuration such that it reverts to a 2-keto compound when the phosphate is removed. The product is pyruvate, not the 3-carbonyl semi-aldehyde of malonate.

THE OXIDATION OF PYRUVATE

The Pyruvate Dehydrogenase Complex

Pyruvate is generated from glucose in the soluble cytoplasm of the cell, but it can diffuse into the mitochondria, which have a pyruvate dehydrogenase complex

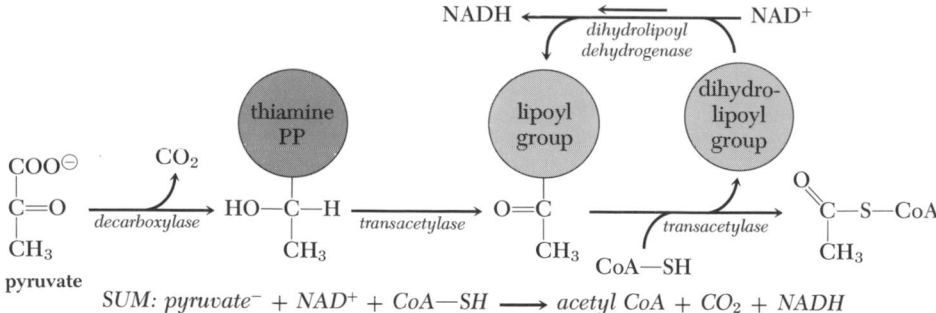

$$SUM:\ pyruvate^- + NAD^+ + CoA{-}SH \longrightarrow acetyl\ CoA + CO_2 + NADH$$

Figure 13-8 The oxidative decarboxylation of pyruvate by the pyruvate dehydrogenase complex follows the same kind of path seen with α-ketoglutarate dehydrogenase. The initial decarboxylation results in the attachment of an ethylol group to the decarboxylase peptide. The group is transferred to a lipoyllysyl residue on a transacetylase peptide, thereby becoming oxidized to an S-acetyl group. The acetyl group is transferred to coenzyme A, forming acetyl coenzyme A and releasing the now-reduced lipoyl peptide. The dihydrolipoyl peptide is re-oxidized by NAD through catalysis by a dihydrolipoyl dehydrogenase, which is a flavoprotein (not shown).

on the inner membrane. Pyruvate is an α-ketocarboxylate, and it is oxidized by the same type of mechanism as is α-ketoglutarate (p. 218). It is decarboxylated to form an *ethylol* group on thiamin pyrophosphate, and this group is oxidized by a lipoyllysyl residue to form an *acetyl*dihydrolipoyllysyl residue on a second protein. Finally, the acetyl group is transferred to coenzyme A, and the dihydrolipoyllysyl group is reoxidized by a flavoprotein with the generation of NADH (Fig. 13-8).

This is the termination of the entire sequence from glucose to acetyl coenzyme A, which can enter the citric acid cycle of the mitochondrion in which it is generated and be oxidized to CO_2 and water to complete the combustion of glucose.

The Effect of Thiamine Deficiency on the Oxidation of Pyruvate

Thiamine must be supplied in the diet as a precursor for thiamine pyrophosphate. (If you guess that the coenzyme is formed by a transfer of the pyrophosphate group from ATP, leaving AMP, you are absolutely right. There is a small amount of a *thiamine pyrophosphokinase* in animal tissues to catalyze this reaction.) If the supply of thiamine is restricted, then one or more enzymes requiring thiamine pyrophosphate will also be deficient. It is worth a moment to consider the nature of this deficiency.

There is a potential competition among the various tissues in the body for any compound required by more than one of them, whether the requirement be large or small. A particular kind of cell may have an advantage because of an ability to absorb the compound at lower concentrations than can other kinds of cells. It may have a disadvantage because the compound turns over rapidly, that is to say that it either decomposes or diffuses out of the cell rapidly. In the case of a co-enzyme, there may be a competition among various apoenzymes within a particular cell for the cofactor. One enzyme may have a lower affinity for the coenzyme, so that it binds a lesser portion of a limited supply.

In essence, there are a large number of variables influencing the character of a particular nutritional deficiency in metabolic terms. Beyond this, there are further variables in manifestations of the deficiency as impairment of human functions. An individual can lose a substantial portion of the function of some tissues and still go through the motions of living, perhaps for years. It is this reserve capacity that makes it possible for a surgeon to remove a kidney, or a lobe of the liver, or a lung, and so on. However, some parts of the central nervous system cannot be dissected so lightly. Pain, paralysis, or loss of sensory function can be seriously disabling. Hence the aphorism, "There are no benign tumors of the brain," and it is probably equally accurate to say that there are few harmless biochemical impairments of the brain. As a result, a nutritional deficiency may be causing a substantial loss of function of other tissues before critical symptoms appear from a disturbance of metabolism in the brain, which may be minor in a quantitative sense.

All of these factors—those determining the vulnerability of particular tissues, and those determining the apparent site of disability—are well displayed in human deficiencies of thiamine. We have seen that thiamine pyrophosphate is required for the oxidation of pyruvate during the metabolism of glucose, and it is also required for the oxidation of α-ketoglutarate in the citric acid cycle, which oxidizes acetyl coenzyme A derived from fats and amino acids as well as from glucose. These processes are important to many tissues. We shall see later that the coenzyme is also necessary in other metabolic routes.

Yet, all of the evidence we have at hand links the deficiency primarily to disturbances of carbohydrate metabolism, especially in the brain. For example, the thiamine requirement of a human is dependent upon carbohydrate intake. On typical mixed diets in this country, an adult will need about 1.5 micrograms per gram of carbohydrate ingested. This would appear to indicate that the thiamine deficiency is primarily being manifested in carbohydrate metabolism, rather than in other routes, and this in turn indicates that pyruvate dehydrogenase loses thiamine pyrophosphate more readily than does α-ketoglutarate dehydrogenase of the citric acid cycle, which is also required for fat metabolism.

This interpretation is supported by studies on the concentration of pyruvate and α-ketoglutarate in the blood of normal and of thiamine-deficient individuals during fasting, and following ingestion of 100 grams of glucose (data from Metabolism, *14*: 141 [1965]):

	Normal		Thiamine-deficient	
	Fasting	After glucose	Fasting	After glucose
(pyruvate)	31 μM	42 μM	49 μM	115 μM
(α-ketoglutarate)	6 μM	7 μM	11 μM	14 μM

These data show that the concentration of α-ketoglutarate in the blood of thiamine-deficient individuals is elevated by a smaller fraction than is the concentration of pyruvate, both before and after eating glucose.

There have been, and still are, ample opportunities for studying the gross effects of these enzymatic disabilities in humans. This may seem strange in view of the fact that thiamine pyrophosphate is an obligatory coenzyme for all of the organisms from which natural human foods are derived, and therefore ought to be a constant constituent of the diet. Indeed, a primary thiamine deficiency, as opposed to a deficiency secondary to a starvation diet, would be a rare thing were it not for two human traits. The first of these is a dislike of coarse food. Hence, humans developed the technique of removing the hard outer layers from seeds, leaving the soft, starchy interior for cooking directly or making into flour. Unfortunately, most of the thiamine pyrophosphate is in the cells of the outer layers. Therefore, those people heavily dependent upon seeds for food are liable to a deficiency of thiamine, and this is most acute in the rice-consuming areas of the Orient. The deprivation is not complete, so that the deficiency develops relatively slowly in many cases, and the resultant illness is called *beri-beri*.

The second human trait resulting in thiamine deficiency is the desire to diminish unpleasant stimulations through the use of ethanol as a depressant. In plain English, a drunk may live only with his bottle, and eat so little food that an acute deficiency of thiamine rapidly develops. The manifestations differ from those of beri-beri, and the illness is known as *Wernicke's encephalopathy*. In addition to its occurrence in alcoholics, it was also well known in Japanese prisoner camps during the Second World War.

The first symptoms of beri-beri are usually abnormal sensations in the limbs. The early disturbance in nervous function may grow worse, finally resulting in paralysis and wasting of the limbs (dry beri-beri). This is consistent with the major dependence of the nervous system on glucose metabolism as a source of energy. However, the nerves of other individuals may successfully compete for the limited thiamine supply, and the cardiovascular system may be caught short. This may be manifested by congestive heart failure with an *edema*—a seepage of liquid into the tissues so that they become puffy (wet beri-beri). There may be a combination of neural and cardiac symptoms, in which the neurological symptoms can be a life-saver by keeping the patient bed-ridden so that he doesn't overload his damaged heart.

Infantile beri-beri is a leading cause of death in some areas, causing vomiting and a peculiar aphonia—a soundless crying.

The more sudden deficiency of Wernicke's disease primarily appears as defects of the central nervous system, with acute mental disturbances and failures of motor control, but even in these cases, the mental impairment sometimes masks effects on the circulation that can result in sudden cardiac failure.

Here, then, are two clinical entities with deprivation of thiamine as a common fundamental cause, but with the tissue predominantly affected depending upon the rate of deprivation. We have dwelt upon this deficiency at some length to illustrate how difficult it is to recognize all of the factors that influence the rate of a single enzymatic reaction in a real animal, and at the same time to see that these factors are concrete, potentially measurable differences between tissues and between enzymes. Unfortunately, most of the necessary metabolic information was discovered after the greatest surge of interest in beri-beri, and there is little current effort to make the necessary measurements for a more complete understanding, despite the prevalence of the condition.

Arsenite and Pyruvate Dehydrogenase

Trivalent arsenic, in addition to its formerly more common use by suicides and murderers, comes in contact with humans through pigments and insecticides. Its toxicity results from its strong affinity for thiol groups:

$$R-As{=}O \; + \; \genfrac{}{}{0pt}{}{HS-R'}{HS-R'} \quad \longrightarrow \quad R-As\genfrac{}{}{0pt}{}{S-R'}{S-R'} \; + \; H_2O$$

and the association is stronger if a pair of thiol groups exists on the same compound. For example, arsenite accumulates in keratin owing to the number of sulfhydryl groups involved in the formation of this protein. Enough hair may resist dissolution in an otherwise well-aged corpse to permit proof of arsenic poisoning decades after the event.

The affinity of arsenite for pairs of sulfhydryl groups favors association with the dihydrolipoyl residue formed as an intermediate by the α-ketocarboxylate dehydrogenases. Enzymes containing lipoate appear to be the locus for the lethal effects because the arsenite chelate cannot be reoxidized by dihydrolipoate dehydrogenase. Indeed, arsenite poisoning mimics the effects of thiamine deficiency on the nervous system. The accumulation of pyruvate in the blood following administration of arsenite also buttresses the interpretation of its primary action as an inhibitor of pyruvate dehydrogenase and similar enzymes. This property is used experimentally to block metabolic routes selectively at points where lipoyl enzymes are required.

Mercuric ions also complex very tightly with thiol groups—so tightly that strongly associated complexes can be formed even when the two thiol groups are on separate molecules. (This property also accounts for the extreme insolubility of mercuric sulfide.) The ability to combine tightly with other thiols may account for the fact that mercuric poisoning has more general systemic effects than does arsenite poisoning, although measurements of pyruvate concentration in mercuric poisoning evidently have not been made, and the different effects of the metals may be due to varying absorption by tissues.

Treatment of poisoning by either arsenite or mercuric ions uses 2,3-dimercaptopropanol, also known as British Anti-Lewisite, BAL, because it was originally developed as an antidote for that arsenical war gas. BAL, with its adjacent sulfhydryl groups, can compete with dihydrolipoyl residues for binding with the metallic ions, forming a soluble chelate that is excreted in the urine:

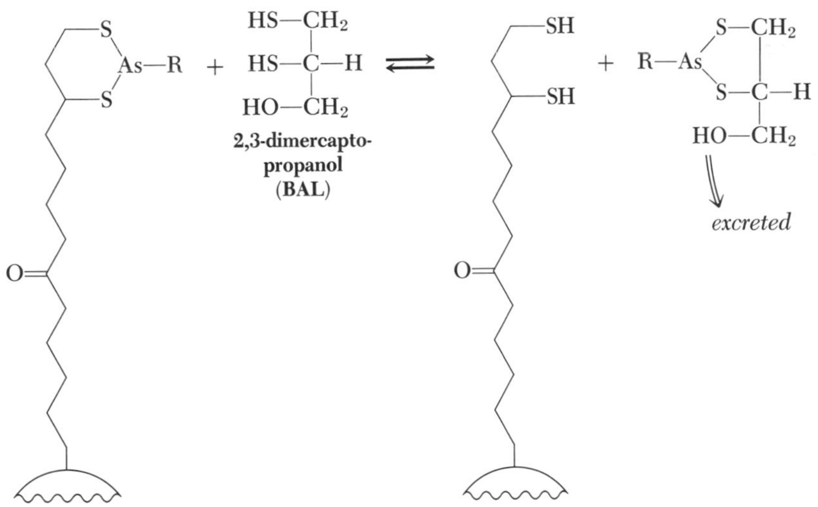

dihydrolipoyl-
arsenite chelate
on enzyme

2,3-dimercapto-
propanol
(BAL)

excreted

A practical point that ought to be kept firmly in mind by everyone is that the stability of the complexes of these ions makes them cumulative poisons, and this is especially true of mercuric ion. They can only be eliminated slowly, so that repetitious administration may cause an accumulation to toxic levels. Metallic mercury is slowly converted to mercuric ions in the body, and spilled mercury can create a hazardous vapor pressure in a room. Failure to recover spillage carefully is a well-known cause of poisoning in laboratory workers. The number of children harmed by its use as a plaything is unknown.

THE YIELD OF HIGH-ENERGY PHOSPHATE FROM THE OXIDATION OF GLUCOSE

In striking the balance for the yield of ATP from the complete oxidation of one mole of glucose, it is important to remember that two moles of triose phosphates are metabolized from each mole of glucose, so that the stoichiometry for all of the reactions beginning with the oxidation of glyceraldehyde-3-phosphate must be multiplied by two. Otherwise, the problem is only one of adding all of the reactions we have considered:

1. Subtract one high-energy phosphate for the phosphorylation of glucose: -1
2. Subtract one high-energy phosphate for the phosphorylation of fructose-6-phosphate: -1
3. Add two high-energy phosphates for each of the two pairs of electrons transported from glyceraldehyde-3-phosphate to mitochondria *via* NADH and the glycerol phosphate shuttle: $+4$
4. Add high-energy phosphate for phosphorylation of ADP by 2 moles of 1,3-diphosphoglycerate: $+2$
5. Add high-energy phosphate for phosphorylation of ADP by 2 moles of phospho-*enol*-pyruvate: $+2$
6. Add three high-energy phosphates for the reoxidation by the mitochondrial electron transport system of each of the two moles of NADH formed by oxidation of pyruvate: $+6$
7. Add 12 high-energy phosphates for the complete oxidation of each of the 2 moles of acetyl coenzyme A in the citric acid cycle: $+24$

TOTAL: $+36$

We therefore arrive at an overall stoichiometry, verifiable by addition of the individual reactions:

$$\text{glucose} + 6\,O_2 + 36\,ADP^{3-} + 36\,P_i^{2-} + 36\,H^+ \longrightarrow 6\,CO_2 + 36\,ATP^{4-} + 42\,H_2O$$

and the P : O ratio for this process is 3.0. The generation of a given quantity of ATP by oxidizing glucose consumes slightly less oxygen than it would by oxidizing fatty acids (P : O ratio $\sim$ 2.8)

Recapitulation of types of reactions

1. Some aldoses and ketoses, or their derivatives, are interconvertible by isomerases:

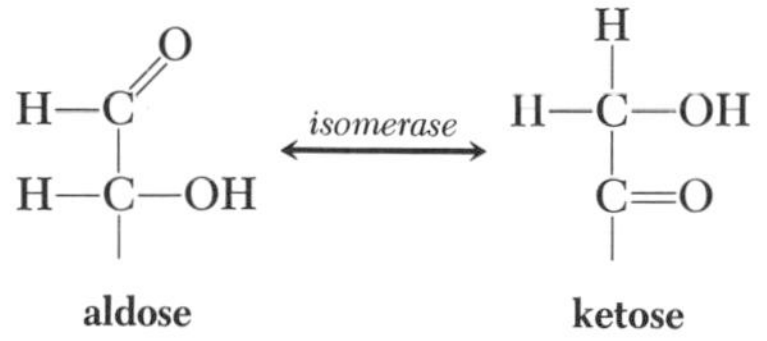

aldose   ketose

Examples: The reactions catalyzed by phosphoglucoisomerase and by triose phosphate isomerase.

2. 3-Hydroxycarbonyl compounds may be formed, or cleaved, by aldolases:

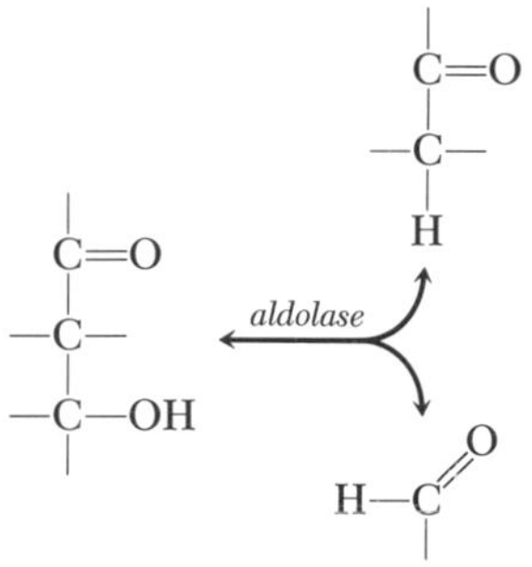

Example: The reaction catalyzed by phosphofructoaldolase.

3. Aldehydes may be oxidized to form acyl phosphates:

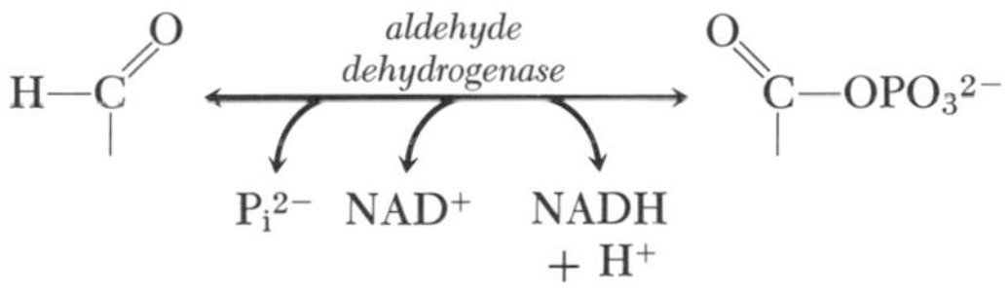

P_i^{2-} NAD^+ NADH + H^+

Example: The reaction catalyzed by glyceraldehyde-3-phosphate dehydrogenase.

4. Phosphates may be transferred from one carbon to another on the same molecule:

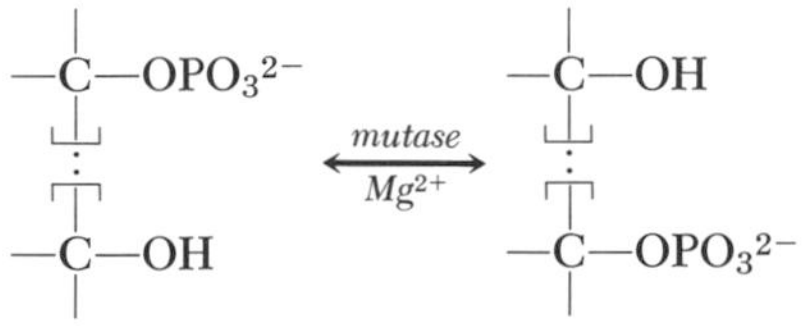

Examples: The reactions catalyzed by phosphoglycerate 2,3-mutase and phosphoglycerate 1,3-mutase.

5. ATP may be generated by transfer of a phosphate from an acyl phosphate:

$$\underset{\displaystyle |}{\overset{\displaystyle O}{\overset{\displaystyle \|}{C}}}-OPO_3{}^{2-} \quad \xleftrightarrow[\substack{\nearrow \quad \searrow \\ ADP^{3-} \quad ATP^{4-}}]{\substack{acyl\ kinase \\ Mg^{2+}}} \quad \underset{\displaystyle |}{COO^{\ominus}}$$

Example: The transfer of phosphate from 1,3 diphosphoglycerate to ADP catalyzed by 3-phosphoglycerate 1-kinase.

6. ATP may be generated by transfer of a phosphate from an enol phosphate:

$$\underset{\displaystyle |}{\overset{\displaystyle |}{\underset{\displaystyle C}{\overset{\displaystyle C}{\|}}}}-OPO_3{}^{2-} \quad \xleftrightarrow[\substack{\nearrow \quad \searrow \\ ADP^{3-} \quad ATP^{4-}}]{\substack{H^+ \nwarrow \\ kinase \\ Mg^{2+}}} \quad \underset{\displaystyle |}{\overset{\displaystyle |}{\underset{\displaystyle C=O}{\overset{\displaystyle -C-H}{}}}}$$

Example: The transfer of phosphate from phospho-*enol*-pyruvate to ADP catalyzed by enolase.

7. A glycol monophosphate may be dehydrated to form the phosphate ester of an enol:

$$\underset{\displaystyle H-\overset{\displaystyle |}{\underset{\displaystyle |}{C}}-OPO_3{}^{2-}}{\overset{\displaystyle -\overset{\displaystyle |}{\underset{\displaystyle |}{C}}-OH}{}} \quad \xleftrightarrow[\substack{\searrow \\ H_2O}]{\substack{enolase \\ Mg^{2+}}} \quad \underset{\displaystyle C-OPO_3{}^{2-}}{\overset{\displaystyle \overset{\displaystyle |}{\underset{\displaystyle \|}{C}}-}{}}$$

Example: The dehydration of 2-phosphoglycerate catalyzed by enolase.

Further reading

Axelrod, B.: *Glycolysis.* In Greenberg, D. E., ed.: *Metabolic Pathways,* Vol. 1 (p. 112). Academic Press (1967).

Crane, R. K.: *Na+-dependent Transport in the Intestine and Other Animal Tissues.* Federation Proc., *24:* 1005 (1965).

Hirschhorn, N., et al.: *Decrease in Net Stool Output in Cholera during Intestinal Perfusion with Glucose-containing Solutions.* New Eng. J. Med., *279:* 176 (1968).

Morgan, H. E., et al.: *Factors Affecting Glucose Transport in Heart Muscle and Erythrocytes.* Federation Proc., *24:* 1040 (1965).

Grossbard, L., and R. T. Schimke: *Multiple Hexokinases of Rat Tissues.* J. Biol. Chem., *241:* 3546 (1966).

Kemp, R. G., and E. G. Krebs: *Binding of Metabolites by Phosphofructokinase.* Biochemistry, *6:* 423 (1967).

Hochster, R. M., and J. H. Quastel, eds.: *Metabolic Inhibitors—A Comprehensive Treatise.* Academic Press (1963). See especially Chapters 20 and 21.

Beaton, G. H., and E. W. McHenry, eds.: *Nutrition—A Comprehensive Treatise.* Academic Press (1964). See especially Chapters 2 and 5.

Scrutton, M. C., and M. F. Utter: *The Regulation of Glycolysis and Gluconeogenesis in Animal Tissues.* Ann. Rev. Biochem., *37:* 249 (1968).

Dickens, F., P. J. Randle, and W. J. Whelan: *Carbohydrate Metabolism and Its Disorders,* Vols. 1 and 2. Academic Press (1968).

Bailey, J. M., P. H. Fishman, and P. G. Pentchev: *Studies on Mutarotases.* J. Biol. Chem., *243:* 4827 (1968). Discussion of biologically active glucose anomers.

GLYCOLYSIS AND THE UTILIZATION OF LACTATE

CHAPTER 14

ARGUMENT

The mechanical energy expended in contracting muscles is derived from the chemical energy of ATP hydrolysis, and the rate at which a muscle can work depends upon the rate at which it can form ATP. Survival of an animal frequently hinges upon its ability to contract muscles more rapidly and strongly than could be sustained by the ATP generated through oxidative phosphorylation. These sudden short-term bursts of activity draw upon more rapidly available sources of high-energy phosphate. One that is readily available is a store of high-energy phosphate in the form of creatine phosphate, which is directly in equilibrium with ATP. Another involves the cleavage of glucose into two moles of lactate with the concomitant formation of two moles of ATP in the cytosol, a process known as glycolysis. Glycolysis results from the accumulation of pyruvate and of electrons in the form of NADH in the cytosol at a faster rate than they can be disposed of by mitochondria, so the equilibrium of a lactate dehydrogenase reaction is displaced to form lactate from pyruvate.

Lactate diffuses into the blood and is slowly disposed of by the liver, which converts a part of it to glucose in a process known as gluconeogenesis. Gluconeogenesis essentially involves the introduction of additional high-energy phosphate to force a reversal of glycolysis. This is accomplished by a pair of reactions utilizing two moles of high-energy phosphate for converting pyruvate to phospho-*enol*-pyruvate and by the direct hydrolysis of hexose phosphate esters. The overall cycle in which glucose is cleaved to

263

lactate in muscles and part of the lactate is converted back to glucose by the liver is known as the Cori cycle. The Cori cycle generates only 37 per cent of the ATP that could be made by the complete oxidation of the amount of glucose consumed in the cycle.

Glycolysis is an anaerobic process in that it can proceed in the absence of oxygen. The term is misleading with skeletal muscles, which usually have an accelerated oxygen consumption caused by the same events that accelerate lactate production, but many microorganisms use glycolysis or related processes as sources of high-energy phosphate, thereby enabling the organisms to survive in the absence of air.

Glycolysis literally means the splitting of glucose. Some tissues, notably the skeletal muscles, convert a part of the absorbed glucose to L-lactate according to the formal stoichiometry:

$$C_6H_{12}O_6 \longrightarrow 2 \; HO-\overset{\displaystyle COO^{\ominus}}{\underset{\displaystyle CH_3}{\overset{|}{\underset{|}{C}}}}-H + 2 \; H^+$$

D-glucose   L-lactate

The important features of this process are that it consumes no O_2 and that it produces H^+.

We saw in the previous chapter that the complete oxidation of glucose involves a production of pyruvate by reactions in the cytosol and the combustion of pyruvate by reactions in mitochondria. The division between cellular compartments adds a complication to the regulation of the rate of glucose metabolism, because there is nothing in the formal nature of the reactions to prevent pyruvate from being produced at a faster rate in the cytosol than it can be oxidized in the mitochondria. This potential imbalance is a reality in skeletal muscles. The muscles of an exercising individual produce pyruvate in excess of the capacity to oxidize the compound, and the disposition of the excess is a significant part of carbohydrate metabolism.

It is well established that the cleavage of ATP to ADP and P_i provides the energy for muscular contraction, although the mechanism of the process is still unknown. The rate at which a muscle can work depends upon the rate at which the ADP produced in the work can be converted back to ATP. Sudden spurts of violent contraction require the ability to deliver large quantities of high-energy phosphate in a hurry, whereas a muscle capable of repeated contraction over long periods of time has a capacity for equally long-term generation of ATP without the continued accumulation of intermediate metabolites.

Skeletal muscles are built to accommodate the kind of load to which they are subjected, and vary in composition within an animal, as well as from animal to animal. A muscle able to work over long periods depends upon the complete oxidation of glucose and fatty acids. It contains more mitochondria to sustain a constantly high rate of oxidation of acetyl coenzyme A and has a higher content of myoglobin to deliver the oxygen required by the electron transport chain. Muscles used for short bursts of very heavy activity utilize ATP at a faster rate than could be sustained by mitochondrial oxidations. These muscles depend more upon readily available stores of high-energy phosphate and upon the generation of ATP from the conversion of glucose to pyruvate in the soluble cytoplasm, and have fewer mitochondria.

Extreme examples of these differing types are evident on every platter of cooked chicken. Chickens constantly use their legs to run around in search of food, but they only rarely fly, and when they do take to the air, the muscles moving the wings must contract rapidly and powerfully to get the heavy carcass off the ground. Therefore, the leg muscles are dark meat, rich in heme proteins, whereas the breast muscles are white meat, deficient in heme proteins and dependent on non-oxidative reactions as ready sources of high-energy phosphate. This does not imply an all-or-none division. All muscles utilize mitochondrial reactions and consume oxygen for the production of ATP, and these oxidations are accelerated upon demand. Likewise, all muscles have inherent stores of metabolites for making ATP without consuming oxygen. The differences between muscles come in the relative balance of the maximum possible rate of generation of ATP by the two processes.

SHORT-TERM GENERATION OF ATP

There are two ways in which an immediate supply of ATP can be obtained without a concomitant consumption of oxygen. One is through the use of stored high-energy phosphate that can be transferred to ADP, and the other is through the conversion of glucose to lactate with accompanying substrate-level phosphorylation of ADP.

Creatine Phosphate

Skeletal muscles store high-energy phosphate in the form of the phosphoric amide of creatine, creatine phosphate, commonly called *phosphocreatine:*

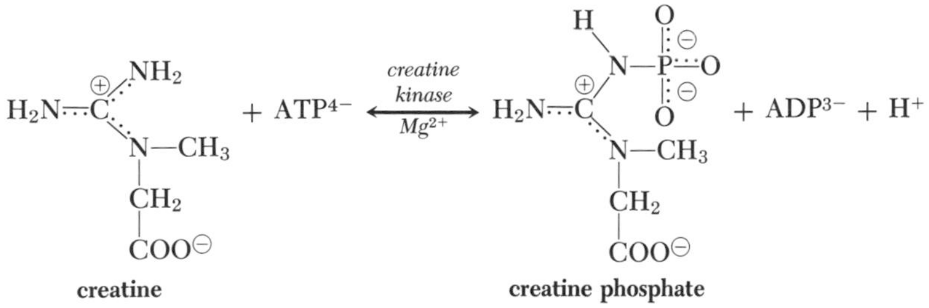

$$\text{creatine} + \text{ATP}^{4-} \underset{Mg^{2+}}{\overset{\textit{creatine kinase}}{\rightleftharpoons}} \text{creatine phosphate} + \text{ADP}^{3-} + \text{H}^+$$

$$K'_{eq} = 0.05 \ (30°, \text{pH } 7.4, 0.002 \text{ M Mg}^{2+})$$

Creatine is a substituted guanidine, which is phosphorylated by ATP in a reaction catalyzed by creatine kinase. The reaction is freely reversible, and the position of equilibrium is such that in muscles of median composition, neither extremely rich nor extremely poor in mitochondria, about two-thirds of the 40 micromoles of creatine in a gram of tissue will be in the form of creatine phosphate at rest, whereas nine-tenths of the total 7 micromoles of adenosine phosphates will be in the form of ATP.

Creatine kinase, and therefore phosphocreatine, is in the same cellular compartment with the ATP immediately available for the contractile proteins. When contraction occurs, the ATP concentration falls and the ADP concentration rises as a result of the utilization of high-energy phosphate for mechanical work. These changes in concentration unbalance the previous equilibrium of the creatine kinase reaction, and phosphate will be transferred to ADP from phosphocreatine, making more ATP for use in contraction and liberating creatine.

Since the regeneration of ATP by this mechanism involves only a single enzymatic reaction and a readily available substrate, it is rapid and permits immediate strong contraction.

Most invertebrates use other guanidinium compounds for high-energy phosphate storage. *Phosphoarginine* is most common throughout the phyla, but some annelids phosphorylate *glycocyamine, taurocyamine,* or *lombricine*:

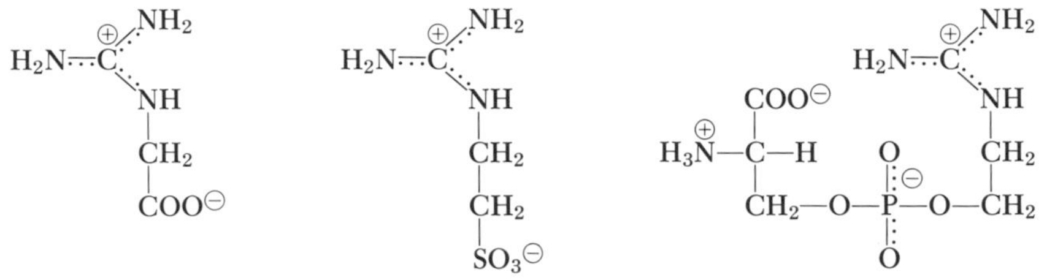

glycocyamine   taurocyamine   lombricine

Lactate Formation

Let us consider more carefully the simple stoichiometric requirements of pyruvate formation, neglecting modifications of rate through allosteric and similar controls. We saw that the conversion of glucose to pyruvate consumes NAD, ADP, and P_i, and produces NADH and ATP. If all of the NAD is converted to NADH, or if all of the ADP is converted to ATP, the absence of these essential reactants will prevent further conversion of glucose to pyruvate.

Regulation of the rate of pyruvate production by the availability of ADP provides an automatic adjustment of glucose metabolism to produce the high-energy phosphate being utilized in muscular contraction or in other metabolic processes. However, an accumulation of ADP cannot accelerate the Embden-Meyerhof pathway unless the NADH produced in that pathway is somehow re-oxidized to NAD.

In the resting muscle, and in most other tissues consuming glucose, the glycerol phosphate shuttle can oxidize NADH in the cytosol and transport the electrons into mitochondria fast enough to keep pace with the pyruvate being produced in response to ATP demand. ATP is utilized in these tissues for protein synthesis, maintenance of concentration gradients across membranes, and so on, and this utilization results in the formation of ADP. The ADP accelerates the conversion of glucose to pyruvate, but not so much that more NADH is being produced than can be handled by the glycerophosphate shuttle, and more pyruvate is being produced than can be oxidized in the mitochondria of these tissues.

A working skeletal muscle is a different story. We must never lose sight of the tremendous metabolic load that can be imposed by muscular contraction—a load that may dwarf the requirements at rest. For example, a young man in good physical condition suddenly springing into action at the maximum rate he can sustain for a period of five minutes will increase his oxygen consumption from around a quarter of a liter per minute to 4 or more liters per minute—a 16-fold, or greater, increase. Practically all of the extra oxygen is being used to generate more high-energy phosphate in his skeletal muscles and heart in response to the stimulus represented by the sudden increase in ADP concentration in his skeletal muscles.

What happens to glucose metabolism when the ADP concentration is kept high by repeated muscular contraction? ADP is no longer limiting for the formation of pyruvate. The rising ADP concentration will also accelerate the electron transport chain in mitochondria, causing an increased rate of oxidation both of pyruvate, through the citric acid cycle, and of glycerophosphate, which is bringing electrons from the cytosol.

If these two processes continued to operate at a balanced rate, glucose would be completely oxidized to CO_2 and water, but at an increased rate. However, skeletal muscles are built with a greater concentration of the enzymes responsible for pyruvate production in the cytosol than of the enzymes in mitochondria catalyzing the oxidation of pyruvate. In other words, the production of pyruvate can swamp the oxidative processes. Even though the oxygen consumption has increased many-fold, the concentration of pyruvate still rises, and part of it diffuses out into the blood.

The concentration of NADH also rises in the cytosol because the glycerophosphate shuttle is not rapid enough to reoxidize the nucleotide as fast as it forms. This would act as a limit on glucose metabolism, were it not for the fact that NADH can be utilized as a reducing agent within the soluble cytoplasm. We have already noted that the equilibrium position of alcohol dehydrogenases favors the reduction of carbonyl compounds to the alcohol, and this circumstance can be used to dispose of any excess NADH. The cytosol of skeletal muscle contains an active *lactate dehydrogenase,* catalyzing the reaction:

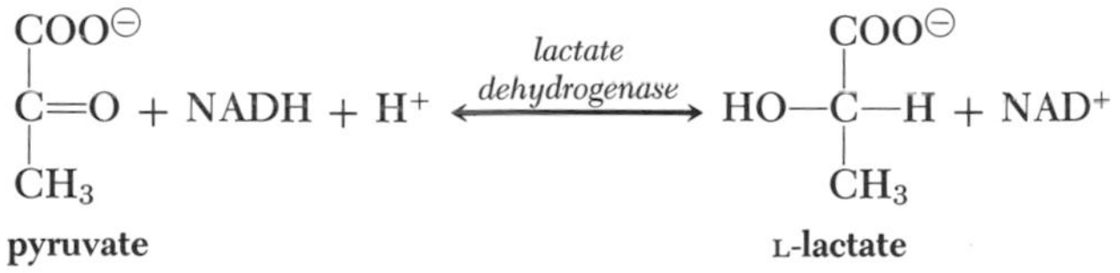

$$\underset{\text{pyruvate}}{\overset{\displaystyle COO^{\ominus}}{\underset{\displaystyle CH_3}{\overset{\displaystyle |}{\underset{\displaystyle |}{C=O}}}}} + NADH + H^+ \underset{\xrightarrow{\hspace{1cm}}}{\overset{lactate}{\overset{dehydrogenase}{\rightleftharpoons}}} \underset{\text{L-lactate}}{\overset{\displaystyle COO^{\ominus}}{\underset{\displaystyle CH_3}{\overset{\displaystyle |}{\underset{\displaystyle |}{HO-C-H}}}}} + NAD^+$$

Pyruvate and NADH cannot continue to accumulate in the cytosol, because this enzyme will catalyze their reaction, producing lactate that diffuses out into the blood and at the same time producing NAD to be re-utilized in the Embden-Meyerhof pathway:

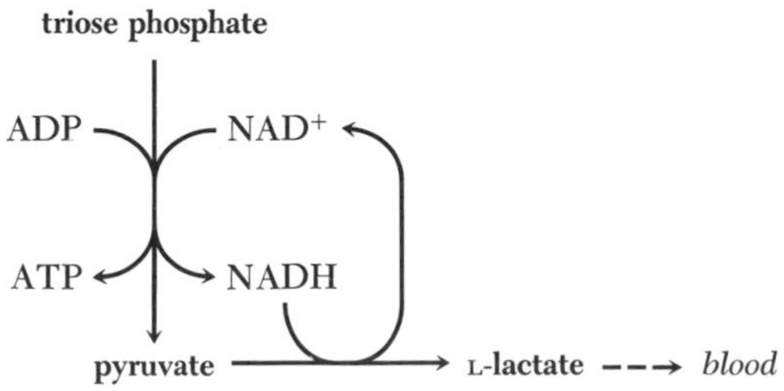

triose phosphate

ADP — NAD$^+$

ATP — NADH

pyruvate → L-lactate $- - →$ *blood*

The amount of lactate going into the blood is greater than the amount of pyruvate also escaping the tissue, and the relative amounts of the two reflects the ratio of concentrations of NADH and NAD.

Stoichiometry of Lactate Production

When the lactate dehydrogenase reaction is added to the stoichiometry for the production of pyruvate (p. 260), the overall reaction is:

$$\text{glucose} + 2\ ADP^{3-} + 2\ P_i^{2-} \longrightarrow 2\ (\text{lactate})^- + 2\ H_2O + 2\ ATP^{4-}$$

Glycolysis therefore represents the cleavage of glucose into two moles of lactate with the concomitant production of two moles of ATP.

Glycolysis does not involve a consumption of oxygen, and is therefore an *anaerobic* process—one that can occur in the absence of air. Skeletal muscles removed from an organism do indeed produce lactate under a nitrogen atmosphere, but it is quite incorrect to conclude from this that skeletal muscles can be regarded as anaerobic organs. We have noted that sustained muscular contractions involve a massive increase in the utilization of oxygen, and lactate production is only a small part of muscle metabolism in most animals.

An accompanying production of H^+ is another feature of lactate formation with functional importance. Glycolysis in itself does not involve production or consumption of H^+. However, glycolysis occurs because of an increased cleavage of ATP during muscular contraction:

$$2\ ATP^{4-} + 2\ H_2O \longrightarrow 2\ ADP^{3-} + 2\ P_i^{2-} + 2\ H^+$$

and when this is added to the glycolytic reaction, the overall economy for muscular contraction plus glycolysis becomes:

$$\text{glucose} \longrightarrow 2(\text{lactate})^- + 2\ H^+$$

The result is a drop in the pH of the blood during rapid glycolysis. We noted in Chapter 4 that the *Bohr effect* in hemoglobin causes an increased release of oxygen when the pH falls. An increased production of lactate in a working muscle therefore will cause an increased release of oxygen to that muscle. Lactate production represents a removal of electrons from triose phosphates at a rate greater than the rate of transfer to oxygen, and the Bohr effect serves as a mechanism for minimizing a deficiency of the supply of oxygen as the cause of the unbalance.

Quantitative Significance of Lactate Production

How much does lactate production contribute to muscular contraction? We can get an idea from measurements made on young men working at a strenuous rate such that they are not completely exhausted in 20 minutes, but close to it. For example, an 83 kg man working so that his oxygen consumption was increased by 3.6 liters per minute accumulated 0.5 mole of lactate during this period, mostly during the first few minutes. One mole of ATP is produced per mole of lactate formed, so glycolysis accounted for the production of 0.5 mole of ATP during this period.

The average ATP utilization by his muscles can be estimated from the oxygen consumption. The complete oxidation of a mole of glucose consumes 134.4 liters of oxygen (6 × 22.4), and produces 36 moles of ATP with perfect coupling of oxidation and phosphorylation, so the ATP production is approximately 0.267 moles per liter of oxygen consumed. We therefore arrive at a rate of ATP production due to oxidative phosphorylation of $0.267 × 3.6 = 0.96$ moles per minute above the resting level.

From this we see that lactate production in this man only represented the formation of an amount of ATP that could be generated in 0.5 minute by oxidative phosphorylation. The same man at rest would have about 1 mole of high-energy phosphate as phosphocreatine and ATP in all of his muscles, equivalent to another minute of work if he could use all of it.

With these figures in mind, the picture is clear. Stored high-energy phosphate sustains strong contraction for a period of less than a minute in full emergency effort. Glycolysis can add energy equivalent to another 30 seconds of effort. (It could be two to four minutes if *all* pyruvate were converted to lactate.) Beyond that, the rate of work is limited by the rate of oxidative phosphorylation, again underlining the fact that skeletal muscles are aerobic organs in most cases, and lactate production is primarily useful for prompt, but limited, responses.

The picture is reinforced by data obtained with runners. Figure 14-1 is a plot of the world record performances in races as of 1966, in which the rate of running

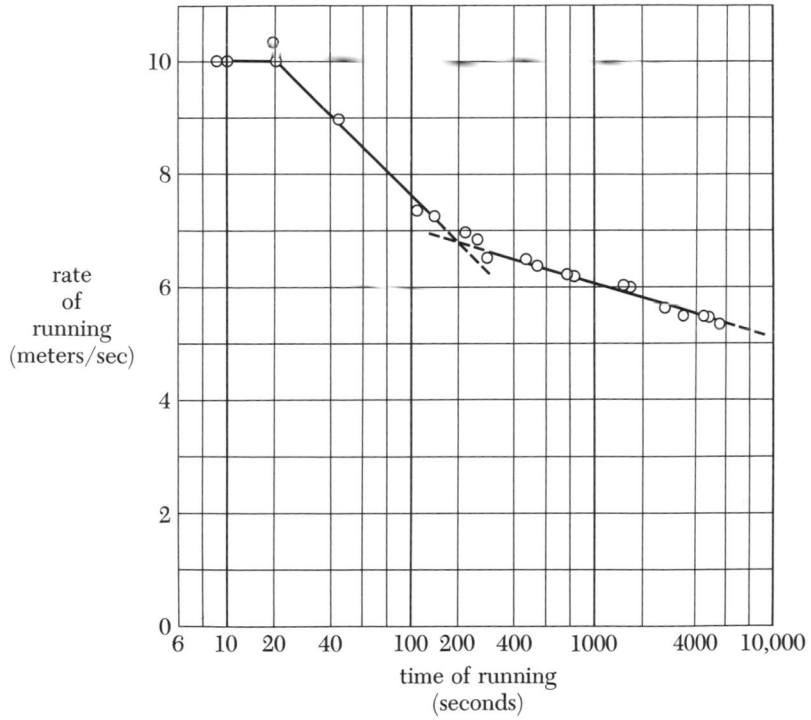

Figure 14-1 World record track performances, plotted so as to show the intensity of work as a function of the time for which it can be sustained. The data suggest that humans working at their maximum exhaust some source of energy in 20 seconds, and there is some additional source that can be mobilized for only 200 seconds.

is compared with the total time of running. Performance in the sprints, lasting for 10 to 20 seconds, is considerably better than in the longer races. Measurements show that the concentration of blood lactate rises little, and may even fall, in these brief exertions. Sprinters may even hold their breath for the period of exercise. Both of these observations are consistent with the view that the main source of energy for these short-term maximum efforts is the immediately available store of high-energy phosphate, mostly as creatine phosphate. As the high-energy phosphate store is depleted, lactate production and oxidative phosphorylation together provide energy in response to the rising ADP concentration. Measurements of the rate of appearance of lactate support the view that this route is largely exhausted in two to three minutes, although there is a small subsequent formation of lactate from the glucose that is continually, although slowly, being returned to the muscles. ATP is also being produced by oxidative phosphorylation during this period. Inspection of the plot shows that there is indeed a break in the trend of the data near 200 seconds of running, which would be consistent with a nearly total dependence on oxidative phosphorylation as the source of ATP for any effort after that time.

The Exclusion of Glycerol Phosphate as an End Product

The reduction of dihydroxyacetone phosphate by NADH in the cytosol is a part of the glycerol phosphate shuttle. Superficially, it appears that this reaction could also be utilized in skeletal muscle to dispose of excess NADH. This would involve the oxidation of one mole of glyceraldehyde-3-phosphate to eventually form pyruvate, and the reduction of one mole of dihydroxyacetone phosphate to form glycerol-3-phosphate (Fig. 14-2). Reactions such as this in which the oxidation of half of the substrate is counter-balanced by reduction of the other half are called *dismutations.*

There is no net production of ATP in this reaction, because the two moles produced in the formation of the one mole of pyruvate replace the two moles earlier used in converting glucose to fructose-1,6-diphosphate. The reaction would be useless as an energy source, but this is not an explanation of why in fact it is not a significant process in muscle even though the necessary enzymes are all present.

One reason that glycerol-3-phosphate doesn't accumulate hinges on kinetics. The reactions from triose phosphate to lactate are accelerated by ADP. Other things being equal, the process consuming the most ADP will be favored, and lactate production consumes two moles per mole of triose handled, whereas the dismutation consumes none. This is more readily seen if one imagines that all of the glucose is indeed converted to a mixture of pyruvate and glycerol-3-phosphate. The enzymes are all present to convert this mixture into lactate, and such a conversion will consume ADP and therefore be accelerated by a rising concentration of ADP.

The latter possibility also reflects the second reason that lactate is formed rather than a mixture of pyruvate and glycerol-3-phosphate. The negative standard free energy change for lactate formation is some 22,000 calories per mole greater than the change for the dismutation, even with the production of two high-energy phosphate bonds.

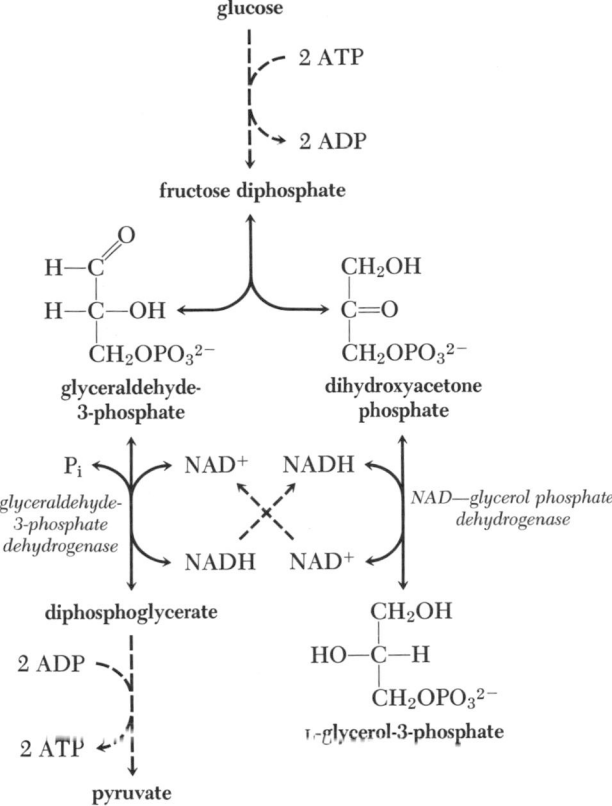

Figure 14-2   The dismutation of glucose to pyruvate and glycerol-3-phosphate. The electrons released as NADH in forming pyruvate are utilized to reduce dihydroxyacetone phosphate. The enzymes for this sequence are present in muscle and it is theoretically possible. However, there are additional reactions resulting in even greater free energy loss that prevent accumulation of glycerol phosphate in mammalian muscle.

Glycolysis and Related Processes in Other Organisms

Glycolysis (since it can produce high-energy phosphate in the complete absence of oxygen) and similar mechanisms are used by microorganisms evolved to fill ecological niches in which there is a supply of carbon compounds and little oxygen. These organisms are properly named *anaerobes;* they may be *obligatory,* and use only anaerobic pathways, or *facultative,* and be able to switch from aerobic to anaerobic metabolism when needed. The overall anaerobic processes of microorganisms are called *fermentations.*

The metabolic pathways of anaerobes may be quite complex even though the overall stoichiometry must balance without a consumption of oxygen, and they serve to remind us that the primitive earth probably had no oxygen in its atmosphere. The earliest organisms may well have used similar pathways to generate energy.

Some microorganisms produce lactate as the primary product of glucose metabolism, and the *Lactobacilli* are often cited as examples, but the most widely known fermentation is the cleavage of glucose to form ethanol and CO_2, especially predominant in yeasts. Yeasts use the same reactions of the Embden-Meyerhof pathway found in animals to produce pyruvate. Pyruvate is decarboxylated in yeasts by an enzyme utilizing thiamine pyrophosphate as a coenzyme. The mechanism is the same as in the pyruvate dehydrogenase reaction of animals, except that there are no associated dehydrogenases. The intermediate hydroxyethyl thiamine derivative decomposes to liberate free *acetaldehyde* into the medium, which is then reduced by NADH to *ethanol:*

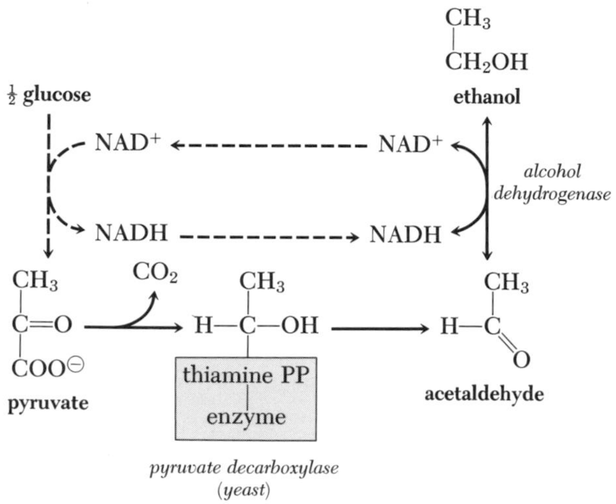

$$\frac{1}{2}\ \text{glucose} \quad\quad\quad\quad \overset{\displaystyle CH_3}{\underset{}{\overset{|}{CH_2OH}}}\ \ \text{ethanol}$$

pyruvate decarboxylase
(yeast)

There are many other and more complicated fermentations known in microorganisms, producing mixtures of products such as butanol, butyrate, glycerol, acetone, and acetate. Hydrogen gas is also a frequent product.

THE UTILIZATION OF LACTATE: GLUCONEOGENESIS

Lactate introduced into the blood from skeletal muscles is metabolized by other tissues, and is not excreted. Some organs, such as the heart, may at times use lactate from the blood as a fuel, oxidizing it to CO_2 and H_2O, but most of the lactate is converted back to glucose by the liver in a process known as *gluconeogenesis*—the formation of glucose from noncarbohydrate precursors. We shall later see that some of the reactions by which glucose is made from lactate are also used to form glucose from amino acids.

Description of the Route

All tissues contain a lactate dehydrogenase, and we noted that the ratio of lactate to pyruvate at equilibrium depends upon the ratio of NADH to NAD. The heart has sufficient mitochondria to handle electron flow under moderate loads and maintains a high enough concentration of NAD for the oxidation of lactate to pyruvate rather than the production of lactate, although this is not always the case.

Similarly, the liver is an organ of relatively constant metabolic rate, and never has the sudden massive ADP load that promotes glycolysis in muscles. The liver also has several routes for removing pyruvate, so that the concentration of this metabolite is kept at relatively low levels. Finally, the liver utilizes NADH in the cytosol by other processes in addition to the glycerol phosphate shuttle. The combination of high [lactate] / [pyruvate] and low [NADH] / [NAD+] favors the oxidation of lactate to pyruvate in that organ.

Even when a low blood concentration of lactate might favor formation of the compound, the liver produces little, because the lactate dehydrogenase of liver differs from the enzyme of muscle. The liver isozyme is constructed so as to be inhibited by lactate itself when pyruvate is the substrate.

The liver, like other tissues, can oxidize pyruvate to acetyl coenzyme A and carry the acetyl coenzyme A through the citric acid cycle. This happens to some extent, but much of the pyruvate is used to produce glucose. Now, glycolysis is just the reverse process, the conversion of glucose to pyruvate with the accompanying production of two moles of high-energy phosphate. We saw that the thermodynamics of glycolysis favor the formation of pyruvate so much that the reverse reaction could not be forced by any reasonable concentration of its end products. The reactions of gluconeogenesis must differ from those of glycolysis in some way that couples them to the utilization of even more high-energy phosphate. In other words, more high-energy phosphate must be consumed in converting pyruvate to glucose than is produced by converting glucose to pyruvate.

The effective introduction of more high-energy phosphate for gluconeogenesis is accomplished in two ways. Firstly, pyruvate is converted to phospho-*enol*-pyruvate by a new pair of reactions requiring two moles of high-energy phosphate. Secondly, the hexokinase and phosphofructokinase reactions, with their equilibria completely unfavorable for glucose formation, are replaced by *phosphatases* catalyzing the hydrolytic removal of the respective phosphate groups.

Phospho-*enol*-pyruvate is converted to pyruvate during glycolysis by the pyruvate kinase reaction. This step is replaced in gluconeogenesis by a *carboxylation* of pyruvate to form oxaloacetate, which is then *decarboxylated* by a second enzyme to form phospho-*enol*-pyruvate (Fig. 14-3). Both the carboxylation and the decarboxylation require high-energy phosphate, so that two high-energy phosphate bonds are broken to create the one in phospho-*enol*-pyruvate, whereas there is only a one-to-one exchange in glycolysis. This raises the concentration of phospho-*enol*-pyruvate sufficiently in the liver to cause a reversal of the same reactions by which fructose-1,6-diphosphate is converted to phospho-*enol*-pyruvate in skeletal muscle during glycolysis.

The two phosphate groups are removed by a specific *fructose-1,6-diphosphatase* and a specific *glucose-6-phosphatase*, first forming fructose-6-phosphate, which is then converted to glucose-6-phosphate by the same phosphoglucoisomerase reaction used in glycolysis, and finally producing free glucose, which can diffuse out of the liver into the blood (Fig. 14-4).

The special enzymes involved in gluconeogenesis are sufficiently important to justify detailed consideration. We shall see that they are involved in amino acid metabolism and fatty acid synthesis, as well as the Cori cycle.

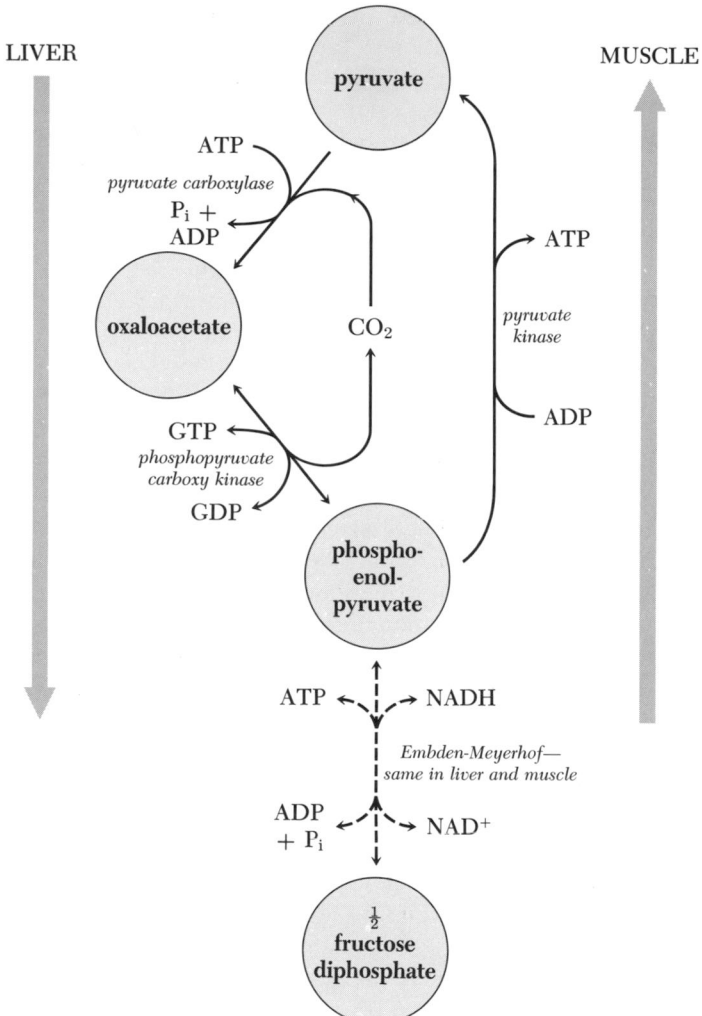

Figure 14-3 The flow of carbons between fructosediphosphate and pyruvate is in opposite directions in the muscle and liver during the Cori cycle. The steps between phospho-*enol*-pyruvate and fructose-diphosphate are the same in both tissues (*bottom center*). CO_2 is added to pyruvate and then removed from the resultant oxaloacetate (*upper left*) in the liver, with each step utilizing high-energy phosphate. The expenditure of two moles of high-energy phosphate drives the equilibrium toward phosphopyruvate. Skeletal muscle lacks the carboxylase enzymes and utilizes only the pyruvate kinase reaction, which catalyzes the transfer of a single mole of high-energy phosphate. The equilibrium of this reaction favors pyruvate formation.

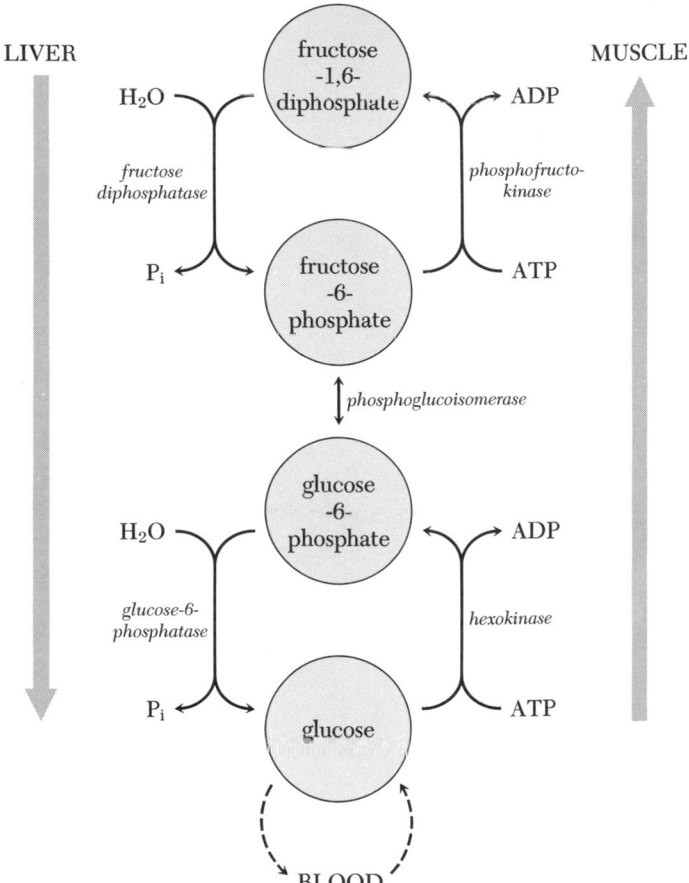

Figure 14-4 The interconversion of fructosediphosphate and glucose occurs by different routes in liver and muscle. The liver contains phosphatases catalyzing simple hydrolyses of fructosediphosphate and glucose-6-phosphate. These are irreversible reactions and therefore cause the conversion of fructose-diphosphate to glucose, which diffuses into the blood. The skeletal muscle lacks glucose-6-phosphatase and has little fructosediphosphatase; it only utilizes the kinases, which catalyze irreversible phosphorylations by ATP. (The liver also has the kinases, but they only function during times of large glucose supply.) This figure is an extension of Figure 14-3.

Pyruvate Carboxylase

Pyruvate carboxylase is a mitochondrial enzyme catalyzing the reaction:

$$
\begin{array}{ccc}
\text{CH}_3 & \text{CO}_2 & \text{COO}^\ominus \\
| & & | \\
\text{C}=\text{O} & \xrightarrow[\quad \text{Mg}^{2+},\ \text{Mn}^{2+} \quad]{\textit{pyruvate carboxylase}} & \text{CH}_2 \\
| & & | \\
\text{COO}^\ominus & & \text{C}=\text{O} \\
 & \text{ATP}^{4-}\ \text{H}_2\text{O} \qquad 2\,\text{H}^+\ \text{ADP}^{3-} & | \\
 & + \text{P}_i^{2-} & \text{COO}^\ominus \\
\textbf{pyruvate} & & \textbf{oxaloacetate}
\end{array}
$$

Carboxylation reactions of this kind involve the participation of *biotin*, a coenzyme attached to lysyl residues of the specific enzymes. Biotin serves as a carrier for the "active" CO_2 that is transferred as a substrate (Fig. 14-5).

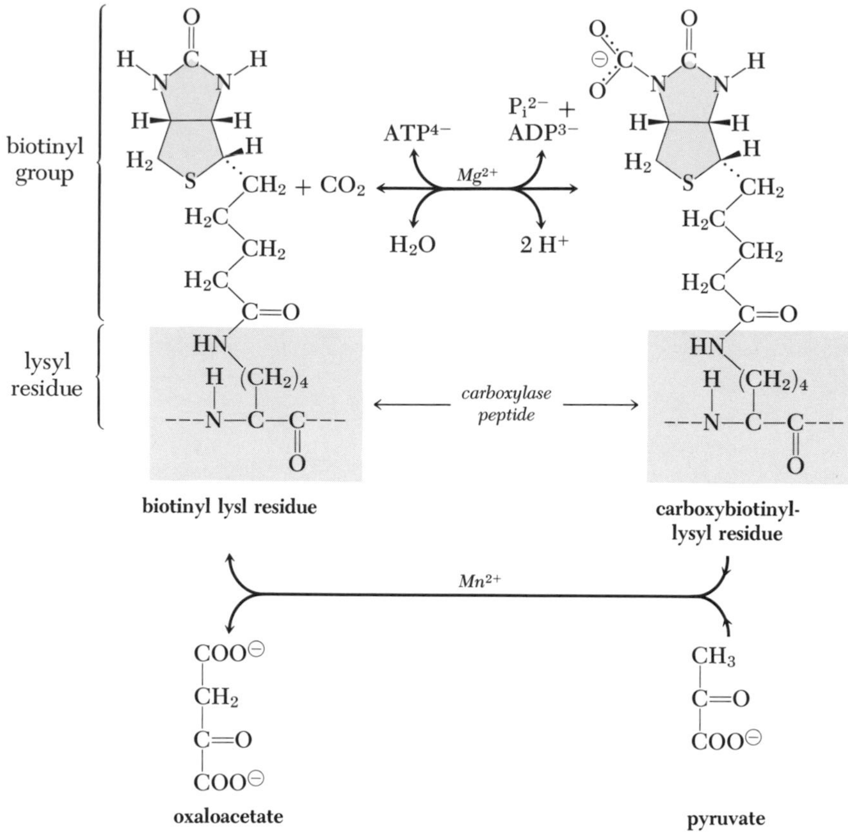

Figure 14-5 The mechanism of pyruvate carboxylase involves a biotinyl group attached to a lysyl residue. The biotinyl group is carboxylated at the expense of high-energy phosphate. The carboxyl group can then be transferred to pyruvate without further expenditure of ATP.

Pyruvate carboxylase is unusual in requiring two metallic ions: magnesium ion is required for the carboxylation of the biotinyllysyl residue and manganous ion is required for binding the keto compounds to the active site of the enzyme.

The most important property of pyruvate carboxylase from the standpoint of regulation of metabolism is an absolute requirement for the presence of *acetyl coenzyme A*, which does not participate in the reaction in any way, but functions solely as a modulating activator. Such an absolute requirement for a modifier is unusual, and it appears to be one of the key mechanisms of control in hepatic metabolism. Pyruvate can diffuse in and out of mitochondria, but acetyl coenzyme A cannot. The localization of the carboxylase step within mitochondria and its sensitivity to acetyl coenzyme A makes gluconeogenesis, which is mainly a process of the cytosol, sensitive to changes in the oxidative metabolism.

Let us pursue this important point. There are two purposes for synthesizing oxaloacetate in the routes we have discussed. One is to provide a carrier of carbons in the citric acid cycle and the other is to participate in gluconeogenesis. An accumulation of acetyl coenzyme A can be used as a useful signal in both functions. If the accumulation results from insufficient oxaloacetate for full operation of the citric acid cycle, it will activate pyruvate carboxylase so as to remedy the deficit. On the other hand, accumulating acetyl coenzyme A may represent a surfeit of substrates for oxidation, with requirements for high-energy phosphate being fulfilled by the initial oxidations of pyruvate or fatty acids so that ADP is not available for a more rapid citric acid cycle. In this case, it is quite appropriate and economical for more glucose to be made by the liver from the excess pyruvate, *via* oxaloacetate, so that it will be used by other organs.

The reasoning in the two cases is summarized in the following:

Case 1: There is a deficit of oxaloacetate.

    a. The deficit slows down the formation of citrate to remove acetyl coenzyme A *via* the citric acid cycle.

    b. The accumulation of acetyl coenzyme A accelerates the pyruvate carboxylase reaction so as to relieve the deficit and permit faster removal of acetyl coenzyme A.

Case 2: There is a surfeit of substrates.

    a. The oxidation of fatty acids, and of pyruvate, forms acetyl coenzyme A.

    b. These oxidations create high-energy phosphate in the mitochondria.

    c. The high-energy phosphate is enough in itself to meet most of the demand of the tissue for ATP.

    d. Therefore, the electron carriers are already largely reduced, and the citric acid cycle will not proceed very rapidly, causing an accumulation of acetyl coenzyme A.

    e. The accumulating acetyl coenzyme A will accelerate the pyruvate carboxylase, causing the formation of oxaloacetate. The accumulation of oxaloacetate, together with the abundance of high-energy phosphate, will accelerate gluconeogenesis.

    f. The accelerated gluconeogenesis removes part of the excess substrates, and will utilize what would otherwise be excess high-energy phosphate for this purpose.

Phospho-*enol*-pyruvate Carboxykinase

This enzyme catalyzes the interconversion of oxaloacetate and phospho-*enol*-pyruvate:

$$K' = 1.7 \,(\text{pH } 7,\, 25°)$$

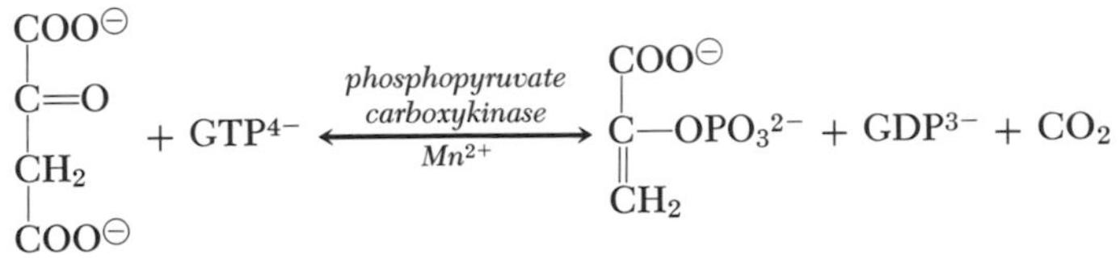

The enzyme differs from pyruvate carboxylase in mechanism; the reaction does not involve biotin, and CO_2 is not "activated." GTP evidently acts to cleave oxaloacetate, with the liberation of CO_2 and the formation of some nucleotide-pyruvate intermediate on the enzyme surface.

The reaction also differs in function. It involves the guanosine nucleotides, and this fact takes our discussion into a largely unexplored realm. The degree to which the guanosine nucleotides are segregated in cells is unknown. In some species, most of the carboxykinase is located in the soluble cytoplasm of the liver, whereas in others it is distributed more evenly between mitochondria and cytosol. Its occurrence in mitochondria can be rationalized as enabling GTP generated by the cleavage of succinyl coenzyme A in the citric acid cycle to be used for phospho-*enol*-pyruvate formation. Gluconeogenesis would then be directly driven by the citric acid cycle. However, in animals such as the rat, in which little carboxykinase occurs in mitochondria, the control would depend upon the rate of appearance of GTP in the soluble cytoplasm. The factors regulating this appearance are not known, except that there is apparently no intermingling of the guanosine nucleotides between the two cellular compartments, so that the generation of GTP in the soluble cytoplasm would depend upon transfer of phosphate from ATP, which can cross the mitochondrial membranes.

Fructose-1,6-diphosphatase

The phospho-*enol*-pyruvate formed by the action of the carboxykinase can be converted to fructose diphosphate through the ordinary glycolytic reactions, by way of phosphoglycerate and the triose phosphates. This overall sequence:

$$2 \text{ (phosphopyruvate)}^{3-} + 2 \text{ NADH} + 2 \text{ ATP}^{4-} + 2 \text{ H}^+ + 2 \text{ H}_2\text{O} \rightleftharpoons$$
$$\text{(fructosediphosphate)}^{4-} + 2 \text{ P}_i^{2-} + 2 \text{ ADP}^{3-} + 2 \text{ NAD}^+$$

has an apparent equilibrium constant near 0.1 M, which is sufficient to make the equilibrium concentration of fructose diphosphate of the order of the concentration of the phosphopyruvate, or higher. (There is considerable uncertainty in the numerical value of the equilibrium constant over this large number of reactions.) No important control other than the concentrations of the reactants and products is known in this sequence, which may go either way, depending upon the availability of glucose or of lactate to the tissue.

Fructosediphosphate is then hydrolyzed to form fructose-6-phosphate in an essentially irreversible reaction catalyzed by a specific enzyme found in the soluble cytoplasm of liver:

$$
\begin{array}{c}
\text{CH}_2\text{OPO}_3{}^{2-} \\
| \\
\text{C}=\text{O} \\
| \\
\text{HO}-\text{C}-\text{H} \\
| \\
\text{H}-\text{C}-\text{OH} \\
| \\
\text{H}-\text{C}-\text{OH} \\
| \\
\text{CH}_2\text{OPO}_3{}^{2-} \\
\text{D-fructose-} \\
\text{1,6-diphosphate}
\end{array}
\quad
\xrightarrow[\text{fructose diphosphatase}]{\begin{array}{c}\text{H}_2\text{O}\quad\text{HOPO}_3{}^{2-}\\ \searrow_{\text{Mg}^{2+}}\nearrow\end{array}}
\quad
\begin{array}{c}
\text{CH}_2\text{OH} \\
| \\
\text{C}=\text{O} \\
| \\
\text{HO}-\text{C}-\text{H} \\
| \\
\text{H}-\text{C}-\text{OH} \\
| \\
\text{H}-\text{C}-\text{OH} \\
| \\
\text{CH}_2\text{OPO}_3{}^{2-} \\
\text{D-fructose-} \\
\text{6-phosphate}
\end{array}
$$

The enzyme is also active in kidney, but only 0.1 as much is found in skeletal muscle, and even this low activity may be due to other cells—connective tissue, the vascular

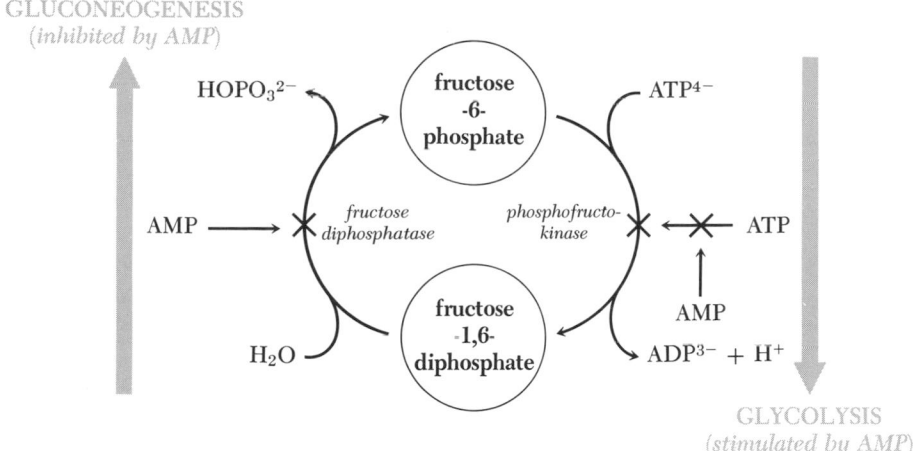

Figure 14-6   The liver contains both phosphofructokinase and fructose diphosphatase. The combined action of these enzymes would result in a constant hydrolysis of ATP. The contrary effects of AMP on the two enzymes prevent both from being simultaneously active. AMP inhibits the phosphatase, whereas it relieves the inhibition of the kinase by ATP.

bed—rather than the contractile fibers themselves. In any event, the difference between these tissues makes the skeletal muscle ineffective as a site of glu-coneogenesis compared to the liver and kidney.

The hydrolysis of fructosediphosphate, like its formation in the reaction cata-lyzed by phosphofructokinase (p. 246) is an important control point. Phosphofructo-kinase controls glycolysis, whereas the phosphatase controls gluconeogenesis.

AMP inhibits fructosediphosphatase. Therefore, any failure of the liver to maintain the adenosine nucleotides in the form of ATP will tend to slow down gluconeogenesis, and conserve the supply of pyruvate for oxidation in the citric acid cycle so as to maintain the high-energy phosphate concentration.

We already saw that AMP also relieves the inhibition of phosphofructokinase by ATP, thereby stimulating the formation of pyruvate from glucose-6-phosphate. AMP is therefore acting in a contrary way on this opposing pair of reactions, and this in itself is an important safety device for tissues containing both enzymes (such as the liver), because the pair, if uncontrolled, would be continually hydrolyzing ATP without accomplishing any useful result (Fig. 14-6).

Glucose-6-phosphatase

Glucose-6-phosphatase catalyzes the hydrolysis of glucose-6-phosphate:

$$2\text{-}O_3P\text{—}O\text{—}CH_2 \quad + H_2O \xrightarrow{\text{glucose-6-phosphatase}} \quad HO\text{—}CH_2 \quad + HOPO_3^{2-}$$

D-glucose-6-phosphate   D-glucose

Like the other enzymes of gluconeogenesis, glucose-6-phosphatase is present in the liver and kidneys, but is scant in skeletal muscle. It is the terminal enzyme of the sequence forming glucose, which can then diffuse into the blood for use by other organs.

Despite its importance, we have a tantalizing lack of information on this enzyme because it is located on the endoplasmic reticulum of the cell, and there are the same problems in studying it as those that are encountered with the enzymes on the cristal membranes of mitochondria. It is known to be inhibited by glucose, and this would superficially appear to be an important kind of regulation—preventing the formation of more glucose when the concentration is already elevated in the blood—but the K for glucose is nearly 0.1 M, so this effect would only appear at abnormally high concentrations of the sugar.

Beyond this, preparations of the enzyme also actively catalyze the hydrolysis of inorganic pyrophosphate and the transfer of phosphate from pyrophosphate or various nucleotide polyphosphates to glucose. It is not known whether these latter reactions are physiological functions or artifacts occurring after the disruption of the endoplasmic reticulum. In short, we cannot give a precise picture of the factors regulating the enzymatic reaction.

However, we shall later see that this is one of the enzymes whose synthesis is unusually responsive to environmental changes, so it may be that the principal regulation comes through changes in the amount of enzyme present rather than through effects on the activity of the protein itself.

The Stoichiometry of Gluconeogenesis and the Cori Cycle

The overall equation for the formation of glucose from lactate according to the reactions outlined above is:

$$2 \text{ (lactate)}^- + 4 \text{ ATP}^{4-} + 2 \text{ GTP}^{4-} + 6 \text{ H}_2\text{O} \longrightarrow$$
$$\text{glucose} + 4 \text{ ADP}^{3-} + 2 \text{ GDP}^{3-} + 6 \text{ P}_i^{2-} + 4 \text{ H}^+$$

Six moles of high-energy phosphate are consumed for each mole of glucose produced, and these must be generated by other processes in the liver. For ease of comparison, let us assume they are generated by the oxidation of part of the lactate to CO_2 and water, although they may in fact be generated by the oxidation of fatty acids or other substrates.

If lactate is being completely oxidized, then the NADH generated in the soluble cytoplasm by the lactate dehydrogenase will be reoxidized by the glycerophosphate shuttle, just as it is in the complete oxidation of glucose (p. 253), and there will be two high-energy phosphates generated per NADH oxidized, or four per pair of lactate molecules converted to pyruvate.

Earlier, we showed that the complete oxidation of the resultant pyruvate in mitochondria results in the formation of 15 high-energy phosphate bonds per molecule of pyruvate, or 30 per pair. There are therefore a total of 34 high-energy phosphate bonds created per pair of lactate molecules completely oxidized to CO_2 and water. Now, if we compare the two routes for lactate utilization:

$$2 \text{ lactate} \longrightarrow \text{glucose:} \qquad \sim P = -6$$
$$2 \text{ lactate} \longrightarrow CO_2, H_2O: \quad \sim P = +34$$

we see that the complete oxidation of 6 out of every 40 pairs of molecules of lactate will generate enough high-energy phosphate to convert the remaining 34 pairs to glucose (6×34 moles produced $= 34 \times 6$ moles consumed). In other words, 85 per cent of the lactate will be converted to glucose, and 15 per cent will be oxidized to CO_2 and H_2O to maintain a constant high-energy phosphate balance in the liver, if this is the only metabolite being utilized.

This is an important concept to grasp, so let us put it in another way. Lactate presented to the liver will be converted to glucose by the processes of gluconeogenesis. This conversion represents a drain on high-energy phosphate. The processes of oxidative phosphorylation will be accelerated to replace the high-energy phosphate, and oxidation will continue to an extent sufficient to replace the drain caused by gluconeogenesis. If lactate is the only available substrate, it will be partially oxidized. Oxidation of 15 per cent of the total is sufficient to provide the high-energy phosphate utilized in converting the remaining 85 per cent to glucose.

This provides us with a basis for a preliminary assessment of the overall economy of lactate production. Let us outline the balance, beginning with the conversion of 100 millimoles of glucose to 200 millimoles of lactate in rapidly contracting skeletal muscles:

Skeletal muscle: 100 millimoles of glucose yield 200 millimoles of lactate plus 200 millimoles of ATP.

Liver: 200 millimoles of lactate arrive in the blood from muscles.
    a. 30 millimoles are oxidized to CO_2, producing 510 millimoles of ATP.
    b. 170 millimoles are converted to 85 millimoles of glucose, utilizing 510 millimoles of ATP.

Skeletal muscle: 85 millimoles of glucose arrive in the blood from the liver.
    a. This replaces all but 15 millimoles of the original glucose converted to lactate.
    b. Therefore, the 200 millimoles of ATP originally gained represent the loss of 15 millimoles of glucose, or an overall production of $200/15 = 13.3$ moles of ATP per mole of glucose consumed.

Now, complete oxidation of glucose in the skeletal muscle produces 36 moles of ATP per mole of glucose consumed. Therefore, the relative efficiency of lactate production as a source of ATP compared to the oxidative route is $13.3/36 = 0.37$. Put another way, 63 per cent of the potential ATP is sacrificed in order to have a rapid, emergency production by glycolysis. In total amounts, this loss is really not very great. We have already noted that lactate production accounts for less than 30 seconds of full-scale ATP utilization. Even in our civilized times, being able to go all-out for a brief period is sometimes life-saving, and the waste of what is at most another minute's worth of ATP isn't a great price.

Recapitulation of types of reactions

1. Carbon dioxide may be attached to a biotinyl group on an enzyme with a concomitant cleavage of ATP, followed by transfer of the resultant carboxylate group to another compound with an active methylene carbon:

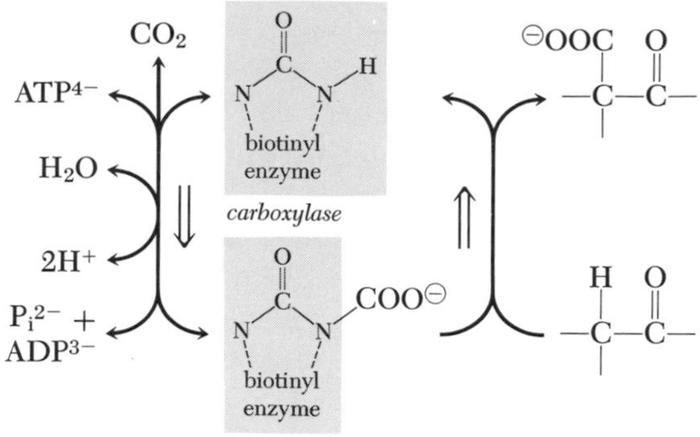

Example: The carboxylation of pyruvate to form oxaloacetate.

2. Carbon dioxide may be attached directly to an enol phosphate ester, with a concomitant transfer of the high-energy phosphate:

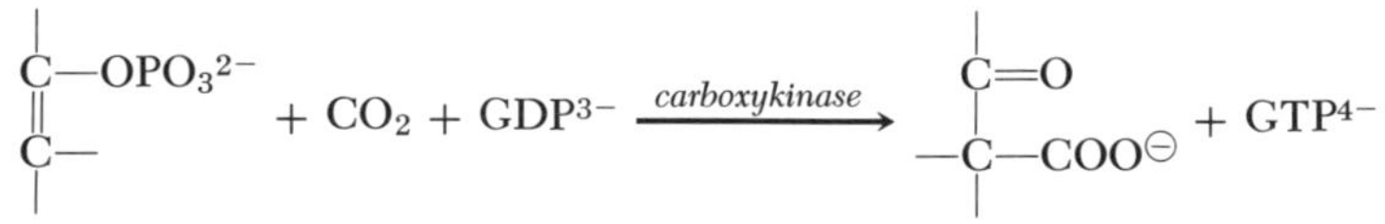

$$\begin{array}{c} | \\ C-OPO_3^{2-} \\ || \\ C- \\ | \end{array} + CO_2 + GDP^{3-} \xrightarrow{carboxykinase} \begin{array}{c} | \\ C=O \\ | \\ -C-COO^{\ominus} \\ | \end{array} + GTP^{4-}$$

Example: The reversible carboxylation of phospho-*enol*-pyruvate to form oxaloacetate and GTP.

3. Phosphate esters may be hydrolyzed:

$$-\overset{|}{\underset{|}{C}}-OPO_3^{2-} + H_2O \xrightarrow{phosphatase} -\overset{|}{\underset{|}{C}}-OH + HOPO_3^{2-}$$

Examples: The specific hydrolyses catalyzed by fructose-1,6-diphosphatase and glucose-6-phosphatase.

Further reading

Chemistry and Metabolism of L- *and* D-*lactic acids.* Ann. N. Y. Acad. Sci., *199:* 851–1165 (1965). A valuable collection of a number of articles.

Bueding, E., and E. Farber: *Comparative Biochemistry of Glycolysis.* In Florkin, M., and H. S. Mason, eds.: *Comparative Biochemistry,* Vol. 1 (p. 411). Academic Press (1960).

Wood, W. A.: *Fermentation of Carbohydrates.* In Gunsalus, I. C. and R. Y. Stanier, eds.: *The Bacteria,* Vol. 2 (p. 59). Academic Press (1961).

Dunn, A., M. Chenoweth, and L. D. Schaeffer: *Estimation of Glucose Turnover and the Cori Cycle Using Glucose-6-t-[14]C.* Biochemistry, *6:* 6 (1967).

Morrison, J. F., and A. H. Ennor: *N-Phosphorylated Guanidines.* In Boyer, P. D., H. Lardy, and K. Myrbäck, eds.: *The Enzymes,* Vol. 2 (p. 89). Academic Press (1960).

Mildvan, A. S., M. C. Scrutton, and M. F. Utter: *Pyruvate Carboxylase VII. A Possible Role for Tightly Bound Manganese.* J. Biol. Chem., *242:* 3488 (1966).

Wood, H. G., J. J. Davis, and H. Lochmüller: *The Equilibria of Reactions Catalyzed by Carboxytransphosphorylase, Carboxykinase and Pyruvate Carboxylase and the Synthesis of Phosphoenolpyruvate.* J. Biol. Chem., *241:* 5692 (1966).

Weber, G., ed.: *Advances in Enzyme Regulation.* Pergamon (1964). Pertinent discussions appear in several volumes; volume 2 contains discussions of lactate dehydrogenase, regulation of pyruvate carboxylase, and gluconeogenesis in general.

Scrutton, M. C., and M. F. Utter: *The Regulation of Glycolysis and Gluconeogenesis in Animal Tissues.* Ann. Rev. Biochem., *37:* 249 (1968).

PART III

Storage of the Major Fuels

NON-REDUCTIVE STORAGE: GLYCOGEN

CHAPTER 15

ARGUMENT

Glycogen is a polymer used to supply glucose in intervals between meals and at times of heavy muscular work. It is composed of glucosyl residues linked mainly between carbons 1 and 4, but with branches formed at intervals by additional linkages between carbons 1 and 6, so that the entire molecule has a tree-like outline. The main stores of glycogen are in the skeletal muscles and the liver. The storage of glucose involves the extension of the glycogen branches by a transfer of glucosyl units from uridine diphosphate glucose, which in turn is formed from glucose-6-phosphate *via* glucose-1-phosphate. Branches are formed by a transfer of segments of 1→4 chains onto C-6 of adjacent chains.

The glucosyl units are recovered for metabolism by cleavage with inorganic phosphate to produce glucose-1-phosphate, catalyzed by phosphorylase. Extensive breakdown of glycogen also requires rearrangement of the chains and removal of the 1→6 branches by other enzymes.

Since glycogen is both formed from, and broken down to, glucose-1-phosphate, the enzymes catalyzing these pathways are tightly regulated, and in an intricate way. The regulation hinges upon an alteration of catalytic activity through the phosphorylation of seryl residues in the responsible enzymes. In skeletal muscle, such phosphorylations are triggered by the hormone, adrenaline, arriving from the adrenal medulla in response to an exciting stimulus and causing an enzyme, adenyl cyclase, to produce a cyclic nucleotide, cyclic-AMP. Cyclic-AMP is an activator for protein kinases, and its appearance eventually results in the phosphorylation of the enzyme adding glucosyl residues, glycogen synthase, and the enzyme removing glucosyl residues, phosphorylase. The change has opposite effects on the activities of these enzymes, with phosphorylated glycogen synthase being inactive and phosphorylated phosphorylase being active, so the result of adrenaline arrival is to stop glycogen synthesis and accelerate glycogen breakdown in the muscle. When adrenaline is removed, as it is by rapid destructive processes, the entire chain of events is reversed through the action of phosphatases that cleave the phosphoproteins, so glycogen synthase now becomes active and phosphorylase inactive. This restores the resting state in which glycogen can be formed without being broken down.

Similar transformations occur with the enzymes in liver, except that adenyl cyclase in that organ is activated by the hormone, glucagon, secreted from the pancreas in response to falling blood glucose concentrations. This causes phosphorolysis of glycogen and release of the glucosyl units as free glucose into the blood. The pancreas therefore regulates blood glucose concentration by elaborating insulin at times of high concentration to promote uptake by skeletal muscle and other tissues, and by elaborating glucagon at times of falling concentration to promote liberation of glucose from the liver to maintain the level.

Many genetic abnormalities of glycogen metabolism are known in humans, and affect the ability to store or to break down glycogen in either the muscles or the liver. Some of these defects permit survival into adulthood with a reasonable ability to do light work, and this is an indication that glycogen is not an obligatory fuel for muscles.

The ability of an organism to interrupt its food intake—to eat meals rather than repeated small snacks and to sleep long intervals—depends upon a capacity to store transient excesses of absorbed compounds for later use. The demand for high-energy phosphate is not transient, and carbon compounds supplying the processes by which it is generated must always be available from internal stores. Most organisms, plant or animal, have among their stores some polymer of a carbohydrate, a *polysaccharide*. Most commonly, this is a polymer of glucose, *starch* in the case of plants, and *glycogen* in animals. These are homopolysaccharides—polymers made from one kind of sugar.

THE NATURE OF GLYCOGEN

In a formal sense, glycogen contains glucosyl residues joined by ether linkages, which are made by the loss of water between hydroxyl groups. The linkage always involves the hemiacetal hydroxyl on C-1. Most of the bonds are formed to C-4 of the neighboring residue (Fig. 15-1 A), and the kind of polymer known as an *amylose* chain contains nothing but these 1→4 glucosidic bonds, and is therefore a 1→4 glucan. Starches contain a high proportion of amylose, which has the ability to form a bright blue complex with iodine.

However, glycogen does not have long amylose chains, because of the presence of many branches, and gives a less intense red-violet color with iodine. These branches are formed by additional glucosidic bonds with the hydroxyl group on C-6 (Fig. 15-1 B), and the interior of the molecule has 1→6 bonds at an average spacing of four 1→4 linked residues. The resultant highly branched arrangement is sketched in cross-section in Figure 15-1 C.

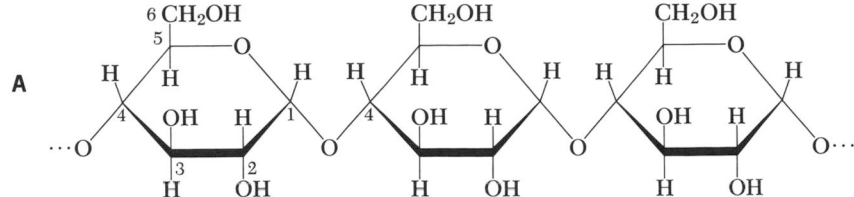

A. Glycogen contains linear amylose chains, made by linking carbons 1 and 4 of glucose through oxygen.

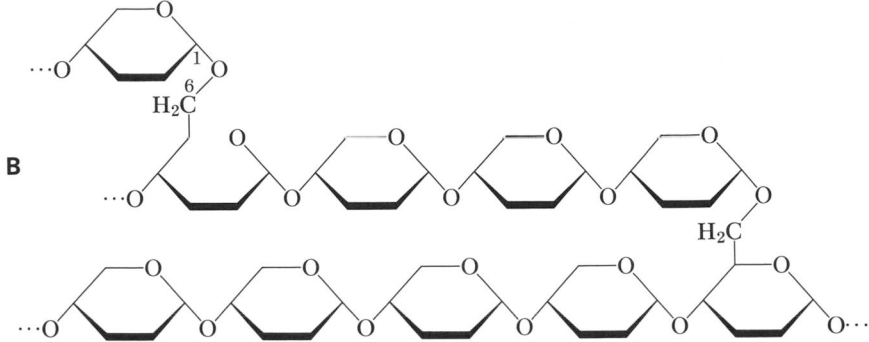

B. Branches occur in the amylose chains where carbon 6 of a residue is also linked through oxygen to the C-1 terminal of another chain segment.

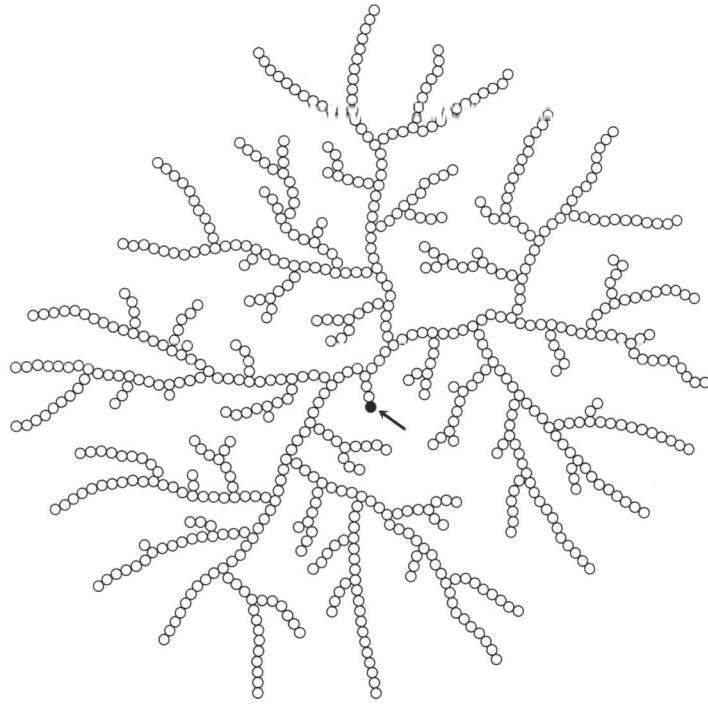

C. A cross-section through glycogen showing the tree-like structure created by branched amylose chains. The short inner segments are actually part of branches extending above and below the cross-section. The circles represent glucosyl residues; only one residue (arrow) doesn't have C-1 attached to an ether linkage so that it may assume an open-chain form and mutarotate.

Figure 15-1 The structure of glycogen.

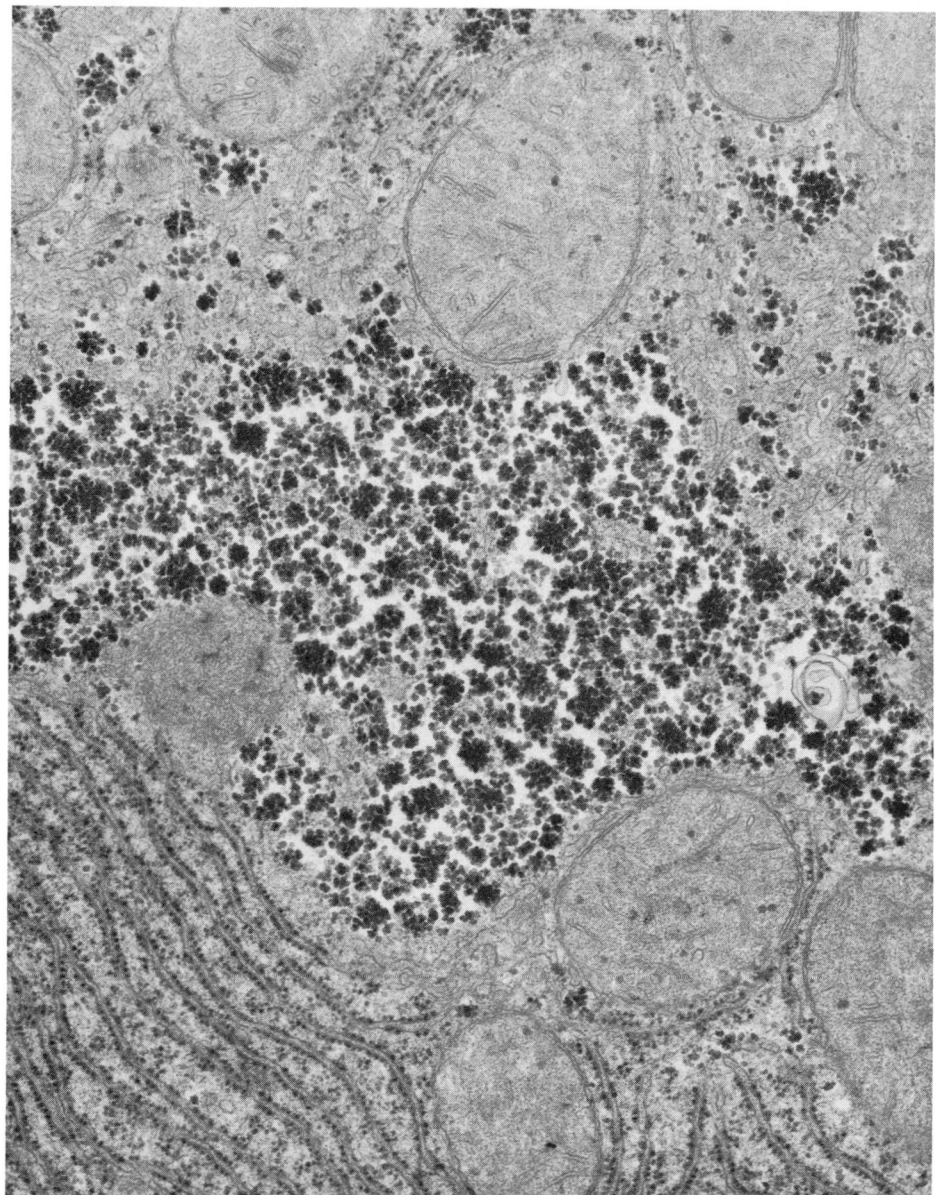

Figure 15-2   Dense clumps of glycogen fill the center of this view of the liver of a rat fed a high-carbohydrate diet. The large organelles are mitochondria. Rough endoplasmic reticulum and associated ribosomes are at the lower left; smooth endoplasmic reticulum is at the upper right. Magnification 31,000×. (Courtesy of Dr. Robert R. Cardell, Jr.)

Glycogen appears as discrete particles in cells (Fig. 15-2), evidently composed of aggregates of molecules with molecular weights ranging around 10^7, of which about two-thirds is water (the many hydroxyl groups make the molecule quite polar). The particles also contain the enzymes concerned with the formation and breakdown of glycogen, but the spatial arrangement of the protein constituents in relation to the polysaccharide is unknown.

The amount of glycogen varies widely among different tissues. Liver and muscles contain the largest stores, but the amount depends upon the nature of the diet and the activity of the animal. We shall mainly be concerned with these two tissues. Other tissues contain glycogen, which is important to their economy, but it is usually of lower concentration and a small fraction of the total in the body. In addition to the natural variability, the measurement of glycogen content is notoriously subject to error, partly because of the difficulties in quantitative separation and partly because of the rapid decomposition of the compound in excised tissue samples. The values for glycogen content found in various handbook tables are frequently an uncertain guide.

In general, the content of glycogen in the liver will vary greatly with diet, while that in the muscles will be less affected. On the other hand, exercise rapidly depletes the glycogen of the muscles. The effect of diet is well illustrated by data obtained with rabbits that had been starved for 96 hours or fed a standard commercial diet supplemented by force-feeding sucrose through a stomach tube (Data for intermediate conditions are also given in the original source: Bull. J. Hopkins Hosp., Vol. 61, p. 349 [1937]):

Treatment	Glycogen Content (g/kg of tissue)*	
	Liver	Skeletal muscle
Starved 96 hours	16.5	5.2
Force-fed sucrose	115	13

*Dry weight of glycogen obtained per kilogram of fresh tissue.

Reliable data for human tissues are especially scarce because fresh tissue specimens from healthy, unanesthetized individuals are not available through ordinary laboratory practices. (There is a high negative correlation between skill with the switchblade and skill in glycogen analysis. An extension of this phenomenon no doubt explains the apparent absence of useable information from the experiments performed on prisoners in the German concentration camps during the Second World War.) However, there have been measurements made on biopsy specimens from the liver and skeletal muscles of unanesthetized humans after an overnight fast, the so-called basal condition, in which the liver content is relatively low, and the results range as follows (Clinical Sci., Vol. 7, p. 287 [1949]):

Tissue	Glycogen Content (g/kg of tissue)
Liver	10 to 40
Pectoralis major*	11 to 39
Gastrocnemius*	8 to 22

*The pectoralis major runs from the chest to the upper arm. The gastrocnemius is in the calf and connects the upper leg to the Achille's tendon.

The range is consistent with information on the total glycogen store available for exercise, which we shall treat later (p. 524). It may be seen from both of the above sets of data that the liver has a potentially higher concentration of glycogen than does skeletal muscle. We can estimate that a nominal man on an average diet will have, at rest in the daytime, something near 50 grams of glycogen in a kilogram of liver and 15 grams of glycogen in a kilogram of muscle.

However, a man weighing 70 kilograms will have about 30 kilograms of skeletal muscle, but only 1.6 kilograms of liver. Therefore, the major part of his glycogen store will be in the muscles—a total of 450 grams, compared to 80 grams in the liver. A glucosyl residue in glycogen, $C_6H_{10}O_5$, has a formula weight of 162, so the total store in both tissues will be on the order of 3.3 moles of glucose.

MECHANISM OF GLYCOGEN STORAGE

Glycogen is mainly formed by enlarging existing molecules, using glucosyl residues from a nucleotide, *uridine diphosphate glucose (UDP-glucose)*, which in turn is derived from glucose-6-phosphate. The overall process is conveniently thought of in three stages: (1) the conversion of glucose-6-phosphate to UDP-glucose; (2) the transfer of glucosyl units from UDP-glucose onto terminal glycogen chains, so as to make a growing amylose chain composed of 1→4 linkages; and (3) the creation of branches by shifting a portion of the chain onto the C-6 hydroxyl group of an adjacent chain.

The formation of UDP-glucose proceeds in a straightforward way. It is the first carbon of glucose that will later be attached to glycogen, so the sequence begins by a freely reversible mutase reaction, in which the phosphate group of glucose-6-phosphate is transferred to form *glucose-1-phosphate* (Fig. 15-3).

Like the phosphoglycerate mutase (p. 254), phosphoglucomutase utilizes the corresponding diphosphate compound in small concentrations as an intermediate. There is a small amount of a kinase present that catalyzes the phosphorylation of glucose-1-phosphate by ATP to form the glucose-1,6-diphosphate necessary to insure maximum activity of the phosphoglucomutase. This is a trivial reaction in quantitative terms compared to the mutase reaction itself because so little of the diphosphate is necessary.

Glucose-1-phosphate then reacts with UTP to form UDP-glucose and inorganic pyrophosphate. As is the case with other pyrophosphorylase reactions, the constant removal of the pyrophosphate through hydrolysis by inorganic pyrophosphatase pulls an otherwise reversible reaction in the direction of UDP-glucose formation. The UTP utilized in this reaction is generated by a *nucleoside diphosphokinase* reaction: $ATP^{4-} + UDP^{3-} \rightleftharpoons ADP^{3-} + UTP^{4-}$, so that the ultimate source of the high-energy phosphate is ATP. The concentration of UTP probably controls the rate of UDP-glucose formation, because the K_M for UTP with UDP-glucose pyrophosphorylase is 2×10^{-4} M, and the reported concentration of UTP in the soluble cytoplasm of liver is near 3×10^{-4} M.

UDP-glucose acts as a donor of glucosyl residues for extending the terminal branches of glycogen in a reaction catalyzed by *glycogen synthase*. The reaction is specific for the hydroxyl group on C-4 of the glycogen residues, so the new glucosyl residue simply extends the 1→4 chain (Fig. 15-4). Since the nature of

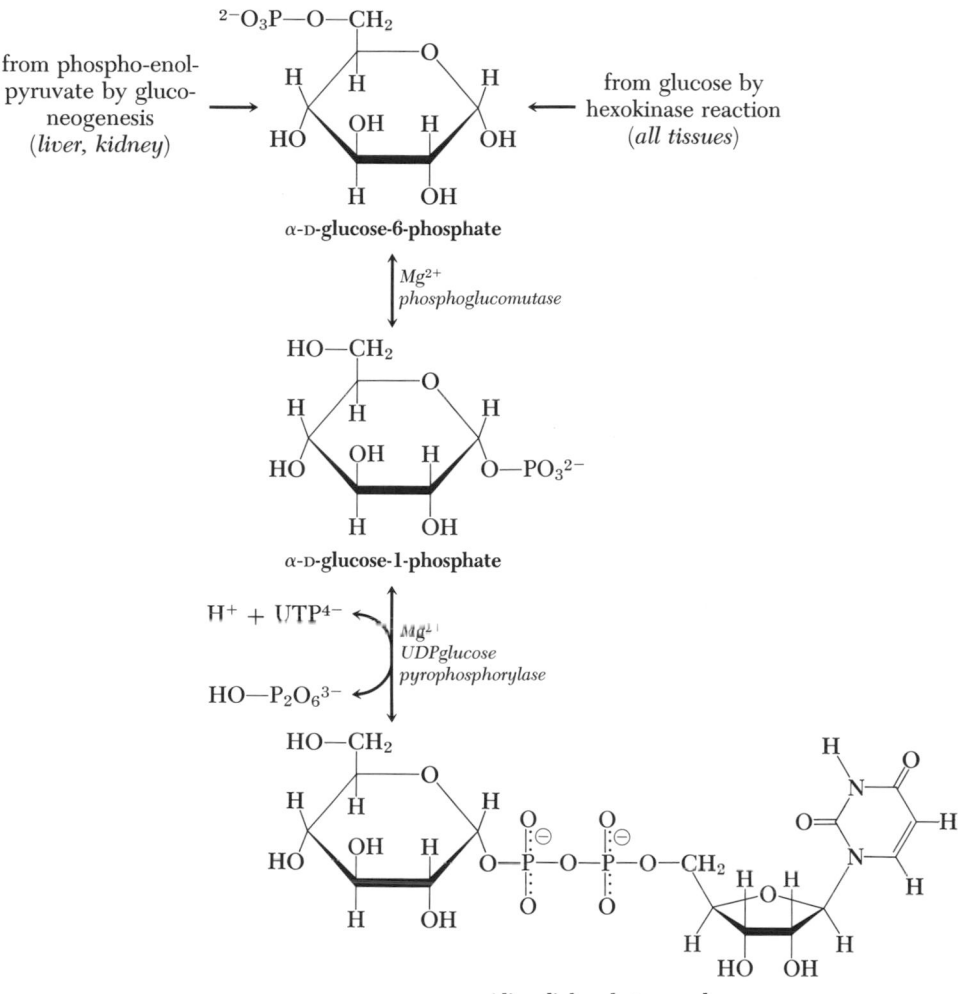

uridine diphosphate-α-D-glucose
(**UDPglucose**)

Figure 15-3   Excess glucose-6-phosphate is converted to UDP-glucose for storage as glycogen in all tissues. The excess may be formed from circulating glucose by the hexokinase reaction in all; the liver and kidney also form glucose-6-phosphate from phosphopyruvate by gluconeogenesis. The conversion involves a shift of the phosphate group to carbon 1 by a mutase reaction. UTP then reacts with the phosphate, transferring the UMP moiety to the sugar phosphate, and liberating pyrophosphate derived from UTP. The entire sequence is reversible with isolated enzymes, but is pulled toward UDP-glucose formation in tissues by hydrolysis of inorganic pyrophosphate.

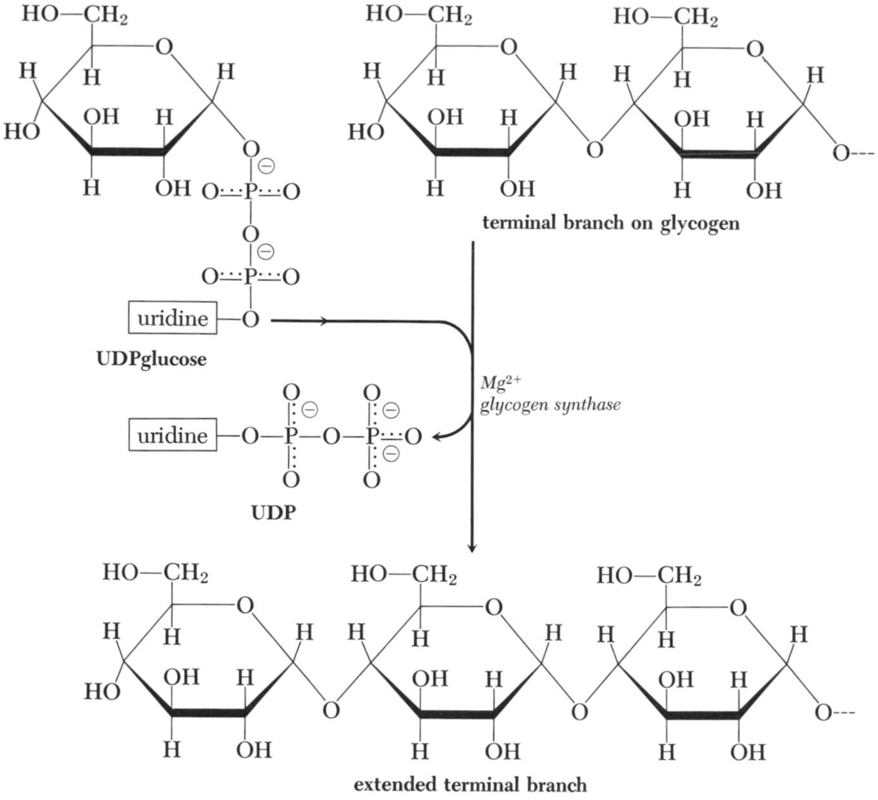

Figure 15-4 Terminal amylose chains on glycogen are extended by transfer of additional glucosyl residues from the phosphate of UDP-glucose to the free hydroxyl group on carbon 4 of the chain. The extended branch can react again in the same way with another molecule of UDP-glucose, successively adding more glucosyl residues.

the chain is not changed by the extension, the glycogen synthase reaction can be repeated to add another glycosyl residue from UDP-glucose, and this can be continued indefinitely. A molecule of UDP is released for each glucosyl residue added.

If nothing else happened, the result of the glycogen synthase reaction would be long amylose chains composed only of 1→4 bonds. However, there is a branching enzyme, a *glycosyl 4:6-transferase*, which catalyzes the transfer of segments of amylose chain onto the C-6 hydroxyl of neighboring chains (Fig. 15-5). This enzyme moves a block of seven residues from a chain at least 11 residues in length, and transfers it onto an amylose chain at a point four residues removed from the nearest branch. (Transfer of seven residues is favored, but specificity is not absolute.) The new branch is therefore seven residues in length, and the remaining stub from which it was removed is at least four residues in length, but more commonly six to nine residues long. 1→6 Glucosides have about 1000 calories less standard free energy content per mole than do 1→4 glucosides, so the equilibrium of reaction favors branching.

The new branch and the remaining stub from which it was obtained can grow in length by addition of more 1→4 glucosyl residues from UDP-glucose through the action of glycogen synthase. When enough have been added to extend them to 11 residues beyond the branch, a further transfer by the branching enzyme may create still another branch, which can grow in length. What limits the size of the molecule is not known. It may be that the branches become so crowded on the periphery that there isn't room to make any more. Some believe that when this happens, some terminal chains continue to extend by 1→4 linkages until they are well clear of the bulk of the molecule, when branching begins again to form a nucleus of a new glycogen molecule. In this picture, a glycogen particle consists of spheres of highly-branched molecules actually bound together by a single thread of amylose chain. Another possibility is that the size of the glycogen molecule is limited by the arrangement of enzymes on the surface, so that they cannot remain in active proximity after the sphere grows too large. There is too little information to justify either speculation. What is certain is that skeletal muscle does not accumulate glycogen indefinitely, even with a high concentration of blood glucose. The content of glycogen in the liver can rise markedly under heavy dietary loading, but even in this tissue glucose is absorbed slowly enough with more usual conditions that an increased fraction of the excess supply is used for making fat (next chapter) after the glycogen content rises above 50 to 60 grams per kilogram of tissue.

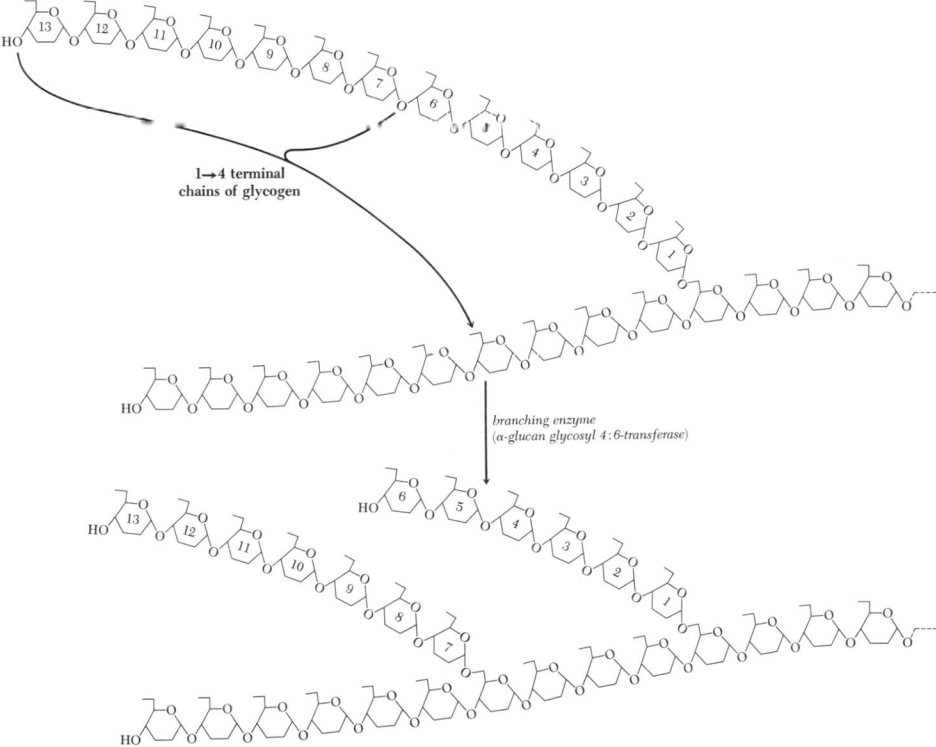

Figure 15-5 Branches are created in glycogen by transferring seven-residue segments of amylose terminal chains to hydroxyl groups on carbon 6 of a glucosyl residue that is four residues removed from an existing branch. A terminal branch must be at least 11 residues long before a segment is transferred from it.

UTILIZATION OF GLYCOGEN

Glycogen is mainly degraded by a simple phosphorolysis of the 1→4 glucosidic bonds to form glucose-1-phosphate, accompanied by reactions to remove the 1→6 branches. The primary reaction is catalyzed by the enzyme, *phosphorylase*, cleaving the terminal 1→4 glucosidic bonds with inorganic phosphate so as to shorten the chain by one residue (Fig. 15-6). The reaction is freely reversible; the equilibrium ratio of P_i to glucose-1-phosphate concentrations is near 3.5. However, the concentration of P_i in the liver is some 30-fold greater than the concentration of glucose-1-phosphate, so the reaction always proceeds in the direction of formation of glucose-1-phosphate.

Phosphorylase contains *pyridoxal phosphate* as a coenzyme. We previously discussed this compound in connection with the mechanism of transamination (p. 116). Although it is linked to a lysyl residue on the phosphorylase peptides, as it is in the transaminases, the function of the coenzyme in the phosphorylase reaction is not clear. The aldehyde group evidently does not participate in the reaction, because the enzyme can be reduced by sodium borohydride without losing activity, and this causes the conversion of the lysyl-pyridoxal Schiff's base to a stable secondary amine.

When a glucosyl residue is removed by phosphorolysis, the remaining branch on the glycogen molecule can be attacked again to remove another residue, followed by further attack on the remainder. However, phosphorylase by itself will not catalyze the removal of all of the terminal residues. In anthropomorphic terms, it won't go near the branches. Expressed more mechanistically, the specificity of the enzyme is such that the bond attacked must be at least four residues removed from a 1→6 branch, and this enzyme alone will not make the outer branches any shorter than four glucosyl residues. The function of this specificity is not clear. Glycogen completely sheared by phosphorylase to the four-residue stubs is called the phosphorylase limit *dextrin*, or Φ-dextrin. Dextrin is a term borrowed from starch chemistry, describing partially hydrolyzed glucose homopolymers, and it is widely used for describing intermediates obtained during chemical investigation of structure, but this does not imply the occurrence of these discrete types of molecules in cells.

Further degradation of glycogen requires a second enzyme, a glucosyl transferase akin to the branching enzyme functioning during glycogen synthesis, but differing in that it transfers three glucosyl residues from a branch onto a chain terminus so as to lengthen it (Fig. 15-7). Suppose that phosphorylase has adjacent chains shortened as much as possible at a branch. Each will be four glucosyl residues in length. The *glycosyl 4:4 transferase* will catalyze the removal of three of the residues from the stub of the branch, leaving a single residue attached to C-6, and transfer them to the end of the other stub, which now has seven residues in 1→4 linkage, and can again be attacked by phosphorylase.

The final reaction in the breakdown is the hydrolysis of the single residue remaining on C-6 at the branch to yield free glucose. The reaction is catalyzed by *amylo-1→6-glucosidase*, which is absolutely specific for a 1→6 linkage to a single residue, thereby protecting longer branches on glycogen from attack by this enzyme. Such attacks would cause internal disruption of the molecule.

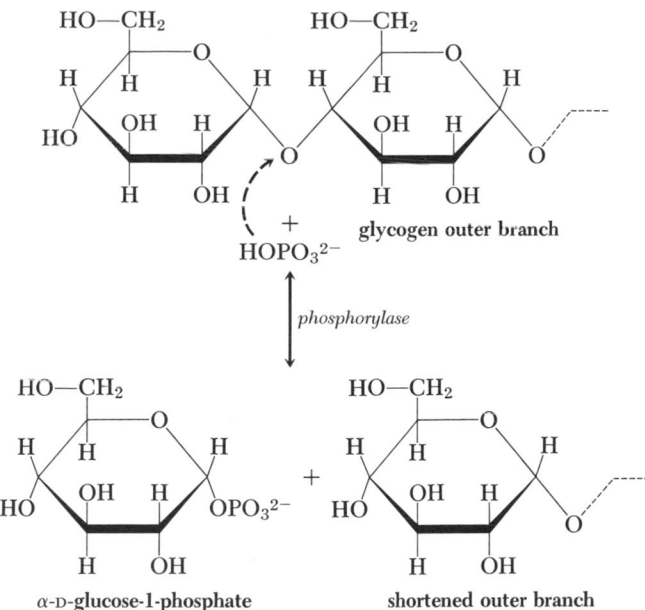

Figure 15-6 The recovery of the glucosyl residues stored as glycogen begins with phosphorolysis of terminal glucosidic bonds to form glucose-1-phosphate. Glucose-1-phosphate is in equilibrium with glucose-6-phosphate (not shown) and can supply the Embden-Meyerhof pathway in all cells or be used to replenish the blood glucose from the liver.

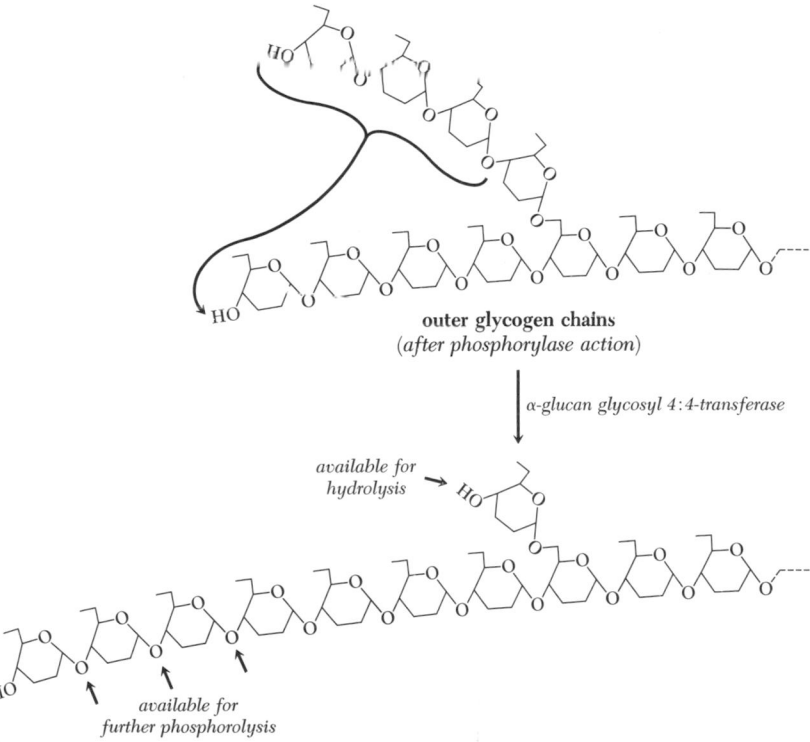

Figure 15-7 Phosphorylase removes glucosyl residues no closer than four residues from a branch point in glycogen. Another enzyme, a glycosyl 4:4-transferase, catalyzes the transfer of three residues from the stub of a branch to the C-4 hydroxyl group of an amylose stub. The 1→6-linked residue is now exposed to hydrolysis by amylo-1,6-glucosidase, and the extended 1→4-amylose chain can again be attacked by phosphorylase until the three added glucosyl residues are removed.

297

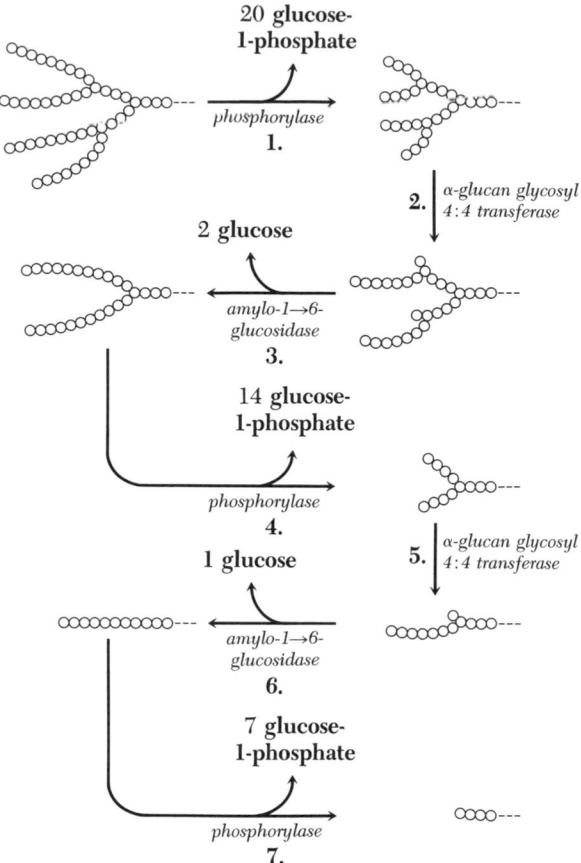

Figure 15-8   Summary of the sequence of glycogen breakdown. The diagram shows the fate of four outer chains and the next two branches from which they arise. Each chain is shown with nine residues beyond the branch.

1. Successive action of phosphorylase strips all but four residues from each branch, forming 20 molecules of glucose-1-phosphate.

2. Segments of three residues are removed from the C-4 oxygen of each branched residue (that is, the residue linked 1→6) and transferred to the C-4 oxygen at the end of the other stubs. Where there were four residues in each stub beyond the branch, there is now a single 1→6 residue, and seven residues in the 1→4 chain.

3. The branched residues are now removed by hydrolysis, liberating one molecule of free glucose from each branch for a total of two. The remainder has a single branch point with 11 residues in each branch.

4. Phosphorylase can again attack the terminal branches until only four residues remain in each, liberating 14 molecules of glucose-1-phosphate. (The action of phosphorylase and of the glucosidase can be interspersed, rather than occurring in the simple sequence given here, but the effect is the same.)

5. The 4:4-transferase again catalyzes the movement of three glucosyl residues from the 1→6 stub to the 1→4 stub.

6. The one remaining 1→6-linked residue is removed by hydrolysis.

7. Phosphorylase can now remove seven residues as glucose-1-phosphate before coming within four residues of the next branch point.

SUM: Out of the 44 glucosyl residues removed, three have appeared as free glucose and 41 as glucose-1-phosphate.

Removal of the single branched residue leaves a straight chain of 1→4 linked residues down to the next branch, permitting the removal of more residues as glucose-1-phosphate by the action of phosphorylase before reaching its limit of four residues next to a branch. What is the result? The whole sequence causes a continual stripping of glucosyl residues as glucose-1-phosphate, except for the single residues at branches that are removed as free glucose. The degree of branching is such that 11 to 14 molecules of glucose-1-phosphate are formed for each molecule of free glucose released.

CONTROL OF GLYCOGEN METABOLISM

The reactions for the formation of glycogen and its breakdown taken together (Fig. 15-9) have a potential for a profitless cycling of glucosyl residues, accomplishing nothing but a continuous breakdown of high-energy phosphate. On the face of it, glucosyl residues would be moved from glucose-1-phosphate to glycogen at the expense of UTP, and therefore of ATP, and moved back again by the action of phosphorylase. The situation is analogous to the simultaneous presence of phosphofructokinase and fructosediphosphatase in the liver, and it might be predicted that there is also some control preventing the processes of glycogen synthesis and glycogen breakdown from operating simultaneously. This is indeed the case, but the control is considerably more involved than anything we have previously considered.

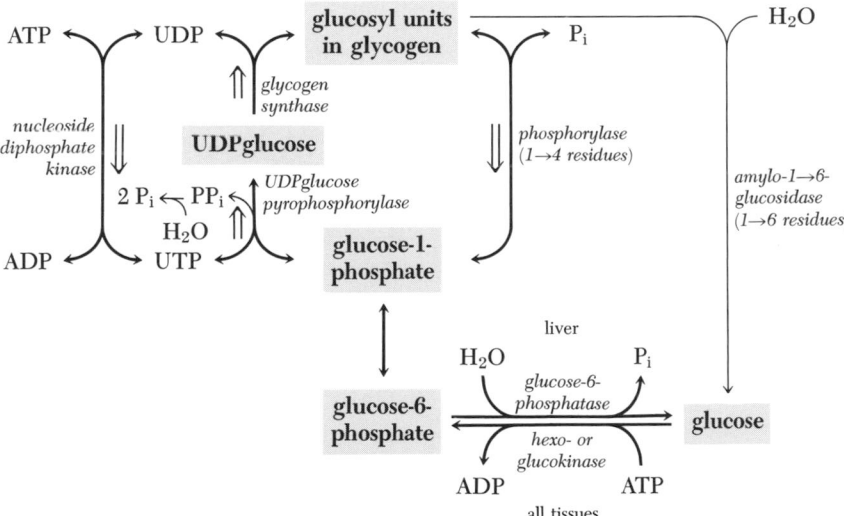

Figure 15-9   Summary of the overall process of glycogen formation and breakdown. Glycogen is formed from glucose-1-phosphate at the ultimate expense of ATP (*upper left*). Glycogen is broken down mostly to glucose-1-phosphate by simple phosphorolysis (*center right*), although a small part appears directly as glucose (*far right*). The net result of this collection of reactions, if all operated at once, would be the hydrolysis of ATP to ADP and P_i.

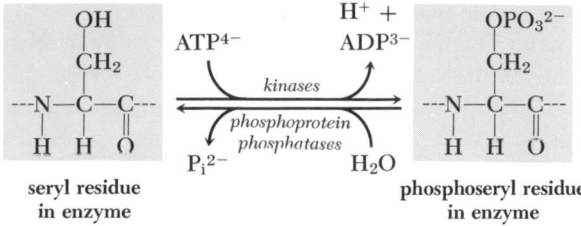

seryl residue
in enzyme

phosphoseryl residue
in enzyme

Figure 15-10   Enzymes are sometimes regulated by being phosphorylated or dephosphorylated through the action of other enzymes. A regulating kinase catalyzes the transfer of phosphate from ATP to seryl residues on the affected enzyme. The phosphate groups are removed by hydrolysis catalyzed by a phosphatase. The modified enzyme may be more or less active than the free form.

The primary control involves changes in glycogen synthase and phosphorylase such that the activity of one enzyme is repressed while the other is stimulated. The novelty lies in the nature of the changes in the enzymes, which are due to actual covalent chemical reactions, rather than to association with allosteric modifiers. The enzymes themselves act as substrates for kinases causing a phosphorylation of seryl residues or for phosphatases removing the phosphate groups by hydrolysis (Fig. 15-10). Here we have enzymes built to act on enzymes and thereby modify their activity.

Control of Phosphorylase in Muscle

Let us first consider the case of the phosphorylases found in skeletal muscle. These enzymes are composed of peptide chains having one seryl group that can be phosphorylated and one pyridoxal phosphate. When these seryl groups are free, two peptide chains are associated, and this form of the enzyme is known as *phosphorylase b*. Phosphorylase b is the substrate for a specific *phosphorylase kinase*, which occurs in much lower concentration than the phosphorylase. The kinase catalyzes the transfer of a phosphate group from ATP to each of the two seryl residues in phosphorylase b. (The semantics can be tricky—phosphorylase kinase catalyzes the phosphorylation of phosphorylase.) The presence of these phosphate ester groups changes the character of the peptides so that they now associate into molecules of four chains each. In other words, two of the dimers, phosphorylase b, associate into tetramers when their seryl groups are phosphorylated, and this tetramer is known as *phosphorylase a* (Fig. 15-11).

The regulation of phosphorylase hinges on the different properties of the two forms of the enzyme. The phosphorylated tetramer, phosphorylase a, will catalyze the formation of glucose-1-phosphate from glycogen without the presence of any activator. The free dimer, phosphorylase b, is completely inactive *unless* AMP is present, and the greater the amount of AMP, the more active it becomes with a given amount of substrates, because AMP lowers the K_M for both the glycogen end groups and for inorganic phosphate.

This now brings us to the second important phase of the control mechanism.

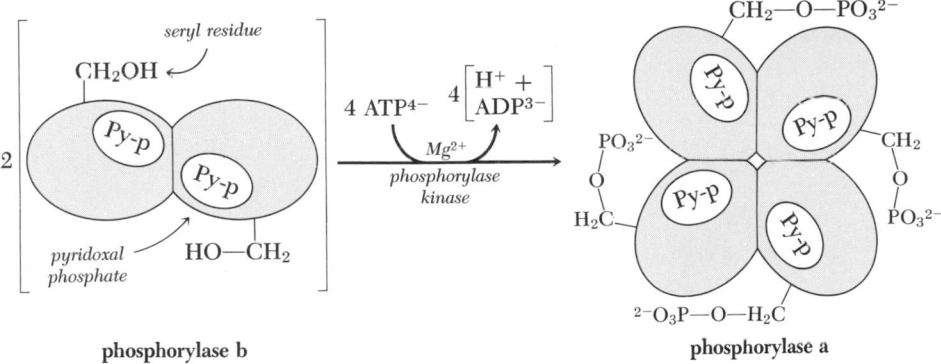

phosphorylase b   **phosphorylase a**

Figure 15-11   Phosphorylase b is a dimer, with each peptide unit binding pyridoxal phosphate and having a seryl group that reacts with ATP. When phosphorylase kinase catalyzes the phosphorylation of these groups, the presence of the phosphate causes the most stable form to be a tetramer, made by combining a pair of dimers. The pyridoxal phosphate is believed to act to stabilize these particular forms.

What governs the activity of phosphorylase kinase? What causes the formation of the tetramer that is independently active in breaking down glycogen? Let us digress for a moment. The rapid mobilization of glucose-1-phosphate, and therefore of glucose 6-phosphate, from glycogen is imperative at times when severe and sudden muscular exertion is begun. The signal for such exertion does not come from changes in concentration of compounds in the blood, it comes from the nervous system. At the same time, the nervous system excites the adrenal medulla, causing this tissue to release *adrenaline* into the blood stream:

$$OH$$

$$OH$$

$$HO—C—H$$

$$CH_2$$

$$N$$

$$H \quad CH_3$$

D-adrenaline
(epinephrine)

(Adrenaline is usually called *epinephrine* by American physiologists and pharmacologists, largely because the former used to be a trade name. However, the compound is named adrenaline in *Chemical Abstracts,* which is the authoritative source for chemical nomenclature in this country, in the literature of other countries, and in the usage of other educated men.) The adrenaline that is produced comes in contact with an enzyme, *adenyl cyclase,* present in the plasma membrane of cells of the skeletal muscles and heart, which catalyzes an intramolecular condensation of ATP to produce cyclic adenosine-3',5'-monophosphate, or *cyclic AMP*:

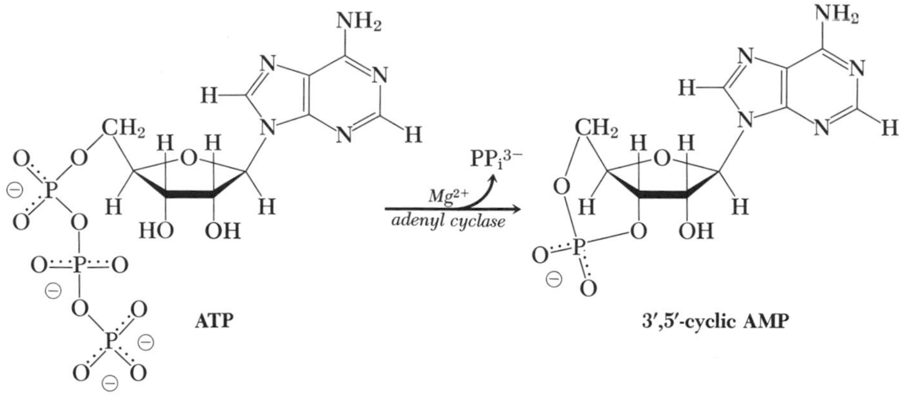

(We shall see that cyclic AMP is involved in the regulation of a number of processes. Right now, we are concerned only with glycogen metabolism.)

Cyclic AMP is an activator of one or more protein kinases in muscles, catalyzing the phosphorylation of seryl groups at the expense of ATP; the kinases have only recently been discovered and their exact character and specificity is unknown. One of them catalyzes the phosphorylation of phosphorylase kinase, which is necessary before this enzyme has catalytic activity and can in turn cause the formation of phosphorylase a.

The stimulation of the adrenal medulla by a signal to flee or fight therefore triggers a complex series of enzymatic activations in skeletal and cardiac muscle that finally results in a rapid breakdown of glycogen; thereby providing a ready source of glucose-1-phosphate for conversion to glucose-6-phosphate and thence to pyruvate. These activating events are summarized in Figure 15-12.

An important part of this arrangement is the cascade effect:

1. A small amount of adenyl cyclase and its stimulating adrenaline creates much more cyclic AMP.

2. The cyclic AMP activates a protein kinase, which can catalyze the phosphorylation of much more phosphorylase kinase.

3. Phosphorylase kinase catalyzes the conversion of phosphorylase b to phosphorylase a. The molar concentration of the kinase is only about 0.1 that of phosphorylase peptides.

4. Phosphorylase a will catalyze the formation of orders of magnitude more glucose-1-phosphate.

Each step in this sequence is magnified by the following reaction, and the result is that a minute amount of adrenaline has a very large effect—so large that 500 μg is a potent dose for humans. The blood concentration of adrenaline during normal function lies in the nanomolar range. The compound is therefore a very powerful *hormone,* a compound elaborated by one tissue in small concentrations to affect the activity of other tissues.

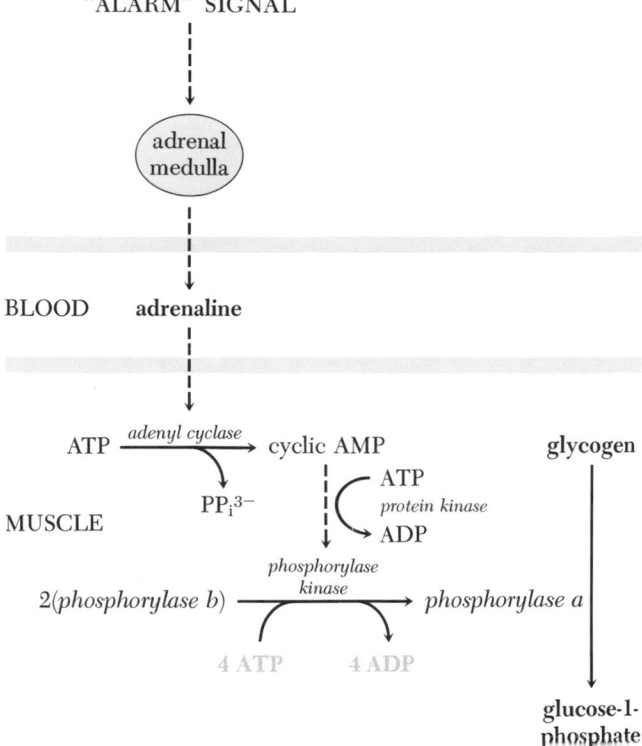

Figure 15-12   The adrenal medulla is stimulated to release adrenaline into the blood by signals from the central nervous system. Adrenaline enters the muscles and activates adenyl cyclase, which catalyzes the formation of 3′,5′-cyclic-AMP. Cyclic AMP triggers, among other things, the action of a protein kinase that phosphorylates still another kinase. The latter kinase catalyzes the phosphorylation of phosphorylase, which results in the formation of phosphorylase a, the active form of the enzyme. Phosphorylase a, in turn, catalyzes the phosphorolysis of glycogen. This Rube Goldberg sequence is designed so that a little adrenaline has a large effect.

When the stimulus is removed and the skeletal muscles can safely relax and the heart slows down, another sequence of events occurs to restore the original state of the enzymes. Adrenaline is rapidly removed by a sequence of reactions we shall consider later, so rapidly that its half-life in blood is only 10 seconds or so. This removes the stimulation of adenyl cyclase, so that cyclic AMP is no longer formed, and the existing compound is hydrolyzed by a phosphodiesterase to yield ordinary 5′-AMP:

The removal of cyclic AMP causes the protein kinases to lose their activity, so that active phosphorylase kinase cannot be formed. The phosphoseryl groups of both phosphorylase kinase and phosphorylase a are hydrolyzed by the phosphoprotein phosphatase mentioned above, and all of the enzymes are in their original resting form.

Two points ought to be made. First, the recovery processes are slower than the activations. If they weren't, the combination of enzymes would be operating in the kind of ATP-hydrolyzing cycle that the regulation of glycogen metabolism is designed to avoid. Any combination of kinases and phosphatases of equal activity and affecting the same metabolites is clearly wasteful. Secondly, the effect of the activating system on ATP balance is small. The ATP utilized is negligible in comparison to the amount generated by the major metabolic pathways, and the ADP or AMP produced by the activating system is not sufficient in itself to have important regulatory effects on the rate of glycolysis or oxidative phosphorylation.

Now, let us consider phosphorylase b, which is only active in the presence of AMP. Even at rest, muscles utilize ATP to maintain tone, and therefore require oxidative phosphorylation. The muscles may be used at more sedate occupations without particular stimulation of the adrenal medulla. It would not do to have a supply of glycogen sitting inviolate within the muscle if the supply of other substrates is inadequate to sustain even modest demands for high-energy phosphate. This is prevented through an activation of phosphorylase b, the "resting" phosphorylase, by AMP. Remember that ADP and AMP are in equilibrium through the adenylate kinase reaction, 2 ADP $\rightleftharpoons$ ATP + AMP, so the concentration of AMP will rise whenever the concentration of ADP rises and the concentration of ATP falls. Therefore, phosphorylase b will increase in activity and produce more glucose-1-phosphate from glycogen whenever the supply of other substrates is unable to cope with the demand for oxidative phosphorylation.

Control of Phosphorylase in Liver

The control of phosphorylase activity in the liver resembles the sequence in skeletal muscle, but with differences appropriate to the function of the organ.

First, liver phosphorylase is a dimer of two peptide chains, and does not associate further when seryl residues are phosphorylated. Secondly, the non-phosphorylated form is not affected by AMP, so that the ability to break down liver glycogen hinges only upon the action of the liver phosphorylase kinase. This kinase is like the muscle enzyme in requiring the presence of cyclic AMP for activation. However, the liver adenyl cyclase that generates the necessary cyclic AMP is not as sensitive to adrenaline as is the enzyme in skeletal muscle. Instead, it is activated by another hormone, *glucagon*, which has no effect on the enzyme in skeletal muscle. Glucagon is a peptide elaborated by the pancreatic islets in response to *lowered* concentrations of glucose in the blood—the contrary of the signal for insulin release by the pancreas. Low blood glucose therefore promotes the formation of cyclic AMP in the liver through glucagon release, thereby triggering activation of the liver phosphorylase by the phosphorylase kinase and increasing the breakdown of liver glycogen to form glucose-6-phosphate *via* glucose-1-phosphate. This increases the liberation of free glucose into the bloodstream by way of the glucose-6-phosphatase reaction, and maintains the blood glucose concentration at the expense of glycogen in the liver.

Control of Glycogen Synthase

Glycogen synthase also exists in forms interconvertible by a kinase and phosphoprotein phosphatase. There is not as clear a picture of the differences in the two forms of the enzyme as there is with phosphorylase, but the effects of the changes are clear. The non-phosphorylated enzyme is the independently active form, designated the *I form*. When the enzyme is phosphorylated, it is only fully active when glucose-6-phosphate is also present as an activator. This is the dependent, or *D form*.

The kinase catalyzing the phosphorylation of the enzyme is activated by cyclic AMP, and will therefore be operating at the same time that the phosphorylase kinase is in the active state. However, the conversion of glycogen synthase to the phosphorylated form is changing it from an enzyme independently extending glycogen chains to one that extends the chains only when the glucose-6-phosphate concentration is elevated. The effect of glucose-6-phosphate is dramatic, with the V_{max} for the phosphorylated enzyme being raised 43-fold by a saturating concentration of the activator (0.01 M glucose-6-phosphate). The apparent K_M for the activator is 5×10^{-4} M.

This means that stimuli that increase the breakdown of glycogen through the formation of phosphorylase a will simultaneously shut off the formation of glycogen by changing glycogen synthase to the dependent form, with the important proviso that this change prevents the formation of glycogen only if glucose-6-phosphate is being removed (by conversion to pyruvate in the case of the muscles, or by conversion to blood glucose in the case of the liver). If the signal was a false alarm, which may be particularly frequent in the case of adrenaline release, unused glucose-6-phosphate can apparently be converted back to glycogen, because the excess of the compound will keep the glycogen synthase active, even though the enzyme has been phosphorylated in response to the alarm signal.

In most cases, the increase in adrenaline does precede active muscular contraction or the increase in glucagon does signal a low blood glucose level, and these other circumstances will cause glucose-6-phosphate to be removed at a faster rate than it was before elaboration of the hormones was provoked. In these cases with little accumulation of glucose-6-phosphate, cyclic AMP acts to shut off the synthesis of glycogen and accelerate its breakdown. Later removal of phosphate from the enzymes by phosphoprotein phosphatases during the recovery phase shuts off the breakdown of glycogen and accelerates its synthesis (Fig. 15-13).

GENETIC DEFECTS IN GLYCOGEN METABOLISM

A number of humans have been discovered who lack one of the enzymes concerned with glycogen metabolism. None of the defects is common, so they are

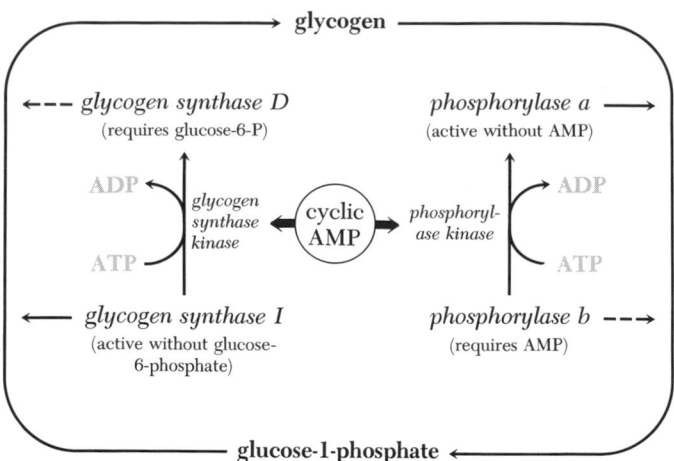

Figure 15-13   Whether glycogen shall be formed from glucose-1-phosphate or degraded to make glucose-1-phosphate is mainly determined by the presence or absence of cyclic AMP. Without cyclic AMP, the critical enzymes are in non-phosphorylated forms; glycogen synthase is independently active (*solid arrow*), while phosphorylase b is inactive (*dashed arrow*) unless a rise in 5′-AMP concentration signals a deficit of high-energy phosphate. Formation of cyclic AMP results in activation of the kinases that cause transformation of the enzymes to their phosphorylated forms; the resultant glycogen synthase D is inactive unless there is an accumulation of glucose-6-phosphate, but phosphorylase a is now independently active in catalyzing the breakdown of glycogen.

not of quantitative importance in the catalogue of human ills, but their occurrence is of great value in assessing mammalian carbohydrate metabolism. For example, a decade ago it was believed that glycogen was synthesized by phosphorylase because it catalyzes a reaction readily reversible in the test tube. A patient was discovered who was weak and who developed severe pain in his muscles after modest exercise, and he was shown to have high concentrations of glycogen and an *absence of phosphorylase* in his skeletal muscles. This proved that he was not synthesizing glycogen by the phosphorylase reaction, and provided strong supporting evidence for the newly discovered UDP-glucose pathway that we have discussed. (The condition is known as *McArdle's disease*, after its discoverer. A psychiatrist engagingly confesses [Ann. Internal Med., Vol. 62, p. 412 (1965)] that he thought McArdle's patient was displaying classic hysteria owing to an unhappy childhood. No comment is offered on the contribution of the phosphorylase deficiency to the unhappy childhood.)

McArdle's disease tells us something else. Impairment of glycogen utilization in skeletal muscles is not fatal. The impairment is very real; there isn't any unsuspected route for utilizing the polysaccharide, and this is easily demonstrated by shutting off the blood supply to an arm with a tourniquet and measuring the lactate concentration. Clenching of the fist causes a sharp rise in a normal individual, but not in one with McArdle's disease. Hence, we see that skeletal muscles can perform to some extent when supported only by the oxidation of substrates other than carbohydrate. However, the importance of carbohydrate metabolism for full efficiency is shown by the weakness of the patients and is supported by the uncommon occurrence of the genetic lesion in the population—this is a mutation that is evidently rapidly eliminated.

Deficiencies have also been described in the *debranching enzyme,* amylo-1,6-glucosidase, causing an accumulation of abnormal glycogen with too many branches, and of the *branching enzyme,* α-glucan glycosyl 4:6-transferase, causing accumulation of glycogen with too few branches, which stains blue with iodine.

This picture of glycogen metabolism as a convenience for full efficiency of muscle rather than an absolute necessity is reinforced by the discovery of individuals with a deficiency of *phosphofructokinase* in skeletal muscles, who have the same clinical picture found in McArdle's disease and who also accumulate glycogen because of inability to utilize it. These individuals can't even utilize glucose taken up from the blood stream directly, and their existence forecloses any interpretation of the survival of those with phosphorylase deficiency being due to an increased ability of skeletal muscle to utilize blood glucose.

To cap the story, a few have been found who lack *glycogen synthase* in both liver and muscle. Glycogen was absent, and yet the individuals survived into infancy.

Many known defects cause an accumulation of glycogen in the liver. As we have seen, the liver store is used to stabilize the blood concentration. Individuals can survive, although precariously, without this mechanism. The first kind of glycogen storage disease recognized, *von Gierke's disease,* was shown to be due to an absence of *glucose-6-phosphatase* in the liver, and this defect not only prevents the later discharge of glucose taken up by the liver in times of high blood concentration, but also of the glucosyl residues created by gluconeogenesis. The liver becomes very large, as do the kidneys, which also have the enzymes of gluconeogenesis. This genetic defect is dangerous because there is no means of maintaining the glucose supply for the brain between meals or for converting lactate to circulating glucose, and individuals with it frequently die in infancy.

Deficiencies of *phosphorylase* and of *phosphorylase kinase* in the liver are also known, both of which cause enlargement owing to the accumulation of glycogen. Even though liver glycogen cannot be utilized, the consequences are not as drastic as they are in the glucose-6-phosphatase deficiency, because the liver can still create glucose from pyruvate.

The most puzzling of the glycogen storage diseases to the biochemist is *Pompe's disease,* in which glycogen accumulates, but the only enzyme known to be missing is a *glucosidase* found in lysosomes. This enzyme catalyzes the direct hydrolysis of glucosyl residues from glycogen, forming free glucose rather than glucose-1-phosphate. The condition is interesting because we know little about the functions of lysosomes. They are particulate structures in the cell (see frontispiece) loaded with a battery of hydrolytic enzymes: peptidases to attack proteins, esterases to attack lipids, phosphatases for phosphate esters, etc., many of which are active only at relatively acidic pH values. The lysosomes are believed to behave as scavengers—engulfing and digesting damaged structures in the cell. In Pompe's disease, many tissues become loaded with glycogen. The level in the heart and skeletal muscles may rise to 100 grams of glycogen per kilogram of tissue. Does this imply that a major part of the glycogen is normally hydrolyzed by lysosomes? It seems very unlikely, particularly in light of the other glycogen storage diseases in which the lysosomes have a normal complement of glucosidase. Perhaps glycogen particles do become damaged occasionally. The attached enzymes might be denatured so that the polysaccharide becomes metabolically inert. In such a case, the lysosome may serve to remove the particle, and a failure to do so may result in an accumulation of these inactive granules. This is a very speculative interpretation, but not impossibly far-fetched. (The lysosome need not be able to distinguish between functional and damaged constituents of a cell. It may engulf all at a rate that is relatively slow compared to the rate of the metabolic processes utilized to replace them, but fast compared to the rate of damage, so that the concentration of damaged constituents is kept low without utilizing more than a fraction of the metabolic capacity of the cell for replacement.)

THE EFFICIENCY OF GLYCOGEN STORAGE

When an organ such as skeletal muscle converts blood glucose to glycogen, two high-energy phosphates are expended per glucosyl residue stored, according to the balance:

(1)   glucose + ATP^{4-} $\longrightarrow$ (glucose-6-phosphate)$^{2-}$ + ADP^{3-} + H^+

(2)   (glucose-6-phosphate)$^{2-}$ $\longrightarrow$ (glucose-1-phosphate)$^{2-}$

(3)   (glucose-1-phosphate)$^{2-}$ + UTP^{4-} + H^+ $\longrightarrow$ (UDP-glucose)$^{2-}$ + PP_i^{3-}

(4)   (UDP-glucose)$^{2-}$ + glycogen $\longrightarrow$ UDP^{3-} + glucosyl-glycogen + H^+

(5) PP_i^{3-} + H_2O $\longrightarrow$ 2 P_i^{2-} + H^+

(6) UDP^{3-} + ATP^{4-} $\longrightarrow$ UTP^{4-} + ADP^{3-}

SUM: glucose + 2 ATP^{4-} + glycogen + H_2O $\longrightarrow$

glucosyl-glycogen + 2 ADP^{3-} + 2 P_i^{2-} + H^+

When the tissue later utilizes the glucosyl residues, over 90 per cent of them will be directly converted to glucose-1-phosphate without any expenditure of high-energy phosphate, and complete oxidation of this compound will yield 37 high-energy phosphates per mole.

Now, if glucose is the only compound metabolized by the tissue during storage, 36 high-energy phosphates can be created per mole of the compound taken up and oxidized. This is sufficient to store 18 moles of glucose as glycogen, or 94.7 per cent of the total glucose (18/19, with the other 1/19 going to CO_2 and H_2O to provide the energy). This 94.7 per cent of the extra glucose available that is now stored as glycogen can later provide $0.947 \times 37 = 35$ moles of high-energy phosphate per mole of the original glucose, instead of the 36 moles that could have been generated by immediate oxidation without forming glycogen. In other words, the storage of glycogen and its later use in the same tissue will yield 35/36, or 97 per cent of the potential high-energy phosphate in the excess glucose, if the tissue has perfect coupling of oxidation and phosphorylation. The loss of useable energy is therefore very small compared to the advantages obtained by storage.

Recapitulation of types of reactions

1. Phosphate esters may react with nucleoside triphosphates to form an esterified nucleoside diphosphate:

$$H^+ + (\text{nucleoside triphosphate})^{4-} + R\text{—}CH_2OPO_3^{2-} \longrightarrow$$

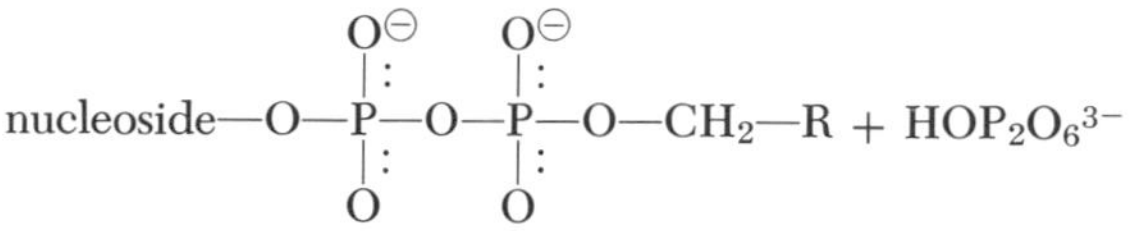

$$\text{nucleoside—O—}\overset{\overset{\displaystyle O^{\ominus}}{|}}{\underset{\underset{\displaystyle O}{|}}{P}}\text{—O—}\overset{\overset{\displaystyle O^{\ominus}}{|}}{\underset{\underset{\displaystyle O}{|}}{P}}\text{—O—}CH_2\text{—}R + HOP_2O_6^{3-}$$

Example: The formation of uridinediphosphate glucose from glucose-1-phosphate and UTP.

2. The alkyl group esterified with nucleoside diphosphates may be transferred to oxygen atoms of other compounds:

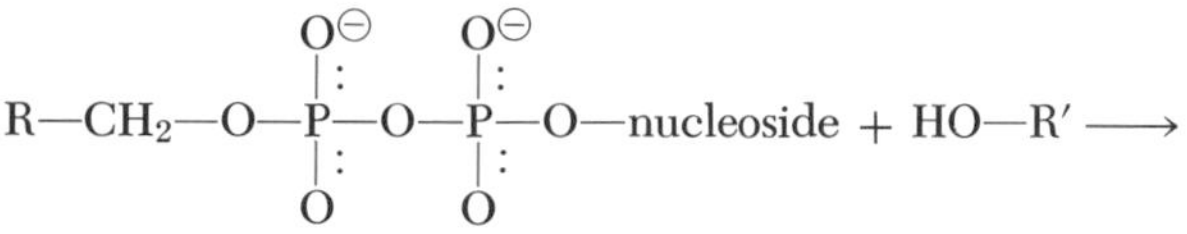

$$R\text{—}CH_2\text{—O—}\overset{\overset{\displaystyle O^{\ominus}}{|}}{\underset{\underset{\displaystyle O}{|}}{P}}\text{—O—}\overset{\overset{\displaystyle O^{\ominus}}{|}}{\underset{\underset{\displaystyle O}{|}}{P}}\text{—O—nucleoside} + HO\text{—}R' \longrightarrow$$

$$R\text{—}CH_2\text{—O—}R' + (\text{nucleoside diphosphate})^{3-} + H^+$$

Example: The transfer of the glucosyl residue from uridine diphosphate glucose to the end of glycogen chains by glycogen synthase.

3. Glucosidic bonds may be cleaved by inorganic phosphate:

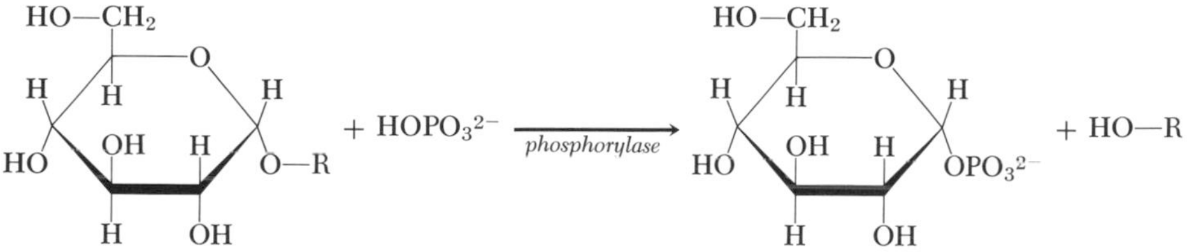

Example: Glycogen phosphorylase reaction.

4. The sugar residues in glycosides may be transferred from one chain to another.
See p. 295 for an illustration of this reaction as exemplified by the glycogen branching enzyme, and p. 297 for an illustration as exemplified by the 4:4 transferase.

5. Glycosides may be hydrolyzed:

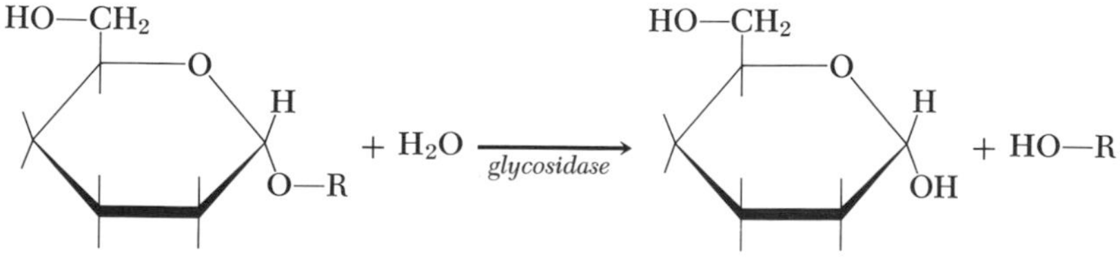

Examples: The hydrolysis of 1→6 residues by amylo-1,6-glucosidase; the hydrolysis of glycogen by the lysosomal glucosidase.

6. Seryl groups in proteins may be phosphorylated:

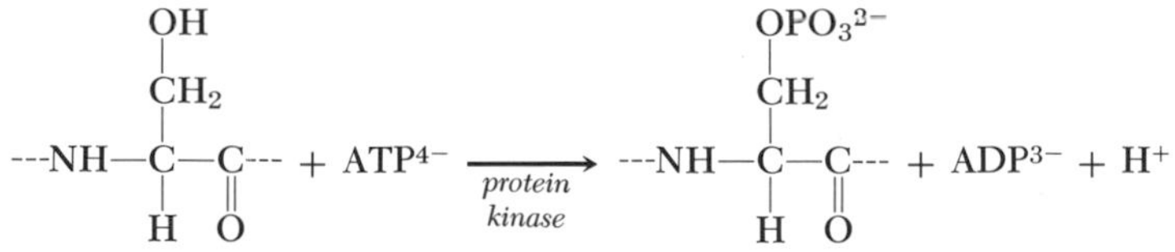

Examples: The phosphorylation of phosphorylase b by ATP, catalyzed by phosphorylase kinase; the phosphorylation of glycogen synthase by ATP, catalyzed by glycogen synthase kinase.

7. Phosphoseryl groups in proteins may be hydrolyzed:

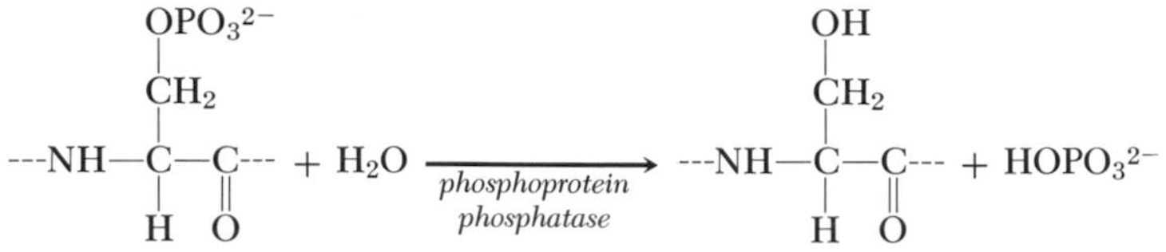

Examples: The dephosphorylation of phosphorylase a; the dephosphorylation of glycogen synthase D.

8. Nucleotide triphosphates may react intramolecularly to form cyclic mononucleotides:

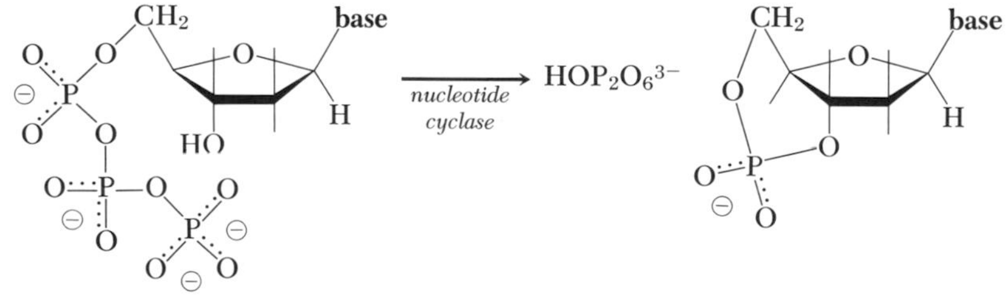

Example: The formation of 3′,5′-cyclic-AMP in the adenyl cyclase reaction.

9. Cyclic nucleotides may be hydrolyzed to form the open mononucleotide:

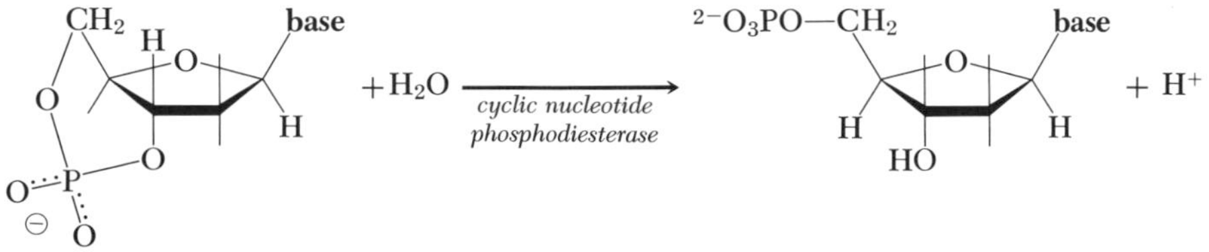

Example: the hydrolysis of cyclic-AMP by a cyclic nucleotide phosphodiesterase to form 5′-AMP.

Further reading

Hassid, W. Z.: *Biosynthesis of Complex Saccharides*. In Greenberg, D. M. ed.: *Metabolic Pathways*, 3rd ed., Vol. 1 (p. 309). Academic Press (1967).

Whelan, W. J., and M. P. Cameron, eds.: *Control of Glycogen Metabolism*. Ciba Found. Symposium, Little, Brown (1964). A collection of articles on then-current problems by several authorities.

Metabolic Effects of Catecholamines. Pharmacol. Rev., *18*: 145–315 (1966). Another collection of articles by authorities.

Steinitz, K.: *Laboratory Diagnosis of Glycogen Diseases*. Adv. Clinical Chem., *9*: 227 (1967). This excellent review contains a summary of the glycogen storage diseases and their relationship to the metabolic pathway in addition to the material implied by the title.

Brown, D. B., and B. I. Brown: *Action of a Muscle Branching Enzyme on Polysaccharides Enlarged from UDP-[^{14}C] glucose*. Biochem. Biophys. Acta, *130*: 263 (1966).

Davis, C. H., *et al.*: *Interrelationships among Glycogen Phosphorylase Isozymes*. J. Biol. Chem., *242*: 4824 (1967).

Walsh, D. A., J. P. Perkins, and E. G. Krebs: *An Adenosine 3′,5′-monophosphate-dependent Protein Kinase from Rabbit Skeletal Muscle*. J. Biol. Chem., *243*: 3763 (1968). See also pages 2200 and 2209 in the same volume.

Hug, G., W. K. Schubert, and G. Chuck: *Phosphorylase Kinase of the Liver: Deficiency in a Girl with Increased Hepatic Glycogen*. Science, *153*: 1534 (1966).

Dickens, F., P. J. Randle, and W. J. Whelan, eds.: *Carbohydrate Metabolism and Its Disorders*, Vols. 1 and 2. Academic Press (1968).

Larner, J., *et al.*: *Hormonal and Non-Hormonal Control of Glycogen Synthesis*. Adv. Enz. Regul., *6*: 409 (1968).

Note: Chapter 23 of this text contains a discussion of cyclic AMP in a larger context and appropriate references.

REDUCTIVE STORAGE: FATS

CHAPTER 16

ARGUMENT

Droplets of triglyceride accumulated within the cells of adipose tissue are the greatest reserve of oxidizable fuel in man. The stored triglycerides are in part derived from fats in the diet, with the constituent fatty acids being modified only as is required to attain a liquid character and a reasonably large potential yield of energy in the fat. With many diets, it is glucose that is available in excess of immediate metabolic demands, and fat is laid down from this surplus carbohydrate supply.

The conversion of glucose into fatty acids by adipose tissue involves the same formation of pyruvate by the Embden-Meyerhof pathway seen in other tissues, and the same mitochondrial oxidative decarboxylation of pyruvate to produce acetyl coenzyme A. The carbon skeleton of the fatty acids is built up from these acetyl units. The energy necessary for the condensation by which 2-carbon units are added to the growing chain is supplied by an ATP-driven carboxylation followed by decarboxylation. The enzyme complex catalyzing fatty acid synthesis contains the same phosphopantetheine group found in coenzyme A, but it is attached to a peptide.

The conversion of acetyl groups into the hydrocarbon chain of fatty acids also involves reductions, and the necessary electrons are transferred through a phosphorylated form of NAD, nicotinamide adenine dinucleotide phosphate, or NADP. Some of the electrons arise from the Embden-Meyerhof pathway, and are transferred from NADH to NADP by the use of a pair of malate dehydrogenases specific for the respective nucleotides. The remainder of the electrons arise from a sequential oxidation of glucose-6-phosphate by a pair of NADP-specific dehydrogenases, which catalyze the initial steps of a sequence of reactions known as the pentose phosphate pathway. The entire process for synthesis of fatty acids from glucose hinges on the use of citrate as a means of transporting acetyl groups into the cytosol, with the simultaneous use of the resultant oxaloacetate as a carrier for transferring electrons to NADP.

The primary product of fatty acid synthesis is palmitate, but acids of longer chain length and containing double bonds are necessary for the creation of fat with the requisite physical properties. The carbon chain can be lengthened through the addition of further acetyl groups within mitochondria. Double bonds can be added, but only at positions between the second and tenth carbons, by oxidations with molecular oxygen, catalyzed by enzymes of a type known as mixed function oxidases. These oxidases simultaneously cause the oxidation of NADPH as a means of "activating" molecular oxygen.

Triglycerides are made by a sequence of reactions in which the coenzyme A esters of various saturated and unsaturated fatty acids are added to a glyceryl group supplied in the form of glycerol-3-phosphate.

Triglycerides are transported through the blood in a variety of combinations with protein and phospholipid, which add sufficient hydrophilic character to stabilize aqueous dispersions. Fat absorbed from the intestine appears as chylomicra, which are droplets of triglyceride with a thin surface film of protein and phospholipid, whereas fat elaborated from the liver appears as various lipoproteins containing lesser proportions of triglyceride.

The triglycerides of adipose tissue are mobilized for use through a hydrolysis triggered by adrenaline (or other agents) in a mechanism mediated by cyclic AMP. The resultant fatty acids complex with serum albumin for transport through the blood to the tissues where they are oxidized.

Fat is the greatest reserve of oxidizable compounds in man. The median quantities of triglycerides in the whole body that can be utilized without extensive tissue destruction are 7.7 kilograms at age 25 in American males and 13.2 kilograms at age 45. Women have even more, with 10.0 and 18.6 kilograms at ages 25 and 45, respectively. These are the amounts of true stores, and do not include the structural lipids found within cell membranes or around nerves.

Most people have a vague idea of the character of fat and give it an equally diffuse anatomical role. Let us take a moment to discuss the sites of deposition so that we can more fully appreciate the biochemistry of these tissues. The fat depot of mammals, including man, is made of globules of triglycerides contained within specialized cells, and the droplets may comprise as much as 90 per cent of the mass of the cells. The cells are formed by differentiation of the reticulum surrounding blood vessels and of the associated endothelium of the capillary bed, so they are in a sense part of the reticulo-endothelial system, which forms blood cells. Indeed, one of the principal sources of blood cells, the bone marrow, undergoes partial transition into a fat depot, so there are red marrow and white marrow. In most parts of the body, the separation of function is more sharp, and there is a discrete *adipose tissue*. Adipose tissue is widely distributed, even in muscles, but it is especially prevalent under the skin, around deep blood vessels, and in the abdominal cavity. It is a relatively late evolutionary development. For example, sharks and related cartilaginous fish store a large amount of lipid, but it is deposited in the liver, which may be as loaded with fat as is the adipose tissue of mammals.

The bony fish, such as the salmon, store lipid in and between the fibers of skeletal muscle.

Adipose tissue appears relatively formless and is difficult to handle experimentally because of its high content of fat, but it is in fact a well-organized tissue with a rich blood supply, nerve endings for stimulation, and an active metabolism appropriate for its important function as an internal larder. Adipose tissue can become the largest in the body, comprising half or more of the total mass of some individuals. Humans can become tubs of lard. Such people are objects of humor, disdain, or concern in our society, but in societies subject to famine they may be happily living on their own fat while burying the last of their formerly trim companions.

The fats being stored represent carbon compounds ingested in excess of the amount required to meet demands for oxidative phosphorylation and for replenishing the glycogen stores. As we saw earlier, the bulk of a triglyceride molecule is made of fatty acid residues, with the glycerol moiety being a small part of the total number of atoms, so that the major quantitative aspect of fat storage is the accumulation of long-chain acyl residues. These may be derived directly from dietary fat or they may be synthesized from acetyl coenzyme A, which in turn can be formed from glucose or amino acids. In short, an excess of any of the major nutrients can be stored as acyl residues in triglycerides. Let us first consider how fatty acid chains are formed from acetyl coenzyme A, and then look at the combination of the chains, from all sources, into triglycerides.

THE SYNTHESIS OF ACYL RESIDUES

Source of the Carbon Atoms

Acetyl coenzyme A is the source of the carbon atoms used to make acyl residues. The conversion of acetyl groups to fatty acids is in a sense the reversal of the processes of oxidation of the fatty acids in which electrons are removed for oxidative phosphorylation, and electrons must be added back in order to form the alkyl chains. The way in which this is accomplished is in some ways reminiscent of the the reverse of the oxidative reactions, but the two processes are quite separate, both in mechanism and in physical location. The oxidation of acyl residues occurs in mitochondria, as we saw, but synthesis occurs in the soluble cytoplasm, and this location imposes a further requirement—the transfer of acetyl groups from the mitochondria in which they appear to the soluble cytoplasm. Furthermore, synthesis of fatty acids for storage is mainly confined to adipose tissue and the liver, whereas oxidation principally occurs in muscles.

We might surmise, and shall later see, that there is some mechanism of control preventing the simultaneous breakdown of fat through mitochondrial oxidations and synthesis of fat in the soluble cytoplasm, so as to prevent purposeless recycling. If the acyl residues are excluded as sources of acetyl coenzyme A, the remaining source is pyruvate, and the oxidation of pyruvate is a mitochondrial process. Acetyl coenzyme A cannot diffuse freely from mitochondria to the soluble cytoplasm, so the acetyl groups must be carried between these cellular compartments in other forms. The major form for transport is as citrate, which can pass between these compartments.

Citrate is, of course, formed from oxaloacetate and acetyl coenzyme A as the first step in the citric acid cycle of mitochondria. A leakage of citrate implies that the cycle is being robbed, but this is not so. In the first place, fat synthesis implies a surfeit of substrates—a time of rest when the capacity to generate high-energy phosphate is in excess of the demand for other processes. In the second place, the demand for complete oxidation of substrates through the citric acid cycle in adipose tissue and liver, where fat synthesis mainly occurs, is not nearly as great as it is in muscles.

Citrate appearing in the cytosol from the mitochondria is cleaved into oxaloacetate and acetyl coenzyme A (Fig. 16-1).

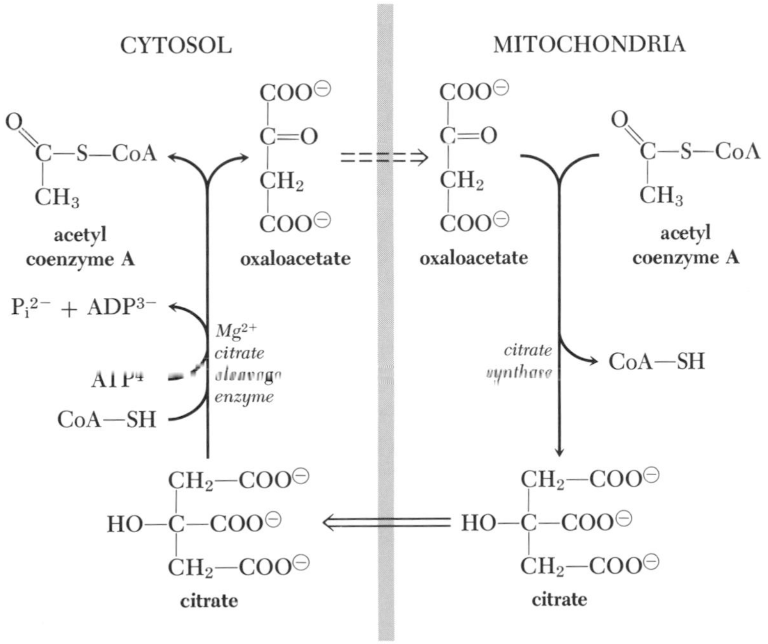

Figure 16-1 Citrate can act as a carrier of acetyl groups from the mitochondria to the cytosol because it is freely diffusible through the membranes. It is formed by the citrate synthase reaction and cleaved by an enzyme catalyzing the simultaneous hydrolysis of ATP, which drives the reaction. Oxaloacetate does not return as such to the mitochondria, but there are reactions by which it is replaced, implied by the dashed arrow.

The energy necessary to form the thiol ester bond is supplied through a simultaneous cleavage of ATP. As the figure shows, the carbons of oxaloacetate are acting in this case as a carrier of the carbons of acetyl coenzyme A from mitochondria to the cytosol, and the transport costs one mole of high-energy phosphate per mole of acetyl group. (We shall later see that oxaloacetate as such need not always diffuse into the mitochondrion, but the effect is as if it did.)

Synthesis of the Carbon Chain

In fatty acid synthesis we are dealing with a situation reminiscent of gluconeogenesis: Steps involving carboxylation and decarboxylation are introduced, and the concomitant loss of high-energy phosphate drives what would otherwise be unfavorable reactions (in this case, condensation and reduction rather than cleavage and oxidation of the fatty acid chain).

The initial two carbons of what is to be a fatty acid are provided by the acetyl group of acetyl coenzyme A. The remainder of the carbons are supplied in the form of *malonyl coenzyme A*, which is made by carboxylation of acetyl coenzyme A in the cytosol:

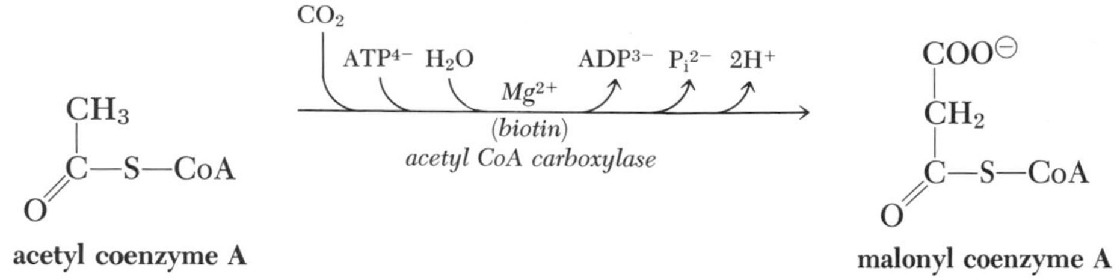

acetyl coenzyme A   malonyl coenzyme A

This is a straightforward carboxylation of the type in which CO_2 is activated by attachment to a biotinyl residue on the *acetyl CoA carboxylase*, with energy supplied through hydrolysis of ATP.

During the reactions in which the hydrocarbon chain of the fatty acids is formed, the growing acyl residues are attached to the enzyme complex catalyzing the synthesis. This *fatty acid synthase* contains a residue of *phosphopantetheine*, the same structure found in coenzyme A, but attached in this case to a seryl residue of the synthase peptide (Fig. 16-2). The enzyme has its own acyl carrier built in to serve the same function during fatty acid synthesis that coenzyme A does during fatty acid oxidation. The reactive part of the molecule is identical in the two cases, but coenzyme A has phosphopantetheine attached to a 3′,5′-*bis*-phosphoadenosyl group and is free to migrate from one enzyme to another.

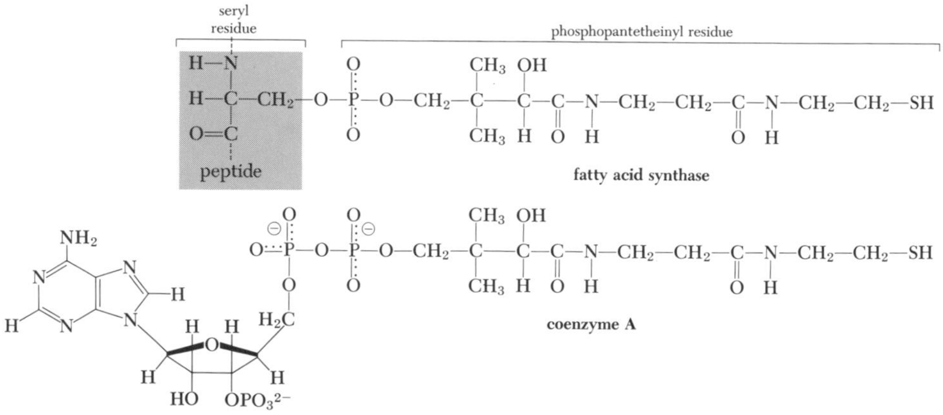

Figure 16-2 A phosphopantetheinyl residue is a part of fatty acid synthase and of coenzyme A. The sulfhydryl group of the residue is used for carrying acyl groups in both cases.

The long chain of the phosphopantetheinyl group enables its sulfhydryl group to sweep through large arcs, and the growing fatty acid is attached to this group for contact with active sites catalyzing the necessary reductions and dehydrations. (A separate peptide unit, the *acyl carrier protein*, is used by some microorganisms for this purpose, but it has not been possible to dissect such a separate peptide from the vertebrate enzyme complexes.)

The fatty acid synthase also makes use of another sulfhydryl group provided by a cysteinyl residue in its peptides. This group is used as a sort of fixed storage site for acyl groups, contrasting with the mobile site on the phosphopantetheinyl residue.

Synthesis begins with the transfer of an acetyl group from coenzyme A onto the phosphopantetheinyl group attached to fatty acid synthase (Fig. 16-3). (This

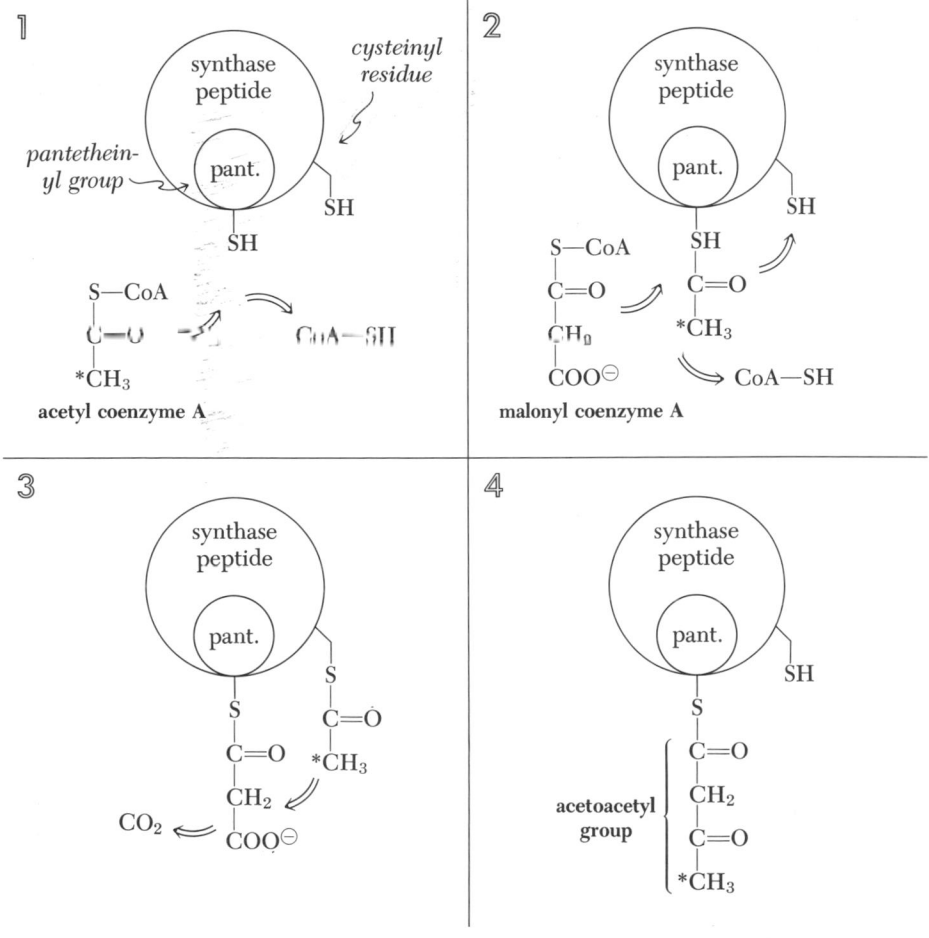

Figure 16-3 The chain-lengthening process in fatty acid synthesis. The figure shows the initial addition of 2-carbon units, but the same mechanism is involved for later additions.
  1. A priming acetyl group is attached to the pantetheinyl group with release of coenzyme A.
  2. The acetyl group moves to a cysteinyl residue, and its place is taken by a malonyl group.
  3. The acetyl group condenses with the methylene carbon of the malonyl group. CO_2 is liberated.
  4. The result is an acetoacetyl group attached to the pantetheinyl group on the enzyme.

and the following transfers may involve the intermediate formation of esters of a seryl group at the transferase active site.) The acetyl group is then moved onto a cysteinyl residue and a malonyl group is transferred onto the phosphopantetheinyl group. The acetyl and malonyl groups are now both attached to the synthase and the stage is set for condensation. In effect, the acetyl group displaces CO_2 from the malonyl group, forming a carbon-to-carbon bond so that there is now the 4-carbon acetoacetyl group attached to phosphopantetheine on the enzyme.

The remainder of the transformations involve the reduction of the 3-keto compound to the saturated hydrocarbon chain in the reverse of the sort of route seen in fatty acid oxidation. However, the oxidative mechanism employs a flavoprotein at one stage because the thermodynamics of the reaction requires this stronger oxidizing agent (p. 219). This wouldn't do for the reverse reaction in which the requirement is for a stronger reducing agent, and it turns out that a reduced nicotinamide nucleotide is the electron donor for both intermediate steps. The particular nucleotide is not NADH itself, but a derivative containing an additional phosphate esterified to the 2'-carbon of the adenosyl moiety:

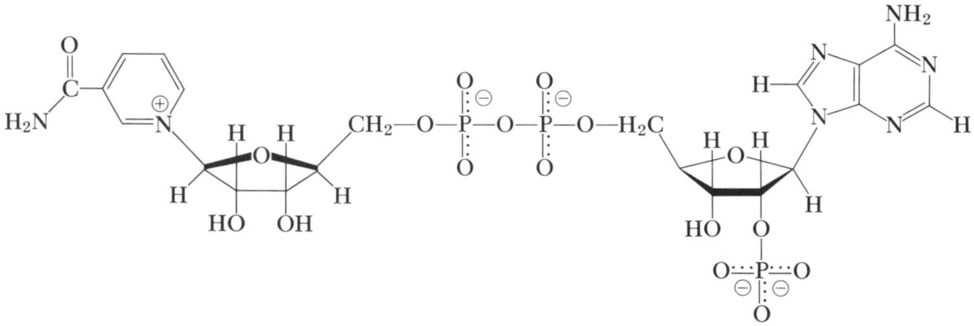

nicotinamide adenine dinucleotide phosphate
(NADP)

The oxidized form has the trivial name of *nicotinamide adenine dinucleotide phosphate (NADP)*, and the reduced form is abbreviated *NADPH*, analogous to NADH. The older name, still used by many, is *triphosphopyridine nucleotide (TPN)*.

The actual sequence of changes by which the acetoacetyl group is reduced is quite similar to a reversal of the steps in oxidation of fatty acids (Fig. 16-4). However, the group remains attached to the phosphopantetheinyl residue on the fatty acid synthase throughout the intermediate steps. The enzyme complex presumably has different sites with the necessary catalytic activities, all within reach of the phosphopantetheinyl chain as it swings about.

The figure shows another distinctive characteristic of fatty acid synthesis. Synthesis involves the D-3-hydroxyacyl compound, which is dehydrated to the *trans*-enoyl derivative. The reaction is an exception to the usual equilibration of L and *trans* isomers.

The product of the initial set of reductions is the *butyryl* derivative of the phosphopantetheinyl group on the enzyme. Now, the butyryl group can be displaced onto the neighboring cysteinyl group by an incoming malonyl group in the

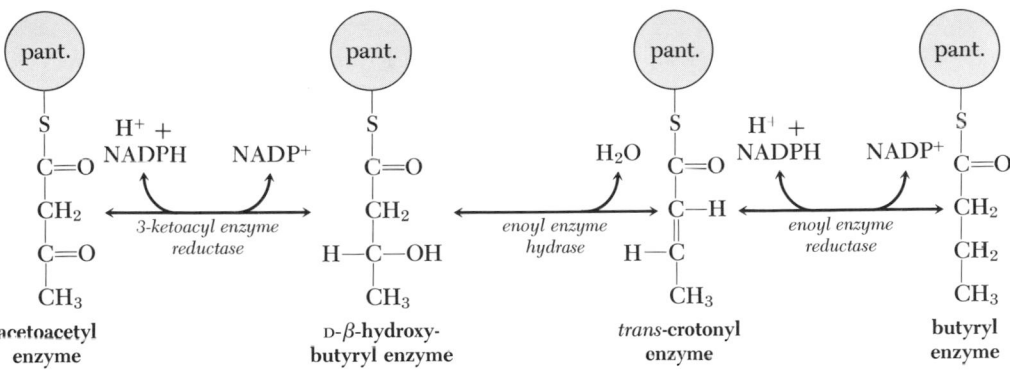

Figure 16-4 The reduction of 3-ketoacyl groups, in this case the acetoacetyl group, by fatty acid synthase occurs while the group is attached to the phosphopantetheinyl sulfhydryl group. The reactions resemble the reversal of fatty acid oxidation, but they all occur on the same enzyme complex.

same way that an acetyl group was displaced in the initial steps of synthesis (Fig. 16-3). The butyryl group will then be transferred onto the methylene carbon of the malonyl group, displacing CO_2 in the same way. The resultant 3-ketohexanoyl group will go through the same cycle of reductions by NADPH to form the saturated 6-carbon hexanoyl group. This in turn can be displaced by a new malonyl group, and so on, making a growing fatty acid chain in which all but the terminal pair of carbons have passed through malonyl coenzyme A.

The chain grows on the fatty acid synthase until seven pairs of carbon atoms have been added from malonyl coenzyme A onto the original pair from acetyl coenzyme A, thereby producing a 16-carbon *palmitoyl* residue. The final step is the release of the chain as free palmitate. It is not known how chain lengthening is limited to the 16-carbon saturated fatty acid, or how the residue is released from the protein.

In any event, the stoichiometry for the formation of palmitate from acetyl coenzyme A, allowing for the intermediate formation of seven moles of malonyl coenzyme A, becomes:

$$(1) \quad 7 \text{ (acetyl—S—CoA)} + 7 \text{ } CO_2 + 7 \text{ } ATP^{4-} + 7 \text{ } H_2O \longrightarrow$$
$$7 \text{ (malonyl—S—CoA)}^- + 7 \text{ } ADP^{3-} + 7 \text{ } P_i^{2-} + 14 \text{ } H^+$$

$$(2) \quad \text{(acetyl—S—CoA)} + 7 \text{ (malonyl—S—CoA)}^- + 14 \text{ NADPH} + 20 \text{ } H^+ \longrightarrow$$
$$\text{(palmitate)}^- + 7 \text{ } CO_2 + 14 \text{ NADP}^+ + 8 \text{ CoA—SH} + 6 \text{ } H_2O$$

$$\text{SUM: } (3) \quad 8 \text{ (acetyl—S—CoA)} + 7 \text{ } ATP^{4-} + 14 \text{ NADPH} + 6 \text{ } H^+ + H_2O \longrightarrow$$
$$\text{(palmitate)}^- + 7 \text{ } ADP^{3-} + 7 \text{ } P_i^{2-} + 8 \text{ CoA—SH} + 14 \text{ NADP}^+$$

Before leaving the stoichiometry, let us consider the utilization of NADPH. NAD and NADP have very similar standard free energies of oxidation and reduction. Why, then, does the cell contain two nucleotides capable of reacting in the same way? The answer is that the nucleotides can be distinguished through enzyme specificity and used in separately controlled sequences involving the same sort of reaction. In general, NADP is used as an electron carrier for reductive syntheses, such as the formation of fatty acids we are now considering, whereas NAD is used more in the processes of energy production. This is not a hard and fast rule, but it is frequently true.

The essential point is that the ratio of concentrations of the oxidized and reduced forms of these two nucleotides can be maintained at different levels by their participation in different reactions. The ratio is approximately 4:1 in favor of the oxidized form of NAD in the liver, whereas it is approximately 3:1 in favor of the reduced form of NADP. Other things being equal, these differences will have the following results: Dehydrogenases with a specificity for NAD will tend to oxidize the substrate and form NADH, whereas dehydrogenases with a specificity for NADP will tend to reduce the substrate and remove NADPH.

Sources of Electrons: NADP-Specific Isocitrate and Malate Dehydrogenases

Since NADPH is the source of the electrons used in fatty acid synthesis, it follows that it must be generated by other reactions in which NADP oxidizes substrates. None of the major oxidative pathways we have discussed to this point involves a transfer of electrons from substrates to NADP. However, there are three major potential sources of NADPH in the soluble cytoplasm.

The soluble cytoplasm contains an *isocitrate dehydrogenase* specific for NADP that oxidizes isocitrate to α-ketoglutarate and CO_2 with an accompanying formation of NADPH. The reaction is identical to the oxidation of isocitrate in the citric acid cycle (p. 213) except for the different nicotinamide nucleotide. (There is also an NADP-specific isocitrate dehydrogenase in mitochondria used to generate NADPH in that organelle, but we are now concerned with the cytosol.) We may as well admit to begin with that we are including this reaction in the discussion for sake of completeness; it is not a significant source of NADPH in adipose tissue, and its quantitative importance in other organs has not been delineated. This ought not be taken as discounting the possible significance of the reaction in the metabolic economy; the enzyme has a high activity in many tissues, and citrate is a circulating metabolite, as well as being an intermediate in the citric acid cycle and a frequent dietary constituent. We simply don't know enough to say why NADPH is sometimes generated in the cytosol of tissues by this particular reaction.

The second potential source of NADPH in the cytosol is a reaction catalyzed by a *malate dehydrogenase,* in which NADP is the oxidizing agent and there is a simultaneous decarboxylation:

$$\begin{array}{ccc}
\overset{\displaystyle COO^{\ominus}}{\underset{\displaystyle \underset{\displaystyle COO^{\ominus}}{\overset{\displaystyle |}{\underset{\displaystyle |}{CH_2}}}}{\overset{\displaystyle |}{HO-C-H}} & \xrightarrow[\substack{NADP\text{—}malate\\ dehydrogenase}]{NADP^+ \quad NADPH} & \overset{\displaystyle COO^{\ominus}}{\underset{\displaystyle CH_3}{\overset{\displaystyle |}{\underset{\displaystyle |}{C=O}}}} + CO_2 \\
\text{L-malate} & & \text{pyruvate}
\end{array}$$

We already pointed out that 3-ketocarboxylates are easily decarboxylated, and that reactions producing them may, or may not, involve the simultaneous loss of CO_2. The NAD-coupled malate dehydrogenase of mitochondria produces oxaloacetate without decarboxylation, but the NADP-coupled enzyme of the cytosol causes the simultaneous decarboxylation to pyruvate.

This reaction is indeed an important source of NADPH for fatty acid synthesis in adipose tissue whenever glucose is being used as the source of the acetyl coenzyme A from which fatty acids are formed. It is part of an ingenious scheme in which electron transfer utilizes as a carrier the same oxaloacetate transporting acetyl groups into the cytosol in the form of citrate (Fig. 16-5).

The transformation of glucose into fatty acids involves the oxidation of glucose to acetyl coenzyme A and the reduction of acetyl coenzyme A to fatty acids. Now, the conversion of glucose to acetyl coenzyme A includes the production of pyruvate in the cytosol and an associated production of NADH, followed by the oxidation of pyruvate in mitochondria. Let us analyze what happens, step by step:

1. The citric acid cycle is relatively slow in adipose tissues. Pyruvate diffusing into mitochondria can in part be oxidized to acetyl coenzyme A, and in part carboxylated to form oxaloacetate. These two compounds condense to form citrate, of which a little is oxidized and the balance diffuses into the cytosol.

2. Citrate is cleaved in the cytosol to form oxaloacetate and acetyl coenzyme A. The acetyl coenzyme A is utilized for fatty acid synthesis.

3. The cytosol also contains an *NAD-specific* malate dehydrogenase; we have already seen that the conversion of glucose to pyruvate causes NADH to accumulate in the cytosol. The combination of NADH from the Embden-Meyerhof pathway

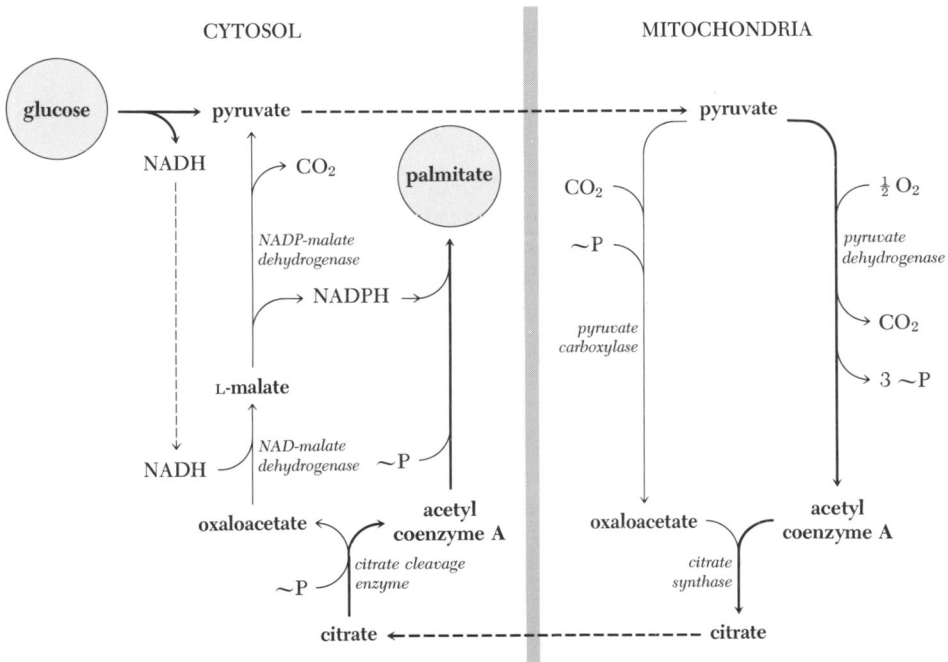

Figure 16-5 The use of oxaloacetate as a carrier for electrons as well as for acetyl groups. See the text for details.

and oxaloacetate from citrate cleavage will cause their conversion to NAD and malate in the presence of the dehydrogenase. The equilibrium favors this conversion.

4. The malate being formed is a substrate for the *NADP-specific* malate dehydrogenase also in the cytosol, which catalyzes the oxidative decarboxylation of malate to form pyruvate, CO_2, and NADPH. The combination of oxidation and decarboxylation in a single reaction makes the equilibrium favor the formation of NADPH, whereas the oxidation without decarboxylation catalyzed by NAD-specific malate dehydrogenase favors the utilization of NADH and the formation of NAD.

5. The pyruvate produced in the cytosol moves into the mitochondria and replaces the pyruvate utilized to form oxaloacetate, so the only pyruvate disappearing is that which is oxidized to acetyl coenzyme A.

The net effect of this sequence is to transfer the electrons in the form of NADH taken from glucose during pyruvate formation to NADP, rather than using them to generate high-energy phosphate or to form lactate. It could not function if electrons were being moved to mitochondria *via* the glycerophosphate shuttle, but this pathway is indeed slow in adipose tissue. The stoichiometry for the transfer shows that one high-energy phosphate is expended for each NADPH formed:

$$(4) \quad NADH + (oxaloacetate)^{2-} + H^+ \longrightarrow (malate)^{2-} + NAD^+$$
$$(5) \quad (malate)^{2-} + NADP^+ \longrightarrow (pyruvate)^- + NADPH + CO_2$$
$$(6) \quad (pyruvate)^- + CO_2 + ATP^{4-} + H_2O \longrightarrow$$
$$(oxaloacetate)^{2-} + ADP^{3-} + P_i^{2-} + 2\ H^+$$

$$\text{SUM: } (7) \quad NADP^+ + NADH + ATP^{4-} + H_2O \longrightarrow$$
$$NADPH + NAD^+ + ADP^{3-} + P_i^{2-} + H^+$$

Sources of Electrons:
The Pentose Phosphate Pathway

NADPH is also generated in the cytosol of many cells through a sequence of reactions for the metabolism of glucose-6-phosphate quite different from the Embden-Meyerhof pathway we have hitherto emphasized. This route is variously known as the *pentose shunt*, the *hexosemonophosphate pathway*, and other permutations of these words. Only a small fraction of the carbons of glucose pass through this route, even in adipose tissue, but it is a major source of NADPH and also provides a mechanism for creating the pentose phosphates for the synthesis of nucleotides, and is involved in plant photosynthesis. The general idea of the pentose phosphate pathway is very simple—glucose-6-phosphate undergoes two successive oxidations by NADP, the final one being an oxidative decarboxylation to form a pentose phosphate. The remainder of the reactions are concerned with a transformation of the pentose phosphate into triose and hexose phosphates that can be re-used.

Glucose-6-phosphate contains a number of hydroxyl groups, and could in principle be the substrate for a variety of dehydrogenases. In fact, only one is of importance in mammalian tissues, and this *glucose-6-phosphate dehydrogenase* catalyzes the oxidation of the β-anomer by NADP (Fig. 16-6).

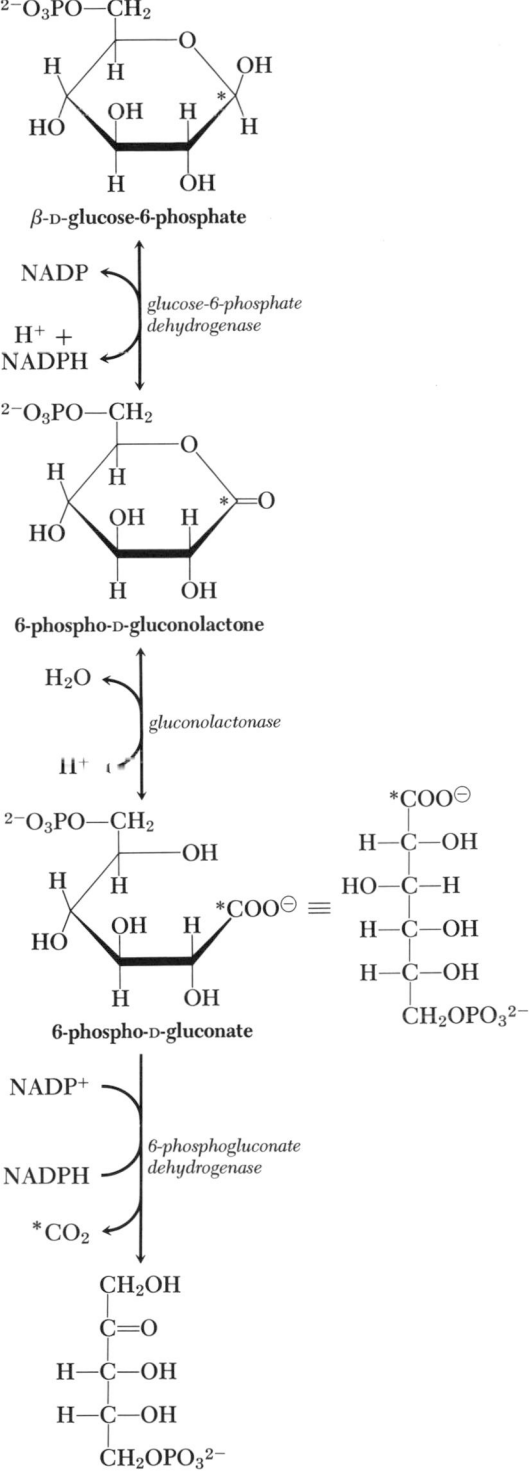

Figure 16-6   The pentose phosphate pathway begins with a pair of oxidations of glucose-6-phosphate by NADP. The first of these produces the lactone of 6-phosphogluconate, which is hydrolyzed to the free compound. 6-Phosphogluconate is then oxidatively decarboxylated.

The oxidation is a conversion of an alcohol group to a carbonyl group, but the substrate is really a potential aldehyde in the form of its hemiacetal and the product is a potential acid in the form of its lactone, or inner ester. The ester bond of this 6-phosphogluconolactone is hydrolyzed by a *gluconolactonase* to form *6-phospho-D-gluconate*. (See p. 717 for a discussion of the nomenclature of the acidic derivatives of sugars.)

6-Phosphogluconate is a substrate for a second dehydrogenase, and its third carbon is oxidized by NADP, with a simultaneous decarboxylation (Fig. 16-6). The reaction is a typical oxidative decarboxylation of a 3-hydroxycarboxylate without release of the intermediate 3-keto compound, and is therefore of the same type as the reactions catalyzed by isocitrate dehydrogenase and NADP-malate dehydrogenase. The product D-*ribulose-5-phosphate* is the ketose isomer of D-ribose-5-phosphate.

This sequence of reactions produces two moles of NADPH for each mole of CO_2 produced from glucose-6-phosphate in the pentose phosphate pathway. This is the crucial part of the overall economy of the pathway: One carbon of glucose appears as CO_2 for each *pair* of NADPH generated. The pentose phosphate, representing five-sixths of the original glucose, is recovered for further use by the remaining reactions of the pathway, which we shall now consider.

Disposition of the Pentose Phosphates

In handling the pentose phosphates, the problem solved by evolution is essentially one of shuffling the carbons around so as to wind up with hexoses and trioses. The redistribution is accomplished mainly through the action of two enzymes, a *transketolase* catalyzing the transfer of 2-carbon units from one sugar to another, and a *transaldolase* catalyzing the transfer of 3-carbon units.

Transketolase contains *thiamine pyrophosphate* as a coenzyme, which is used to bind the intermediate aldehyde much as it is in the pyruvate or α-ketoglutarate dehydrogenase reactions (p. 215). The first two carbons of a ketose phosphate are transferred to an aldose phosphate with intermediate carriage on the thiamine pyrophosphate (Fig. 16 7).

As Figure 16-7 indicates, the preferred substrates for the enzyme are ketoses with a D-*threo* configuration next to the carbonyl group, that is, with a D-configuration on the fourth carbon, and an L-configuration on the third. Since the second carbon of the aldose remaining after removal of two carbons is derived from the fourth carbon of the ketose, it follows that the aldose phosphate substrates must have the D-configuration of the carbon adjacent to the aldehyde group. (Remember that enzymes catalyze reactions in either direction, and products must be able to act as substrates.)

An aldose phosphate and a ketose phosphate capable of being substrates for transketolase are formed from D-ribulose-5-phosphate through a pair of reactions (Fig. 16-8, top).

The isomerization to D-ribose-5-phosphate is catalyzed by a straightforward *phosphoribose isomerase* analogous to the phosphoglucose isomerase and triose phosphate isomerase seen in the Embden-Meyerhof pathway. (This reaction also provides a source of the pentose for nucleotide formation.) The formation of D-xylulose-5-phosphate by an *epimerization,* or inversion of configuration of one

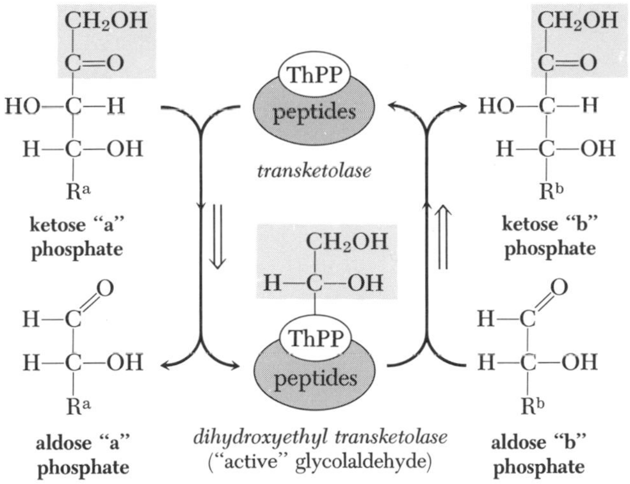

Figure 16-7   The action of transketolase. A ketose phosphate ("a") reacts with thiamine pyrophosphate on the enzyme; the first two carbons are transferred to the coenzyme, leaving the remainder of "a" as an aldose phosphate. The dihydroxyethyl group is then transferred from thiamine pyrophosphate to a second aldose phosphate ("b"), converting it to a new ketose phosphate.

of the carbons, is something we have not encountered before, and is a kind of reaction frequently encountered in carbohydrate metabolism for the conversion of one kind of sugar into another. The particular *ribulose phosphate 3-epimerase* used here will only catalyze the inversion of the third carbon.

Since ribose-5-phosphate and xylulose-5-phosphate are appropriate substrates, transketolase catalyzes the transfer of a 2-carbon unit from the ketose to the aldose (Fig. 16-8, middle). The 3-carbon remainder of the ketose is the familiar glyceraldehyde-3-phosphate, but the only known function for the 7-carbon sedoheptulose-7-phosphate in animals is its participation as an intermediate in the pentose phosphate pathway. (Before the discovery of its metabolic importance, sedoheptulose was regarded as a trivial oddity; its only known occurrence was as a polymer in plants of the genus *Sedum*.)

Transaldolase catalyzes the transfer of a dihydroxyacetone unit from a ketose to an aldose, and its specificity is such that the products of the transketolase reaction we just mentioned are the preferred substrates (Fig. 16-8, bottom). (Transaldolase does not utilize thiamine pyrophosphate as a coenzyme, because the group being transferred is a ketone, not an aldehyde. The nature of the intermediates involved is unknown.)

The combination of the transketolase and transaldolase reactions accomplishes the conversion of six of the 10 carbon atoms in a pair of pentose phosphates to fructose-6-phosphate, which is directly interconvertible with glucose-6-phosphate through the phosphoglucose isomerase reaction. The remaining four carbons are in the form of D-erythrose-4-phosphate. This aldose phosphate can also act as a substrate for the transketolase reaction, accepting a pair of carbons from xylulose-5-phosphate (Fig. 16-9).

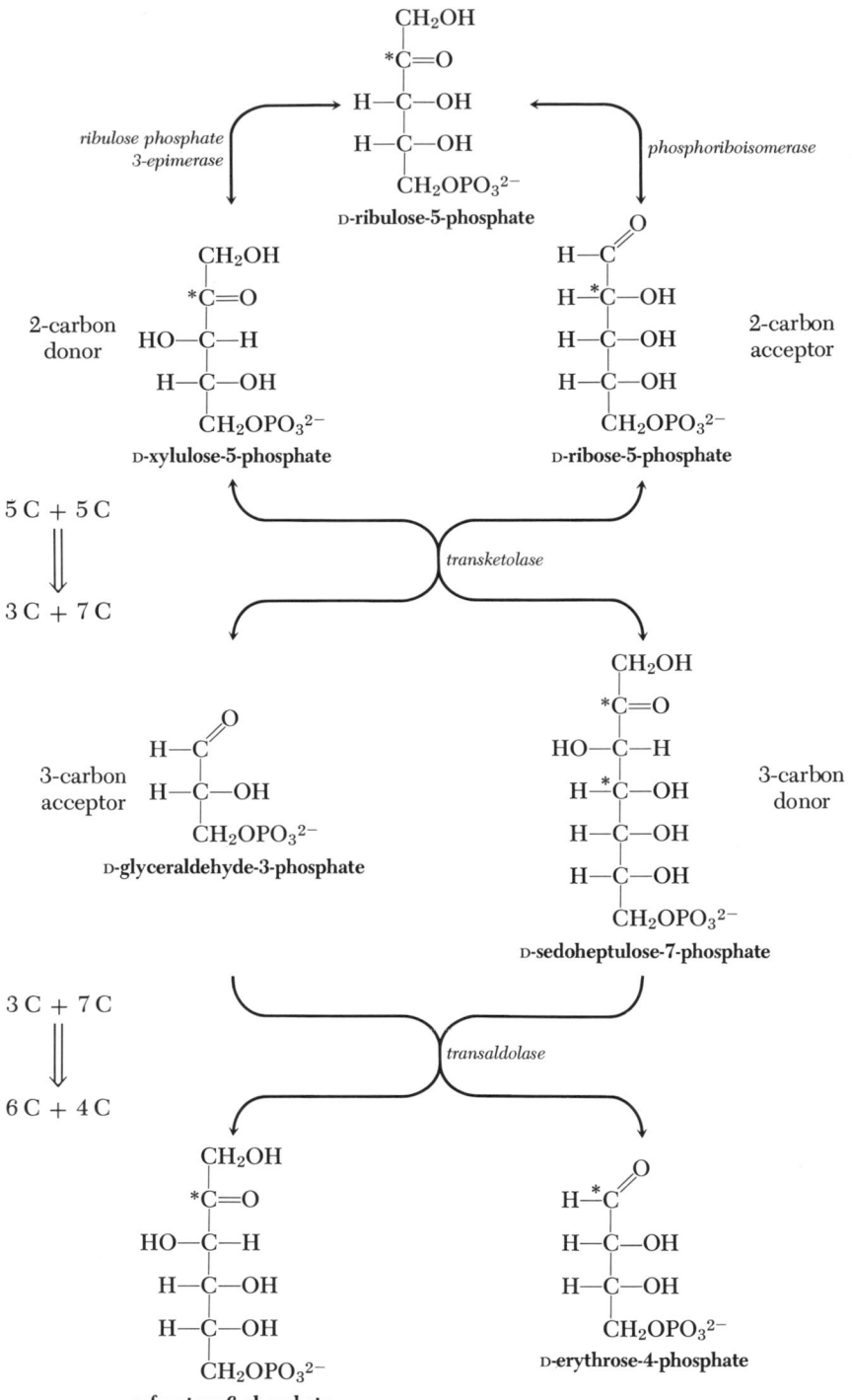

Figure 16-8 Ribulose-5-phosphate is converted into a pair of substrates for transketolase by the action of an epimerase and an isomerase. The 3-carbon and 7-carbon sugar phosphates formed by transketolase are substrates for transaldolase, which transfers a 3-carbon unit. The result is the recovery of six out of 10 carbons in a pair of pentose phosphate molecules as a molecule of fructose-6-phosphate. The fate of the remaining four carbons is shown in the next figure.

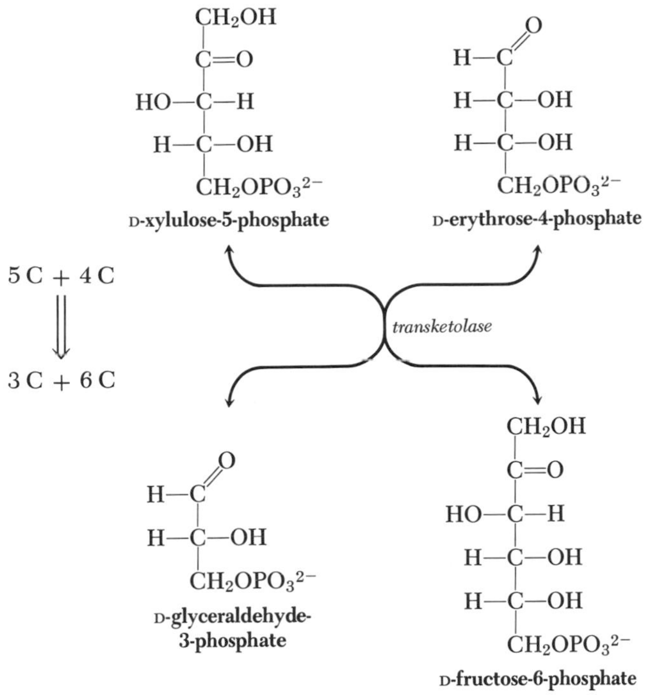

CH₂OH

$$\begin{array}{c} CH_2OH \\ | \\ C=O \\ | \\ HO-C-H \\ | \\ H-C-OH \\ | \\ CH_2OPO_3{}^{2-} \end{array}$$

D-**xylulose-5-phosphate**

$$\begin{array}{c} H-C \overset{O}{\nearrow} \\ | \\ H-C-OH \\ | \\ H-C-OH \\ | \\ CH_2OPO_3{}^{2-} \end{array}$$

D-**erythrose-4-phosphate**

5 C + 4 C

transketolase

3 C + 6 C

$$\begin{array}{c} H-C \overset{O}{\nearrow} \\ | \\ H-C-OH \\ | \\ CH_2OPO_3{}^{2-} \end{array}$$

D-**glyceraldehyde-**
3-phosphate

$$\begin{array}{c} CH_2OH \\ | \\ C=O \\ | \\ HO-C-H \\ | \\ H-C-OH \\ | \\ H-C-OH \\ | \\ CH_2OPO_3{}^{2-} \end{array}$$

D-**fructose-6-phosphate**

Figure 16-9 The final transformation of the pentose phosphate pathways involves the reaction between an additional molecule of pentose phosphate and the erythrose phosphate remaining from previous transformations. The action of transketolase causes the formation of glyceraldehyde phosphate and fructose phosphate, representing final recovery of the original pentose phosphate as intermediates of the Embden-Meyerhof pathway.

The result is the formation of hexose phosphate and triose phosphate, completing the conversion of pentose phosphate into these sugar esters.

What do we now have? The transketolase reaction has been used twice and the transaldolase reaction once, and if we sum up the number of carbons involved, we have this:

$$\begin{array}{llr} 5\,C + 5\,C \longrightarrow 3\,C + 7\,C & \text{(transketolase)} \\ 3\,C + 7\,C \longrightarrow 6\,C + 4\,C & \text{(transaldolase)} \\ 5\,C + 4\,C \longrightarrow 3\,C + 6\,C & \text{(transketolase)} \\ \hline \end{array}$$

SUM: $5\,C + 5\,C + 5\,C \longrightarrow 6\,C + 6\,C + 3\,C$

In other words, three moles of pentose phosphates are converted to two moles of hexose phosphate and one mole of triose phosphate. The reactions all balance without change in water, hydrogen ion, or other reactants.

Considering the entire process starting with glucose-6-phosphate, for each six molecules oxidized, six molecules of ribulose-5-phosphate will be produced according to the following stoichiometry:

(7)   6 (glucose-6-phosphate)$^{2-}$ + 12 NADP$^+$ + 6 H$_2$O $\longrightarrow$
      6 (ribulose-5-phosphate)$^{2-}$ + 6 CO$_2$ + 12 NADPH + 12 H$^+$

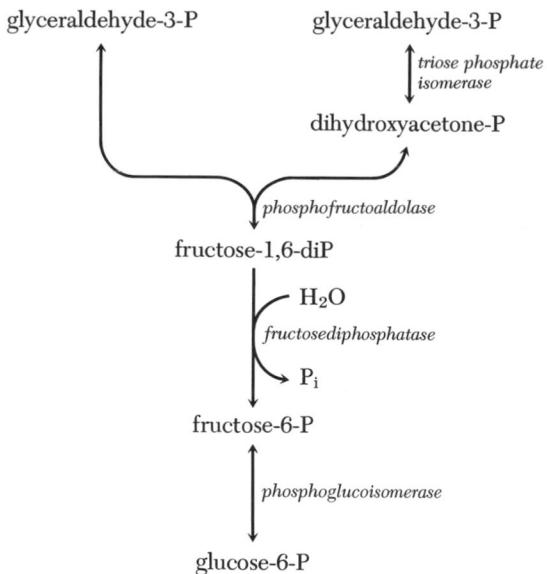

Figure 16-10   Cells containing fructosediphosphatase may recover glyceraldehyde-3-phosphate from the pentose phosphate pathway as glucose-6-phosphate.

If we now add the stoichiometry for the pentose phosphate disposal, in which two moles of hexose phosphate and one mole of triose phosphate are formed for each three moles of pentose phosphate handled:

$$(8) \quad 6 \text{ (ribulose-5-P)}^{2-} \longrightarrow$$
$$4 \text{ (glucose-6-P)}^{2-} + 2 \text{ (glyceraldehyde-3-P)}^{2-} \text{ we have:}$$
$$\text{SUM: (9)} \quad 2 \text{ (glucose-6-P)}^{2-} + 12 \text{ NADP}^{+} + 6 \text{ H}_2\text{O} \longrightarrow$$
$$2 \text{ (glyceraldehyde-3-P)}^{2-} + 6 \text{ CO}_2 + 12 \text{ NADPH} + 12 \text{ H}^{+}$$

The net effect of this process is the same as if one mole of glucose-6-phosphate had been oxidized to CO_2 and H_2O by NADP while another mole was being converted to a pair of triose phosphates.

This is the stoichiometry involved in fatty acid synthesis, but we ought to note that tissues with an active fructosediphosphatase may have a completely closed cycle, because the glyceraldehyde phosphate is always interconvertible with dihydroxyacetone phosphate, and this pair of triose phosphates can be converted to fructose-1,6-diphosphate (Fig. 16-10). If the fructose-1,6-diphosphate is hydrolyzed to fructose-6-phosphate, which is in equilibrium with glucose-6-phosphate, one of the two moles of glucose-6-phosphate in equation (9) cancels, and we are left with:

$$\text{SUM: (10)} \quad \text{(glucose-6-P)}^{2-} + 12 \text{ NADP}^{+} + 6 \text{ H}_2\text{O} \longrightarrow 6 \text{ CO}_2 + 12 \text{ NADPH} + 12 \text{ H}^{+}$$

The stoichiometry proves what is not so obvious from the individual reactions: The pentose phosphate pathway is a means of oxidizing glucose to CO_2 and H_2O, using only NADP as an oxidizing agent. Some microorganisms employ this route as a major part of their metabolism for generating energy, but in animals it is used only as a source of NADPH for syntheses and of the pentose phosphates.

Even though the triosephosphates are not converted back to glucose-6-phosphate in adipose tissue, the net effect is still the complete oxidation of a portion of the glucose-6-phosphate. The remainder is converted to glyceraldehyde-3-phosphate, which can be carried through the Embden-Meyerhof pathway to pyruvate, so that the overall balance of equation (9) is the same as would be obtained if half of the glucose-6-phosphate were completely oxidized in the pentose phosphate pathway and the other half were carried through the regular Embden-Meyerhof pathway to pyruvate. This is an important part of the route for converting glucose to fatty acids in adipose tissue.

It ought to be noted, although we shall return to the point later, that the reactions discussed here also provide a means for converting pentose phosphates into hexose phosphates, so that a sugar such as ribose in the diet can be converted to glucose.

The Overall Stoichiometry of
Fatty Acid Synthesis

Let us look at the overall stoichiometry of fatty acid synthesis in adipose tissue according to our formulation of the routes. The result is a beautiful demonstration that metabolism is an array of reactions ordered both in character of the intermediates and in anatomical location of the enzymes, and susceptible to rational analysis. However, analysis is not easy and requires a step-by-step approach.

There are five processes to be balanced (neglecting H_2O and H^+ for the moment):

1. Formation of palmitate:

Cytosol

$$8 \text{ acetyl—S—CoA} + 7 \text{ ATP} + 14 \text{ NADPH} \longrightarrow$$
$$\text{palmitate} + 7 \text{ (ADP} + P_i) + 14 \text{ NADP}$$

2. Metabolism of glucose by Embden-Meyerhof pathway:

Cytosol

$$\text{glucose} + 2 \text{ (ADP} + P_i) + 2 \text{ NAD} \longrightarrow 2 \text{ pyruvate} + 2 \text{ ATP} + 2 \text{ NADH}$$

Mitochondria

$$2 \text{ pyruvate} + O_2 + 6 \text{ (ADP} + P_i) \longrightarrow 2 \text{ acetyl—S—CoA} + 6 \text{ ATP} + 2 \text{ CO}_2$$

3. Transfer of electrons by malate dehydrogenase pathways:

Cytosol

$$\text{oxaloacetate} + \text{NADH} \longrightarrow \text{malate} + \text{NAD}$$
$$\text{malate} + \text{NADP} \longrightarrow \text{pyruvate} + CO_2 + \text{NADPH}$$

Mitochondria

$$\text{pyruvate} + CO_2 + \text{ATP} \longrightarrow \text{oxaloacetate} + \text{(ADP} + P_i)$$

4. Transfer of acetyl coenzyme A *via* citrate:

Cytosol

$$\text{citrate} + \text{CoA—SH} + \text{ATP} \longrightarrow \text{oxaloacetate} + \text{acetyl—S—CoA} + \text{(ADP} + P_i)$$

Mitochondria

$$\text{acetyl—S—CoA} + \text{oxaloacetate} \longrightarrow \text{citrate} + \text{CoA—SH}$$

5. Metabolism of glucose by pentose phosphate pathway:

Cytosol

glucose + ATP + 6 NADP $\longrightarrow$

glyceraldehyde—3—P + 3 CO_2 + 6 NADPH + ADP

glyceraldehyde—3—P + NAD + 2 ADP + P_i $\longrightarrow$ pyruvate + NADH + 2 ATP

Mitochondria

pyruvate + $\frac{1}{2}$ O_2 + 3(ADP + P_i) + CoA—SH $\longrightarrow$

acetyl—S—CoA + CO_2 + 3 ATP

The essential question is, how much of the NADPH will be produced by the malate transfer process and how much through the pentose phosphate pathway? If the system is perfect, the two routes ought to balance in such a way that glucose and oxygen will disappear and palmitate and CO_2 will accumulate without any change in the concentrations of NADH or NADPH.

This can be solved fairly simply. Eight moles of acetyl coenzyme A must be supplied for each mole of palmitate formed. Let the number of moles formed from triose phosphate produced in the Embden-Meyerhof pathway be E, and the number produced in the pentose phosphate pathway be P; P + E = 8. For each produced in the Embden-Meyerhof pathway, there is a concomitant formation of one mole of NADPH by transfer from NADH. For each produced in the pentose phosphate pathway, there are seven moles of NADPH also formed—six in the pathway itself and one more by transfer from NADH. Therefore, 7P + E = 14, to account for the NADPH used in palmitate synthesis. When we solve these two simultaneous equations, we find that P = 1, and E = 7. Therefore, only one mole of glucose will pass through the pentose phosphate pathway producing one mole of triose phosphate, and three and a half moles will pass through the Embden-Meyerhof pathway, producing seven moles of triose phosphate for each mole of palmitate synthesized.

Let us now cast a table of balance and see what we have (positive sign for production, negative for consumption):

Process	Glucose	O_2	CO_2	cytosol NADH	cytosol NADPH	Acetyl Coenzyme A	ATP
Palmitate synthesis	0	0	0	0	−14	−8	−7
Embden-Meyerhof	−3.5	−3.5	+7	+7	0	+7	+28
Pentose phosphate	−1	−0.5	+4	+1	+6	+1	+4
Acetyl transfer	0	0	0	0	0	0	−8
Electron transfer	0	0	0	−8	+8	0	−8
SUM	−4.5	−4.0	+11	0	0	0	+9

How about that! Everything works out and there is even a small production of high-energy phosphate. Essentially, the balance states that adipose tissue consumes 27 carbon atoms as glucose to store 16 carbon atoms as palmitate, with the remaining 11 appearing as CO_2. It does so in such a way that the eight moles of acetyl coenzyme A that must be transported from mitochondria to cytosol exactly equals the number of pairs of electrons that must be transferred from NADH to NADP in the cytosol. The citrate moving out of mitochondria supplies both needs in the cytosol, and pyruvate returning replaces the oxaloacetate used in making citrate, so there is no need for a diffusion of oxaloacetate across the mitochondrial membrane. The entire process is shown as a flow sheet in Figure 16-11.

We have emphasized repeatedly that logical coherence is a necessary part of any explanation of metabolic pathways, but intellectual satisfaction must be buttressed by measurement of actual circumstances. This is technically very difficult to do for routes as complicated as those we are discussing in a complete tissue because the errors are fairly large. However, such information as we have is entirely consistent with the proposed summation. For example, there are estimates of the relative fraction of glucose being metabolized by various routes during rapid fat synthesis in adipose tissue. The actual results with values predicted by the above summary are:

	Found	Predicted
Fraction of glucose metabolized by pentose phosphate pathway	0.25	0.22
Fraction of total CO_2 produced by pentose phosphate pathway	0.33	0.27
Ratio of carbons in fatty acids to carbons in CO_2	1.25	1.45

The observations are amazingly close to the predictions, considering all of the other processes that occur in adipose tissue, even though the major function is to store fat.

We have other confirmatory observations. The predicted process makes no use of the citric acid cycle, and adipose tissue using glucose as the primary fuel indeed has a very slow citric acid cycle. (This also confirms that the conversion of glucose to fatty acids produces, rather than consumes, high-energy phosphate.)

Adipose tissue can synthesize fatty acids from acetate. What changes would be predicted? According to our summation, if acetyl coenzyme A is not being produced from glucose, then the source of ATP and of over half of the NADPH is lost. Observation shows that the pentose phosphate pathway and the citric acid cycle do indeed accelerate markedly to convert acetate to palmitate, thereby supplying the necessary additional NADPH and high-energy phosphate.

The agreement between prediction and observation, along with the logical simplicity, encourages a belief that our rationale and the actual events are nearly alike, and the formal equation for synthesis of fatty acids from glucose in adipose tissue is:

$$(11) \quad 4.5 \text{ glucose} + 4 \text{ O}_2 + 9 \text{ (ADP}^{3-} + \text{P}_i^{2-}) + 8 \text{ H}^+ \longrightarrow$$
$$(\text{palmitate})^- + 11 \text{ CO}_2 + 9 \text{ ATP}^{4-} + 20 \text{ H}_2\text{O}$$

Lengthening of the Carbon Chain

Many of the acyl residues of triglycerides contain more than 16 carbons, and are created by an extension of the palmitoyl chain in mitochondria. Before this can occur, the palmitoyl group must be transferred from the soluble cytoplasm where it is synthesized. It is first converted to the coenzyme A ester, thereby expending two of the high-energy phosphates created during synthesis, and then to the carnitine derivative, which can cross the mitochondrial membrane (Fig. 16-12).

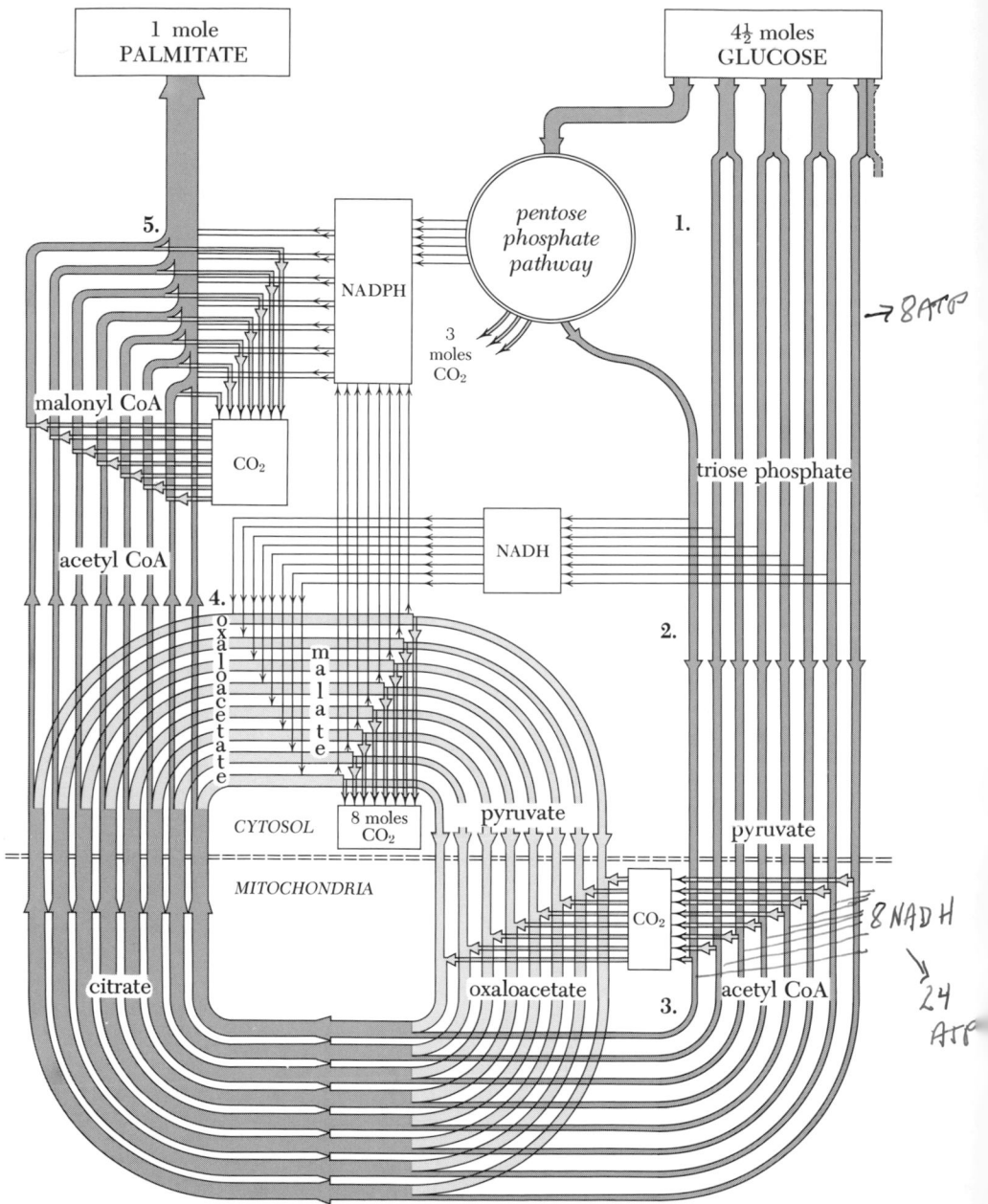

Figure 16-11 *See legend on opposite page.*

Figure 16-11   Flow sheet for the synthesis of palmitate from glucose. The flow of carbons appearing in palmitate is shown by heavy shading. Lighter shading indicates the recycling of oxaloacetate *via* pyruvate from the cytosol to the mitochondria. Transfers of electrons are indicated by narrow arrows.

1. Four and a half moles of glucose are required to make 1 mole of palmitate; one of these passes through the pentose phosphate pathway to generate 6 moles of NADPH. The remainder are converted directly to triose phosphates by the Embden-Meyerhof pathway.

2. The 8 moles of triose phosphates created in step 1 are converted to pyruvate by the Embden-Meyerhof pathway, with the formation of 8 moles of NADH.

3. Eight moles of pyruvate are oxidized to acetyl coenzyme A in the mitochondria, and the acetyl coenzyme A combines with oxaloacetate to form citrate (8 moles).

4. The citrate passes from the mitochondria into the cytosol, where it is cleaved to acetyl coenzyme A and oxaloacetate. The oxaloacetate is reduced to malate, using the 8 moles of NADH formed in step 2. The malate is then oxidized to pyruvate by the NADP-coupled dehydrogenase to form an additional 8 moles of NADPH. The pyruvate diffuses back into the mitochondria, where it is carboxylated to regenerate oxaloacetate consumed in step 3.

5. The 8 moles of acetyl coenzyme A transported into the cytosol as citrate are carboxylated (7 moles) and condensed to form palmitate. The necessary reductions consume the NADPH generated in steps 1 and 4.

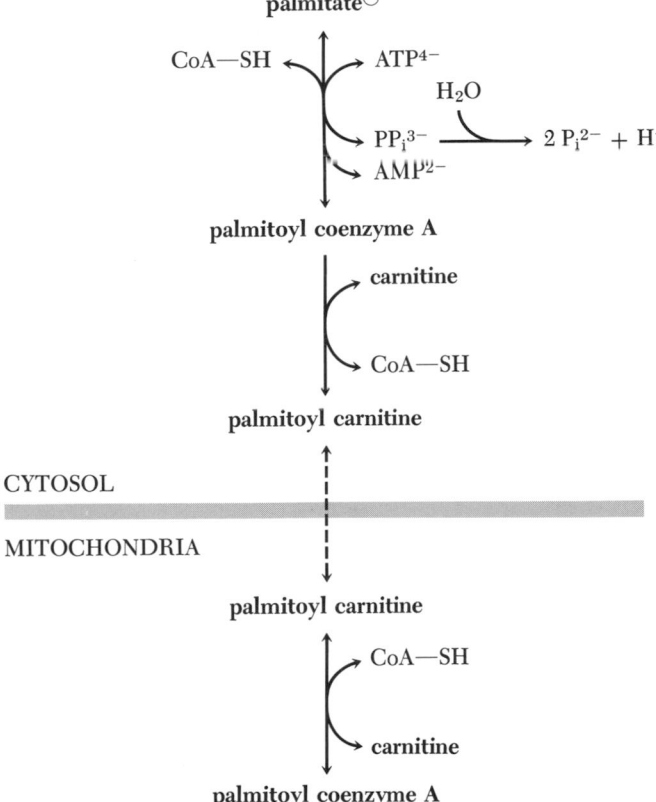

Figure 16-12   The transport of palmitate from the cytosol where it is synthesized to the mitochondria for chain lengthening and introduction of double bonds involves transport as the carnitine ester, formed by exchange after the coenzyme A ester is synthesized.

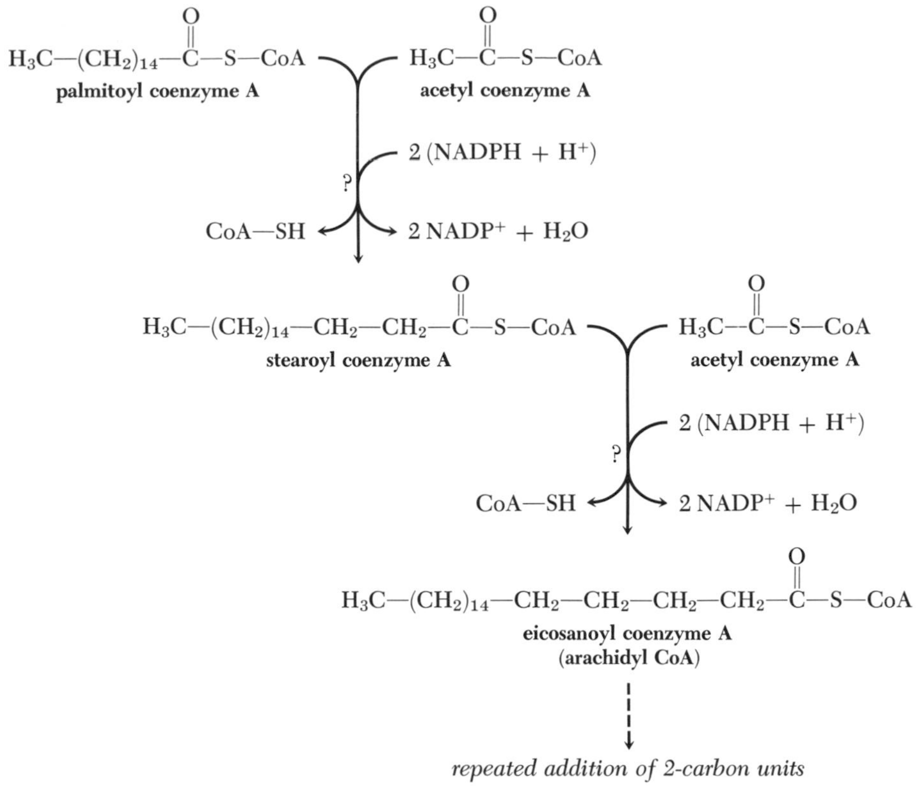

Figure 16-13   Successive addition of acetyl units to the palmitoyl residue in mitochondria create longer chain fatty acids for incorporation into phospholipids and other structural components.

Long-chain acyl coenzyme A compounds can add two carbons within mitochondria by condensing with acetyl coenzyme A, followed by reduction with two moles of NADPH. The responsible enzymes have not been characterized, and it is not certain how this system operates without the driving force of malonyl coenzyme A formation and decarboxylation, which creates the favorable equilibrium for palmitate formation in the cytosol. In any event, palmitoyl coenzyme A can be converted to stearoyl coenzyme A (octadecanoyl coenzyme A), which in turn can add another two carbons to form eicosanoyl coenzyme A, and so on (Fig. 16-13). This makes it possible to have a mixture of acyl residues available, with varying numbers of pairs of carbon atoms. The process also explains why most of the naturally occurring fatty acids do have an even number of carbon atoms.

Creation of Double Bonds

We have already noted that some fatty acids have double bonds in the chain. The predominant residue in human adipose tissue is the singly unsaturated oleoyl group, with 18 carbons and a double bond in the middle of the chain. This double bond can be formed by oxidizing stearoyl coenzyme A, but the way in which the oxidation is accomplished is different than anything we have seen before, because the enzyme system utilizes molecular oxygen and is present in the endoplasmic reticulum of the cell. (What is known as the endoplasmic reticulum is broken up

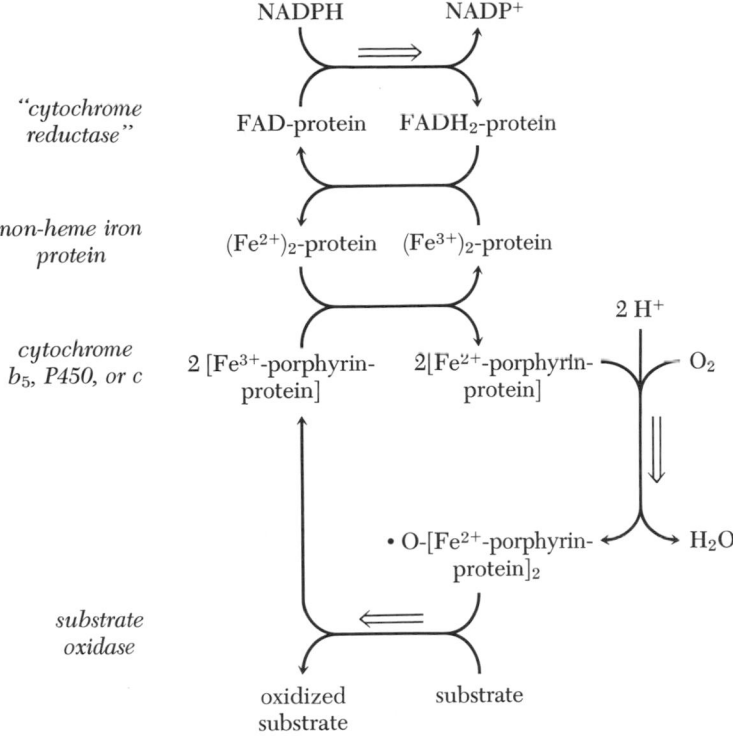

Figure 16-14 The mixed function oxidases utilize NADPH to activate molecular oxygen by half-reduction, leaving a single atom available for attack on the substrate. The exact mechanism is not clear, and the sketch shows a speculative possibility.

into the *microsome* fraction when cells are fractionated by differential centrifugation for laboratory study of its enzymes.)

The endoplasmic reticulum contains a complicated system for electron transport, less well understood than that of mitochondria. The oxidations accomplished by this system are not used for energy production, but for the modification of structural components or for the degradation of metabolites occurring in relatively low concentrations. In short, oxidation is not coupled to phosphorylation in this system. Many of the oxidations are accomplished by "activated" forms of molecular oxygen created by the partial reduction of oxygen by NADPH or NADH. A general scheme for this sort of system is shown in Figure 16-14. In fact, there are several similar systems in the reticulum. The particular examples at hand are less well explored than others, but evidence indicates that both NADH and NADPH are utilized, and that a cytochrome of the b type, cytochrome b_5, is also involved.

An enzyme in the reticulum oxidizes the C-9–C-10 bond of acyl coenzyme A according to this equation:

(stearoyl coenzyme A)

$$\text{or} \quad + \text{NADPH} + O_2 + H^+ \longrightarrow$$

(palmitoyl coenzyme A)

(oleoyl coenzyme A)

$$\text{or} \quad + \text{NADP}^+ + 2H_2O$$

(palmitoleoyl coenzyme A)

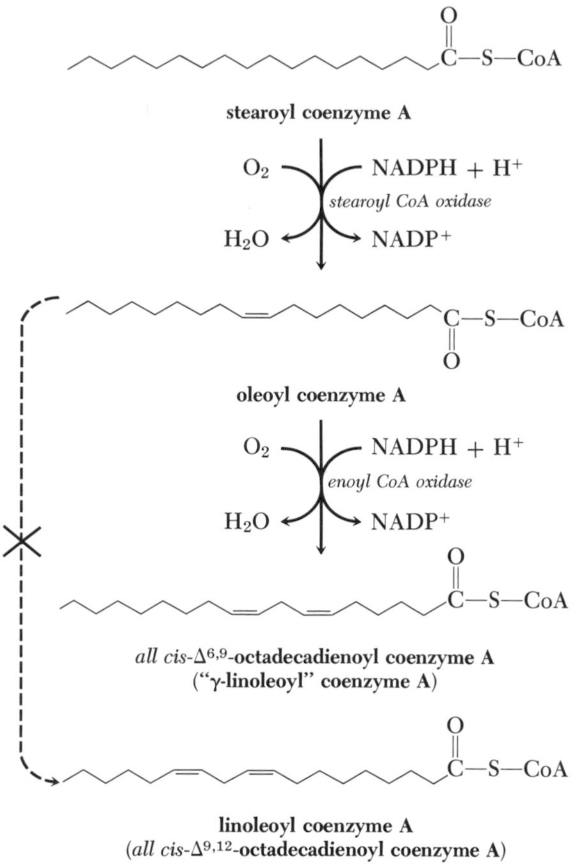

Figure 16-15   Double bonds can be introduced into fatty acids by oxidation at C-9 and at successive 3-carbon intervals toward the carboxyl end. The mammalian enzyme cannot introduce double bonds beyond C-9, and therefore cannot form linoleate. The oxidations occur with the coenzyme A esters of the fatty acids.

Palmitoleate is cis-Δ^9-hexadecenoate, in other words, palmitate with a double bond after C-9, just as oleate is stearate with a double bond after C-9. This enzyme is therefore responsible for producing the bulk of the unsaturated fatty acid in man.

However, there are additional unsaturated fatty acids, some of which must be supplied in the diet. There is no enzyme system in mammals capable of introducing a double bond beyond the ninth carbon atom of a fatty acid chain, but there is an additional oxidase in the endoplasmic reticulum that can form double bonds at three-carbon intervals toward the carboxyl end of the chain, and it does so by the same sort of mechanism as that shown for the formation of oleoyl coenzyme A (Fig. 16-15). As Figure 16-15 indicates, formation of linoleoyl coenzyme A is not possible because of the limited specificity of the enoyl coenzyme A oxidase.

Fatty acids such as linoleic (all cis-$\Delta^{9,12}$-octadecadienoic) and linolenic (all cis-$\Delta^{9,12,15}$-octadecatrienoic), which are essential in the diet, provide the building blocks for a variety of other polyunsaturated acids, created by adding on more 2-carbon units through the chain lengthening enzymes and forming more double

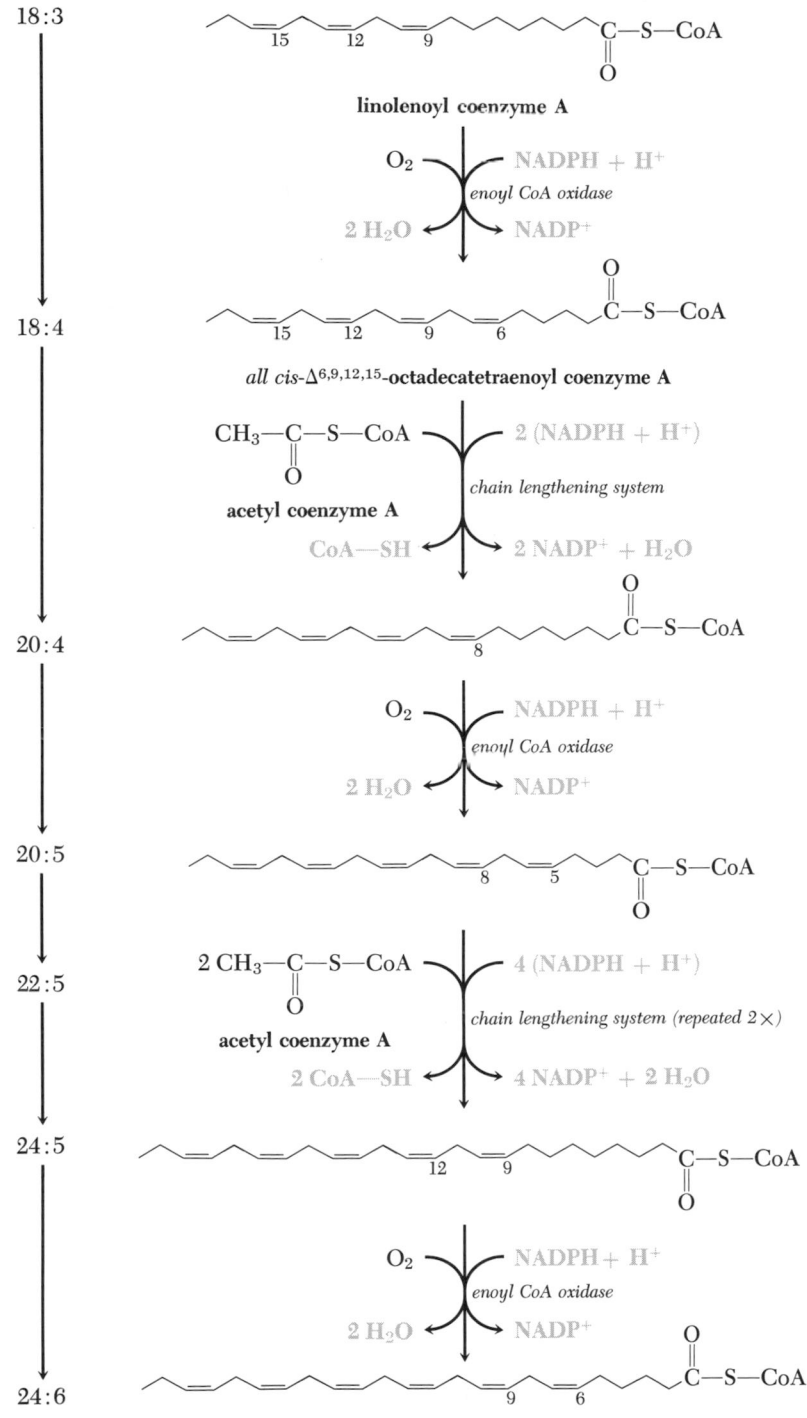

18:3

linolenoyl coenzyme A

O$_2$ ⟶ NADPH + H$^+$
enoyl CoA oxidase
2 H$_2$O ⟵ NADP$^+$

18:4

all cis-Δ^{6,9,12,15}-**octadecatetraenoyl coenzyme A**

CH$_3$—C—S—CoA ⟶ 2 (NADPH + H$^+$)
chain lengthening system
acetyl coenzyme A
CoA—SH ⟵ 2 NADP$^+$ + H$_2$O

20:4

O$_2$ ⟶ NADPH + H$^+$
enoyl CoA oxidase
2 H$_2$O ⟵ NADP$^+$

20:5

2 CH$_3$—C—S—CoA ⟶ 4 (NADPH + H$^+$)
chain lengthening system (repeated 2×)
acetyl coenzyme A
2 CoA—SH ⟵ 4 NADP$^+$ + 2 H$_2$O

22:5

24:5

O$_2$ ⟶ NADPH + H$^+$
enoyl CoA oxidase
2 H$_2$O ⟵ NADP$^+$

24:6

all cis-Δ^{6,9,12,15,18,21}-**tetracosahexenoyl coenzyme A**

Figure 16-16   An example of how a combination of chain-lengthening and oxidation can produce a polyunsaturated fatty acid from dietary linoleate (as the coenzyme A ester).

bonds toward the carboxyl group by the enoyl coenzyme A oxidase. Figure 16-16 shows an example of a route by which 24-carbon chains with six double bonds, fairly abundant in some structural lipids, can be made from linolenate supplied in the diet.

THE FORMATION OF TRIGLYCERIDES

Having at hand the necessary mixture of fatty acids in the form of the coenzyme A esters, triglycerides are formed by a succession of transfers of the acyl residues onto the glycerol skeleton provided originally in the form of L-glycerol-3-phosphate (Fig. 16-17). The glycerol phosphate is available through the reduction of dihydroxyacetone phosphate. We mentioned this reaction as being a part of the shuttle for transferring electrons from the soluble cytoplasm to mitochondria, especially in muscle, but it also acts as a source of the glyceryl group in triglyceride formation. Note, however, that the contribution of glycerol by glucose to stored fat is quantitatively a small fraction of the total mass. A typical triglyceride will have a total of 55 carbon atoms, with 52 of them provided by the fatty acid residues and only three provided by the glyceryl group.

It might as well be admitted at once that there is much to be learned about the specificity of the enzymes concerned in triglyceride synthesis in various tissues. There is a preference for incorporating an unsaturated residue such as the oleoyl residue indicated in the diagram on carbon 2 of glycerophosphate. The enzymes are a part of the endoplasmic reticulum in most tissues, so the difficulty of handling this membrane material is added to the experimental difficulty of using the only slightly soluble lipids as substrates for investigative purposes. In addition, there is an unknown amount of rearrangement of residues between glycerides after the molecules are formed. Be that as it may, we have information on the final result. In the triglycerides of human adipose tissue, the percentage of the residues that are unsaturated is about 40 per cent on C-1 of the glycerol chain, 75 per cent on C-2, and 60 per cent on C-3.

This is not completely a reflection of the relative synthesis of the various types of fatty acid. It ought not to be forgotten that dietary fatty acids may be incorporated into triglycerides without change. In fact, the composition of the diet partly determines the character of the stored fat. This is shown strikingly by experiments in which individuals were fed corn oil, linseed oil, or coconut oil for periods of a year or more, and the fatty acid composition of their subcutaneous adipose tissue was compared with that of individuals eating the usual random American diet. The fatty acids of these oils and the resultant changes in the adipose tissue are shown in Table 16-1.

Figure 16-17 The formation of a triglyceride involves the transfer of successive acyl groups from coenzyme A esters onto glycerol derived from dihydroxyacetone phosphate. The first two additions are made on phosphorylated glycerol. The phosphate group is then hydrolyzed to form the free diglyceride, making the hydroxyl group available for transfer of the final acyl group. The particular residues shown are examples; several kinds of residues may be used.

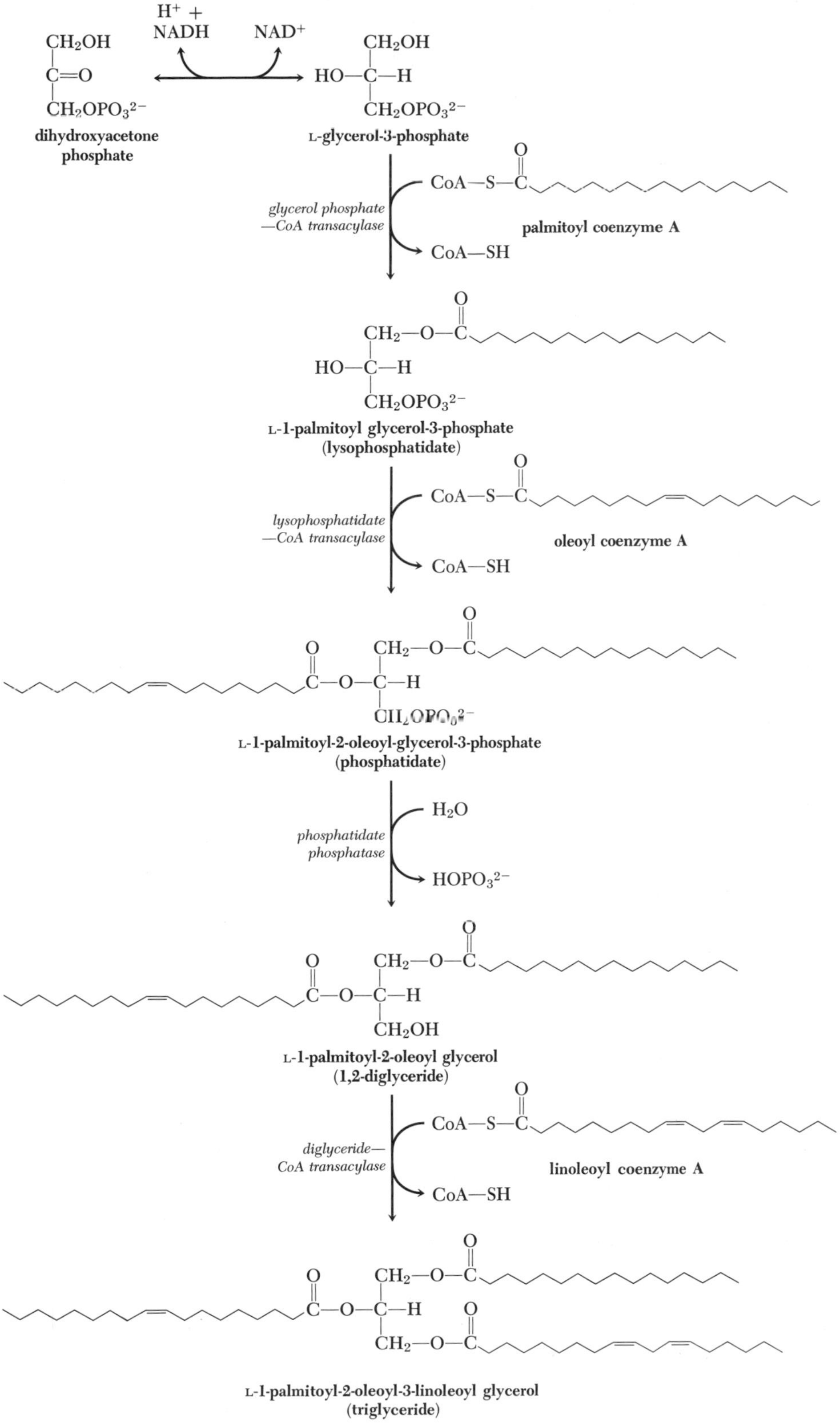

Figure 16-17 *See legend on opposite page.*

TABLE 16-1. COMPOSITION OF TRIGLYCERIDES IN PLANT OILS AND IN ADIPOSE TISSUE OF HUMANS EATING THE OILS[*]

			PERCENTAGE OF TOTAL FATTY ACID RESIDUES				
Fatty Acid	Humans on Random American Diet	Corn Oil	Humans with 40% of Energy Source as Corn Oil for 3 Years	Linseed Oil	Humans Eating 83 g of Linseed Oil Daily for 1 Year	Coconut Oil	Humans Eating 60 g of Coconut Oil for 1.5 Years
8:0						8	
10:0						10	
12:0	0.7		0.1		0.1	44	14.5
14:0	3.3		0.7		1.2	13	13.9
16:0	19.5	12	15.3	6	14.7	6	17.5
16:1	6.9		2.2		5.8		7.6
18:0	4.2	2	2.2	4	5.4	2	2.7
18:1	46.3	29	32.1	22	.35.5	14	30.4
18:2	11.4	58	45.2	15	20.5	2	9.3
18:3	0.4			52	13.7		0.1

[*] Data for adipose tissue from J. Hirsch in *Handbook of Physiology*, p. 148, section 5, American Physiological Society (1965).

Several things emerge from these data. First, the data with a random diet show that the adipose tissue isn't very fancy in its synthesis of fatty acids. The palmitoyl and oleyl residues make up nearly two-thirds of the total, and linoleate from the diet makes up the next most abundant residue. There is little chain lengthening beyond 18 carbon atoms and little introduction of additional double bonds beyond the first needed to supply the oleoyl and palmitoleoyl residues. The fact that a diet high in 18:2 or 18:3 fatty acids leads to a storage of these residues in adipose tissue suggests that there is no inherent screen against polyunsaturated fatty acids. Eating coconut oil, with its abundance of short chain residues, does lead to an increased storage of 12:0 and 14:0 residues, but most of the 8:0 and 10:0 residues found in the food have evidently been modified or disposed of elsewhere.

It turns out that such modifications of dietary fatty acids as are necessary are made by other tissues, especially the liver. The liver utilizes the carbons of short-chain fatty acids and turns out the stock 16- and 18-carbon chains for storage in adipose tissue. The liver also forms the long-chain polyunsaturated compounds needed for making structural lipids.

In addition to changes in composition caused by diet, adipose tissue can be modified in one other important way. The amount of unsaturated fatty acid residues, primarily oleoyl, increases with exposure to cold. A similar variation is found between species of normally extreme habitats. The apparent reason for this is to maintain the lipid in a liquid state. Introduction of double bonds lowers the melting point of a triglyceride. For example, trioleoyl glycerol melts at $-5°C$, whereas, tristearoyl glycerol melts at $54°C$. The more double bonds, the colder exposed skin can become without crystallizing the subcutaneous triglycerides.

The same effect could be achieved by storing shorter chain saturated fatty acids, for example, 8:0, the triglyceride of which melts around $3°C$; this stratagem is indeed employed by many microorganisms, some of which do not possess any unsaturated fatty acids and obtain the necessary consistency in their lipids by these shorter chains. (Of course, we are talking about structural lipids rather than adipose tissue here, but the principle is the same.) When we think on it, though, we can see that these shorter chain acids have less potential per carbon atom for ATP production upon later oxidation—no oxidations are required to convert acetate, $2:0$, to acetyl coenzyme A, only one pair to convert butyrate, $4:0$, and so on. Probably the 16- and 18-carbon acids we store, with a judicious sprinkling of single double bonds, represent the most efficient later yield of ATP consistent with proper melting point at our internal temperatures.

THE MOBILIZATION AND TRANSPORT OF LIPID

Neither the triglycerides nor the free fatty acids are soluble enough to be transported in the blood stream as free compounds. Free fatty acids are transported mainly in combination with *serum albumin*—the most abundant protein in blood plasma, constituting about 4 out of 7 grams of protein in each 100 milliliters. The amount transported at one time is not great, ranging around 2 milligrams of fatty acid per gram of albumin, but the turnover is very rapid, so that some 25 grams of fatty acid are shuttled from one tissue of another every hour through the blood.

Triglycerides are transferred as droplets stabilized by the presence of proteins, phospholipids and cholesterol esters—compounds that we mentioned earlier in connection with the lipid of mitochondrial membranes. These droplets may be of molecular size or so large as to give a milky appearance to blood plasma.

Chylomicrons

Fatty acids from the food are converted into triglycerides by the intestinal mucosa. In some unknown way, the mucosa forms the triglycerides into an emulsion stabilized by the presence of small amounts of protein (~ 1.5 per cent), phospholipids (~ 9 per cent), and cholesterol (~ 7 per cent). The protein is contributed by the mucosa itself and the phospholipid by the liver *via* the bloodstream. This is of some importance, as we shall develop later, because the absorption of fat and its transport can be impaired by defects in protein synthesis or in phospholipid synthesis. The droplets formed by the process range around 1 micron in diameter, and are elaborated into the lymphatic drainage of the gut, appearing in the venous circulation through the thoracic duct. These are the particles that cause a milky plasma after a meal rich in fat.

Chylomicrons may supply several tissues. Half of them disappear from the blood in humans each five to 15 minutes. The liver removes some, perhaps most in a person who has fasted. In a well-fed animal, the adipose tissue appears to take up the bulk. Uptake by tissues involves a hydrolysis of the constituent triglycerides by *lipoprotein lipase*—an esterase—to form the free fatty acid anions and glycerol. The target tissue, if it is the liver, may oxidize the fatty acid, or if it is either the liver or the adipose tissue, convert it back to a triglyceride. Humans are known who have a congenital deficiency of lipoprotein lipase, and chylomicrons may persist in their blood for more than 14 hours after a fatty meal, confirming our belief in the importance of this enzyme for utilization of triglycerides by tissues. (Children so afflicted get xanthomas, yellow bumps rich in lipid, on their skin. Treatment is simply avoiding a fatty diet.)

Plasma Lipoproteins

The liver produces three types of lipoproteins found in plasma. They are usually designated according to electrophoretic mobility, *α-lipoprotein* moving fastest, *β-lipoprotein* next most rapidly, and a *pre-β-lipoprotein* least rapidly (chylomicrons do not migrate). The α- and β-lipoproteins contain distinct types of peptide chains, but the pre-β-lipoprotein contains a mixture of the two types. These proteins may also be separated in the ultracentrifuge. The α-lipoprotein is about half peptide and half lipid (30 per cent phospholipid and 18 per cent cholesterol and cholesterol esters). The β-lipoprotein with only one-quarter peptide, 22 per cent phospholipid, 43 per cent cholesterol and its esters, and 10 per cent triglycerides, is of lower density and floats to the top of the container upon centrifugation. The pre-β-lipoprotein is even less dense, with 50 to 80 per cent triglyceride, and floats more rapidly. It is this protein with which we shall be primarily concerned, since it is the form in which triglycerides are transported from the liver to other tissues. (The function of the other lipoproteins is still obscure.)

We have seen that the liver can synthesize fatty acids from acetyl coenzyme A, which in turn can be derived from carbohydrate or from short chain fatty acids. These fatty acids, together with those obtained from the circulation, can be remade into triglycerides, which are elaborated from the organ as the pre-β-lipoprotein complex. The liver therefore acts as a clearing-house for the fatty acids, degrading some and building others so as to create the characteristic chain length for the triglycerides used by other tissues.

The triglycerides circulating in combination with lipoprotein may be taken up by adipose tissue or used directly by skeletal muscle. In either case, they are hydrolyzed by lipoprotein lipase to the component fatty acids before passing into the interior of the cells.

Release of Stored Triglycerides

The globule of stored triglyceride within an adipose tissue cell is not attacked by lipoprotein lipase, which is used only to hydrolyze the triglycerides of chylomicrons or lipoprotein being brought to the cell for storage. Breakdown of the stored material depends upon the action of other lipases within the cell. It is evident that breakdown must be under the same sort of control as that exerted on the breakdown of glycogen—it must not proceed until there is a demand. All of the evidence at hand indicates that control is maintained through a lipase that is activated by *3',5'-cyclic AMP*, the same compound that triggers the activation of phosphorylase in liver and skeletal muscle, but serving a quite different function in adipose tissue.

The formation of cyclic-AMP in adipose tissue is stimulated by adrenaline as it also is in muscles. Therefore, the alarm signal represented by the elaboration of this hormone causes the release of free fatty acids from the triglyceride stores for use by the skeletal muscles. However, this signal can be overridden by the presence of insulin. Since the pancreas secretes insulin in response to a high concentration of glucose in the blood, its effect on the adipose tissue presumably serves to delay the elaboration of fatty acids in those individuals who have built up a store of glycogen in skeletal muscles available for emergency use.

The formation of cyclic-AMP in adipose tissue is also stimulated by *glucagon* and by *adrenocorticotrophic hormone (ACTH)*. Glucagon is secreted in response to low blood glucose levels and presumably triggers the formation of free fatty acids from adipose tissue to act as oxidizable substrates for skeletal muscle when the glycogen store is low. ACTH is a peptide secreted by the pituitary in response to a variety of stresses, emotional and physical. It is better known for its effect on the adrenal cortex—causing increased formation of steroid hormones, the effects of which we shall treat later—but it also causes an increased output of fatty acids by adipose tissue. Again, the physiological value is apparently one of having increased amounts of substrate available for use during a time of stress.

REGULATION OF FATTY ACID SYNTHESIS

The synthesis of fatty acids from glucose implies an excess of the substrate over metabolic demands, and the process is regulated to insure that synthesis proceeds only under those circumstances. The major mechanism of control involves the first reaction peculiar to the formation of fatty acids, the carboxylation of acetyl coenzyme A to form malonyl coenzyme A.

Acetyl CoA carboxylase exists as an inactive monomer and an active polymer in the form of filaments containing 10 to 20 of the monomer units. The association of the monomers necessary for activity is promoted by citrate. Citrate therefore stimulates fatty acid synthesis by causing an accelerated production of malonyl coenzyme A. However, malonyl coenzyme A competes for citrate and prevents the aggregation of enzyme necessary for its own production. This effect of malonyl coenzyme A serves to limit its irreversible formation if there is some defect in the later reactions of fatty acid synthesis preventing the utilization of malonyl coenzyme A from keeping pace with production. Such a defect might be, for example, an inadequate supply of NADPH, which in turn would be a result of a low supply of glucose.

The importance of the activation by citrate is not as well documented by actual demonstrations of changes in concentration of this compound with variations in substrate supply as we might like, but the effect fits what might rationally be expected. Since citrate is derived from oxaloacetate and acetyl coenzyme A, both of which are produced from pyruvate, an increased supply of glucose might well result in a rising concentration of citrate, which would in turn cause more rapid conversion of acetyl coenzyme A to malonyl coenzyme A, and thence to fatty acids. Now, acetyl coenzyme A inhibits its own production from pyruvate by the pyruvate dehydrogenase reaction, so increased removal of this compound would also increase its formation from pyruvate. We have already discussed the controls by which increased utilization of pyruvate is reflected in its increased formation from glucose (Chapter 13).

There is still an additional control on acetyl coenzyme A carboxylase: It is *inhibited* by *acyl coenzyme A* compounds, such as palmitoyl coenzyme A. The concentration of these compounds would be expected to rise if the capacity of the enzyme system forming triglycerides is exceeded by the supply of fatty acids, and such overloading occurs when fatty acids are being supplied in large amounts from the diet. When this happens, the control mechanism prevents the conversion of glucose to fatty acids. Adipose tissue has an especial requirement for the conservation of glucose because it is completely dependent upon glucose as a source of glycerophosphate used in making triglycerides.

Liver has a *glycerokinase* catalyzing the phosphorylation of free glycerol, and can use glycerol-3-phosphate produced by the reaction to make triglycerides or can convert it to dihydroxyacetone-3-phosphate for disposal by any of the main routes of carbohydrate metabolism:

LIVER

LIVER OR ADIPOSE TISSUE

$$
\begin{array}{c}
\text{CH}_2\text{OH} \\
|\\
\text{HO}-\text{C}-\text{H} \\
|\\
\text{CH}_2\text{OH} \\
\text{glycerol}
\end{array}
\quad
\xrightarrow[\text{glycerokinase}]{\overset{\text{H}^+ +}{\overset{\text{ATP}^{4-}\ \text{ADP}^{3-}}{\underset{Mg^{2+}}{\frown}}}}
\quad
\begin{array}{c}
\text{CH}_2\text{OH} \\
|\\
\text{HO}-\text{C}-\text{H} \\
|\\
\text{CH}_2\text{OPO}_3{}^{2-} \\
\text{L-glycerol-3-phosphate}
\end{array}
\quad
\underset{\substack{\text{glycerol phosphate}\\ \text{dehydrogenase}}}{\overset{\overset{\text{H}^+ +}{\text{NAD}^+\quad \text{NADH}}}{\rightleftharpoons}}
\quad
\begin{array}{c}
\text{CH}_2\text{OH} \\
|\\
\text{C}=\text{O} \\
|\\
\text{CH}_2\text{OPO}_3{}^{2-} \\
\text{dihydroxyacetone} \\
\text{phosphate}
\end{array}
$$

Adipose tissue has no such kinase and cannot recover glycerol released upon hydrolysis of triglycerides. The tissue must have some glucose available as a source of glycerol-3-phosphate even when the fatty acids are supplied from the diet.

THE EFFICIENCY OF STORAGE OF FAT

From the information at hand, we can estimate the efficiency of triglyceride storage. When the source of the fatty acids is dietary fat, the cost of storage is relatively low, with a maximum of two high-energy phosphates expended per transfer across plasma membranes to form triglycerides. There may be only two such transfers, one in the intestinal mucosa and another in adipose tissue. Since the oxidation of palmitate results in the creation of some 129 high-energy phosphate bonds, the storage may be costing as little as 3 per cent of the potential energy, which is a low price for the advantages obtained.

What is the cost of storing carbohydrates as triglycerides? A complete calculation allowing for the unsaturated fatty acids and the glycerol portion is tedious, and the result is not much different from that obtained with the simpler example of one fatty acid residue, so let us make a first approximation on that basis.

According to equation (11) (p. 331), the consumption of 4.5 moles of glucose is necessary for the production of one mole of palmitoyl coenzyme A and a surplus of seven moles of high-energy phosphate in adipose tissue. If the 4.5 moles of glucose were oxidized directly, $4.5 \times 36 = 162$ moles of high-energy phosphate would have been produced, rather than the seven actually formed. Let us assume that the seven moles are necessary to maintain adipose tissue; they certainly can't be transferred to another tissue. We then have an efficiency for the later generation of high-energy phosphate from palmitate of $129/162 = 0.80$ compared to the potential generation from the glucose utilized. This is lower than the 97 per cent recovery calculated for storage of glucose as glycogen. Why, then, is excess carbohydrate stored as fat?

The answer lies in the physical properties of the two types of storage materials. Anhydrous glycogen weighs about 162 grams per mole of glucosyl residue, which yields a maximum of 37 moles of high-energy phosphate upon later use. The organism is therefore carrying some 4.4 grams, dry weight, for each potential mole of high-energy phosphate represented by glycogen. However, glycogen isn't anhydrous in the cells, and the additional water in its structure will raise the weight to around 13 grams per mole of high-energy phosphate.

A typical triglyceride, 1-palmitoyl-2,3-dioleoyl glycerol has a molecular weight of 859 and will yield 437 moles of high-energy phosphate upon later oxidation (129 from palmitate, 144 from each oleate, and 20 from glycerol). Only 1.9 grams of anhydrous triglyceride are carried per mole of potential ATP. Even if one assumes that an additional third of the anhydrous weight must be added to allow for water and the other cellular components of adipose tissue, the additional burden is only 2.5 grams per mole of high-energy phosphate compared to the 13 with glycogen storage.

Put another way, our median 25-year-old American male with his 7.7 kilograms of triglyceride would be walking around with an additional 32 kilograms (70 pounds) of body weight if his fat store were replaced with glycogen of equal potential for high-energy phosphate production.

The efficiencies of glycogen and triglyceride storage are really not as high on a long-term basis as the estimates we have derived, because there is a constant breakdown and resynthesis of these stores. Estimates of the magnitude of turnover vary, with some running as high as a turnover of 10 per cent of the triglycerides per day in the median male. Opposed to this are data showing that it takes several weeks to change the composition of adipose tissue in response to changes in dietary fatty acids. Even the highest estimate of turnover only represents an expenditure of around 1 per cent of the potential ATP yield per day, which doesn't seem to be an exorbitant carrying charge in an active man.

BROWN ADIPOSE TISSUE

Animals have another type of adipose tissue that has a high content of mitochondria, giving it a distinctly brown hue. Adult humans have little of the tissue, but it is prominent in newborn, who have it in the upper torso and the neck. It is also common in adults of animals that hibernate. The tissue acts as a means of producing heat, which is quite different from the function of ordinary adipose tissue as a reservoir of potential energy.

Larger animals appear to be limited in the rate of metabolism by the difficulty in disposing of heat. George Gamow has estimated that an elephant would be literally red-hot if its metabolic rate equalled that of a mouse. The point here is that the limitations on the elephant's metabolism come not from some inherent theoretical limit of the metabolic scheme, but from a built-in regulation adjusting its energy metabolism to its capacity for heat disposal. However, infants of a given species, including humans, have a high ratio of surface area to tissue mass as compared to adults, and are more vulnerable to cold. They are equipped with brown adipose tissue in which metabolism is deliberately inefficient, so that the rate of oxidation, and therefore the rate of heat production, is not geared to the utilization of high-energy phosphate by other metabolic processes. The mitochondria of brown adipose tissue apparently have a total lack of coupling between oxidation and phosphorylation in the electron transport chain. They do produce small amounts of high-energy phosphate by substrate level phosphorylations, such as the single mole created from each mole of succinyl CoA in the citric acid cycle.

The result is that the rate of oxidation is no longer influenced by the concentration of ATP and ADP in mitochondria. What actually does control it is unknown, but it is probably the rate of breakdown of triglycerides, which in turn is regulated by neural or endocrine stimulation much as in ordinary white adipose tissue. There may be other factors, because it has been found that mitochondria of the brown adipose tissue from adult rats that have been acclimated to a cold environment consume oxygen at about twice the rate of mitochondria from the same tissue in rats maintained at more equable temperature, but the mitochondria from both sets of animals do not have oxidative phosphorylation. (Adult rats grow brown adipose tissue under these circumstances—perhaps humans do also.)

This tissue, then, is an example of a lipid store used for emergency maintenance of body temperature when the ordinary metabolic processes plus the increased metabolism caused by shivering are inadequate. The fact that it exists makes belief in the efficiency of the metabolism of other tissues much stronger. If they were inefficient, their metabolism could also be used for heat production, and there would be no need for a specialized brown adipose tissue.

Recapitulation of types of reactions

1. Acetyl coenzyme A may be formed by cleavage of 3-hydroxycarboxylates, with the thiol ester generated at the expense of ATP:

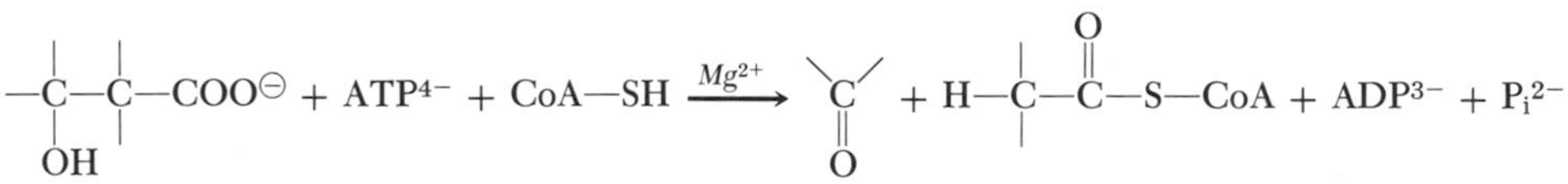

$$-\overset{|}{\underset{OH}{C}}-\overset{|}{C}-COO^{\ominus} + ATP^{4-} + CoA-SH \xrightarrow{Mg^{2+}} \overset{}{\underset{O}{C}} + H-\overset{|}{C}-\overset{O}{\overset{||}{C}}-S-CoA + ADP^{3-} + P_i^{2-}$$

Example: The cleavage of citrate to oxaloacetate and acetyl coenzyme A.

2. C-C condensations may be driven by simultaneous decarboxylations:

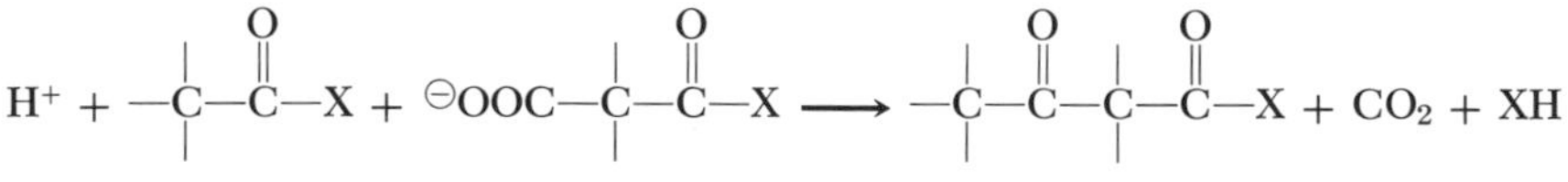

$$H^+ + -\overset{|}{C}-\overset{O}{\overset{||}{C}}-X + {}^{\ominus}OOC-\overset{|}{C}-\overset{O}{\overset{||}{C}}-X \longrightarrow -\overset{|}{C}-\overset{O}{\overset{||}{C}}-\overset{|}{C}-\overset{O}{\overset{||}{C}}-X + CO_2 + XH$$

Example: The condensation of malonyl groups carried by acyl carrier protein with acyl groups on other molecules of acyl carrier protein.

3. Unsaturated carbon chains may be reduced by NADPH:

$$-C{=}C- + \text{NADPH} + \text{H}^+ \longrightarrow -\underset{\text{H}}{\underset{|}{C}}-\underset{\text{H}}{\underset{|}{C}}- + \text{NADP}^+$$

Example: The reduction of Δ^2-enoyl acyl carrier protein.

4. Aldopyranoses may be oxidized to lactones:

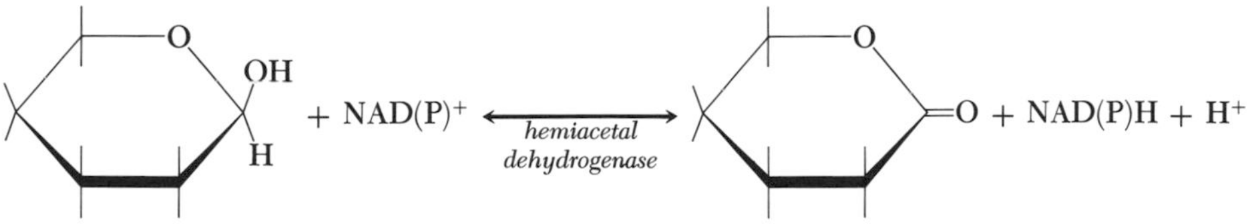

$+ \text{NAD(P)}^+ \xrightleftharpoons[\textit{dehydrogenase}]{\textit{hemiacetal}} \quad {=}\text{O} + \text{NAD(P)H} + \text{H}^+$

Example: The oxidation of glucose-6-phosphate by NADP.

5. Lactones may be hydrolyzed:

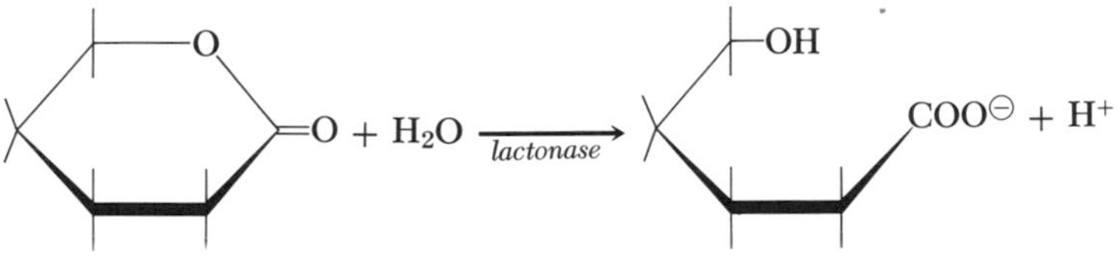

${=}\text{O} + \text{H}_2\text{O} \xrightarrow{\textit{lactonase}} \quad \text{COO}^\ominus + \text{H}^+$

Example: The hydrolysis of 6-phosphogluconolactone.

6. Two-carbon units may be transferred from 2-ketoses to aldoses:

$$
\begin{array}{l}
\text{CH}_2\text{OH} \\
|\\
\text{C}{=}\text{O} \\
|\\
\text{CH}{-}\text{OH} + \text{H}{-}\text{C}{\overset{\text{O}}{\big\diagup}} \\
|\\
\text{R} \qquad\qquad \text{R}'
\end{array}
\xrightleftharpoons[]{\textit{transketolase}}
\begin{array}{l}
\qquad\qquad\quad \text{CH}_2\text{OH} \\
\qquad\qquad\quad |\\
\qquad\qquad\quad \text{C}{=}\text{O} \\
\qquad\qquad\quad |\\
\text{H}{-}\text{C}{\overset{\text{O}}{\big\diagup}} + \text{CH}{-}\text{OH} \\
\qquad\qquad\quad |\\
\text{R} \qquad\quad \text{R}'
\end{array}
$$

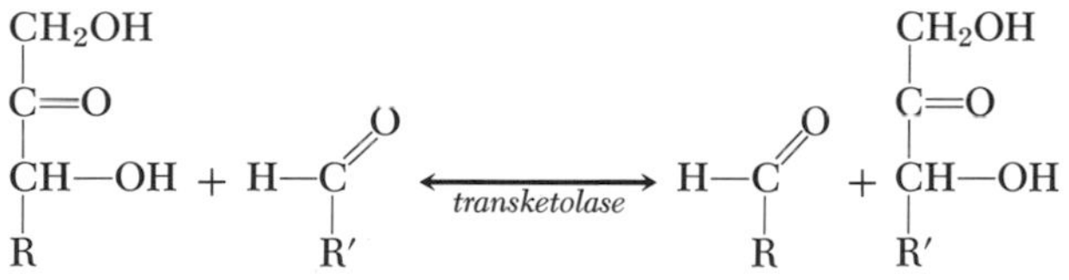

Example: The transketolase reaction.

7. Three-carbon units may be transferred from 2-ketoses to aldoses:

$$
\begin{array}{l}
\text{CH}_2\text{OH} \\
|\\
\text{C}{=}\text{O} \\
|\\
\text{CH}{-}\text{OH} \\
|\\
\text{CH}{-}\text{OH} + \text{H}{-}\text{C}{\overset{\text{O}}{\big\diagup}} \\
|\\
\text{R} \qquad\qquad \text{R}'
\end{array}
\xrightleftharpoons[]{\textit{transaldolase}}
\begin{array}{l}
\qquad\qquad\quad \text{CH}_2\text{OH} \\
\qquad\qquad\quad |\\
\qquad\qquad\quad \text{C}{=}\text{O} \\
\qquad\qquad\quad |\\
\qquad\qquad\quad \text{CH}{-}\text{OH} \\
\qquad\qquad\quad |\\
\text{H}{-}\text{C}{\overset{\text{O}}{\big\diagup}} + \text{CH}{-}\text{OH} \\
\qquad\qquad\quad |\\
\text{R} \qquad\quad \text{R}'
\end{array}
$$

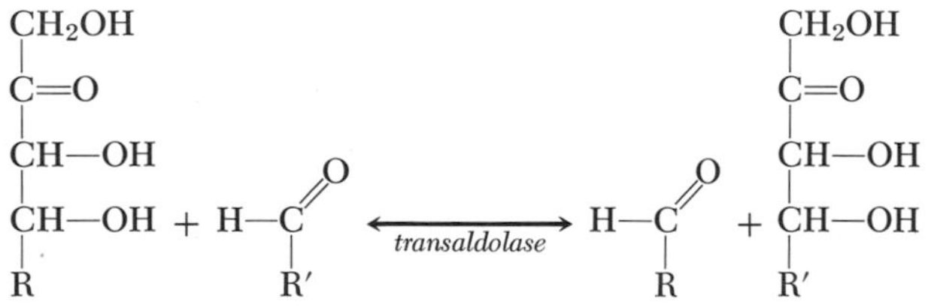

Example: The transaldolase reaction.

8. The configuration of an optically active alcohol may be inverted:

$$H—\overset{\displaystyle R}{\underset{\displaystyle R'}{C}}—OH \xrightleftharpoons[epimerase]{} HO—\overset{\displaystyle R}{\underset{\displaystyle R'}{C}}—H$$

Example: The interconversion of D-ribulose-5-phosphate and D-xylulose-5-phosphate.

9. Saturated hydrocarbon groups may be oxidized by molecular oxygen, with the simultaneous oxidation of NADPH:

$$\overset{\displaystyle |}{\underset{\displaystyle |}{\overset{CH_2}{CH_2}}} + O_2 + NADPH + H^+ \xrightarrow{oxidase} \overset{\displaystyle |}{\underset{\displaystyle |}{\overset{CH}{CH}}} + NADP^+ + 2\,H_2O$$

or

$$\overset{|}{CH_2} + O_2 + NADPH + H^+ \xrightarrow{hydroxylase} \overset{|}{CH}—OH + NADP^+ + H_2O$$

Example: The formation of unsaturated fatty acid esters of coenzyme A (the first reaction above—the second reaction is illustrated in a later chapter).

10. Acyl coenzyme A and alcohols may react to form oxy-esters:

$$R—\overset{\displaystyle O}{\overset{\|}{C}}—S—CoA + HO—R' \xrightarrow{transacylase} R—\overset{\displaystyle O}{\overset{\|}{C}}—O—R' + HS—CoA$$

Example: The formation of triglycerides.

Further reading

Masoro, E. J.: *Physiological Chemistry of Lipids in Mammals.* W. B. Saunders (1968).
Adipose Tissue Metabolism and Obesity. Ann. N. Y. Acad. Sci., *131:* 1–683 (1965).
Renold, A. E., and G. F. Cahill, Jr.: *Adipose Tissue.* In Field, J., ed.: *Handbook of Physiology.* Williams and Wilkins (1965).
Kinsell, L. W., ed.: *Adipose Tissue as an Organ.* Charles C Thomas (1962).
Rodahl, K., and B. Issekutz, eds.: *Fat as a Tissue.* McGraw-Hill (1964).
Olson, J. A.: *Lipid Metabolism.* Ann. Rev. Biochem., *35:* 559 (1966).
Shapiro, B.: *Lipid Metabolism.* Ann. Rev. Biochem., *36:* 247 (1967).
Lindgren, F. T., and A. V. Nichols: *Structure and Function of Human Serum Lipoproteins.* In Putnam, F. W., ed.: *The Plasma Proteins.* Academic Press (1960).
Smith, R. E., J. C. Roberts, and K. J. Hittelman: *Non-phosphorylating Respiration of Mitochondria from Brown Adipose Tissue.* Science, *154:* 653 (1966).
Lynen, F.: *The Role of Biotin-Dependent Carboxylations in Biosynthetic Reactions.* Biochem. J., *102:* 381 (1967). Includes excellent discussion of fatty acid synthesis.
Chesterton, C. J., P. H. W. Butterworth, and J. W. Porter: *Sites of Binding of Acetate, Malonate and Acetoacetate to the Pigeon Liver Fatty Acid Synthetase.* Arch. Biochem. Biophys., *126:* 864 (1968).
Frederickson, D. S., R. I. Levy, and R. S. Lees: *Fat Transport in Lipoproteins–An Integrated Approach to Mechanisms and Disorders.* New Eng. J. Med., *276:* 34, 94, 148, 215, and 273, (1967). Classic review.
Flatt, J. P., and E. G. Ball: *Studies on the Metabolism of Adipose Tissue XIX.* J. Biol. Chem., *241:* 2862 (1966).
Katz, J., and R. Rognstad: *The Metabolism of Tritiated Glucose by Rat Adipose Tissue.* J. Biol. Chem., *241:* 3600 (1966).
Stoll, E., et al.: *Liver Acetyl CoA Carboxylase: Activation of Model Reactions by Tricarboxylic Acids.* Proc. Natl. Acad. Sci., *60:* 986 (1968).

PART IV

The Nitrogen Economy

AMINO ACIDS AS SOURCES OF ENERGY

CHAPTER 17

ARGUMENT

All of the amino acids that occur in proteins also are found in free form in the blood, the extracellular fluid, and within cells. This pool of free amino acids is augmented by the supply from the diet, and by the continual breakdown of tissue proteins. It is depleted by the utilization of amino acids for the synthesis of proteins and for the production of other nitrogenous constituents of the body. Any excess of supply over utilization is oxidized, providing high-energy phosphate.

Most of the nitrogen of the amino acids is converted to urea for excretion in the urine. Urea contains two nitrogen atoms, one derived from ammonium ion and the other from aspartate. Amino acid metabolism therefore includes reactions by which the nitrogen derived from a mixture of amino acids can be equally partitioned between ammonium ion and aspartate. Some of the amino acids contain reactive groups enabling a direct deamination to form ammonium ions. Most undergo transamination so that their nitrogen appears in the form of glutamate. Glutamate may transaminate with oxaloacetate to create the aspartate necessary for urea synthesis, but part of it is oxidized by NAD, which causes a release of ammonium ions in amounts necessary to match the total quantity of aspartate.

The removal of nitrogen atoms, which occurs early in the course of metabolism of most amino acids, leaves carbon skeletons. These organic compounds can be completely oxidized with a concomitant production of high-energy phosphate. However, deamination and other early steps of amino acid metabolism occur in the liver, whereas the major demand for high-energy phosphate is in the peripheral tissues. The liver therefore converts the carbon skeletons into compounds that can be transported in the blood and utilized by other tissues. This enables the energy of oxidation of amino acids to be available generally in the body without the necessity for duplicating the enzymes involved in nitrogen disposal.

For example, part of the carbon from many amino acids is converted to pyruvate or to intermediates of the citric acid cycle, all of which can be converted to oxaloacetate. Glucose can be made from oxaloacetate and liberated into the blood. Part of the carbon from other amino acids is converted to acetyl coenzyme A or other intermediates of fatty acid oxidation. The liver converts acetyl coenzyme A to free acetoacetate, or the corresponding reduced compound, D-3-hydroxybutyrate. These 3-oxybutyrates also diffuse into the blood, and can be metabolized by the skeletal muscles or the brain through conversion to acetoacetyl coenzyme A and subsequent oxidation in the citric acid cycle. The overall effect is the conversion of amino acids to a mixture of glucose and acetyl units that can be utilized by the regular routes of carbohydrate and fat metabolism.

The transformations by the liver involve some reactions not encountered in glucose or fat metabolism. Methylmalonyl coenzyme A is formed from some amino acids. This 4-carbon branched chain compound is rearranged to succinyl coenzyme A in a reaction requiring the presence of cobamide coenzyme. Cobamide coenzyme is a complex porphyrin-like compound containing cobalt and derived from vitamin B_{12}. Another distinctive set of reactions involves the degradation of the aromatic rings found in phenylalanine, tyrosine and tryptophan. These include hydroxylases utilizing molecular oxygen for a simultaneous substitution of oxygen on the rings and oxidation of tetrahydrobiopterin. Tetrahydrobiopterin is related to the vitamin, folic acid, and cannot be synthesized by mammalian tissues. The aromatic rings are broken by oxidases, which introduce both atoms of molecular oxygen to cleave one of the C—C bonds.

Some of the amino acids contain sulfur, and the routes of their metabolism include reactions for converting the sulfur to inorganic sulfate, which is excreted.

We opened our discussion of biochemistry by examining the amino acids as constituents of proteins. The use of amino acids as precursors of proteins is the central fact dominating any consideration of amino acid metabolism, and we must constantly keep it in mind.

Even so, it is easy to show that the maintenance of a supply of amino acids from which proteins can be made isn't a simple process. We eat proteins, which are hydrolyzed to the constituent amino acids in the gut. At the same time, tissue proteins are also being hydrolyzed to form amino acids, which mix with those derived from food as an *amino acid pool* in the extracellular fluid upon which all tissues may draw for their requirements.

Part of the amino acid pool is used to rebuild tissue proteins, but the total amount of protein remains relatively constant in most adults from day to day, which means that the quantity of amino acids used to make tissue proteins is no greater than the quantity obtained by the breakdown of tissue proteins. Therefore, the usual adult will have a surplus of amino acids equivalent to the amount he has ingested. Amino acids aren't excreted in significant quantities, and the surplus must be disposed of in other ways. In short, there is an active metabolism of amino acids that constantly must be reconciled with maintenance of a supply for protein synthesis.

Tissues draw upon the excess amino acid pool for the synthesis of other nitrogenous constituents, such as the nucleic acids and porphyrins. These compounds are constantly broken down and re-made, as are the proteins themselves. However, the demand for this purpose is not enough to account for more than a small fraction of the amino acid intake in common diets.

The bulk of the excess amino acid intake is degraded through oxidative pathways associated with the formation of high-energy phosphate: Amino acids are used as fuel. How important a fuel are they? Usually less significant than fat or carbohydrate, but sometimes of major importance. The extent of the contribution to the energy supply obviously depends upon the composition of the diet. The rib roast eaten by an affluent human contains protein and fat to the extent of one-quarter each of its cooked weight, with virtually no carbohydrate. The halibut or fried chicken more likely to be on most tables has three to six times as much protein as fat in the original tissues, but the fat will be increased in the final food to some extent through the cooking oil used in its preparation.

Considering the total food intake, most humans obtain only one-tenth of their high-energy phosphate by oxidizing amino acids, but purely carnivorous animals generate nearly one-half of their supply in this way.

Since there are some 20 amino acids commonly metabolized during the utilization of proteins, the description of protein metabolism inevitably becomes more complicated than the description of the major routes of carbohydrate and fat metabolism, and it is tempting to skip over the subject lightly with a few superficial generalities, but the temptation ought to be resisted. The metabolism of amino acids, including the oxidative degradation and the use as precursors for nucleic acids and other nitrogenous constituents, is so complicated because it is so important to the life of the cell. We are dealing here with the maintenance and construction of the machinery that is driven by the energy obtained from combustion of fats and carbohydrates.

Furthermore, aberrations of the nitrogen economy and its deliberate manipulation through drugs include many of the most pressing of current medical problems. We shall also see that maintenance of nitrogen economy in the entire population is also among our most pressing social problems.

GENERAL DESCRIPTION OF
AMINO ACID DEGRADATION

Let us begin, then, with a discussion of the oxidative degradation of amino acids, recognizing that we are looking at only part of a larger whole that will gradually be revealed in the later chapters.

The metabolism of many of the amino acids begins with the removal of the nitrogen atom, which is then handled by separate metabolic pathways. Most of the excess nitrogen is excreted in the urine as *urea*. Urea contains two nitrogen atoms, one derived from *ammonia* and the other from *aspartate*. Part of our story is therefore a description of the processes of *deamination* of amino acids and of converting the nitrogen in equivalent amounts to ammonia and aspartate for urea synthesis.

The other part of amino acid degradation is the disposition of the carbon skeletons. This is the part in which most of the high-energy phosphate is generated by mitochondrial electron transfers. However, the metabolism of the carbon skeletons begins in the liver, and that organ has a relatively small capacity for oxidation. Hence, the liver converts the carbon skeletons to *glucose* by the reactions of gluconeogenesis, and to either *fatty acids* by fatty acid synthesis from acetyl coenzyme A or to *ketone bodies*. The ketone bodies are *acetoacetate* and *3-hydroxybutyrate*, which can be metabolized by the muscles. The result is the transformation of the amino acids by the liver into standard fuels which can be utilized by the other tissues for generation of high-energy phosphate. This part of our story, then, is a description of the routes by which the carbon skeletons are converted to phospho-*enol*-pyruvate as a precursor of glucose and to acetyl coenzyme A as a precursor of either the fatty acids or the ketone bodies.

DEAMINATION AND TRANSAMINATION

There are two common mechanisms for removing the amino group from amino acids. In many cases, the group is transferred to another carbon skeleton by *transamination* to form a new amino acid. The new amino acid is usually L-glutamate. Glutamate can be oxidized to liberate ammonia, which is then available for urea synthesis, or its nitrogen may be transferred again by transamination to form L-aspartate, the other precursor of urea. (We discussed the mechanism of *glutamate-aspartate transaminase* at some length in Chapter 6. Other transaminases act in the same way, utilizing *pyridoxal phosphate* as a coenzyme.)

The amino group of some amino acids is removed directly as ammonia. This can only be accomplished with those compounds containing other substituent groups to make the direct deamination energetically feasible.

The Transaminases and Glutamate Dehydrogenase

Much of the nitrogen of the amino acids sooner or later appears in the form of glutamate because of the action of transaminases. Glutamate, in turn, is subject to *oxidative deamination* in a reaction catalyzed by *glutamate dehydrogenase*:

$$
\begin{array}{ccc}
\text{COO}^{\ominus} & & \text{COO}^{\ominus} \\
| & & | \\
\overset{\oplus}{\text{H}_3\text{N}}-\text{C}-\text{H} & \text{H}_2\text{O} \qquad \text{H}^+ & \text{C}=\text{O} \\
| & \qquad\qquad & | \\
\text{CH}_2 + \text{NAD}^+ \;\rightleftarrows\; & \underset{\substack{glutamate \\ dehydrogenase}}{} & \text{CH}_2 + \text{NH}_4^+ + \text{NADH} \\
| & & | \\
\text{CH}_2 & & \text{CH}_2 \\
| & & | \\
\text{COO}^{\ominus} & & \text{COO}^{\ominus} \\
\text{L-glutamate} & & \alpha\text{-ketoglutarate}
\end{array}
$$

Glutamate dehydrogenase is present in the mitochondria of most, if not all, tissues. The enzyme will catalyze the reaction with either NAD or NADP. We are now primarily concerned with the oxidation of glutamate and the accompanying release of ammonia, but ought to note that the reaction provides a capability for creating glutamate from ammonia by a reversal of the reaction using NADPH.

The transaminations by which amino acids and glutamate are equilibrated are catalyzed by a number of enzymes, each specific for one or a few amino acids. The combination of the action of these enzymes with that of glutamate dehydrogenase has the same effect as if there were individual dehydrogenases for the other amino acids. For example, consider the fate of accumulating alanine in the presence of glutamate-alanine transaminase and glutamate dehydrogenase (Fig. 17-1). Alanine reacts with α-ketoglutarate to form glutamate and pyruvate. The resultant glutamate is then oxidized by NAD, liberating ammonium ion and regenerating the α-ketoglutarate originally used for transamination. The net effect is the oxidation of alanine by NAD to form pyruvate, ammonium ion, and NADH. Any other amino acid that will transaminate with α-ketoglutarate may be deaminated in the same way so as to provide the necessary ammonia for urea synthesis.

This elegantly simple system has an obvious advantage. The equilibrium constants of the transamination reactions are near unity, so their rates will depend upon

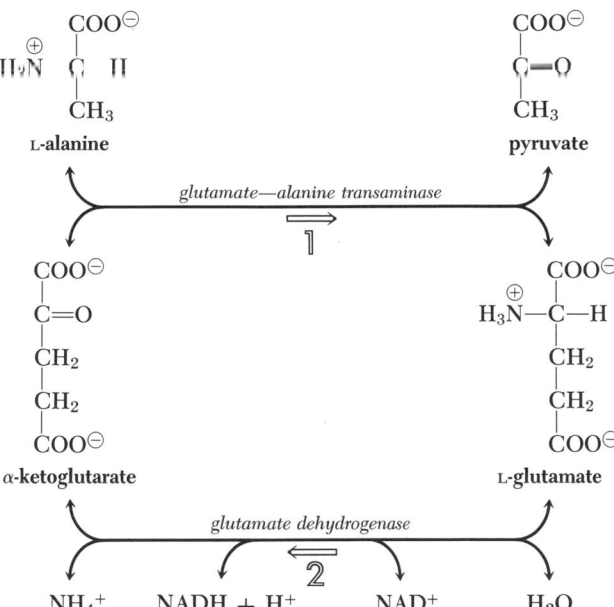

Figure 17-1 The concerted action of glutamate-alanine transaminase and glutamate dehydrogenase achieves the same result as the action of a hypothetical alanine dehydrogenase.

1. The amino group of alanine is transferred to α-ketoglutarate, forming pyruvate and glutamate.

2. The resultant glutamate is oxidized by NAD, with the nitrogen originally in alanine now appearing as NH_4^+. The α-ketoglutarate consumed in the first step is now regenerated and there is no net change in the concentration of this compound. The overall reaction involves only the disappearance of alanine and NAD, with the appearance of pyruvate, NADH and NH_4^+.

the rate of removal of the products of the reactions. One of the products usually is glutamate, so the rate of ammonia formation hinges in part upon the glutamate dehydrogenase reaction, which is under complex control by modulators. Due to the nature of this system, this control also affects the rate of ammonia production from all of the amino acids transaminating with α-ketoglutarate.

Since so much amino acid nitrogen is funnelled through the glutamate dehydrogenase reaction, the enzyme is a potential point of control in the metabolism of many amino acids. The activity of the enzyme is altered by a variety of compounds, but the critical effect appears to be a potent non-competitive *inhibition* by *guanosine triphosphate*. This may well provide negative feedback so as to adjust the utilization of amino acids as sources of energy to demand. As we have seen, part of the energy captured in the operation of the citric acid cycle appears as GTP during the conversion of succinyl coenzyme A to succinate. Succinyl coenzyme A is formed by the oxidation of α-ketoglutarate, which is also a product of the glutamate dehydrogenase reaction. Now, if GTP is not being utilized rapidly, its concentration will rise and inhibit the conversion of glutamate to α-ketoglutarate, thereby preventing the utilization of the carbon skeleton of glutamate to feed the citric acid cycle. Prevention of the oxidation of glutamate will in turn tend to make its concentration rise and slow down its production through the readily reversible transaminase reactions.

The inhibition of glutamate dehydrogenase by GTP can be overridden in two ways. Firstly, *ADP activates* the enzyme. An ADP accumulation represents an increased demand for high-energy phosphate, and activation of glutamate dehydrogenase supplies more substrate for the citric acid cycle. Secondly, several *amino acids*, of which leucine is a prominent example, *activate* the enzyme at some site quite independent of the nucleotide-binding site. The effect here is to cause increased degradation of the amino acids through more rapid disposal of their nitrogen whenever they begin to accumulate, and this will occur even if the total demand for high-energy phosphate is not very great at the time of their accumulation.

Finally, glutamate dehydrogenase is inhibited by NADH above and beyond the effect of the nucleotide as a reactant, which is to say that the enzyme has modifier sites, in addition to catalytic sites, for binding NADH. Therefore, the oxidation of glutamate will tend to slow down when electron transport is swamped by other substrates.

These various effects are too complex to analyze in detail with the information we have at hand, and there are many vexing anomalies with which students of nitrogen metabolism are wrestling, but the general trend of the effects seems to be quite clear. Glutamate is oxidized faster when the need for energy is greater or when amino acids are accumulating for disposal.

Direct Deamination:
Serine and Threonine Dehydratases

Compounds containing amino groups, such as amino acids, could in theory lose ammonia by a direct deamination analogous to the dehydration of an alcohol, or by a hydrolytic deamination to produce an alcohol:

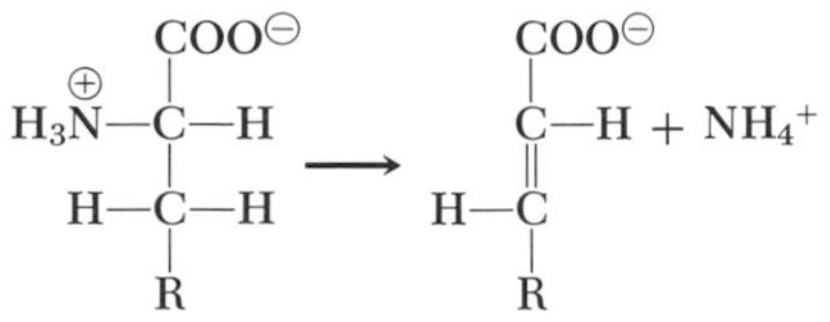

$$\begin{array}{ccc}
\overset{\ominus}{C}OO^{\ominus} & & COO^{\ominus} \\
\overset{\oplus}{H_3N}-\overset{|}{C}-H & & \overset{||}{C}-H + NH_4^+ \\
H-\overset{|}{C}-H & & H-\overset{|}{C} \\
\overset{|}{R} & & \overset{|}{R}
\end{array}$$

The position for equilibrium of such reactions with simple amines is unfavorable, and they are only feasible in aqueous solution when the character of the products is such as to form a resonating conjugated system or to undergo further reaction with additional liberation of free energy. Therefore, only a few of the amino acids are deaminated directly without oxidation. Among these are serine and threonine, which we shall consider now as examples.

Serine and threonine have hydroxyl groups in addition to the ammonium group on their carbon chains. In effect, the carbon skeleton of these amino acids is already more oxidized than that of many amino acids. They are somewhat analogous to glycerate, with its adjacent hydroxyl group, and we saw in the discussion of the Embden-Meyerhof pathway that the dehydration of glycerate to form pyruvate liberates enough free energy to sustain the formation of a high-energy phosphate. Similarly, the deamination of serine and threonine to form the corresponding 2-keto compounds liberates enough free energy to be a feasible reaction—these deaminations are, in effect, irreversible:

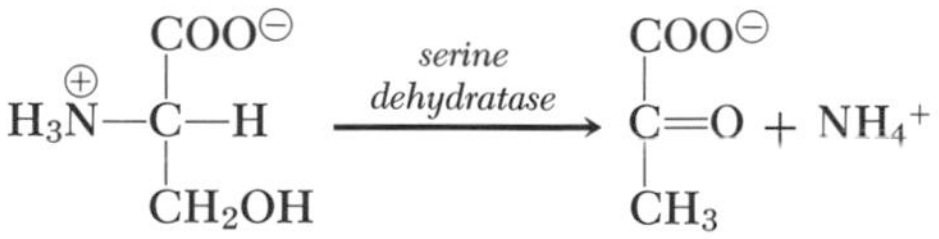

$$\begin{array}{ccc}
COO^{\ominus} & \xrightarrow{\text{serine dehydratase}} & COO^{\ominus} \\
\overset{\oplus}{H_3N}-C-H & & C=O + NH_4^+ \\
CH_2OH & & CH_3 \\
\text{L-serine} & & \text{pyruvate}
\end{array}$$

$$\begin{array}{ccc}
COO^{\ominus} & \xrightarrow{\text{threonine dehydratase}} & COO^{\ominus} \\
\overset{\oplus}{H_3N}-C-H & & C=O + NH_4^+ \\
H-C-OH & & CH_2 \\
CH_3 & & CH_3 \\
\text{L-threonine} & & \text{2-ketobutyrate}
\end{array}$$

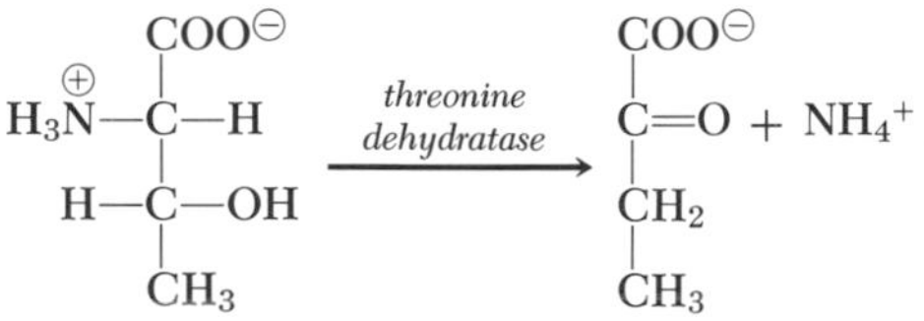

The enzymes are called dehydratases, rather than deaminases, because the reaction proceeds by the initial loss of the elements of water, although water is later added back during the loss of ammonia from an unstable intermediate. They are additional examples of enzymes containing *pyridoxal phosphate* as a coenzyme, which forms a Schiff's base with the amino acids, much as it does on the transaminases (Fig. 17-2). We shall see that this versatile coenzyme plays many functions in nitrogen metabolism.

DISPOSAL OF NITROGEN—UREA SYNTHESIS

One of the fundamental facts about animals is that they are intolerant of even modest concentrations of ammonium ion in the cellular environment. The reason

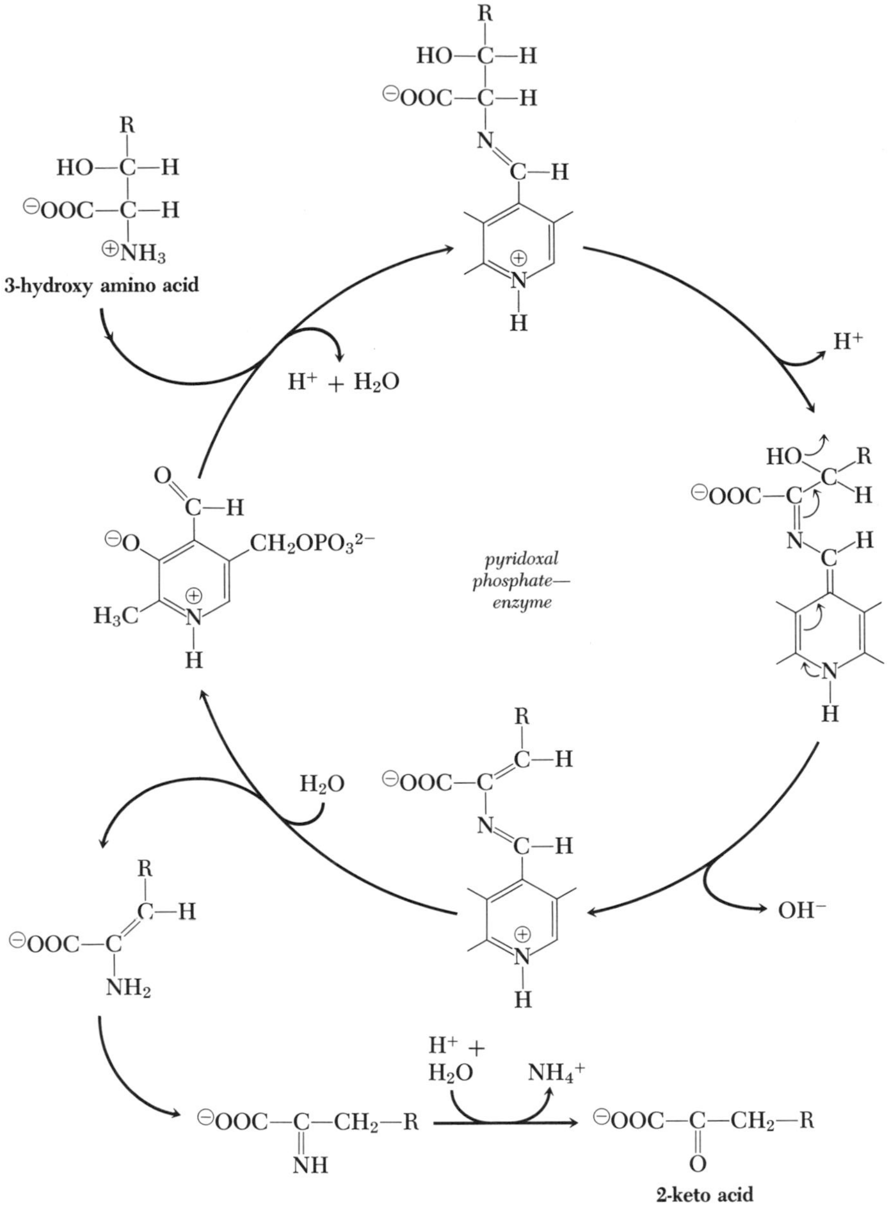

Figure 17-2 The deamination of hydroxy amino acids by dehydratases involves a combination of the amino group with the aldehyde group of pyridoxal phosphate on the enzyme to form an imine, or Schiff's base (*top center*). A series of electron shifts results in the elimination of the water, leaving an unsaturated imine (*bottom center*). Hydrolysis of the imine bond regenerates the aldehyde group of the coenzyme and liberates an unstable unsaturated amine, which spontaneously rearranges and hydrolyzes to yield NH_4^+ and the 2-keto analogue of the original amino acid.

for this sensitivity is not entirely clear, but it probably is a result of the kinds of mechanisms that have been developed for regulating the ionic environment of cells.

The higher animals may well be coping with mechanisms designed for a different environment. Simple organisms living in water have no problem with nitrogen disposal. Ammonia diffuses out freely and is thereby diluted to a very low concentration. The first organisms probably developed ionic gradients and the basic metabolic processes of amino acid metabolism to fit this circumstance. Adaptation for more efficient disposal of nitrogen only became necessary with the development of larger size and an enclosed circulation with impermeable skin, which were necessary preludes for movement from marine to terrestrial environments.

In any event, modern animals of the higher phyla have efficient means of maintaining the ammonium ion concentration at very low levels—4×10^{-5} M is the upper limit for the normal range in human blood. Many animals, including the mammals, accomplish this by converting ammonia to urea. (It is common to speak of ammonia and ammonium ion somewhat interchangeably—they are in equilibrium. At the pH of blood, only 1 per cent of the total is present as the base, so it is better to express concentrations as ammonium ion concentrations. Ammonia is the form that passes through cell membranes. The particular ionic form used in many enzymatic reactions has not been determined.)

Now, urea is made by a simple hydrolysis of the amino acid, *arginine*, catalyzed by the enzyme, *arginase*:

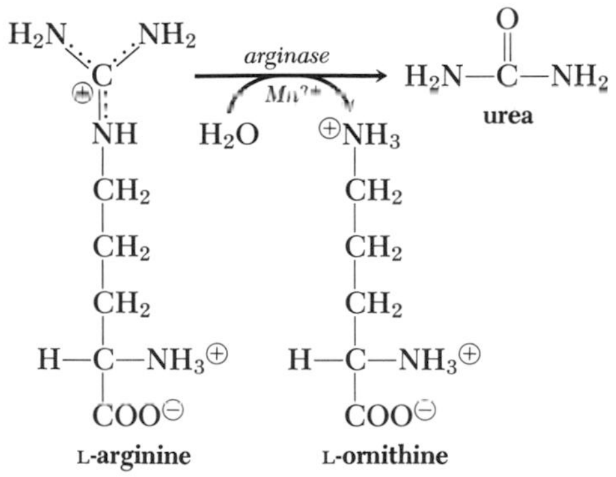

The other product of the hydrolysis is the dibasic amino acid, *ornithine*. (Ornithine is the next lower homologue of lysine, but it is not used for protein formation.)

Urea synthesis therefore requires some mechanism for producing arginine. This is done by a cyclical process in which the ornithine portion of the molecule is used over and over again, acting only as a carrier for the carbon atom and two nitrogen atoms released as urea. This is analogous to the continual reuse of the oxaloacetate portion of citrate in the citric acid cycle.

We have considered the citric acid cycle earlier because of its central role in metabolism. Historically, the urea cycle became known earlier, being described by H. A. Krebs (who later discovered the citric acid cycle) in the late 1930's, and this rationale of metabolic processes in terms of a cycle was one of the major milestones in modern biochemistry.

As we mentioned earlier, one of the nitrogen atoms in urea is derived from ammonia and the other from aspartate. The carbon is obtained from CO_2. The

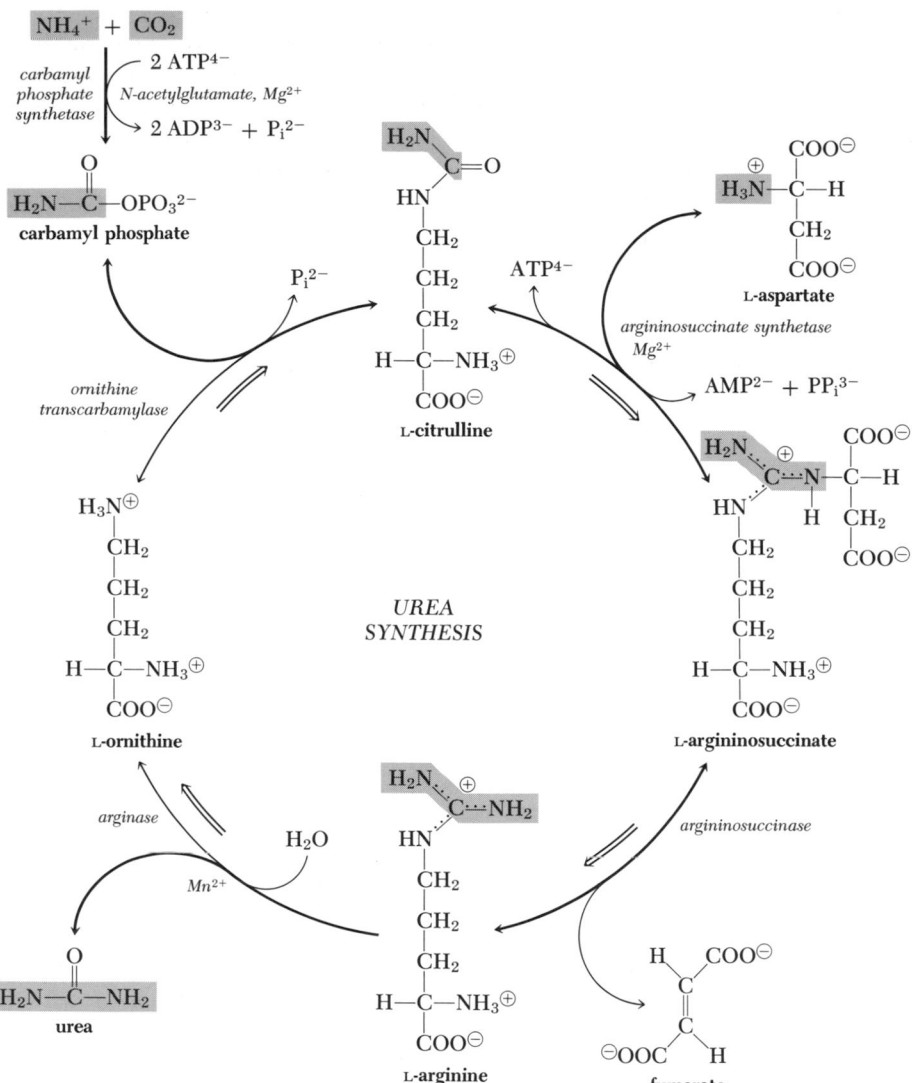

Figure 17-3   Urea synthesis. One of the nitrogen atoms and the carbon atom of urea come from NH_4^+ and CO_2 (*top left*). The other nitrogen atom comes from aspartate (*top right*). The amino acid, ornithine, acts as a carrier for the groups, and it is converted to arginine. Arginine is hydrolyzed to form urea, with the liberation of the ornithine carrier. The entire process is driven by the ultimate cleavage of four high-energy phosphate bonds, including the hydrolysis of PP_i.

complete process of urea synthesis is shown diagrammatically in Figure 17-3, and the reactions can be summarized in tabular form:

1. $NH_4^+ + CO_2 + 2\ ATP^{4-} + H_2O \longrightarrow$
$$\text{(carbamyl phosphate)}^{2-} + 2\ ADP^{3-} + P_i^{2-} + 3\ H^+$$

2. $\text{(carbamyl phosphate)}^{2-} + \text{(ornithine)}^+ \longrightarrow \text{(citrulline)} + P_i^{2-} + H^+$

3. $\text{(citrulline)} + \text{(aspartate)}^- + ATP^{4-} \longrightarrow$
$$\text{(argininosuccinate)}^- + H^+ + AMP^{2-} + PP_i^{3-}$$

4. $\text{(argininosuccinate)}^- \longrightarrow \text{(arginine)}^+ + \text{(fumarate)}^{2-}$

5. $\text{(arginine)}^+ + H_2O \longrightarrow \text{(urea)} + \text{(ornithine)}^+$

SUM: $NH_4^+ + CO_2 + 3\ ATP^{4-} + 2\ H_2O + \text{(aspartate)}^- \longrightarrow$
$$\text{(urea)} + 2\ ADP^{3-} + 2\ P_i^{2-} + AMP^{2-} + PP_i^{3-} + \text{(fumarate)}^- + 5\ H^+$$

The intermediate amino acids in the cycle, *citrulline* and *argininosuccinate*, like ornithine, are not ordinarily found in proteins.

The enzymes of urea synthesis are distributed between the mitochondria and the soluble cytoplasm, and we shall see that this is an important feature, aiding in the transport of carbon atoms for gluconeogenesis. The formation of carbamyl phosphate and the transfer of the carbamyl group to ornithine, producing citrulline, occur in mitochondria. (The mitochondrial membrane is permeable to amino acids and to ammonia.) The transfer of nitrogen from aspartate to citrulline and the hydrolysis of arginine occur in the soluble cytoplasm. Fumarate therefore is released in the cytosol, which also contains the enzymes necessary for converting fumarate to glucose.

The mechanistic novelties in the cycle are the use of *carbamyl phosphate* and *aspartate* as nitrogen donors. We shall see that carbamyl phosphate is also used in the formation of pyrimidines, but the compound is made for this purpose in the cytosol and by an entirely different reaction.

It is energetically feasible to make carbamyl phosphate by the expenditure of only one high-energy phosphate rather than the two actually used, but the reaction would be reversible if that were the case. It is so critical for the higher organisms to maintain very low ammonia concentrations that two moles of high-energy phosphate are expended to drive the reaction far in the direction of ammonia uptake.

The carbamyl phosphate synthetase used in urea synthesis requires the presence of N-acetyl-L-glutamate in low concentrations as a coenzyme. The function of the compound is not clear.

We shall also later see reactions in which aspartate is used as a nitrogen donor during the formation of purine nucleotides. These reactions proceed in a manner analogous to the formation of arginine, with the formation of intermediates similar to argininosuccinate that are cleaved to release fumarate, leaving the nitrogen on the purine ring.

Glutamine

Ammonia is produced in most, if not all, of the tissues of the body, but its transformation into urea mainly occurs in the liver. However, the concentration of ammonia is kept below a toxic level by conversion to glutamine, which acts as a temporary storage for transport. (Glutamine also is, of course, one of the amino acids incorporated into protein, and the same conversion provides a supply for protein synthesis.)

The formation of glutamine is a simple combination of ammonium ion and glutamate at the expense of one high-energy phosphate:

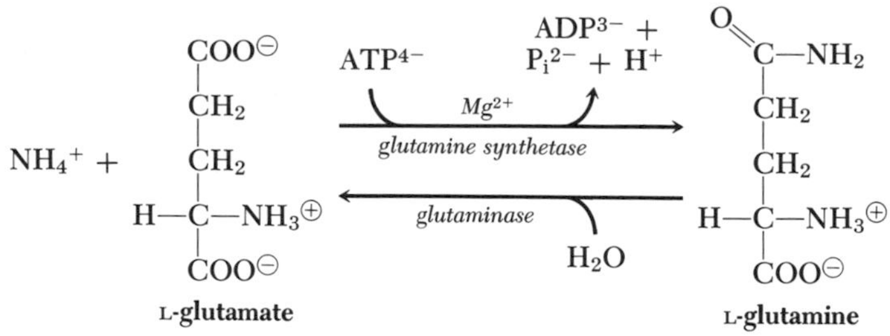

The compound can diffuse through the plasma membrane of cells into the blood. The liver and kidney are especially rich in *glutaminases*, which catalyze the hydrolysis of glutamine into its constituent ammonium ion and glutamate within those tissues, making the ammonium ion available for urea synthesis in the liver or for direct excretion in the kidney. The latter will be considered later (p. 620).

The formation of glutamine is also used for transient storage of part of the nitrogen in the diet. There is an immediate rise in the glutamine concentration of the blood following ingestion of ammonium salts or proteins, and the concentration only falls as urea synthesis proceeds to dispose of the excess nitrogen permanently.

Glutamine has still another metabolic role that we shall encounter repeatedly. It serves as a donor of an amide group for the formation of a variety of compounds, including pyrimidines, purines, and amino sugars found in structural polysaccharides. Interference with the metabolism of glutamine will therefore be reflected in disturbances of a variety of metabolic pathways.

Ammonia Toxicity—Hepatic Coma and Genetic Defects

The combination of temporary storage of ammonia as glutamine and permanent disposal as urea is more than adequate to sustain the organism handling moderate loads of amino acids. Over three-quarters of the liver can be removed in experimental animals without a toxic accumulation of ammonia because the animals eat little and the increase in breakdown of tissue proteins is not too great.

However, the capacity to cope with ammonia may fall to a dangerously low level if there is extensive loss of functional liver. Degeneration of the liver arises from a number of causes, but perhaps the most common is too much alcohol coupled with too little of other dietary constituents, leading to *hepatic cirrhosis*. The ammonia concentration may rise high enough to cause coma in such cases.

There have been various reports of improvement through the administration of glutamate or of ornithine to bolster the inadequate rates of ammonia removal in such cases, but the consequences of the hepatic degeneration are usually so manifold that little can be done other than to diminish the sources of the ammonia.

The easiest way to do this is to cut down on the protein intake, but many of the individuals so affected have already been on an inadequate protein diet, so this is dangerous. Some help is obtained by giving antibiotics that kill the intestinal flora and thereby eliminate the sudden peaks of ammonia in the portal circulation after meals. This does not solve the overall problem, because half of the nitrogen, as we have seen, must pass through ammonia on its way to urea, and elimination of all ammonia formation would also eliminate all urea formation.

The mechanism by which ammonia upsets brain function is not clear. There is the possibility that ammonium ion is directly antagonistic to neural transmission. Another suggestion has been that ammonia prevents the normal function of the citric acid cycle in brain by causing a constant removal of α-ketoglutarate to form glutamate and then glutamine. Such a drain would deplete the oxaloacetate carrier so that acetyl CoA could not be handled. Since the brain is dependent upon the action of the citric acid cycle for most of its energy, an impairment of the cycle would indeed be expected to be disabling.

Since the loss of functional tissue in hepatic cirrhosis has multiple effects, there is always ground for suspecting something other than the ammonia accumulation as being the primary cause of the interference with nervous activity. Very convincing evidence for the especially deleterious effect of ammonia is provided by known cases of genetic defects in urea synthesis. First, it must be emphasized that there is no known case of a living person who has a total absence of any enzyme concerned with urea synthesis. The defects are such that the activity of the enzymes concerned is less than normal, but as we have emphasized, there is considerable reserve capacity for urea formation, and even though one of the enzymes may have only 5 per cent of its normal activity, this is still sufficient to convert much of the ingested amino nitrogen to urea.

The known defects, and the enzymes shown to be deficient, are as follows:

1. *Hyperammonemia:* Two established cases with deficiencies of ornithine transcarbamylase and carbamyl phosphate synthetase. Which of the two is the primary defect is unknown, although the ornithine transcarbamylase has the lowest fraction of the normal activity.

2. *Citrullinemia:* Two known cases with deficiencies of argininosuccinate synthetase.

3. *Argininosuccinic aciduria:* Twelve known cases with deficiencies of argininosuccinase.

The severity of the defects decreases in the order shown, which is also the order of occurrence of the reactions in urea synthesis. All are characterized by impairment of the central nervous system and by elevated ammonia concentrations in the blood following ingestion of protein. Symptoms of neural disability in all are related to the concentration of ammonia.

METABOLISM OF THE CARBON SKELETONS
OF THE AMINO ACIDS

Now that we have in mind the general routes by which nitrogen from the amino acids is handled, let us turn to the disposition of the carbon skeletons. The peripheral tissues of the body have a capacity for handling some of the amino acids, but the major part of the preliminary reactions occurs in the liver, sometimes supplanted to an important extent by the kidneys. As we noted, these reactions involve the formation of phospho-*enol*-pyruvate, which can be converted to glucose, or acetyl coenzyme A, which can be converted to fatty acids or the ketone bodies.

Formation of Phospho-*enol*-pyruvate

The metabolism of lactate, which we discussed in Chapter 14, involves its oxidation to pyruvate. In times of a plenitude of glucose, pyruvate may be oxidized to acetyl coenzyme A and stored as fatty acids. More often, the compound is converted to glucose by the reactions of gluconeogenesis, which successively involve conversions to oxaloacetate and phospho-*enol*-pyruvate, followed by reversal of the reactions of the Embden-Meyerhof pathway.

Therefore, any amino acid that is metabolized to form *pyruvate* or *oxaloacetate* is a *glucogenic* amino acid, because these compounds are direct precursors of glucose. Most of the amino acids are glucogenic.

It is easy to see that *alanine* and *aspartate* are of this type, because simple transamination of these amino acids results in the formation of pyruvate and oxaloacetate, respectively. It isn't so obvious that amino acids metabolized by routes forming *any* component of the citric acid cycle are also glucogenic, because they can be converted to oxaloacetate. This point deserves detailed consideration.

When excess *glutamate* is metabolized, the carbon skeleton appears as α-keto-*glutarate*. We shall see that carbon skeletons from several amino acids are partially converted to *succinyl coenzyme* A. Components of the citric acid cycle, such as citrate and malate, are also commonly present in the diet. They sometimes constitute more than 1 per cent of the wet weight of many fruits. The citric acid cycle by itself does *not* provide a route for the complete combustion of any of these compounds, because it does not consume oxaloacetate.

When additional quantities of intermediates in the cycle appear in a cell, either from the diet or from the metabolism of amino acids, they represent the effective addition of additional amounts of the oxaloacetate carrier. Such compounds will undergo the reactions of the cycle until oxaloacetate is produced. Oxaloacetate will tend to accumulate because it is neither destroyed nor produced by the *complete* citric acid cycle.

This is an important point, so let us approach it from another direction. Mitochondria require a certain concentration of oxaloacetate to carry acetyl groups through the citric acid cycle, and the oxaloacetate is used over and over. If a compound is inserted in the cycle, it will cause the formation of more oxaloacetate, above and beyond the concentration used for carrying acetyl groups, and the excess oxaloacetate is removed by other routes.

Since oxaloacetate is a direct precursor of phospho-*enol*-pyruvate, it follows that three of the carbons of any intermediate of the citric acid cycle can be converted to glucose by the liver or kidneys. The remaining carbons appear as CO_2 (Fig. 17-4).

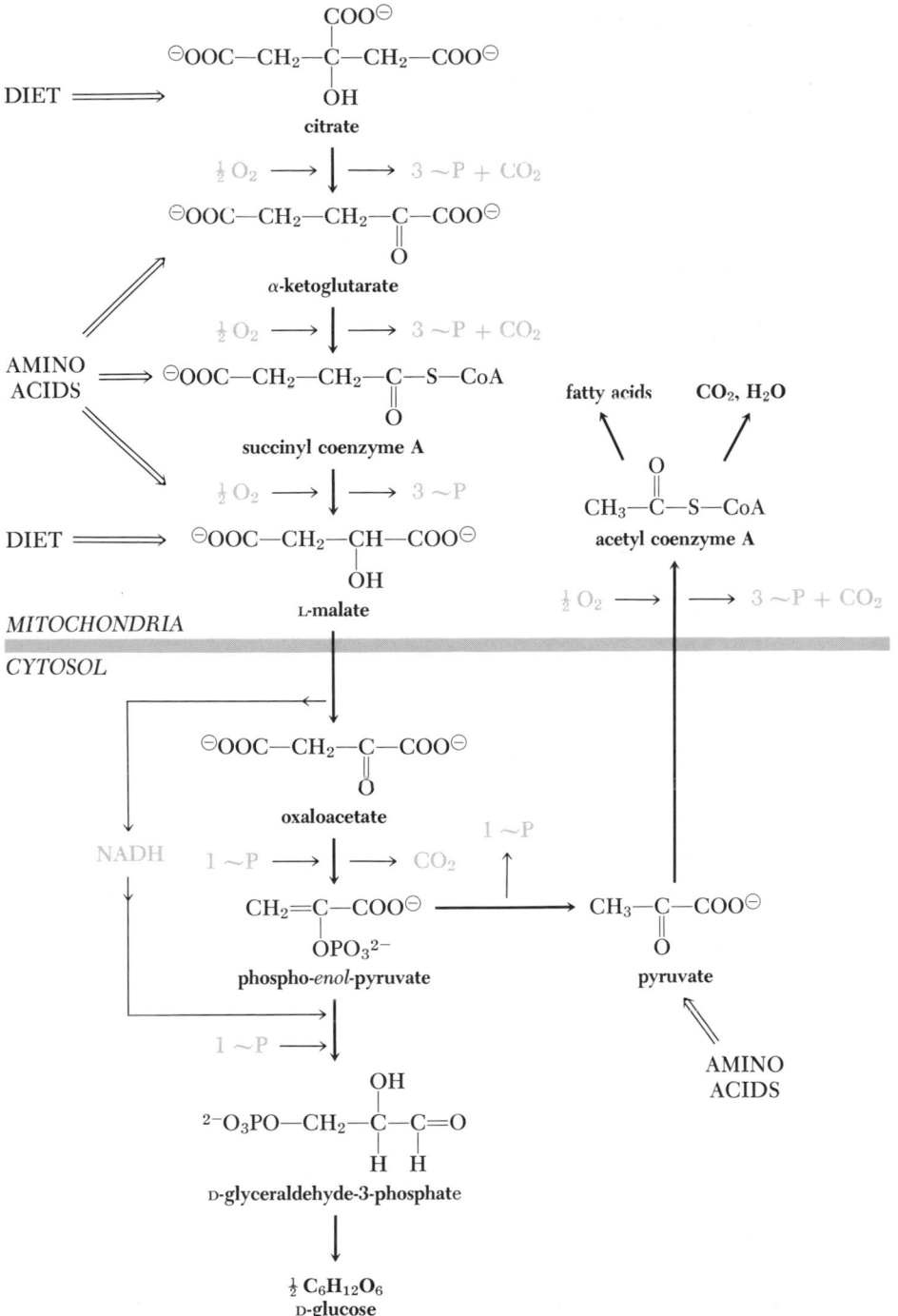

Figure 17-4 Components of the citric acid cycle may arise from the metabolism of amino acids or may be ingested in the diet. They undergo the reactions of the cycle in mitochondria. Further metabolism involves the conversion in the cytosol of malate to phospho-*enol*-pyruvate, which may be disposed of in various ways, depending upon conditions. It may be used as a precursor of glucose by the reactions of gluconeogenesis (*bottom*), or it may be converted to pyruvate at times when gluconeogenesis is not occurring. Pyruvate, which is also a direct product of the metabolism of other amino acids, returns to the mitochondria for oxidation to acetyl coenzyme A under these conditions. Acetyl coenzyme A, in turn, may be used in the synthesis of fatty acids when there is a surfeit of fuel, or it may be oxidized to CO_2 and H_2O. If the latter occurs, the original component of the cycle will have been completely oxidized in the liver.

Formation of Succinyl Coenzyme A
from Propionyl Coenzyme A

It was mentioned in the discussion of fatty acid oxidation that the uncommon compounds with odd numbers of carbon atoms gave rise to the coenzyme A ester of the 3-carbon acid, propionate. Propionyl coenzyme A is converted to succinyl coenzyme A for disposal. The compound (or an intermediate of its conversion) is also the product of metabolism of several amino acids, including *threonine, valine, isoleucine,* and *methionine.* The route is therefore of general significance.

Consider the metabolism of threonine. We already saw that ammonia is removed by the action of threonine dehydratase, leaving *2-ketobutyrate.* The first reaction of this compound is predictable (Fig. 17-5). It is a typical 2-ketocarboxylate, the next higher homologue of pyruvate, and undergoes the same kind of oxidative decarboxylation seen with pyruvate and α-ketoglutarate. The products are CO_2 and *propionyl coenzyme A.*

Before going further, we ought to note that both the deamination and the oxidative decarboxylation are irreversible reactions. Threonine cannot be created from the products of its metabolism and it is an *essential amino acid* in the diet.

The route for metabolism of propionyl coenzyme A is summarized in Figure 17-6. The first reaction is catalyzed by *propionyl CoA carboxylase,* which contains biotin and acts by the same sort of mechanism seen with acetyl CoA carboxylase and pyruvate carboxylase. The product is *methylmalonyl coenzyme A,* which is an optically active compound. A racemase exists to catalyze the equilibration of the two isomers. The new isomer is the substrate for the next reaction in the sequence. (It isn't at all clear why different isomers are used in this sequence.)

The final reaction is the rearrangement of (R)-methylmalonyl coenzyme A to succinyl coenzyme A, catalyzed by a mutase. Since succinyl coenzyme A is converted to oxaloacetate, it follows that threonine and all other compounds giving rise to propionyl coenzyme A or methylmalonyl coenzyme A are glucogenic.

We have not seen anything comparable to the mutase reaction heretofore. It involves a coenzyme derived from vitamin B_{12}, and let us digress to consider this important compound.

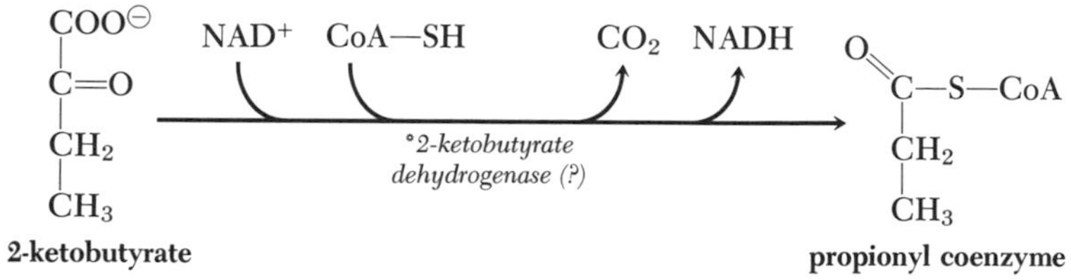

2-ketobutyrate   propionyl coenzyme A

* *Note: Enzyme complex believed to include decarboxylase containing thiamine pyrophosphate, trans-acylase containing lipoate, and dihydrolipoyl dehydrogenase containing FAD.*

Figure 17-5   Oxidative decarboxylation of 2-ketobutyrate. Some believe the reaction is catalyzed by the pyruvate dehydrogenase complex; others vote for a more specific enzyme acting by a mechanism similar to that of pyruvate or α-ketoglutarate dehydrogenases.

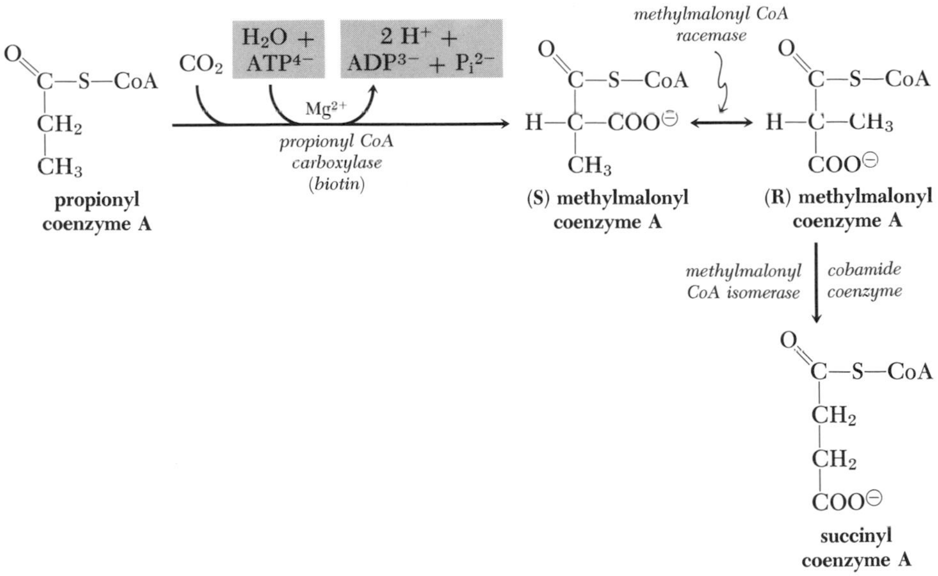

Figure 17-6 Propionyl coenzyme A is metabolized by a carboxylation followed by rearrangements producing succinyl coenzyme A.

Vitamin B_{12} and Cobalamin Coenzymes

The coenzyme for methylmalonyl CoA mutase is 5' *deoxyadenosylcobalamin* (Fig. 17-7)—a very complicated molecule containing cobalt bound to nitrogen or carbon on all six coordination positions. The salient part of the molecule is the large tetrapyrrole ring surrounding the cobalt, which resembles a porphyrin ring superficially, but differs in being more saturated and in lacking one methylene bridge. This is known as the corrin ring. The *cobalamins* are *corrinoid* compounds.

Cobalamin coenzyme has two unusual chemical features. First, it contains a metal-carbon bond, the only known biological example of this linkage. Secondly, the cobalt is present at a univalent oxidation state. The mechanism of action of the coenzyme is still unknown, but is believed to involve rupture of one of the cobalt coordination bonds by the substrate.

What is commonly referred to as *vitamin B_{12}* is the same as the cobalamin coenzyme, except that the deoxyadenosyl group is replaced by a cyanide ion, and the valence of the cobalt is +3, so that there is a net charge of +1 remaining on the metal. This is *cyanocobalamin,* which is the form of the compound that was originally isolated during the search for the vitamin. There is some doubt as to whether cyanocobalamin occurs as such in a natural diet, because the natural coenzyme is very susceptible to oxidation and disruption in the presence of cyanide. Although we ordinarily think of cyanide as a dangerous poison, it is produced by many microorganisms in small amounts, and the affinity of the cobalamins for cyanide is so great that cyanocobalamin may be created during the handling of the microbial cultures from which it is isolated. Most of the cobalamin in animal tissues is present as the coenzyme, which is therefore the natural dietary form of the vitamin.

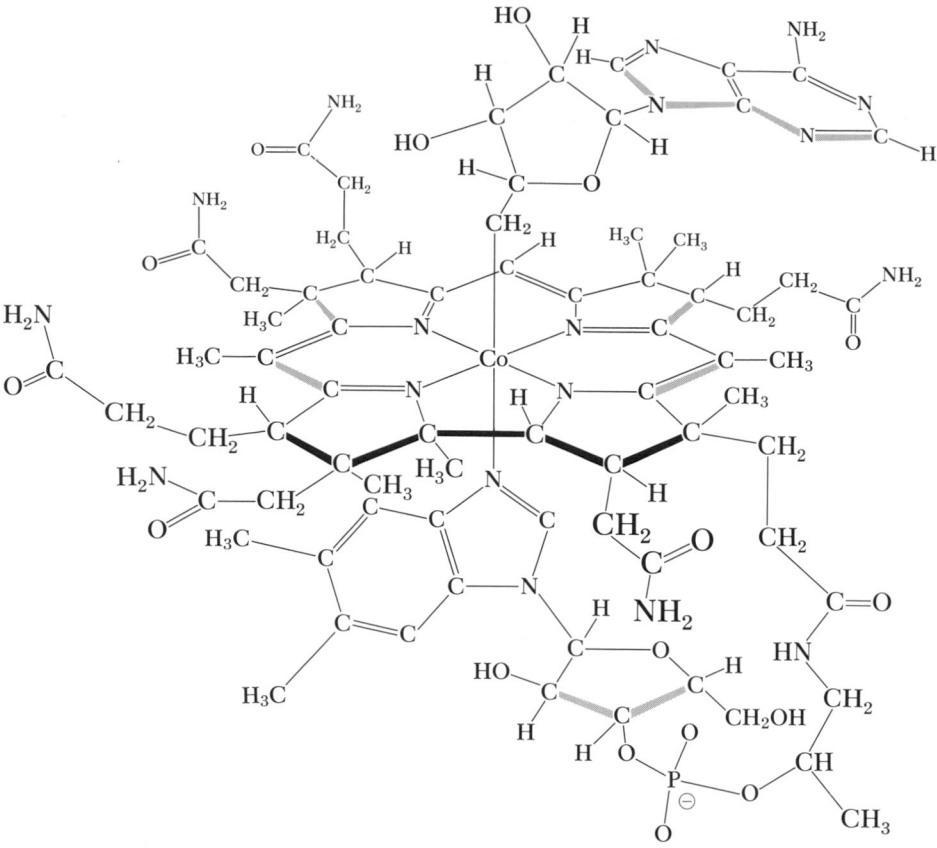

Figure 17-7   The structure of cobamide coenzyme. The large ring seen on edge around the central cobalt atom, resembling a porphyrin ring with one methylene group missing, is the *corrin* ring. The corrin-cobalt complex, complete with the various side chains on the ring and including the ribosyl group attached through a 3'-phosphate diester bond (*bottom right*), is *cobamide*. The further combination with a dimethylbenzimidazole ring, viewed face-on below the corrin ring and with three of substituent hydrogen atoms removed for clarity, constitutes a *cobalamin*. The final coordination position of cobalt is usually occupied by another group, in this case 5'-deoxyadenosine, combined through a carbon atom; cobamide coenzyme is *5'-deoxyadenosylcobalamin*. Commercially available vitamin B_{12} has a cyanide group attached and is *cyanocobalamin*. The figure is an idealization of a true projection of the spatial position of the atoms.

The search for the vitamin was originally spurred because it is deficient in the tissues of humans suffering from *pernicious anemia*. The name implies the low concentration of hemoglobin resulting from the condition, but the effects also include serious disturbances of the central nervous system that result in abnormal sensation, motion, or thought. (The condition shows how widespread the ramifications can be from a disturbance of seemingly isolated parts of amino acid metabolism.)

Curiously, pernicious anemia turned out to be a genetic defect of the stomach, rather than a dietary deficiency disease. The absorption of dietary cobalamins depends upon the formation by the gastric mucosa of a carbohydrate-rich protein, known as the *intrinsic factor*. People with pernicious anemia do not make this protein. Before purified cyanocobalamin was available, treatment consisted of eating hog stomach preparations to supply the missing intrinsic factor and liver to supply the vitamin (the *extrinsic factor*).

The amount of cobalamin required in the diet is very small—a few micrograms a day for a normal individual. Even those with pernicious anemia can absorb enough if the oral dose is raised to the level of 1 milligram or so. The cobalamins therefore represent some of the most potent biological agents known. The formation of these compounds is the only known biological function of cobalt, but it is a critical one for animals, and this brings us to a striking paradox. It is almost a truism that the metabolic processes of plants and animals proceed with the same kinds of reactions, with the important exception of photosynthesis. It is therefore startling to discover that the yeasts, the green algae, and all of the higher plants have no need for cobalt and contain no cobalamins, despite the fact that these organisms deal with the same kinds of compounds metabolized by cobalamin-dependent reactions in animals. This drastic evolutionary parting of the metabolic way has little direct consequences over most of the world. However, there are some areas, particularly in Australia, in which the soils have a very low cobalt content. Plants grow, despite their low cobalt content, but some animals that eat the plants don't.

The need for cobalt is especially pronounced in ruminants, and it is the proclivity of the Australians to herd sheep that exposed the unsuspected deficiency in their land. Cud-chewing animals depend upon microbial fermentations in their rumen to break down cellulose in the diet. (Cellulose is a glucose polysaccharide, but unlike starch and glycogen, it is made of β-glucosyl residues, and is not attacked by the usual battery of enzymes in mammalian intestines.) The fermentations produce acetate and butyrate, which can serve as sources of fatty acids, but they also convert a large part of the carbohydrate to propionate, and this is the major source of glucose for the animals.

The conversion of propionate to glucose depends upon the same sequence of reactions we have been discussing, with the addition of a thiokinase reaction to form propionyl coenzyme A from the free compound. Therefore, the ruminants have a rapid propionate metabolism, including an active methylmalonyl mutase and its necessary cobalamin coenzyme. If they ingest enough cobalt, the microorganisms that digest carbohydrate will also synthesize the requisite cobalamin for them.

Man is not dependent upon this metabolic route for most of his glucose supply, so inability to handle propionyl coenzyme A accounts for only part of the effects of pernicious anemia. (We shall see other reactions requiring cobalamin coenzyme.) However, the failure does provide a means of assessing lack of cobalamin in a practical way, because a failure to convert methylmalonyl CoA to succinyl CoA causes the appearance of methylmalonate in the urine. (Thiolases catalyzing the hydrolysis of coenzyme A esters are known, and presumably serve to remove excesses.) Figure 17-8 shows the sensitivity of this excretion as an index. Data are plotted from an individual with pernicious anemia who was treated with cyanocobalamin orally. Normal function of hematopoietic tissues was promptly restored, as is shown by the sharp rise in reticulocyte count, which later declined as complete maturation of new erythrocytes proceeded. The increase in reticulocytes was also accompanied by a sharp rise in the concentration of hemoglobin, as would be expected. The level of cobalamin compounds in the blood rose somewhat later, but methylmalonate was still detectable in the urine, and it was only after 240 days that it declined to unmeasurable levels.

Formation of the Oxybutyrates

Now that we have considered the formation of precursors of glucose from amino acids, let us turn to acetyl coenzyme A and other intermediates of fatty acid

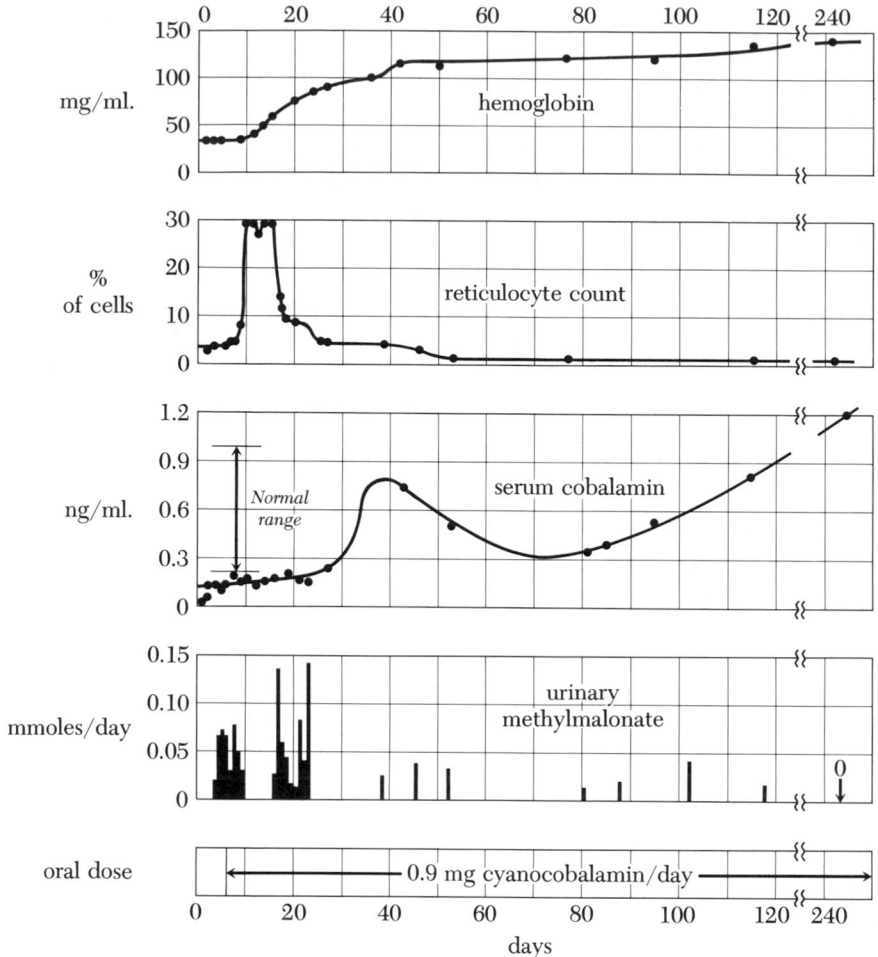

Figure 17-8   The response of a patient with pernicious anemia to the oral administration of cyanocobalamin. Adapted from S. B. Kahn, *et al.*, in J. Lab. Clin. Med., Vol. *66:* p. 75 (1965).

metabolism derived from amino acids. We have previously emphasized that glucose cannot be synthesized from acetyl coenzyme A. It must be converted to forms other than glucose for transport to the peripheral tissues. An obvious possibility is the synthesis of fatty acids followed by transport from the liver as triglycerides associated with lipoproteins. This indeed occurs at times when there is a temporary excess glucose supply, with suppression of gluconeogenesis in the liver and conversion of glucose to fatty acids in the adipose tissue.

There is another route for the transport of acetyl units from the liver and kidneys that we have not discussed in earlier chapters. It becomes increasingly important as the total glucose supply diminishes and fatty acids are mobilized. Indeed, acetyl units from amino acids are mixed with acetyl units derived from fatty acids in the liver under these conditions. The liver and kidneys utilize acetyl

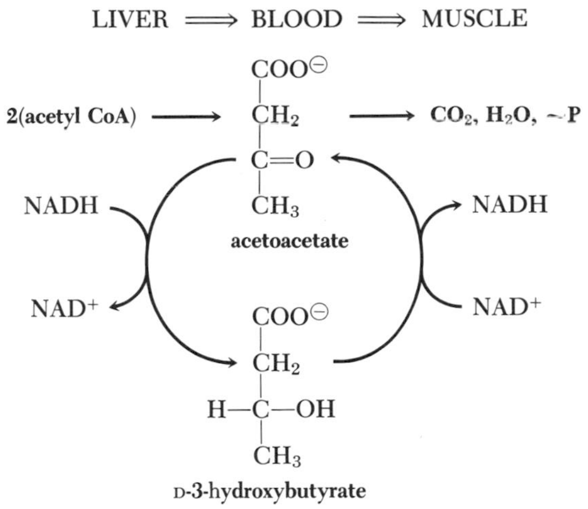

LIVER $\Longrightarrow$ BLOOD $\Longrightarrow$ MUSCLE

Figure 17-9 Both liver and muscle contain a 3-hydroxybutyrate dehydrogenase, catalyzing the interconversion of acetoacetate and D-3-hydroxybutyrate. Both of the compounds appear in the blood, although the hydroxybutyrate can only be formed from acetyl coenzyme A *via* acetoacetate in the liver and can only be oxidized *via* acetoacetate in the muscles.

groups by the mechanism outlined below to produce *acetoacetate* (3-oxobutyrate in systematic nomenclature) and D-*3-hydroxybutyrate*, which diffuse into the blood. These oxybutyrates are actively metabolized by muscles, and sometimes by the brain, to provide acetyl coenzyme A for the citric acid cycle of those tissues. They therefore comprise another type of oxidizable compound that can be elaborated by the liver for use in the peripheral tissues (Fig. 17-9).

Acetoacetate spontaneously decarboxylates to form acetone. This is a slow reaction at physiological pH values, and the amount of acetone formed is a small fraction of the total acetoacetate produced. Diabetes and starvation, for reasons we shall discuss in later chapters, cause an accumulation of acetoacetate, and the resultant rise in the concentration of acetone may be sufficient to make its odor detectable in the breath. This is part of the reason acetoacetate, 3-hydroxybutyrate, and acetone were collectively called the *"ketone bodies"* by early investigators. The term is still in use. However, acetone is a minor product in quantitative terms, and we shall mainly be concerned with the oxybutyrates.

The concentrations of acetoacetate and 3-hydroxybutyrate tend to rise and fall together. Combustion of the amount produced by the liver and kidneys of a well-fed individual on a carbohydrate-rich diet may represent only a few per cent of the oxidative metabolism of his muscles, and the compounds are utilized so efficiently that the blood concentration is maintained in the range of 10 to 30 micromolar for each compound, with acetoacetate usually being somewhat lower in concentration than 3-hydroxybutyrate. However, conditions causing greater dependence on the metabolism of fats and proteins result in an increased production of these compounds so that the blood concentration rises into the millimolar range. (There is great variation, but 2 mM acetoacetate and 8 mM D-3-hydroxybutyrate would be typical values for the blood of a starved obese person.) An individual with an accumulation in this range is said to be in a state of *ketosis,* and the increased concentration in the blood is denoted as a *ketonemia.*

Acetoacetate is produced in the liver and kidneys by a simple two-step process

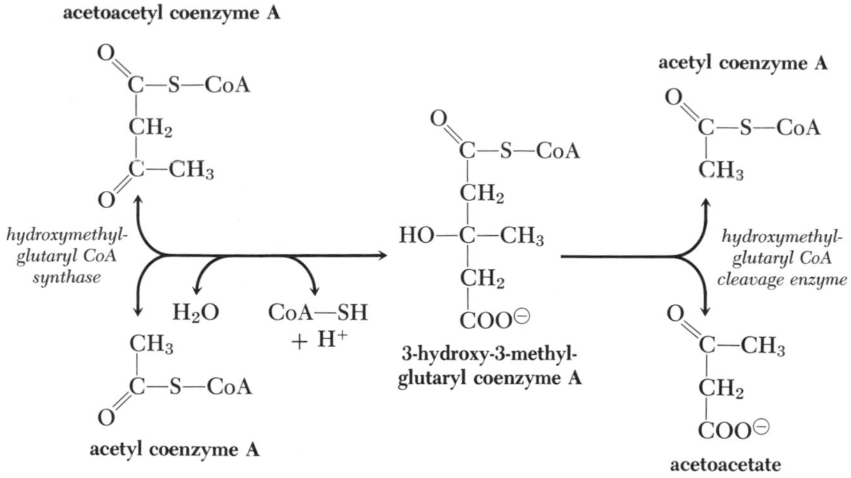

Figure 17-10   The formation of acetoacetate from acetyl coenzyme A involves the formation and cleavage of 3-hydroxy-3-methylglutaryl coenzyme A. Since acetoacetyl coenzyme A is in equilibrium with acetyl coenzyme A by the thiolase reaction (not shown), this route represents the total synthesis of acetoacetate from acetyl coenzyme A.

(Fig. 17-10) in which acetyl coenzyme A and acetoacetyl coenzyme A condense to form *3-hydroxy-3-methylglutaryl coenzyme A* with the loss of one molecule of coenzyme A; followed by cleavage of the 6-carbon coenzyme A ester to yield free acetoacetate and acetyl coenzyme A. If we think of acetoacetyl coenzyme A as being analogous to oxaloacetate, we see that the condensation reaction is exactly analogous to the formation of citrate in the first step of the citric acid cycle (p. 211). However, it is catalyzed by a quite different synthase (sometimes named *β-hydroxy-β-methylglutaryl CoA thiolase,* since the reaction is more reversible than the formation of citrate). The cleavage is different than the cleavage of citrate in the cytosol. First, cleavage occurs at a different point than condensation so that the acetyl moiety released has the thiol ester group and is therefore derived from the original acetoacetyl coenzyme A rather than from the original acetyl coenzyme A. The acetyl group derived from acetyl coenzyme A upon condensation becomes a part of the 4-carbon acetoacetate that forms by cleavage. Secondly, the coenzyme A ester of acetoacetate is not regenerated, so the reaction is irreversible without the use of a high-energy phosphate to drive it. Even so, this sequence represents a mechanistic variant of the formation and cleavage of citrate as a means of transporting carbons.

What is the effective result? Acetyl coenzyme A and acetoacetyl coenzyme A are in equilibrium because of the reaction catalyzed by acetoacetyl coenzyme A thiolase, which we have considered as part of fatty acid oxidation (p. 230). This means that *all* of the carbons of 3-hydroxy-3-methylglutaryl coenzyme A can be formed from acetyl coenzyme A. Therefore, carbon atoms of amino acids that are converted to acetyl coenzyme A or other intermediates of fatty acid oxidation, as well as the carbon atoms of the fatty acids themselves, may be used to form acetoacetate in the liver and kidneys. Such amino acids are said to be *ketogenic.*

The formation of acetoacetate appears to occur in both the cytosol and the mitochondria, but the major production is in the latter organelles. The mitochondrial cristae also contain a *3-hydroxybutyrate dehydrogenase*, a typical alcohol dehydrogenase catalyzing the reaction:

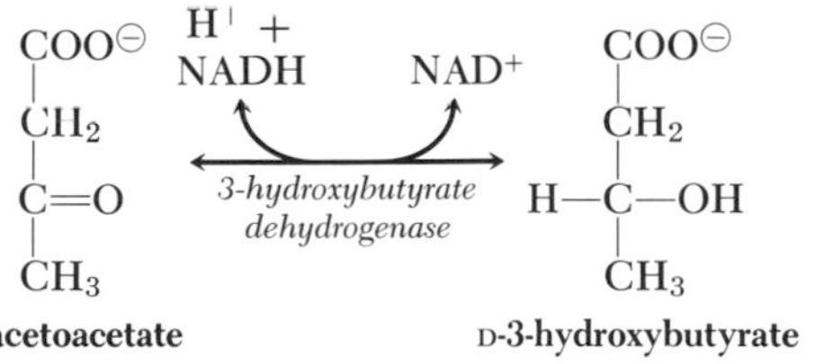

acetoacetate   D-3-hydroxybutyrate

The enzyme causes a conversion of part of the acetoacetate to D-3-hydroxybutyrate, and the amount of conversion is determined by the ratio of NADH to NAD maintained within the mitochondria. Indeed, the relative amounts of the two oxybutyrates produced is used as an index of the state of reduction of NAD in mitochondria. The formation of hydroxybutyrate is, in effect, withdrawing electrons from mitochondria to make a more reduced substrate, and later oxidation of the compound in peripheral tissues can produce more high-energy phosphate than does the oxidation of acetoacetate.

The 3-oxybutyrates appearing in the blood diffuse into the skeletal and cardiac muscles. The mitochondria of these tissues also contain an NAD-coupled 3-hydroxybutyrate dehydrogenase catalyzing the conversion of this compound to acetoacetate. However, the muscle mitochondria also have another enzyme, *acetoacetate—succinate CoA transferase*, that catalyzes the transfer of coenzyme A from succinyl coenzyme A to acetoacetate. The reaction is reversible, but the constant removal of the resultant acetoacetyl coenzyme A to form acetyl coenzyme A makes it proceed in one direction physiologically. The uptake of acetoacetate catalyzed by this enzyme is, in effect, at the expense of one mole of high-energy phosphate. This is so because it involves the conversion of succinyl coenzyme A to succinate, and GTP would otherwise be obtained by this conversion through the succinyl CoA synthetase reaction in the citric acid cycle (p. 218).

In sum, the muscles have a mechanism for converting D-3-hydroxybutyrate to acetoacetate, and the carbons of acetoacetate obtained in this way and by direct diffusion from the blood are injected into the citric acid cycle in the form of acetyl coenzyme A. Two moles of acetyl coenzyme A are obtained from one mole of acetoacetate (Fig. 17-11, p. 375).

METABOLISM OF SPECIFIC AMINO ACIDS

We have shown in a general way how nitrogen is removed from some of the amino acids and converted to urea. We have discussed the formation of glucose and of fatty acids or oxybutyrates from the carbon skeletons. The unique part of the metabolism of each amino acid therefore becomes the process by which the precursors of glucose or of acetoacetate are formed. Let us now consider these peculiar steps for each amino acid whose metabolism can be completely described in these terms, beginning with those whose metabolism has already been outlined. (There are a few amino acids directly involved in more complicated processes, and the metabolism of these is treated in the following chapter.)

Alanine

Nitrogen: Transferred to α-ketoglutarate, forming glutamate.
Carbons: Appear as pyruvate.
Transamination of alanine directly produces pyruvate:

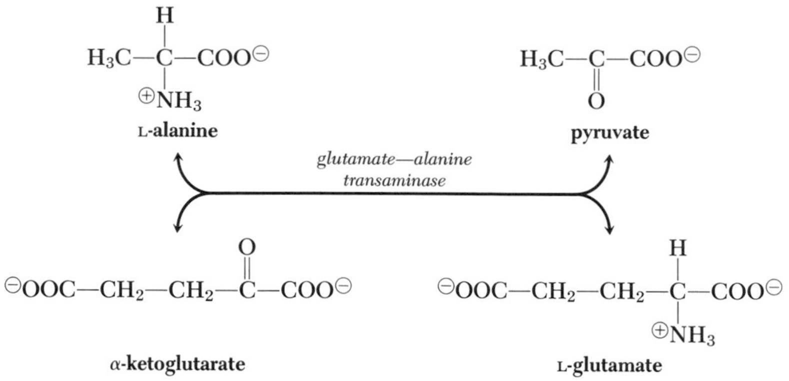

All three of the carbons of this amino acid therefore can be converted to glucose. The metabolism of its nitrogen atom is essentially the metabolism of glutamate nitrogen.

Aspartate

Nitrogen: May be used directly in formation of urea, or transferred to α-ketoglutarate, forming glutamate.
Carbons: Appear directly as oxaloacetate or fumarate.
One of the steps in urea synthesis directly utilizes the nitrogen atom of asparatate, leaving the carbon skeleton as fumarate. However, part of a large excess of aspartate may undergo transamination in the cytosol with the carbon skeleton appearing as oxaloacetate:

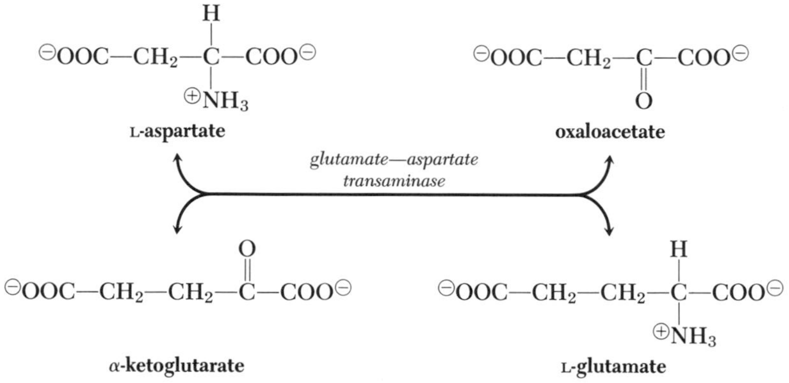

The result is the same in either case because fumarate from urea synthesis is converted to oxaloacetate by the fumarase and NAD-malate dehydrogenase also present in the cytosol. Three of the carbons of aspartate therefore appear as glucose, with one appearing as CO_2.

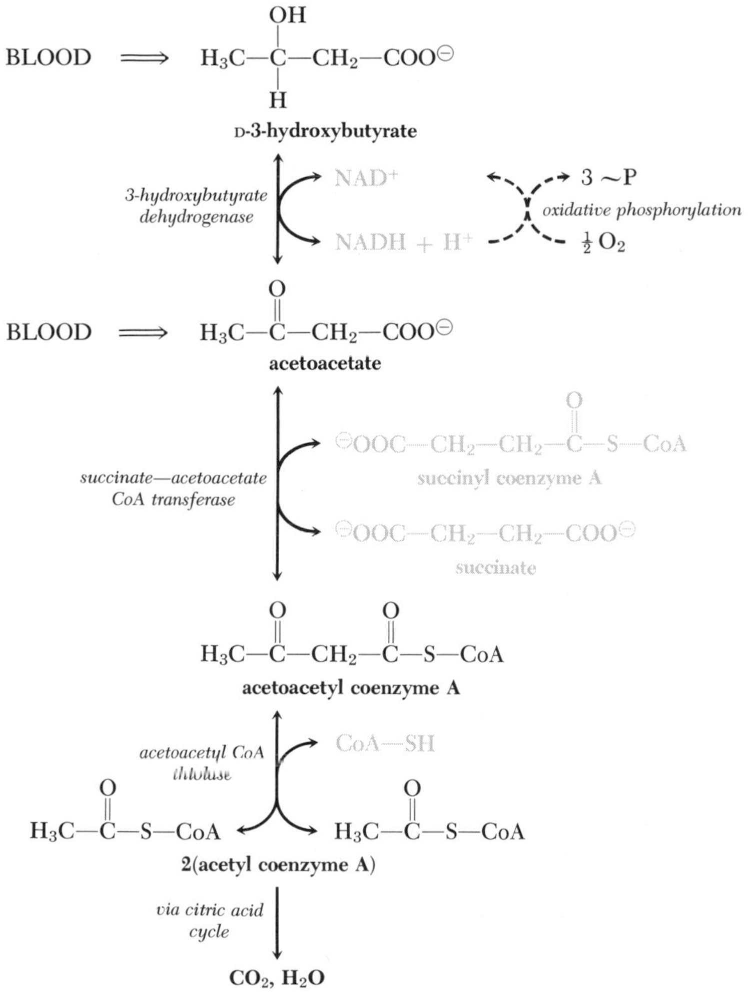

Figure 17-11 Skeletal and heart muscle convert the oxybutyrates arriving from the blood to acetyl coenzyme A for complete combustion in the citric acid cycle. D-3-Hydroxybutyrate is oxidized to acetoacetate in the mitochondria. Acetoacetate obtained both from this reaction and from the blood is converted to acetoacetyl coenzyme A by transfer of a coenzyme A residue from succinyl coenzyme A.

Glutamate

Nitrogen: May be used to form aspartate for incorporation into urea.

Carbons: Appear directly as α-ketoglutarate.

The α-ketoglutarate produced by either the glutamate dehydrogenase or transaminase reactions is oxidized by the enzymes of the citric acid cycle to oxaloacetate. Therefore, three of the five carbons of glutamate may be converted to glucose, with the other two appearing as CO_2.

Serine

Nitrogen: Direct deamination produces NH_4^+.

Carbons: Appear directly as pyruvate.

Since serine dehydratase causes the direct conversion of serine to ammonium ion and pyruvate, all three carbons may be converted to glucose.

Threonine

Nitrogen: Direct deamination produces NH_4^+.
Carbons: Converted to succinyl coenzyme A *via* propionyl coenzyme A.

The formation of propionyl coenzyme A and its conversion to succinyl coenzyme A enables three of the carbons of threonine to be converted to glucose; the fourth appears as CO_2.

Branched-Chain Amino Acids—Leucine, Isoleucine, and Valine

Nitrogen: Transferred from all to α-ketoglutarate, forming glutamate.
Carbons: Leu: Converted to 3-hydroxy-3-methylglutaryl coenzyme A, the precursor of acetoacetate.

          Ile: Converted to equal amounts of succinyl coenzyme A (*via* propionyl coenzyme A) and acetyl coenzyme A.

          Val: Converted to succinyl coenzyme A and CO_2 *via* methylmalonyl coenzyme A.

All of the branched-chain amino acids transaminate with α-ketoglutarate. The next reaction for each is oxidative decarboxylation of the resultant 2-ketocarboxylates to form coenzyme A esters (Fig. 17-12).

The oxidative decarboxylations are all irreversible, and there is no means of creating the branched-chain 2-ketocarboxylates by other metabolic routes in animals. Leucine, isoleucine, and valine are *essential amino acids,* and must be supplied in the diet.

Some genetic defects in the enzymes catalyzing the first two steps in branched-chain amino acid metabolism are illuminating. For example, a single enzyme in heart muscle catalyzes the transamination of all three amino acids. However, the transaminations in the liver are catalyzed by three different enzymes. The quantitative importance of the liver compared to the heart in handling the amino acids is shown by a rare genetic defect in humans causing *hypervalinemia,* in which there is an accumulation of valine without an accompanying accumulation of leucine or isoleucine. If the heart, and very likely other muscles, were handling a major part of the load, either all three or none of the amino acids should accumulate, since a single enzyme handles all.

The *oxidative decarboxylations* are also catalyzed by different enzymes, one specific for 2-ketoisocaproate and 2-keto-3-methylvalerate derived from leucine and isoleucine, respectively, while the other enzyme is specific for 2-ketoisovalerate derived from valine. Some humans lack the former enzyme, and have *maple syrup urine disease.* The primary effect is the accumulation of the 2-keto analogues of leucine and isoleucine. However, these compounds act as inhibitors of the dehydrogenases attacking 2-ketoisovalerate and 2-ketobutyrate, so these products of valine and threonine metabolism also accumulate.

The name of the disease is derived from the odor of the urine from affected

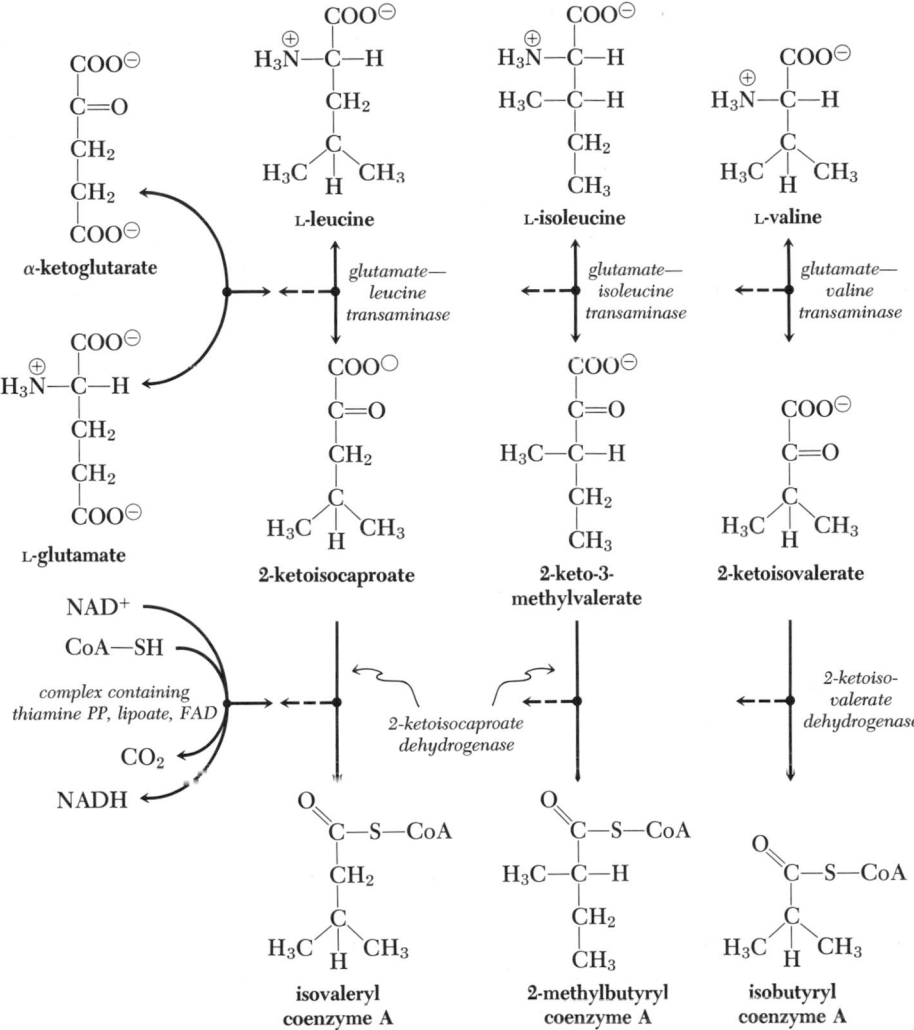

Figure 17-12 Leucine, isoleucine and valine are metabolized by similar initial steps, beginning with a transamination with α-ketoglutarate, catalyzed by a separate enzyme for each amino acid. The resultant 2-ketocarboxylates are oxidatively decarboxylated by enzyme complexes resembling pyruvate dehydrogenase, forming coenzyme A esters of branched-chain fatty acids containing one less carbon atom than the original amino acid. The succeeding steps are detailed in the next three figures.

infants, which connoisseurs say is similar to that of certain vintages of maple syrup. Since no other condition is known to create a urine of even remotely similar aroma, pediatricians are not faced with the necessity for such acute olfactory discrimination. (Besides, the condition is rather rare.) The odor is actually the result of polymerization of 2-hydroxybutyrate derived as a by-product of 2-ketobutyrate.

The name has a certain puckishness and probably won't survive in these prosaic times, but the consequences of the condition are grim. Affected infants develop severe neurological damage within a few days after birth and die within several weeks unless treated. The damage evidently is a result of toxicity to the brain of the accumulating 2-keto compounds. Treatment involves rigid dietary restrictions to reduce the intake of the amino acids to the minimal level consistent with growth, and this precludes the use of anything approaching the varied American diet with its variety of protein sources.

Let us go on now to consider the further metabolism of the carbon skeletons. The coenzyme A esters produced by oxidative decarboxylation are typical fatty acid metabolites, except for the branches in the hydrocarbon chain, and they undergo the same kind of oxidation by flavoproteins seen with other acyl coenzyme A compounds. Until recently, there was no reason to believe that the necessary dehydrogenases differed from the enzymes used in fatty acid metabolism. However, a brother and sister have been discovered with apparently normal fatty acid metabolism who excrete isovalerate, but not isobutyrate or 2-methylbutyrate, and the existence of these two individuals implies that isovaleryl coenzyme A is metabolized by a different enzyme than those acting on the short-chain fatty acyl coenzyme A compounds or the corresponding branched-chain esters derived from valine and isoleucine.

The metabolism of the *leucine* carbon skeleton also deviates from that of the others in several later details, so let us consider the fate of isovaleryl coenzyme A separately. The predicted initial oxidation results in the formation of the corresponding enoyl coenzyme A ester (Fig. 17-13). However, the ordinary hydration followed by oxidation to yield a 3-ketoacyl coenzyme A compound that would be expected with ordinary fatty acid derivatives cannot occur because of the branch on the third carbon.

A different route has evolved, in which the compound is carboxylated by a typical biotin-containing carboxylase at the expense of a high-energy phosphate. The result is the coenzyme A ester of an unsaturated dicarboxylic acid. This is then hydrated to form *3-hydroxy-3-methylglutaryl coenzyme A*, which is also the intermediate used in the production of acetoacetate by the liver (p. 372). Therefore, the compound will be cleaved in the liver to yield acetyl coenzyme A and acetoacetate by routes we have already considered. Leucine, therefore, is unlike the amino acids considered earlier. It is a *ketogenic,* rather than a glucogenic amino acid, and all six carbons may appear as acetoacetate.

The branched-chain acyl coenzyme A compounds derived from *valine* and *isoleucine* can, and do, undergo the standard sequence of oxidation, hydration and oxidation seen with ordinary fatty acid derivatives, with the qualification that the valine derivative loses its coenzyme A along the route (Fig. 17-14 A, B). The methylacetoacetyl coenzyme A derived from isoleucine is cleaved just as acetoacetyl coenzyme A is, except that one of the products of the cleavage is propionyl coenzyme A rather than acetyl coenzyme A (Fig. 17-14 B). The route of metabolism of the semialdehyde of methylmalonate, which is derived from valine, is less certain. It evidently is oxidized to methylmalonate, and then activated to form the corresponding coenzyme A ester.

Since both *isoleucine* and *valine* give rise to the formation of one mole of succinyl coenzyme A *via* methylmalonyl coenzyme A, three of the carbons of these amino acids may appear as glucose. Isoleucine also forms a mole of acetyl coenzyme A, and two of its carbons may appear as acetoacetate *via* that compound. There is also a net production of one mole of CO_2 from isoleucine and two from valine.

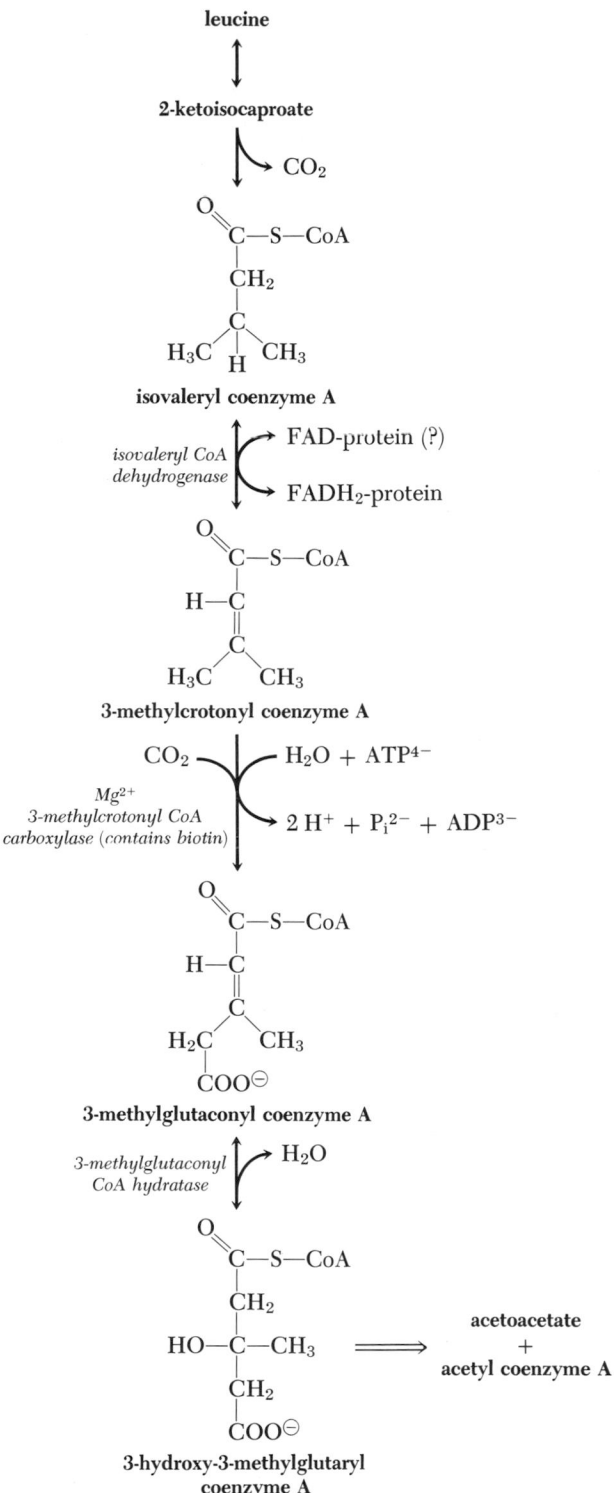

Figure 17-13 Isovaleryl coenzyme A obtained by the metabolism of leucine is converted to 3-hydroxy-3-methylglutaryl coenzyme A, the same compound that is an intermediate in the formation of acetoacetate from acetyl coenzyme A. Since the acetyl coenzyme A formed by the cleavage of this compound can be recycled to make more of the intermediate, the carbon skeleton of leucine can be completely converted to acetoacetate by the liver.

379

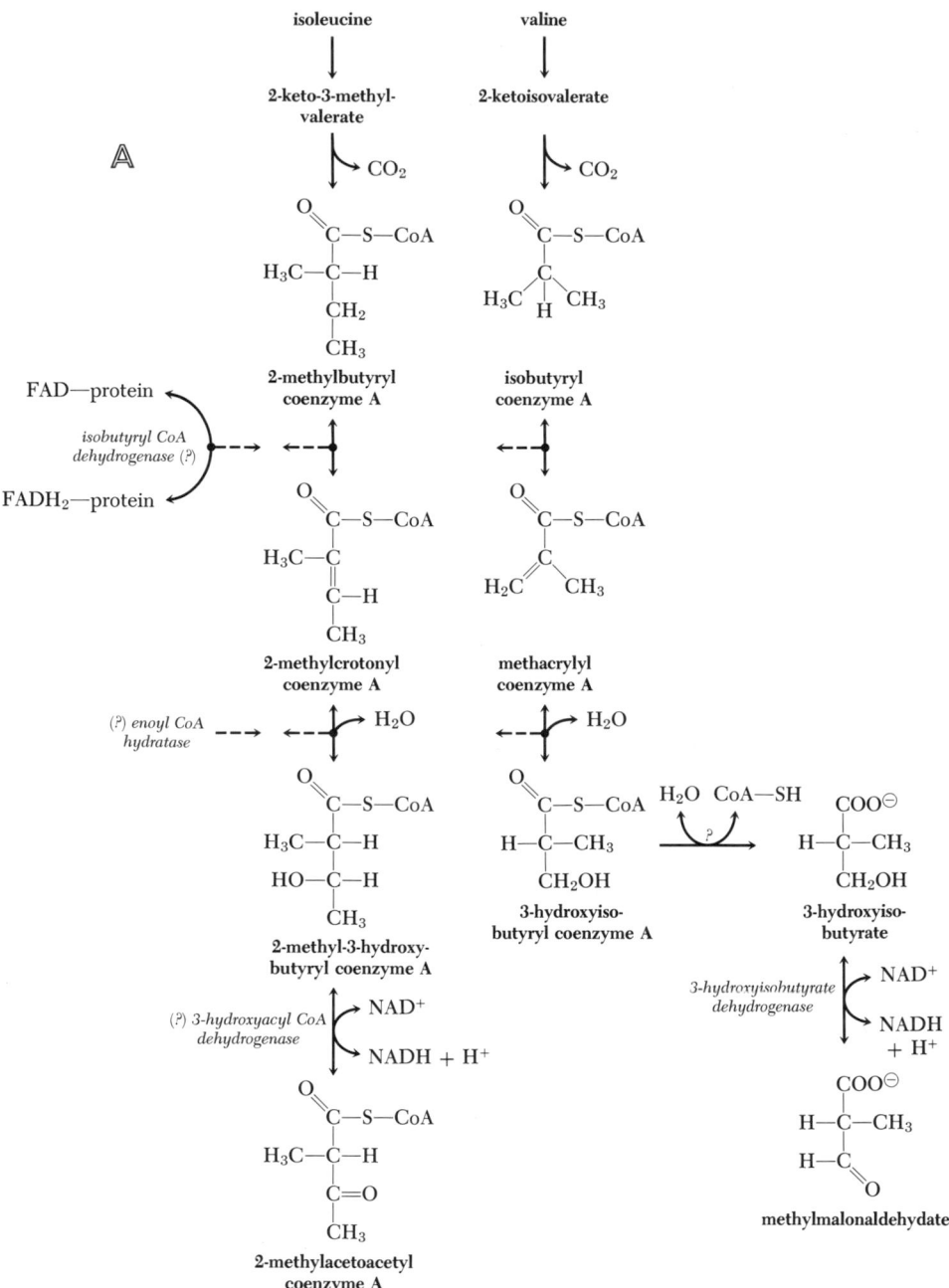

Figure 17-14   The metabolism of the branched-chain coenzyme A esters obtained from isoleucine and valine begins with an oxidation and a hydration like those of ordinary fatty acid metabolism (part A). The further metabolism of the isoleucine skeleton continues the resemblance, with oxidation of the 3-hydroxy derivative to a 3-keto compound, followed by thiolysis to form acetyl coenzyme A and propionyl coenzyme A. However, the 3-hydroxyisobutyryl coenzyme A obtained from valine is converted to methylmalonyl coenzyme A by a route believed to involve the oxidation of the free compound, first to an aldehyde and then to a carboxylate.

Illustration continues on opposite page.

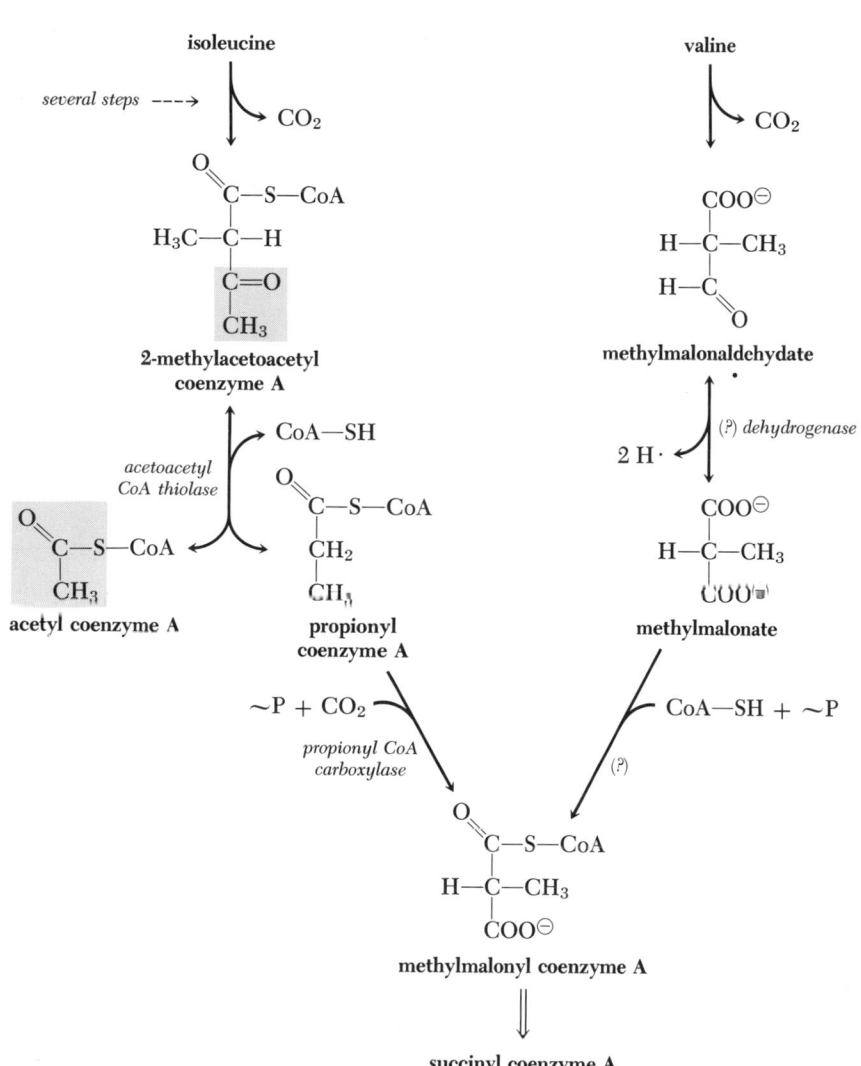

Figure 17-14   *Continued.*

Cysteine and Cystine

Nitrogen: Transferred to α-ketoglutarate, forming glutamate.
Carbons: Converted to pyruvate.
Sulfur: Appears as sulfate.

When we discussed the formation of proteins, we noted that disulfide bonds, representing cystine, are interconverted with pairs of sulfhydryl groups by exchange reactions (p. 95). This is a more general type of reaction within cells, so that the cystine present in dietary protein or within cellular proteins undergoing degradation is interconvertible with cysteine, which is the form that enters the major pathways for metabolism. Cells contain a particular tripeptide, *glutathione*, which is the active compound in disulfide-sulfhydryl exchanges (Fig. 17-15). Reduced glutathione can be regenerated from the oxidized compound by the action of a glutathione reductase, which uses NADPH as the reducing agent.

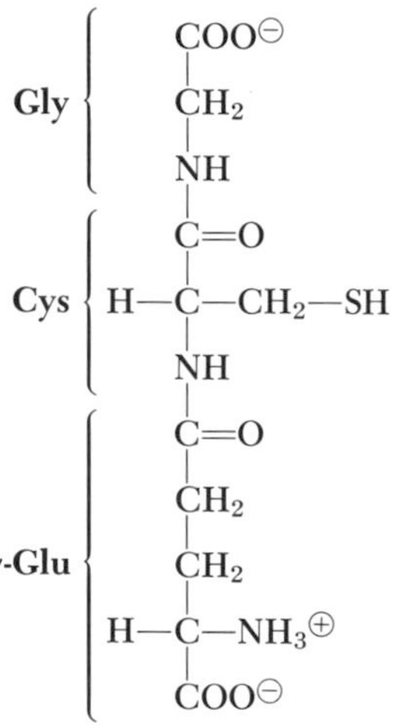

$$\text{a. } 2(GSH) + R-S-S-R' \xrightarrow[\text{exchange enzymes}]{\text{various -SH}}$$

$$G-S-S-G + R-SH + R'-SH$$

$$\text{b. } G-S-S-G + NADPH + H^+ \xrightarrow[\text{(FAD?)}]{\text{glutathione--NADPH} \atop \text{reductase}}$$

$$2(GSH) + NADP^+$$

Gly: COO$^{\ominus}$ — CH$_2$ — NH

Cys: C=O — H—C—CH$_2$—SH — NH

γ-Glu: C=O — CH$_2$ — CH$_2$ — H—C—NH$_3$$^{\oplus}$ — COO$^{\ominus}$

glutathione (GSH)
(γ-glutamylcysteinylglycine)

Figure 17-15 Glutathione is a widely distributed tripeptide used for the maintenance of sulfhydryl groups in proteins (reaction a). The necessary electrons are obtained from NADPH (reaction b).

Parenthetically, glutathione is formed by specific enzymes, not by the general processes of peptide formation. γ-Glutamyl cysteine is formed by one enzyme at the expense of ATP, and this dipeptide is conjugated with glycine at the expense of a second mole of ATP.

In any event, cystine enters metabolism as cysteine. Metabolism of cysteine produces pyruvate and inorganic sulfate as products, but there is still uncertainty over the route. In the soluble cytoplasm, cysteine is oxidized to cysteine sulfinate (Fig. 17-16), followed by transamination with α-ketoglutarate to form the corresponding keto compound, which releases SO_2 as sulfite, thereby yielding pyruvate. The sulfite is then oxidized by a *sulfite oxidase* to form inorganic sulfate. This oxidase is found on the endoplasmic reticulum, and the route of electron transfer is still unknown, except that it evidently involves one of the unusual cytochromes found in that organelle. The importance of sulfite oxidase is shown by the discovery of a human infant born without the enzyme and with such severe neurological impairment as to be almost functionally brainless.

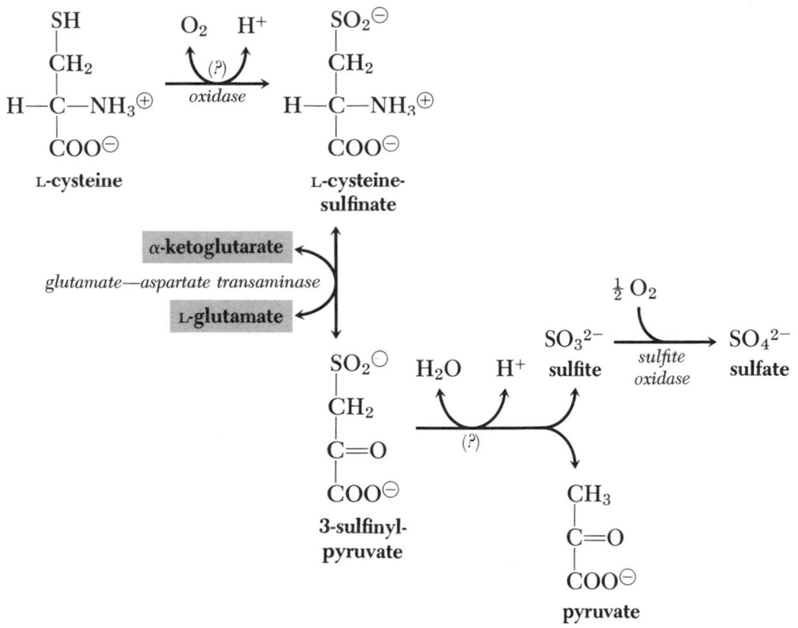

Figure 17-16   The metabolism of cysteine involves successive oxidations of the sulfur to sulfate, with an intervening removal of the nitrogen by transamination and the conversion of the carbon skeleton to pyruvate.

However, there is evidence that the major pathway of cysteine breakdown occurs in mitochondria, and that cysteine sulfinate is not an intermediate. The end products are still inorganic sulfate and pyruvate, so the stoichiometry of the overall route is the same, except that the nature of the enzymatic steps is unknown.

The two important points emerging from what is known are that the carbon skeleton of cysteine is introduced directly into the pathway of carbohydrate metabolism at the pyruvate stage with all three carbons convertible to glucose, and that the sulfur is mainly transformed to inorganic sulfate. Cysteine therefore provides a source of sulfate, which, as we shall see, is introduced into several structural components, especially complex polysaccharides (Chapter 24).

Now that the major route for cysteine metabolism is firmly in mind, it is probably safe to note that sulfur metabolism is considerably more complex and more poorly understood than has been implied. A variety of possible reactions have been reported, and the purpose of the routes is not at all clear in most cases. For example, a minor fraction of the sulfur of the body is excreted in the urine as thiosulfate and thiocyanate. These compounds apparently arise by the route shown in Figure 17-17. The carbon skeleton still appears as pyruvate in these routes, but the fate of the sulfur differs. We shall later mention some routes in which the carbon skeleton, as well as the sulfur, is used for the formation of other cellular constituents in small amounts.

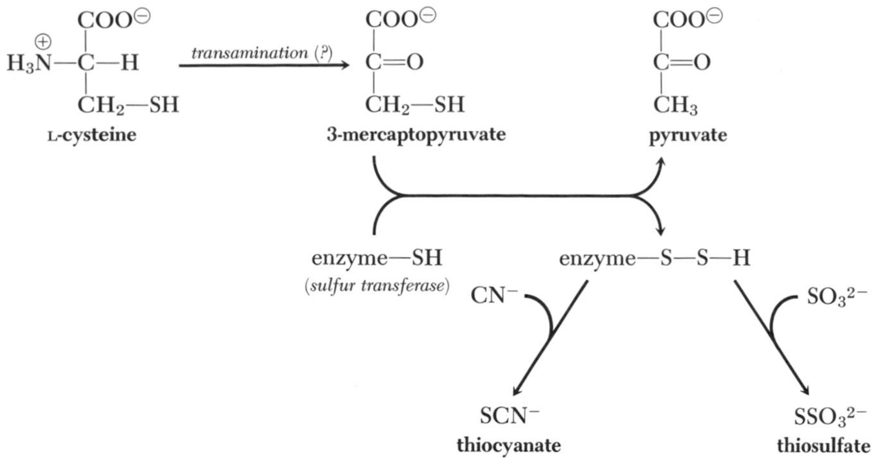

Figure 17-17 A minor route of cysteine metabolism involves an initial transamination without oxidation of the sulfur, forming 3-mercaptopyruvate. Another enzyme reacts with mercaptopyruvate to form pyruvate, with the sulfur remaining on the enzyme as a hydrodisulfide group. This group will react with a variety of anions, including cyanide and sulfite. Such reactions account for the trace amounts of thio compounds normally occurring.

Phenylalanine and Tyrosine

Nitrogen: Transferred to α-ketoglutarate, producing glutamate.

Carbons: Appear as acetoacetate and fumarate.

The aromatic amino acids have a metabolism as complex as that of the sulfur amino acids, but the major pathway is clearly defined. Phenylalanine is oxidized to form tyrosine, and tyrosine is further metabolized, with the ultimate formation of fumarate and acetoacetate. The sequence of oxidations by which the aromatic ring is broken is catalyzed by a number of enzymes localized in the soluble cytoplasm and the endoplasmic reticulum. These reactions therefore do not lead to the generation of high-energy phosphate. In fact, many of them are catalyzed by mixed-function oxidases and consume NADPH along with molecular oxygen. Of course, the fumarate and the acetoacetate eventually produced go into the regular metabolism, and since a mole of each is produced, the potential production of high-energy phosphate in the whole body is considerable, with three carbons convertible to glucose, four appearing as acetoacetate, and two appearing as CO_2.

The conversion of phenylalanine to tyrosine involves an oxidation by *phenylalanine hydroxylase.* It is an irreversible reaction, and *phenylalanine is an essential amino acid.* This reaction consumes molecular oxygen and at the same time causes the oxidation of a type of cofactor we have not seen before, *tetrahydrobiopterin* (Fig. 17-18). The mechanism of the reaction is unknown, but the cofactor presumably "activates" oxygen in some way typical of mixed-function oxidases. In any event, the reduced pterin must be regenerated by a second enzyme at the expense of NADPH.

We shall consider other pteridine coenzymes, which are *folic acid* derivatives, in some detail in the next chapter. The biological significance of the compounds to animals was first noted in an odd way. The pterins were known to occur in animal pigments, for example

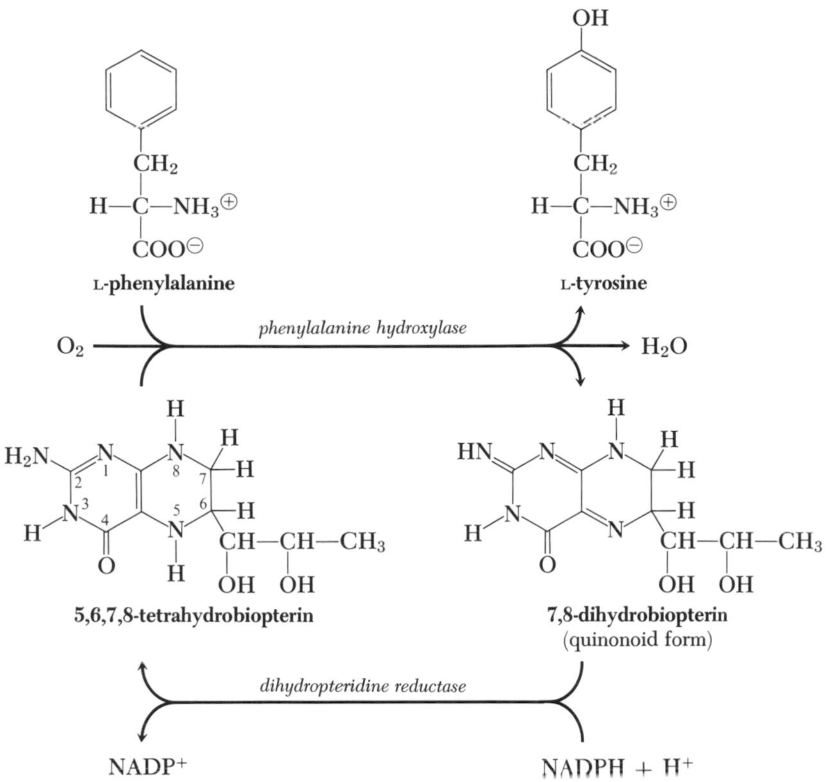

Figure 17-18   The conversion of phenylalanine to tyrosine by a mixed-function oxidase that uses a coenzyme, tetrahydrobiopterin, to activate molecular oxygen.

on the wings of butterflies. The basis for their chemistry was laid back in the late 1800's, but the subject remained stagnant until the 1930's, when one of the compounds, xanthopterin, was synthesized. This compound was shown to alleviate an anemia produced in rats fed on goat milk—a curious observation. More significantly, salmon in Washington hatcheries fed on diets in which yeast was used as a vitamin supplement also became anemic, but this anemia was corrected by feeding small amounts of xanthopterin. Obviously, the fish needed the insect pigment. With this background, it was much easier to connect the pteridines with the vitamin folic acid, which was isolated in the 1940's, and to show that it was chemically related. So much attention was concentrated on folate derivatives that the simpler pterins were neglected, and it was only in the last few years that the use of these compounds in oxidation-reduction reactions, such as the oxidation of phenylalanine, was demonstrated.

The major pathway for the metabolism of *tyrosine* begins with a transamination with α-ketoglutarate, catalyzed by a specific transaminase (Fig. 17-19). The resultant *p*-hydroxyphenylpyruvate is oxidized by another unusual oxidase, which contains ferrous iron and causes the addition of a full molecule of O_2. The enzymatic steps have not been characterized, but the result is a hydroxylation of the ring, an oxidative decarboxylation of the side chain, and a shift of the acetyl group on the ring. The resultant 3,4-dihydroxyphenylacetate has the trivial name

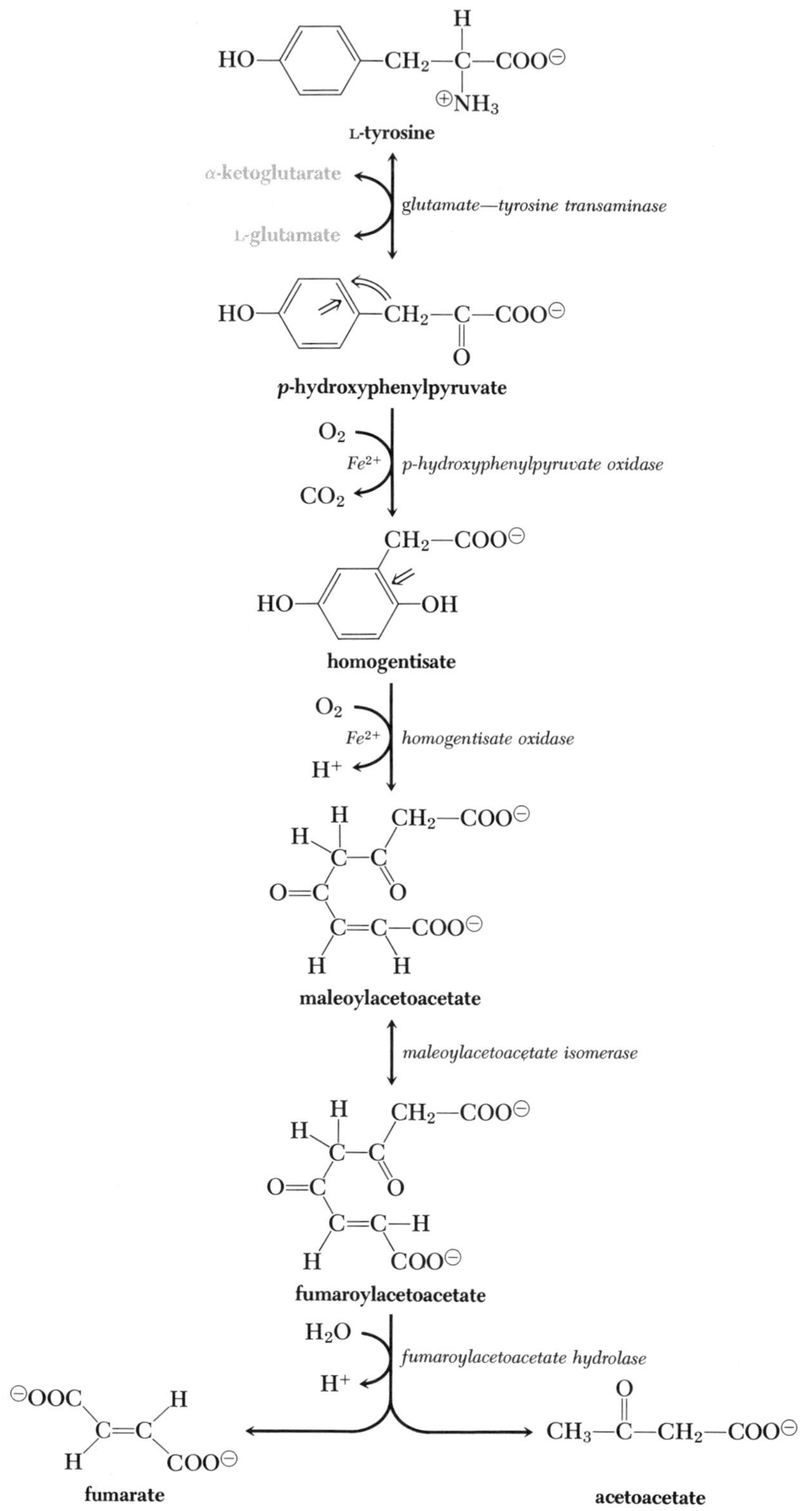

Figure 17-19   The major route for the metabolism of tyrosine.

of *homogentisate*. Homogentisate is attacked by still another oxidase containing ferrous iron, which cleaves the ring to form the *cis*-unsaturated compound, C-maleoylacetoacetate. An isomerase converts this to the *trans*-compound, C-fumaroylacetoacetate, and this is hydrolyzed to yield fumarate and acetoacetate.

This is a remarkably complex series of oxidations, parting from the normal types of enzymes we have encountered in other connections. However, rupture of an aromatic ring so as to yield the same metabolites encountered in fat and carbohydrate metabolism is in itself somewhat of a stunt. Does the fact that this can take place explain why proteins have phenylalanine rather than phenylglycine or phenylaminobutyrate? Who knows? Furthermore, we shall see that similar kinds of reactions are used to make several other kinds of aromatic compounds in small quantities, so the type of reaction has been evolved for more general purposes than the particular sequence we are considering.

A number of genetic defects in the metabolism of aromatic amino acids are known in humans. The most common one is the absence of phenylalanine hydroxylase. Phenylalanine accumulates in the blood and tissues. Part of it is transaminated, forming phenylpyruvate. This compound, as well as phenylalanine, is excreted in the urine. Hence the condition is called *phenylketonuria*. It is also known as phenylpyruvic oligophrenia, because the condition results in early neurological damage preventing normal intellectual development. Treatment consists of early avoidance of high concentrations of phenylalanine in the diet, but the problem is the same as in maple syrup urine disease. The body cannot synthesize phenylalanine, so some must be present to build normal proteins; it is difficult to gain a proper balance, and requires an expensive diet.

Phenylketonuria is unusually common for a genetic defect. The full-blown condition is only seen in homozygotes, yet it occurs once in each 10,000 births, which means that the heterozygotes, detectable only by deliberately loading them with phenylalanine and testing for blood levels of phenylalanine and phenylpyruvate, must be about one in 100 of the population. Such a prevalence makes one immediately suspicious that the prevalence is due to a balanced polymorphism, similar to the hemoglobinopathies. But under what circumstances could this condition possibly be advantageous? We don't know. Perhaps the conservation of phenylalanine had some advantage on some peculiar ancestral diet. Perhaps it is still advantageous in some regions of the world where food is frequently in short supply.

People with phenylketonuria also excrete some other aromatic compounds (Fig. 17-20), representing aberrations of the normal process of metabolism due to the high concentration of phenylpyruvate. The compound is reduced by a phenyllactate dehydrogenase in the liver, forming phenyllactate. Phenylpyruvate is also oxidized in the same way as is the normal p-hydroxyphenylpyruvate, except that the product is o-hydroxyphenylacetate, rather than the 2,5-dihydroxyphenylacetate that is homogentisate. The monohydroxy compound is not further metabolized, and appears in the urine.

The suspicion that the prevalence of phenylketonuria may reflect some occasional advantage to the heterozygotes is deepened by the fact that the other known genetic abnormalities of tyrosine metabolism are much less common, despite the fact that their consequences appear less grave.

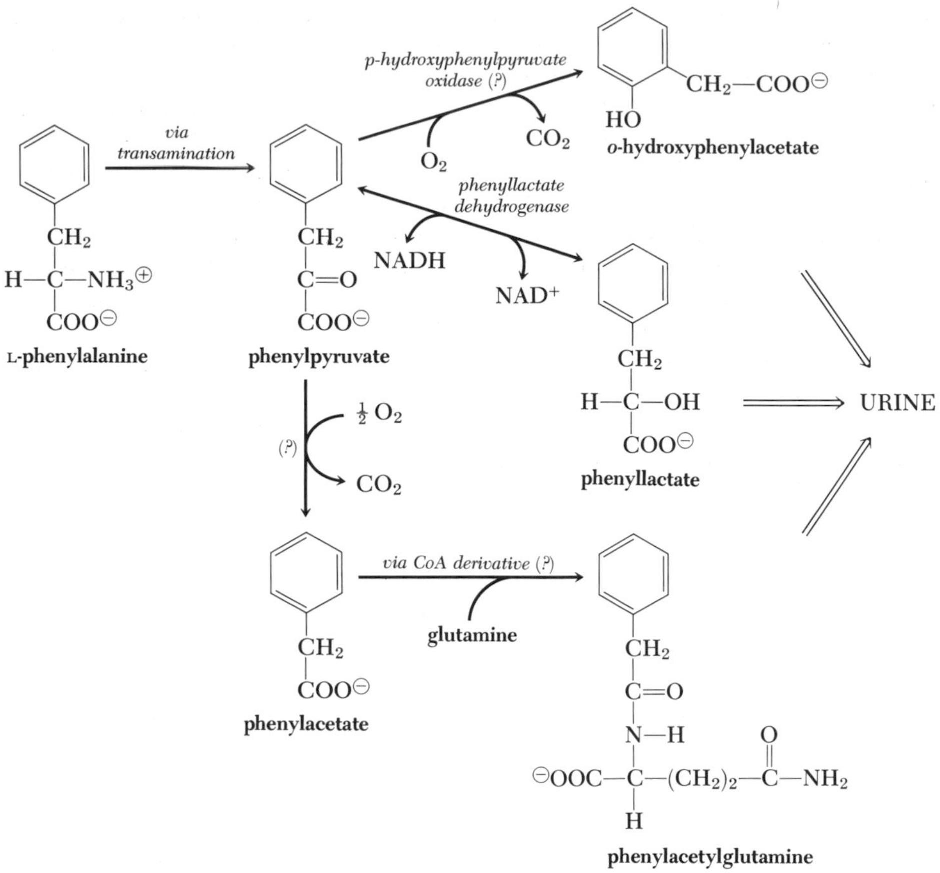

Figure 17-20   Ordinarily trivial routes of phenylalanine metabolism become more important in patients lacking phenylalanine hydroxylase, and a variety of products appear in the urine in increased amounts.

A single clear-cut case of tyrosinosis was reported by Grace Medes in 1932. The facts are consistent with the absence of p-hydroxyphenylpyruvate oxidase in this individual, because she excreted both tyrosine and the keto acid. The excretion of both increased with phenylalanine feeding, but homogentisate was metabolized at the normal rate. No obvious deleterious effects of the condition were observed. The case is of historical interest in that the metabolism of the aromatic acids was not well understood until painstaking studies were made of the metabolism of this one human. The compounds that did and didn't appear in the urine now appear obvious in light of our knowledge, but much of that knowledge is a result of just those measurements. This was an unusual patient, because there have been recent reports of a few more cases around the world who have been in serious difficulty in early childhood, with the enzyme defect causing kidney damage and early death. The probability of mutation ought to be as high for this enzyme as it is for phenylalanine hydroxylase, and the rarity of its genetic deletion reflects the absence of any advantage created by it for the heterozygote.

Finally, in about 1 in 200,000 births, infants are found whose urine turns black on standing due to the homogentisate they are secreting because they lack homogentisate oxidase. Homogentisate is a substituted hydroquinone, and these diphenols are notoriously susceptible to auto-oxidation, forming a mixture of highly-colored products. The condition is known as *alcaptonuria*, and individuals with it live until well into reproductive age with no difficulty other than whatever esthetic offense the darkening urine may represent. Many in their fourth or fifth decade will develop arthritis. The degeneration of the connective tissue in the joints is apparently associated with a deposition of pigment, presumably resulting from further oxidation of homogentisate in cartilage.

What do these genetic defects teach us? The active life possible with alcaptonuria shows that the oxidation of aromatic amino acids is not imperative for energy production. We have to stretch our imagination to visualize circumstances in which the few grams of available metabolite represented by these compounds might tip the balance. Those with phenylketonuria or tyrosinosis are in more difficulty, but their greater problems evidently arise from the accumulation of metabolites, which in themselves disturb metabolic processes, rather than from a failure to produce some needed compounds or an adequate supply of high-energy phosphate.

Tryptophan

Nitrogen: One appears as alanine, and one as NH_4^+.

Carbons: Three appear as alanine, four as crotonyl coenzyme A, one as formate, and three as CO_2.

Tryptophan is usually the least abundant of the amino acids in the diet, and is not a major substrate for the generation of high-energy phosphate. However, the unusual indole ring that it contains is used as a precursor for a variety of cellular components, as we shall see in Chapter 19. The formation of these substances is usually satisfied with a small fraction of the dietary consumption. The balance of the carbon skeleton is metabolized to CO_2 by way of alanine made from the side-chain and crotonyl CoA formed from the ring, which can be converted to glucose and acetoacetate, respectively. Metabolism begins with oxidation by O_2, catalyzed by *tryptophan pyrrolase*, an enzyme containing heme. This reaction opens the indole ring, and succeeding reactions oxidize the phenyl ring and open it, with the intermediate removal of carbons as formate and alanine (Fig. 17-21). The oxidations are of the same sort encountered in the metabolism of phenylalanine and tyrosine. We have already considered the disposal of alanine. We shall consider the fate of formate more completely in the next chapter. Suffice it to say now that it can be oxidized to CO_2 and H_2O, but the nature of the responsible enzyme mammalian tissues has not been settled.

The remaining reactions by which the carbons are converted to CO_2 and crotonyl CoA, which is an intermediate of fatty acid metabolism, involve different compounds than those we have seen, but no particular novelties in the types of reactions (Fig. 17-22).

It ought to be noted that the processes for cleaving the indole ring are irreversible, like those involved in opening the phenyl ring of tyrosine. There is no mechanism for forming the indole ring in mammals, and *tryptophan is an essential amino acid* that must be supplied in the diet.

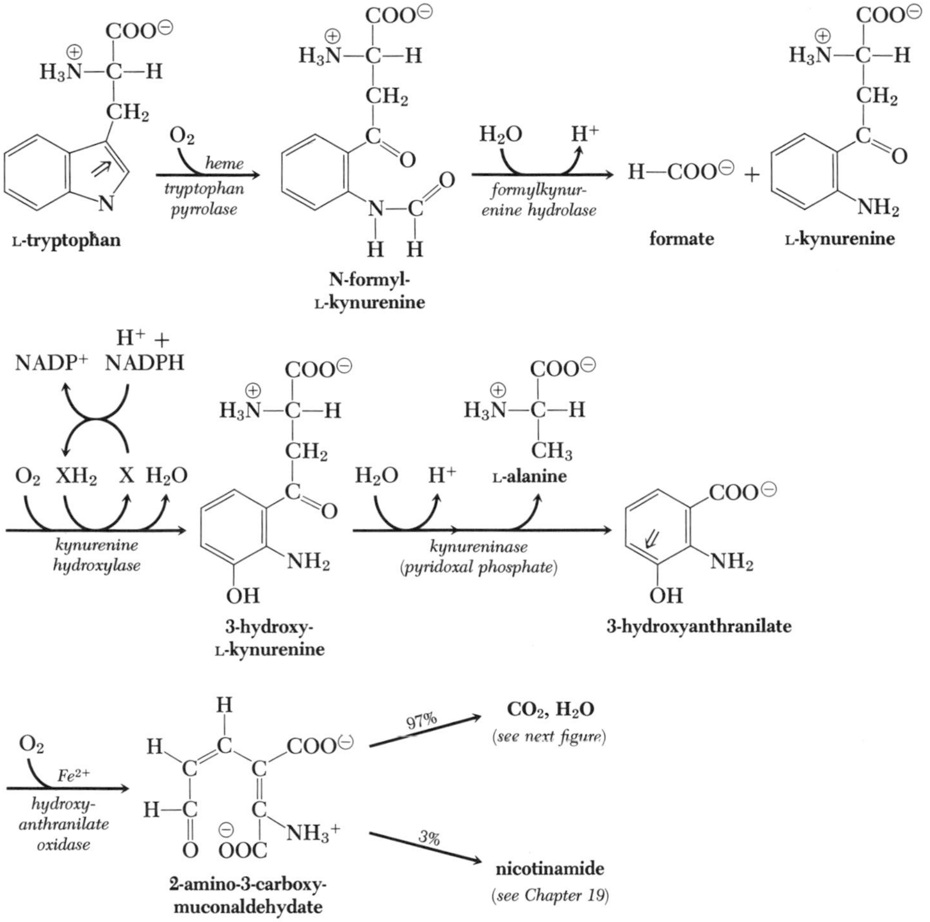

Figure 17-21   The route for the complete metabolism of tryptophan involves oxidases that cleave the two rings, with the concurrent formation of formate and of alanine. Part of the final intermediate shown here may be used to form nicotinate, but most is further metabolized, as shown in the next figure.

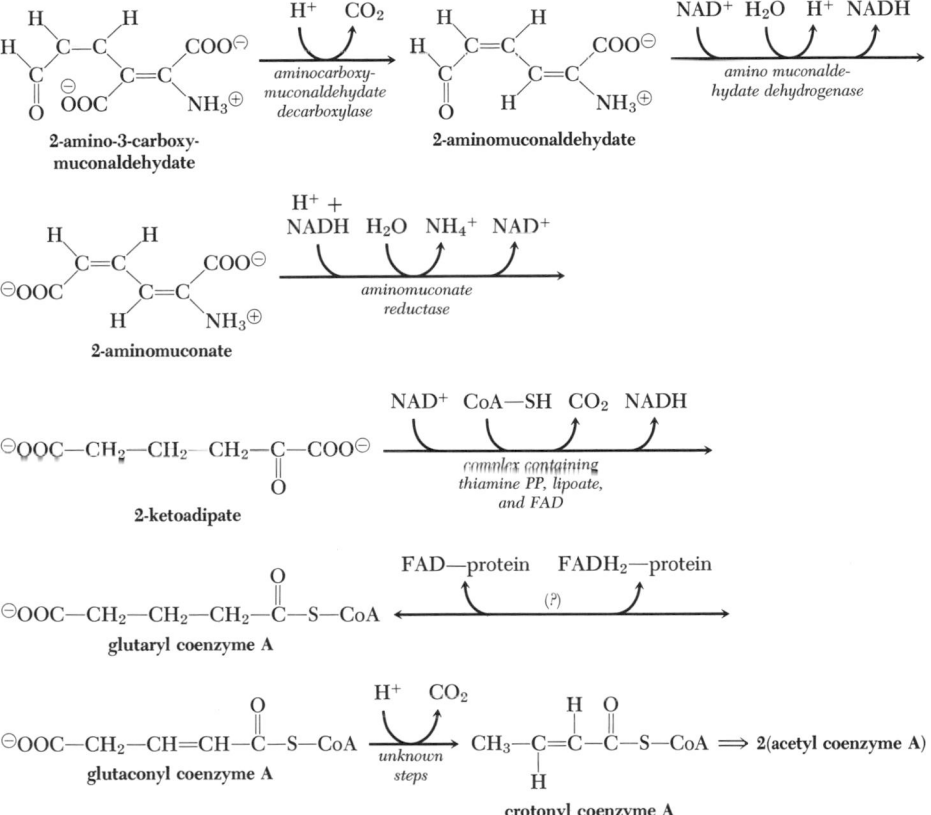

Figure 17-22   The final steps in the metabolism of tryptophan result in the liberation of NH_4^+ and the formation of crotonyl coenzyme A, which is a normal intermediate of fatty acid metabolism. The final three steps are also involved in the metabolism of lysine.

Lysine

Nitrogen: Both transferred to α-ketoglutarate forming glutamate.
Carbons: Two appear as CO_2 and four as crotonyl coenzyme A.

Lysine, like tryptophan, is metabolized *via* 2-ketoadipate, eventually forming crotonyl CoA. However, lysine is a relatively abundant amino acid. The amount consumed on most diets is comparable to the intake of the branched-chain amino acids.

Despite this relative abundance, the metabolism of lysine is still unsettled, except that it is clear the free amino acid must be modified before it can be a substrate for a transaminase, and that it is an *essential amino acid.* Apparently there are two routes in mammals (Fig. 17-23), one of which involves blocking the terminal ammonium group with an acetyl group, so as to make the α-ammonium group susceptible to transamination.

The other route involves a condensation with α-ketoglutarate and a reduction to form the compound, *saccharopine,* which is then cleaved with an oxidation on the opposite side of the nitrogen bridge so as to release glutamate and the semialdehyde of 2-aminoadipate. The semialdehyde is then oxidized to form 2-aminoadipate. This is the next higher homologue in the dicarboxylate amino acid series; it does not occur in proteins, but it will transaminate, forming 2-ketoadipate.

The route from acetylated lysine also arrives at 2-aminoadipate, but through the heterocyclic compound, *pipecolate.* According to available data, this route accounts for perhaps one-quarter of the total degradation of lysine, the balance probably passing through saccharopine, which is metabolized in liver mitochondria.

From an experimental standpoint, lysine is unusual in that the nitrogen of the amino acid will not equilibrate with the nitrogen of ammonia or other amino acids. An essential amino acid, such as valine, that is metabolized by an initial transamination will be constantly equilibrated with its corresponding 2-keto acid, even though the *net* reaction may be proceeding in the direction of breakdown of the amino acid. Therefore, nitrogen from glutamate will be appearing in the amino acid. This is demonstrated by the use of N^{15} as a tracer. Unfortunately, it is common to misstate the experimental inertness of lysine as implying a metabolic inertness compared to the other amino acids. We know this is not so. We eat comparable amounts of lysine in the diet, and it does not accumulate, which means that it must be metabolized at a rate comparable to that of the other essential amino acids.

The essentiality of lysine does raise some question about the proposed route through saccharopine for its degradation. As is implied in Figure 17-23, the nature of many of the steps has not been characterized, but most of the reactions between 2-ketoadipate and lysine ought to be reversible through this route, in contrast to the route through pipecolate, in which the hydrolysis of the terminal acetyl group is irreversible. The further metabolism of 2-ketoadipate begins with an irreversible oxidative decarboxylation, so there would be no possibility of making lysine from

---

Figure 17-23   Proposed routes for the metabolism of lysine. All of the reactions have been demonstrated, but the nature of some of the individual steps is not clearly defined and there is disagreement on the relative importance of the two routes shown. In any event, the carbon skeleton appears as 2-ketoadipate, which is metabolized as shown in the previous figure.

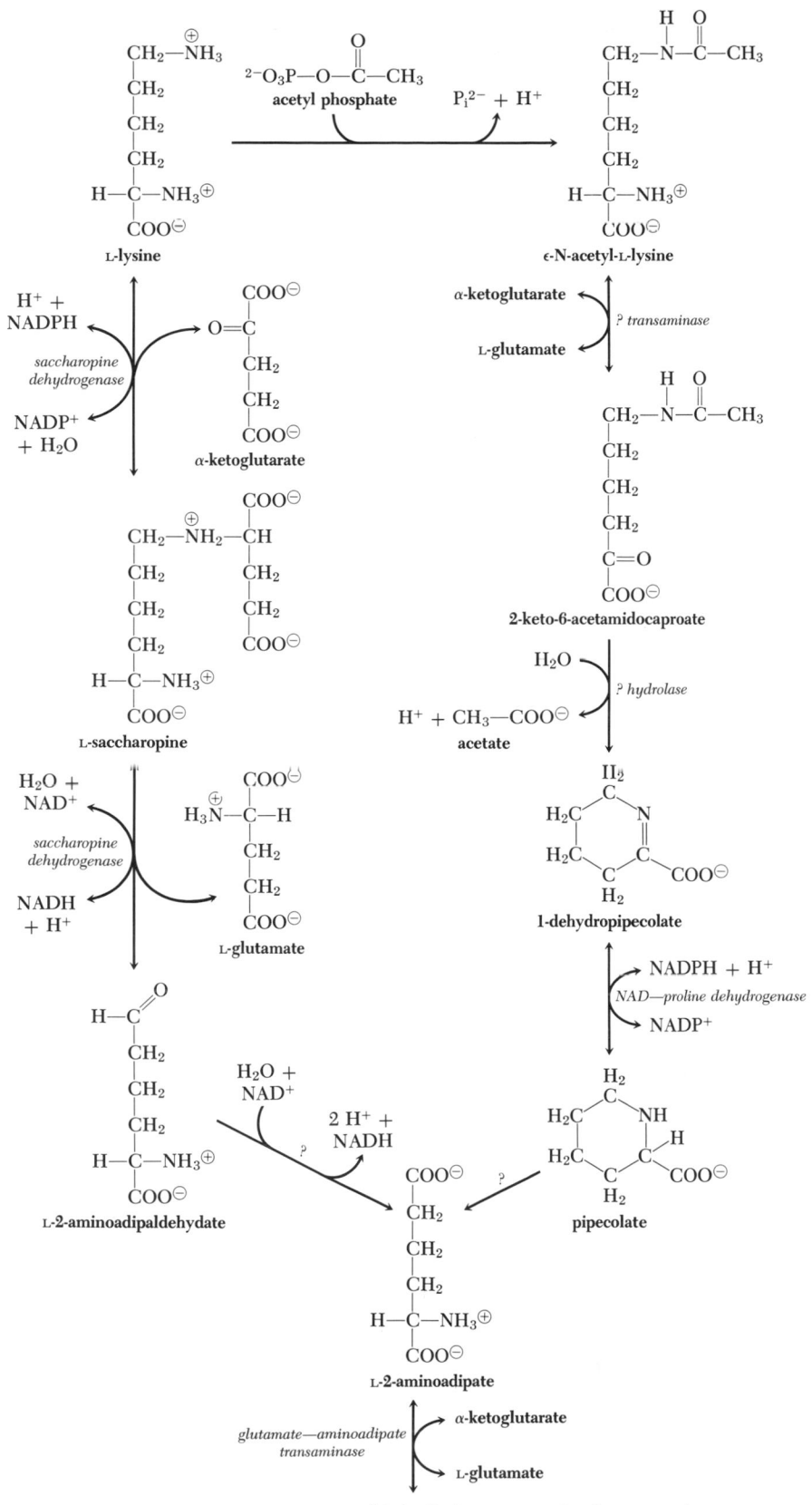

L-lysine

ε-N-acetyl-L-lysine

saccharopine dehydrogenase

α-ketoglutarate

L-saccharopine

2-keto-6-acetamidocaproate

saccharopine dehydrogenase

L-glutamate

1-dehydropipecolate

NAD—proline dehydrogenase

L-2-aminoadipaldehyde

pipecolate

L-2-aminoadipate

glutamate—aminoadipate transaminase

2-ketoadipate ⟹ crotonyl coenzyme A

393

glucose or fatty acids. However, it ought to be possible to make lysine from 2-aminoadipate, and therefore from 2-ketoadipate, which in turn can be formed from tryptophan. In fact, lysine cannot be replaced by tryptophan or by 2-amino-adipate. Either one of the oxidations in the saccharopine route involves a stronger oxidizing agent than NAD, or there is some further reaction coupled to it, or the evidence for the route is wrong after all.

Proline

Nitrogen and carbon skeleton: Appear as glutamate.

The transformation of L-proline to L-glutamate is simple. Proline is oxidized by a cytochrome-linked enzyme in mitochondria, presumably a flavoprotein, forming an unsaturated ring. Simple hydrolysis of the ring at the double bond would yield the semialdehyde of glutamate, and this reaction does spontaneously occur. However, the open ring form does not appear to a significant extent as an intermediate, because pyrroline carboxylate is directly oxidized to yield free glutamate in a reversible reaction (Fig. 17-24).

A stronger oxidizing agent is required to convert proline to pyrroline carboxylate because the equilibrium of the same reaction with nicotinamide nucleotides is strongly in favor of proline. In fact, an enzyme exists in the cytoplasm of the cell to catalyze the reduction of pyrroline carboxylate by NADPH, thereby providing a means for synthesizing proline, since the pyrroline carboxylate can be supplied from glutamate. Proline therefore need not be supplied in the diet. When dietary proline is supplied, its fate is that of glutamate, with three carbons convertible to glucose.

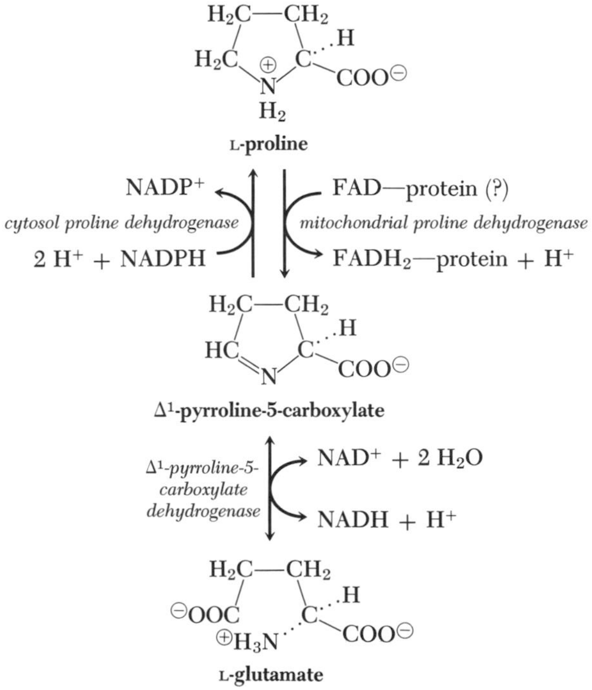

Figure 17-24   Proline may be converted to, or formed from, glutamate *via* an intermediate pyrroline-5-carboxylate. Proline is irreversibly oxidized by a mitochondrial flavoprotein or irreversibly formed through reduction of the pyrroline-carboxylate by NADPH. The different redox potential of the flavin and of the nicotinamide nucleotide account for the different position of equilibrium in the two cases. (Hydroxyproline is metabolized in a similar way, with the formation of 4-hydroxyglutamate. The disposition of this compound *via* the corresponding keto compound, 2-keto-4-hydroxyglutarate, is mentioned in the next chapter.)

Arginine

Nitrogen: Two appear directly as urea and one appears in glutamate.
Carbons: One appears directly as urea and five appear in glutamate.

As we have seen, arginine is hydrolyzed to urea and ornithine as a part of the process of urea synthesis. Ornithine is re-used in the urea cycle, but excess arginine derived from the diet or tissue proteins causes the formation of excess ornithine beyond the amount needed to sustain the urea cycle.

Ornithine, unlike lysine, transaminates with α-ketoglutarate:

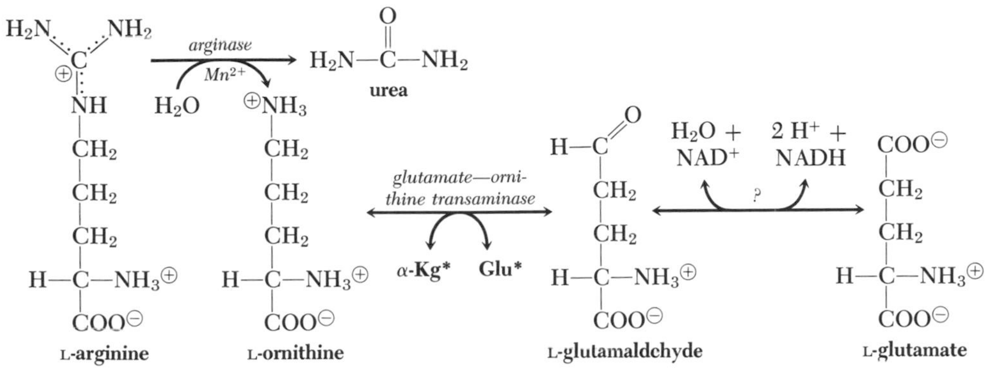

* α-Kg = α-ketoglutarate
Glu = glutamate

The transamination is unusual in that the terminal ammonium group of ornithine, not the α-group, is involved. The product is glutamaldehyde, which is oxidized to glutamate in a reversible reaction. (The aldehyde is the open-chain form of pyrroline carboxylate, the intermediate of proline metabolism. Proline, ornithine and glutamate are readily interconvertible in mammals.)

It follows that three of the carbons of arginine may be converted to glucose.

TOTAL STOICHIOMETRY FOR OXIDATION OF AMINO ACIDS

When protein is being metabolized, whether it came from the diet or from the tissues, all of the amino acids are presented more or less simultaneously, and in unequal amounts. We know from observation that this mixture of amino acids can be handled by some cells, such as those in the liver, and these cells must have a distribution of enzymes between the mitochondria and the cytosol that will make everything come out all right. Let us now tackle the problem of analyzing just how this occurs.

The amino acids from the mixture of proteins in most animal tissues are capable of sustaining the nitrogen economy without any supplementation from other sources. It is therefore a useful first approximation to consider what will happen if all of the amino acids found in something like beef muscle are metabolized together, and at rates proportional to the quantity ingested, so that they all disappear with nothing extra remaining to be metabolized alone. Such an assessment is probably also a rough approximation of what occurs during the constant turnover of amino acids from the individual's own tissues, since the amino acid composition of tissues does not differ greatly among vertebrates in general, and even less among mammals.

Initial Stoichiometry

Let us consider the fate of 1000 millimoles of the mixture of amino acids obtained by hydrolysis of the proteins of beefsteak. The average formula weight of the amino acid residues in these proteins is near 110, so this quantity of amino acids will be obtained from 110 grams of protein, which in turn is the amount in about 600 grams of raw meat. Since some of the amino acids contain more than one nitrogen atom, the total nitrogen content of the protein is near 1390 milliatoms. (The quantity is not known with this precision for real foods, but it is more convenient to calculate with integer values. As a frame of reference it is helpful to know that the quantity we are dealing with is about what would be found in the daily intake of an adult on a protein-rich diet.)

The calculations that follow are based on the following composition (Orr, M. L., and B. K. Watt: *Amino Acid Content of Foods*. Home Economics Research Report No. 4, U.S.D.A. [1957]), given in millimoles of each residue per 1000 millimoles of all residues:

Ala	82.5	His	28.5	Pro	54.5
Arg	47	Ile	51	Ser	51
° Asx	89	Leu	79.5	Thr	47
Cys	13	Lys	76	Trp	7.5
° Glx	131	Met	21	Tyr	24
Gly	105	Phe	32	Val	60.5

amide nitrogen (= Asn + Gln) = 110

° Asx = Asn + Asp; Glx = Gln + Glu.

Using the routes for metabolism developed in the preceding discussion (and anticipating in some cases those discussed in the next chapter), a summation of the initial reactions will give a result like this:

CONSUMED (except initial Asp, Glu)

780 (amino acids)
857 O_2
570 (α-ketoglutarate)

PRODUCED (plus initial Asp, Glu)

nitrogenous compounds	remaining carbons	other products
°47 (urea)	206 (pyruvate)	34 SO_4^{2-}
377 NH_4^+	180 (succinate)	ca. 2500 ~ P
831 (glutamate)	56 (fumarate)	
89 (aspartate)	84 (acetoacetate)	
	200 (3-hydroxybutyrate)	
	401 CO_2	

° Some urea is formed directly from the arginine found in the protein, hence the 47 millimoles of urea appearing at this stage.

The value for high-energy phosphate is uncertain because of lack of knowledge of the precise route of metabolism of a few of the amino acids. The correct maximum yield is not likely to differ by more than 10 per cent of the value given, and this uncertainty does not affect the general principles being illustrated.

The balance that has been struck through simple, although tedious, arithmetic makes the metabolism of the 20 amino acids look much simpler. The problem now becomes one of analyzing the way in which urea is formed from the listed nitrogenous compounds, and at the same time showing how glucose may be made from pyruvate and the intermediates of the citric acid cycle. Although these processes occur together, it is convenient to consider them separately and then combine the results.

Disposal of Nitrogen

In order to form urea, there must be equal amounts of NH_4^+ and of aspartate available. The initial reactions summarized in the preceding produce 1296 milliatoms of nitrogen in the form of glutamate, aspartate, and NH_4^+, but the bulk is as glutamate (830 millimoles). The nitrogen must be redistributed so as to make a total of 648 millimoles each of NH_4^+ and aspartate:

	glutamate	aspartate	NH_4^+	total amino N
required	0	648	648	1296
at hand	830	89	377	1296
necessary change	− 830	+ 559	+ 271	1296

These adjustments are easily made through the action of glutamate dehydrogenase and glutamate-aspartate transaminase in mitochondria:

1. $\quad$ 271 (glutamate) + 136 $O_2 \longrightarrow$ 271 NH_4^+ + 271 (α-ketoglutarate) + 813 ~ P
2. $\quad$ 559 (glutamate) + 559 (oxaloacetate) $\longrightarrow$
$$559 \text{ (aspartate)} + 559 \text{ (α-ketoglutarate)}$$

So far so good. This gets rid of the glutamate and produces the necessary precursors for urea synthesis. However, it also requires 559 millimoles of oxaloacetate, and where is this to come from? The answer is that oxaloacetate can be made from the intermediates of the citric acid cycle that also are formed from the amino acids. We could use any of these for the purpose at hand and eventually arrive at the same conclusion, but to make it easier let us assume that all of the pyruvate, fumarate, and succinate is used to provide oxaloacetate in the mitochondria, with the remainder of oxaloacetate made by oxidizing part of the α-ketoglutarate:

3. $\quad$ 206 (pyruvate) + 56 (fumarate) + 180 (succinate) + 117 (α-ketoglutarate) +
$$89 \text{ CO}_2 + 383 \text{ O}_2 \longrightarrow 559 \text{ (oxaloacetate)} + 1915 \sim P$$

(CO_2 is being consumed to carboxylate pyruvate, with only part provided by the oxidation of α-ketoglutarate.)

We have now definitely provided everything for the synthesis of urea:

4. $\quad$ 648 NH_4^+ + 648 (aspartate) + 648 CO_2 + 2592 ~P $\longrightarrow$
$$648 \text{ (urea)} + 648 \text{ (fumarate)}$$

Of course, this process involves the formation of citrulline in mitochondria, which, together with the aspartate formed in the preceding reactions, moves into the cytosol where arginine and fumarate are made, followed by hydrolysis of the arginine into urea and ornithine. The ornithine diffuses back into mitochondria to complete the cycle, as outlined earlier in the discussion of urea synthesis.

If we now add the original stoichiometry to reactions (1) through (4), we have this situation:

CONSUMED

1000 (amino acids)
1240 O_2
336 CO_2

PRODUCED

nitrogenous compounds	*remaining carbons*	*other products*
695 (urea)	144 (α-ketoglutarate)	34 SO_4^{2-}
	648 (fumarate)	2600 $\sim$ P
	84 (acetoacetate)	
	200 (3-hydroxybutyrate)	

Disposal of Carbons

Urea, acetoacetate, and 3-hydroxybutyrate diffuse into the blood, with the urea being excreted by the kidney and the 3-oxybutyrates being utilized by peripheral tissues. The remaining carbons are present as 143 millimoles of α-ketoglutarate and 648 millimoles of fumarate, both of which are precursors of glucose. Now, α-ketoglutarate can be oxidized within mitochondria to form malate, which can diffuse into the soluble cytoplasm:

5.   144 (α-ketoglutarate) + 144 O_2 $\longrightarrow$ 144 (malate) + 864 $\sim$ P + 144 CO_2

Since the cytosol contains fumarase, malate and fumarate are effectively equilibrated, and the result is that there is now a sum of 792 millimoles of malate available in the cytosol for gluconeogenesis.

Malate is an ideal precursor for glucose because the transformation involves no net transfer of electrons from or to other compounds. This is so because the oxidation of malate produces the NADH that is later utilized in converting phosphopyruvate to triose phosphate. Therefore, the final reaction completing our stoichiometry for metabolism of beefsteak proteins is:

6.   792 (malate) + 1588$\sim$P $\longrightarrow$ 396 (glucose) + 791 CO_2

Total Stoichiometry

When we include reactions (5) and (6) in the stoichiometry, the final result is:

1000 (amino acids) + 1383 O_2 $\longrightarrow$ 695 (urea) + 396 (glucose)
+ 84 (acetoacetate) + 200 (3-hydroxybutyrate) +
34 SO_4^{2-} + 600 CO_2 + ca 1900 $\sim$P

(It ought to be made clear that the liver produces varying proportions of acetoacetate and 3-hydroxybutyrate in response to changing nutritional circumstances.

The 84 : 200 ratio used here, which was selected for ease in calculation, is about in the middle of the observed range.)

The important thing about a stoichiometry based on such complicated sets of reactions is its validity. Does this equation have any relevance to real events oc curring in the intact animal? There are some experimental findings that it helps explain, but there is also a direct test available through old observations. When it became clear that glucose could be formed from part of the amino acids of ingested protein, a direct measurement of the conversion was made in the following way: Animals were caused to excrete glucose in the urine, either by removing the pancreas so as to create diabetes or by treating them with *phlorhizin*, which prevents the reabsorption of glucose from the filtered plasma in the kidney. (Phlorhizin is a polyphenolic glycoside found in the bark of apples, cherries, and other rosaceous plants.) When properly performed, the experiments would result in nearly quanti- tative recovery of any glucose formed. The amount of amino acids metabolized to produce the glucose could be measured by analyzing the excretion of nitrogen. The results were expressed as a *D : N ratio*, the weight of glucose (dextrose) formed per weight of nitrogen excreted.

Now, the theoretical D : N ratio, according to our stoichiometry, would be 396×180 grams of glucose per $695 \times 2 \times 14$ grams of excreted N, which calcu- lates out as 3.66. The actual observed values on animals fed meat were 3.63 in dogs, 3.68 in humans treated with phlorhizin, and 3.63 to 3.73 in humans with diabetes. This agreement with theory is almost too good to be true!

Another test is through the relative value of protein as a fuel, as compared to fats and carbohydrates. This will be discussed more extensively in Chapter 22, when we shall use the theoretical balance in reasoning about the total metabolic economy. Since we have information on the amounts of glucose and of the 3-oxybutyrates produced, we can also estimate the potential energy available to the peripheral tissues from amino acid metabolism. Another important finding is that amino acid metabolism in the liver produces high-energy phosphate, even though gluconeogenesis is occurring. However, the P : O ratio is only 0.69, which means that the consumption of oxygen in the liver for amino acid metabolism produces less ATP for use in chemical syntheses than it does in other tissues in which most of the carbon passes through the complete citric acid cycle.

We could have made these important points by simply stating the final stoi- chiometry without wading through the details of its development. However, there are two reasons for understanding *how* the liver achieves the particular result. First, by successfully tackling something as complex as amino acid metabolism, we gain confidence in a rational approach to an understanding of the entire metabolic economy. One of the most exciting challenges to biochemists today is the correlation of our knowledge of what used to be regarded as independent metabolic processes. Secondly, when we get everything laid out, we see that the entire metabolism of amino acids forms an integrated whole that is not at all obvious from the final stoichiometry or from consideration of the individual processes.

Fig. 17-25 summarizes the entire process, and there are a number of striking things that are apparent:

1. Over half of the total nitrogen in the amino acids passes through glutamate. Two-thirds of the glutamate is used to form aspartate by transamination, and only one-third is oxidatively deaminated by the glutamate dehydrogenase reaction.

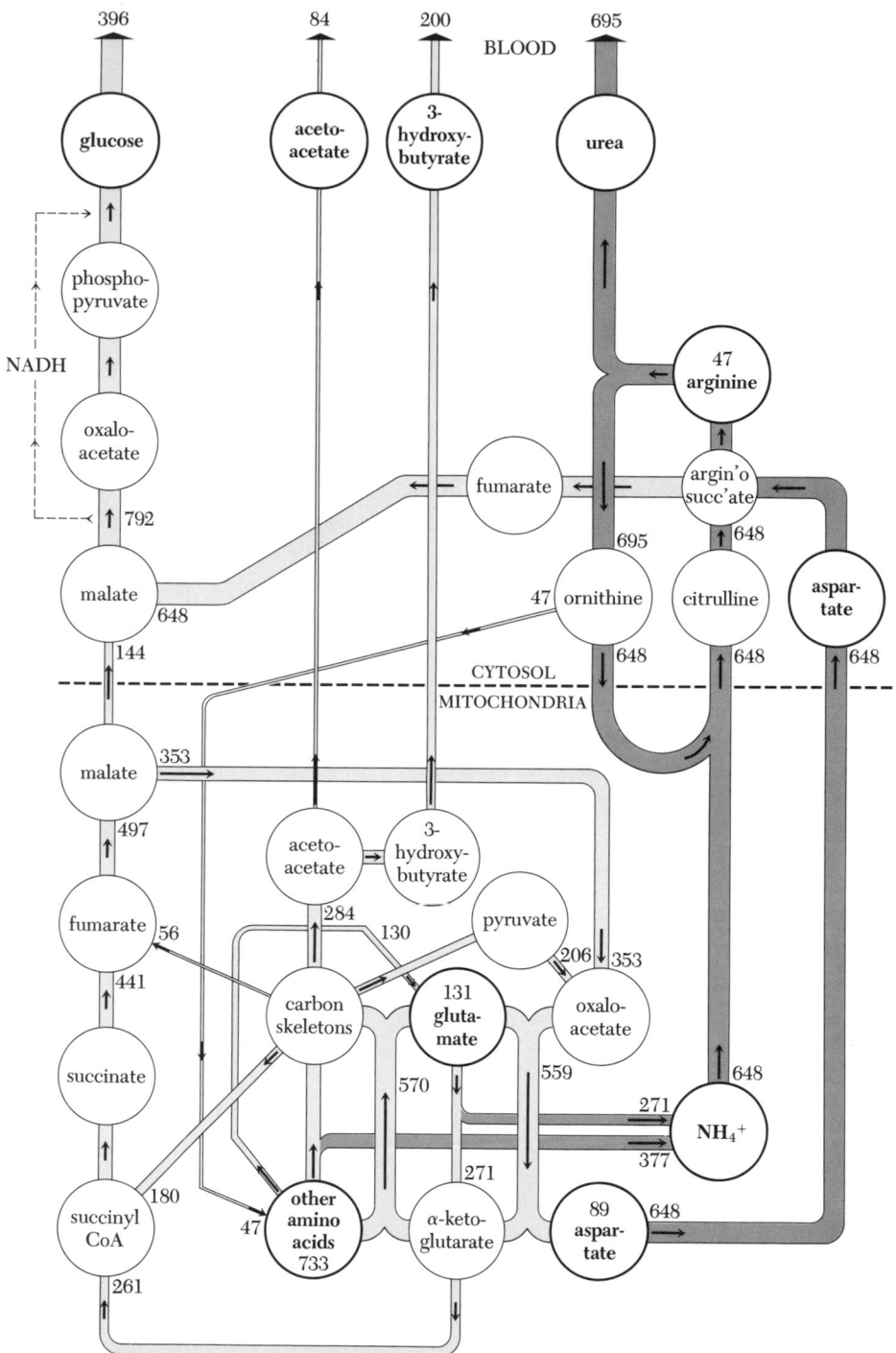

Figure 17-25 *See legend on opposite page.*

2. The reactions of the citric acid cycle are an integral part of the overall mechanism of nitrogen disposal in the liver. This puts an extra load on the terminal reactions of the cycle, particularly those following succinyl coenzyme A. Put another way, amino acid metabolism pre-empts part of the capacity of the liver to oxidize acetyl coenzyme A through the complete citric acid cycle, and this probably accounts for the evolution of the release of acetyl groups from the liver in the form of the 3-oxybutyrates to be oxidized in other tissues.

3. Aspartate is the form in which most of the carbons used for gluconeogenesis are transferred from the mitochondria to the cytosol. This transport is a function for aspartate in addition to its provision of half the nitrogen for urea synthesis. We now see an excellent reason for the division of the reactions of urea synthesis between the mitochondria and cytosol and the use of aspartate as the final nitrogen donor. Urea synthesis and gluconeogenesis from amino acids are tied together in a single process.

4. Only a small fraction of the carbons must cross the mitochondrial membranes in the form of malate. Experiments with isolated mitochondria show that this compound can cross, although there is some reason to believe that it is an active process requiring the expenditure of high-energy phosphate.

The amount that is involved is so small that it raises an interesting point. What would have to be changed so that no malate crossed the mitochondrial membrane, and all of the carbons for gluconeogenesis were carried in the form of aspartate? Essentially, it would involve a shift in the composition of the amino acid mixture so that there is more ammonium ion available directly. It so happens that amino acids are partially deaminated in other tissues and by the microorganisms of the gut during digestion. We don't have enough information to be able to say that the effect is such as to eliminate the necessity for moving carbons as malate, but it is a provocative possibility.

One built-in balance does become evident, however. The use of glutamine as a means of transporting nitrogen from other tissues to the liver provides a balanced supply of aspartate and ammonium ion for urea synthesis. The ammonium ion appears directly upon hydrolysis of glutamine, and the aspartate is created from the remaining glutamate. The latter conversion involves transamination with oxaloacetate, and replacement of the oxaloacetate by oxidation of α-ketoglutarate through the citric acid cycle reactions. This is a part of the scheme shown in Figure 17-25,

---

Figure 17-25 Outline of the metabolism in the liver of a mixture of amino acids similar to that derived from beef muscle as it occurs when gluconeogenesis is proceeding at a maximum rate. The original amino acids and the products, other than CO_2, are shown in bold-face. The numbers appearing within the circles are the millimoles of the amino acid originally present per mole of total amino acid. The numbers alongside the arrows indicate the millimoles of intermediate undergoing the particular transformations per mole of total amino acid. The width of the arrow also indicates this number, and not the relative quantity of carbon atoms involved (except in the case of glucose, where the arrow is not shrunk in half to match the combination of two moles of triose phosphates). The pathway of the nitrogen atoms is indicated by heavy shading—some of these transformations also involve disposition of the carbon atoms. Light shading indicates transformations involving only carbon atoms.

Two striking features shown by the diagram are the heavy utilization of aspartate for transporting carbon skeletons for gluconeogenesis as well as nitrogen for urea synthesis from the mitochondria to the cytosol, and the utilization of the latter steps of the citric acid cycle, but not the complete cycle.

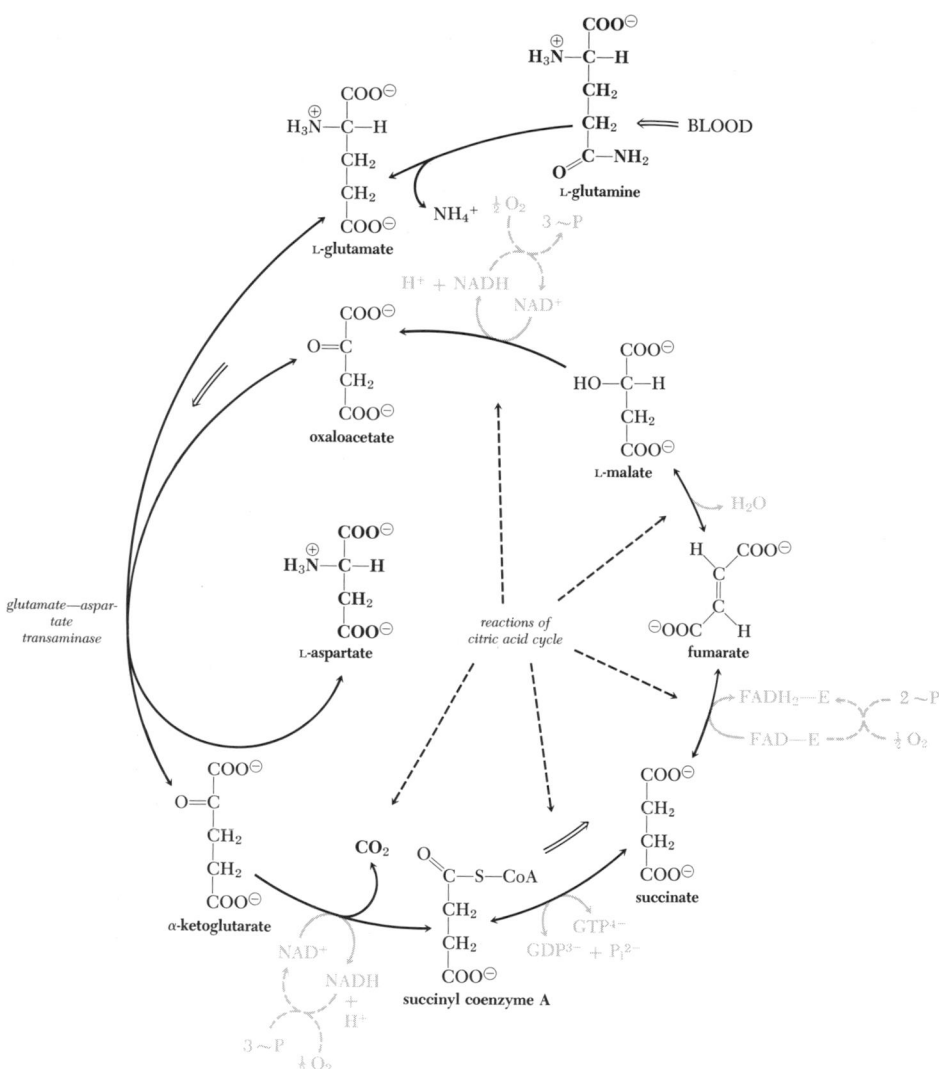

Figure 17-26   The metabolism of glutamine in the liver. The advantage of glutamine as a carrier of nitrogen from other tissues may lie in the ability to convert it to equimolar amounts of NH_4^+ and aspartate for urea synthesis, with the intermediate conversions of the carbon skeleton providing sufficient high-energy phosphate both for urea synthesis and for the conversion of the resultant fumarate to glucose.

and is detailed separately in Figure 17-26. It may well be that this is one of the reasons glutamine is used as a nitrogen carrier. (Asparagine would also provide equivalent amounts of aspartate and ammonium ion, but without the production of high-energy phosphate to maintain urea synthesis and gluconeogenesis within itself, whereas high-energy phosphate is produced by the oxidations involved in the formation of aspartate from glutamate.)

D-AMINO ACIDS

The coding for protein synthesis is directed specifically to the incorporation of L-amino acids into peptide chains. However, amino acids of the D-configuration occur fairly widely in plants and microorganisms, and are apparently used to make small peptides that will resist metabolism, therefore being useful in cell walls and as inhibitors to other organisms. Many of the antibiotics are peptides of this kind. For example, *Bacillus brevis* produces a variety of cyclic peptides lumped under the name *gramicidin*. Among these is gramicidin J_1, which contains one mole each of D-ornithine, L-valine, L-ornithine, D-phenylalanine, D-leucine, L-phenylalanine, and L-proline, in that order, with the L-proline combined with the initial D-ornithine to complete the closed peptide ring. (Gramicidin is used largely as a topical antibiotic.) The occurrence of the D-amino acids in animal tissues is evidently very limited. The only clear-cut example is the existence of D-serine in some invertebrates.

Be that as it may, D-amino acids will be presented to animals, either in ingested plants and microorganisms or by synthesis in the flora of the gut, and these compounds are metabolized. The route is through a simple oxidation by a flavoprotein, *D-amino acid oxidase*, found in the liver and kidneys, which catalyzes an oxidative deamination (Fig. 17-27). The 2-ketocarboxylate formed by the reaction may be degraded through the regular routes or may be converted to the L-amino acid by transamination with glutamate, in which case there has been an effective conversion of the D-form to the L-isomer. This may be a significant route when the supply

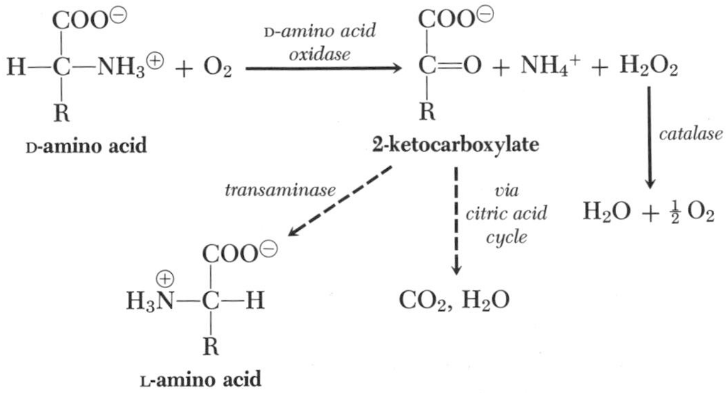

Figure 17-27   D-Amino acids are oxidized to NH_4^+ and the corresponding 2-keto compound. Molecular oxygen is consumed, with one atom appearing as hydrogen peroxide, which may be cleaved by the action of catalase, as shown, or may be used as an oxidizing agent by various peroxidases. The keto-carboxylate is identical to the intermediate of metabolism of L-amino acids, which may be formed from it in many cases, or it may be oxidized.

of amino acids is low, because animals lacking some essential amino acids (especially phenylalanine, tyrosine, tryptophan, and methionine) will grow if the D-isomer of the missing amino acid is fed.

D-amino acid oxidase is interesting in another connection. It is an example of a metallo-flavoprotein that produces hydrogen peroxide as a product, probably through a sequence involving one-electron reductions of flavoprotein pairs:

$$\text{protein-(FAD)}_2 + \text{substrate} \longrightarrow \text{oxidized substrate} + \text{protein-(FADH} \cdot)_2$$
$$\text{protein-(FADH} \cdot)_2 + O_2 \longrightarrow \text{protein-(FAD)}_2 + H_2O_2$$

Hydrogen peroxide is used as an oxidizing agent in a minor class of reactions catalyzed by enzymes known as peroxidases, of which we shall encounter some examples later. More importantly, it is decomposed to water and oxygen by a ubiquitously distributed enzyme, *catalase*, which is a heme protein:

$$H_2O_2 \longrightarrow H_2O + 1/2\ O_2$$

This sort of sequence produces no high-energy phosphate and is mainly of value for eliminating compounds, or converting them to useable metabolites.

There are also L-amino acid oxidases of similar mechanism known, but they have a limited occurrence in animal tissues. One ought to be mentioned because of its widespread use as an analytical and preparative tool. This is the L-amino acid oxidase occurring in snake venoms. It is not clear what its especial value is to the snakes, since the lethal action of the venoms is due more to the lecithinases and neurotoxins they contain, but it is especially useful to biochemists, because the venom is a relatively concentrated source of the enzymatic activity with far fewer contaminating enzymes than are found in most tissues.

AMINO ACID ANTAGONISTS

Amino acid molecules, like those of other metabolites, may be chemically modified in such a way that they will bind to enzymes, but be incapable of performing all of the functions of the natural molecule and therefore be inhibitors. There has been much interest in antagonists for the amino acids in particular, because of the hope of finding a compound that would block the metabolism of malignant tumors without affecting the host. The hope has not been realized.

With some modifications, the mimicry is very effective, and the amino acid analogue will go through several steps of the regular metabolic route. For example, p-*fluorophenylalanine* so closely resembles ordinary phenylalanine that it is incorporated extensively into proteins of animals to which it is administered. These proteins frequently retain their functional characteristics, but at least some enzymes have lower than normal activity, and the cumulative impairment of function is such that 1 mg of fluorophenylalanine per gram, dry weight, of diet is toxic.

Similarly, *ethionine*, the S-ethyl analogue of methionine, is extensively incorporated into proteins, but the compound is toxic.

Amino acid antagonism has had an unexpected effect on the flour milling industry. It used to be a practice to treat flour with nitrogen trichloride in order

p-fluorophenylalanine 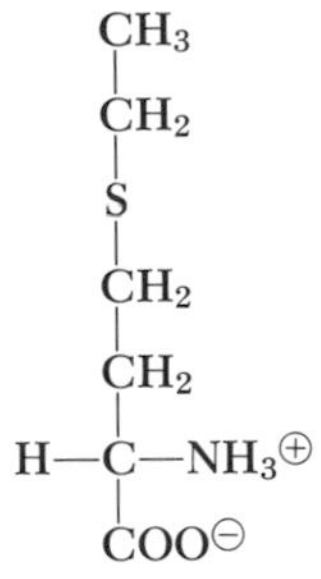   ethionine 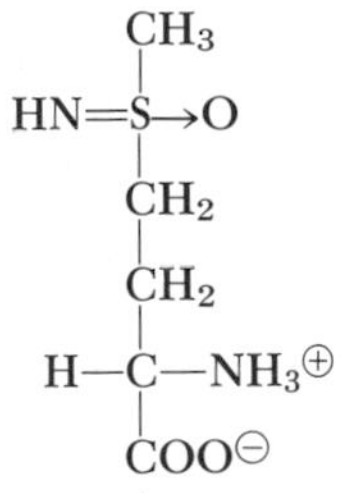   methionine sulfoximine

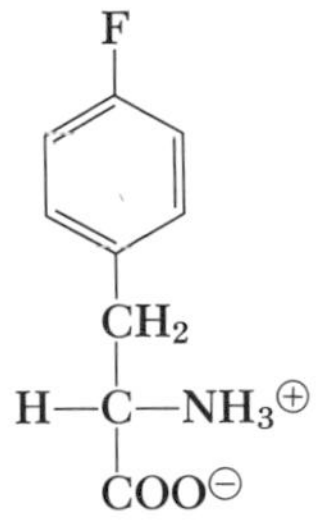

to bleach it to the pristine whiteness associated with cleanliness and gracious living by housewives. In 1946 it was shown that this treatment was responsible for running fits in dogs, a condition of puzzling etiology known for some decades before that, which occurred in dogs that were fed flour products. Not long thereafter, it was found that nitrogen trichloride reacted with methionine residues in the flour to form methionine sulfoximine. This compound is not only an antagonist of methionine, but it also inhibits glutamine synthetase, and the latter effect may well account for the neurological effects of the compound.

Treatment of flour for human consumption with nitrogen trichloride was quietly dropped, and current sociological problems cannot be blamed on this cause. It is not known whether the practice did cause occasional running fits or other aberrations in human behavior in the past.

Recapitulation of types of reactions

1. Amines, including amino acids, may be deaminated through an oxidation by NAD:

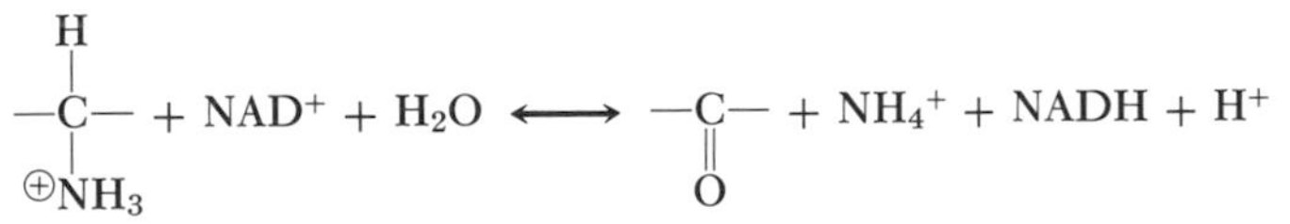

$$-\overset{\overset{\text{H}}{|}}{\underset{\underset{\oplus}{\text{NH}_3}}{\text{C}}}- \;+\; NAD^+ + H_2O \;\longleftrightarrow\; -\overset{\text{C}}{\underset{\overset{||}{\text{O}}}{}}- \;+\; NH_4^+ + NADH + H^+$$

Example: The reaction catalyzed by glutamate dehydrogenase.

2. 3-Hydroxy amines may be deaminated by an initial dehydration, followed by hydrolytic cleavage of the resulting imine:

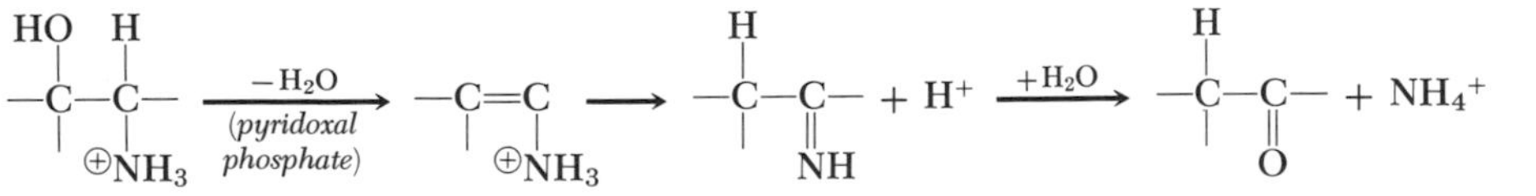

$$-\overset{\overset{\text{HO}}{|}}{\text{C}}-\overset{\overset{\text{H}}{|}}{\underset{\underset{\oplus}{\text{NH}_3}}{\text{C}}}- \xrightarrow[\substack{(pyridoxal \\ phosphate)}]{-H_2O} -\text{C}=\underset{\overset{|}{\underset{\oplus}{\text{NH}_3}}}{\text{C}} \longrightarrow -\overset{\overset{\text{H}}{|}}{\text{C}}-\underset{\overset{||}{\text{NH}}}{\text{C}}- + H^+ \xrightarrow{+H_2O} -\overset{\overset{\text{H}}{|}}{\text{C}}-\underset{\overset{||}{\text{O}}}{\text{C}}- + NH_4^+$$

Example: The reactions catalyzed by serine dehydratase and threonine dehydratase.

3. Coenzyme A may be transferred from one acyl group to another:

$$R-\overset{\overset{\displaystyle O}{\|}}{C}-S-CoA + R'-COO^{\ominus} \longleftrightarrow R-COO^{\ominus} + R'-\overset{\overset{\displaystyle O}{\|}}{C}-S-CoA$$

Example: The reaction catalyzed by acetoacetate-succinate CoA transferase.

4. The carbon skeletons of some compounds may be rearranged in reactions involving cobamide coenzymes:

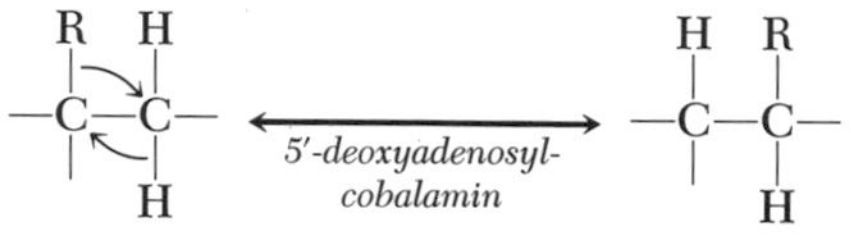

Example: The interconversion of methylmalonyl coenzyme A and succinyl coenzyme A.

5. Mercaptans may be oxidized to sulfinates:

$$R-SH + O_2 \xrightarrow{(?)} R-SO_2{}^{\ominus} + H^+$$

Example: The oxidation of cysteine to cysteine sulfinate.

6. Sulfinates may be oxidized to sulfonates (or sulfite to sulfate):

$$R-SO_2{}^{\ominus} + \tfrac{1}{2}O_2 \xrightarrow{(?)} R-SO_3{}^{\ominus}$$

Example: The oxidation of sulfite to sulfate by sulfite oxidase.

7. Aromatic rings may be oxidized to create hydroxyl groups with the simultaneous oxidation of other electron acceptors:

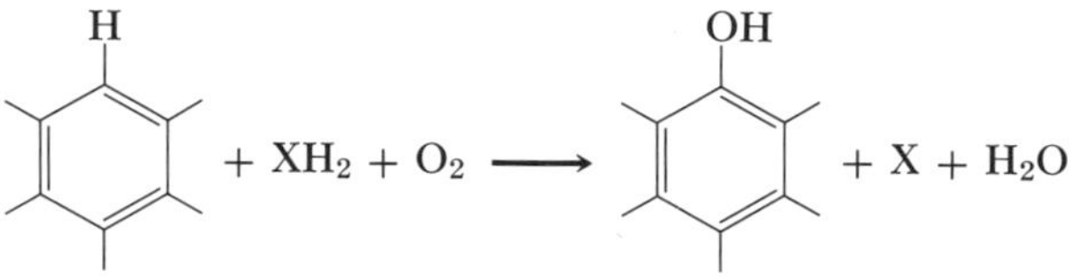

X = dihydrobiopterin in some, but not all, hydroxylations

Examples: The oxidation of phenylalanine to tyrosine catalyzed by phenylalanine hydroxylase; the oxidation of kynurenine by kynurenine hydroxylase.

8. Aromatic rings may be opened by oxidations utilizing a complete molecule of oxygen:

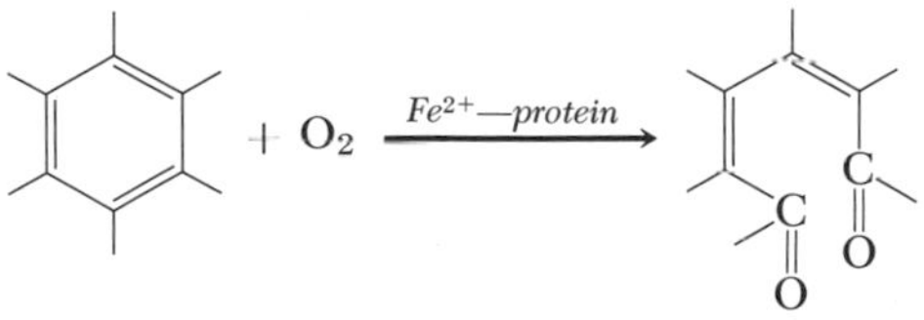

Examples: The oxidation of homogentisate to form maleoylacetoacetate; the oxidation of tryptophan by tryptophan pyrrolase; the oxidation of hydroxyanthranilate by hydroxy-anthranilate oxidase.

9. The nitrogen of aspartate may be transferred to carbonyl groups through the intermediate formation of an imino compound, followed by cleavage to release fumarate:

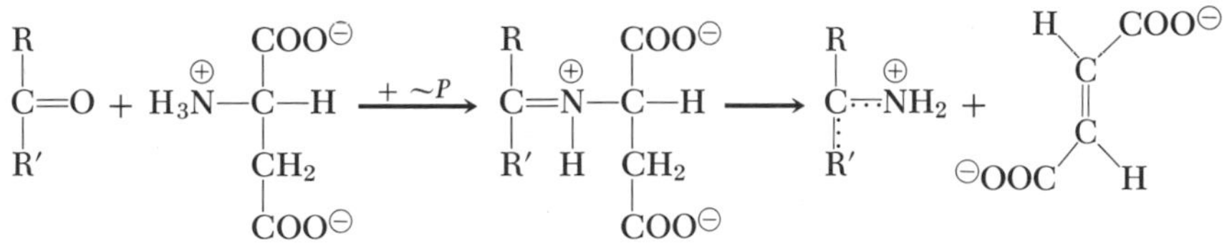

Example: The conversion of citrulline to arginine.

10. Guanidinium compounds may be hydrolyzed to form urea and an amine:

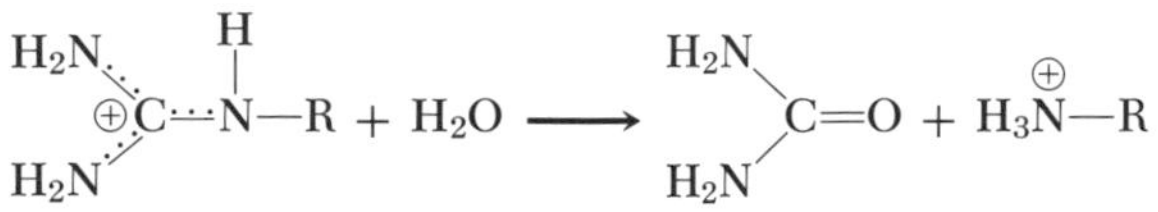

Example: The hydrolysis of arginine to urea and ornithine.

11. Amines, including amino acids, may be oxidized by flavoproteins to form ammonia, followed by reaction of the reduced flavoprotein with oxygen to form hydrogen peroxide:

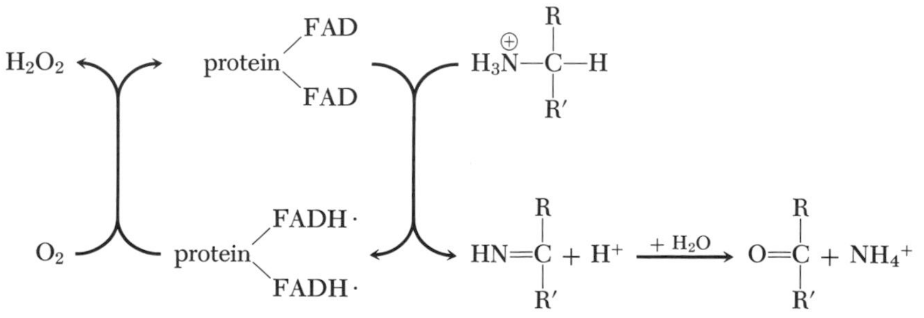

Example: The oxidation of D-amino acids catalyzed by D-amino acid oxidase.

Further reading

Meister, A.: *Biochemistry of the Amino Acids,* 2nd ed., Vols. 1 and 2. Academic Press (1965). *The monograph on the subject*—an extraordinary individual accomplishment.

Munro, H. N., and J. B. Allison, eds: *Mammalian Protein Metabolism,* Vols. 1 and 2. Academic Press (1964).

Greenberg, D. M., ed.: *Metabolic Pathways,* Vol. 3. Academic Press (1969).

Hsia, D. Y-Y., ed.: *Symposium on Treatment of Amino Acid Disorders.* Am. J. Diseases Child., *113:* 1 (1967). Complete issue on genetic abnormalities of amino acid metabolism.

Tanaka, K., M. A. Budd, M. L. Efron, and K. J. Isselbacher: *Isovaleric Acidemia: A New Genetic Defect of Leucine Metabolism.* Proc. Natl. Acad. Sci., *56:* 236 (1966).

Kahn, S. B., et al.: *Methylmalonic Excretion, A Sensitive Indicator of Vitamin B_{12} Deficiency in Man.* J. Lab. Clin. Med., *66:* 75 (1965).

Sauer, F., and J. D. Erfle: *On the Mechanism of Acetoacetate Synthesis by Guinea Pig Liver Fractions.* J. Biol. Chem., *241:* 30 (1966).

Connelly, J. L., D. J. Danner, and J. A. Bowden: *Branched Chain α-Keto Acid Metabolism, I. and II.* J. Biol. Chem., *243:* 1198 and 3526 (1968).

Cohen, P. P., and G. W. Brown, Jr.: *Ammonia Metabolism and Urea Biosynthesis.* In Florkin, M., and H. S. Mason, eds.: *Comparative Biochemistry,* Vol. 2 (p. 161). Academic Press (1960).

Lusk, G.: *The Science of Nutrition,* 4th ed. W. B. Saunders (1931). The title is misleading. This is a superb summary of early work on the metabolic economy.

MacKenzie, I. L., and R. M. Donaldson, Jr.: *Vitamin B_{12} Absorption and the Intestinal Cell Surface.* Federation Proc., *28:* 41 (1969).

Ronzio, R. A., and A. Meister: *Phosphorylation of Methionine Sulfoximine by Glutamine Synthetase.* Proc. Natl. Acad. Sci., *59:* 164 (1968).

Besrat, A., C. E. Polan, and L. M. Henderson: *Mammalian Metabolism of Glutaric Acid.* J. Biol. Chem., *244:* 1461 (1969).

Owen, E. E., et al.: *Liver and Kidney Metabolism During Prolonged Starvation.* J. Clin. Invest., *48:* 574 (1969). Includes discussion of role of the oxybutyrates in brain metabolism.

Fellman, J. H., et al.: *Soluble and Mitochondrial Forms of Tyrosine Aminotransferase—Relationship to Human Tyrosinemia.* Biochemistry, *8:* 615 (1969).

THE ONE-CARBON POOL

CHAPTER 18

ARGUMENT

Single-carbon groups that are more reduced than carboxylate groups are transferred for such diverse purposes as the formation of purine rings and the methylation of amines or alcohols. These one-carbon units are handled mainly in two ways. Methylidyne and methylene groups, which correspond to the oxidation level of formate and formaldehyde, respectively, appear as bridges between two nitrogen atoms of the coenzyme, tetrahydrofolate. Tetrahydrofolate is a reduced pteridine compound with the primary function of acting as a carrier for one-carbon units. Although free formate and formaldehyde will react with the coenzyme to form active one-carbon donors, they are usually not as important a source as is the amino acid, serine. The terminal carbon of serine can be transferred to tetrahydrofolate as a methylene carbon, and the remainder of the amino acid is liberated in the form of glycine. Glycine itself is also subject to degradation by routes in which its second carbon becomes a methylene group attached to tetrahydrofolate.

Methyl groups, which correspond to the oxidation state of methanol, are handled in a different way. The primary methyl group donor is S-adenosyl-L-methionine, which is formed by reaction of methionine with ATP, and this compound is a substrate for a variety of enzymes used to methylate amines and alcohols. S-adenosylhomocysteine remains after the transfer of the methyl group, and it is hydrolyzed to adenosine and homocysteine. Homocysteine, which represents all of the methionine molecule except the methyl group, may be degraded by irreversible routes eventually yielding propionyl coenzyme A. Since the metabolism of homocysteine is irreversible, sufficient methionine must be present in the diet to replace the loss of this portion of the molecule.

The two kinds of one-carbon donors, the tetrahydrofolate derivatives

409

and S-adenosylmethionine, are metabolically related. Methylenetetrahydrofolate may be reduced to 5-methyltetrahydrofolate, and the methyl group of this compound transferred to homocysteine, thereby forming methionine. The transfer is catalyzed by an enzyme that also requires cobamide coenzyme. The process serves to conserve the irreplaceable homocysteine portion of the methionine molecule by using it repeatedly as a carrier for methyl groups derived from other sources *via* the one-carbon pool.

There is another relationship between the one-carbon donors. A large portion of the methyl groups in the body occur as choline in phospholipids, and significant amounts of choline also are ingested in most diets. When the supply of choline is deficient, it can be synthesized at the expense of methyl groups derived from methionine, but when an excess of choline is present, it can be used to replenish both the methionine and the tetrahydrofolate one-carbon pools. The alcoholic group of choline is oxidized to the corresponding carboxylate, which is betaine. One of the methyl groups on betaine may be transferred to homocysteine, creating methionine, and the other two methyl groups are oxidized to formaldehyde, which may react with tetrahydrofolate.

One of the major uses of methyl groups is the formation of creatine. The carbon skeleton and the amidinio group of creatine are provided by glycine and arginine, respectively, but the N-methyl group is derived from methionine. Much of the creatine is present as phosphocreatine in muscles. Phosphocreatine undergoes a slow, spontaneous cyclization yielding the metabolically inert compound, creatinine, which is excreted in the urine. Creatinine formation proceeds at a nearly constant rate, and represents a continual drain on the one-carbon pool. The methyl groups lost in this way must be replaced by dietary choline or methionine, or be formed *via* 5-methyltetrahydrofolate.

We have already seen many examples of the utilization or formation of CO_2 through carboxylations or decarboxylations. Carbon dioxide is the most oxidized form of a single carbon atom. However, single carbon atoms are also produced and transferred in less fully oxidized states. These more reduced one-carbon units are used to introduce carbon atoms into the purine ring and to create methyl groups on compounds such as choline, creatine, and the methylated bases found in nucleic acids. The *one-carbon pool* describes the group of compounds, partially interconvertible, acting as sources or intermediate carriers of these single-carbon groups.

We shall first discuss the coenzyme, *tetrahydrofolate*, that carries one-carbon units in various oxidation states. Understanding its function will permit us to complete the discussion of the metabolism of *serine, glycine* and *histidine,* and to show how these amino acids act as sources of one-carbon units. (The important utilization of tetrahydrofolate transport in synthesizing the bases of nucleic acids will be considered in Chapter 20.) Then we shall turn to the transfer of methyl groups, partially derived from the tetrahydrofolate compounds, but mainly supplied by dietary *methionine* and *choline*.

Tetrahydrofolate

Tetrahydrofolate has the specific function of carrying one-carbon units. It is a reduced pteridine like tetrahydrobiopterin (p. 385), but it is more complex, having within it residues of *p*-aminobenzoic acid and glutamate (Fig. 18-1). The portion without the glutamyl residue is tetrahydropteroic acid, so the compound is more systematically named as *tetrahydropteroylglutamate*. The two names for the compound are respectively abbreviated as $H_4folate$ and $H_4PteGlu$ in the current literature (THF or THFA in the older literature). The reduced form is readily oxidized upon exposure to air, forming folate, which is the usual commercial reagent and dietary form, but enzymes exist to catalyze its reduction back to the active coenzyme (Fig. 18-1, bottom).

Mammals cannot synthesize the pteridine ring, so folate qualifies as a vitamin in the sense that mammalian tissues must be supplied with the compound. However,

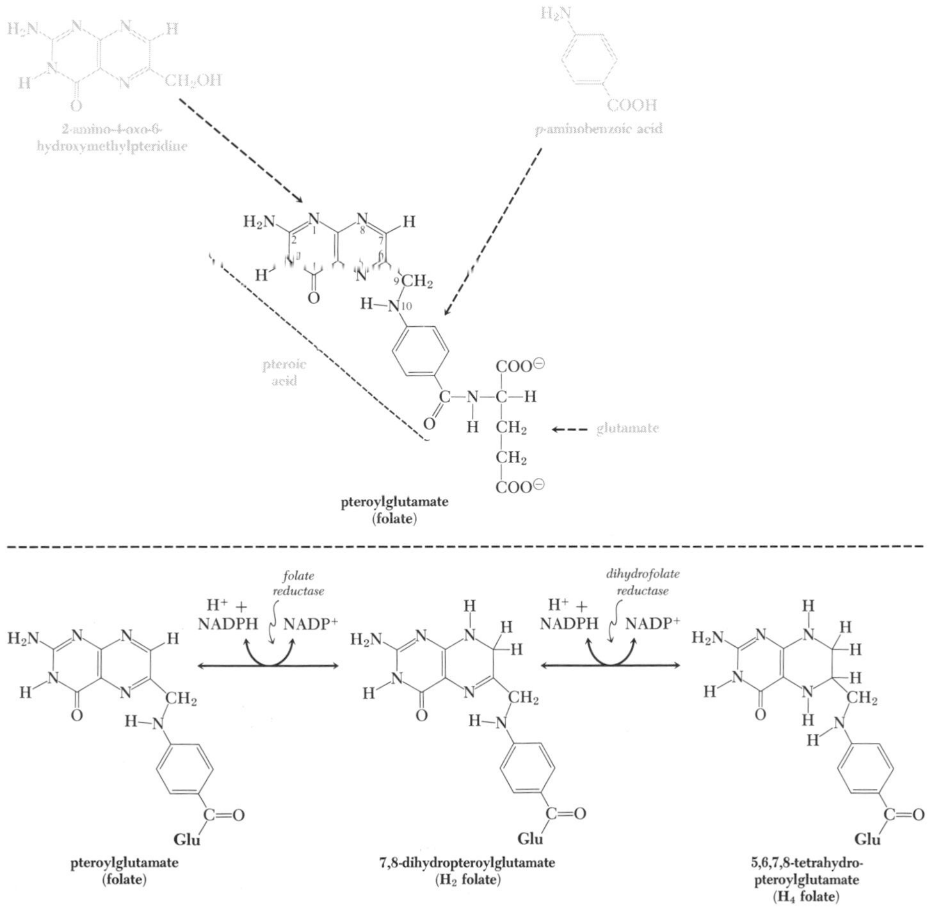

Figure 18-1   The structure of the vitamin, pteroylglutamate, more commonly known as folate, is shown at the top, along with the structures and names of its precursors. The oxidized form usually found in the diet is reduced in two steps to the active coenzyme, tetrahydrofolate (*bottom*).

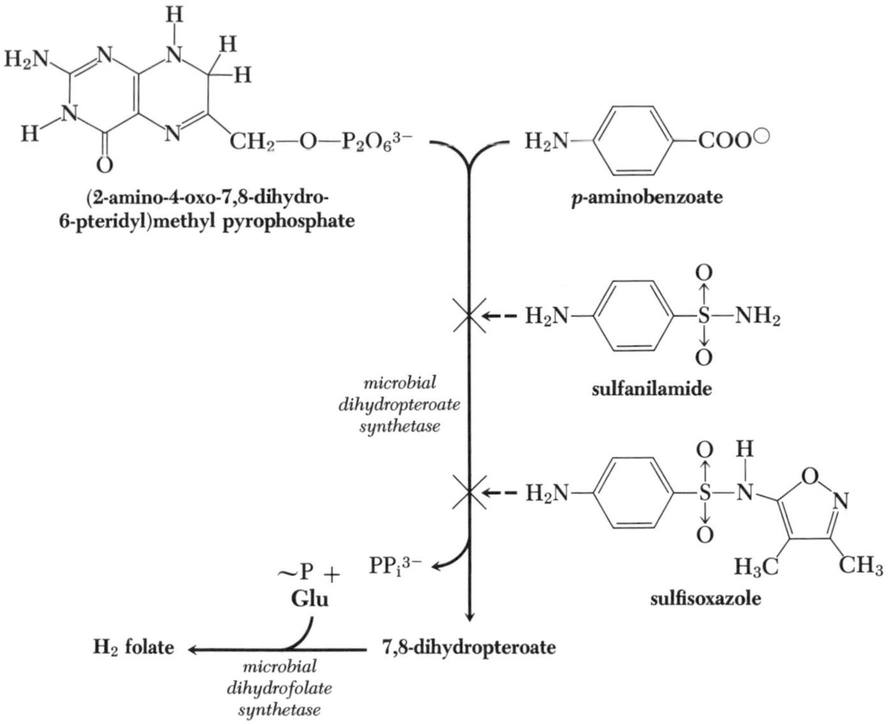

Figure 18-2   The antimicrobial action of sulfonamides depends upon their competitive inhibition of one of the steps in folate synthesis.

an external dietary source is frequently not required because the microorganisms of the intestinal tract readily form the compound. Indeed, creation of a folate deficiency in experimental animals usually requires administering an antibiotic to suppress the intestinal growth of microorganisms as well as feeding them a deficient diet, which is in itself a painstaking procedure owing to the wide distribution of the compound in natural foods.

The ability of some microorganisms to make folate provides an Achilles heel for prevention of their growth. One of the steps in the synthesis is a combination of a pteridine with *p*-aminobenzoate. Substituted sulfonamides (Fig. 18-2) are effective competitive inhibitors for this process. Since humans rely on outside supplies of folate, rather than making their own, they are not vulnerable to this kind of inhibition, and the sulfonamides have relatively low toxicity at concentrations effective in preventing growth of susceptible organisms.

5,10-Methylene-H$_4$folate—"active formaldehyde"

Formaldehyde is a naturally occurring metabolite, which will spontaneously react with tetrahydrofolate to form a methylene bridge between N-5 and N-10 (Fig. 18-3). The process occurs in two steps, with the transient formation of an intermediate 5-hydroxymethyl compound, which spontaneously cyclizes to form 5,10-methylene-H$_4$folate. This compound is at a key point in the one-carbon pool, able

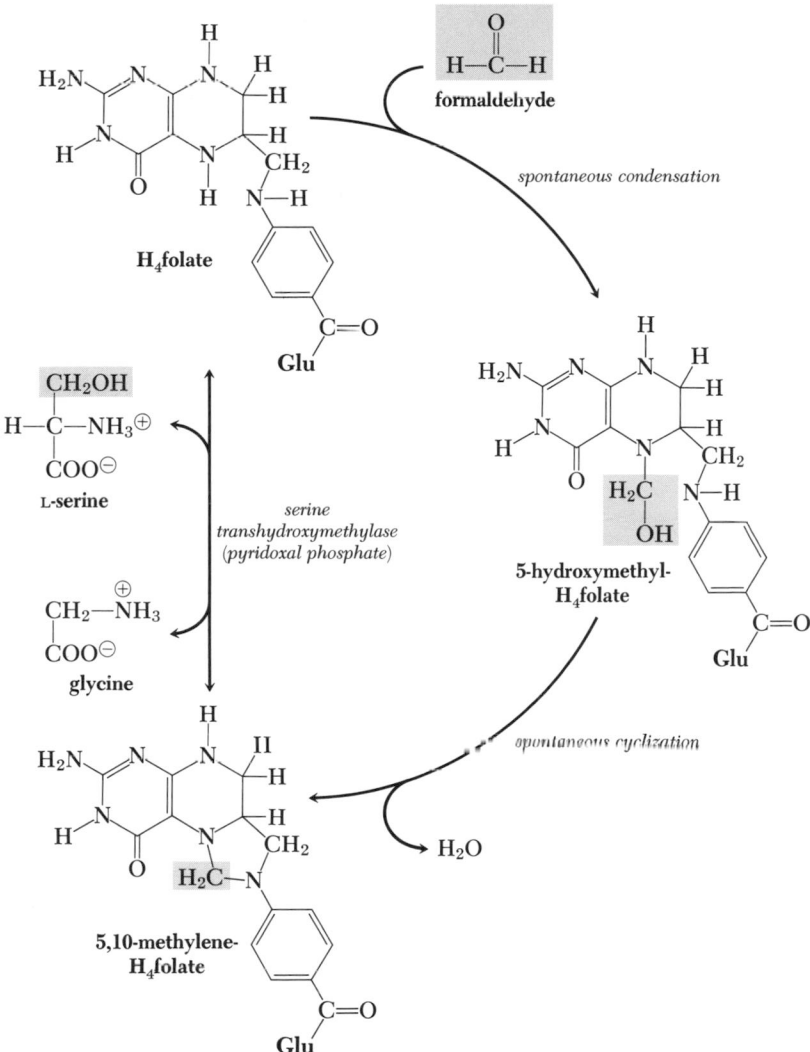

Figure 18-3 Tetrahydrofolate is used as a coenzyme for carrying one-carbon groups. One of the most important intermediates is 5,10-methylene-H₄folate (*bottom*), which may be formed by the reaction of H₄folate with formaldehyde *via* an intermediate hydroxymethyl derivative (*right*). An even more important source of the one-carbon group is the amino acid, serine. One of its carbons may be transferred to H₄folate, leaving glycine as the other product (*left*).

to react so as to form most of the other forms of one-carbon units, and is occasionally called "active formaldehyde," a name persisting from earlier days when the nature of the compound was unknown.

While the formation of methylene-H₄folate from formaldehyde is an important reaction, the major source of the compound is the amino acid, L-*serine*. In a formal sense, serine is equivalent to a condensed form of formaldehyde and glycine, analogous to the hydroxymethyl-H₄folate intermediate shown in Figure 18-3. An enzyme is present in liver, and probably in most tissues, that catalyzes the direct transfer of the hydroxymethyl group of serine to H_4folate, as is shown in the figure, without the intermediate appearance of formaldehyde. The enzyme contains pyridoxal phosphate as a coenzyme.

The Oxidation and Reduction of Methylene-H_4folate

Single carbon units can exist in various oxidation states, which are represented by the series of simplest organic compounds: methane, methanol, formaldehyde, formate, carbon dioxide. These carbon units can, except for methane, appear in other compounds in the same oxidation states by the formal elimination of water. Let us take a moment to summarize the possibilities (Fig. 18-4).

PARENT COMPOUND		CONDENSED FORMS	
CH_4 methane	$\xrightarrow{+ R-H}$ none		
CH_3OH methanol	$\xrightarrow[- H_2O]{+ R-H}$ $R-CH_3$ methyl group	$\xrightarrow{+ R'-H}$ none	
H—C—H (O) formaldehyde	$\xrightarrow{+ R-H}$ $R-CH_2OH$ hydroxymethyl group	$\xrightarrow[- H_2O]{+ R'-H}$ $R-CH_2-R'$ methylene group	
H—C—OH (O) formic acid	$\xrightarrow[- H_2O]{+ R-H}$ R—C—H (O) formyl group	$\xrightarrow[- H_2O]{+ R'-H_2}$ R—C=R' methylidyne group	
HO—C—OH (O) carbonic acid	$\xrightarrow[- H_2O]{+ R-H}$ R—C—OH (O) carboxyl group	$\xrightarrow[- H_2O]{+ R'-H}$ R—C—R' (O) carbonyl group	

Figure 18-4   The above table shows one-carbon compounds of various oxidation states and their formal relationship to various groups found as substituents of natural compounds. There is no direct biological interconversion of methanol and methyl groups. Otherwise, examples may be found of all of the interconversions.

Carbonic acid, the hydrated form of CO_2, is included in the figure for completeness, but compounds at this oxidation state are handled by carboxylation and decarboxylation reactions, not by the reactions of the one-carbon pool. The figure has the general R- designation for substituents on the one-carbon unit, but the actual nature of the groups that can undergo such interchanges is limited by thermodynamics, and in most cases the reacting atoms will be N, O, or S. In tetrahydrofolate, as we have seen, the one-carbon unit is linked to nitrogen atoms.

All of the possible structures of one-carbon groups shown in the figure are found in biological compounds, and all except the carbonic acid derivatives are part of important tetrahydrofolate derivatives. Since each row differs from its neighbor

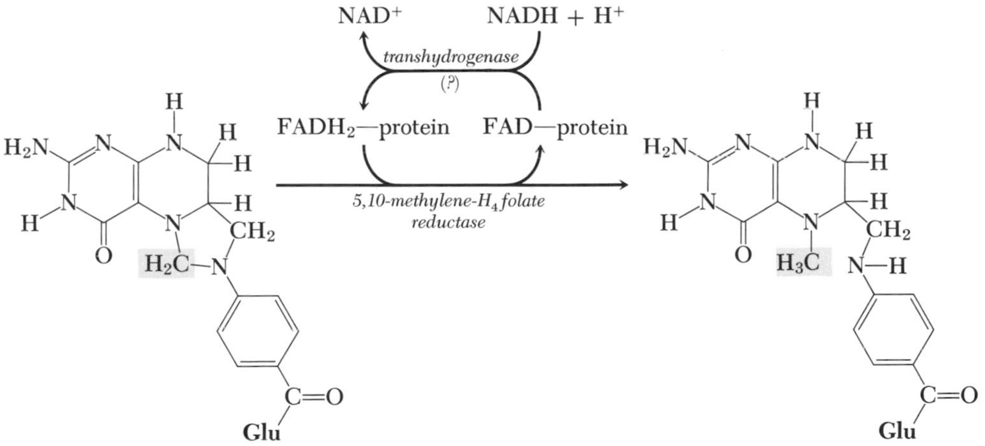

Figure 18-5 Reduction of the one-carbon group in 5,10-methylene-H_4folate to form 5-methyl-H_4folate involves transfer of electrons from NADH through an intermediate flavoprotein.

by the equivalent of two electrons, the rows are interconvertible only by oxidations or reductions, and there are enzymes catalyzing the conversion of methylene-H_4folate to more oxidized or reduced compounds.

One of these enzymes catalyzes the reduction of the compound to form *5-methyltetrahydrofolate* (Fig. 18-5). The enzyme is a flavoprotein, which is unusual among those catalyzing substrate reduction, and the mechanics of electron flow have not been worked out in mammalian systems. In *E. coli*, a favorite microorganism for biochemical studies, a second flavoprotein is used to transfer electrons from NADH to the enzyme, reducing its FAD; it is the reduced FAD that donates electrons to the methylene carbon. In any event, the sequence is essentially irreversible and sufficiently unlike most substrate oxidations and reductions to justify naming the enzyme as a *reductase* rather than as a dehydrogenase.

The oxidation of methylenetetrahydrofolate to *5,10-methylidyne-H_4folate* is a straightforward reversible reaction using NADP as an oxidizing agent. The one-carbon unit remains bound in a ring in the product, but as a methylidyne, rather than a methylene, group (Fig. 18-6). This group is at the oxidation state of formate, and another enzyme catalyzes partial hydrolysis of the ring to make *10-formyl-H_4folate*, which can also be made from free formate at the expense of a mole of high-energy phosphate.

These tetrahydrofolate derivatives are the donors of one-carbon units for a variety of purposes. We shall shortly see that the 5-methyl compound may be used to supply the methyl group of methionine, which in turn is transferred to form a variety of methylated compounds. In Chapter 20, we shall see that the 5,10-methylidyne and the 10-formyl derivatives are both used to supply carbon atoms for synthesis of purine rings. The 5,10-methylene compound is the source of the methyl group for making thymine during DNA synthesis.

Let us turn now to a consideration of the production of these one-carbon units during the metabolism of amino acids.

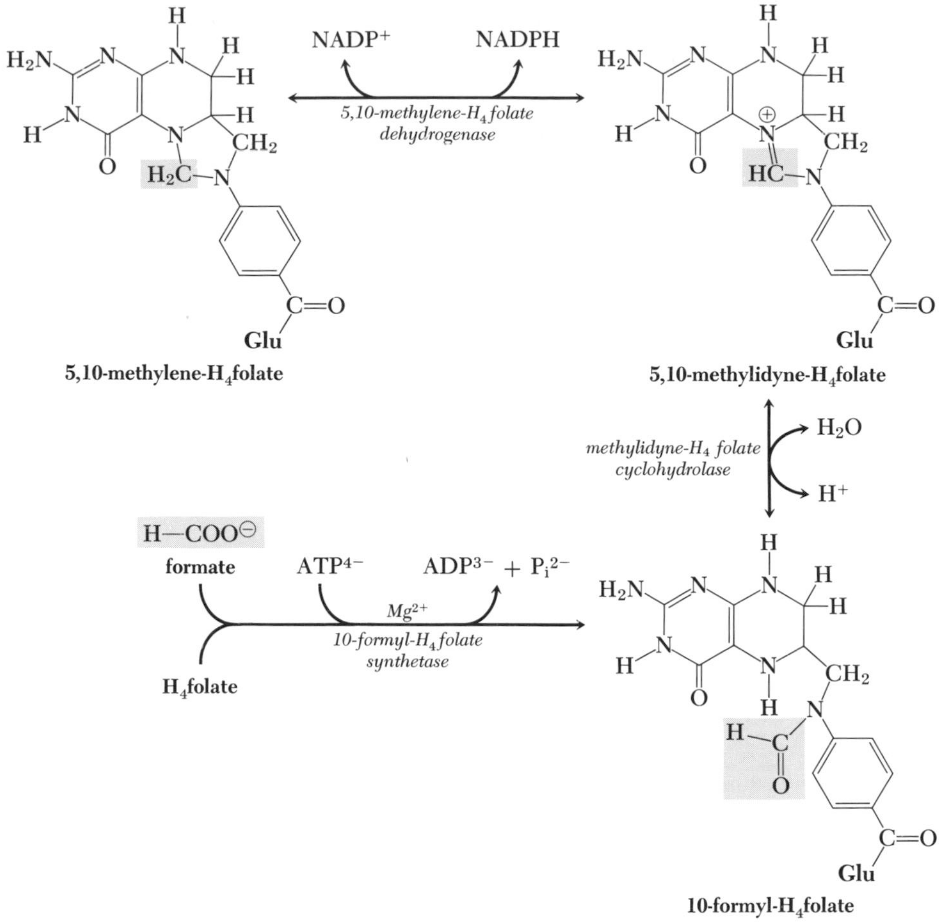

Figure 18-6 5,10-Methylene-H_4folate can be oxidized by NADP to form 5,10-methylidyne-H_4folate in a reversible reaction. The methylidyne compound and 10-formyl-H_4folate are interconvertible by the action of a second enzyme, catalyzing a simple hydrolysis. 10-Formyl-H_4folate can also be formed by the combination of formate with H_4folate at the expense of high-energy phosphate.

PRIMARY SOURCES FOR THE ONE-CARBON POOL

Serine

Nitrogen and two carbons: Appear as glycine.

One carbon: Appears as 5,10-methylene-H_4folate.

We discussed a major route for the degradation of serine—deamination to form pyruvate—in the previous chapter. However, the transfer of one carbon to tetrahydrofolate shown earlier in Figure 18-3 is also a major reaction. Through this route, serine provides a large fraction of one-carbon units and also serves as a precursor of the amino acid, *glycine*, which therefore need not be in the diet.

Serine is a common constituent of dietary proteins, but it also can be synthesized from glucose by two different routes (Fig. 18-7). The existence of independent

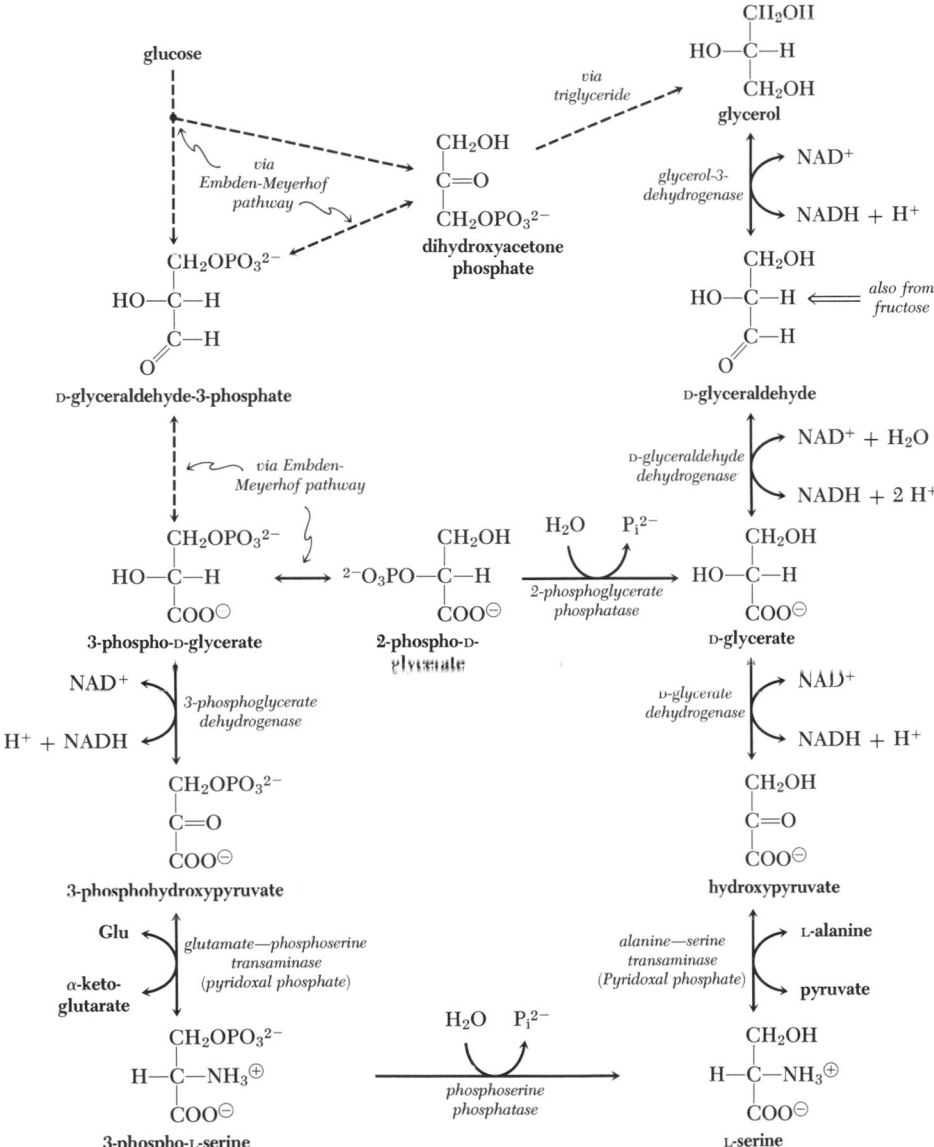

Figure 18-7 Serine may be formed by glucose by two independent routes. One involves phosphorylated derivatives, with the necessary nitrogen obtained from glutamate (*left*). The other involves the non-esterified compounds, and the nitrogen is obtained from alanine (*right*). Since serine is an important donor of one-carbon units by the transhydroxymethylase reaction (Figure 18-3), the routes shown here provide a means for the total synthesis (from glucose and any donor of amino groups) not only of serine, itself, but also of the one-carbon units and of glycine, which is another product of the reaction.

means of forming a common constituent of the diet ought to make us suspect that it is a major metabolic intermediate in addition to being a component of proteins. This is indeed so. The turnover of serine in the amino acid pool is almost as rapid as that of glutamate and aspartate. The continual formation of the amino acid insures a supply of one-carbon units independent of the immediate dietary supply. (We shall also see that serine is used as a precursor of the bases in phospholipids.)

One of the routes for the formation of serine begins with D-glycerate, which can be made *via* glyceraldehyde from the glycerol released by triglyceride breakdown or from fructose by a route to be considered in detail later (p. 633). It also can be formed by direct hydrolysis of 2-phosphoglycerate. Successive oxidations of glyceraldehyde by straightforward dehydrogenase produce *hydroxypyruvate*, which is the 2-keto analogue of serine. A specific transaminase, which uses alanine rather than glutamate, converts hydroxypyruvate to serine.

Similarly, 3-phospho-D-glycerate from the regular Embden-Meyerhof pathway of glucose metabolism can be oxidized by a separate enzyme to phosphohydroxy-pyruvate, which transaminates with glutamate to produce *phosphoserine*. A specific phosphatase catalyzes the hydrolysis of phosphoserine to produce serine. The purpose of having two separate routes is not clear. The relative amounts of serine made by them differs from one species to another and changes during the development of a given individual in a species. We evidently have not realized all of the relationships of serine and phosphoserine, and are therefore missing an important clue.

Since both glycerol and phosphoglycerate are obtained from glucose, glucose can provide all of the carbons of serine, and can contribute in this way to the one-carbon pool.

Glycine

Nitrogen: Route unsettled; part may appear as glutamate by transamination, part as NH_4^+.

Carbons: Route unsettled; part may appear as formate and CO_2 and part as 5,10-methylene-H_4folate and CO_2.

Glycine is the simplest of the amino acids, abundant in most foods, and yet it is a source of embarrassment to mammalian biochemists because of considerable uncertainty about the route of its metabolism. It now appears that the major route involves a reaction only recently demonstrated in animal tissues by which glycine is cleaved to form NH_4^+, CO_2, and 5,10-methylene-H_4folate. The reaction involves the oxidation of glycine after it has combined with pyridoxal phosphate on the enzyme (Fig. 18-8).

From this we see that glycine itself is a source of one-carbon units. Indeed, it is possible for the methylene group derived from glycine to be transferred onto a second molecule of glycine to form serine, since the transhydroxymethylase reaction is freely reversible. This may occur at times when there is an abundant supply of glycine, and serine is being actively degraded to pyruvate. In such a case, three of the carbon atoms of a pair of glycine molecules would appear as serine, with the fourth carbon appearing as CO_2.

There is evidence that a portion of the carbon chain of glycine appears as the corresponding carbonyl compound, *glyoxylate*. This could occur either by an oxidation or a transamination (Fig. 18-9). The appropriate enzymes for both routes

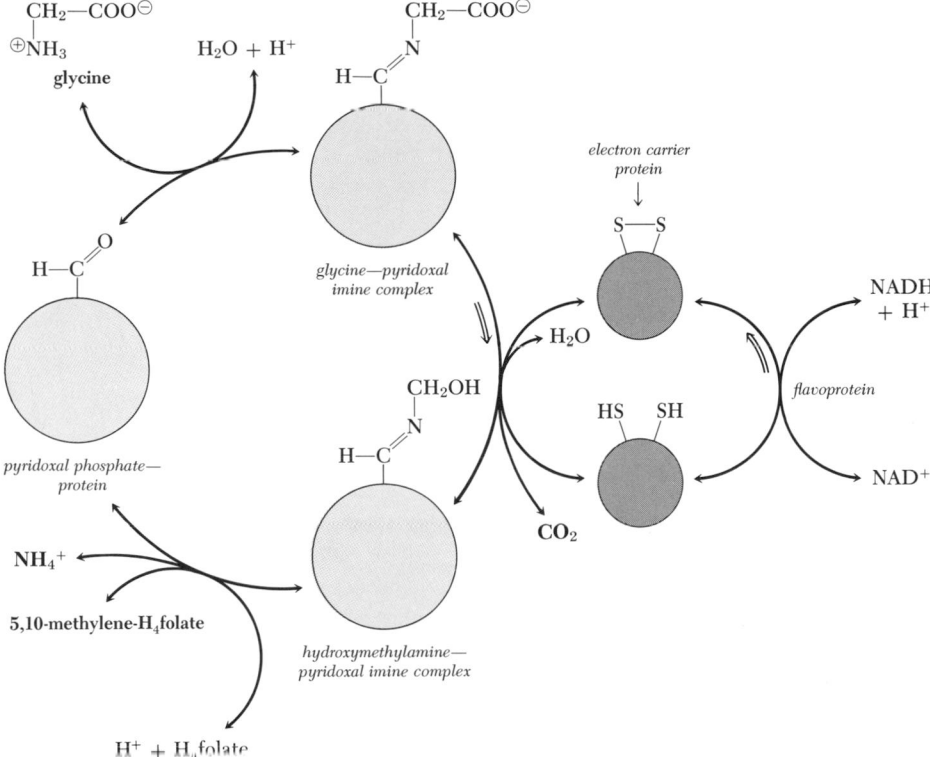

Figure 18-8   Glycine is oxidized in the form of the glycine-pyridoxal imine carried on the enzyme. The electrons are transferred through a disulfide on a second protein. The advantage of this arrangement is that the hydroxymethylamine group formed from glycine is stabilized, and the hydroxymethyl portion can be transferred directly to H_4folate without the intermediate appearance of formaldehyde.

have been demonstrated. However, the activity of glycine oxidase is low, and the transamination reactions appear to be irreversible in the direction of glycine formation, rather than utilization. Glyoxylate is oxidized to formate and CO_2, thereby contributing one of the carbons to the one-carbon pool.

Although we can't give a clear picture of the metabolism of glyoxylate at the moment, the compound ought to be kept in mind. Current investigations suggest that there are additional routes for its metabolism. It is also formed during the metabolism of *hydroxyproline*. (The route of hydroxyproline metabolism is similar to that of proline, resulting in the formation of the 4-hydroxy analogue of α-ketoglutarate. This 2-keto-4-hydroxy glutarate is cleaved by an aldolase to form glyoxylate and pyruvate.)

While we have emphasized the gaps in our knowledge of glycine metabolism, this should not obscure what is clear. Both of the likely routes result in the potential introduction of C-2 of glycine into the one-carbon pool. Both result in the eventual appearance of the nitrogen as ammonia. Glycine and serine are freely interconvertible.

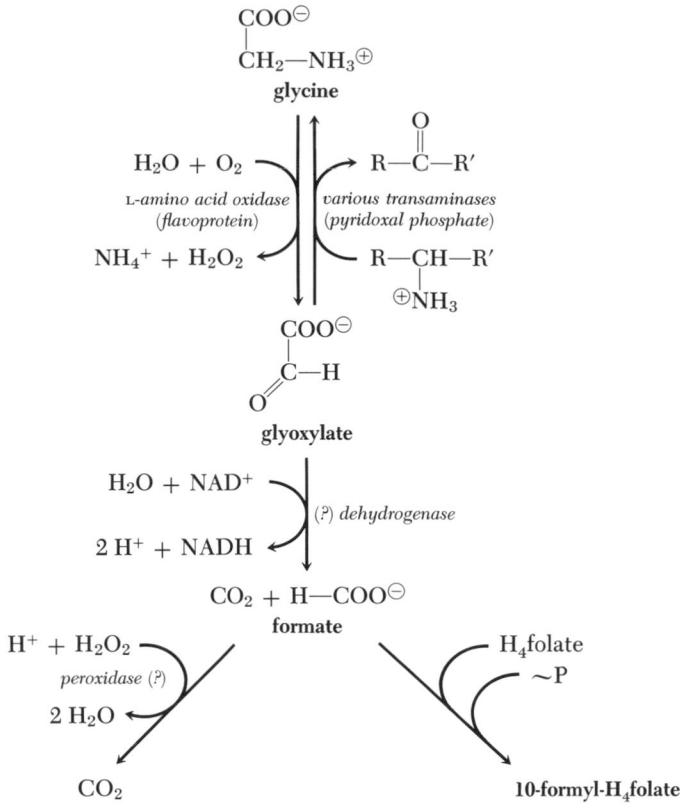

Figure 18-9 Glycine and glyoxylate are metabolically interchangeable, but there is some uncertainty over the routes. Known reactions are shown, but the amount of compound handled by each is debatable.

Histidine

Nitrogens and carbons: One nitrogen and five carbons appear as glutamate, two nitrogen as NH_4^+, and one carbon as methylidyne-H_4folate.

The imidazole ring of histidine contains a methylidyne carbon between two nitrogens, and the route for the degradation of the amino acid involves the transfer of this carbon to H_4folate. (Fig. 18-10). The sequence begins with two unusual reactions. The first, catalyzed by *histidase*, is a removal of ammonia to leave a double bond. This is analogous to a dehydration removing water, but there are not many examples known of this kind of reaction, and most occur in microorganisms. The unsaturated compound, *urocanate,* is excreted to a small extent in the urine, but most is hydrated in a second unusual reaction in which there is a general shift of double bonds, so that the oxygen is added to the imidazole ring, but it is the propionate side chain that becomes saturated. The product is imidazolonepropion-ate, and the ring of this compound now contains an amide aroup, which is cleaved by hydrolysis to form the open chain compound, *formiminoglutamate.*

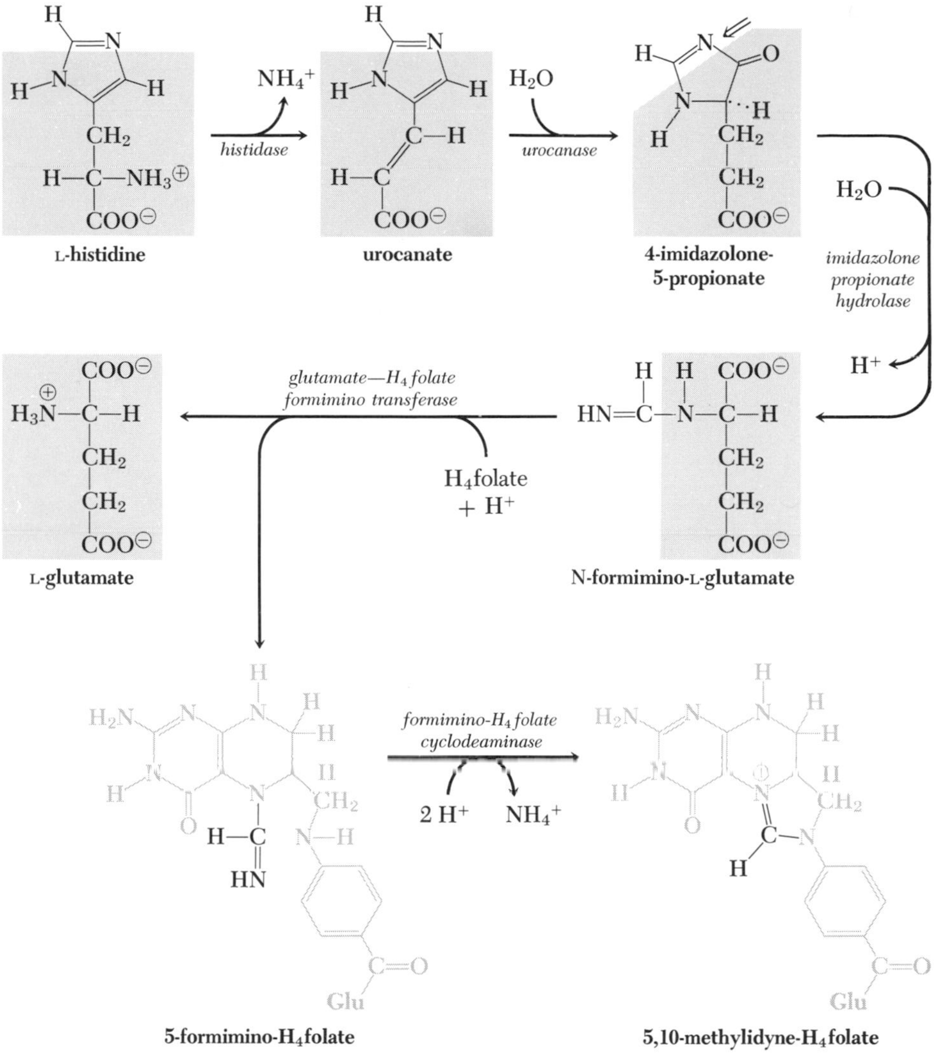

Figure 18-10 The metabolism of histidine. Direct deamination is a feasible first reaction in this case because the product, urocanate, is stabilized by resonance. The resultant shift in electronic structure makes the following hydration and hydrolysis by which the ring is opened feasible. One of the ring carbons, along with a nitrogen, is then transferred to H$_4$folate, with the remainder of the original histidine molecule appearing as glutamate.

The formimino group is transferred to tetrahydrofolate, leaving free glutamate and forming *5-formimino-H$_4$folate*, from which the ammonium group is cleaved during ring closure to produce 5,10-methylidyne-H$_4$folate.

The degradation of histidine therefore involves no oxidative reactions peculiar to the route, and by a combination of group transfers and rearrangements the amino acid is converted intact to glutamate, two molecules of ammonia, and a methylidyne group on tetrahydrofolate. Of course, this means that histidine, in addition to being a donor of a single carbon to the one-carbon pool, is also a *glucogenic* amino acid.

The reactions of its degradation are irreversible, and *histidine is an essential amino acid.*

THE UTILIZATION OF METHYL GROUPS

A substantial part of the metabolism of one-carbon units involves the transfer of methyl groups, primarily to form N-methyl compounds. The fundamental methyl group donor is the amino acid, L-*methionine*. In order to drive these transmethylations to completion, methionine is "activated" by converting it to S-*adenosyl methionine* at the expense of high-energy phosphate (Fig. 18-11).

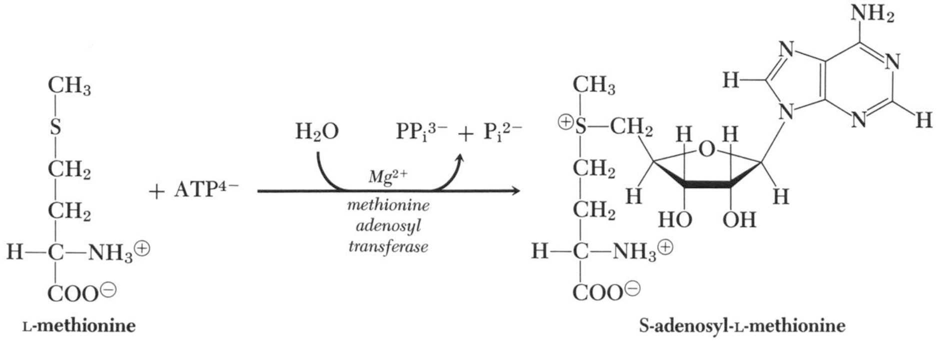

Figure 18-11 Methionine is activated by the transfer of an adenosyl group from ATP to the sulfur atom. The resultant S-adenosylmethionine serves as a donor of methyl groups in forming a variety of compounds. The phosphates of ATP transiently appear as inorganic triphosphate (PPP_i), which is rapidly cleaved to form PP_i and P_i.

The sulfonium group represents a high-energy compound, with a standard free energy of hydrolysis (yielding adenosine, methionine, and H^+) of approximately -6000 calories per mole. Its formation by transfer of an adenosyl group is an unusual reaction. The three phosphate groups of ATP are bound to the enzyme as triphosphate during the reaction, and then cleaved to form pyrophosphate and orthophosphate before release. The reaction is therefore driven by the energy of cleavage of the ester bond between adenosine and the triphosphate and of cleavage of one pyrophosphate link. Of course, the inorganic pyrophosphate formed is maintained at low concentrations by inorganic pyrophosphatase, which also serves to drive the reaction.

Creatine

Synthesis: From glycine, arginine, and methionine.
Degradation: Excreted as creatinine—not recovered.
One of the important functions of S-adenosyl methionine is to donate a methyl group for the formation of creatine, used to store high-energy phosphate in skeletal muscle. Creatine could be named as N-methyl-N-amidinioglycine, which would substantially describe the groups used in forming it but is not very good nomenclature. The amidinio group is supplied by a transfer from arginine to glycine (Fig. 18-12), forming guanidinioacetate and ornithine. (Guanidinioacetate is more ac-

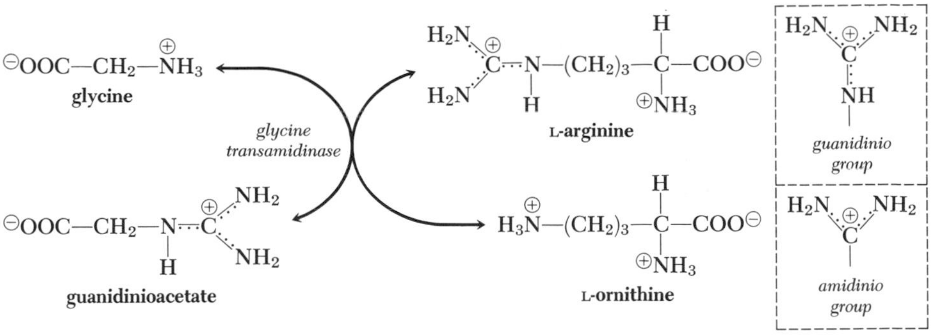

$$^{\ominus}OOC-CH_2-\overset{\oplus}{N}H_3 \quad \text{glycine}$$

glycine transamidinase

$$^{\ominus}OOC-CH_2-\underset{H}{N}\overset{NH_2}{\underset{}{\overset{}{\cdots}C}}\overset{}{\underset{NH_2}{}}$$

guanidinioacetate

L-arginine

L-ornithine

guanidinio group

amidinio group

Figure 18-12   The formation of guanidinioacetate, the precursor of creatine.

cepted nomenclature than amidinioglycine. The structure of these two groups is also shown in the figure.) In a sense, glycine is being used in this reaction to cleave arginine in the way that water is used in the arginase reaction. Guanidine is the nitrogen analogue of urea, and arginine is acting as an amidine donor in both cases, with ornithine being the remaining product.

Finally, the guanidinioacetate, also known trivially as glycocyamine, is methylated to form creatine. S-Adenosylmethionine is the methyl donor, and forms *S-adenosylhomocysteine* as the other product (Fig. 18-13). This is an irreversible reaction under physiological conditions. Much of the liberation of free energy in this reaction comes from the loss of the sulfonium structure. The product, S-adenosylhomocysteine, is a simple thioether, analogous to methionine itself, so a proton is liberated, and the energy of ionization is a part of the total.

At rest, most of the creatine will exist in skeletal muscles as phosphocreatine.

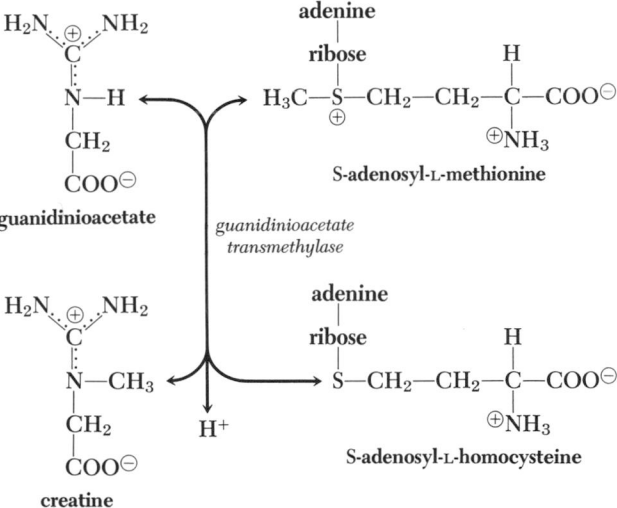

Figure 18-13   Guanidinioacetate is converted to creatine by transfer of a methyl group from S-adenosylmethionine. This reaction is typical of the transmethylases, with S-adenosylhomocysteine being the other product.

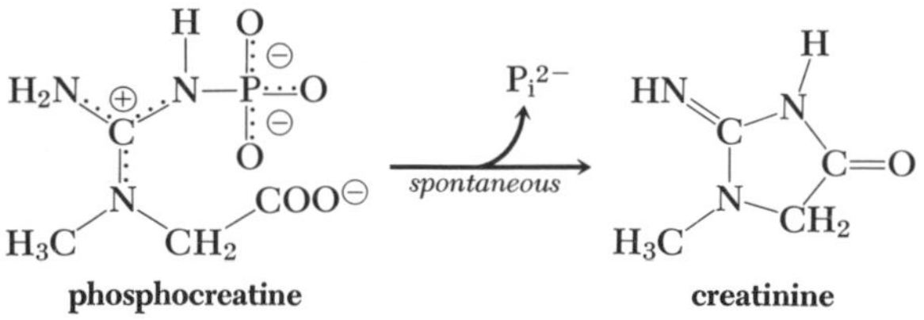

Figure 18-14 Phosphocreatine spontaneously cyclizes with the loss of P_i to form creatinine, which is excreted in the urine.

We pointed out that a man has on the order of 40 micromoles of creatine per gram of muscle, or a total of about 1.1 moles in a 70 kg man. The presence of this large amount of phosphocreatine causes a continual loss of the methyl groups represented by creatine, because there is a constant, although slow, spontaneous cyclization of phosphocreatine to form *creatinine* (Fig. 18-14), which is excreted in the urine. The loss is not trivial—it amounts to 15 millimoles (1.7 grams) per day for an average male adult, and is remarkably constant, depending only on the amount of muscle of other animals eaten and on the amount of creatine in the body. In other words, a constant fraction of the phosphocreatine is broken down each day to form creatinine. The constancy is sometimes used to test the reliability of urine samples—the idea being that a bottle alleged to contain all of a day's urine ought to contain all of a day's creatinine and this can be measured.

The existence of this leak, dependent upon phosphocreatine content and therefore upon skeletal muscle mass, is used to assess the latter value. When muscles degenerate for any reason—from paralysis as in poliomyelitis, or from muscular dystrophy—the creatinine content of the urine falls, owing to the large amount of creatine being lost. Creatinine is ordinarily eliminated rapidly, and any rise in blood concentration is a sensitive indicator of kidney damage. From the standpoint of biochemical economy, creatinine excretion is important mainly as a constant load on the supply of transferable methyl groups. There is no known way of recovering the methyl groups transferred to creatine.

Choline

Supply: Synthesized as phosphatidylcholine from serine and methionine; obtained in diet.

Degradation: One methyl group to methionine, two to formaldehyde, remainder of molecule converted to glycine.

We have previously mentioned two functions for choline. One is the use in nerve conduction as *acetylcholine,* but this is a quantitatively small requirement, imperative as it may be in a qualitative sense. The other is as a constituent of *phosphatidylcholine*—the important phospholipid involved in membrane structure and in the lipoproteins used for triglyceride transport.

Because of its general occurrence in living tissues, choline is constantly supplied by the diet of animals, and is sometimes classified as a vitamin. However, phosphatidylcholine can be synthesized from other dietary constituents—even from glucose and ammonia—and it is misleading to list choline among the essential components in food.

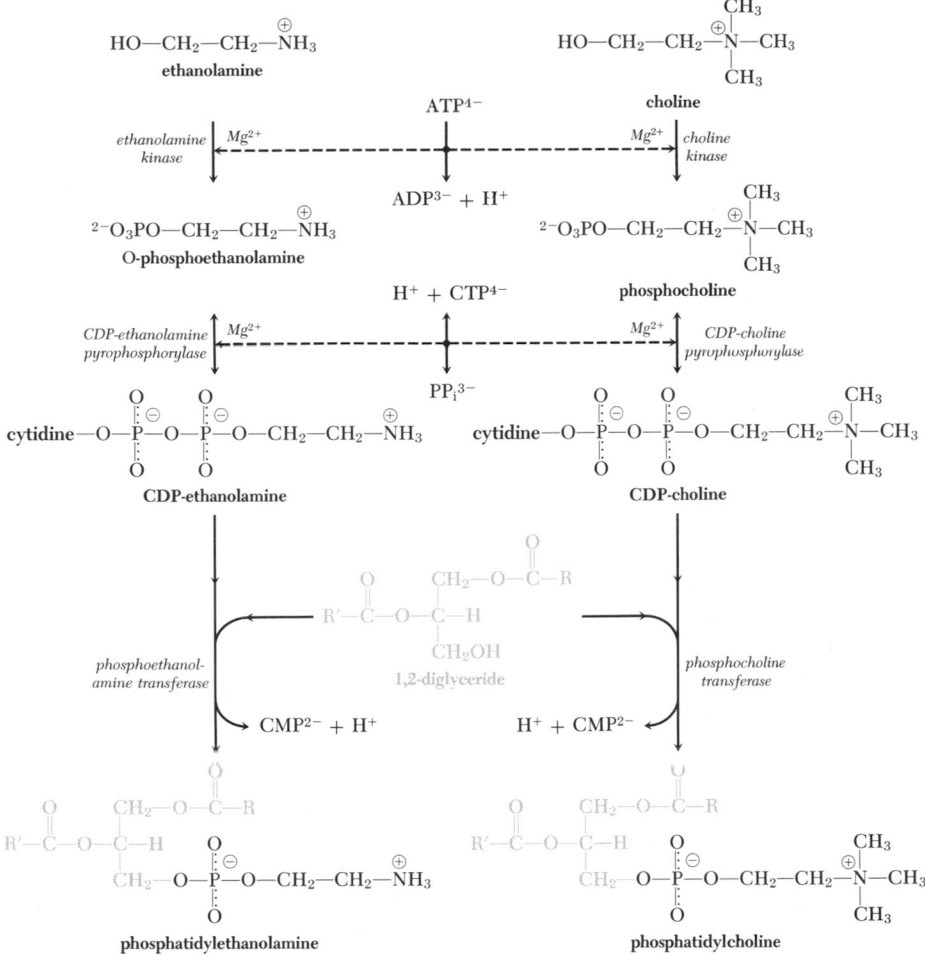

Figure 18-15   Ethanolamine and choline are incorporated into phospholipids by similar sets of reactions involving the intermediate formation of cytidine diphosphate esters. The 1,2-diglycerides to which the phosphorylated bases are attached usually differ from those involved in triglyceride synthesis in that they have longer chain fatty acid residues with more double bonds.

It is probably correct to say that animals on most diets do use the free amines to make phosphatidylethanolamine and phosphatidylcholine, using the bases liberated from the breakdown of phospholipids over again and making up from dietary sources any deficits caused by degradation of the compounds. The route by which these phosphatidylamines are generated begins with the phosphorylation of the hydroxyl group of the free compounds by straightforward kinase reactions (Fig. 18-15), forming *phosphoethanolamine* or *phosphocholine*. (These compounds are not named ethanolamine or choline phosphate because such names also describe the salts of the amines with phosphoric acid.) The ester phosphates then react with cytidine triphosphate to form *cytidine diphosphate choline* and *cytidine diphosphate ethanolamine*. Except for the different nucleotides, these reactions are analogous

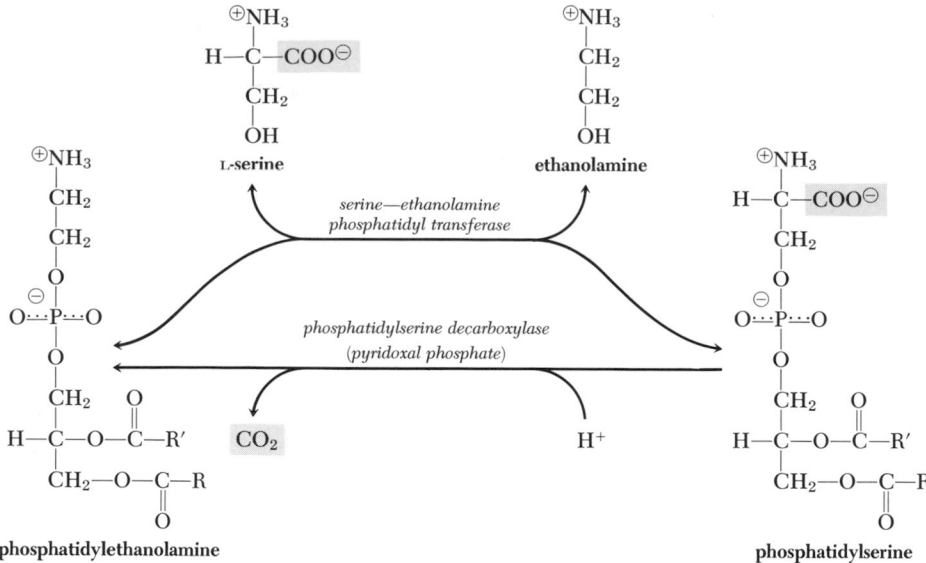

Figure 18-16 The various phosphatidylserines are formed from the corresponding phosphatidyleth-anolamines by an exchange of serine for ethanolamine. (There are many kinds, differing in the nature of the fatty acid residues.) Phosphatidylserine is a normal constituent of phospholipids, but also is a precursor of phosphatidylethanolamine through the loss of the carboxyl group. The enzyme is typical of amino acid decarboxylases in containing pyridoxal phosphate as a coenzyme, and we shall see more examples of this kind of reaction.

to the formation of uridine diphosphate glucose from glucose-1-phosphate and UTP. (With these reactions we have now shown examples of the participation of all of the common nucleoside triphosphates in metabolic reactions, in addition to their use as precursors of RNA. A complete rationalization of the use of cytidine nucleotides in phospholipid formation is not at hand, owing to lack of information on the compartmentalization of the compounds within the cell.)

Finally, the CDP-amines react with free 1,2-diglycerides to form either phosphatidylethanolamines (*cephalins*) or phosphatidylcholines (*lecithins*), and the remainder of the nucleotide portion is released as CMP. This reaction, forming a phosphate ester rather than an ether, therefore differs from the glycogen synthetase reaction in which UDP, not UMP, is released from UDP-glucose. CMP can be converted back to CTP by successive transfers from ATP:

$$\text{CMP} + \text{ATP} \longrightarrow \text{CDP} + \text{ADP} \qquad (CMP\ kinase)$$
$$\text{CDP} + \text{ATP} \longrightarrow \text{CTP} + \text{ADP} \qquad (nucleoside\ diphosphokinase)$$

CMP kinase is specific for the cytidine phosphates and may govern the rate of supply of CTP, although this is pure conjecture.

Should a dietary supply of either ethanolamine or choline fail, routes are available for the formation of both phosphatidyl compounds. The ethanolamine in phospholipids can be created from serine by an ingenious arrangement (Fig. 18-16). The phosphatidyl group in phosphatidylethanolamine is transferred to serine, creating *phosphatidylserine*, which in itself is a minor constituent of phospholipids, and releasing free ethanolamine. A second enzyme catalyzes the decarboxylation of phosphatidyl serine, leaving phosphatidylethanolamine as a product. This is the first example we have seen of a general class of enzymes, the *amino acid decarboxylases*. They are primarily used in mammals for the sort of purpose we see here, the formation of amines that are components of cells, and are therefore in lower concentration than the enzymes used for degradations or major syntheses. All of the amino acid decarboxylases have *pyridoxal phosphate* as a coenzyme.

The sum of these two enzymatic reactions amounts to a net formation of ethanolamine from serine:

1. serine + phosphatidylethanolamine $\longrightarrow$
    phosphatidylserine + ethanolamine
2. phosphatidylserine + H^+ $\longrightarrow$ phosphatidylethanolamine + CO_2

SUM: serine + H^+ $\longrightarrow$ ethanolamine + CO_2

Since serine can be made from glucose and ammonia, it follows that ethanolamine can also be completely synthesized from these compounds, and a dietary supply is not necessary.

The three methyl groups necessary to convert phosphatidylethanolamine to phosphatidylcholine are contributed by S-adenosylmethionine (Fig. 18-17). The

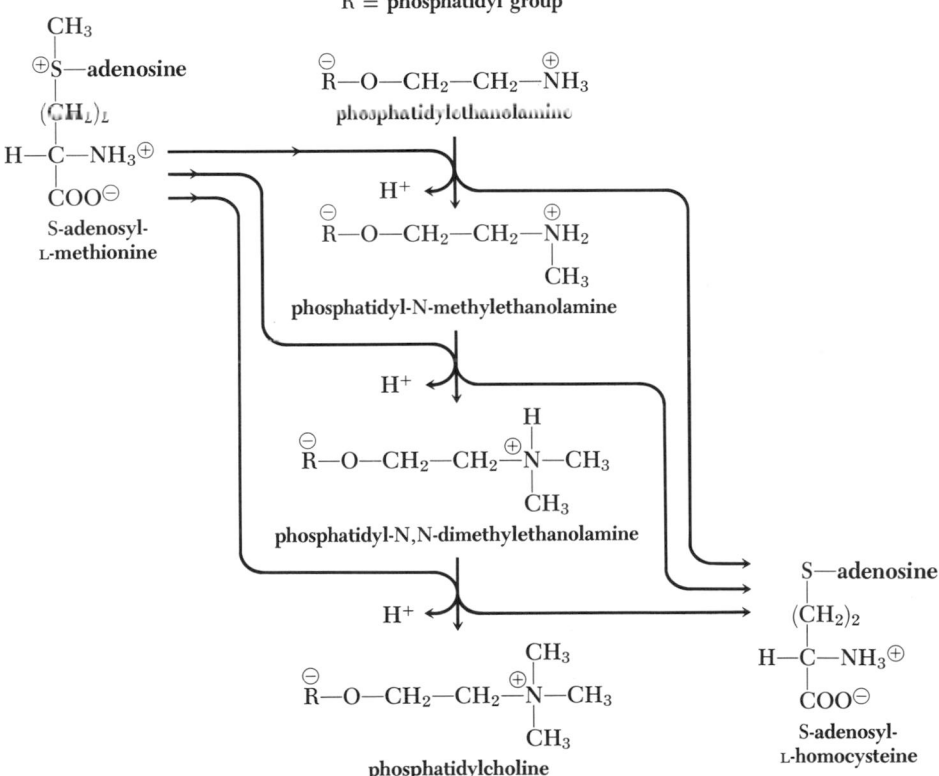

Figure 18-17 A phosphatidylcholine is formed from the corresponding phosphatidylethanolamine by three successive transfers of methyl groups from S-adenosyl-methionine.

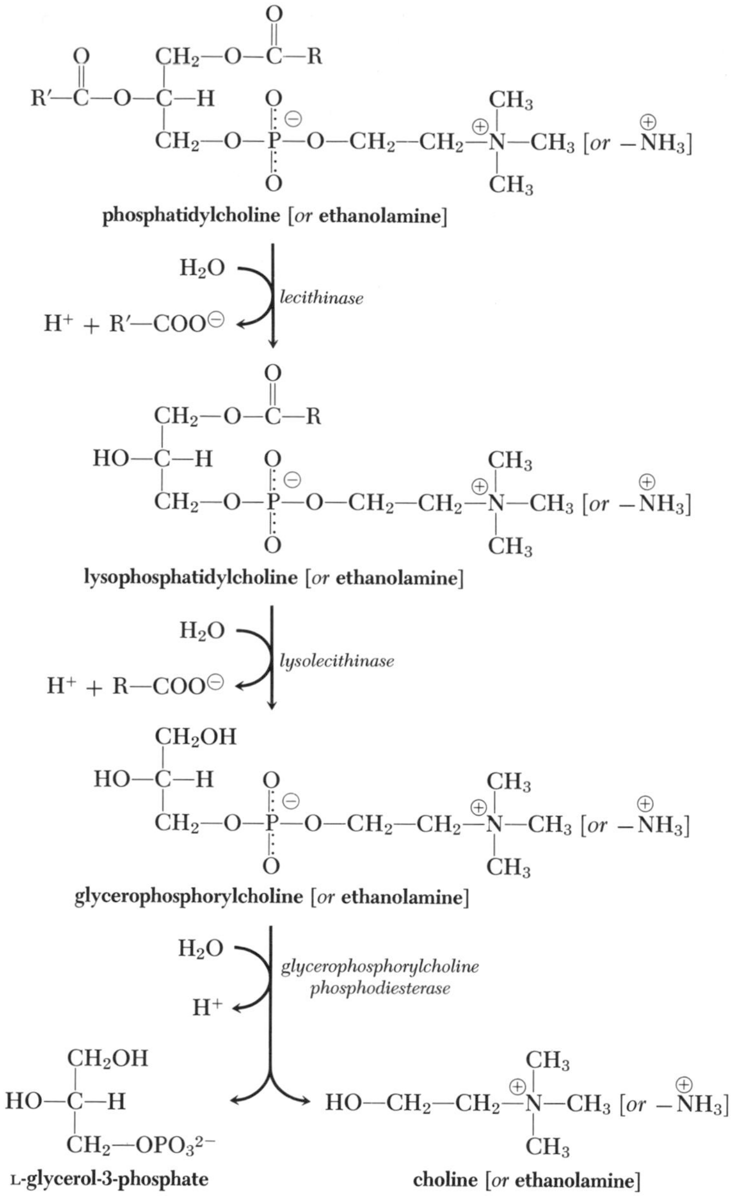

Figure 18-18 The phospholipids are degraded by successive hydrolyses of the fatty acid residues followed by hydrolysis of the base-phosphate bond. The glycerol-3-phosphate and free base may be re-incorporated into other molecules, or may be degraded.

enzymes catalyzing the three steps have not been characterized, but it appears from data with crude tissue preparations that the first methylation is catalyzed by one enzyme, and the next two by another. Whatever the route, three moles of S-adenosylmethionine are utilized to create one mole of phosphatidylcholine, and three moles of S-adenosylhomocysteine are thereby liberated. Both phosphatidyl-ethanolamine and phosphatidylcholine are constantly being hydrolyzed in cells. This is especially true of the phospholipids involved in fat transport. The sequence of hydrolyses (Fig. 18-18) can eventually lead to the formation of free ethanolamine and free choline, which are remade into phospholipids in other tissues or degraded by processes shortly to be considered.

Methylation of RNA

We noted in our discussion of protein synthesis that transfer RNA and ribosomal RNA contain bases other than the usual guanine, adenine, cytosine, and uracil. The distinctive bases frequently are methylated derivatives of the more common compounds (p. 38). All of these methyl derivatives are created by enzymatic transfer of methyl groups from S-adenosylmethionine to the ribonucleates, which originally contain only the non-methylated bases at the time of assembly on DNA templates. Even the thymine found in RNA is made by methylation of uracil, not by direct incorporation of thymidine at the time of nucleic acid formation. The bases to be methylated appear to be located in particular sequences, at least in the case of transfer RNA. There are several different enzymes involved, which have not been separated completely. (Separate enzymes are known to occur for methylating individual bases.) For our purposes now, the important thing to recognize is that the formation of these specific ribonucleates represents a drain upon the supply of methyl groups represented by S-adenosylmethionine.

REGENERATION OF METHIONINE

When the methyl group of methionine is utilized for the various kinds of purposes we have mentioned, the remainder of the molecule remains intact in the form of S-adenosyl homocysteine. A hydrolase exists to cleave this molecule:

$$\text{S-adenosylhomocysteine} + H_2O \longrightarrow \text{homocysteine} + \text{adenosine}$$

Homocysteine can again be methylated to form methionine, provided that appropriate donors are present. To the extent that this occurs, there is no consumption of the homocysteine portion of the methionine molecule.

The primary source for methyl groups used in the regeneration of methionine is choline. However, in order to make the liberation of free energy large enough to sustain the formation of methionine, choline is first oxidized at the alcohol carbon to create the corresponding carboxylate (Fig. 18-19). After this occurs, one of the methyl groups can be recovered as methionine. The two oxidations necessary for the conversion of choline to *betaine* occur in mitochondria and are linked to the electron transport chain, so 5 moles of high-energy phosphate can also be generated. Betaine is named from its discovery in beets, *Beta vulgaris,* and the English, in particular, emphasize this origin by pronouncing the word "beet-ane," but most use the more systematic pronunciation, "bay-tah-een."

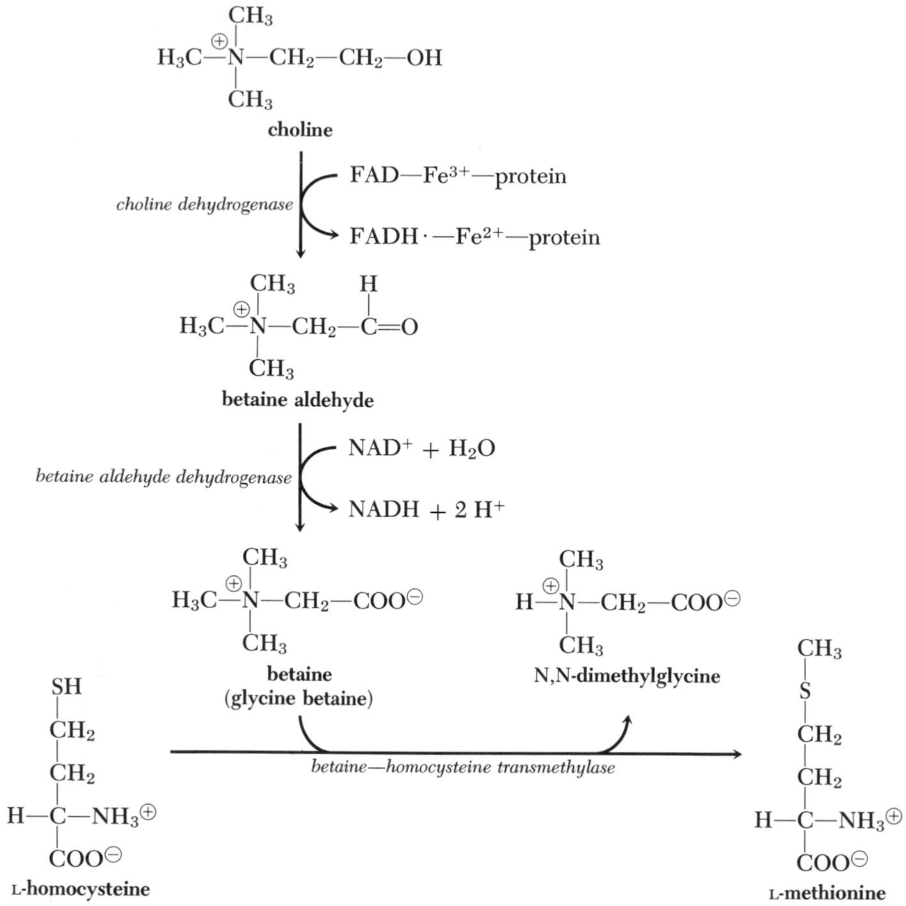

Figure 18-19   Excess choline is degraded in a way permitting direct recovery of one of the methyl groups by transfer to homocysteine.

Betaine is N,N,N-trimethylglycine. Other trimethyl derivatives of amino acids have been discovered (for example, carnitine, p. 228) and are also called betaines, so the original compound would be more exactly designated as glycine betaine, but this is rarely done, and the unqualified term refers to the glycine derivative.

After the methyl group is transferred to homocysteine, the remaining dimethyl glycine no longer has a "high-energy" methyl group, owing to loss of the quaternary nature of the nitrogen. However, the methyl groups are removed by specific flavoprotein oxidases in mitochondria, which convert the methyl groups to formaldehyde (Fig. 18-20). These oxidases are linked to the same electron transfer flavoprotein used to carry electrons from the acyl coenzyme A dehydrogenases in fatty acid oxidation. Each oxidation therefore generates two moles of high-energy phosphate per pair of electrons transferred.

The formaldehyde released from the methyl groups can be oxidized to formate and then to CO_2 in the mitochondria, or can enter the one-carbon pool by combining with tetrahydrofolate.

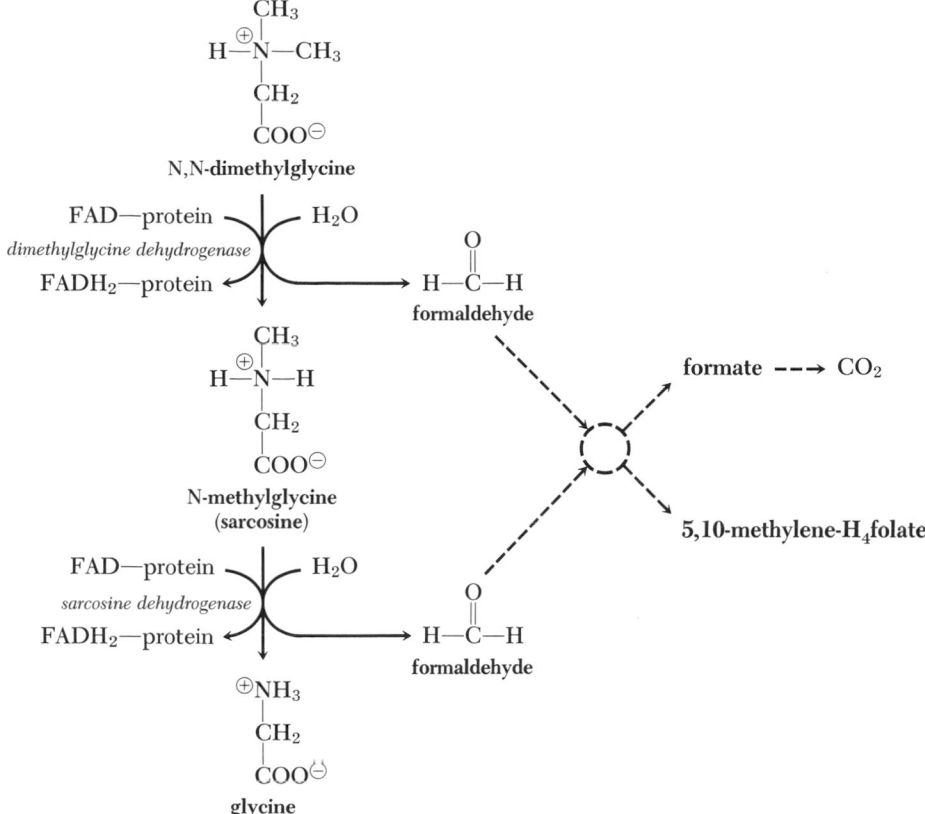

Figure 18-20 The methyl groups of the dimethylglycine resulting from choline metabolism are removed by oxidation to formaldehyde. The formaldehyde may be re-inserted into the one-carbon pool by combination with H_4folate, or may be oxidized to formate and then to CO_2 at times when the H_4folate is saturated with one-carbon units.

As we have seen many times over, the existence of routes for the utilization of a compound to form a product and its regeneration from this product does not imply a wasteful cyclical process. Methyl groups are donated by methionine to form a choline residue in phospholipids, not free choline, and this route will be employed when phospholipid synthesis is necessary and the supply of free choline is not adequate to maintain the lipids. The route for the oxidation we just discussed begins with free choline, and implies an excess of choline in the liver, supplied by degradation of phospholipids elsewhere or by the diet, and it represents a means for recovering the methyl groups for use—one as a direct donor in the form of methionine, and the others as part of the general one-carbon pool.

The one-carbon pool itself can contribute methyl groups to homocysteine, forming methionine. We saw that the one-carbon folate derivatives can be reduced to *5-methyl-H₄folate*. The methyl group on this compound can be transferred to homocysteine by an enzyme that requires the presence of *cobamide coenzyme* (Fig. 18-21).

A reducing agent and S-adenosylmethionine must also be present for the re-

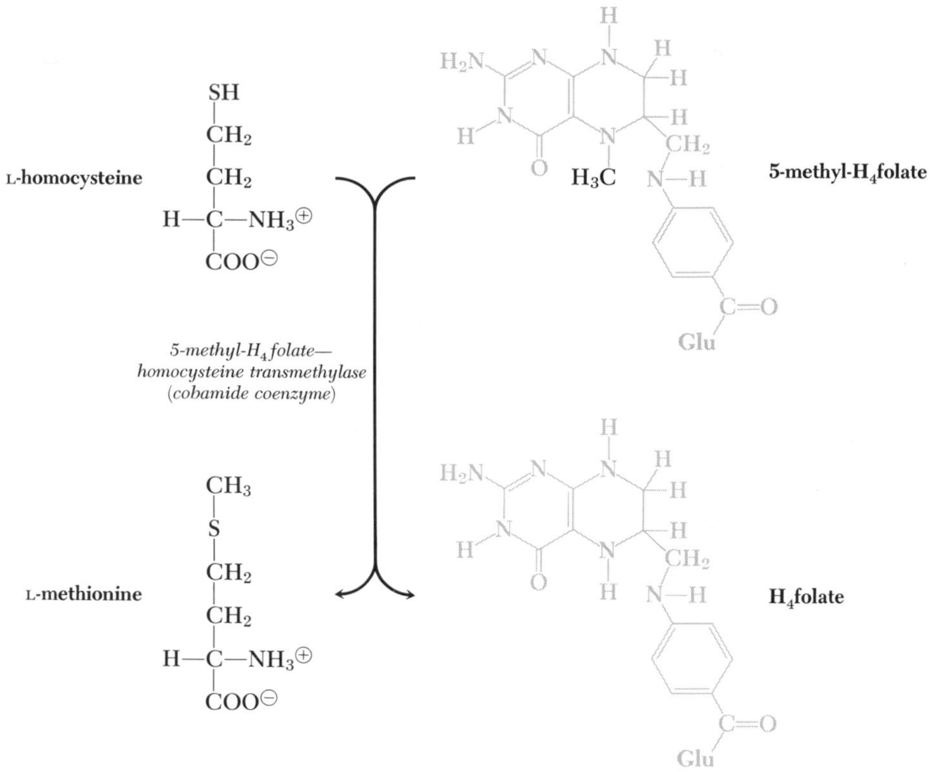

Figure 18-21 Homocysteine can be converted back to methionine by transfer of a methyl group from methyl-H_4folate. This reaction may serve to maintain the supply of methyl groups at times when there is a shortage of dietary methionine and choline. Cobamide coenzyme is required for the reaction.

action to proceed in experimental systems. The role of these compounds in the cell is problematical, because the methyl transfer involves the formation of a cobalt-methyl intermediate of the cobamide coenzyme, in which the cobalt is in a reduced valence state $(+1)$. The reduced coenzyme is notoriously unstable under laboratory conditions, and it is not certain whether the presence of S-adenosylmethionine is necessary to maintain it as the more stable methylated form during the course of an experiment or is necessary for some mechanism not presently understood.

DEGRADATION OF HOMOCYSTEINE

Nitrogen: Released as NH_4^+.
Carbons: Converted to succinyl coenzyme A.
Sulfur: Appears as cysteine.
   Although some homocysteine can be recovered as methionine by methylation reactions, part is constantly lost through degradation by an irreversible route in which the sulfur appears as cysteine and the carbon skeleton as 2-ketobutyrate,

which we have already seen is metabolized *via* propionyl coenzyme A and CO_2 to succinyl coenzyme A. Therefore, methionine is a source of cysteine and methyl groups and is a glucogenic amino acid. *Methionine is an important dietary constituent for three purposes.* Most importantly, it is required as an amino acid for the formation of proteins, and no other dietary component will substitute for this purpose, since the homocysteine chain cannot be made. As a source of methyl groups it is important, but not imperative, since choline in the diet can provide methyl groups, and a further supply is available through the one-carbon pool by the formation of methyl-H_4folate. Finally, it is a source of sulfur available for the formation of cysteine. This relieves the necessity of an absolute requirement for cysteine in the diet, but it means that sufficient methionine must be present to provide an excess sulfur supply, if the content of cysteine is low.

The actual degradation of homocysteine is a simple two-step process involving the formation of *cystathionine* and its cleavage to form cysteine, 2-ketobutyrate and ammonium ion (Fig. 18-22). Both of the enzymes involved have pyridoxal phosphate as a coenzyme.

Congenital defects of both these enzymatic steps are known in humans. *Homocystinuria,* the excretion of the oxidized form of homocysteine, occurs in a congenital absence of cystathionine synthase, which also results in mental retardation. *Cystathioninuria,* the excretion of cystathionine, would be expected to result from a congenital deficiency of cystathioninase, but this has not been discovered. Instead, the known cases result from formation of enzyme with an apparent low affinity for its coenzyme, pyridoxal phosphate. The effect is the same—the loss of ability to cleave cystathionine in the tissues—but it serves to emphasize that lack of an enzymatic activity does not imply that the corresponding protein is not being made in any form. The protein may be present, but the mutation may have changed some critical amino acid residue so that it can no longer catalyze an enzymatic reaction, and this in turn may be due to loss of ability to bind substrates, to bind coenzymes, or to carry out the catalytic mechanism.

Cystathioninuria also is associated with mental defects. When one considers the unusually high content of cystathionine in the brains of normal individuals (0.2 to 0.5 mg per g of tissue, as contrasted with 7 to 8 μg per g in other tissues) and

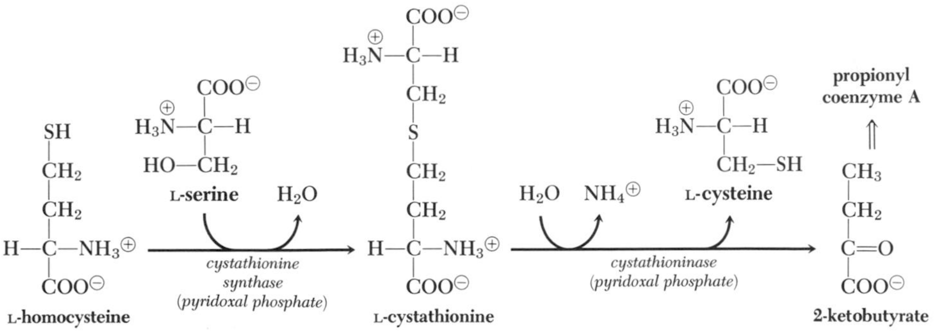

Figure 18-22   The degradation of homocysteine. The effective result is the transfer of the sulfur to serine, forming cysteine, and a deamination. The carbon skeleton appears as 2-ketobutyrate, which is also the product of threonine metabolism.

the disturbances in the nervous system apparently created by inability to make or destroy this amino acid, one suspects it may have some presently unknown function in nerve physiology. Primates have a higher concentration in the brain than do other mammals, whereas birds have much lower level than do mammals. This is not to imply that the term "hypocerebrocystathioninic" could rationally be substituted for bird-brained as an epithet.

Recapitulation of types of reactions

1. Tetrahydrofolate reacts with formaldehyde to incorporate a methylene group:

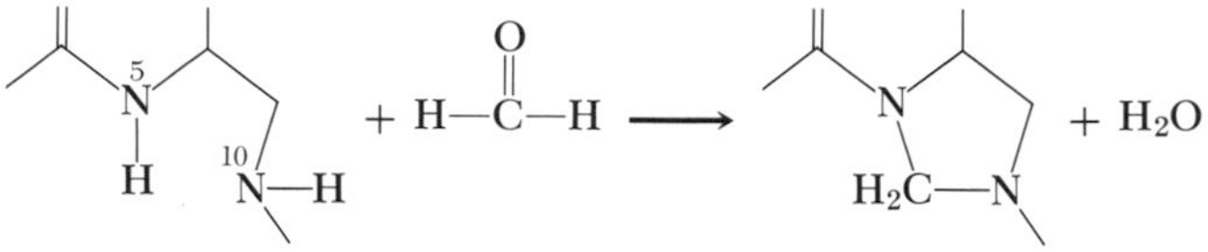

H$_4$folate   formaldehyde   5,10-methylene-H$_4$folate

2. Tetrahydrofolate reacts with serine to incorporate a methylene group and release glycine:

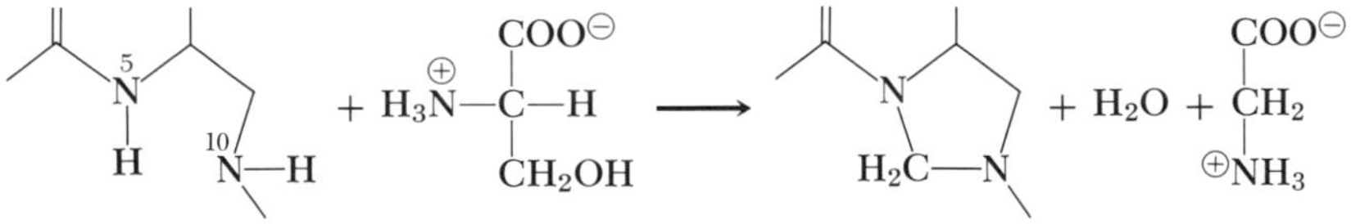

H$_4$folate   L-serine   5,10-methylene-H$_4$folate   glycine

3. 5,10-Methylenetetrahydrofolate may be reduced to 5-methyltetrahydrofolate:

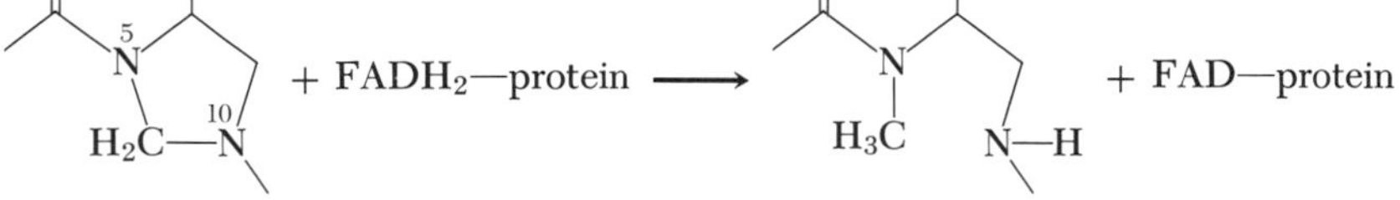

5,10-methylene-H$_4$folate   5-methyl-H$_4$folate

4. 5,10-Methylenetetrahydrofolate may be oxidized to 5,10-methylidynetetrahydrofolate:

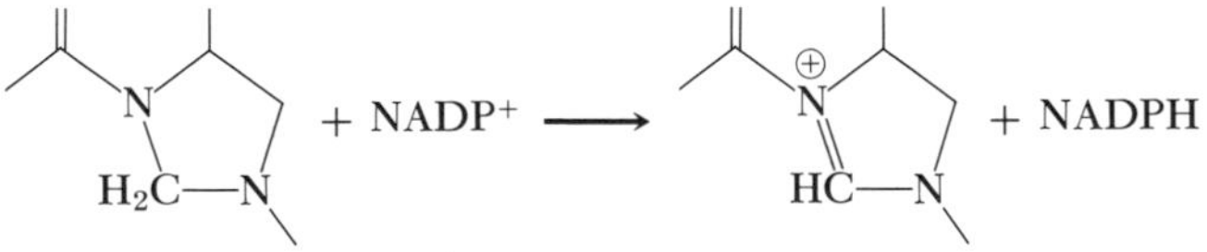

5,10-methylene-H$_4$folate   5,10-methylidyne-H$_4$folate

5. 5,10-Methylidynetetrahydrofolate is interconvertible with 10-formyltetrahydro-folate:

5,10-methylidyne-H$_4$folate   10-formyl-H$_4$folate

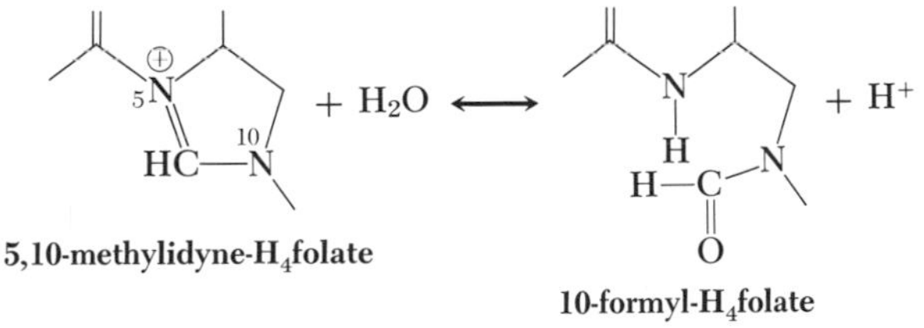

6. Tetrahydrofolate reacts with formate to form 10-formyltetrahydrofolate:

H$_4$folate   formate   10-formyl-H$_4$folate

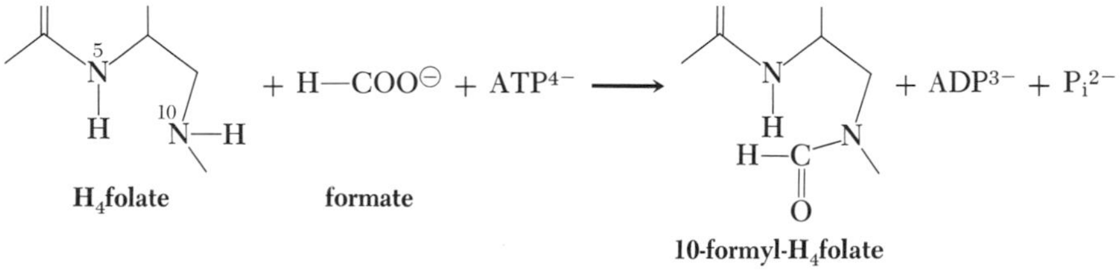

7. Tetrahydrofolate reacts with N-formimino-L-glutamate to incorporate the formimino group and release glutamate:

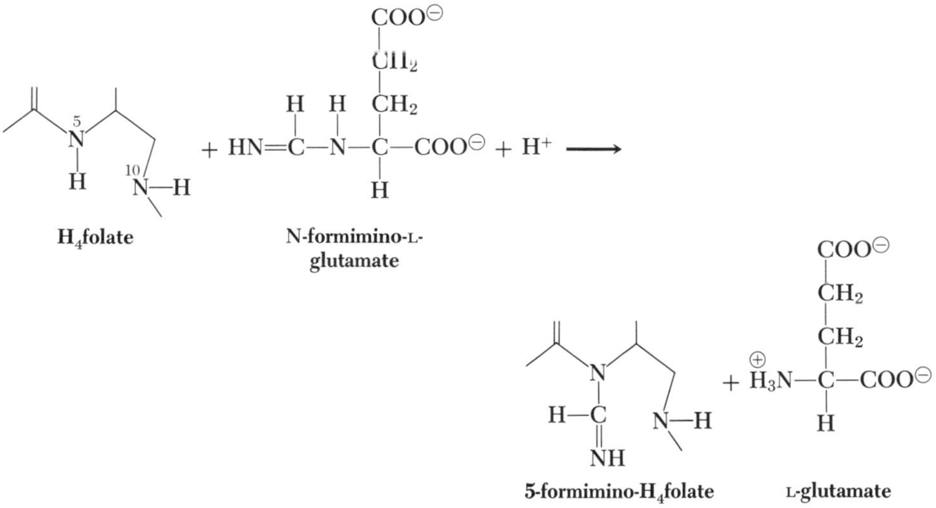

H$_4$folate   N-formimino-L-glutamate   5-formimino-H$_4$folate   L-glutamate

8. 5-Formiminotetrahydrofolate loses ammonium ion and undergoes cyclization to form 5,10-methylidynetetrahydrofolate:

5-formimino-H$_4$folate   5,10-methylidyne-H$_4$folate

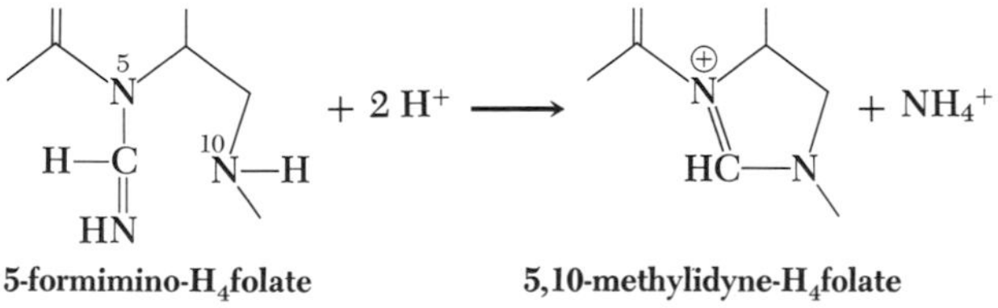

9. Glycine may react with a pyridoxal phosphate containing protein to form a Schiff's base that is oxidatively decarboxylated, followed by transfer of the remaining methylol group to tetrahydrofolate, with release of ammonium ion:

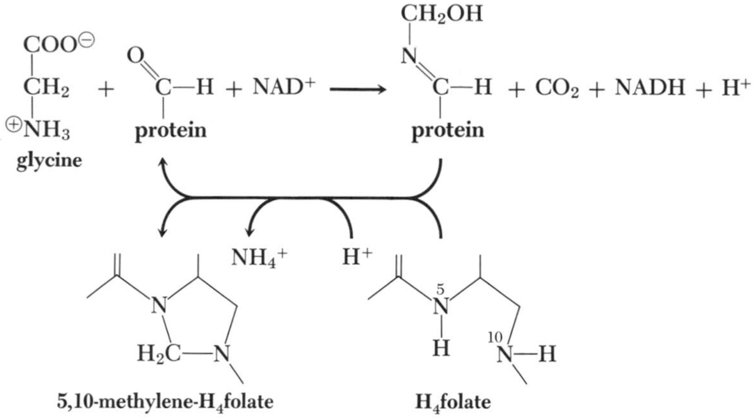

5,10-methylene-H_4folate H_4folate

10. ATP may react with methionine to convert the thioether group to the sulfonium group of S-adenosylmethionine:

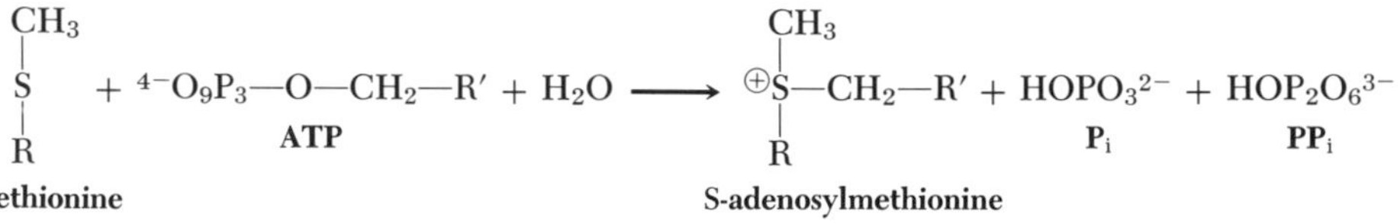

11. The methyl group of S-adenosylmethionine may be transferred to a nucleophilic atom, typically N or O:

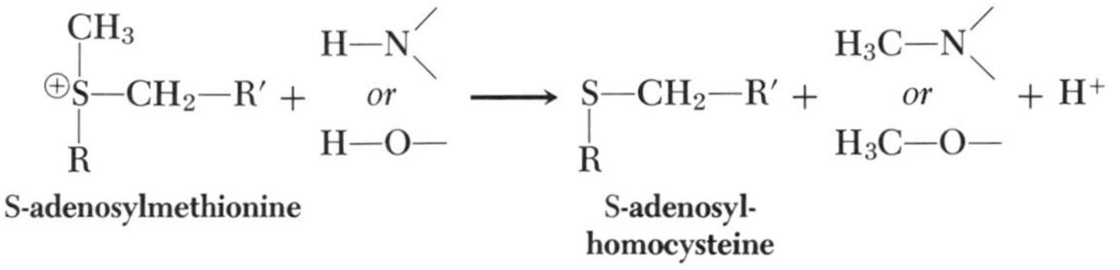

Example: The methylation of guanidinioacetate to form creatine.

12. Thioethers may be hydrolyzed to form mercaptans and alcohols:

$$R—S—R' + H_2O \longrightarrow R—SH + HO—R'$$

Examples: The hydrolysis of S-adenosylhomocysteine; the hydrolysis of cystathionine (also accompanied by deamination).

13. The methyl group of 5-methyltetrahydrofolate may be transferred to homocysteine:

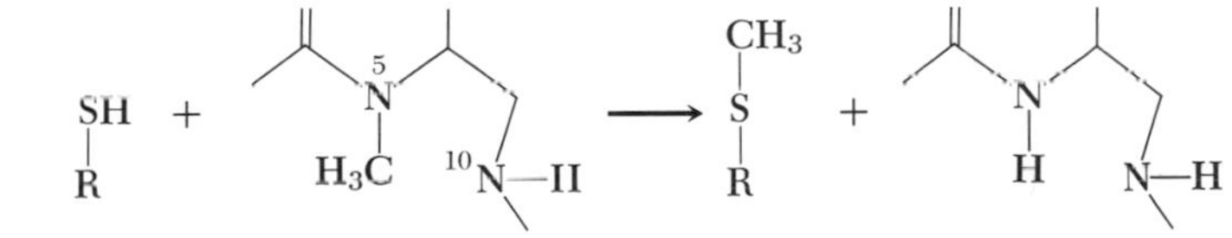

homocysteine   5-methyl-H_4folate   methionine H_4folate

14. A methyl group of betaine may be transferred to homocysteine:

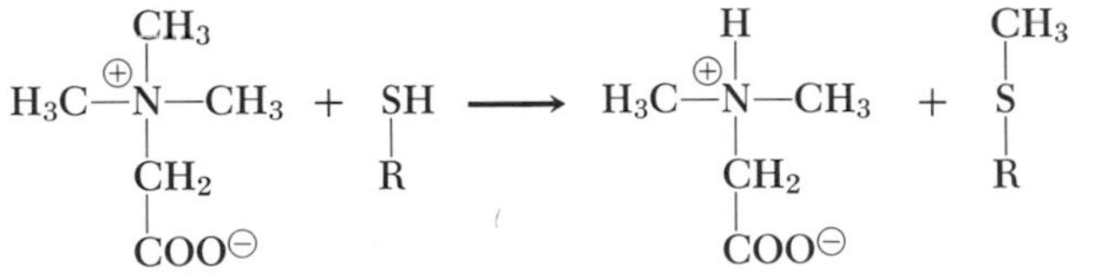

betaine   homocysteine   N,N-dimethylglycine   methionine

15. Mercaptans and alcohols may combine to form thioethers:

$$R\text{—}SH + HO\text{—}R' \xrightarrow{\text{\textit{pyridoxal phosphate}}} R\text{—}S\text{—}R' + H_2O$$

Example: The reaction of homocysteine and serine to form cystathionine.

16. N-methyl groups may be removed by oxidation to form formaldehyde:

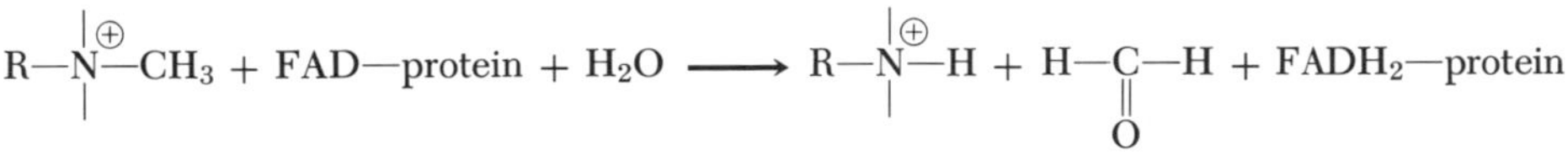

Example: The oxidative demethylation of dimethyl glycine and of sarcosine.

17. Nucleoside diphosphate alkyl esters may be cleaved by alcohols to form phosphodiesters with the release of nucleoside monophosphate:

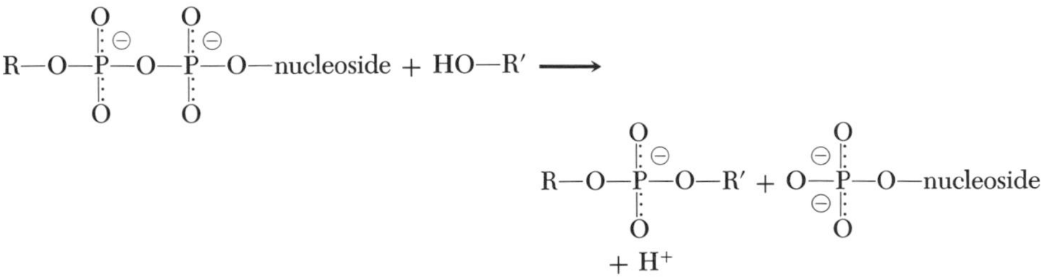

Example: The reactions of cytidine diphosphate choline or cytidine diphosphate ethanolamine with diglycerides to form phosphatidylcholine or phosphatidylethanolamine.

18. Amino acids may be decarboxylated to form amines:

$$
\underset{\text{amino acid}}{\text{R}-\underset{\overset{|}{\text{H}}}{\overset{\overset{\text{COO}^{\ominus}}{|}}{\text{C}}}-\text{NH}_3^{\oplus}} + \text{H}^+ \xrightarrow{\overset{\textit{pyridoxal}}{\textit{phosphate}}} \underset{\text{amine}}{\text{R}-\text{CH}_2-\text{NH}_3^{\oplus}} + \text{CO}_2
$$

Example: The decarboxylation of phosphatidylserine to form phosphatidylethanolamine.

Further reading

Meister, A.: *Biochemistry of the Amino Acids,* 2nd ed. Academic Press (1965).

Greenberg, D. M., ed.: *Metabolic Pathways,* 2nd ed., Vol. 3. Academic Press (1968).

Huennekens, F. M., and M. J. Osborn: *Folic Acid Coenzymes and One-Carbon Metabolism.* Adv. Enzymology, *21:* 369 (1959).

Hitchings, G. H., and J. J. Burchell: *Inhibition of Folate Biosynthesis and Function as a Basis for Chemotherapy.* Adv. Enzymology, *27:* 417 (1965).

Sato, T., Y. Motokawa, H. Kochi, and G. Kikuchi: *Glycine Synthesis by Extracts of Acetone Powder of Rat-liver Mitochondria.* Biochem. Biophys. Res. Comm., *28:* 495 (1967).

Thompson, J. S., and K. E. Richardson: *Isolation and Characterization of an L-Alanine: Glyoxalate Aminotransferase from Human Liver.* J. Biol. Chem., *242:* 3614 (1967).

Shapiro, S. K., and F. Schlenk, eds.: *Transmethylation and Methionine Biosynthesis.* University of Chicago Press (1965).

Hsia, D. Y-Y, ed.: *Symposium on Treatment of Amino Acid Disorders.* Am. J. Diseases Child. *113:* 1 (1967).

Baguley, B. C., and M. Staehelin: *The Specificity of Transfer Ribonucleic Acid Methylases from Rat Liver.* Biochemistry, *7:* 45 (1968).

The following two papers detail some of the recent studies on glyoxylate metabolism: Schlossberg, M. A., et al.: *Isolation and Identification of 5-Hydroxy-4-ketovaleric Acid as a Product of α-Ketoglutarate: Glyoxylate Carboligase.* Biochemistry, *7:* 333 (1968).

Kobes, R. D., and E. E. Dekker: *2-Keto-4-hydroxyglutarate Aldolase of Bovine Liver.* J. Biol. Chem., *244:* 1919 (1969).

AMINES

CHAPTER 19

ARGUMENT

The constituents of the body include a number of compounds containing amine groups in addition to the amino acids found in proteins. Some are of major importance even though they occur in low concentrations and account for only a small fraction of the total nitrogen metabolism. For example, acetylcholine and noradrenaline act as transmitters in particular parts of the nervous system, and their effects are so distinctive that nerves are classed as cholinergic or adrenergic, depending upon which transmitter is involved. Acetylcholine is formed by direct acetylation of choline with acetyl coenzyme A. Noradrenaline is derived from tyrosine by oxidation of the aromatic ring, decarboxylation, and further oxidation of the side chain. In the adrenal medulla, noradrenaline is methylated to produce the hormone, adrenaline. Somewhat related reactions occur in other parts of the nervous system and in the intestinal mucosa and convert tryptophan to serotonin. Serotonin has powerful physiological effects, but the mechanism of its action is uncertain.

Other amino acid decarboxylases exist to form amines. Histamine is made from histidine in mast cells and 4-aminobutyrate is made from glutamate in the central nervous system. The exact function of these amines is unknown. Ornithine is decarboxylated to form putrescine, which is combined with part of the methionine molecule to form spermidine in many tissues. These amines are believed to play a role in membrane formation and nucleic acid replication.

Another side reaction to tryptophan metabolism results in the formation of nicotinate mononucleotide for incorporation in NAD. Since only a fraction of the tryptophan being degraded can be diverted into this pathway, it can only result in complete replacement of the dietary nicotinate requirement when there is an ample supply of tryptophan above that necessary for protein synthesis. Consequently, people whose diets are low in tryptophan are also susceptible to a deficiency of nicotinate.

There are some amino acid derivatives known to occur in relatively high concentrations, but whose function is yet an enigma. For example, N-acetylaspartate is abundant in brain for unknown reasons. Muscles contain high concentrations of carnosine and anserine, which are derivatives of histidine. These compounds are likely to be involved in the contractile process, but we don't know how.

The amino acids are precursors of a wide variety of small molecules containing nitrogen that occur in relatively low concentrations, but have potent physiological actions. Many have been studied intensively because of their function in the nervous system; others are hormones or constituents of coenzymes; still others are known mainly from the effects of excessive concentration with their physiological role still unknown. With the major types of reactions of amino acids fresh in mind, this is a good time to consider the formation of some of these compounds.

AMINES OF THE NERVOUS SYSTEM

Acetylcholine appears to be a general transmitting agent between effector nerves outside the central nervous system, that is, at all synapses in ganglia transmitting signals to tissues. It is formed by a simple reaction between choline and acetyl coenzyme A. We already discussed the role of acetylcholinesterase in destroying the acetylcholine liberated at synapses or at the motor endplate on skeletal muscles when we used it as an example of an enzyme (p. 100).

Other amines are the transmitters at the terminals of nerves that comprise the sympathetic nervous system and that are in part of the central nervous system. The transmitters in sensory neurons are yet to be characterized.

Adrenaline and Noradrenaline

Most of the tyrosine in the body that is not used for protein synthesis is metabolized to CO_2 and urea by the route beginning with transamination (p. 386). A small part of the tyrosine is diverted to make other components in some tissues by pathways beginning with an oxidation. These tissues are derived from the ectodermal layer of an embryo and include some nerves, the skin, and the adrenal medulla. While the routes are trivial in a quantitative sense, they are of major qualitative importance, since they form amines used in nerve transmission, the hormone, *adrenaline*, and the pigment of the skin, *melanin*.

The oxidation of tyrosine in nerves and the adrenal medulla is catalyzed by a *tyrosine hydroxylase*, which resembles phenylalanine hydroxylase (p. 385). It utilizes molecular oxygen, which is activated by tetrahydrobiopterin so that one atom of oxygen goes onto the aromatic ring of the substrate and the other oxidizes the coenzyme (Fig. 19-1). The product is *3,4-dihydroxyphenylalanine*. This amino acid is then decarboxylated by the action of an enzyme typical of amino acid decarboxylases in having pyridoxal phosphate as a cofactor. The amine that is formed is *3,4-dihydroxyphenylethylamine*. Some of the nerves in the central nervous system have especially high concentrations of this compound.

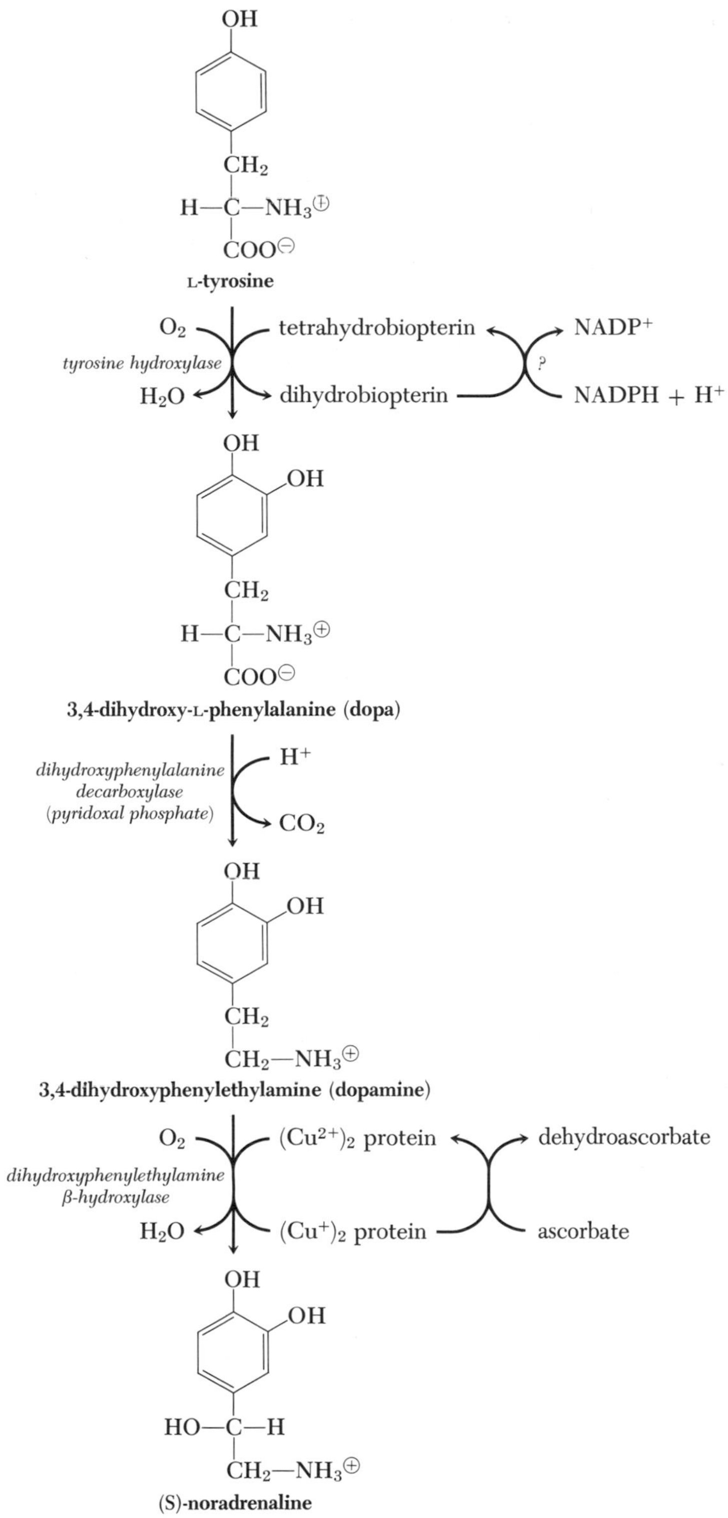

Figure 19-1   The formation of noradrenaline from tyrosine in the adrenergic nerves and the adrenal medulla begins with a hydroxylation involving tetrahydrobiopterin, followed by a decarboxylation. The side-chain hydroxyl group is introduced by another mixed-function oxidase using ascorbate as the second electron donor. Its mechanism is shown in the next figure.

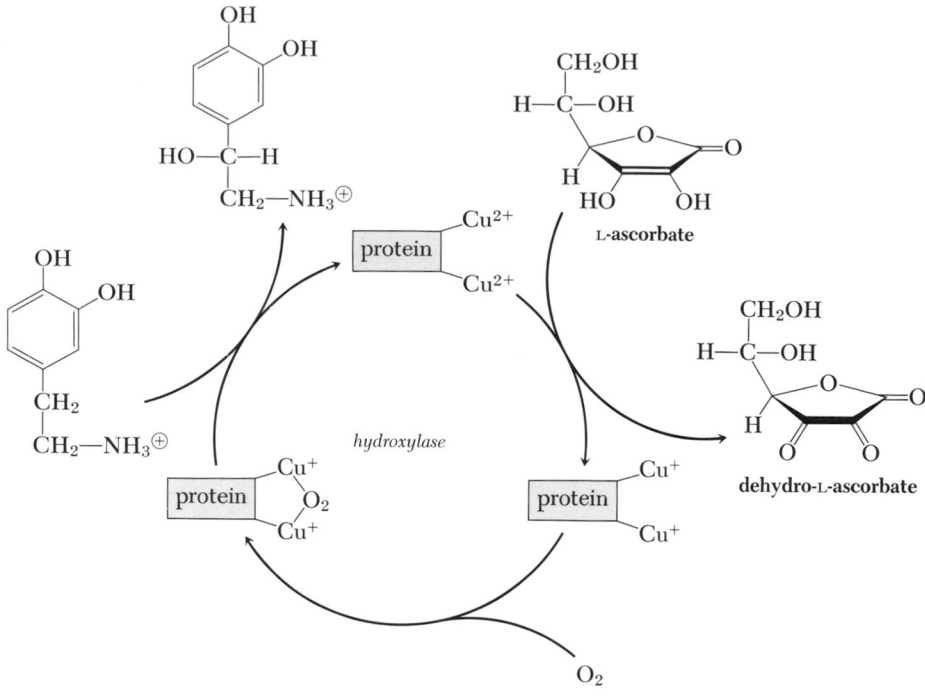

Figure 19-2   The mechanism of dihydroxyphenylethylamine β-hydroxylase involves the reduction of cupric ions on the enzyme by ascorbate, followed by combination of the cuprous enzyme with molecular oxygen. One atom of the activated oxygen is transferred to the substrate and the other appears as water. (The addition and subtraction of the necessary protons on the enzyme during these transformations is not shown.)

In other nerves and the adrenal medulla, dihydroxyphenylethylamine is again oxidized, this time on the β-carbon of the side chain, producing *noradrenaline*. The oxidase is a copper-protein, and is another example of a class of mixed-function oxidases that apparently use *ascorbate* (vitamin C) as a second reducing agent in reactions with molecular oxygen (Fig. 19-2). Noradrenaline is the transmitter substance in sympathetic nerves and at some synapses in the central nervous system. It is also synthesized in the adrenal medulla where it is further methylated to form *adrenaline,* with S-adenosyl methionine as methyl donor (Fig. 19-3).

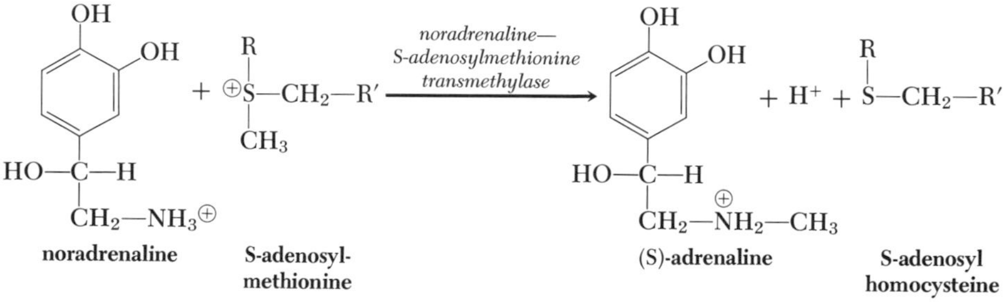

Figure 19-3   The formation of adrenaline by the methylation of noradrenaline.

The enzymes catalyzing the conversion of tyrosine to noradrenaline in nerves are contained in organelles near the synapses. The noradrenaline is bound and stored in particulates associated with these organelles. Evidently the active structure of the synapse contains a small amount of the amine, which is replenished from these particulate stores after discharge during transmission.

As in the case of acetylcholine, it is necessary to remove the amine that is released during discharge so as to restore the synapse to its original state, ready to transmit another impulse. For some reason, the adrenergic nerves are not built to destroy noradrenaline at the site of its action in the way that acetylcholine is removed in the cholinergic nerves. Instead, part of the amine diffuses into the blood and is carried away; another part is taken up again into the storage granules, and this storage in some way involves the expenditure of ATP. The concentration of the amine is kept from rising, either within the nerve or within the blood, by enzymes that catalyze the conversion of the amine to inactive forms.

Within the nerve, any amine diffusing away from the synapse into the remainder of the cell is destroyed by a *monoamine oxidase* of uncharacterized nature in the mitochondria. The oxidase is probably a flavoprotein like the enzyme found in the mitochondria of other tissues. It catalyzes the formation of an aldehyde (Fig. 19-4), which in turn is oxidized to the carboxylate, probably by an NAD-aldehyde dehydrogenase.

The amines in the blood stream are removed by the liver, which has an enzyme transferring methyl groups from S-adenosylmethionine onto one of the phenolic hydroxyl groups. Part of the resultant methoxy compound is excreted in the urine and part is further oxidized. The N-methyl group of O-methyladrenaline can be removed as formaldehyde by an oxidation similar to the reaction of N-methylglycine (p. 431), yielding the O-methyl derivative of noradrenaline. The primary amines are oxidized by amine oxidase to eventually yield the carboxylates, which are also excreted in the urine.

The determination of 3-methoxy-4-hydroxymandelate excretion is a useful guide in following the clinical management of tumors such as pheochromocytoma and neuroblastoma that cause the formation of adrenaline or noradrenaline. (The compound is known clinically as *vanillmandelate* or *vanillylmandelate,* an excruciating example of bastard nomenclature whose origin will be evident in the next paragraph.)

3,4-Dihydroxyphenyl derivatives are widespread in plants. The methoxy compounds frequently have a pleasant odor; for example, eugenol in cloves and vanillin in vanilla are mostly responsible for the flavor of these spices.

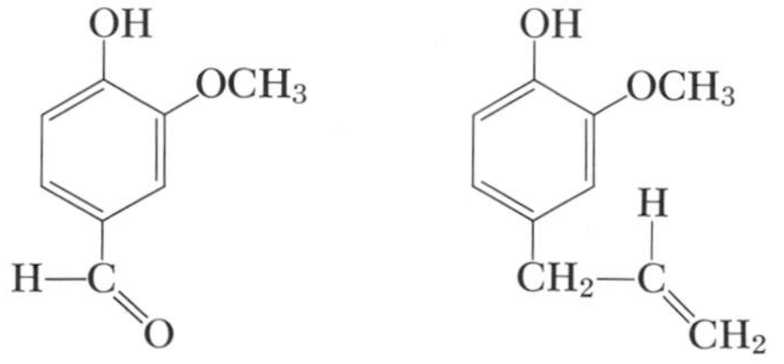

OH
OCH₃

H—C
O

vanillin
(in vanilla)

OH
OCH₃

H
CH₂—C
CH₂

eugenol
(in cloves)

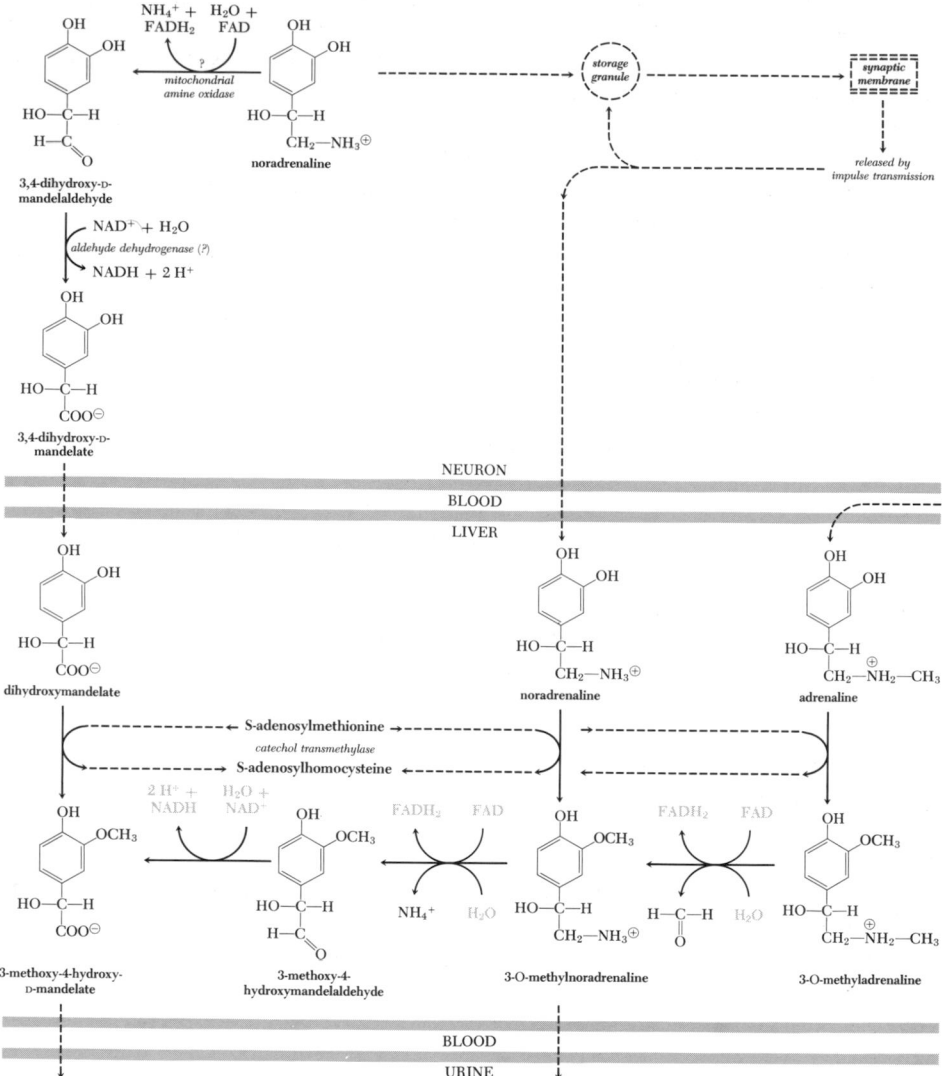

Figure 19-4   The degradation of noradrenaline and adrenaline. Noradrenaline formed in neurons is stored in granules to provide a supply for charging the synapses. Part of the noradrenaline released upon stimulation (*top right*) is stored again in the granules. Part is oxidatively deaminated in the neuron (*top left*), with the resultant aldehyde being further oxidized to 3,4-dihydroxy-D-mandelate, which diffuses into the blood. Another part of the noradrenaline diffuses into the blood. Both of these compounds appearing in the blood, along with adrenaline elaborated by the adrenal medulla, are further metabolized in the liver. Hepatic metabolism begins in each case with the methylation of one of the phenolic groups. The N-methyl group of the adrenaline derivative is then removed by oxidation to formaldehyde, leaving the 3-O-methylnoradrenaline (*bottom center*) also obtained by methylation of noradrenaline. The amino group of this compound is then removed by oxidative deamination, followed by oxidation of the aldehyde, with the ultimate formation of 3-methoxy-4-hydroxy-D-mandelate in all cases. This compound, along with some of the O-methylnoradrenaline, appears in the urine. (There are other possible combinations of the various kinds of reactions shown, leading to the formation of lesser amounts of other products.)

5-Hydroxytryptamine (Serotonin)

5-Hydroxytryptamine is concentrated in some cells of the central nervous system, and is believed to be a transmitter, or at least to directly affect synaptic transmission in those cells. (It is also formed in the intestinal mucosa for an unknown reason.) The route to the amine is like the route to noradrenaline in that it begins by the direct hydroxylation of an aromatic ring, in this case of tryptophan (Fig. 19-5). The enzyme utilizes tetrahydrobiopterin as a coenzyme and hydroxylates the indole ring of tryptophan at C-5. The 5-hydroxytryptophan formed can be acted on by dihydroxyphenylalanine decarboxylase in the neurons to form 5-hydroxy-tryptamine.

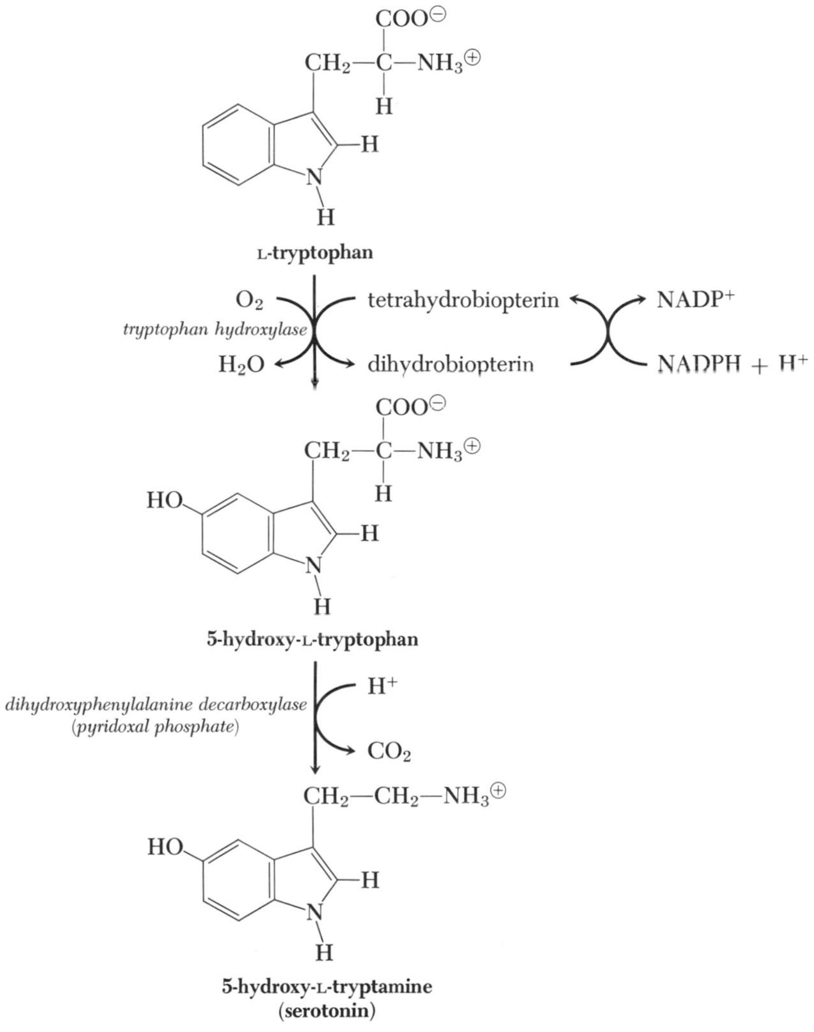

Figure 19-5   The formation of serotonin from tryptophan. Although different enzymes are involved in the first step, this transformation resembles the first two steps in the formation of noradrenaline from tyrosine. Indeed, the same enzyme catalyzes the decarboxylation reaction in both cases.

Hydroxytryptamine is removed by the action of monoamine oxidase and aldehyde dehydrogenase, which convert it to the corresponding carboxylate:

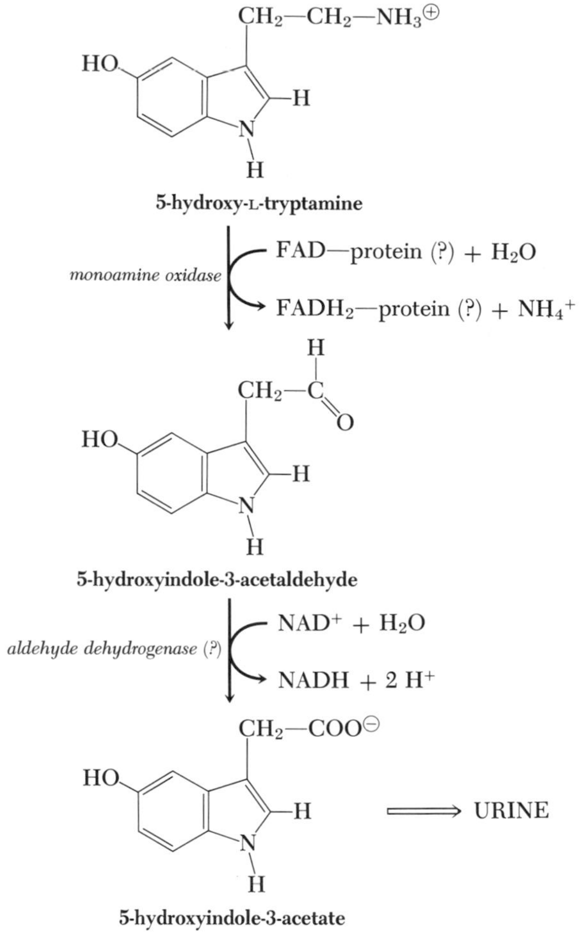

5-hydroxy-L-tryptamine

monoamine oxidase

FAD—protein (?) + H_2O

FADH$_2$—protein (?) + NH_4^+

5-hydroxyindole-3-acetaldehyde

aldehyde dehydrogenase (?)

NAD^+ + H_2O

NADH + 2 H^+

5-hydroxyindole-3-acetate $\Longrightarrow$ URINE

The carboxylate is excreted in the urine.

4-Aminobutyrate

Glutamate is decarboxylated in the brain by a specific enzyme, containing pyridoxal phosphate as a coenzyme:

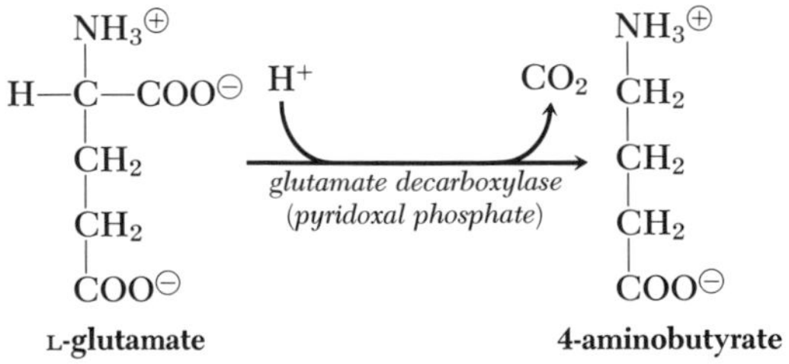

L-glutamate

H^+ → *glutamate decarboxylase (pyridoxal phosphate)* → CO_2

4-aminobutyrate

The resultant 4-aminobutyrate accumulates to a total concentration of 0.8 millimoles per kilogram of tissue, and the local concentration in some neurons is presumed to be even higher. The function of the compound is unknown. It inhibits neural transmission, and there are strong advocates of the belief that it is a physiological inhibitor (but there are equally strong dissenters).

The compound is metabolized at a rapid rate by a route involving an initial transamination followed by an NAD-linked dehydrogenation to succinate:

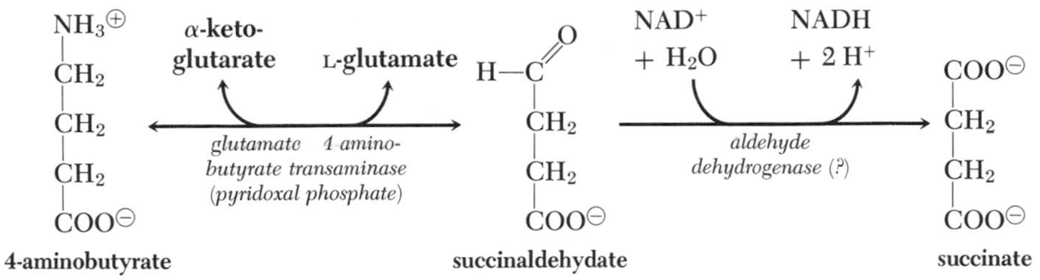

4-aminobutyrate   succinaldehydate   succinate

On the face of it, the route represents a bypass of the α-ketoglutarate dehydrogenase reaction in the citric acid cycle:

1. glutamate $\longrightarrow$ 4-aminobutyrate + CO_2
2. 4-aminobutyrate + α-ketoglutarate $\longrightarrow$ succinaldehydate + glutamate
3. succinaldehydate + NAD^+ $\longrightarrow$ succinate + CO_2 + NADH

SUM: α-ketoglutarate + NAD^+ $\longrightarrow$ succinate + CO_2 + NADH

Much has been made of this as a major route of glucose metabolism in brain, but optimistic estimates allow no more than 0.1 of the α-ketoglutarate to be handled by this route. Beyond that, it seems purposeless as a bypass in the citric acid cycle, because the generation of GTP ordinarily occurring from succinyl coenzyme A would be lost.

Neuropharmacology

Control of the metabolism and function of the amines in the nervous system of humans is the basis of action for a number of the most important drugs available to a clinician, and the study of this action is a large part of pharmacology (Fig. 19-6). In general, drugs may compete for a receptor site for the active amine at the synapse or receptor junction. *Curare* (p. 106) is an example of such a blocking agent effective at neuromuscular junctions. Others have been developed with considerable selectivity for ganglionic sites or adrenergic receptors. *Lysergic acid diethylamide* (LSD) apparently owes its hallucinogenic activity to competition with 5-hydroxytryptamine in the central nervous system.

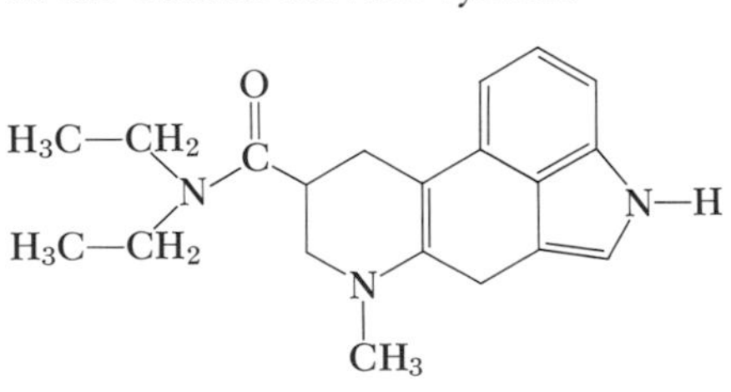

lysergic acid diethylamide (LSD)

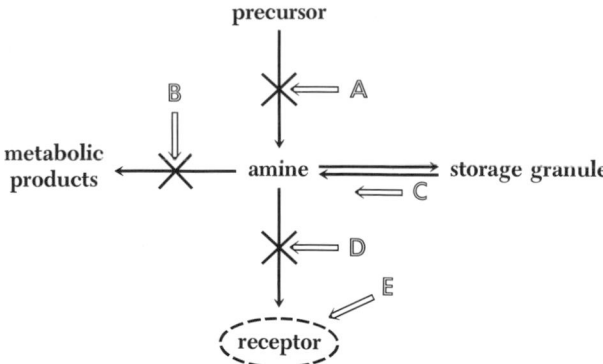

Figure 19-6   A compound may have pharmacological effects by acting on any part of the metabolic processes involving a biologically active amine. (A) A compound may diminish the effects of the amine by preventing its formation. (B) Another compound may exaggerate the effects by preventing the degradation of an amine so that it accumulates. (C) Still another compound may exaggerate the effects by causing increased mobilization of the free compound from storage granules or any other bound form. (D) The action of the amine may be prevented by a compound inhibiting its combination with the receptor site. (E) Finally, some compounds may combine with the receptor and mimic the action of the amine so as to produce the same effects as the natural compound, or to augment its action.

Other drugs may prevent the synthesis of an active amine, or discharge the amines from storage granules. In either case, the amine is not available for recharging synapses at the usual rate, and the sensitivity of response is diminished. The tranquilizer, *reserpine,* acts in the latter way.

The concentration of an amine may be increased by inhibition of its destruction. We already mentioned agents preventing the destruction of acetylcholine (p. 106). Agents inhibiting monoamine oxidase slow the removal of noradrenaline or 5-hydroxytryptamine within neurons, and the increase of the amines also increases the sensitivity of the neurons to stimulation, so that the drugs act as antidepressants. Other stimulating agents, such as *amphetamine,* appear to mimic the active amines and in themselves sensitize neural responses.

Nomenclature

The chemical names for many drugs are unwieldy. Consequently, it is common in the clinical literature to use trivial names that frequently convey little information on chemical structure, or to use abbreviations. This tendency unfortunately carries over to the physiological and pharmacological literature, with the plethora of coined words and abbreviations sometimes obscuring meaning rather than aiding rapid comprehension. For example, LSD is so commonly used for lysergic acid diethylamide that it has passed into the general language, but 2-bromo-D-lysergic acid diethylamide is not abbreviated Br-LSD, but BOL. Even the central nervous system becomes CNS, which is very awkward when the discussion also concerns cyanide ions (CN) or thiocyanate ions (SCN). Biochemists themselves are responsible for *dopa,* meaning 3,4-dihydroxyphenylalanine, and *dopamine,* meaning the corresponding dihydroxyphenylethylamine.

The various dihydroxyphenylalkylamines, such as adrenaline and noradrenaline, are lumped under the term *catecholamine.* The word is derived from catechol, which is a fairly recent shortening of the trivial name, pyrocatechol, or 1,2-dihydroxybenzene (catechol was a more complex aromatic compound from which pyrocatechol was obtained by heating).

Adrenaline and noradrenaline are not the aromatic amines implied by the term catecholamine. However, language does not always obey rules, and much of the modern literature concerned with the action of the compounds uses catecholamine as a generic term for the class of compounds. There is no excuse for including metabolites that are not amines in the term, nor for using it as a substitute for the specific name of a compound, such as adrenaline or noradrenaline, when it is meant in particular.

IMPORTANT AMINES IN OTHER TISSUES

Histamine

Specialized cells, the mast cells, are found in clumps around the blood vessels of loose connective tissue in the skin, in the pleural membrane of the lung, and in the mucosa of the stomach, as well as being scattered about other tissues of the body. These cells have two especially interesting biochemical properties: they form a polysaccharide, *heparin*, which inhibits blood clotting (p. 653), and *histamine*, which is the decarboxylated product of histidine. The cells have the necessary histidine decarboxylase, again with pyridoxal phosphate as a cofactor, catalyzing the formation of this amine (Fig. 19-7).

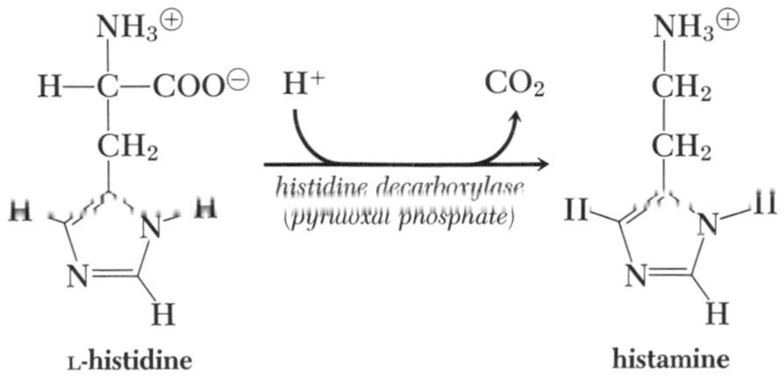

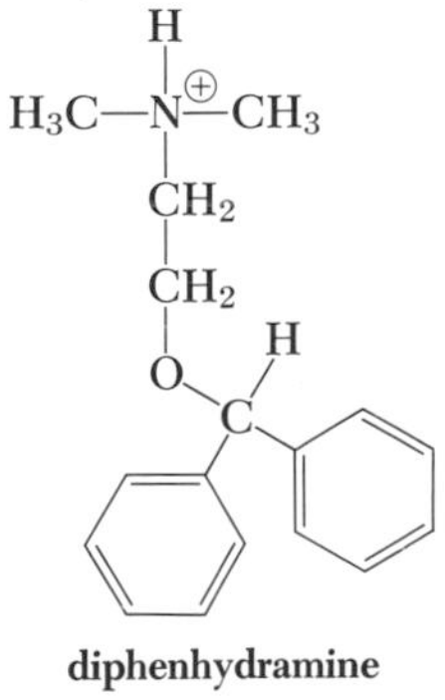

Figure 19-7 Small amounts of histidine are decarboxylated to form histamine. One of the compounds counteracting some of the physiological effects of histamine is shown at the bottom.

Histamine has at least three important physiological effects. (1) It causes an expansion of capillaries (perhaps by a constriction of the small veins leading from them), which results in a loss of fluid through their walls (edema) and a general increase in the volume of the entire vascular bed. The resultant drop in blood

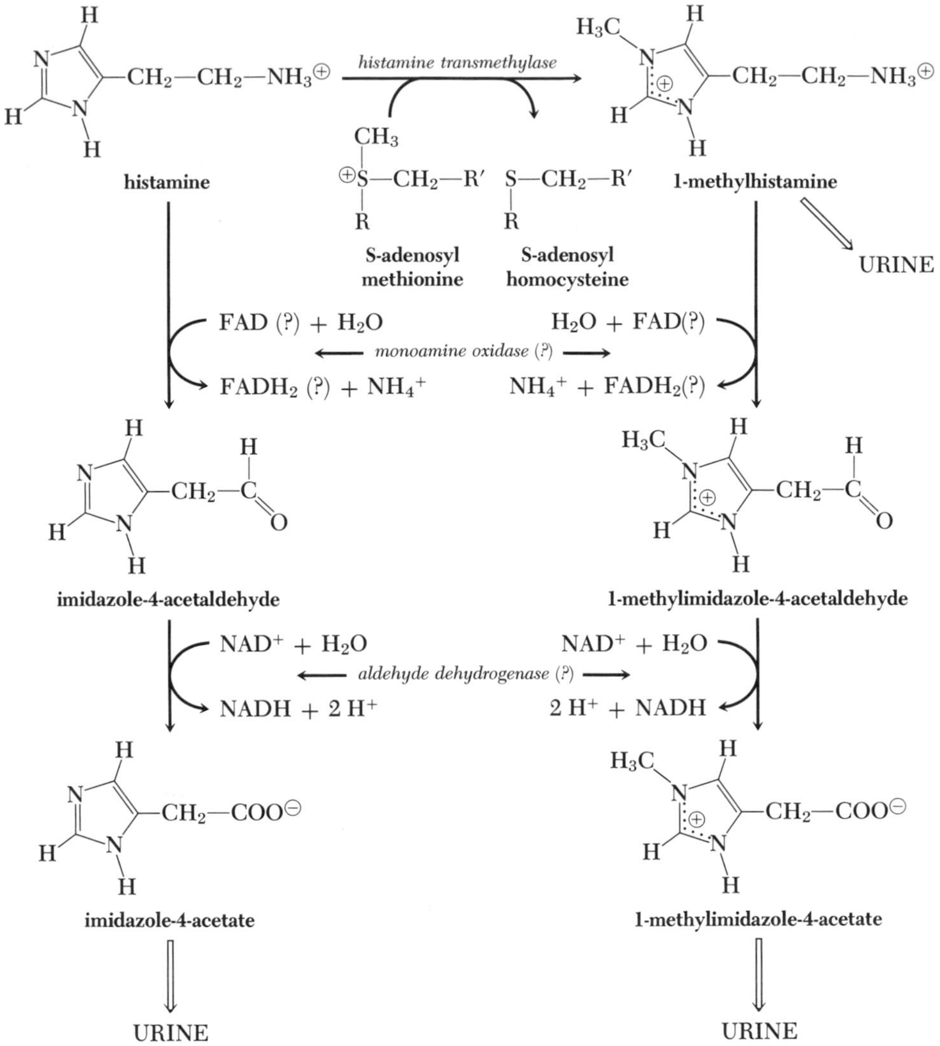

Figure 19-8 Histamine, like many other amines, is removed by oxidative deamination followed by oxidation of the aldehyde. This may occur with the original compound or with a methylated derivative.

pressure may be sufficient to cause shock. (2) It causes a constriction of the bronchial smooth muscles in the lung. (3) Histamine also causes an increase in secretion by the gastric mucosa, particularly of hydrochloric acid. Local release of histamine, particularly from allergic reactions, will cause the development of wheals on the skin from the leaking fluid, and more general release may be fatal because of circulatory failure.

Counteraction of these effects is of sufficient general importance to have led to intensive search for agents that will inhibit the effects of histamine or prevent its release—the *antihistamines,* and the existence of such compounds is generally known through widespread advertising. A typical example is *diphenhydramine* (Benadryl), also shown in Figure 19-7. The resemblance of these compounds to the transmitter amines of the central nervous system is sufficiently close that many of them have side-effects on the brain. Furthermore, while the compounds have valuable use in counteracting the effects of histamine released by hay fever or food allergies, they have no useful effect on the receptors of the gastric mucosa. Agents for preventing the formation of histamine have not been developed.

Histamine, like noradrenaline, may be metabolized by routes involving methylation of the ring or oxidation of the amino group (Fig. 19-8). The methyl derivatives or the imidazoleacetates appear in the urine. The oxidation of the amino group is catalyzed by a relatively non-specific enzyme, diamine oxidase, that also oxidizes a number of aliphatic compounds containing two amino groups. It is not well characterized, and would appear to be a typical flavoprotein oxidase, utilizing molecular oxygen and producing hydrogen peroxide, were it not for the fact that it also contains pyridoxal phosphate, and the only precedent for the association of this coenzyme with an oxidative enzyme is the system that attacks glycine (p. 419).

Putrescine, Spermidine, and Spermine

Putrescine received its ugly name because it was discovered as a product of the action of bacteria in decaying meat, being formed by decarboxylation of ornithine, which in turn arises from bacterial hydrolysis of arginine in the meat. (The product of decarboxylation of lysine was named *cadaverine* for the same reason.)

The polyamino compounds, spermidine and spermine, were named for their discovery in human semen. Indeed, crystals of spermine phosphate in semen were one of the things noted by van Leeuwenhoek with his newly invented microscope, and their presence is still used as a part of the legal identification of suspect stains, but the structure of the compound was only worked out in 1926. It is only in the last decade that it has been realized that these compounds, along with putrescine, are widespread in tissues, and may have important general functions. They appear to occur in association with nucleic acids, as might be expected from their polycationic character, and are therefore under active investigation as possible participants in mechanisms regulating protein synthesis or cell division.

It turns out that the synthesis of the compounds in animals resembles, but is not identical to, the process in microorganisms (Fig. 19-9). Putrescine is formed by the action of a straightforward amino acid decarboxylase on ornithine. However, the aminopropyl groups—one in spermidine and two in spermine—are obtained from S-adenosylmethionine.

We ordinarily think of S-adenosylmethionine only as a methyl group donor. However, Guilio Cantoni noted, and it appears obvious after it has been pointed out, that there is no mechanistic reason that any one of the three substituents on the sulfur atom of S-adenosylmethionine couldn't be the group transferred to another compound. Here we have a case in which it is the amino acid skeleton that is transferred, with an accompanying decarboxylation. (S-adenosylmethionine is decarboxylated by a separate enzyme in some microorganisms, and the remaining aminopropyl group is then transferred, but a single enzyme appears to catalyze both steps in animals.)

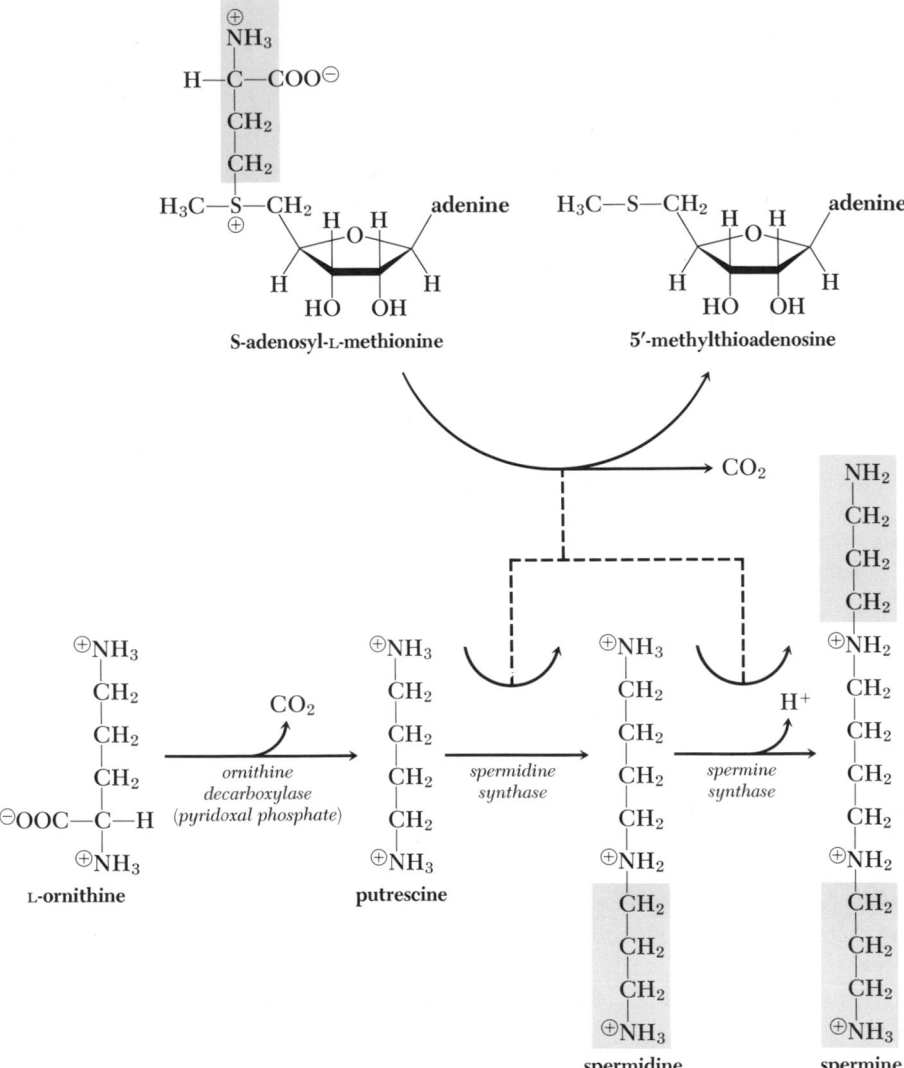

Figure 19-9 The synthesis of putrescine, spermidine, and spermine.

We shall have to wait for further work to define the exact function of many of the amines, but let us turn now to the formation of a nitrogenous compound of well-defined function, the nicotinamide group.

Nicotinamide

Nicotinamide, the all-important active part of NAD and NADP, can be formed by a side-reaction of tryptophan in the liver. The reactions by which this is done diverge from the main route of tryptophan degradation (p. 390) at 2-amino-3-carboxymuconaldehydate. In the main route, the compound is decarboxylated by an enzyme. However, it also spontaneously cyclizes to form *quinolinate* (Fig. 19-10).

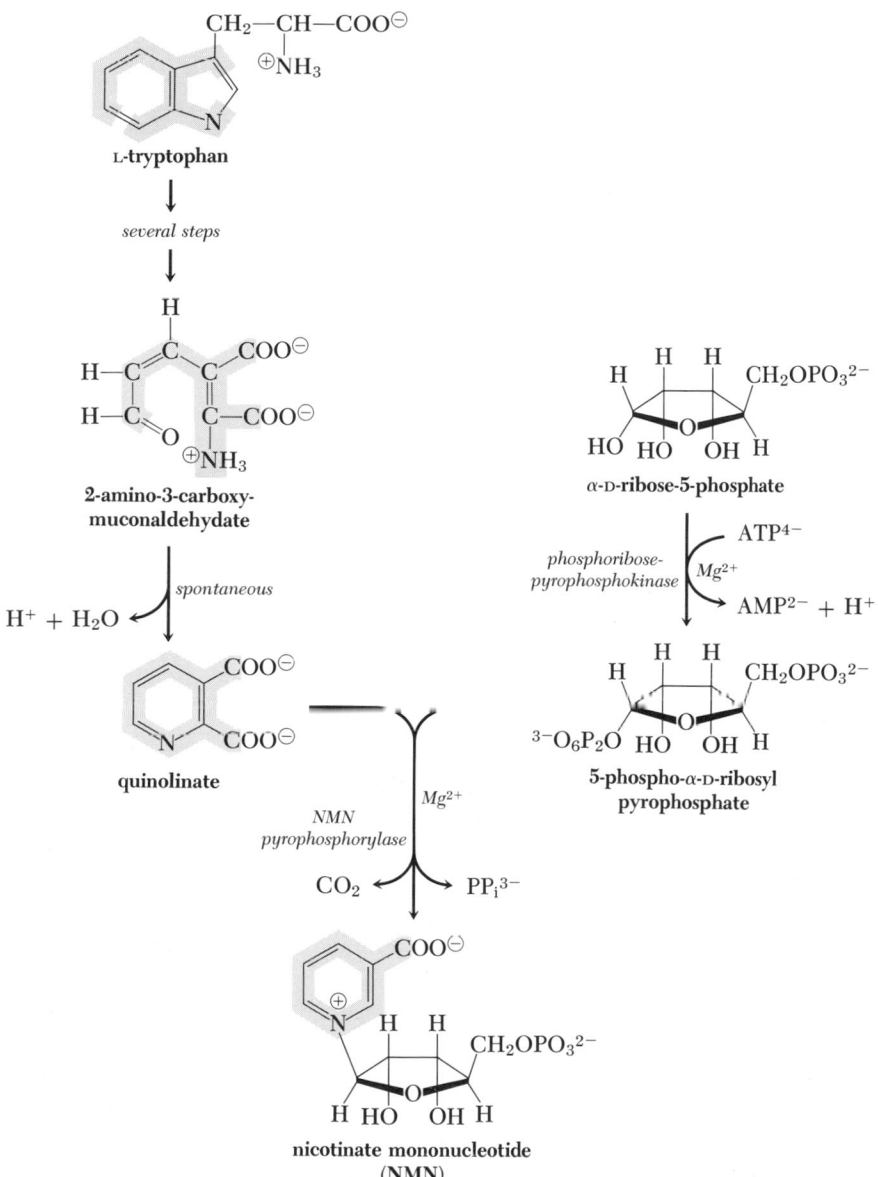

Figure 19-10 One of the intermediates of the normal pathway of tryptophan metabolism, 2-amino-3-carboxymuconaldehydate, spontaneously cyclizes to form quinolinate, which is a precursor for the nicotinate portion of NAD. 5-Phosphoribosylpyrophosphate reacts with quinolinate to form nicotinate mononucleotide, the extra carboxylate group being lost by a simultaneous decarboxylation. The condensation of bases with 5-phosphoribosylpyrophosphate is a general type of reaction for the formation of nucleotides, and the ribosyl compound is produced from ribose-5-phosphate as shown.

Quinolinate condenses with 5-*phosphoribosylpyrophosphate* to form a nucleotide, but CO_2 is lost from the intermediate, so that the product is *nicotinate mononucleotide*. Nucleotides are typically formed by the reaction of a nitrogenous base with phosphoribosylpyrophosphate. This ribosyl compound is the parent for the pentose residue in all of the ribonucleotides, and it is formed from ribose-5-phosphate (p. 324) by a pyrophosphokinase.

Nicotinate mononucleotide condenses with ATP in the reversal of another pyrophosphorylase reaction to form *nicotinate adenine dinucleotide* (deamido-NAD), which then reacts with glutamine as an amide donor to form NAD itself. The transfer of the amide nitrogen requires one high-energy phosphate, and the use of glutamine as a nitrogen donor in this way is found in other reactions we shall encounter. (Recall that the amide nitrogen of glutamine in a sense represents stored ammonia, and the utilization of the compound is a means of taking up ammonia released from amino acids and making other compounds from it without maintaining a toxic pool of the free compound.) NAD may be partially converted to NADP by the reaction of a simple kinase.

It is evident that animals eating a sufficient supply of tryptophan will not require a supply of nicotinamide in the diet, if the reactions described previously can proceed at a rate adequate to maintain the compound. In other words, nicotinamide may not be a vitamin for those animals.

Man and many other animals are able to maintain the supply of NAD and NADP by direct synthesis from tryptophan, without any dietary intake of nicotinate. This does not mean that dietary nicotinate will not be used. An enzyme does exist to catalyze the reaction with phosphoribosylpyrophosphate to form nicotinate mononucleotide directly when nicotinate is available.

The critical point is the concentration of 2-amino-3-carboxymuconaldehydate. It may be too low to maintain adequate nicotinate formation because the supply of tryptophan is small or because it is being decarboxylated rapidly, even at low concentrations, which implies a high concentration of the decarboxylase. This is indeed the circumstance in cats, which have an elevated level of the enzyme compared to other mammals. This may be nothing more than an evolutionary change resulting from a diet purely of meat, with an ample supply of nicotinate provided by the breakdown of the NAD and NADP present in the food.

Although the natural diets of humans are never completely deficient in nicotinate, a diet based extensively on cereals and fruits may contain less than required for the daily turnover of NAD. This is no problem as long as the tryptophan content of the diet is adequate to meet the demands of protein synthesis and still have enough extra passing through the degradative pathway that the bypass *via* quinolinate can supply the missing nicotinate. In human adults, the molar conversion of extra tryptophan to NAD is roughly 3 per cent of the total degraded, so that 60 milligrams of tryptophan yield one milligram of nicotinate as NAD. The daily turnover of NAD is equivalent to about 0.15 milligrams of nicotinate per kilogram of body weight, or 10.5 milligrams total in a 70-kilogram man. It would require, therefore, 630 milligrams of tryptophan per day, above and beyond the minimum amount necessary to maintain protein synthesis, to supply the nicotinate requirement, if there were no other source. (This estimate is probably high, because there is some reason to believe that the efficiency of nicotinate formation from tryptophan may

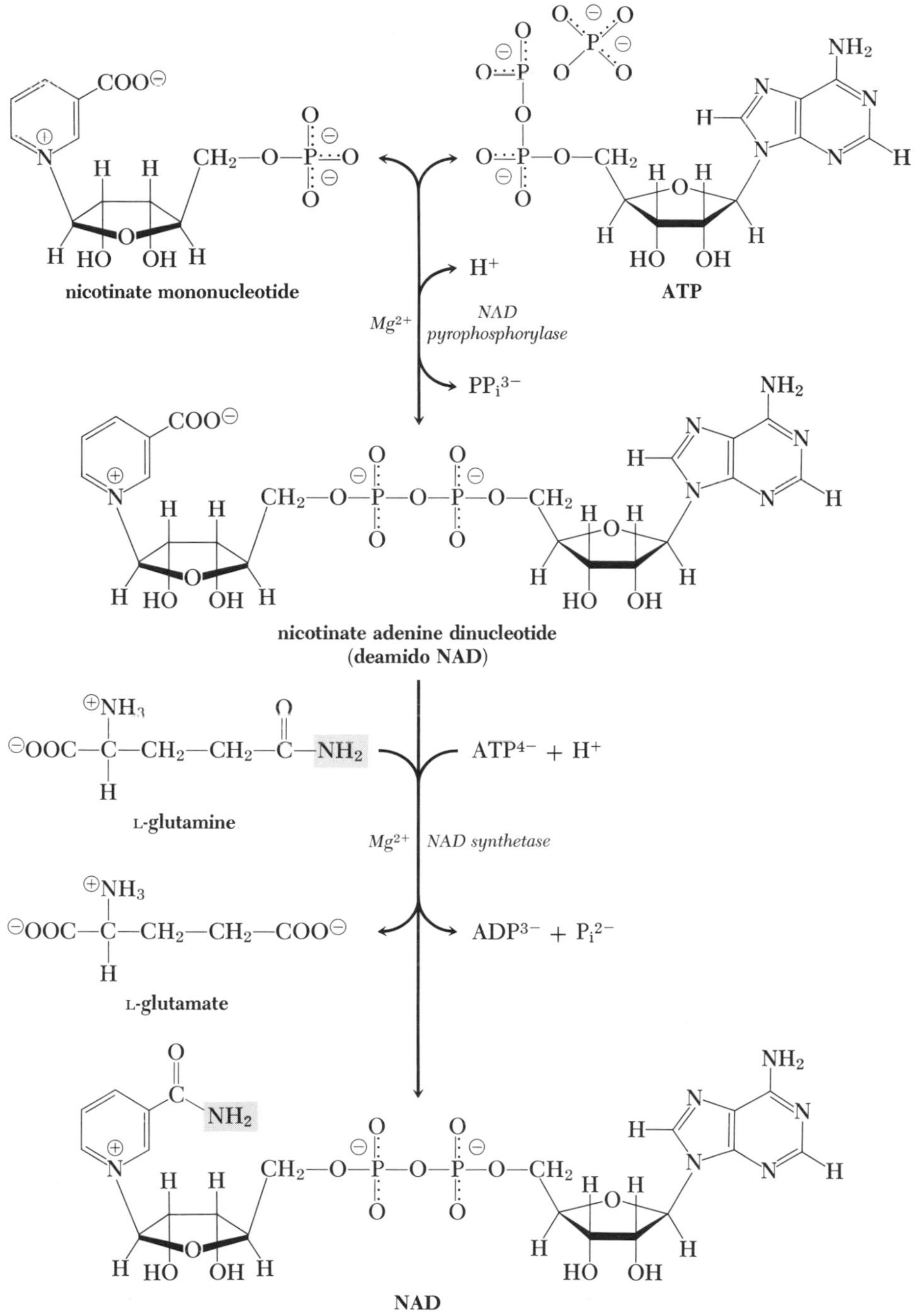

Figure 19-11 The conversion of nicotinate mononucleotide to NAD. The adenosine phosphate portion of the molecule is obtained from ATP and the amide nitrogen by transfer from glutamine.

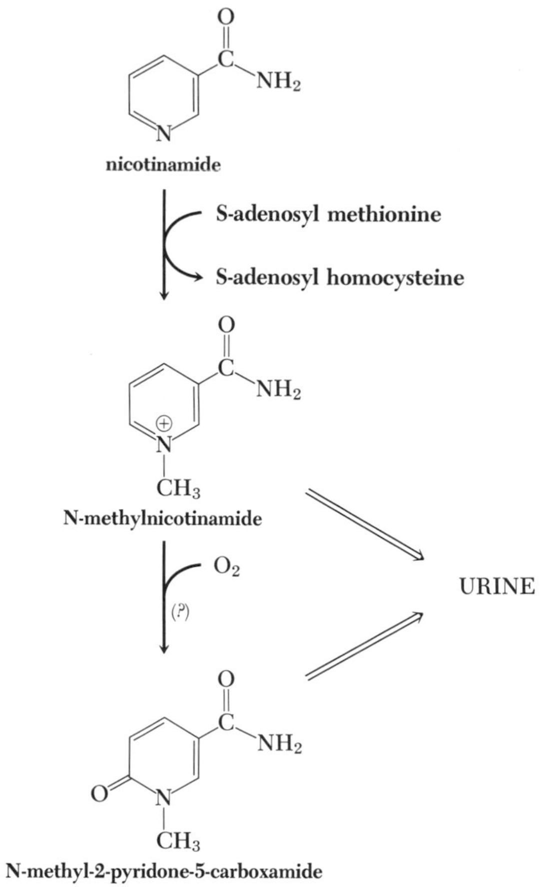

Figure 19-12   Excess nicotinamide is excreted in the form of the N^1-methyl derivative and the corresponding pyridone.

rise with a severe deficiency.) This amount of tryptophan would be supplied by 250 grams of beefsteak, 450 grams of whole oats, or 1330 grams of whole corn.

These figures tell the tale. People subsisting mainly on corn (maize) are subject to tryptophan deficiency. Corn contains somewhat over 10 milligrams of nicotinate per kilogram, but for some reason much of this is not available, and it is not clear whether it is bound in some form resistant to digestion and absorption or whether the corn contains some antagonist to nicotinate. In any event, the combination of low tryptophan and low content of available nicotinate leads to the development of *pellagra* in people on a corn diet. Pellagra is mainly a deficiency of nicotinate, characterized by inflammation of the skin and mucous membranes, along with psychic disturbances common in nutritional disorders. The poor diet necessary to provoke the appearance of pellagra may easily also cause other deficiencies with their symptoms superimposed on the signs of nicotinate deficiency. The lack of tryptophan is an additional deficiency in itself.

The malignancy known as *carcinoid* is an unusual case of nicotinate deficiency. Carcinoid is a proliferation of specialized cells in the appendix or small bowel known as argentaffine cells. These cells have a high capacity for synthesizing 5-hydroxy-tryptamine. When they multiply without normal restraints, the formation of 5-hydroxytryptamine may divert a large fraction of the tryptophan supply, with the major end-product, 5-hydroxyindoleacetate, being excreted in amounts up to 1.5 millimoles per day, contrasted to the normal maximum of 0.05 millimole. The loss of tryptophan is great enough to cause the development of full-blown pellagric dermatitis on top of the other symptoms of the malignancy. The dermatitis can be relieved by giving nicotinate, but this, of course, has no effect on the primary, and lethal, malignancy. (Whether other symptoms of tryptophan deficiency are adding to the patients' discomfort is not known.)

The major end products from the nicotinate of NAD that is broken down in cells are N-methyl nicotinamide (Fig. 19-12) and the oxidized form N-methyl pyridone carboxylate. The methyl group is provided by S-adenosyl methionine. The excretion of these compounds in the urine provides a guide to nicotinate turnover.

Other Examples

Some compounds are embarrassing because they occur at too high concentrations in specialized tissues to ignore them completely, and yet their function is unknown. For example, skeletal muscles contain 20 to 30 millimoles per kilogram of *anserine* and *carnosine*, which are peptides of β-alanine with histidine or

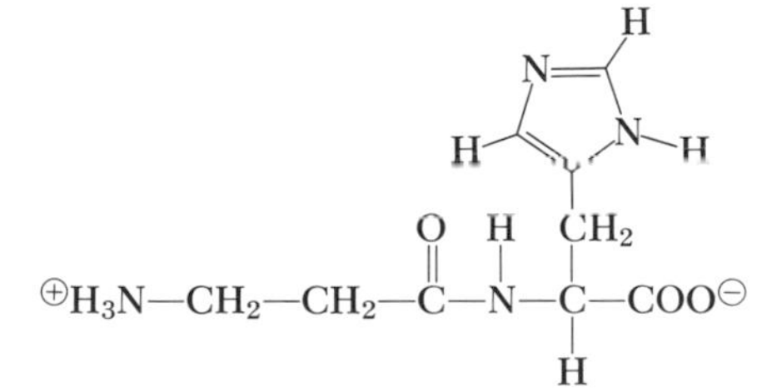

carnosine
(β-alanylhistidine)

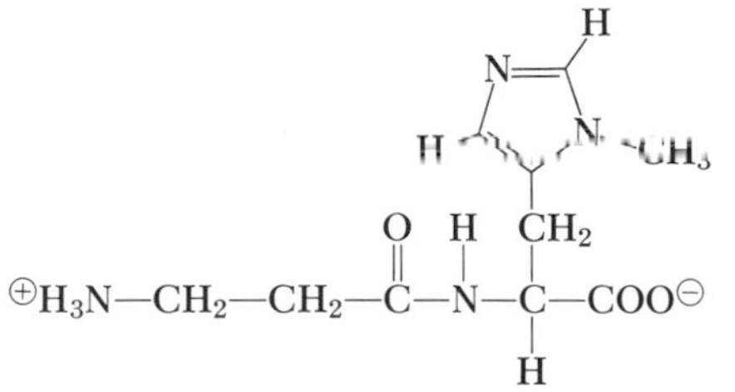

anserine
(β-alanyl-1-methylhistidine)

1-methylhistidine. Carnosine is made from β-alanine and histidine at the expense of ATP, which is cleaved to AMP and PP_i. Anserine is made from carnosine by methylation at the expense of S-adenosylmethionine. This knowledge of the way the compounds are formed doesn't tell us anything about their function. It seems likely that they have something to do with muscular contraction, and there are studies showing some effect on the hydrolysis of ATP, but this is only tantalizing because we don't know the details of the mechanism of muscular contraction.

Another example is N-acetylaspartate, which is present at a concentration of 10 millimoles per kilogram in the brain—a higher concentration than that of

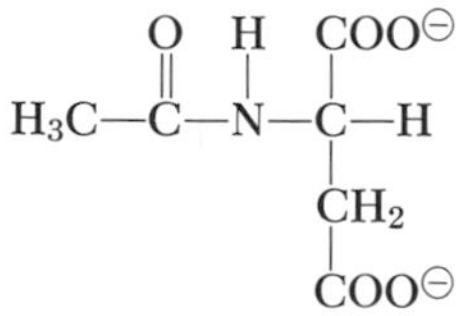

N-acetylaspartate

any other amino acid or derivative, even of glutamate (3 mmole/kg^{-1}), glutamine (6 mmole/kg^{-1}), or aspartate (2.5 mmole/kg^{-1}), which are especially rich in nervous tissue. The only thing known about acetylaspartate is that its acetyl group can be used as a source of acyl residues for making the lipid-rich myelin sheath around some neurons during development, but this does not explain its presence in adults. Indeed, the function of N-acetylamino acids is in general an enigma, despite the known presence of enzymes that make them, both in the free form and in some proteins. (Acetyl coenzyme A is the acetylating agent.) We have mentioned a possible route for the degradation of lysine by way of acetylated derivatives, and the use of acetylglutamate as a coenzyme in the carbamyl phosphate synthetase catalyzing the first reaction of urea synthesis, but if there is a more general function, it is unknown.

Recapitulation of types of reactions

1. Aminoaldehydes may condense to form rings:

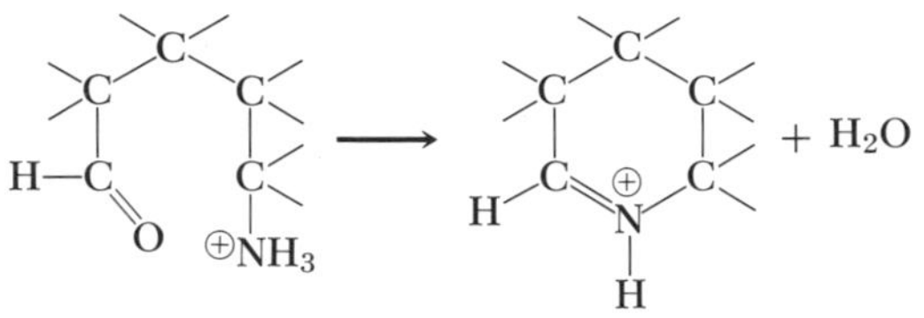

Example: The formation of quinolinate from 2-amino-3-carboxymuconaldehydate. The formation of desmosyl residues in elastin is similar.

2. Nucleotides may be formed by reactions between bases and phosphoribosylpyrophosphate:

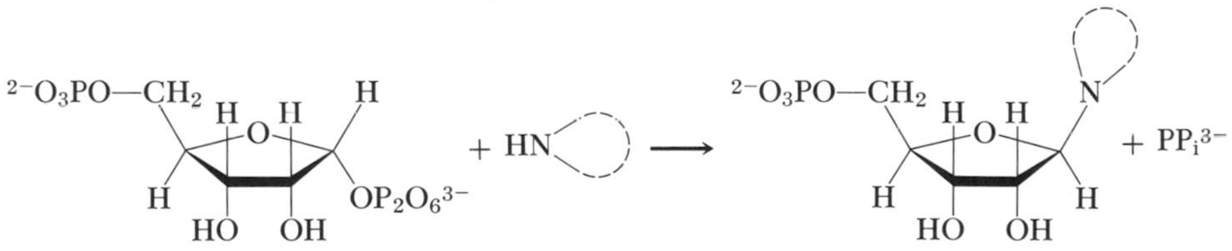

Example: The formation of nicotinate mononucleotide.

3. Pyrophosphate esters may be formed by the reaction of alcohols with ATP, liberating AMP:

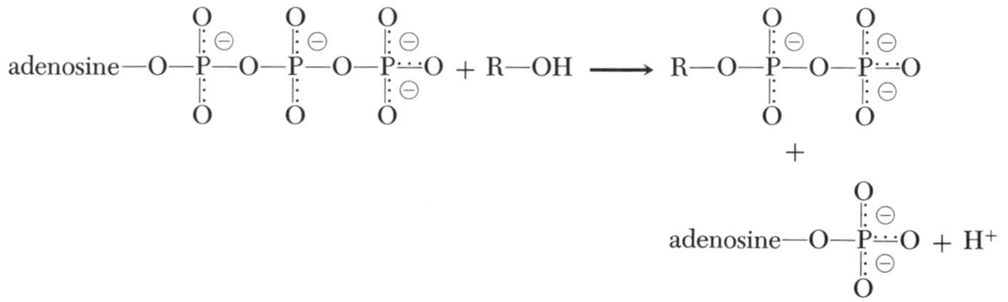

Example: The formation of phosphoribosylpyrophosphate.

4. Glutamine may act as an amide donor:

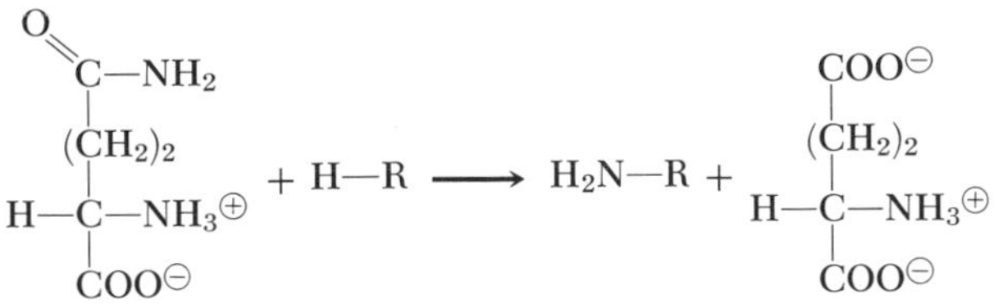

Example: The conversion of nicotinate adenine dinucleotide to NAD.

Further reading

Meister, A.: *Biochemistry of the Amino Acids*, 2nd ed. Academic Press (1965).

Goth, A.: *Medical Pharmacology*, 3d ed. C. V. Mosby (1966).

Goldsmith, G. A.: *The B Vitamins*. In Beaton, G. H., and E. W. McHenry, eds: *Nutrition—A Comprehensive Treatise*, Vol. 2 (p. 166). Academic Press (1964).

Rocha e Silva, M., ed.: *Histamine and Anti-histaminics*. In *Handbook of Experimental Pharmacology*, Vol. 18, part 1. Springer (1966).

Erspamer, V., ed.: *5-Hydroxytryptamine and Related Indolealkylamines*. In *Handbook of Experimental Pharmacology*, Vol. 19. Springer (1966).

von Euler, U. S., S. Rosell, and B. Urnäs, eds.: *Mechanism of Release of Biogenic Amines*. Pergamon Press (1966).

Ijichi, H., A. Ichiyama, and O. Hayaishi: *Studies on the Biosynthesis of Nicotinamide Adenine Dinucleotide*. J. Biol. Chem., *241:* 3701 (1966).

Bridges, J. M., J. B. Gibson, L. W. Loughbridge, and D. A. D. Montgomery: *Carcinoid Syndrome with Pellagrous Dermatitis*. Brit. J. Surgery, *45:* 117 (1957).

Mann, T.: *The Biochemistry of Semen*. John Wiley (1964).

Hagino, Y., et al.: *Metabolism of Pyridinium Precursors of Pyridine Nucleotides in Rat Liver*. J. Biol. Chem., *243:* 4980 (1968).

Avena, R. M., and W. J. Bowen: *Effects of Carnosine and Anserine on Muscle Adenosine Triphosphatase*. J. Biol. Chem., *244:* 1600 (1969).

McEwen, C. M., Jr., et al., and Hellerman, L., and V. G. Erwin: *Mitochondrial Monoamine Oxidase*. J. Biol. Chem., *243:* 5217 and 5234 (1968).

Pegg, A. E., and H. G. Williams-Ashman: *On the Role of S-Adenosylmethionine in the Biosynthesis of Spermidine by Rat Prostate*. J. Biol. Chem., *244:* 682 (1969).

THE TURNOVER OF NUCLEIC ACIDS

CHAPTER 20

ARGUMENT

Nucleic acids and proteins are constantly being made in all of the tissues. All of the constituents of the cells must be formed in the intestinal mucosa, skin, and bone marrow, which have actively dividing cells. Tissues such as the brain, skeletal muscles, and cartilage have few dividing cells, and therefore little need for formation of deoxyribonucleates, but even these cells have a turnover of RNA that requires a supply of nucleotides. Purines and pyrimidines are available in most diets, but animals are not dependent on this source because they can supplement the dietary supply by *de novo* synthesis of the nucleotides. The pentose phosphate portion of the molecules is supplied directly as 5-phosphoribosylpyrophosphate. The purine ring is built up a step at a time on the pentose phosphate, using the amide group of glutamine, a glycine molecule, the nitrogen atom of aspartate, the one-carbon pool, and CO_2 as sources of the constituent atoms. The pyrimidine ring is made from carbamyl phosphate and aspartate and later attached to the ribose phosphate group. The various substituents on the rings that define the respective bases are contributed as the nitrogen of glutamine in the case of guanine and cytosine, and as the nitrogen of aspartate in the case of adenine. The methyl group of thymine is obtained from the one-carbon pool.

Since the formation of nucleotides is necessary for growth, attempts have been made to control cancers by the use of analogues to inhibit some of the synthetic reactions. These analogues include inhibitors of the transfer of amide nitrogen from glutamine, inhibitors of the reduction of dihydrofolate to tetrahydrofolate so as to block proper function of the one-carbon pool, and inhibitors of the enzyme catalyzing synthesis of deoxythymidine phosphate.

The formation of deoxythymidine phosphate involves the transfer of a methylene group from the one-carbon pool, but also requires the reduction of that group to the methyl form found in thymine. The reducing agent is tetrahydrofolate, which becomes dihydrofolate. The production of thymine, which is only required in cells undergoing division, is therefore particularly susceptible to inhibitors of dihydrofolate reduction, as well as of the one-carbon transfer itself, because they prevent the recovery of the dihydrofolate as tetrahydrofolate.

Excess nucleotides, arising both from synthesis and the diet, are degraded. The ribosyl unit is recovered as ribose phosphate. Part of the bases may also be re-incorporated, but part are destroyed. The pyrimidines are eventually converted to ammonia and CO_2, but the purines are deaminated and the ring further oxidized to form uric acid. Uric acid is excreted. However, it has a low solubility, and any tendency toward accumulation, either through excessive production or limited excretion, may lead to formation of crystals. This may occur in the urinary tract, with the development of stones, or in other tissues, leading to gout. Gout is a widespread condition. Another much more rare condition, the Lesch-Nyhan syndrome, is accompanied by excessive formation of uric acid, but the primary defect is the lack of an enzyme necessary for the recovery of guanine or hypoxanthine released during degradation of nucleotides. The Lesch-Nyhan syndrome has attracted widespread interest because patients with it are aggressive toward themselves and others, and it therefore provides a clear-cut example of a direct causal relationship between a distinct personality trait and a metabolic change.

No tissue is completely stable. Nucleic acids and proteins constantly break down and are synthesized in all parts of the body. The attrition of these compounds and the associated requirement for synthesis of replacements is most acute in those tissues in which normal function involves loss of entire cells, such as the skin, the intestinal mucosa, and the blood. The apparatus of cell division is constantly in use in such tissues, and all of the components of the cells are being formed. It is only in dividing cells that the formation of new deoxyribonucleates is required.

Proteins are also made and destroyed in cells that do not divide. Protein synthesis requires ribonucleates, which also are synthesized and then degraded. Messenger RNA is destroyed relatively rapidly, and ribosomal RNA and transfer RNA more slowly. The turnover of protein and of RNA are therefore tied together. The turnover is especially rapid in cells that secrete proteins as a major part of their function. Prominent examples are cells of various parts of the gastrointestinal tract that pour out large volumes of digestive juices, rich in enzymes and mucoproteins. The liver also secretes proteins such as the albumin of the blood, but it has an additional load on the formation of RNA and of protein because of a relatively rapid intracellular turnover, enabling the enzymatic composition of this organ to be adapted to varying metabolic loads.

Even the stable cell population in adult muscles and nerves has an internal turnover of RNA and protein, although at a slower rate than that of most tissues.

We have noted that separate consideration of sequences of reactions in metabolism is a logical device. The limitations of the device, its artificiality, are especially apparent in any attempt to divorce a consideration of the economy of the nucleic acids from the economy of the proteins. Nucleic acids are required for the assembly of amino acids into proteins, and certain amino acids are required for the assembly of the nucleic acids. A change in the economy of one is frequently a direct change in the economy of the other. Therefore, when we discuss sequences of reactions in this chapter that are superficially labelled nucleic acid metabolism, it is critical that their essential interactions with amino acid metabolism not be overlooked.

SYNTHESIS OF NUCLEOTIDES

The pool of nucleoside triphosphates within cells has two functions. One is to act as metabolic intermediates. Much of our discussion of metabolism has concerned the function of the adenine nucleotides in this way, but we have also seen examples of the participation of guanine nucleotides (citric acid cycle and gluconeogenesis), uridine nucleotides (glycogen synthesis), and cytosine nucleotides (phospholipid synthesis). Even if there were no turnover of nucleic acids, we ought to expect a turnover of these nucleotides in line with the general principle that the concentration of metabolic intermediates is usually controlled by regulation of the rates of opposing processes of degradation and synthesis. The second function of the nucleoside triphosphates is to act as precursors for nucleic acids, that is, as substrates for nucleic acid polymerases. We also ought to expect a regulation of the available pool of nucleotides for this purpose; it is self-evident that supplies of all of the constituent nucleotides must be available for the formation of nucleic acids if the organism is to survive.

There is indeed a constant destruction and synthesis of nucleotides. Even though the diet of animals is composed mainly of cells from other organisms and therefore must contain a continual supply of nucleotides, the animals still retain the capacity for making both the purine and pyrimidine nucleotides from amino acids and carbohydrates. This probably attests to the importance of being able to regulate metabolism through destruction of nucleotides at a faster rate than they can be provided in all diets. Experimentation on this point has only scratched the surface of the problem.

Purine Nucleotides

The purine ring found in guanosine and adenosine nucleotides is assembled piece by piece on ribose phosphate, so that the products of synthesis are the finished nucleotides, not the free purines. The origin of the ring skeleton is shown in Figure 20-1. One of the four nitrogen atoms and two of the carbons come from an intact *glycine* molecule. Another nitrogen is contributed by *aspartate*, and two more by *glutamine*. Two of the carbons come from the *one-carbon pool* as tetrahydrofolate derivatives, and the remaining carbon is added as CO_2. The initial product when

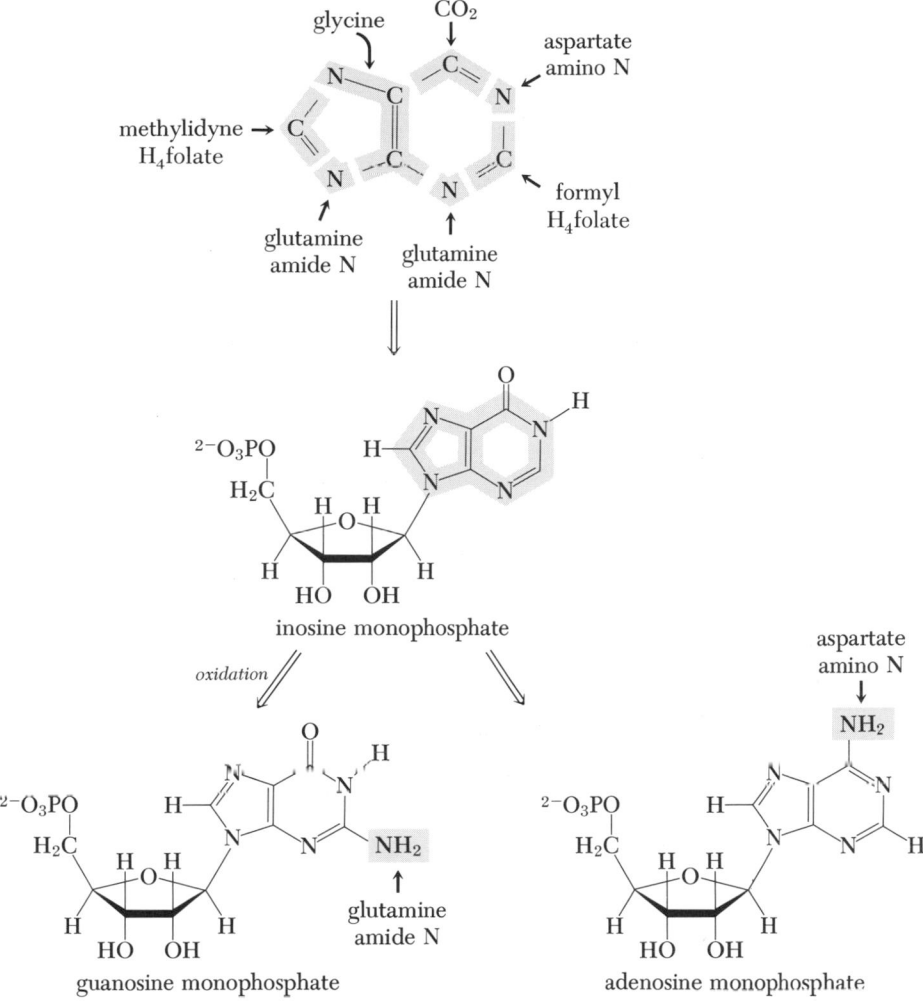

Figure 20-1   Sources of the atoms of purines. The ring is assembled on a ribose-5-phosphate residue obtained from 5-phosphoribosylpyrophosphate (not shown), forming inosine monophosphate as a common precursor of both GMP and AMP. The formation of GMP involves an oxidation and the transfer of the substituent amino group from glutamine. AMP is formed by transfer of the substituent amino group from aspartate.

the ring is assembled is *inosine monophosphate—a hypoxanthine* nucleotide that ought to be called hypoxanthosine monophosphate.

Inosine monophosphate (IMP) is the precursor of both guanosine and adenosine monophosphates. Simple replacement of the oxygen by an amino group derived from aspartate makes the adenine nucleotide. Oxidation of the ring and replacement of the added oxygen by the amide group of glutamine make the guanine nucleotide.

The steps by which this assembly is accomplished are detailed on the following pages.

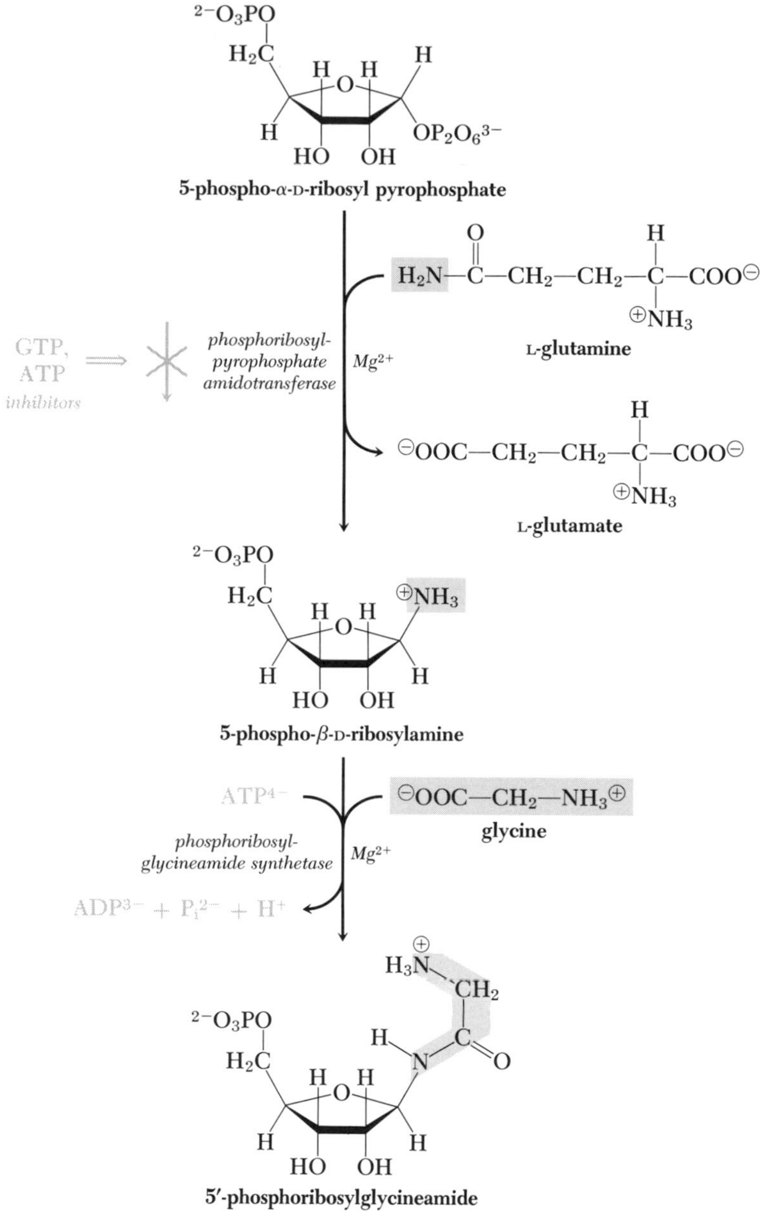

Synthesis continues on facing page

Figure 20-2 The synthesis of purine nucleotides I. See text.

The initial reaction is a transfer of the amide group of glutamine to phosphoribosylpyrophosphate forming *phosphoribosylamine,* and liberating inorganic pyrophosphate. Since the ribosylpyrophosphate has the α-configuration, and the ribosylamine the β-configuration, there is an inversion of configuration during the

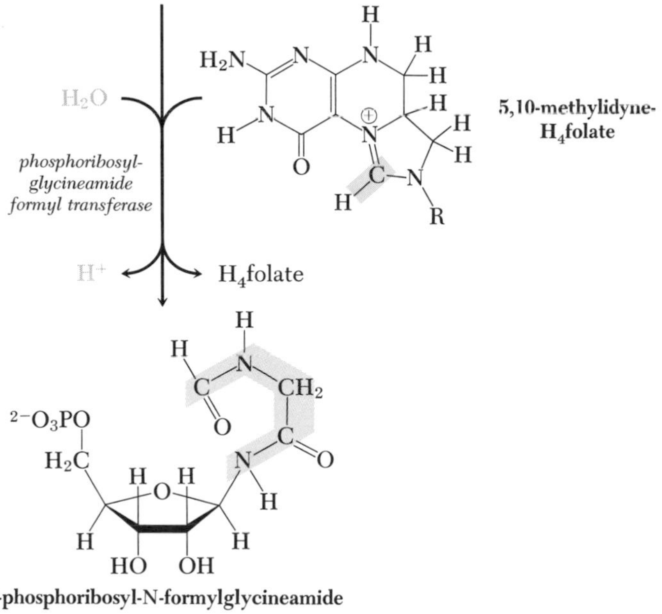

Figure 20-2 *Continued*

reaction. The remainder of the purine ring is built around this added nitrogen, so the products will have the characteristic β-ribosyl configuration found in the nucleotides.

The reaction is irreversible, but is the *rate-controlling reaction* for purine synthesis. The transamidase enzyme is inhibited by both guanosine and adenosine phosphates, and more so by combinations of the two than either alone. Since these are the end products of the synthetic scheme, the inhibition at this initial reaction serves to regulate the sum of the concentration of the purine nucleotides in the cell. Intermediates do not accumulate because the reactions following phosphoribosylamine formation go faster than the formation of the compound. The supply of substrates is probably always adequate for this reaction.

A glycyl group is then attached intact to the amino group on ribose by a straightforward synthetase reaction. ATP is cleaved to drive the reaction reversibly to the formation of phosphoribosylglycineamide. The rate of the reaction is probably always limited by the supply of the ribosyl compound, and not by the supply of ATP or glycine. The reaction is reversible because of the similar free energies of hydrolysis of the aminoacyl amide and ATP, but the product is constantly removed by the following reaction.

A formyl group is then added to the free amino group of the glycyl residue. This completes the atoms necessary for the five-membered ring of the purines. The formyl donor is 5,10-methylidyne-H_4folate in the one-carbon pool. A direct lack of one-carbon units at the formate level probably never impedes this reaction. A deficiency of folate may sometimes affect the supply of substrate, but there is no clear-cut evidence for or against impairment.

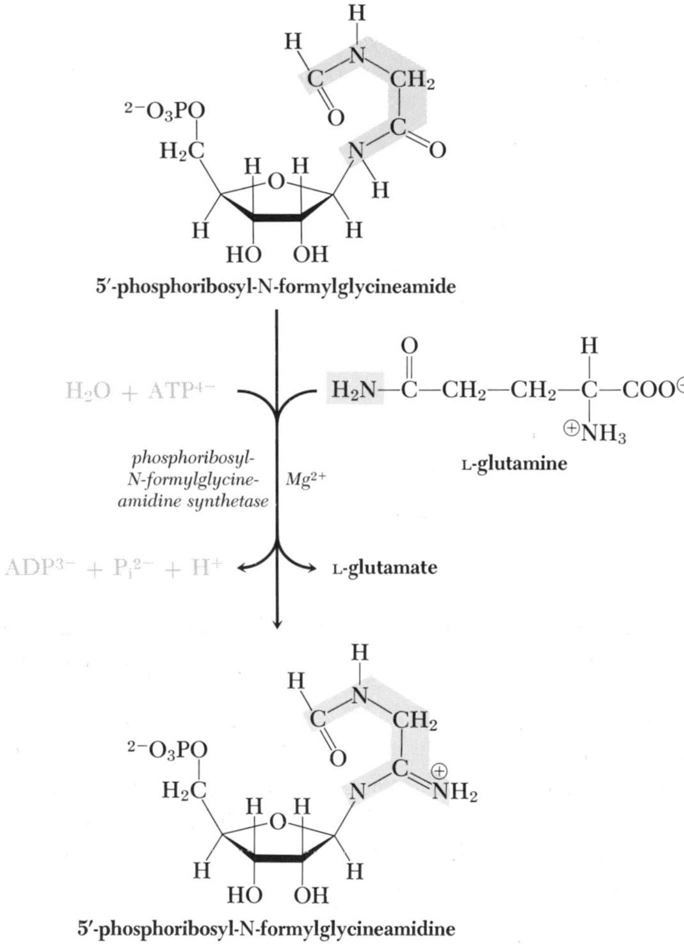

5′-phosphoribosyl-N-formylglycineamide

$H_2O + ATP^{4-}$

phosphoribosyl-
N-formylglycine- Mg^{2+}
amidine synthetase

$ADP^{3-} + P_i^{2-} + H^+$

L-glutamine

L-glutamate

5′-phosphoribosyl-N-formylglycineamidine

Synthesis continues on facing page

Figure 20-3   The synthesis of purine nucleotides II. See text.

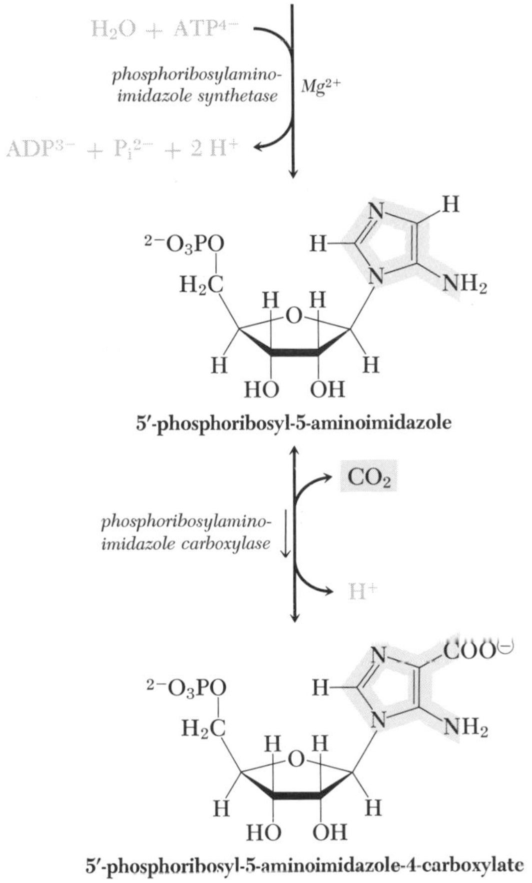

5'-phosphoribosyl-5-aminoimidazole

5'-phosphoribosyl-5-aminoimidazole-4-carboxylate

Figure 20-3 *Continued*

A further nitrogen is next transferred from the amide group of glutamine, and the reaction is driven by the hydrolysis of ATP. Since the free energy of hydrolysis of ordinary amides, in contrast to aminoacyl amides, is less than the free energy of hydrolysis of ATP, the reaction is essentially irreversible.

The five-membered imidazole ring of the purine is then formed in a reaction driven by ATP. The amino group added in the preceding reaction is now sticking out as a handle on which to build the remainder of the purine molecule.

In the next step, CO_2 adds to the imidazole ring. The carboxylase catalyzing this reaction has no biotin, pyridoxal phosphate, or thiamine pyrophosphate attached, and therefore differs in mechanism from most of the enzymes catalyzing addition or removal of carboxylate groups. The reaction is reversible, but the aminoimidazole carboxylate is constantly removed by the following reaction.

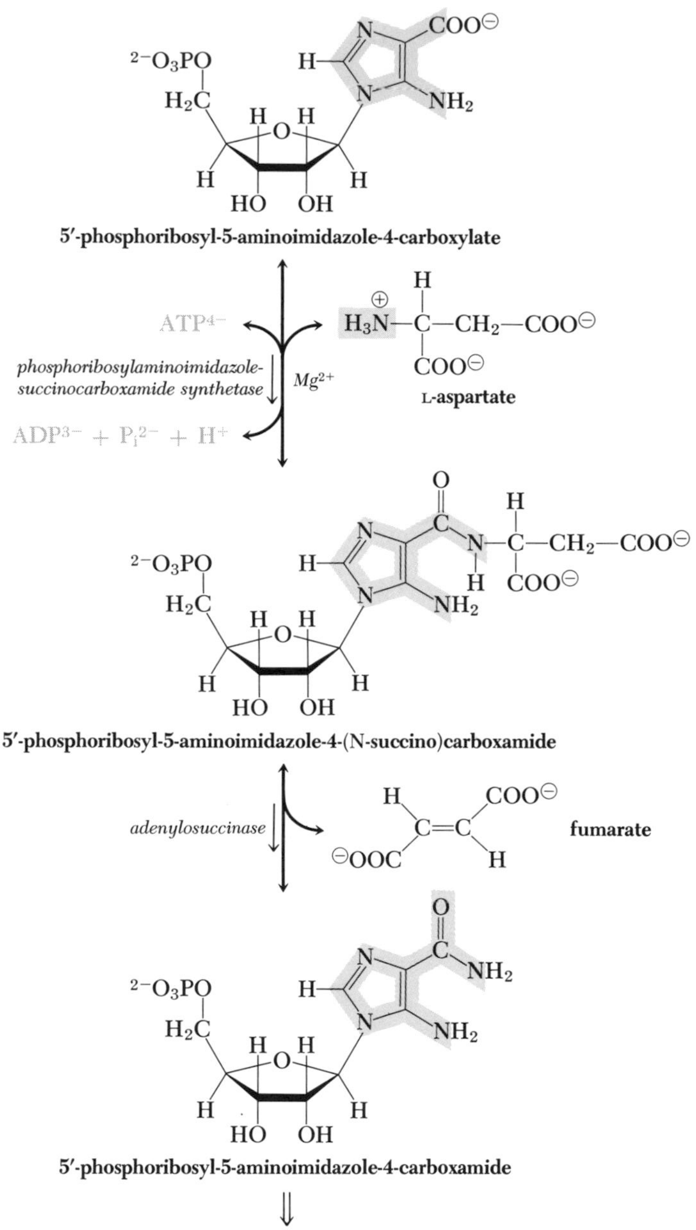

5'-phosphoribosyl-5-aminoimidazole-4-carboxylate

ATP^{4-}

phosphoribosylaminoimidazole-succinocarboxamide synthetase

Mg^{2+}

$H_3N^{\oplus}-\overset{H}{\underset{COO^{\ominus}}{C}}-CH_2-COO^{\ominus}$

L-aspartate

ADP^{3-} + P$_i^{2-}$ + H$^+$

5'-phosphoribosyl-5-aminoimidazole-4-(N-succino)carboxamide

adenylosuccinase

$^{\ominus}OOC-C\overset{H}{=}C\overset{COO^{\ominus}}{\underset{H}{}}$ fumarate

5'-phosphoribosyl-5-aminoimidazole-4-carboxamide

⇓

Synthesis continues on facing page

Figure 20-4 The synthesis of purine nucleotides III. See text.

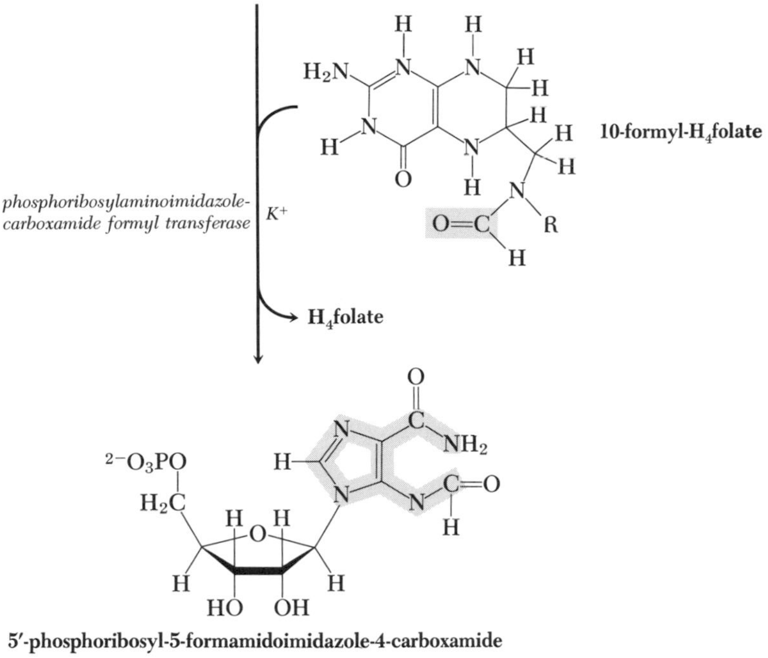

Figure 20-4   *Continued*

In the next pair of reversible reactions, the amino group of aspartate is transferred to form an amide of the carboxyl group created by the preceding reaction.
The first step is the formation of the amide between the aminoimidazole carboxylate
and the amino group of aspartate, using ATP as an energy donor. Again, the reaction
is reversible because an aminoacyl amide is the product. The product then is cleaved
on the other side of the connecting nitrogen atom to liberate fumarate and leave
the aminoimidazole carboxamide as a product. The pair of reactions is analogous
to the mechanism in urea synthesis by which the nitrogen of aspartate is transferred
to citrulline, forming arginine (p. 360), with an intermediate N-succino compound
formed in both cases. (However, ATP is cleaved to form pyrophosphate in
arginine synthesis, with the constant removal of pyrophosphate insuring that the
reaction will be drawn toward arginine.)

Then, the final atom necessary for the purine ring is transferred from
10-formyl-tetrahydrofolate by an irreversible reaction. This reaction, like the
earlier transfer of a methylidyne group, is probably not susceptible to any deficiency
of formyl groups in the one-carbon pool, but is perhaps susceptible to a deficiency
of tetrahydrofolate for carrying the groups. There are conflicting reports of increased
amounts of aminoimidazolecarboxamide appearing in the urine of humans who lack
sufficient tetrahydrofolate. Folate deficiency is known to impair this reaction in some
lower species, but not in others.

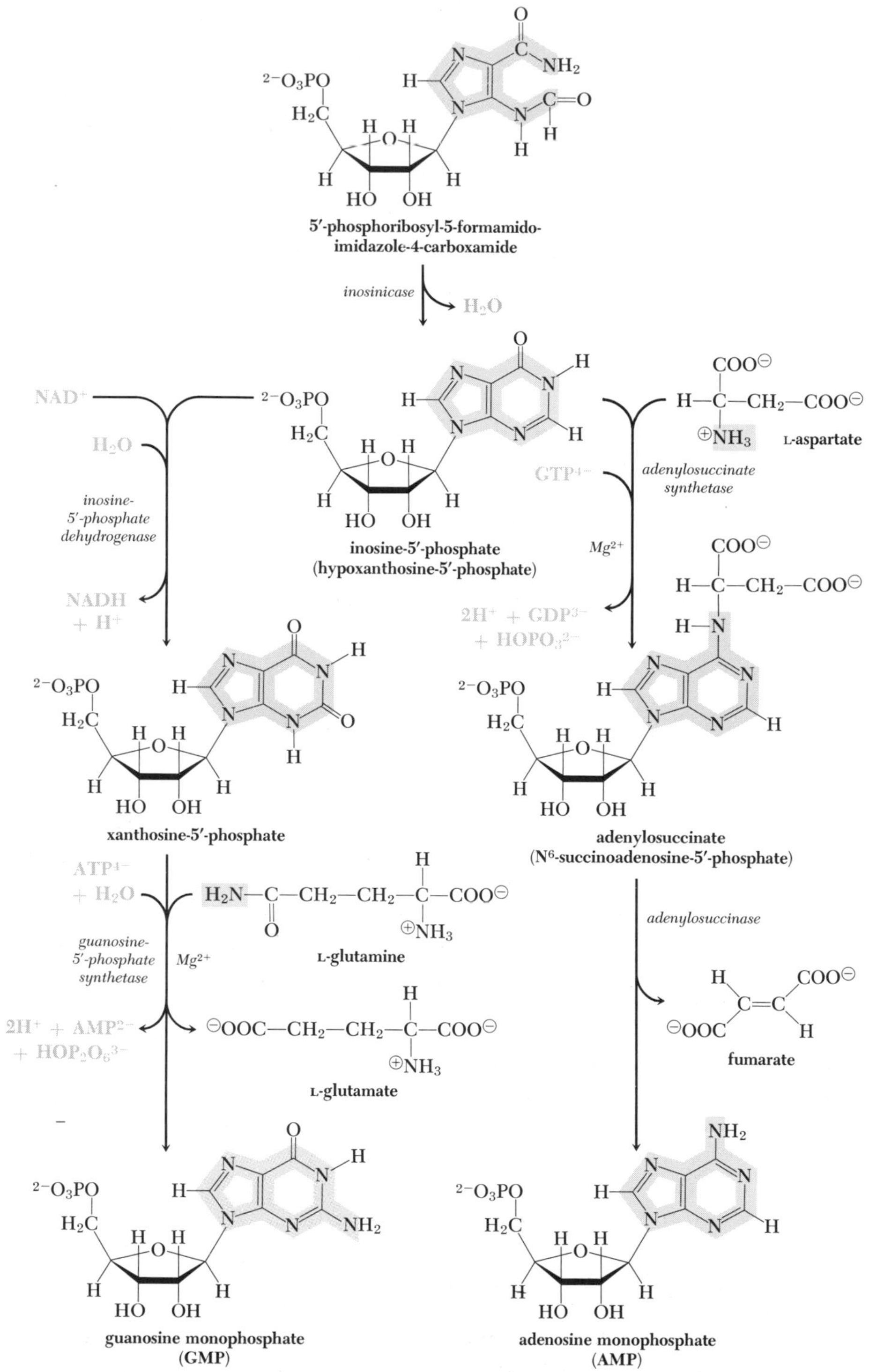

Figure 20-5   The synthesis of purine nucleotides, concluded. See text.

With the final atom for the purine in place, the ring is formed by an enzyme catalyzing the simple removal of the elements of water to produce the purine nucleotide, inosine-5′-phosphate.

The route to *adenosine-5′-monophosphate* (Figure 20-5, *right*) involves the transfer of the amino group of asparate *via* the intermediate N-succino compound, adenylosuccinate, in a way completely analogous to the transfer shown in Figure 20-4. The same enzyme cleaves fumarate in this and the previous set of reactions. However, the synthetase that forms adenylosuccinate differs in utilizing GTP rather than ATP as an energy source. The significance of this difference is not clear. Bacteria have control systems by which the individual purine nucleotides inhibit the first enzyme involved in their formation from inosine monophosphate. Thus, AMP inhibits adenylosuccinate synthetase in these organisms. It is possible that the inhibition depends upon the substrate nucleotide that is used as an energy source (GTP) being different than the ultimate product (AMP). However, this is pure speculation, and the control is yet to be demonstrated in mammalian systems.

Guanine is a more oxidized purine than is adenine. The first step in the formation of *guanosine monophosphate* is therefore the oxidation of inosine-5′-phosphate by NAD to form *xanthosine-5′-phosphate,* the nucleotide of xanthine (2,6-dioxopurine). Guanosine monophosphate is then formed by a transfer of the amide group of glutamine, with the reaction driven by ATP. This reaction differs from the earlier ones in which glutamine is an amide donor in that ATP is cleaved to AMP and inorganic pyrophosphate, at least in bacterial systems. The reaction is irreversible, even without removal of inorganic pyrophosphate, so the functional value of the loss of the extra high-energy phosphate is not clear.

These two sets of reactions complete the formation of the two purine nucleotides required in metabolism and for the formation of ribonucleates on DNA templates by the RNA polymerase reaction (p. 31). As we noted earlier, the methylated purines found in transfer RNA and ribosomal RNA are made by methylation of the completed polynucleotide (p. 429), not by reactions with the free mononucleotides.

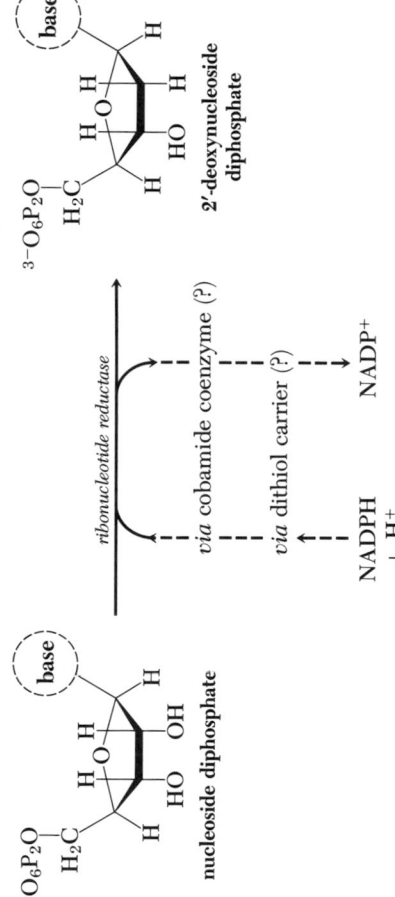

Figure 20-6 The 2'-deoxynucleotides required for the synthesis of DNA are obtained by reduction of ribonucleotides with NADPH. Although the reaction in mammals may require the presence of cobamide coenzyme, as is true of some microorganisms, it now seems more likely that only a small protein with a reactive pair of sulfhydryl groups is used for electron transfer. Such a protein, thioredoxin, is used for this purpose in plants and other microorganisms.

Purine Deoxynucleotides

The deoxyguanosine and deoxyadenosine nucleotides required for synthesis of DNA are formed by reduction of the corresponding ribonucleotides. (There is some discussion as to whether the diphosphates or the triphosphates are substrates for the reducing enzyme. In the absence of conclusive evidence, we shall go along with those who believe that the diphosphates, GDP and ADP, are used.) The mechanism of the process is not completely worked out in mammals. The primary reducing agent appears to be NADPH, but the electrons are transferred through one or more intermediate carriers before they are used to reduce the ribosyl residue (Fig. 20-6). The reduction in some microorganisms requires the presence of cobamide coenzyme. However, plants and most microorganisms have no cobamide coenzyme, and yet they make DNA. In some of these species, a small protein, *thioredoxin*, has been shown to be an electron carrier for ribonucleotide reduction. It contains a cystinyl residue that accepts electrons, probably from NADPH initially, thereby being converted to a pair of cysteinyl groups that are the active reducing agents.

An active enzyme preparation can be made from the cobamide-dependent organisms that will also use reduced thioredoxin, but this doesn't mean much, because a variety of dithiols, including reduced lipoate and synthetic compounds, will also act in the same way. In these organisms, the cobamide coenzyme probably acts as an electron carrier in itself, with the cobalt being reduced and oxidized during the course of the reaction.

The evidence for the participation of cobamide coenzyme, or some related compound, in mammalian ribonucleotide reduction hinges largely on observed failures of mitosis and nucleic acid formation in animals with a cobalamin deficiency, but these failures may be secondary to general metabolic disturbances. It now seems likely that the cobamide coenzyme is not involved.

Whatever the route for formation of deoxyribonucleotides may be, the overall process is closely regulated. Accumulation of these nucleotides in the absence of DNA formation would serve no useful purpose. The regulatory mechanisms hinge upon an inhibition of the reductases by the end-product nucleotides. All of the interrelationships have not been worked out, but it now seems that the enzymes are controlled in such a way as to produce a balanced mixture of the required deoxyribonucleotides. For example, a rise in the concentration of deoxyATP inhibits the reduction of all of the other nucleotides, and the production of this compound is stimulated by deoxyGTP and deoxyTTP, so that if these two compounds are not being removed, they will shut off their own production by *accelerating* the formation of the inhibitory dATP.

Pyrimidine Nucleotides

The pyrimidine ring, unlike the purine ring, is formed as a free base that is later attached to a ribosyl phosphate to make nucleotides. The elements of the ring are contributed completely by *glutamine, CO_2,* and *aspartate.*

The process begins by a reaction between carbamyl phosphate and aspartate to form *carbamylaspartate.* The formation of carbamyl phosphate is also the initial step in urea synthesis (p. 360). We saw that urea formation largely occurs in the liver, but many tissues are able to make the pyrimidines for use in nucleic acid synthesis. Carbamyl phosphate is formed for this purpose by a different enzyme, only recently discovered. The process differs in several ways from the reaction used for making urea. The reaction occurs in the cytosol. Acetyl glutamate is not a required cofactor. Glutamine, rather than ammonia, probably is the direct nitrogen donor. The content of the enzyme is quite low because the demand for pyrimidine synthesis is orders of magnitude lower than the demand for urea synthesis. Most importantly of all, the enzyme is inhibited by UTP, which therefore regulates the initial step in its own production.

The simple removal of water from carbamylaspartate forms a reduced pyrimidine, *dihydroorotate,* which is then oxidized to orotate.

Orotate reacts with phosphoribosylpyrophosphate to form the nucleotide, *orotidine-5'-phosphate,* liberating pyrophosphate. As with other pyrophosphorylases, the removal of the inorganic pyrophosphate drives the reaction toward nucleotide formation.

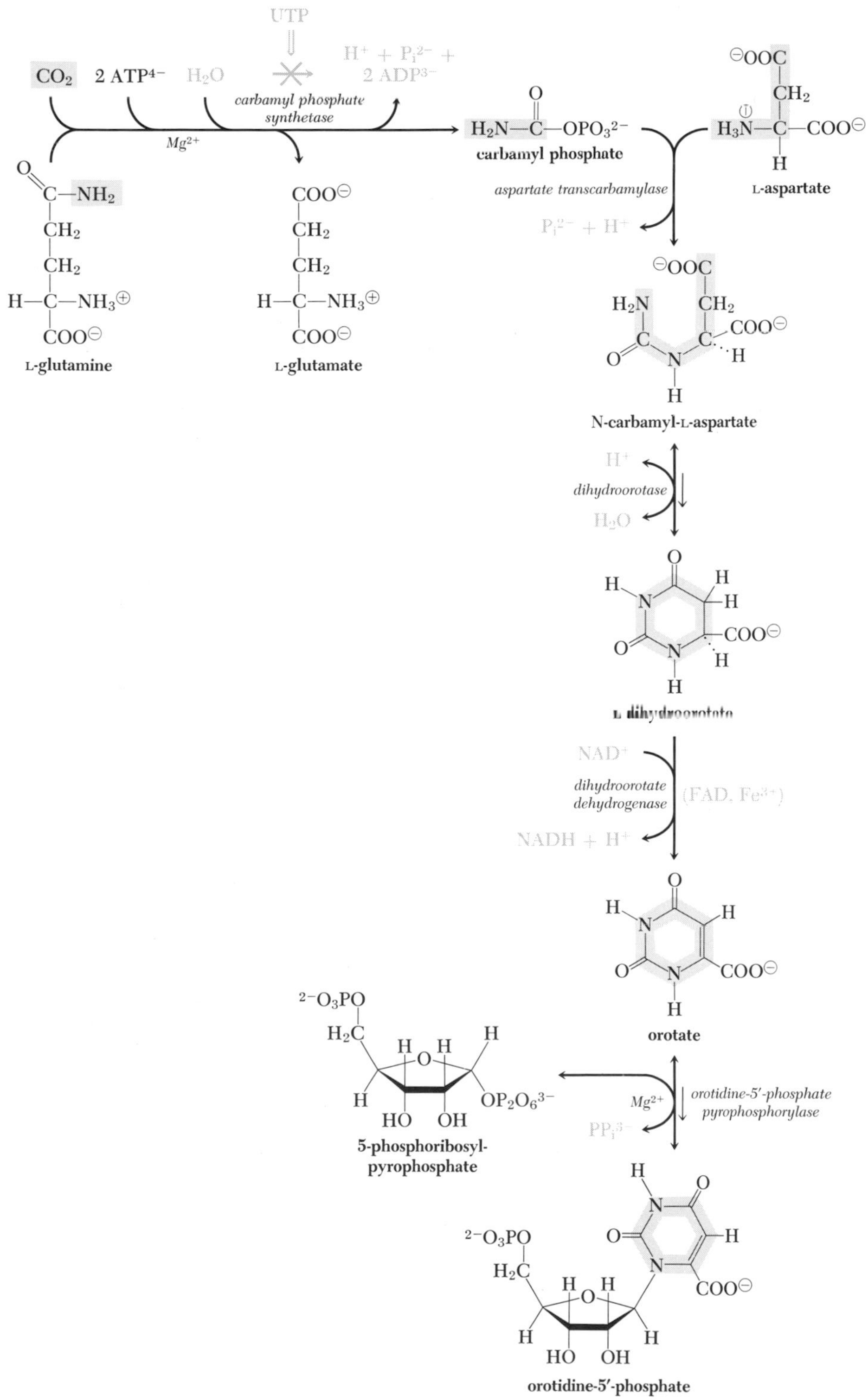

Figure 20-7 The synthesis of pyrimidine nucleotides I. See text.

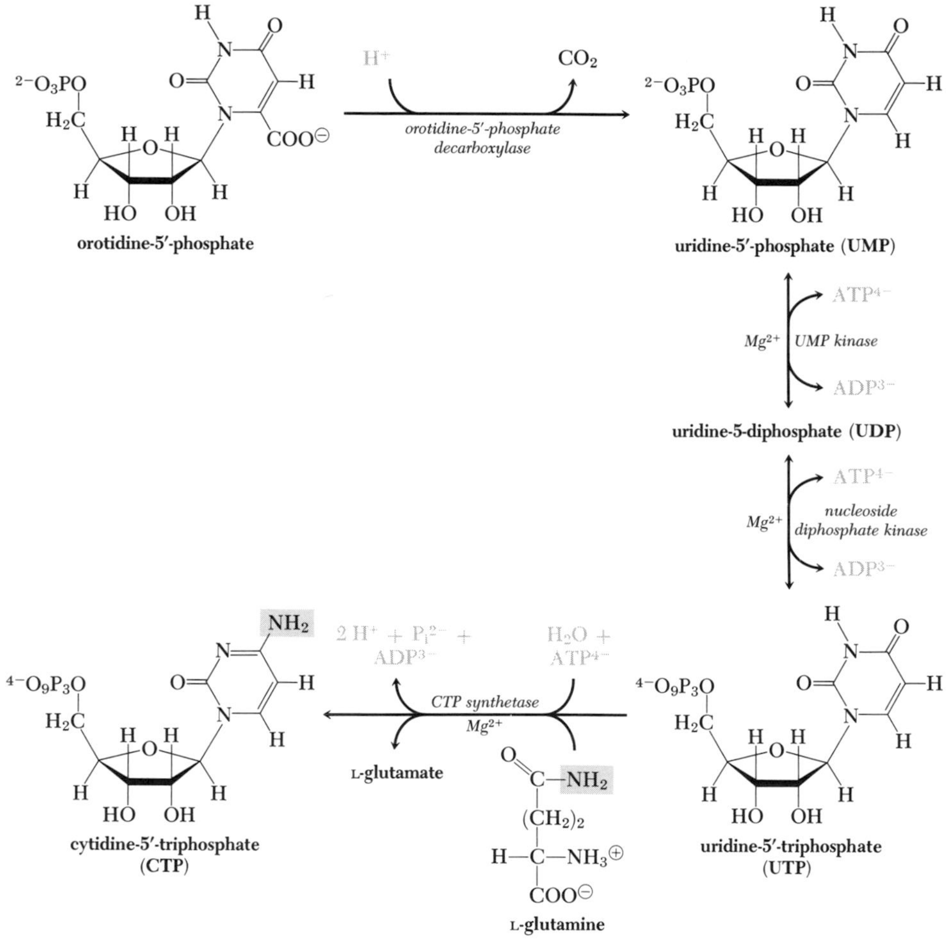

Figure 20-8   The synthesis of pyrimidine nucleotides, concluded. See text.

Orotidine-5'-phosphate is converted to *uridine-5'-phosphate* by a simple decarboxylation. The enzyme catalyzing the reaction has no known cofactors.

Uridine monophosphate, like the other nucleoside phosphates, is converted to its diphosphate by a specific kinase utilizing ATP, and the diphosphate is converted to the triphosphate by a general nucleoside diphosphokinase, also at the expense of ATP. Uridine triphosphate can be used for nucleic acid synthesis or in the formation of metabolic intermediates such as uridine diphosphate glucose, but part of it is consumed to make *cytidine triphosphate.*

Uridine triphosphate is converted to cytidine triphosphate by a transfer of the amide group of glutamine, driven by the cleavage of ATP. This reaction is similar to those using glutamine in the formation of purine nucleotides.

Pyrimidine nucleoside diphosphates, like those containing purines, are reduced upon demand to create 2'-deoxynucleotides as precursors for DNA. The same enzyme system appears to catalyze the reduction of both the purine and pyrimidine nucleotides. The precursor of the deoxythymidine nucleotides is dUMP, and the relationships involved are shown in Figure 20-9. DeoxyTMP is formed from dUMP by an enzyme named *thymidylate synthetase.* The substituent methyl group is obtained by transfer from 5,10-methylene-H_4folate. If the methylene group was transferred without anything else happening, the result would be the appearance of a hydroxymethyl group. In order to make a methyl group, there must be a simultaneous reduction, and the reducing agent in this reaction is the tetrahydrofolate that is also serving as a one-carbon carrier. Tetrahydrofolate contributes two electrons and is converted to *dihydrofolate.* We already saw that the conversion of dietary folate to the active tetrahydrofolate involves two reductions with dihydrofolate as an intermediate (p. 411). We now see that the second of these reductions must be used for the recovery of tetrahydrofolate from dihydrofolate in cells producing DNA in addition to its use in reducing dietary folate.

All of this may seem straightforward, but it contains an enigma. We know that it is a uracil, not a thymine, nucleotide that is reduced to the corresponding deoxynucleotide. This is true with all organisms investigated, including those using the cobamide-dependent reduction and those not requiring cobamide; those reducing UTP and those reducing UDP. Now, the organisms also contain the kinases necessary for interconverting dUMP, dUDP and dUTP. Here lies the problem, because dUTP also will act as a substrate for DNA polymerase, at least under experimental conditions. Why, then, aren't deoxyuridine residues incorporated in the DNA of dividing cells?

Some have tried to rationalize the problem by invoking another well-established route for the formation of the dUMP precursor that excludes the formation of dUDP or dUTP. This route, also shown in Figure 20-9, involves the hydrolytic deamination of dCMP. This route permits the formation of deoxythymidine residues from deoxyuridine formed from deoxycytidine—all as the monophosphates. These reactions indeed exist. However, the rationalization assumes that dividing cells cannot reduce UDP to dUDP and cannot phosphorylate dUMP to dUTP. Both of these assumptions seem to be contrary to the observed behavior of the necessary enzymes, and we are left without an adequate explanation for the exclusion of uracil residues from DNA.

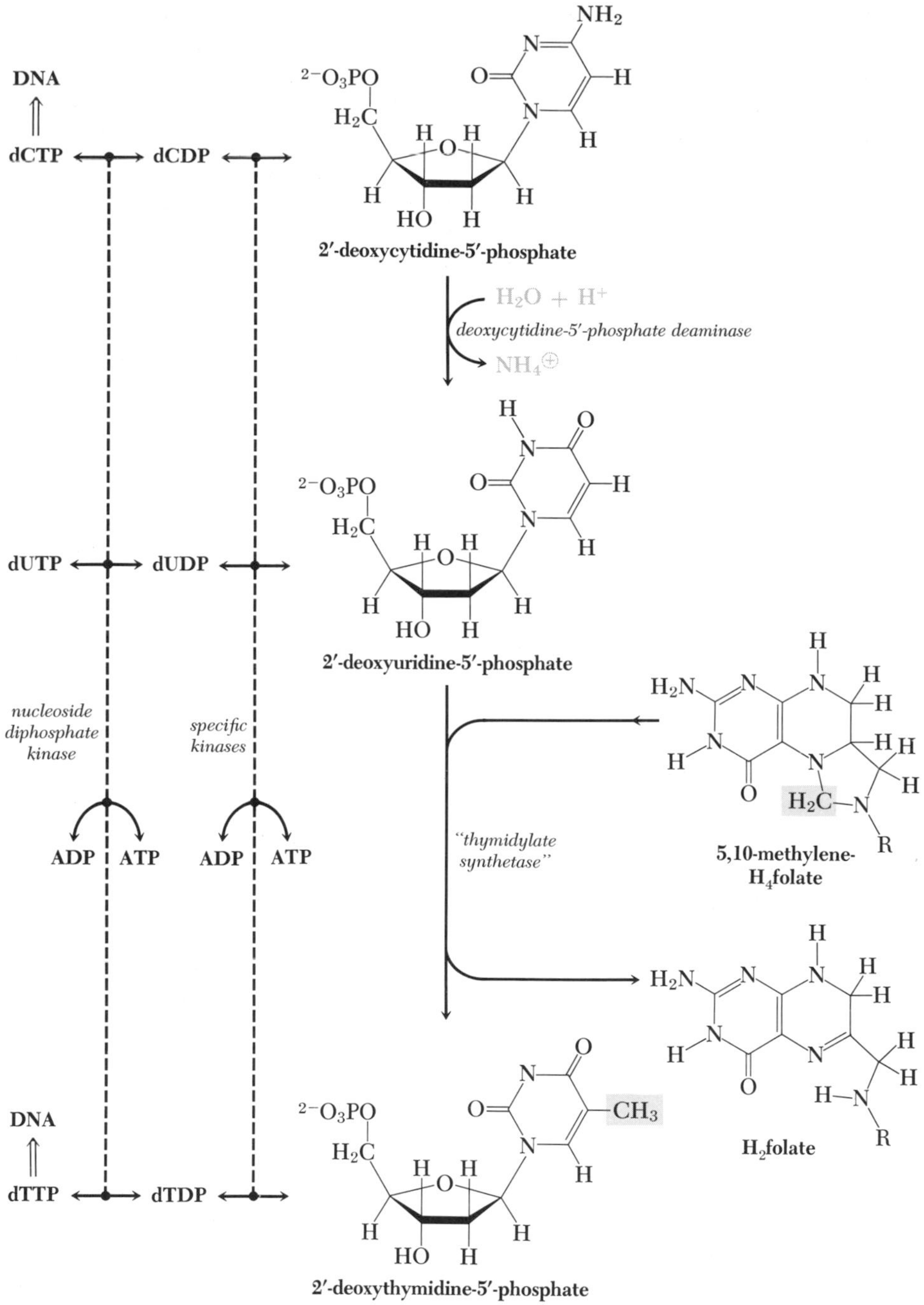

DNA

dCTP ⟷ dCDP ⟷

$^{2-}O_3PO$
H_2C

2'-deoxycytidine-5'-phosphate

$H_2O + H^+$

deoxycytidine-5'-phosphate deaminase

$NH_4^{\oplus}$

$^{2-}O_3PO$
H_2C

2'-deoxyuridine-5'-phosphate

dUTP ⟷ dUDP ⟷

nucleoside diphosphate kinase

specific kinases

ADP | ATP   ADP | ATP

"thymidylate synthetase"

H_2N

H_2C-N

5,10-methylene-H_4folate

H_2N

$H-N$

H_2folate

DNA

dTTP ⟷ dTDP ⟷

$^{2-}O_3PO$
H_2C

CH₃ → CH_3

2'-deoxythymidine-5'-phosphate

Figure 20-9   Deoxythymidine found in DNA is formed as dTMP (*bottom*) from dUMP by transfer of a methylene group from H₄folate. The electrons of H₄folate are simultaneously used to reduce the methylene group to a methyl group. The pteridine compound in this case is serving the usual function of a one-carbon carrier and the function of an electron donor seen with tetrahydrobiopterin in some mixed-function oxidase reactions.

The source of the dUMP is still in doubt. It may arise by deamination of dCMP (*top center*), or there may be a reduction of UDP to dUDP, which is in equilibrium with dUMP. All of the reactions shown have been demonstrated in experimental preparations and it is not known why dUTP is not also used for DNA formation.

Inhibitors of Nucleotide Synthesis

Cancers are made of cells that continue to divide indefinitely. Since cell division requires a net synthesis of nucleic acids, there has been considerable effort made to find compounds that will selectively inhibit the formation of nucleic acids and check the uncontrolled growth of cancer.

The least successful attempts involve the inhibition of the general types of reactions used in the formation of the purine and pyrimidine rings. For example, nitrogen is transferred twice from glutamine in the formation of the purine ring, again in forming GMP from XMP, and again in forming CTP from UTP. Certain diazo compounds, such as *azaserine* and *6-diazo-5-oxo-norleucine*:

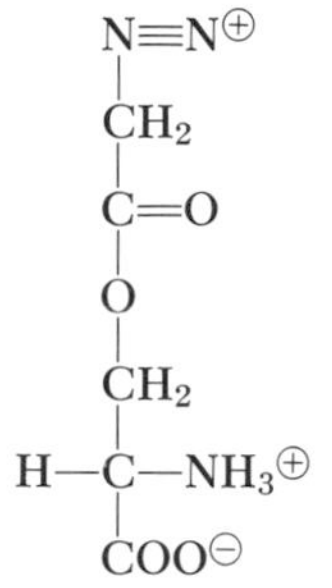

$$N{\equiv}N^{\oplus}$$
$$|$$
$$CH_2$$
$$|$$
$$C{=}O$$
$$|$$
$$O$$
$$|$$
$$CH_2$$
$$|$$
$$H{-}C{-}NH_3^{\oplus}$$
$$|$$
$$COO^{\ominus}$$

O-(2-diazoacetyl)-L-serine
(azaserine)

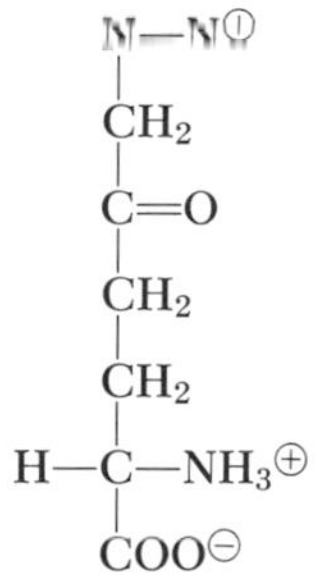

$$N{-}N^{\oplus}$$
$$|$$
$$CH_2$$
$$|$$
$$C{=}O$$
$$|$$
$$CH_2$$
$$|$$
$$CH_2$$
$$|$$
$$H{-}C{-}NH_3^{\oplus}$$
$$|$$
$$COO^{\ominus}$$

6-diazo-5-oxo-L-norleucine

are potent inhibitors of most of these reactions. However, these same reactions are necessary for the maintenance of normal cells, including those that are not dividing. The result is that concentrations of the glutamine analogues that are effective in suppressing cancer growth are dangerously toxic, and these compounds are not useful drugs.

Another general class of reactions used for nucleotide synthesis is the transfer of one-carbon units from tetrahydrofolate derivatives. Such transfers occur twice during the formation of purine rings and once in the formation of deoxythymidine monophosphate. As we saw, the formation of dTMP is a special case in that it involves a simultaneous reduction of tetrahydrofolate to dihydrofolate. There is at least a possibility that a compound inhibiting the reduction of dihydrofolate back to tetrahydrofolate would be selectively toxic to dividing cells, since these are the only cells forming deoxynucleotides. Two such compounds are commercially available, *aminopterin* and *methotrexate*. These competitively inhibit the dehydrogenases

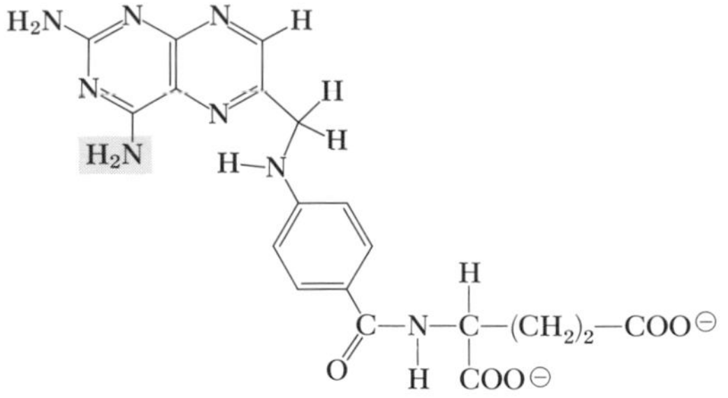

aminopterin
(4-deoxy-4-amino folate)

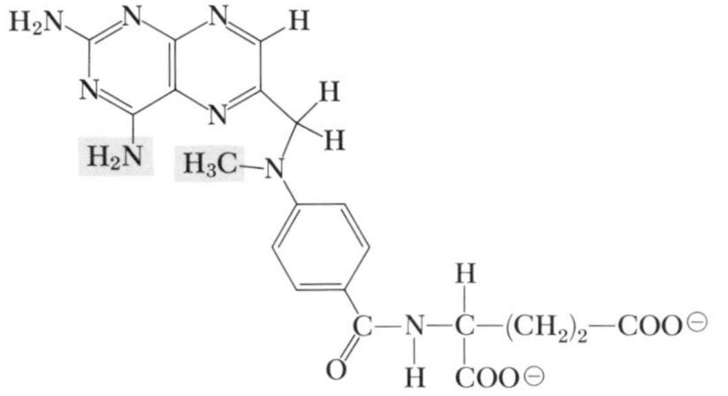

methotrexate
(4-deoxy-4-amino-10-methylfolate)

catalyzing the reduction of folate to dihydrofolate and dihydrofolate to tetrahydrofolate by NADPH. Methotrexate in particular is in common use today to control certain types of cancer, especially choriocarcinomas and childhood leukemias. However, this drug also has a general toxicity because it affects all rapidly dividing cells, such as those in the intestinal mucosa, and prevents the replenishment of tetrahydrofolate from the oxidized form of the vitamin in all tissues. Therefore, it is not possible to give the drug in sufficiently large quantities to kill all cancer cells and effect a real cure of leukemias, but choriocarcinomas are frequently cured.

The more generally successful drugs are those that are analogues of the purine and pyrimidine bases themselves. These include, for example, *6-mercaptopurine* and *5-fluorouracil*. 6-Mercaptopurine is used in adult leukemias, and it has its

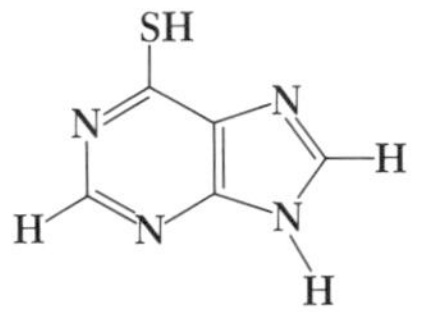

6-mercaptopurine

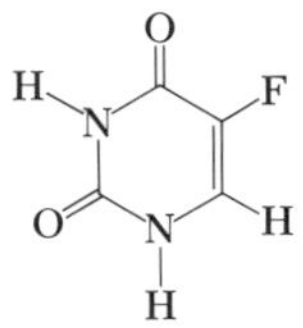

5-fluorouracil

effect because it is incorporated into nucleic acids (as thioguanosine), thereby preventing normal protein synthesis, and also because its nucleotides inhibit the formation of phosphoribosylamine and the conversion of inosine monophosphate to adenosine monophosphate. 5-Fluorouracil, which is of some value in obtaining remission with the more prevalent carcinomas, is also incorporated into RNA, but its major effect comes from the formation of deoxyfluorouridine monophosphate, which is a potent inhibitor of thymidylate synthetase, and therefore prevents normal DNA formation in dividing cells.

There is yet another class of drugs, the alkylating agents, that have their effect through reactions introducing alkyl substituents on the purines of DNA itself. These, of course, do not interfere with the metabolism of nucleotides, and are mentioned here only for completeness. Like the other chemotherapeutic agents, alkylating agents cannot be administered in doses high enough to cure cancers because of their toxic effects on normal tissues, but they can prolong useful life in a number of patients.

DEGRADATION OF NUCLEOTIDES

The pool of free nucleotides in a cell is constantly depleted by degradative processes that serve to maintain concentration through the balance of synthesis and destruction. The breakdown of nucleic acids increases the concentration of free nucleotides and the rate of their destruction. Complete destruction of deoxyribonucleates is not consistent with maintenance of normal cellular function, and does not occur until the entire cell is destroyed. However, the ribonucleates, especially the messenger ribonucleates, are degraded and replaced during normal function. Consequently, the enzymes attacking ribonucleates are more widely distributed, and in a higher concentration, than those affecting deoxyribonucleates. (There is some evidence that deoxyribonucleases may function at a low rate to enable repair of damaged DNA within a cell—presumably by removing abnormal nucleotides created by radiation damage or other causes, thereby permitting insertion of the proper nucleotide. This process is slow compared to other metabolic activities, and therefore difficult to study, so that the details have not been described.)

Degradation of nucleic acids begins with partial hydrolysis (Fig. 20-10) catalyzed by *ribonucleases* or *deoxyribonucleases*. We already mentioned the ribonuclease of the pancreatic juice (p. 106), which hydrolyzes 5'-phosphate ester bonds, liberating 3'-phosphoribosyl residues. The pancreatic enzyme is concerned with the digestion of dietary nucleic acids, and differs from the intracellular ribonucleases, which are found in lysosomes and mainly attack the 3'-bonds of ribonucleates so as to liberate the metabolically active 5'-phosphoribosyl linkage. The effect of the lysosomal nucleases is to cleave the molecule into smaller polynucleotide fragments, with the liberation of only a few free nucleotides. Further hydrolysis to liberate all of the mononucleotides is catalyzed by less specific *phosphodiesterases*.

The combined action of these enzymes puts all of the constituents of nucleic acids into the general metabolic pool of nucleotides.

The phosphate group of nucleotides is liberated by hydrolysis, catalyzed by *nucleotidases*. The nucleosides formed are attacked by *phosphorylases*, with liberation of the free bases and the recovery of the sugar as the 1-phosphate ester. This ester is equilibrated with the 5-phosphate ester by a *phosphoribomutase*, which is analogous in action to phosphoglucomutase and requires the presence of ribose-1,5-diphosphate in small amounts.

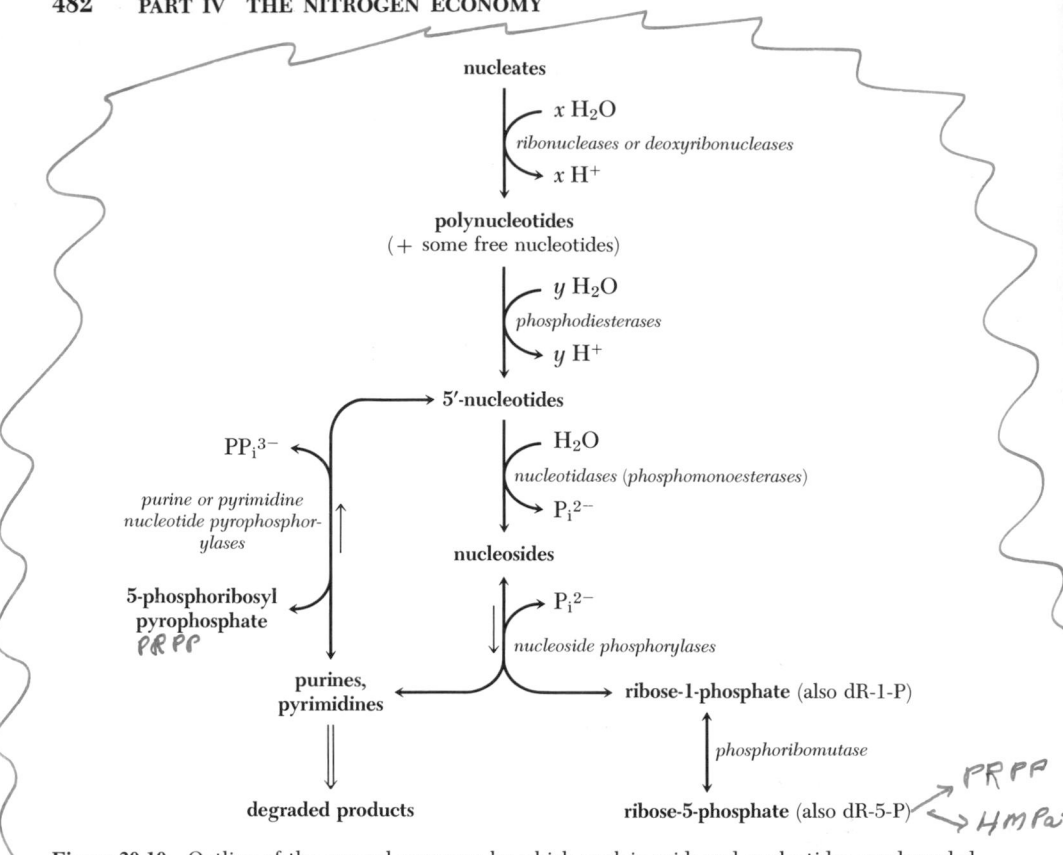

Figure 20-10   Outline of the general processes by which nucleic acids and nucleotides are degraded. The purines and pyrimidines arising in this way or obtained from the diet may be re-incorporated into nucleotides by reactions with phosphoribosylpyrophosphate (*lower left*), or they may be further degraded. The ribose portion appears as ribose-5-phosphate, which may be used to make phosphoribosylpyrophosphate or metabolized by the reactions of the pentose phosphate pathway.

Part of the bases may be recovered by reaction with phosphoribosyl pyrophosphate to reform nucleotides. This represents a *salvage route* whereby purines liberated by nucleic acid destruction can be recovered in other cells without the necessity for complete synthesis. Two *pyrophosphorylases* are involved. One is specific for *adenine* and the other will react with *hypoxanthine* or *guanine*.

Degradation of Purines (Fig. 20-11)

The major purine nucleotide, adenosine-5′-monophosphate, is degraded so as to liberate the base by the pathway outlined before, with one important deviation. The nucleotide is first deaminated by a simple hydrolysis to produce *inosine phosphate*. Of course, inosine phosphate is also the precursor for both adenosine and guanosine phosphates, and could be converted to either, but at least part of the compound is attacked by nucleoside phosphorylase and thereby carried through the transitions to the nucleoside, inosine, and on to the free base, *hypoxanthine*. There is very likely something in the function of AMP deamination that we do not understand. *Adenylate deaminase* is present in quite high concentration in

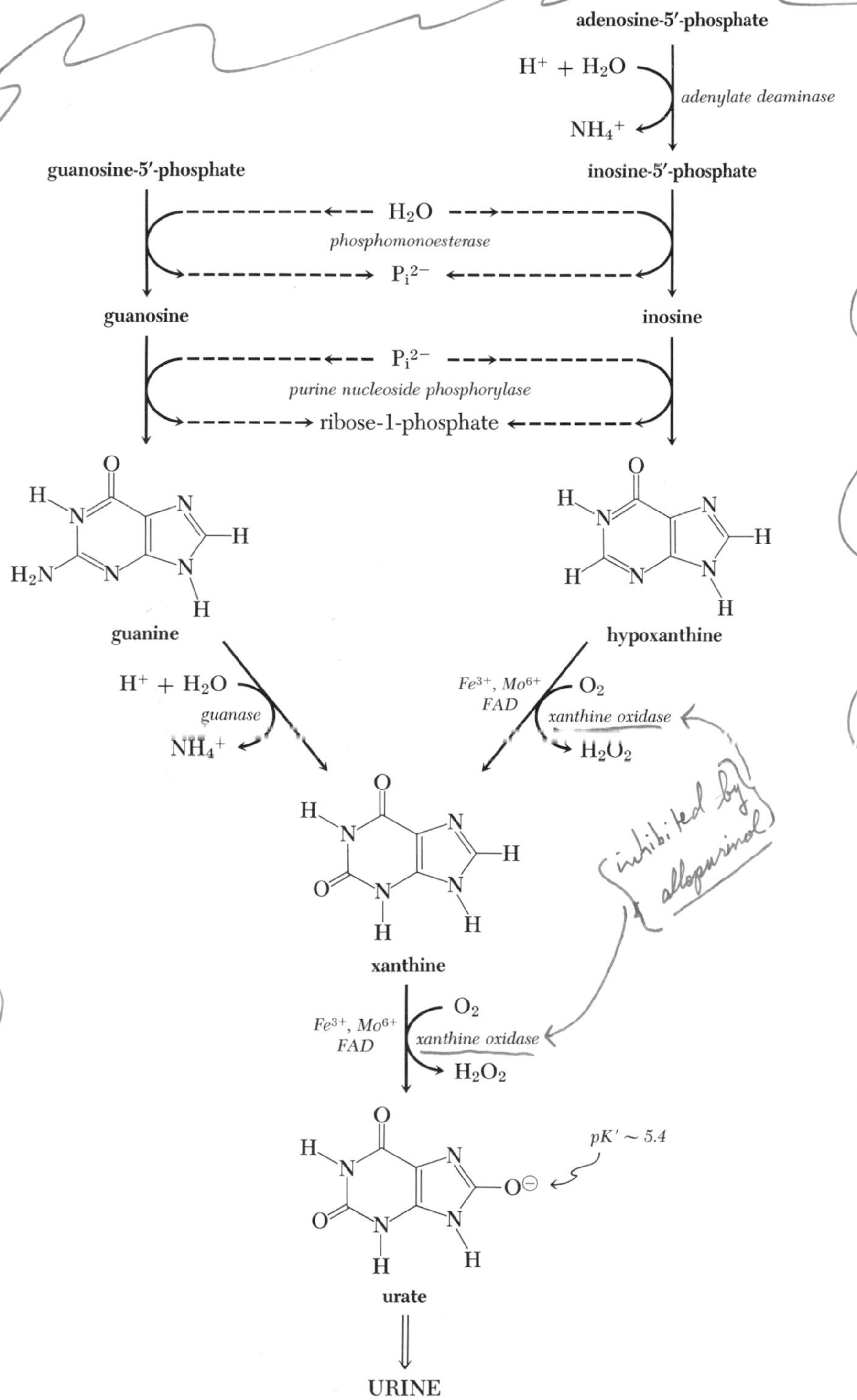

adenosine-5'-phosphate

$H^+ + H_2O$

adenylate deaminase

NH_4^+

guanosine-5'-phosphate   inosine-5'-phosphate

H_2O

phosphomonoesterase

P_i^{2-}

guanosine   inosine

P_i^{2-}

purine nucleoside phosphorylase

ribose-1-phosphate

guanine   hypoxanthine

$H^+ + H_2O$ Fe^{3+}, Mo^{6+} O_2
                                               FAD
guanase   *xanthine oxidase*

NH_4^+ H_2O_2

(inhibited by allopurinol)

xanthine

                              Fe^{3+}, Mo^{6+} O_2
                                     FAD
                                  xanthine oxidase
                                                      H_2O_2

$pK' \sim 5.4$

$-O^{\ominus}$

urate

URINE

Figure 20-11 The conversion of the purines to urate. A deamination and single oxidation of guanine derived from GMP forms urate. AMP is deaminated as a nucleotide with the eventual liberation of hypoxanthine; two oxidations are required to convert hypoxanthine to urate.

skeletal muscle, and it doesn't seem rational to have so much of an enzyme in that tissue if its only purpose is to start nucleotides on the road to destruction. Whatever other purpose it may serve, the enzyme also is necessary for the removal of adenine nucleotides, which is what we are presently considering.

Hypoxanthine is oxidized to *xanthine* by molecular oxygen. The enzyme catalyzing the reaction, xanthine oxidase, is an unusual flavoprotein. It occurs in the soluble cytoplasm of the cell, and yet it also contains non-heme iron and *molybdenum*. The presence of molybdenum in this, as well as the related aldehyde oxidase also present in the cytoplasm, makes it necessary to have traces of the inorganic element in the diet. The molybdenum, ordinarily hexavalent, is reduced to the pentavalent compound during the passage of electrons to oxygen. Oxygen is reduced to hydrogen peroxide as the other product.

Guanine is at the same oxidation level as xanthine, so simple hydrolytic deamination of the free base liberated by degradation of its nucleotides will form xanthine. Xanthine therefore is an intermediate in the degradation of both major purines. Xanthine is also a substrate for xanthine oxidase, which catalyzes the oxidation to *uric acid*, present in the body as urate ion.

Urate is cleared efficiently from the blood stream by the kidney, and is excreted in the urine as the terminal product of metabolism of the purine ring. (Remember that the substituent amino groups of adenine and guanine nucleotides eventually appear as ammonia, so the purines also make a small contribution to the general nitrogen pool.) The first ionization of uric acid has a pK' near 5.4. In an acidic urine, as much as half of the excreted compound may be in the form of the undissociated acid, which has a quite low solubility (about 0.1 gram per liter of pure water at 38°C). Consequently, uric acid may precipitate in the urine. This is ordinarily not significant because the urine contains too many constituents for the ready growth of good crystals, but if the solubility of other compounds, especially calcium salts, is exceeded in the urine, uric acid may co-precipitate with them. The result is a slow, but steady, aggregation of troublesome stones, which may form in either the kidneys or the bladder. However, uric acid precipitation is usually a contributing factor to, rather than a primary cause of, stone formation in the kidneys, where calcium oxalate and phosphate stones predominate.

The amount of urate excreted is an index of purine turnover. It is not an absolute guide, because part of the urate in the blood passes into the gut in gastric juice and bile, where it decomposes further with a liberation of ammonia that is reabsorbed to eventually appear as urea rather than urate. The decomposition is at least partially the result of metabolism by microorganisms, but urate is also known to be unstable in duodenal fluid free of microorganisms. After allowance for the intestinal loss from all causes (20 per cent to 50 per cent of the total urate formed), urate excretion in the urine reflects the presence of purine nucleotides above and beyond the amount required to maintain normal cellular concentrations. Assuming that an individual is an adult with relatively constant body composition, the extra purine nucleotides represent the sum of the amount synthesized by the routes outlined earlier and the amount ingested in the diet. Most American adults put out 2 to 4 millimoles (0.3 to 0.6 gram) of urate per day.

In addition to the major purines, adenine and guanine, methylated purines are also produced, but in lesser amounts, from the degradation of transfer and ribosomal

RNA, and these compounds must be metabolized. It appears that most purines substituted with a methyl group on a ring nitrogen atom are excreted unchanged. The fate of those with a methyl group on a side-chain nitrogen atom is unknown. If such methylamino groups are hydrolyzed in the same way as are the unsubstituted amino groups, free methylamine (or dimethylamine, in the case of di-substituted purines) will be liberated. It may be that this accounts for the small amounts of methylamines normally found in urine.

The low solubility of uric acid is used to advantage by the birds and reptiles. These animals convert the major part of excess nitrogen to uric acid through the pathway of purine synthesis, rather than to urea through the pathway of arginine synthesis. These animals reproduce by laying closed eggs containing a limited amount of water. Urea is relatively innocuous, but its concentration would become intolerably high in the egg. However, uric acid crystallizes out within the egg during development of the embryo, and its concentration in the tissue fluids cannot rise to excessive levels. There may be further advantages in adult life. The elimination of urea requires the concomitant elimination of a fair amount of water to maintain a reasonably dilute urine. Adult birds and reptiles excrete a urine that is a pasty white mass of uric acid crystals, containing much less water per mole of excreted nitrogen than does mammalian urine. This is an obvious advantage to animals in a desert environment, and even more of an advantage to a flying animal that must lift all of the water it is later going to excrete.

Gout and the Lesch-Nyhan Syndrome

The upper limit for the "normal" concentration of urate in the blood is about 0.35 millimolar (60 μg per ml). Sodium urate is soluble to the extent of about 5 millimolar in water at 38°C. Despite this apparent margin of safety, crystals of sodium urate form in the joints and kidneys of some individuals. The deposits in the joints provoke a painful inflammatory condition known as gout. (For some reason, the joint at the base of the big toe is especially susceptible.) The mechanism for deposition is unknown. Some people, including about one out of 20 normal adult males, have elevated concentrations of urate without the appearance of gout. However, 19 out of 20 gouty people have urate concentrations above 0.35 millimolar. These patients are usually sexually mature males, and number some 800,000 people in the United States. There evidently is an association between gout and elevated urate concentrations, but it appears that gout requires some additional factor favoring urate deposition. Given an inherent inability to prevent local concentration of the compound in the joints or other susceptible locations, a gouty attack may be precipitated by anything that elevates the concentration of urate. (Anyone is likely to have a transient episode of gout if the urate concentration becomes high enough. Secondary gout of this sort may be a result of failure of elimination of urate in the kidney, or of a transitory rapid breakdown of tissues so as to load the system with an unusual quantity of nucleotides.) Many of the patients with primary gout have a deficient feedback control so that the initial step in purine synthesis, the formation of phosphoribosylamine, proceeds at an excessive rate.

The classic treatment for gout is administration of colchicine, which interrupts mitosis of leucocytes and prevents development of inflammation. Current rationale is that inflammation is a consequence of phagocytosis of urate crystals by the leucocytes that invade the affected area. (Colchicine interferes with cell division in other tissues, and it has been used to provoke the formation of tetraploids in plant breeding. The possibility of unproven deleterious effects in humans is deemed less dangerous than the well-known effects of uncontrolled gout by most physicians.) Compounds promoting the excretion of urate by the kidney were later introduced as a means of controlling gout. These are outside the scope of our present discussion.

Still more recently, a competitive inhibitor of xanthine oxidase has been developed to control gout by preventing the formation of urate so that excess purines are excreted as hypoxanthine and xanthine. This compound, *allopurinol*, differs from hypoxanthine in the distribution of ring nitrogens. The rationale for its use is that hypoxanthine and xanthine won't precipitate in the joints. Hypoxanthine is more

allopurinol

soluble than urate, although xanthine is not. However, hereditary deficiencies of xanthine oxidase are known, and people with this condition have no apparent difficulties with deposition of xanthine in their tissues, although they do have an increased propensity for the appearance of xanthine stones in the urinary tract. In addition to its more favorable solubility characteristics, hypoxanthine is also used to some extent in the salvage route for re-forming nucleotides, whereas urate is not. Remaking nucleotides won't diminish the necessity for ultimate disposal, but may help to dampen transient fluctuations in concentration. Allopurinol is not a completely safe drug, and it remains to be seen whether its disadvantages are outweighed by its ability to control gout, which is a potentially serious condition.

Some 60 people are known to be afflicted with a condition named the *Lesch-Nyhan syndrome*, which is caused by a hereditary lack of GMP pyrophosphorylase. This X-linked recessive disease is peculiarly horrifying because children afflicted with it mutilate themselves. Characteristically, they will bite off the tips of their fingers, or bite their lips if their hands are protected. The aggressive attacks also may be directed against others as biting, or in later years as the use of obscene language or gestures. The pathos of the affliction is accentuated by the tendency of the children to be very likeable and open, quick to laugh and capable of warm affection. They sometimes are terrified of their own aggression, so that they scream in fright when restraints are removed, and yet they compulsively begin snapping at their fingers.

Here is an enzymatic deficiency that causes symptoms which, considered by themselves, might well be labelled as psychic in origin. The Lesch-Nyhan syndrome has attracted wide attention because of the potentialities it appears to have for relating organic constitution to personality.

The syndrome is also of great interest to students of purine metabolism because of the questions it raises. The missing enzyme is responsible for the salvage of free guanine and hypoxanthine by converting them to the corresponding nucleoside monophosphates. It is especially active in the brain, which has over 10-fold more enzyme than does the liver. The corresponding enzyme catalyzing the salvage of adenine is only a third as active in brain as it is in liver. (The relative rates of the GMP and AMP pyrophosphorylases are almost $1:3$ in the liver and $20:1$ in the brain basal ganglia.) This appears to indicate some especial metabolic role of the guanine nucleotides in the nervous system that may or may not be connected with the symptoms caused by lack of the salvage enzyme. However, this does not explain another striking effect of the Lesch-Nyhan syndrome. The patients excrete large quantities of uric acid—more per unit of body mass than is seen in any other condition. Their synthesis of purines is extraordinarily rapid. (Of course, they also become gouty as an additional tribulation.) This effect cannot be rationalized as a metabolic disturbance in the nervous system. There have been efforts to explain it on the basis of a failure to maintain sufficient levels of GMP for feedback control of purine synthesis in the liver and other tissues, but this is not very satisfying. In short, the existence of the Lesch-Nyhan syndrome serves to expose our lack of knowledge about some fundamental aspects of regulation of purine metabolism.

Gout and the Lesch-Nyhan syndrome also raise a large evolutionary question. Why do the primates excrete uric acid? Most animals degrade the compound further, and avoid the problems created by its low solubility. The notion has been in existence for a long time that gouty people tend to be of higher quality than ordinary folk, perhaps more intelligent. A sound study to test this point has not been made. However, an attempt to correlate the blood urate concentrations of some 800 Army draftees with their scores on classification tests has been made. There appeared to be a significant, although small, correlation. The existence of so many humans with gout, and an even larger number with high urate concentrations without symptoms, does raise the possibility of some evolutionary advantage in the presence of urate that balances the disadvantages of greater mortality from crystal deposition. Gouty people are alleged to be more irritable even before pain appears. We might ask, only half in jest, if the aggressiveness of the patients with Lesch-Nyhan syndrome is an extreme manifestation of a purine-dependent personality trait that sometimes has its advantages in a milder form. (No data are presently available on the urate concentrations in social activists.)

Breakdown of Pyrimidines (Fig. 20-12)

The pyrimidine nucleotides, like the purine compounds, are converted to nucleosides, and then undergo phosphorolysis to release the free pyrimidines. The figure shows a simple hydrolytic deamination of free cytosine to form uracil, but in fact the deamination may occur at the nucleoside or nucleotide stages.

Uracil and thymine are metabolized by the same routes, beginning with a reduction by NADPH to the dihydropyrimidines, followed by a hydrolytic cleavage of the ring. This is, in a sense, the reversal of the synthesis of uracil in which the ring is closed by dehydration of a carbamyl amino acid and then oxidized—the difference being that the synthetic mechanism involves the carboxylate analogues, dihydroorotate and orotate.

The two products are *carbamyl* derivatives of the corresponding β-amino acids,

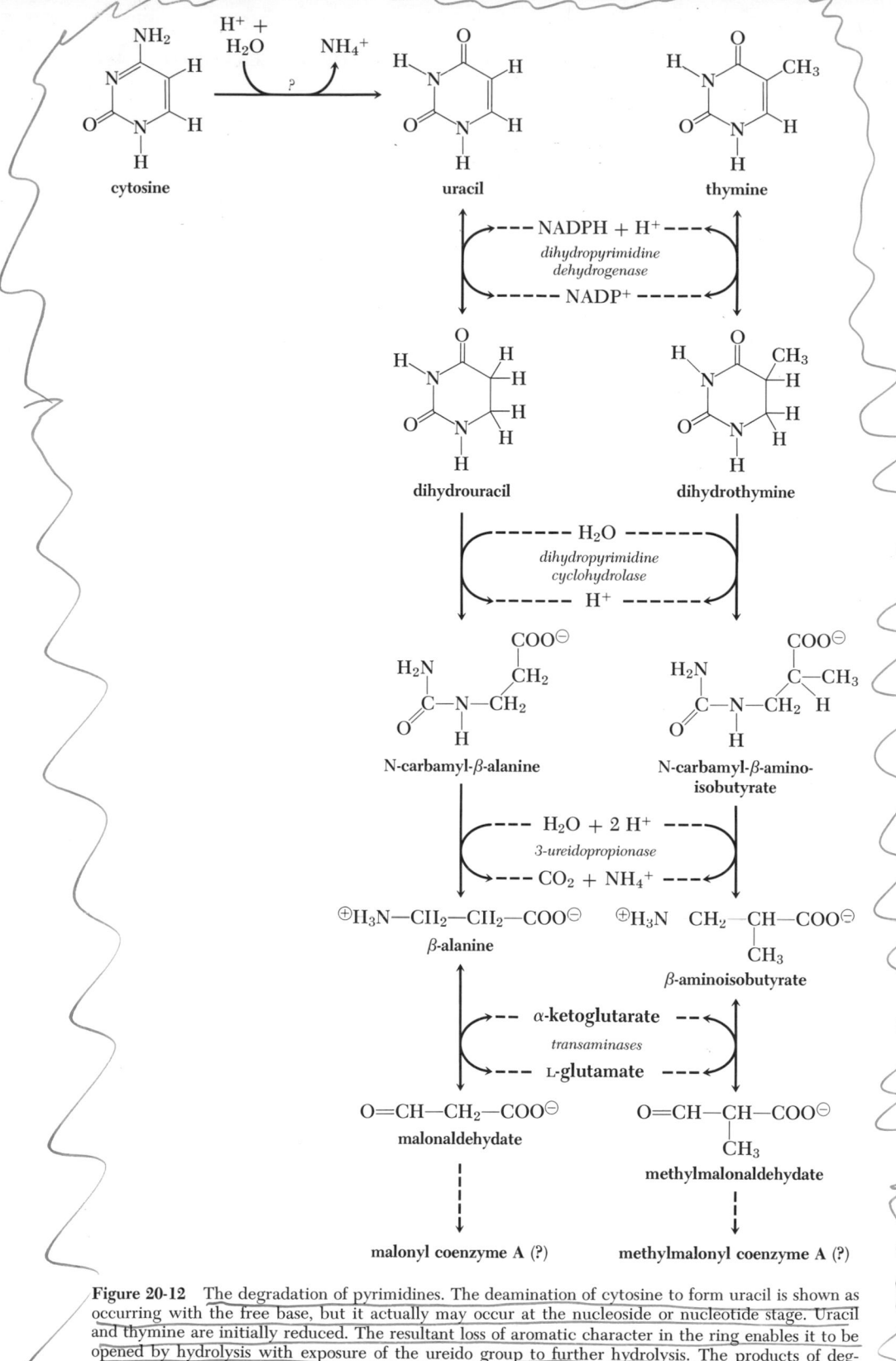

Figure 20-12 The degradation of pyrimidines. The deamination of cytosine to form uracil is shown as occurring with the free base, but it actually may occur at the nucleoside or nucleotide stage. Uracil and thymine are initially reduced. The resultant loss of aromatic character in the ring enables it to be opened by hydrolysis with exposure of the ureido group to further hydrolysis. The products of degradation can enter the general metabolic routes and there are no characteristic excretory products from pyrimidine metabolism.

β-*alanine*, and β-*aminoisobutyrate*. The carbamyl group is removed by hydrolysis, releasing the free amino acids and ammonia and CO_2. The amino acids transaminate with α-ketoglutarate to form semialdehydes of malonate and methylmalonate, which we encountered earlier in the metabolism of valine, and are presumably oxidized to eventually yield malonyl or methylmalonyl coenzyme A.

From this scheme, it is evident that pyrimidine metabolism will result in the formation of urea and CO_2 as the end products, with nothing excreted to differentiate the turnover of these compounds from that of the amino acids.

REGULATION OF NUCLEOTIDE TURNOVER

From the information at hand, it is apparent that there are two parts to the regulation of nucleotide turnover. One is the regulation of the formation and destruction of the free nucleotides and the other is the regulation of the formation and destruction of the nucleic acids.

The regulation of free nucleotide metabolism appears to result from feedback controls that we have already mentioned—the inhibition of phosphoribosylamine formation by the purine nucleotides and the inhibition of the carbamyl phosphate synthetase used in pyrimidine synthesis by the pyrimidine nucleotides. Some organisms use a different mechanism for regulation of the pyrimidines. The pyrimidine nucleotides inhibit the asparate transcarbamylase in these organisms, which do not have the distinctive carbamyl phosphate synthetase used for urea synthesis in mammals. We discussed this as an example of allosteric control (p 139). It is not known whether the breakdown of the nucleotides is or isn't under feedback control. It is entirely possible that it isn't, with the nucleotidases providing a relatively constant slow leak in the pool and concentrations in the pool maintained by varying the synthesis of nucleotides.

The regulation of nucleic acid turnover is a much more complicated, and less well-defined, phenomenon. The rate of deoxyribonucleate turnover is essentially the rate of cell turnover, and we cannot as yet define the factors regulating the rate of cell division or precipitating the destruction of a cell. The lifetime of cells varies enormously, from the one and a half day existence of cells in the mucosa of the small intestine through the one week lifetime in the rectal mucosa and the 120 day lifetime of erythrocytes to the indefinite lifetime of neurons. Of course, the mature mammalian erythrocyte does not contain DNA, but the loss of circulating erythrocytes requires their replacement, and the replacement involves maturation of erythroblasts with an accompanying loss of DNA.

The turnover of the ribonucleates may be regulated in a manner similar to that suggested for the free nucleotides, with a relatively constant loss caused by hydrolysis (although the rate need not be constant from one kind of ribonucleate to another) and a varying rate of synthesis. Considerable evidence is accumulating that variations in the expression of genetic character are largely due to regulation of the rate of synthesis of ribonucleates on the DNA templates of the nucleus. All cells of an organism have the same DNA content, and yet they differ greatly in the proteins synthesized, which in turn determine their character. (It has been found, for example, that a change in the character of the glucose-6-phosphate dehydrogenase in the bread mold, *Neurospora crassa*, causes a change in the morphology of the entire organism.) The protein content of single cells changes in time, responding to the varying character of chemicals diffusing in from the environment, which range from hormones in minute concentration to primary energy sources in relatively high concentrations.

Exploration of the mechanism of these controls of ribonucleate synthesis has only scratched the surface. There is growing belief that the primary blocking agents for ribonucleate synthesis are low molecular weight proteins known as *histones*. These proteins contain high concentrations of either arginine or lysine, and therefore carry a number of positive charges facilitating combination with the negative charges scattered along the backbone of nucleic acids. Experimental evidence suggests that natural combination of histones with newly formed RNA on the DNA molecules in the nucleus prevents further replication of RNA, and that this property is used to suppress the formation of particular proteins.

The histones are known to be altered in three ways within the cell. One is by acetylation of the amino group of lysyl residues, using acetyl coenzyme A, another by methylation of the amino group, using S-adenosylmethionine, and a third by phosphorylation of seryl residues, and it is suggested that these reactions are used to modify the combination of the histones.

However, the modification of genetic expression—the changes in the rate of synthesis of particular kinds of RNA—probably involves the formation of proteins whose only function is to augment or diminish the replicative process. This may or may not involve interaction with the histones. (There are only a few kinds of histones found in particular tissues, and they do not appear to have sufficient variety to permit selective combination with specific sites on the DNA molecules. However, it is possible that the histone is always the blocking agent, but specificity is imparted by other proteins or ribonucleates interacting with the histone-DNA combination at particular loci.) It is certainly true that the content of DNA in highly differentiated organisms is larger than the amount required to form any reasonable projection of the total number of enzymes, structural proteins, transfer RNA, and ribosomal RNA found in those cells. Most of the remaining DNA is believed to function in a regulatory way, and it is at least possible that this regulation is brought about through the synthesis of specific controlling proteins or ribonucleates in low concentrations.

We ought not to forget that mitochondria may have independent controls. Mitochondria have a life of their own within the cell; they have their own apparatus for replication and protein synthesis, including DNA and RNA, so that the cell actually has a dual heredity—one derived from the nucleus, which controls the bulk of protein and nucleic acid turnover, and another derived from mitochondria. This cytoplasmic inheritance may be exclusively through the maternal line. Spermatozoa do have mitochondria at the base of the head, but it is questionable whether they survive the process of implantation into the ovum.

The DNA of mitochondria is a single circular strand, like that of many microorganisms, and this fact leads some to believe that mitochondria are descendants of a symbiotic microorganism living in primitive cells, somewhat akin to the algae found today alive and thriving in the cells of some invertebrates.

Evidence to date indicates that formation of membrane constituents is directed by the mitochondrial DNA, but that the internal soluble proteins are formed extramitochondrially under the direction of nuclear DNA and later incorporated somehow into the mitochondria.

Recapitulation of types of reactions

1. The amide group of glutamine may be transferred to other compounds:

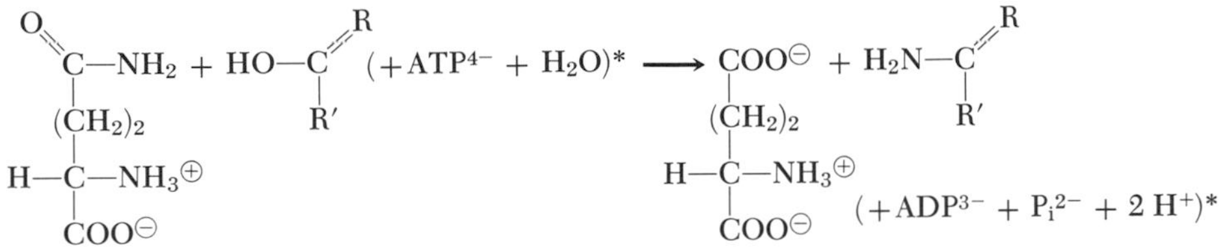

$$
\begin{array}{c}
\text{O} \\
\backslash\text{C}-\text{NH}_2 + \text{HO}-\text{C}^{\diagdown R} \ (+\text{ATP}^{4-} + \text{H}_2\text{O})^* \longrightarrow \text{COO}^{\ominus} + \text{H}_2\text{N}-\text{C}^{\diagdown R} \\
| \qquad\qquad\qquad R' \qquad\qquad\qquad\qquad\qquad | \qquad\qquad R' \\
(\text{CH}_2)_2 \qquad\qquad\qquad\qquad\qquad\qquad\qquad (\text{CH}_2)_2 \\
| \qquad\qquad\qquad\qquad\qquad\qquad\qquad\qquad\qquad | \\
\text{H}-\text{C}-\text{NH}_3^{\oplus} \qquad\qquad\qquad\qquad \text{H}-\text{C}-\text{NH}_3^{\oplus} \\
| \qquad\qquad\qquad\qquad\qquad\qquad\qquad\qquad\qquad | \qquad (+\text{ADP}^{3-} + \text{P}_i^{2-} + 2\,\text{H}^+)^* \\
\text{COO}^{\ominus} \qquad\qquad\qquad\qquad\qquad\qquad \text{COO}^{\ominus}
\end{array}
$$

*ATP not always required

Examples: The formation of phosphoribosylamine*; the formation of phosphoribosylglycineamidine; the formation of GMP; the formation of CTP from UTP; the formation of carbamyl phosphate in the cytosol.

2. Ribonucleotides may be reduced to 2'-deoxyribonucleotides with electrons transferred either through thioredoxin or cobamide coenzyme from NADPH:

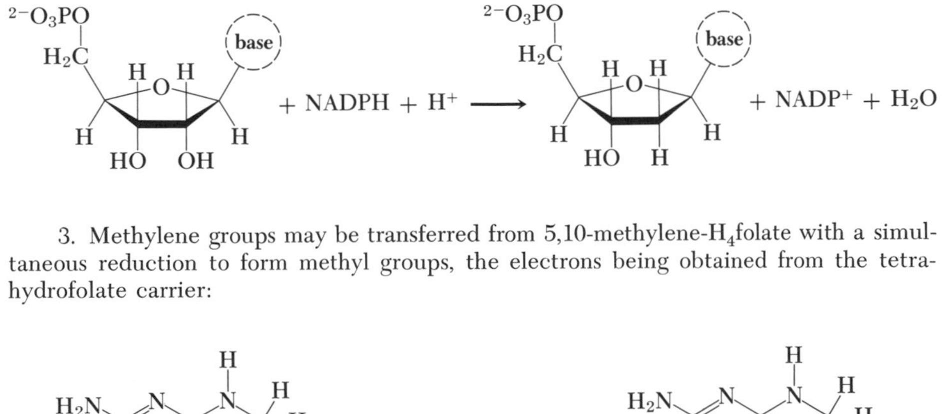

3. Methylene groups may be transferred from 5,10-methylene-H$_4$folate with a simultaneous reduction to form methyl groups, the electrons being obtained from the tetrahydrofolate carrier:

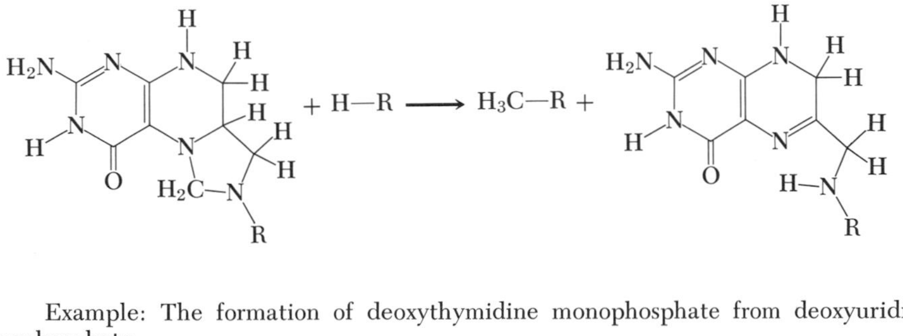

Example: The formation of deoxythymidine monophosphate from deoxyuridine monophosphate.

Further reading

Chargaff, E., and J. N. Davidson, eds.: *The Nucleic Acids*. Academic Press (1960). A multi-volume work. See especially Vol. III: J. M. Buchanan: *Biosynthesis of Purine Nucleotides* (p. 303) and R. E. Handschumacher and A. D. Welch: *Agents Which Influence Nucleic Acid Metabolism* (p. 453).

Davidson, J. N., and W. E. Cohn, eds.: *Progress in Nucleic Acid Research*. Academic Press. Annual volumes containing many articles of interest.

Moore, E. C., and R. B. Hurlbert: *Regulation of Mammalian Deoxyribonucleotide Biosynthesis by Nucleotides as Activators and Inhibitors*. J. Biol. Chem., 241: 4802 (1966).

Goulian, M., and W. S. Beck: *Cobamide-dependent Ribonucleotide Reductase from Lactobacillus leishmannii*. J. Biol. Chem., 241: 4233 (1966).

Tatibana, M., and K. Ito: *Carbamyl Phosphate Synthetase of the Hematopoietic Mouse Spleen and the Control of Pyrimidine Biosynthesis*. Biochem. Biophys. Res. Comm., 26: 221 (1967).

Anderson, E. P., and R. W. Brockman: *Biochemical Effects of Duazomycin A in the Mouse Cell Neoplasm 70429*. Biochem. Pharmacol., 12: 1335 (1963). The title of this paper does not reveal that it contains an appraisal of differential inhibition of amide group transfer by azaserine and diazooxo-norleucine.

Heidelberger, C.: *Cancer Chemotherapy with Purine and Pyrimidine Analogues*. Ann. Rev. Pharmacol., 7: 101 (1967).

Bland, J. H., ed.: *Seminars on the Lesch-Nyhan Syndrome*. Federation Proc., 27: 1019 (1968). A fascinating set of papers.

Wyngaarden, J. B.: *Gout*. Stanbury, J. B., J. B. Wyngaarden, and D. S. Frederickson, eds.: *The Metabolic Basis of Inherited Disease*. McGraw-Hill (1966).

Hager, S. E., and M. E. Jones: *A Glutamine-dependent Enzyme for the Synthesis of Carbamyl Phosphate for Pyrimidine Biosynthesis in Fetal Rat Liver*. J. Biol. Chem., 242: 5674 (1967).

Lonsdale, K.: *Human Stones*. Science, 159: 1199 (1968).

Livingston, D., Crawford, E. J., and M. Friedkin: *Studies with Tetrahydrohomofolate and Thymidylate Synthetase from Amethopterin-resistant Mouse Leukemia Cells*. Biochemistry, 7: 2814 (1968).

Hill, D. L., and L. L. Bennett, Jr.: *Purification and Properties of 5-Phosphoribosyl Pyrophosphate Amidotransferase from Adenocarcinoma 755 Cells*. Biochemistry, 8: 122 (1969).

THE TURNOVER OF PORPHYRINS

CHAPTER 21

ARGUMENT

The turnover of hemoproteins also involves the turnover of porphyrins. Various types of porphyrins are known; they are given names on the basis of the nature of the side-chain substituents and numbers on the basis of the order of arrangement of the substituents. All of the porphyrins in the hemoproteins can be synthesized by a sequence of reactions beginning with a condensation of succinyl coenzyme A and glycine. The result is uroporphyrinogen III, which has acetate and propionate side chains arranged in the same order as the side chains in protoporphyrin IX. Protoporphyrin IX is made from uroporphyrinogen III by a sequence of decarboxylations and oxidations of the side chains.

The porphyrin ring is degraded through reactions that break the porphyrin ring, and form the linear tetrapyrroles, biliverdin and bilirubin. Bilirubin is excreted in the bile after conjugation with glucuronate residues, which are the carboxylate analogue of glucosyl residues. When a pathological condition such as cellular damage to the liver or obstruction of the bile duct interferes with this process, bilirubin accumulates, and the patient develops a jaundice from the yellow color of the compound. Excreted bilirubin is further converted to other yellow bile pigments, especially urobilins and stercobilins, which are mainly responsible for the color of feces. Part of the pigments is reabsorbed and excreted by the kidney, thereby also coloring the urine.

Hemoprotein turnover also involves a turnover of iron. The absorption of iron is partially limited by the quantity already available. This is the only control over the quantity of iron in the body, because there are no regular routes for its excretion. Iron is transported through the plasma membrane of cells in the ferrous state. However, it is stored inside cells, particularly in the intestinal mucosa and the liver, as a ferric chelate of a particular protein, apoferritin. Transport through the blood also occurs in the ferric state in combination with a protein, transferrin. When the amount of iron exceeds the capacity of storage proteins, an excess is deposited in various tissues, especially the liver and spleen, as a complex precipitate known as hemosiderin. A large accumulation of hemosiderin frequently accompanies damage to the liver from a variety of causes.

The turnover of the hemoproteins involves not only the destruction and formation of peptides, but also the destruction and formation of the associated porphyrins and a turnover of the ligated iron. These processes are qualitatively important because of the central role of the cytochromes in metabolism, but more is known about the quantitatively important turnover of hemoglobin, which is present in much larger amounts than are the other hemoproteins. The synthesis and degradation of protoporphyrin IX in hemoglobin is a significant part of the nitrogen economy, and concomitantly involves a major part of the economy of iron in the organism.

Nomenclature of the Porphyrins

Before going further, it is necessary to say something about the classification of porphyrins. The basic unit of porphyrins is the pyrrole ring, with four of these linked to form the large porphyrin ring. Each of the four pyrrole rings may have two side chains attached, and these side chains may differ among the four pyrrole groups. The name of the porphyrin indicates the kinds of side chains and the number of pyrrole groups in the porphyrin having the particular kinds; the name does not indicate the order in which the particular pyrrole groups are linked in the porphyrin. Some of the common kinds of porphyrins contain the pyrrole groups shown in Figure 21-1, and their composition may be summarized as follows:

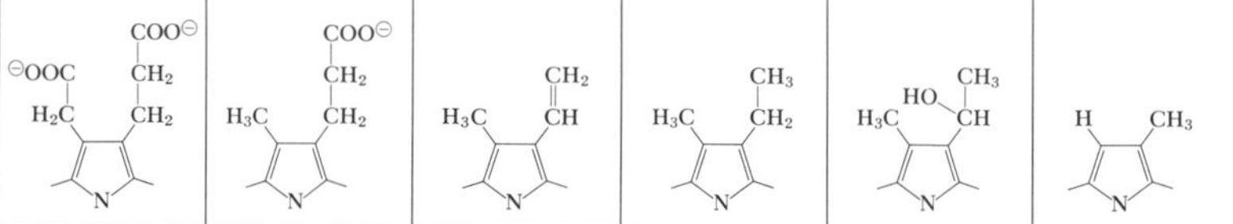

uroporphyrin	4	—	—	—	—	—
coproporphyrin	—	4	—	—	—	—
protoporphyrin	—	2	2	—	—	—
etioporphyrin	—	—	—	4	—	—
hematoporphyrin	—	2	—	—	2	—
mesoporphyrin	—	2	—	2	—	—
deuteroporphyrin	—	2	—	—	—	2

Figure 21-1   Types of porphyrins. The porphyrins are named according to the kinds of pyrroles as defined by the nature of the side chains. The structures of the common constituent pyrroles are given at the top with the number of each kind occurring in particular porphyrins listed below.

uroporphyrins—each pyrrole group has an *acetate* and a *propionate* side chain.

coproporphyrins—each pyrrole group has a *methyl* and a *propionate* side chain.

protoporphyrins—each of two pyrrole groups has a *methyl* and a *propionate* side chain; each of the other two has a *methyl* and a *vinyl* side chain.

etioporphyrins—each pyrrole group has a *methyl* and an *ethyl* side chain.

hematoporphyrins—each of two pyrrole groups has a *methyl* and a *hydroxyethyl* side chain; each of the other two has a *methyl* and a *propionate* side chain.

mesoporphyrins—each of two pyrrole groups has a *methyl* and an *ethyl* side chain, each of the other two has a *methyl* and a *propionate* side chain.

deuteroporphyrins—each of two pyrrole groups has a single *methyl* side chain with a hydrogen atom at the other position; each of the other groups has a *methyl* and a *propionate* side chain.

It is not hard to visualize sequences of decarboxylations, oxidations, hydrations, and reductions by which all of these could be formed from the parent uroporphyrins listed first.

The porphyrins of a given name can vary among themselves in the order in which the pyrrole rings are put together. Suppose we designate the acetate and propionate groups in the uroporphyrins as A and P. There are four possible ways of combining the pyrroles bearing these groups so as to make a porphyrin:

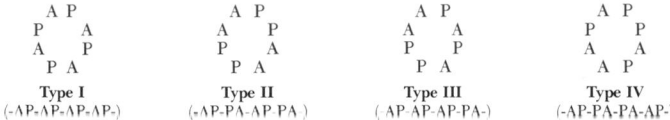

A P	A P	A P	A P
P A	A P	A A	P P
A P	P A	P P	A A
P A	P A	P A	A P
Type I	Type II	Type III	Type IV
(-AP-AP-AP-AP-)	(-AP-PA-AP-PA-)	(-AP-AP-AP-PA-)	(-AP-PA-PA-AP-)

Each of these four uroporphyrins is designated by a Roman numeral. (There are only four uroporphyrins because any other reversal of pyrrole groups beyond those shown is superimposable on one of the four by turning the ring over.)

Now, if two of the groups in uroporphyrin are changed into a third kind of group, which is the circumstance seen in protoporphyrins, then there are 15 possible combinations. Emil Fischer° wrote down the 15 possibilities, and showed that the porphyrin in hemoglobin had the same arrangement as the ninth he had tabulated. Hence, the porphyrin in heme is designated as protoporphyrin IX.

All natural porphyrins are derived from uroporphyrin I, in which there is a regular alternating sequence of groups, as might be expected if the pyrroles are combined head-to-tail, and from uroporphyrin III, which represents a reversal—an isomerization—of one of the pyrrole groups.

The *porphyrinogens* are porphyrins in which the bridge atoms between pyrrole rings are in the reduced, or methylene, state, whereas these atoms are in the methylidyne state in porphyrins.

Porphyrin Synthesis

Animals can make porphyrins. The carbon and nitrogen atoms are completely supplied by succinyl coenzyme A and glycine (Fig. 21-2). The first step in the process

---

°Emil Fischer (1852-1919) was one of the truly great biochemists. He is responsible for much of our fundamental knowledge of proteins, carbohydrates, nucleic acids, and porphyrins.

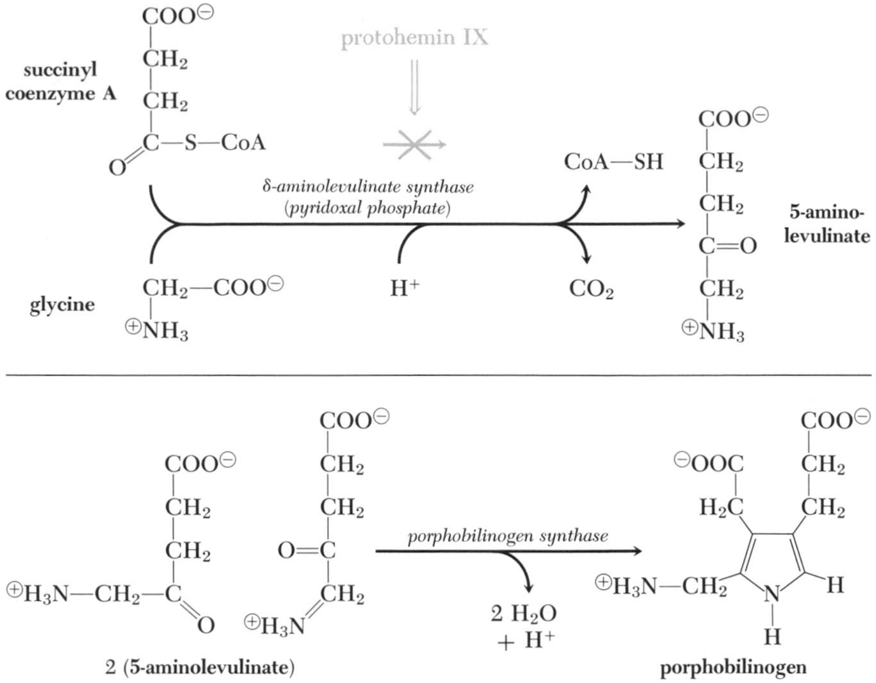

Figure 21-2   5-Aminolevulinate, the initial intermediate in porphyrin synthesis, is formed by the condensation of succinyl coenzyme A and glycine with a concomitant decarboxylation. The synthase catalyzing the reaction is inhibited by protohemin IX, which will accumulate if the supply of porphyrins exceeds the supply of apoprotein, thereby preventing wasteful synthesis. The pyrrole, porphobilinogen, is formed by the condensation of two molecules of 5-aminolevulinate.

is a condensation of these freely available precursors to form *5-aminolevulinate,* and this is the rate-controlling step. The parent indole ring is made in the next step by the condensation of two molecules of 5-aminolevulinate to form *porphobilinogen.*

Four molecules of porphobilinogen are then condensed to make *uroporphyrinogen III.* The condensation is believed to occur in two stages (Fig. 21-3). The first step appears to involve the head-to-tail combination of the four porphobilinogen molecules into a linear tetrapyrrole, with the methylene bridges formed by a loss of ammonia catalyzed by *uroporphyrinogen III "synthase."* This linear intermediate is not released, and in the presence of a second protein, a *cosynthase,* further condensation with a loss of ammonia occurs, but with a simultaneous isomerization of one pyrrole group to give the uroporphyrinogen III configuration.

The remaining steps involve the conversion of the acetate side chains to methyl groups by successive decarboxylations, thereby forming *coproporphyrinogen III* (Fig. 21-4), followed by oxidation of the methylene bridges to methylidyne bridges, and oxidative decarboxylation of two of the propionate side chains to form vinyl groups. The sequence of the latter steps is not known.

The primary regulation of porphyrin synthesis appears to be due to an inhibition of 5-aminolevulinate synthase by protohemin IX. The normal product of

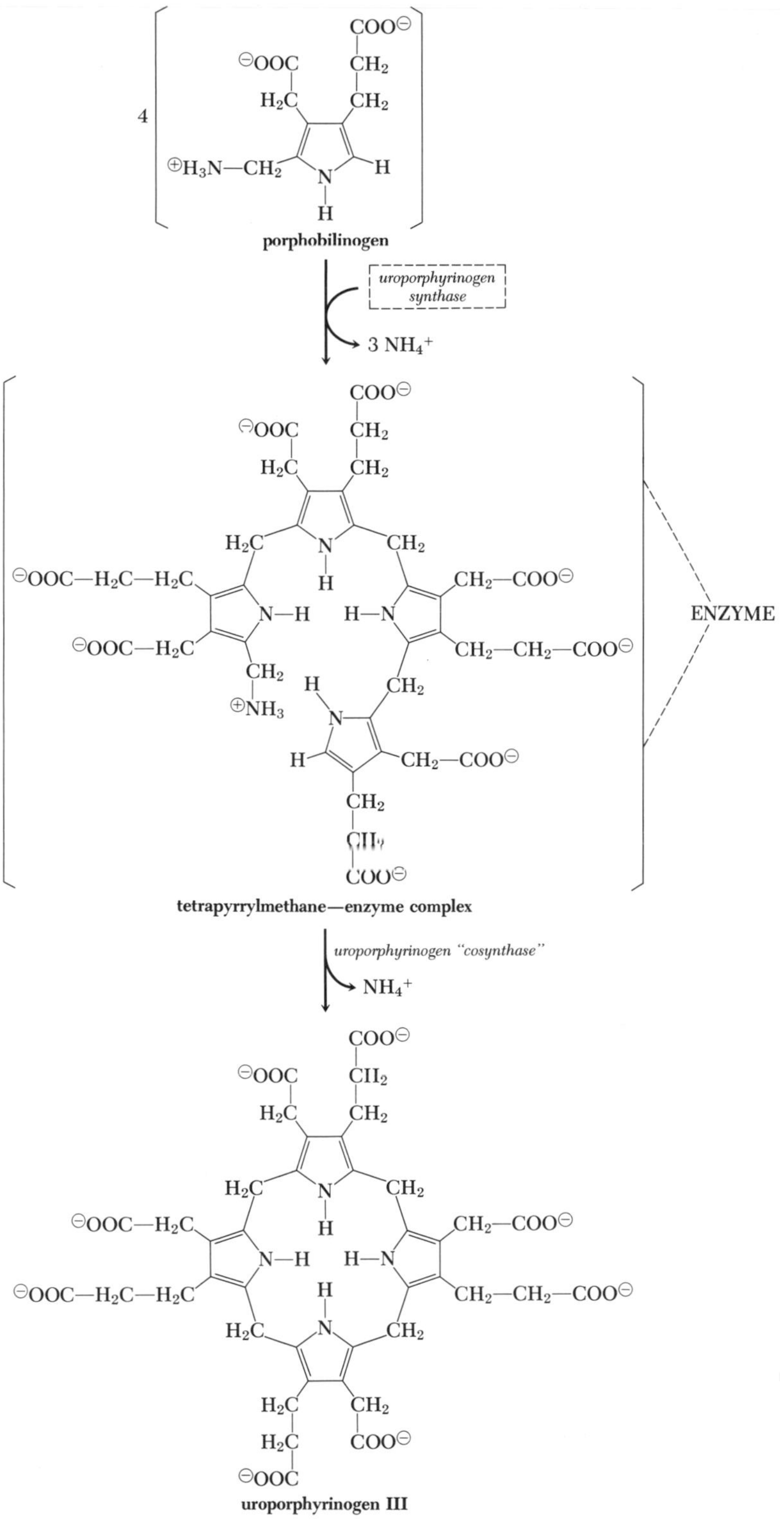

Figure 21-3 Four molecules of porphobilinogen condense to form a linear tetrapyrryl compound on the surface of the enzyme. A second protein, uroporphyrinogen cosynthase, lacks catalytic activity in itself, but it reacts with the tetrapyrryl-synthase complex, in some way causing a reversal of the terminal pyrrol ring and a simultaneous condensation into a porphyrin ring with the loss of ammonia.

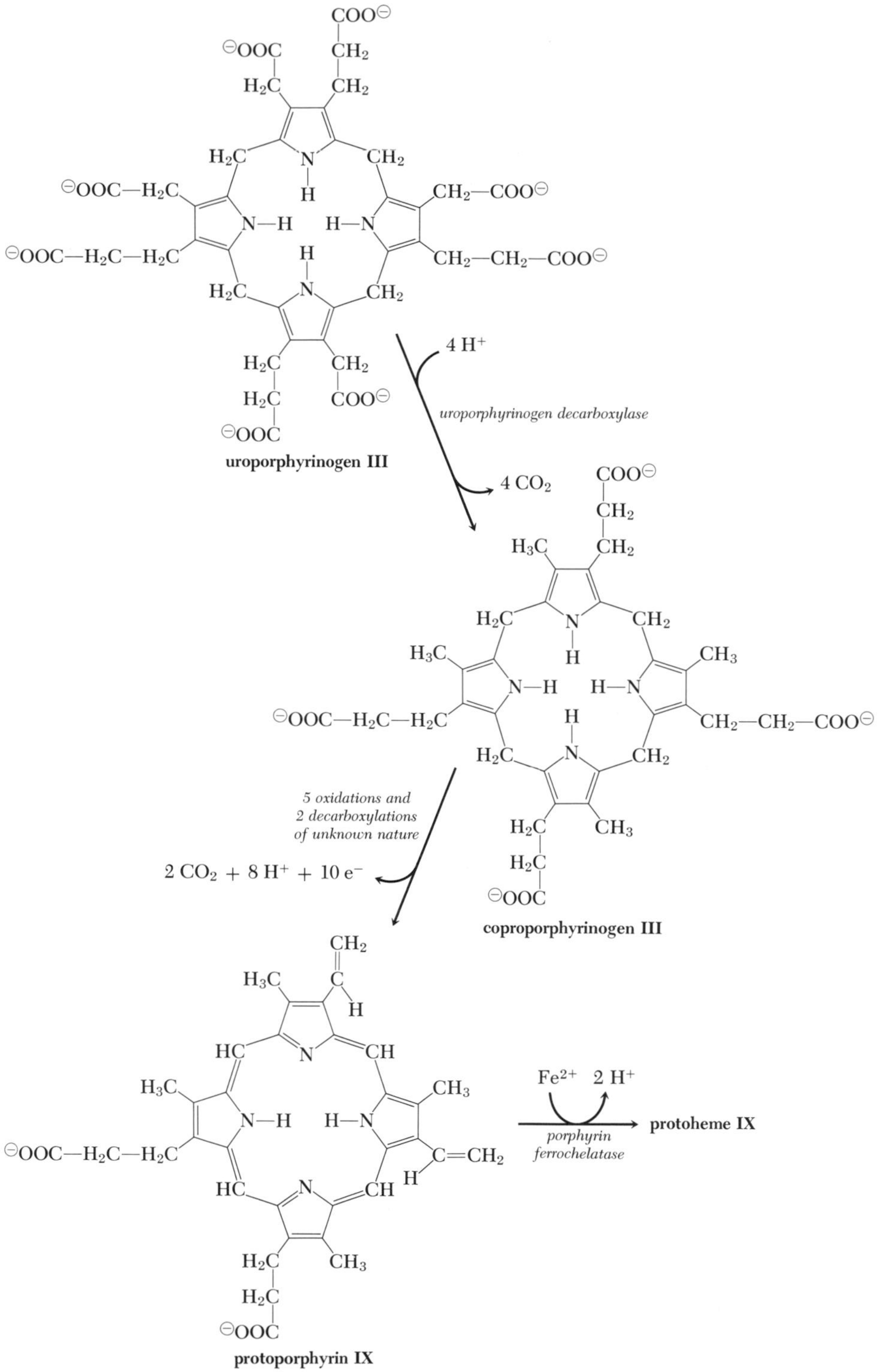

Figure 21-4   The conversion of uroporphyrinogen III to protoporphyrin IX involves straightforward decarboxylations and oxidations for appropriate modification of the side chains. Protoporphyrin IX spontaneously binds ferrous ions, but there is an enzyme to accelerate the chelation.

porphyrin synthesis is protoheme IX, which is quickly bound by globin peptides and protected from oxidation. If the production of the heme exceeds the production of globin, the accumulating free compound will be oxidized spontaneously by molecular oxygen, forming protohemin IX, which inhibits further synthesis of excess porphyrin (Fig. 21-5).

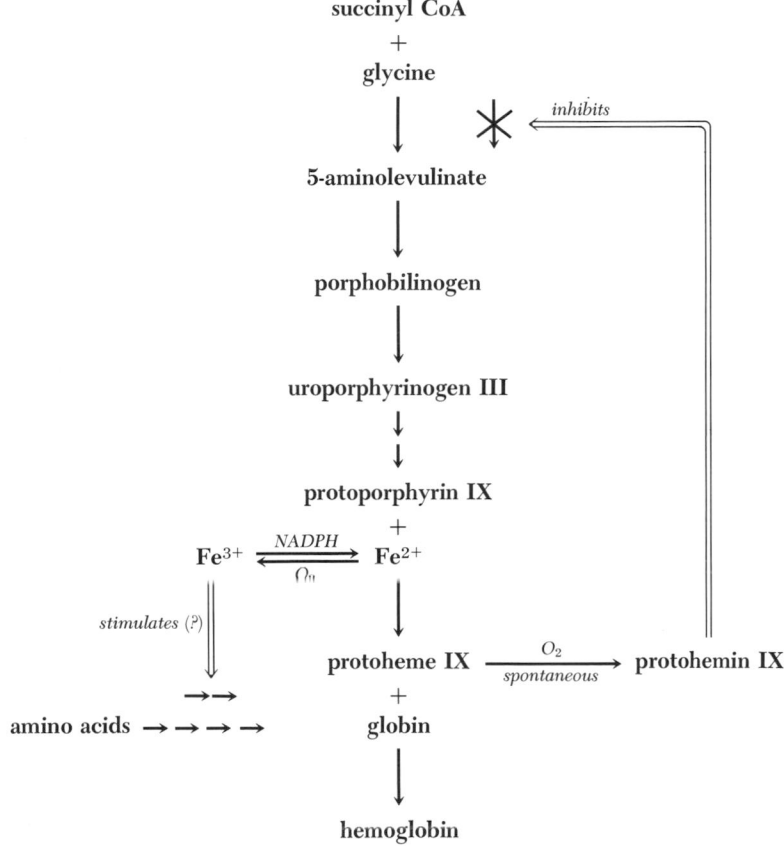

Figure 21-5   Regulation of hemoglobin synthesis according to current concepts. Unless there is globin available to bind newly synthesized protoheme IX, it will be spontaneously oxidized to protohemin IX, which inhibits further formation of porphyrin at the first step peculiar to the process. There is some evidence that the presence of ferric ions stimulates the synthesis of globin. Ferric ions will be present if the supply of iron is adequate to convert all protoporphyrin IX to protoheme IX. If the supply of iron is inadequate, there would be no point in forming more globin. These processes probably regulate the relative balance of the various precursors of hemoglobin, but not the primary adjustments of the total amount of circulating hemoglobin to such factors as environmental oxygen tension.

The combination of protoporphyrin IX with ferrous ions to form heme can occur spontaneously, but there also is an enzyme catalyzing the reaction. Only small amounts of protoporphyrin and free globin accumulate in normal erythroblasts when there is insufficient iron, as is the case in iron-deficient anemias, so there must be additional controls beyond the effect of protohemin IX. There is a possibility that the presence of excess ferric ions is necessary for rapid synthesis of globins, because these ions have been shown to facilitate clustering of ribosomes into polyribosomes.

Such an effect would tie the production of globin to the availability of iron, and there is some evidence that the availability of globin in turn is necessary for the activation of the synthesis of porphyrins.

The Porphyrias

Porphyrias are conditions in which, as the name implies, porphyrins accumulate. The production of porphyrins by erythroblasts and by other tissues, especially the liver, is under separate genetic control. Therefore, porphyrias may be caused by defects in porphyrin synthesis in either kind of tissue, and there are erythropoietic porphyrias and hepatic porphyrias.

Since the primary regulation of the production of porphyrins occurs by modulation of the activity of 5-aminolevulinate synthase, we might expect most of the causes of overproduction to be a result of some failure in this regulation. This indeed appears to be the case in both the erythropoietic and hepatic porphyrias. A failure of the synthase to respond to protohemin IX might be due to a change in the nature of the enzyme itself or to a change in some later process so that protohemin IX does not accumulate. Both possibilities appear to be represented among the porphyrias.

Somewhat fewer than 100 people are known who appear to lack sufficient cosynthase for normal formation of uroporphyrinogen III in their erythroblasts. This recessive mutation does not affect the ability of the cells to condense four molecules of porphobilinogen into a linear tetrapyrrole on the synthase, but prevents the isomerization of one of the pyrroles necessary for normal cyclization into the type III porphyrinogen. The linear tetrapyrrole can, and does, cyclize spontaneously, but the product is the abnormal uroporphyrinogen I. This compound is oxidized spontaneously to uroporphyrin I (Fig. 21-6), which accumulates. Part of the uroporphyrin I is excreted, part is converted to coproporphyrin I, which also is excreted, and part is converted to other type I porphyrins by further metabolism of the side chains. These porphyrins are red to brown, and their accumulation causes a pink color in the teeth, bones, and urine of individuals born with the condition.

Since a major part of the porphyrins formed in the people with this congenital erythropoietic porphyria are being diverted from the normal formation of protoheme IX, the normal regulation by the appearance of excess protohemin IX is not occurring, and the abnormal types of porphyrins are produced at an excessive rate. A major consequence of the accumulation of porphyrins is a sensitivity to light. Exposure to the sun causes blistering.

The other known kinds of porphyrias appear to be due to inherent defects in the regulatory mechanism itself rather than to lack of the modifying protohemin IX. People with one of the conditions, acute intermittent porphyria, have abnormally high activities of 5-aminolevulinate synthase in the liver and other tissues. Consequently, they make excessive amounts of porphobilinogen and 5-aminolevulinate, and somewhat lesser amounts of type III porphyrins, all of which are excreted. Since the porphyrin is of the normal type, less of it accumulates, and the individuals are not photosensitive. The condition appears to be more prevalent in Scandinavia and the British Isles. One in a thousand people in Lapland is afflicted with it, perhaps reflecting the high degree of consanguinity in that area, whereas the incidence is about 15 in a million people in the remainder of northern Europe.

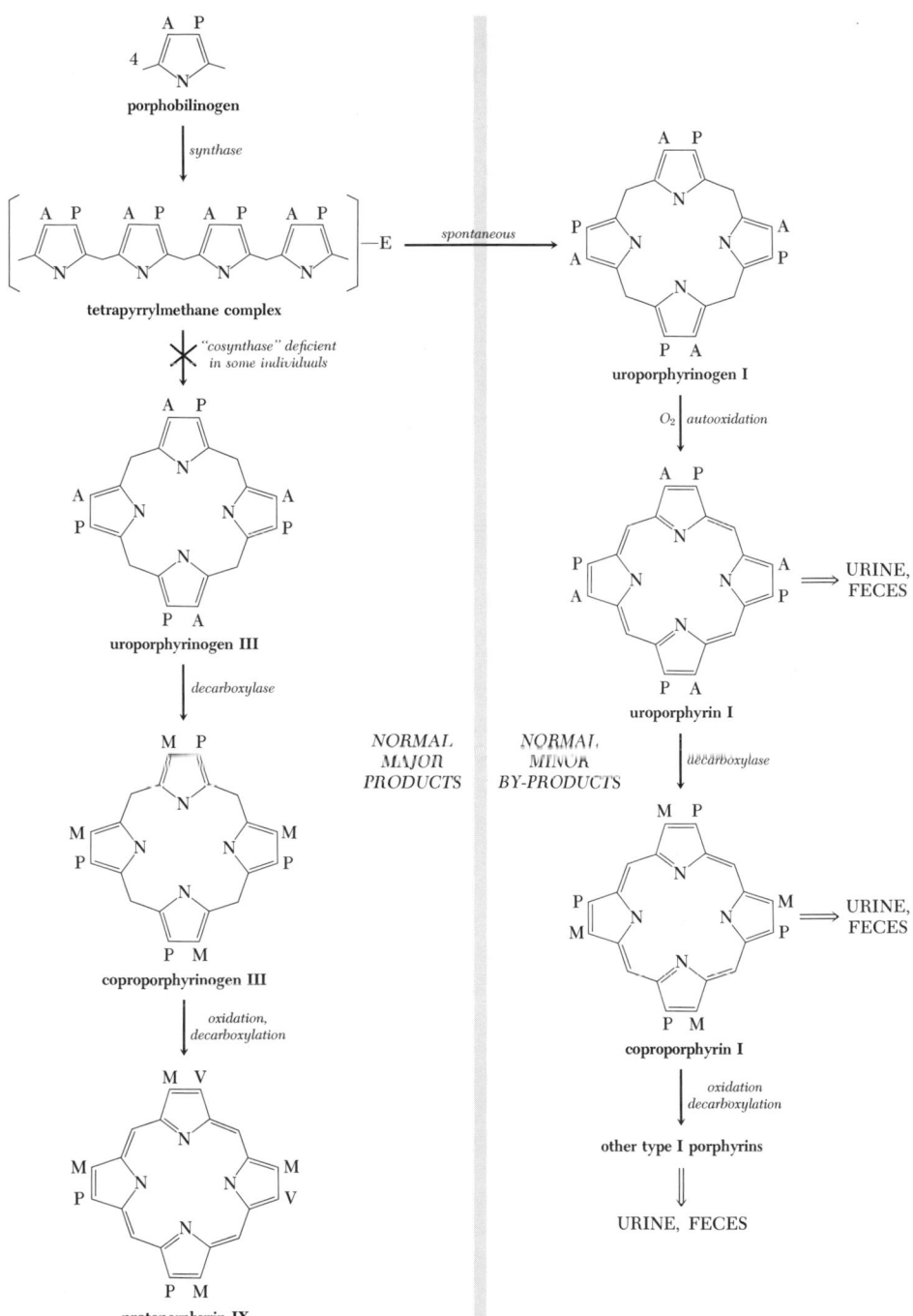

Figure 21-6 The deficiency of uroporphyrinogen III cosynthase found in some individuals prevents the formation of normal amounts of the usual type III porphyrins (*left column*). The resultant accumulation of the tetrapyrrylmethane-enzyme complex accelerates the usually trivial spontaneous cyclization of the intermediate to form uroporphyrinogen I. Uroporphyrinogen is transformed to a succession of type I porphyrins (*right column*) by the enzymes of the normal pathway, but not to a protoporphyrin. The accumulating abnormal porphyrins appear in the urine and feces.

Degradation of Hemoglobin

What exactly causes an erythrocyte to be destroyed is not known, nor is it certain how the destruction is accomplished, except that much of it apparently occurs in the tissues of the reticulo-endothelial system. In any event, the rupture of a red blood cell releases hemoglobin. There is no means of re-incorporating the protein into a new cell, and it is degraded. The globin portion of the molecule is hydrolyzed into the constituent amino acids, but the heme group is handled separately. All of the details of the degradation of heme are not clear, but the general nature of the process is well understood.

After hemoglobin is released from the erythrocyte, it combines with *hapto-globins*, which are proteins in blood plasma with an especial affinity for hemoglobin. The released hemoglobin is susceptible to oxidation of the iron, and a methemo-globin-haptoglobin complex is probably the substrate for the first enzymatic reaction in the degradative process.

Degradation begins with an attack on one of the methylidyne bridges between pyrrole rings, specifically between the two rings carrying vinyl groups (Fig. 21-7). An oxygenase catalyzes the formation of a formyl group, which is removed by an unknown mechanism resulting in a shift of the resonant structure of the tetrapyrrole. The change in the structure decreases the affinity for both ferric ion and globin, which dissociate. The released tetrapyrrole has a green color, and was named *biliverdin* because of this. It and related compounds are responsible for the pigmentation of well-developed bruises.

Biliverdin is reduced by NADPH at the center methylidyne bridge to form *bilirubin*, which is named for its red-brown color. The shift in color comes from the loss of part of the resonance between the two halves of the molecule. Bilirubin is transported in the plasma by attachment to that versatile carry-all, serum albumin, and is taken up by the liver.

Bilirubin is made more soluble in the liver by attaching residues of D-*glucu-ronate* through glycosidic bonds. D-Glucuronate is the 6-carboxylate analogue of glucose, and we shall have more to say about it later when we consider its more important function as a constituent of polysaccharides (p. 578). Suffice it for now to note that UDP-glucuronate is made by the oxidation of UDP-glucose, and its glucuronate residue may be transferred to acceptors such as bilirubin in the presence of appropriate enzymes. The conversion of bilirubin to its glucuronide enables secretion into the bile without the formation of crystals. Even so, there is enough hydrolysis in the bile that free bilirubin is an important constituent of gallstones.

Biliary excretion disposes of bilirubin for all metabolic purposes. However, there are further transitions of esthetic significance. The glucuronyl residues are removed by hydrolysis within the bowel, and the liberated bilirubin undergoes a series of reductions by the microorganisms present, with the formation of *urobilino-gens* and *stercobilinogens*. The methylidyne bridges are reduced in all of these compounds; one or both vinyl groups are reduced in the urobilinogens, and the stercobilinogens have two pyrrole rings also reduced. Part of the urobilinogen is reabsorbed and excreted in the urine. The various bilinogens are colorless, but are spontaneously oxidized to re-form one of the methylidyne bridges upon exposure to oxygen. The partially oxidized compounds are known as the *urobilins* and *stercobilins*. These are the compounds largely responsible for the brown hue of feces

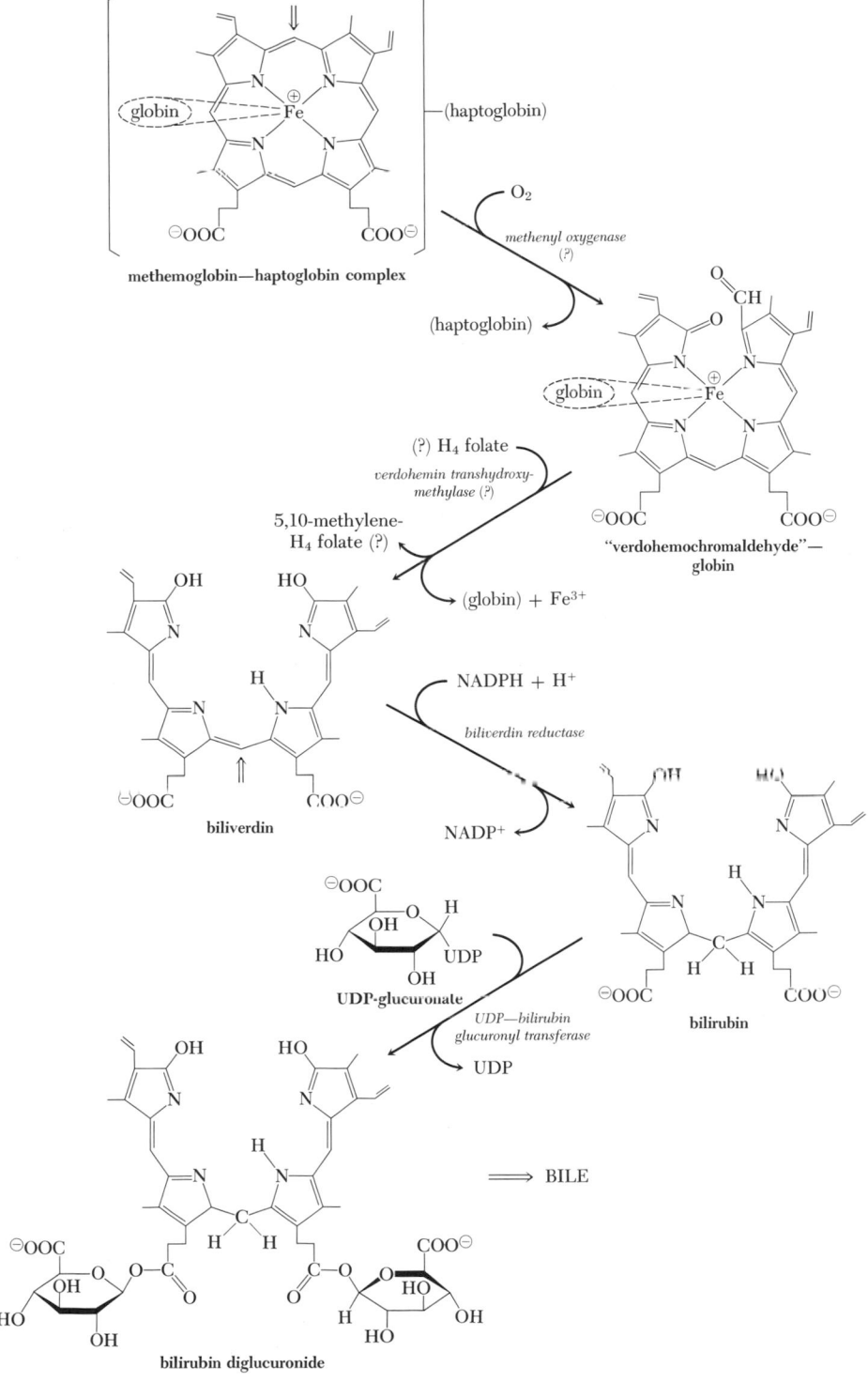

Figure 21-7 Destruction of an erythrocyte results in the spontaneous oxidation of hemoglobin to methemoglobin, which forms a complex with another protein, haptoglobin. Degradation of the porphyrin ring is believed to begin with oxidative scission while the porphyrin is still attached to this complex. After removal of the bridge carbon, the linear tetrapyrrole, biliverdin, appears in free form and is reduced to bilirubin.

503

and the yellow color of urine. Obstruction of the bile duct leads to light-colored feces and yellow skin. The skin is tinted by the bilirubin that normally is converted to urobilin in the feces. Jaundice may also result from hepatic damage, hereditary defects in the formation of bilirubin glucuronide in the liver, or rapid destruction of red blood cells beyond the capacity of the liver to form the glucuronide.

Turnover of Iron

Although it is customary to talk of ferrous and ferric ions in connection with the metabolism of iron, and we shall do so for convenience in indicating the valence state, little of these ions can exist in solutions near pH 7, particularly in the presence of proteins and other compounds with effective ligand groups, because of the high affinity of iron ions at any oxidation state for complexing agents. Even in the absence of other ligands, most of the ions will exist as complexes of water and hydroxide ions, and there is no such absence in tissues or body fluids. This lends considerable complexity to any discussion of iron metabolism because the actual distribution of the metal among potential ligands is unknown, except for the occurrence of a few proteins with a specifically favorable arrangement of ligands for binding iron ions.

The known forms in which the iron pool—as opposed to the specific heme proteins and other iron-containing electron carriers—exists are as follows: Iron is transported in the blood between tissues by combination with specific proteins, *transferrins* (molecular weight equals 74,000), which have a high affinity for two ferric ions. Loading the blood by injection of iron beyond the capacity of transferrin to carry it causes toxic reactions.

Within tissues, ferric ions are stored by combination with another protein, *apoferritin* (M.W. equals 460,000), which can ligate many atoms of the metal per molecule, so that the iron content of *ferritin* approaches one quarter of the total weight. Ferritin forms small granules in the cytoplasm.

As the iron content of a cell represented by ferritin grows, an increased amount of another type of granule, *hemosiderin*, appears. Hemosiderin is a complex of ferric ions with hydroxide ions, various polysaccharides, and proteins, with a third of its weight as iron. It may well be a relatively non-specific precipitate, formed by degradation of ferritin in the lysosomes of cells.

The other fact necessary to understand iron metabolism is that passage of iron through cell membranes occurs in the ferrous oxidation state, but the form of the complex actually crossing the membrane is unknown.

Iron losses from the body are usually low. There is no specific secretory mechanism within the kidney, and transferrin does not appear in the glomerular filtrate, so the amount of iron in the urine is less than 10 μmoles per day. Some is lost in sweat and from desquamation of the skin, but this is usually less than 10 μmoles per day, except that it may be as high as 20 or 30 μmoles with heavy sweating. In all, men lose about 20 μmoles per day with moderate activity. Women lose more through menstruation, ranging from 50 to 15,000 μmoles per period, with an average of 350 μmoles, or somewhat over 12 additional μmoles per day on a long-term basis.

It is obvious that if the absorption of iron exceeds these various small losses, the total iron content of the body will rise with age, and this in fact is a frequent occurrence. The only possible regulation of the quantity of iron in the body, given excessive intake, is a regulation of absorption from the intestine.

Ingested iron salts are reduced to the ferrous state by products of microbial fermentation in the gut, and therefore can pass into the mucosal cells. A portion of the absorbed ferrous ions is oxidized to ferric ions within the mucosa, and is retained there as ferritin. There is some evidence that the remainder combines with another specific carrier, still in the ferrous state, and passes into the blood stream across the mucosa. It was formerly believed that there was an absolute control, under normal conditions, over the amount of iron absorbed because only iron in ferritin was transferred, with the control being maintained by the amount of ferritin in the mucosal cells. When the iron storage capacity of the cells was saturated—thereby indicating a sufficiency of iron within the body—further absorption was believed to be blocked, owing to the use of ferritin as part of the transport mechanism.

Evidence of continued absorption of iron under heavy loading led others to discard the idea of mucosal control completely, and they concluded that iron absorption was dependent only on the amount in the diet and the form in which it appeared at the mucosal surface. (All concede that iron present in the form of insoluble chelates after digestion cannot be absorbed. Such chelates usually represent combination with polyphosphate compounds in the diet. Much of the iron in spinach is not absorbable owing to this fact, and any benefits from the infliction of spinach on the young must be derived from the psychological stresses involved. Nutritionally, it may be more damaging than helpful.)

The truth probably lies between these extreme views. It now appears that the rate of iron absorption is indeed slowed as the amount of iron already present in the body increases, but that this control is not total. If the iron content of the diet increases markedly, there will be increased absorption even though the stores are already adequate. As we shall see, this may be a case where evolution has not yet caught up with the Iron Age.

When ferrous iron appears in the blood stream, it is oxidized to the ferric state by a copper-containing protein, *ceruloplasmin,* named for its sky-blue color. The name *ferroxidase* has been proposed as being more consistent with function. The reduced copper of the enzyme is reoxidized by molecular oxygen, with the production of water. The ferric ions are immediately bound as transferrin and carried to other tissues.

The mode of reduction of transferrin at receptor cells is not known. However, excess iron in these cells is also reoxidized and stored as ferritin, with excesses appearing as hemosiderin, and the deposit of these relatively insoluble stores is especially prominent in the reticulo-endothelial system and the liver.

A continued deficiency of iron in the diet, or excessive losses by bleeding, may deplete the stores to such an extent that normal quantities of hemoglobin cannot be synthesized. We might predict a conservation of necessary iron for the formation of cytochromes and other electron carriers under these conditions, because it might be of evolutionary value to sacrifice part of the oxygen-carrying capacity of the blood rather than sacrifice the essential utilization of oxygen within tissues. This was formerly believed to be true, but there is more recent conflicting evidence that the cytochromes are also lost during iron deficiency.

Many are aware of iron deficiencies, which we shall discuss further in Chapter 27, but the occurrence of excessive absorption of iron is not so well known. This may result from a generalized failure of control with essentially normal intake or from excessive ingestion (or injection) of iron—or both. Given normal metabolism and excessive absorption of iron, the excess will appear as hemosiderin deposits in the liver, and to a lesser extent in the spleen and other tissues. This circumstance always arises from high concentrations of iron in the diet, and there are three main sources. One is dosage with "tonics" or other iron preparations in excessive amounts and sometimes without justification when some modest iron deficiency is suspected or anticipated, as in pregnant women. Traditionally, maiden aunts are blamed for this practice, but many medical practitioners also think like old women in this instance. A second source of iron is through preparation of food or drink in iron containers. This is classically illustrated by the Bantus of South Africa, who prepare meals and make beer in large iron pots so that the average of 50 μmoles originally in the foodstuffs ingested per day is increased to 2000 μmoles or more. Many of these people are ill, and the condition is designated as the *iron pot syndrome*. However, there are more nutritional difficulties than an excessive iron intake involved, and we shall consider it further later. We need not go so far afield, because wines sold in the United States may contain as much as 500 μmoles of iron per liter. Winos also usually have other nutritional problems that complicate any assessment of the contribution of iron intake to their medical problems. Finally, a few areas of the world have extraordinary amounts of iron in water used for drinking, and some people in those areas accumulate excessive stores.

The contribution of hemosiderin deposits (*hemosiderosis*) to cellular damage is still debatable because of the frequent association of high iron intake with other nutritional problems. It seems reasonable to expect that the accumulation of large insoluble granules within liver cells will make them more susceptible to damage, but this has not been clearly proven. In any event, any degeneration of the liver—cirrhosis—associated with excessive iron absorption leads to a more general deposition of hemosiderin throughout the body. This produces a characteristic sign, a discoloration of the skin known as *hemochromatosis* that has been variously described as a metallic slate-gray appearance or a bronzy sheen.

Recapitulation of types of reactions

1. Amino acids may be decarboxylated while simultaneously condensing with a compound containing an active carbonyl group:

$$H^+ + R - \overset{\overset{\displaystyle O}{\|}}{C} - X + H - \overset{\overset{\displaystyle R'}{|}}{\underset{\underset{\displaystyle COO^{\ominus}}{|}}{C}} - NH_3^{\oplus} \xrightarrow{\substack{pyridoxal \\ phosphate}} R - \overset{\overset{\displaystyle O}{\|}}{C} - \overset{\overset{\displaystyle R'}{|}}{\underset{\underset{\displaystyle H}{|}}{C}} - NH_3^{\oplus} + CO_2 + HX$$

Example: The formation of 5-aminolevulinate from glycine and succinyl coenzyme A. This reaction is driven by a simultaneous loss of coenzyme A, but such is not the case with all reactions of this type.

2. The pyrrole ring is formed by condensation of two moles of 5-aminolevulinate:

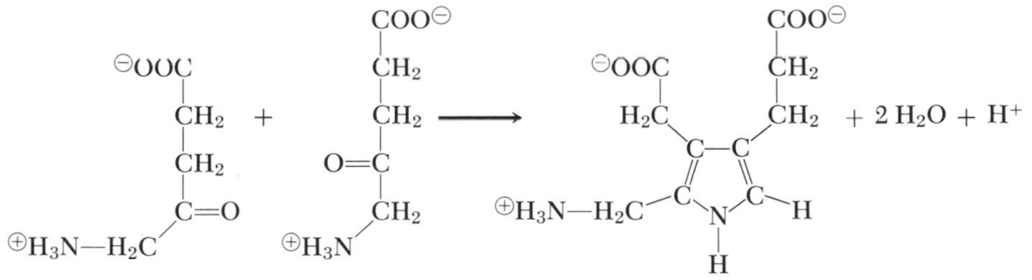

3. Porphobilinogen condenses to form polypyrroles with simultaneous deamination:

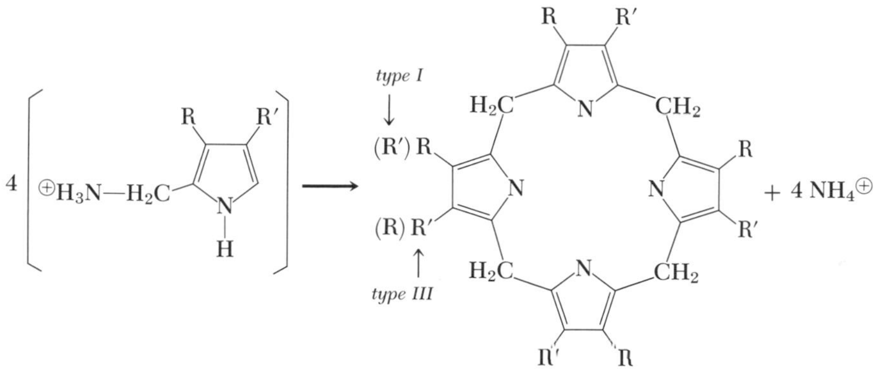

Examples: The formation of uroporphyrinogen III, with a simultaneous isomerization in the presence of a cosynthase; the formation of uroporphyrinogen I in the absence of a cosynthase.

Further reading

Schmid, R.: *The Porphyrias.* Also Pollycove, M.: *Hemochromatosis.* In Stanbury, J. B., J. B. Wyngaarden, and D. S. Frederickson, eds.: *Metabolic Basis of Inherited Disease.* McGraw-Hill (1966).

MacDonald, R.: *Hemochromatosis and Hemosiderosis.* Charles C Thomas (1964).

Singleton, J. W., and L. Laster: *Biliverdin Reductase of Guinea Pig Liver.* J. Biol. Chem., *240:* 4780(1965).

Nakajima, H., et al.: *Studies on Heme α-Methenyl Oxygenase,* I. and II. J. Biol. Chem., *238:* 3784 and 3797 (1963).

Osaki, S., D. A. Johnson, and E. Frieden: *The Possible Significance of the Ferrous Oxidase Activity of Ceruloplasmin in Normal Human Serum.* J. Biol. Chem., *241:* 2746 (1966).

Dowdle, E., et al.: *The Metabolism of [5-14C]-δ-Aminolaevulinic Acid in Normal and Porphyric Human Subjects.* Clin. Sci., *34:* 233 (1968).

Tenhunen, R., H. S. Marver, and R. Schmid: *The Enzymatic Conversion of Heme to Bilirubin by Microsomal Heme Oxygenase.* Proc. Natl. Acad. Sci., *61:* 748 (1968).

Levin, E. Y.: *Uroporphyrinogen III Cosynthetase from Mouse Spleen.* Biochemistry, *7:* 3781 (1968).

PART V

Economy of the Total Body

TOTAL
ENERGY
DEMAND
AND
SUPPLY

CHAPTER 22

ARGUMENT

One of the most important factors in the total metabolic economy is the assurance of sufficient oxidizable substrates from the diet, either directly or after storage as glycogen and triglycerides, to meet the total demand for high-energy phosphate. The amount of high-energy phosphate utilized during the daily life of a human cannot be measured directly with existing techniques. The assessment of the energy requirement therefore must be made indirectly, using techniques devised before the route of energy utilization was known. One method, too cumbersome for routine use, is to place the subject in a calorimeter and measure heat evolution. The other method is to measure the oxygen consumption, carbon dioxide production, and nitrogen excretion, and then to use these values for computing the relative amounts of fat, carbohydrate, and protein metabolized.

When most humans are lying down after an overnight fast, they will be oxidizing relatively constant proportions of the three fuels, mainly fat, so that the oxygen consumption alone can be used as an index of the rate of energy production. This basal metabolic rate (BMR) is useful as a baseline for assessing the total energy demand with varying amounts of physical activity occurring during normal life. It is also used occasionally as a confirmatory tool for diagnosing conditions such as disorders of the thyroid gland.

The methods have limitations. The use of heat production implies a direct relationship between formation of high-energy phosphate and heat of combustion, which is not always true. The use of respiratory exchange hinges on estimating the relative proportion of oxidation of fats and carbohydrates through the respiratory quotient, which is the ratio of carbon dioxide production to oxygen consumption. This is only reliable when the oxidation of fuels is the predominant metabolic process.

Calculations of the stoichiometry of ATP production from the various fuels show that accepted caloric values for the oxidation of fats and carbohydrates are valid comparisons, but this has been established long ago through comparison of the effects of various diets. The major value of the theoretical stoichiometry is in providing better comparison of ingested fuels and stored fuels, which have identical heats of combustion but different effectiveness in the production of ATP upon oxidation, and in arriving at better values for the energy yield from the metabolism of amino acids.

The energy for sustained heavy work is first provided by the oxidation of stored glycogen, followed by the oxidation of triglycerides as the supply of glycogen is exhausted. More oxygen must be consumed to generate equivalent amounts of ATP from the metabolism of triglycerides, so the maximum amount of work that can be performed in a given time declines as the glycogen store disappears.

The metabolism of amino acids usually supplies only a small fraction of the total energy utilized per day. Some humans eat diets in which amino acids provide as much as one-third of the potential energy, but more common diets have at least one-half, and usually more, of the potential energy in the form of carbohydrate. Indeed, the capacity to utilize amino acids is limited by the inability of the liver to handle the quantity that would represent the total daily demand for energy. There is no particular restriction upon the fraction of the energy represented by fats or carbohydrates. Diets rich in either component are tolerated well.

However, those who eat little carbohydrates and the starved who are existing upon their stored triglycerides develop a ketosis, which is usually asymptomatic. This is a result of increased oxidation of fatty acids in the liver, with the consequent diversion of part of the increased acetyl coenzyme A into acetoacetate production. The acetoacetate and 3-hydroxybutyrate formed in this way are used as fuels by the brain, thereby sparing the limited supply of glucose available from gluconeogenesis.

We have at this point completed our detailed consideration of the metabolic routes in which most of the fats, carbohydrates, and proteins are involved. These components of the metabolic economy are interesting in themselves in the way that the individual parts of a finely constructed machine are interesting for the beauty of their design and the elegance of their fabrication, but in order to reach our goal of realizing the contribution biochemistry makes to an understanding of the complete living organism, we must put the components together. When we do this, we ought to have not an intellectual toy but something that will better enable us to deal with reality.

We must be concerned with the mechanisms by which the various tissues are made to operate as parts of a single metabolic unit that is the organism, and we also shall have to say something more about the components used to create the architecture of cells and tissues, but one of the major concerns in assessing the total metabolic economy is inevitably the effect on the economy of the food that is eaten. Here is an area in which we need all of our understanding of the interrelationships of the many metabolic reactions to cope with the simple question, "What is an adequate supply of food?" Even what may have seemed as among the most abstruse of points, the yield of ATP from the oxidation of various fuels, suddenly appears as a real event that determines whether a human dies from starvation or lives. It is this point, the total requirement for ATP, to which we shall first address ourselves in considering the total metabolic economy.

MEASUREMENT OF ENERGY DEMAND AND SUPPLY

One of the simplest and, at the same time, most fundamental approaches to the metabolic economy of an entire organism is to assess its energy exchange with the environment. The evaluation of this basic parameter was begun during the nineteenth century, and it is a tribute to the wide-ranging imagination of the pioneers that the approach they developed has persisted to this date. Knowledge of the biochemical processes was very fragmentary in that century. Metabolic sequences, the idea of high-energy phosphate, and even the amino acid composition of proteins were yet to come. What was known was that organisms oxidized ingested foods and evolved heat, and this knowledge was used to develop the basic concept that the heat of combustion could be used as an index of the energy requirements of the organism and of the relative ability of different foods to meet those requirements.

This was in many ways a bold step. Classical thermodynamics developed as a study of heat engines, and it was clear at an early date that heat could not be used for work in organisms because of their relatively constant temperature. Therefore, heat of combustion would be valuable as a guide only if it had a constant relationship to free energy with all types of nutrients.

Proof that heat exchange was useful as an approximation took much experimentation. It was necessary to build large calorimeters for measuring the heat exchange of animals, thereby providing proof that the heat output for the combustion of foods within an organism was the same as the output obtained by burning the same materials in a bomb calorimeter. Humans, because of their cooperativeness, proved to be the most valuable subjects for this type of experimentation.

The agreement between biological combustion and bomb calorimetry proved to be excellent for fats and carbohydrates. However, there was an important and unsatisfactory correction necessary in the case of proteins because nitrogen is converted to nitrogen gas and oxides of nitrogen in the bomb, whereas it is converted to various other products, mainly urea, in the mammal.

The next step was establishing that heats of combustion did bear some relationship to the relative utilization of the major fuels in the biological system. Perhaps the most reassuring finding was the result of comparing fats and carbohydrates for the maintenance of constant weight. If the heat of combustion of the dietary intake was held constant, it was found that there was little change in weight if fat was substituted for a substantial amount of carbohydrate in the diet. Variation of protein content gave less consistent results, and this is not surprising with the added complications we have noted of the requirements for total nitrogen and essential amino acids. We shall also see that heat of combustion, even after correction for urea output, is an inherently less satisfactory guide to the energy available from utilization of protein.

Despite its limitations, heat of combustion is still used for equating food intake with the energy requirements of the organism. Tabulations of caloric values pervade nutrition and are the basis for innumerable dietary fads.

An important point must be made absolutely clear. The standard unit of heat energy in scientific circles is the *calorie*, which was defined as the amount of heat necessary to raise the temperature of one gram of water by 1° at 15° C. (The calorie is now defined in terms of more easily measured units—either the joule or the watt-hour—for precise use, but this need not affect our considerations.) The heat of combustion of the common foodstuffs lies in the range of several thousand calories per gram. Because of a reluctance to write large numbers, it became the practice to specify the values in *kilocalories*. Someone made the mistake of designating this as a large calorie, spelled with a capital letter (Calories or Cal.). This trivial typographical distinction was soon neglected (what is a thousand-fold among friends?), and we must face the fact that laymen and dieticians mean kilocalories when they say calories. There is no excuse for such license in professional circles. When we say calories, we shall mean it, and 1000 calories will be a kilocalorie, not a Calorie.

The Respiratory Quotient and Indirect Calorimetry

The relative amounts of fats, carbohydrates, and proteins being used for the production of energy can be estimated from three simple measurements: (1) the consumption of oxygen; (2) the production of carbon dioxide; and (3) the excretion of nitrogen. Since nothing more is involved than the collection of urine and feces and the small discomfort of having to breathe through a system in which the disappearance of oxygen and appearance of carbon dioxide can be measured, the monitoring of energy balance in this way has largely replaced the use of cumbersome calorimeters. Let us now analyze how these values can be used.

The first step is to determine how much of the gas exchange—the consumption of oxygen and production of carbon dioxide—results from the metabolism of proteins. If we know the nitrogen excretion and the kinds of proteins being metabolized, the corresponding gas exchange can be calculated from simple stoichiometry. For example, the stoichiometry for the metabolism of some proteins, if all of the nitrogen appears as urea, is as follows:

STOICHIOMETRY OF PROTEIN METABOLISM

	Moles per kilogram of food				
	CO_2	$-O_2$	N	CO_2/N	$-O_2/N$
Beef muscle proteins	37	45	12.6	3.0	3.6
Casein°	37	45	11.2	3.3	4.0
Zein°	39	47	11.4	3.4	4.1

° Casein is the principal protein of milk and zein is the principal protein of corn.

If we assume that an individual is metabolizing a mixture of amino acids approximating the composition of muscle proteins, we can multiply the measured nitrogen excretion by 3.0 and 3.6 to obtain the respective values for carbon dioxide output and oxygen uptake due to the combustion of amino acids. Of course, the values may be in error by 0.1 or more if the mixture of amino acids consumed more closely approximates the composition of casein or zein.

The second step in the calculation is to subtract the values for gas exchange due to protein metabolism, estimated as above, from the total gas exchange. This leaves the values of oxygen consumption and carbon dioxide production due to the metabolism of fats and carbohydrates. Now, the stoichiometry for the metabolism of typical examples of these fuels is as follows:

STOICHIOMETRY
OF FAT AND
CARBOHYDRATE
METABOLISM

	Moles per kilogram of food		
	CO_2	$-O_2$	$CO_2/-O_2$
Starch	37	37	1.00
Sucrose	35	35	1.00
Lactose	35	35	1.00
Corn Oil	65	90	0.72
Pig fat	64	91	0.71

We see right away that the ratio of carbon dioxide production and oxygen consumption is greatly different for the carbohydrates and for fats. This ratio is the *respiratory quotient*, and it is a valuable guide to the relative amounts of oxygen being used for the combustion of fats and of carbohydrates. It may be used both with intact animals and with isolated tissue preparations for this purpose.

Since the respiratory quotient (RQ) for the oxidation of carbohydrates is 1.00, and the value for the oxidation of most natural fats is near 0.71, we can estimate the fraction of the oxygen consumption used for carbohydrate metabolism by the following relationship (after correction for amino acid metabolism):

$$\frac{-O_{2 \text{ (carbohydrate)}}}{-O_{2 \text{ (fat + carbohydrate)}}} = \frac{RQ_{\text{(observed)}} - RQ_{\text{(fats)}}}{RQ_{\text{(carbohydrates)}} - RQ_{\text{(fats)}}} = \frac{RQ_{\text{(obs.)}} - 0.71}{0.29}$$

Knowing the stoichiometry for high-energy phosphate production from fats, carbohydrates, and proteins, we can then use the oxygen consumption for the metabolism of these three kinds of fuel, as calculated by the above methods, and estimate the total energy production in the entire organism (or tissue) from the gas exchange and nitrogen excretion.

The pioneers in the field knew nothing about high-energy phosphate and the mechanisms of utilization of the energy of combustion. Instead, they employed the calculated partition of oxygen consumption among the fuels to estimate the total caloric yield. Hence, the method is termed *indirect calorimetry*, because heat production is calculated from gas exchange rather than by direct observation in a calorimeter. This name gives the implication of lesser reliability, but in fact the measure of oxygen consumption is in many ways a more useful guide to the metabolic economy than is heat evolution.

The factors commonly used in indirect calorimetry are shown in the following table. (The data are given in terms of liters of gas as well as moles, because these units are still in common use in many laboratories.)

	RQ	per gram of food	per mole O_2	per liter O_2
		KILOCALORIES PRODUCED:		
Proteins	0.80	4.32	100	4.46
Fat	0.71	9.46	105	4.69
Starch	1.00	4.18	113	5.05

1 *mole* of urinary N = 372 kcal; 3.00 moles CO_2 produced, and 3.70 moles O_2 consumed.

1 *gram* of urinary N = 26.5 kcal; 4.8 liters CO_2 produced, and 5.9 liters O_2 consumed.

The Basal Metabolic Rate (BMR)

The basal metabolic rate (BMR) is the rate of oxygen consumption, or the calculated equivalent heat production, of an individual lying down and at rest, who is awake and who has had no food for at least 12 hours. This measurement is an important physiological tool because it assesses the energy requirement for maintenance of tissues, response of the nervous system, and any dissipation for heat production through oxidations not coupled to phosphorylations. It is often useful as a guide to thyroid function, since thyroxine has the effect of uncoupling oxidation and phosphorylation, thereby increasing the total oxidation necessary for the production of a given amount of ATP.

The usual determination of BMR is really only a determination of the rate of oxygen consumption. Much of the success of this approach results from the specified conditions, which minimize many potential sources of error. A previously well-fed individual will depend mainly on his fat stores for energy after an overnight fast, with a respiratory quotient near 0.82 if he has been eating the usual American diet. Only 15 per cent of the oxygen consumption will be used for the metabolism of amino acids, thereby minimizing error due to lack of knowledge of the exact composition of the amino acids being metabolized.

It is therefore reasonably safe to assume that most people in the basal condition will be metabolizing a proportion of fuels such that *1 mole of O_2 is equivalent to 108 kcal; 1 liter of O_2 is equivalent to 4.82 kcal.* (The latter value is frequently quoted as 4.8205 kcal per liter, which is a beautiful example of confusing accuracy in arithmetic with experimental precision. One part in 100 is an optimistic expectation for precision in this field.)

The Use of Surface Area

It is common practice to compare the metabolism of individuals on the basis of the area of the body surface. This practice is even extended to such things as the calculation of drug dosages. The advantages and defects of the custom deserve examination.

It ought to be apparent from what we have learned to this point that metabolic processes do not occur predominantly at the body surface; therefore, metabolism and body surface do not have a direct anatomical relationship. However, it was realized over a century ago that mammals have a proportionately slower basal metabolism as their size increases. A large mammal has a greater total metabolism than does the smaller mammal, but the metabolism per unit of body weight decreases with size. Many later studies have shown that the basal metabolism is more nearly proportional to the surface area of the adult animals in various species than it is to the weight of the animals.

We really ought to expect this to be so. As was mentioned in the earlier discussion of brown adipose tissue, there is no inherent reason that the cells of an elephant could not be conducting oxidations at the same rate as do the cells of a mouse. The primary limiting factor is the ability to dispose of the heat generated by metabolism. The primary mechanisms for heat loss for land animals are direct radiation to the environment and evaporation of water on the surface, both of which can be expected to be proportional to the surface area. (There are other mechanisms, such as evaporation through the lungs, so heat loss is not directly proportional to surface area among all animals.)

In short, we are dealing with an evolutionary development in which the metabolic processes of many kinds of animals have been adjusted to conform, at least roughly, to the surface area of the animal rather than to its tissue mass. This is the basis for using the surface area in comparing metabolic rates of species.

The justification for the use of surface area in comparing individuals of one species is considerably more shaky. One would have to argue that the metabolism of individuals somehow adapts to keep their heat output proportional to area, and there is no real evidence that this is so. The evidence at hand suggests that metabolic rate is more nearly proportional to the total amount of protein in the body, and if we could estimate the number of mitochondria, this would probably be the best guide of all. It is well known that, given two men of equal height and weight, the more muscled specimen has the higher metabolic rate. It is known that metabolic rate per unit surface area differs among various groups of people, and differs with age and sex in the same group of people. This is shown in Figure 22-1.

Why, then, aren't metabolic measurements made by simply measuring oxygen consumption per unit time and comparing the results with the average range of values obtained for people of similar age, sex, height, and weight? Perhaps the major reason is the striving for logical unity in science. Since we obviously can't compare the oxygen uptake of a lean person with that of an obese person on the basis of weight—the droplets of triglyceride have no oxygen uptake—we seize on the surface area as the best available means of minimizing individual variations in people with normal thyroid function. This evidently comforts many even though the practice has no biochemical basis and the calculated values still must be compared with empirically determined standards for similar people.

It is exceedingly difficult to measure surface area accurately. Hence, there have been many attempts to estimate this value from more easily measured parameters,

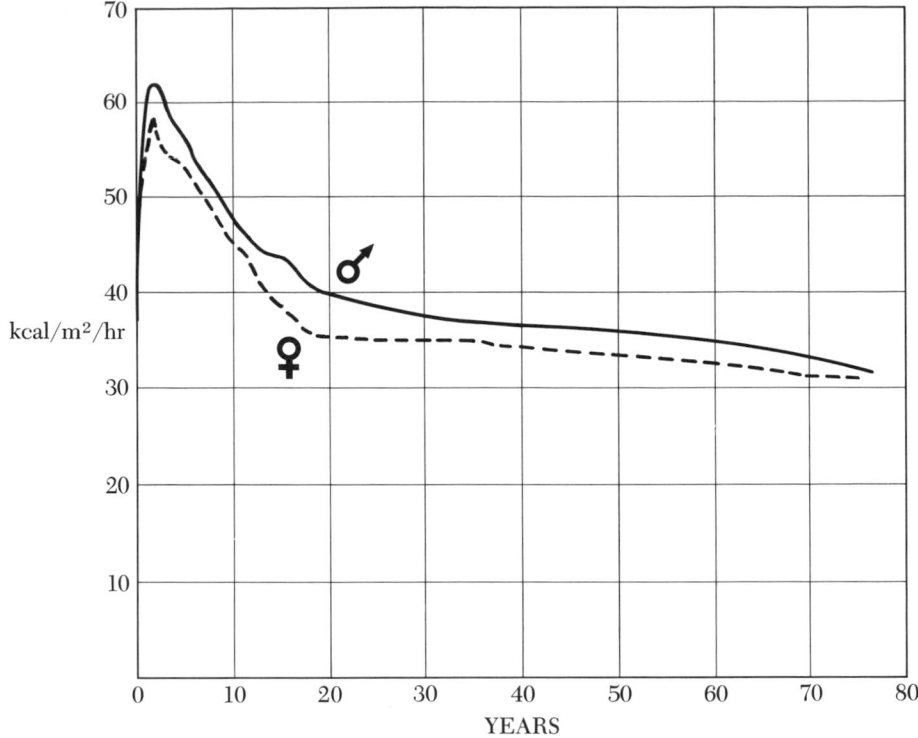

Figure 22-1   The basal metabolic rate of males and females of varying ages, given in terms of square meters of calculated body surface area.

such as the overall length of the body and the body weight. Metabolic laboratories have charts for this purpose, which have been constructed from the relationship:

$$\log A = 0.425 \log W + 0.725 \log H - 2.144$$

in which A = area in square meters; W = weight in kilograms; H = height in centimeters.

One has to be alert for some curious examples of circular logic in the literature on surface area measurements. The object is to determine how closely metabolism is related to surface area. Some proceed to contend that their formulas for determining surface area are better because they result in more constant values of metabolic rate per unit of surface.

The Validity of Calorimetry

There are both technical and theoretical errors in the use of calorimetry for assessment of energy balance. Let us consider some technical aspects first. The weakest aspect of the application of calorimetry is in handling protein metabolism. It is startling to learn that, in this day when the composition and structure of many proteins are known in detail and the fundamental thermochemical parameters have been determined for many thousands of compounds of other types, there are no accurate data for the heats of combustion of a representative selection of proteins. All of the many computations and tabulations of the caloric equivalents of proteins hinge on measurements made in the nineteenth century on samples that appear from the reported stoichiometry to have contained unknown amounts of lipid and carbohydrate.

Furthermore, everything in our approach hinges upon the excretion of nitrogen as a reflection of the metabolism of protein during the period of measurement. Unless the measurements are continued for a long period of time, which is one of the experimental conditions that indirect calorimetry seeks to avoid, the excretion of nitrogen may not be in step with amino acid metabolism. Some reactions may be lagging behind, or glutamine may be accumulating or may be hydrolyzed at an unusual rate. Any transitory change in the rate at which the kidneys remove urea from the circulation can also cause serious error, because the total amount of urea in the body of a normal individual is near half of the total quantity he will excrete in a full day on common mixed diets.

There is an additional complication due to the correction of the heats of combustion of proteins for the heats of combustion of excreted products, and the basis for this correction has not been rigidly re-examined for decades. It turns out that the total feces and urine were arbitrarily assigned as end products of protein metabolism, with the result that the quoted values for the caloric value of protein are low.

It is also assumed in usual practice that the bulk of the excreted nitrogen appears in the urine under basal conditions, so the unpleasant collection and analysis of feces is usually eliminated.

There are further limitations on the value of indirect calorimetry. The measurements are only valid when synthesis of fat from carbohydrates, the Cori cycle, and the interconversion of amino acids are at a minimum, because each of these processes involves gas exchange with a respiratory quotient of its own. For example, we saw that the RQ for the production of fatty acids from glucose is 2.75 (p. 331), and any significant occurrence of this process will completely negate the value of RQ for estimating the relative combustion of fat and carbohydrate in other tissues. (Respiratory quotients greater than 1.00 actually occur in individuals who have recently stuffed themselves with carbohydrates.)

These qualifications mean that the use of calorimetry for the assessment of oxidative metabolism is of value only under two circumstances: when the rate of oxidative metabolism far exceeds the rate of the synthetic processes and when the rate of amino acid metabolism is relatively low compared to the rate of fat and carbohydrate metabolism. The usefulness of the basal metabolic rate is due to the fact that these circumstances are largely achieved under the defined basal conditions. The measurements are also very useful in animals during exercise, at which time the oxidative metabolism far exceeds other processes, and in isolated experimental preparations in which the supply of nutrients can be controlled so as to suppress syntheses.

In sum, there are important technical difficulties in the interpretation of calorimetric measurements, not always evident to those who use them, that ought to be kept in mind, but they still provide a powerful tool for the appraisal of the economy of an intact individual when used correctly.

Let us now turn to a theoretical appraisal of the validity of heats of combustion in the assessment of energy balance. The real energy balance is a result of the formation and utilization of high-energy phosphate, which we cannot measure with present techniques in a living animal. However, we do know the theoretical stoichiometry for the complete oxidation of the various fuels, and we have noted in the preceding chapters various lines of reasoning that suggest that the actual yield of high-energy phosphate is not far from the theoretical value. Even if this isn't so, the proportional yield ought to be about the same for the oxidation of fats, carbohydrates, and proteins, since most of the high-energy phosphate is generated in all cases by the same pathway of oxidative phosphorylation.

However, we have the same complications in this approach as those that appear in calorimetric measurements. We do not have the necessary data for calculating the yield of high-energy phosphate under conditions in which glucose, triglycerides, or amino acids are being partially degraded in one organ, synthesized in another, and finally oxidized in still another. We must assume a metabolic economy in which there is a minimal recycling between organs. As in calorimetry, this condition is most nearly met with starvation, heavy exercise, or in an individual who is somewhat emaciated and eating a diet barely sufficient for the daily demand. Since it is the last condition that is of most practical interest in comparing the relative value of foods as energy sources for most of the people of the world, it may be regarded as the baseline to use for the comparison, in terms of ATP yield or heat combustion.

The following table compares various kinds of foods, both in terms of high-energy phosphate and the classical values for heat derived from complete oxidation. (The yields from glycogen and fat stored within the individual are included for later use. The yield from *ingested* glycogen is the same as that from starch.)

TABLE 22-1. ENERGY YIELD OF COMMON FUELS.

	HIGH-ENERGY PHOSPHATE		HEAT	RATIO
	moles/kg	moles/mole O_2	kcal/mole O_2	kcal/mole ~P
Ingested starch	216	5.8	113	19.5
Ingested fat	502	5.5	105	19.1
Ingested meat protein	215°	4.8	100	20.9
Stored muscle glycogen	228	6.2	113	18.3
Stored fat	510	5.6	105	18.8

° Based on muscle peptides. 15.5 moles ~P per mole N.

The high-energy phosphate yields were computed on the following assumptions:

1. The bulk of the ingested glucose is directly stored as glycogen in skeletal muscles, and completely oxidized in that tissue, so the Cori cycle can be neglected.

2. Ingested triglycerides are hydrolyzed and stored in adipose tissue without modification by the liver.

3. Amino acids are initially metabolized only in the liver. All of the nitrogen is converted to urea. The carbon skeletons are converted to glucose and acetoacetate, which are metabolized in other tissues without intermediate storage.

4. The amount of high-energy phosphate necessary for transport across cell membranes is negligible compared to the total yield from oxidation. This is the most shaky premise, and we can only hope that the relative loss on transport is roughly the same with all types of compounds.

The last column of the table gives the ratio of heat production to high-energy phosphate formation per mole of O_2 utilized. This ratio ought to be the same for all types of foods if heat production is a valid guide for utilizable energy from oxidations. The values for fats and carbohydrates are indeed nearly the same for both the ingested foods and the stored fuels, although they differ for the same food when ingested and when utilized from the body stores.

The value for heat production from proteins obviously gives too high an assessment for the utilizable energy. This is true even though the caloric yield of proteins has been corrected in a way that gives inherently low values, as we discussed earlier. Early experiments on feeding for maintenance of weight showed this discrepancy, which gave rise to an extraordinary rationalization that will be discussed in the next section.

What is the practical result of our re-examination of the caloric values? The usefulness of caloric values for comparing fats and carbohydrates has long been established through experimentation without the aid of our theoretical justification. Indeed, the tables are turned, and this large body of observations generates confidence in the stoichiometric calculation of high-energy phosphate yields for assessing the potential energy of foods. This is important both for immediate use and for its future potentialities. The immediate applications are in adjusting the accepted values for the energy equivalent of protein as a fuel and in making a distinction between the ingested and stored fuels, which have identical heats of combustion but different capacities for supporting work. The future potentialities lie in assessing energy balance under conditions in which processes other than oxidation are proceeding at significant rates and in being able to take advantage of the more detailed analyses of foods now becoming available for computing their nutritional value.

We come, then, to the following conclusions at this point in the discussion:

1. The multitude of tables showing caloric equivalents of foods are useful and sound, provided that the protein content of the food is not unusually high (>0.2 or so of the calculated caloric equivalent). The major objection lies in the implication of using true thermochemical values and of being able to use heat for work.

2. The energy yield from stored glycogen is significantly higher than that of ingested starch, and the "caloric" value of oxygen consumed for the metabolism of glycogen ought to be raised by 7 per cent.

3. The accepted values for the energy yield from meat proteins are too high. If the caloric basis is to be used, and the caloric yield from starch is taken as 4.2 kilocalories per gram, the corresponding yield from meat proteins would be near 300 kcal per mole of N, or 21.5 kcal per g of N. (The current accepted value is 26.5 kcal per g of N. Meat has a relatively high content of glycine, owing to the presence of collagen, and the energy yield per mole of N from glycine is low. It so happens that the fudge factors applied to the original calculation for meat proteins make the value of 26.5 kcal per g of N about right for proteins such as casein and zein, with their larger content of high molecular weight amino acids. This occurred by accident, not design, and illustrates the need for re-evaluation based on amino acid content of proteins as fuels.)

Specific Dynamic Action

The oxygen consumption and heat output of an individual rise upon eating, even if he remains at rest. This elevation in metabolic rate over the basal rate was named the specific dynamic action of foods. It is well established that the specific dynamic action is something above and beyond the result of the exertion of mastication, swallowing, and the increased motility of the gastrointestinal tract. The increased metabolism follows the absorption of the digested foods.

There have been many differing estimates of the magnitude of the specific dynamic action of various foods. These range from near zero for the fats, through 5 to 10 per cent of the total caloric equivalent of ingested carbohydrate, to 20 to 30 per cent of the total caloric equivalent of ingested proteins, when these foodstuffs are tested individually.

The phenomenon is real, and much effort has been expended in attempting to explain it. The problem has been complicated by the inconsistency of the results under varying conditions. For example, the specific dynamic action is greater for foods given to a starving animal than for foods given to an animal that is fed repeatedly. The specific dynamic action of a mixed meal is less than the sum of the effects seen with the constituents given individually.

A compromise in interpretation has been reached by which a specific dynamic action of 10 per cent of the total caloric value is assigned by many to the usual mixed meal. They then rationalize that the immediate output of heat represents something extra, not useful, and additional fuels must be supplied to compensate. Therefore, in computing the necessary food intake for a given individual, it is common to estimate his caloric requirements from his basal metabolic rate, and then add an extra 10 per cent to the intake to compensate for the specific dynamic action.

When we examine the various metabolic pathways previously discussed, it is somewhat difficult to see what all the fuss has been about, because we now have the advantage of being able to balance metabolic processes in terms of high-energy phosphate. It is self-evident that the handling of absorbed nutrients places variable demands on the oxidative metabolism. In the case of fats, we can see that storage in adipose tissue may involve nothing more than re-forming triglycerides, involving the expenditure of approximately 1.6 per cent of the potential $\sim$P yield. Glucose will be converted to glycogen, which involves the immediate expenditure of ATP amounting to 6 per cent of its potential $\sim$P yield. These extra loads on the ATP supply will cause corresponding increases in oxidative phosphorylation, which account for the increased O_2 consumption.

Amino acids, if given to an individual who has fasted, will be converted relatively rapidly to glucose and the oxybutyrates by the liver. According to the stoichiometry we cast for this conversion in Chapter 17, it will involve the consumption of approximately 1.38 moles of O_2 for the production of 1.9 moles of ATP, whereas the production of 1.9 moles of ATP by the oxidation of lactate or of fatty acids would require the consumption of 0.31 to 0.39 moles of O_2, depending upon the process actually involved. This "extra" consumption of one mole of O_2 per mole of mixed amino acids amounts to 20 per cent of the total oxygen consumption for their complete metabolism, and it will be even greater if conditions are such that the amino acid load increases the total oxygen consumption of the liver. It will be less on mixed diets, due to the temporary storage of amino acids in the tissue pools. It is not surprising that the specific dynamic action varies with conditions— the routes and rates of disposal of nutrients depend upon the particular state of the metabolic economy.

Should the specific dynamic action be allowed for in planning diets? Essentially, the phenomenon is nothing more than the initial stages of metabolism of the particular nutrients, and these are included in the total stoichiometry for $\sim$P balance. What is the point, then, of adding an extra 10 per cent to the dietary intake to compensate for specific dynamic action? There wouldn't be any point if we calculated fuel value on the basis of $\sim$P, or used caloric values adjusted to that basis. However, adding an extra 10 per cent to the dietary requirement calculated from the customary values for the caloric equivalent of foods has the same effect as lowering these values so that they are more nearly correct in relation to stored fuels, which we have seen is particularly necessary for ingested protein and carbohydrate.

In summary, the adjustment of dietary intakes for specific dynamic action is one of those lucky accidents. No one has been able to give a sound explanation of the logic for the practice because it isn't logically sound, but making this adjustment fortuitously compensates for discrepancies between true useable energy and heats of combustion. This is especially true in the case of proteins; good correction of the accepted caloric yield can be achieved by assignment of a specific dynamic action of 25 per cent, which is in the range of frequently quoted values.

NATURE OF THE ENERGY SUPPLY

Most animals eat sporadically. The absorption of ingested nutrients may or may not coincide with the times of peak demand for high-energy phosphate, and it certainly does not for those creatures who must hunt their breakfast or for humans who start the day on a cup of coffee. In short, the utilization of stored fuels is a part of the metabolic economy of the well-fed, and not confined to the starved. We shall consider here the utilization of the fuels for strenuous work, the utilization during starvation, and the regeneration of the reserves from ingested fuels.

Utilization During Work

The sequence of utilization of potential substrates for the generation of ATP during heavy activity is simply described: Glycogen is used preferentially, the fats are used next, and the proteins contribute little. Let us now confine ourselves to a description of the events, and consider the biochemical factors governing this sequence when we discuss metabolic regulation in the next chapter.

Let us get a clear picture of the changing fuel economy by studying what happens in a single individual working over an extended period of time. The table on the next page shows what happened in a young man accustomed to physical work who ran intermittently over a period of six hours. His respiratory exchange and nitrogen excretion were measured during this period, and these data enable us to calculate the amount and kind of fuels being utilized.

TABLE 22-2.   THE MARATHON RUNNER.

RATE OF RUNNING	BODY WEIGHT	BLOOD SUGAR	$-O_2$	RQ	~P PRODUCTION total	from glycogen	from fats	fraction from fat	WEIGHT OF FUELS glycogen	fats
km hr⁻¹	kg	mM	moles		moles	moles	moles		g	g
0	59.61	5.6								
11.3	59.32	4.6	2.75	0.97	16.8	15.2	1.6	0.10	67	3
9.3	59.03	4.9	2.34	0.96	14.3	12.5	1.8	0.13	55	4
11.3	58.82	4.8	2.71	0.94	16.4	13.3	3.1	0.19	58	6
9.3	58.65	4.4	2.31	0.88	13.7	8.4	5.3	0.39	37	11
11.3	58.25	4.6	2.76	0.86	16.3	8.8	7.5	0.46	39	15
9.3	58.15	4.5	2.40	0.82	14.0	5.6	8.4	0.60	25	16
11.3	57.93	4.2	2.85	0.82	16.6	6.7	9.9	0.60	29	19
9.3	57.78	4.3	2.44	0.79	14.1	4.2	9.9	0.71	18	20
11.3	57.48	3.8	2.85	0.82	16.6	6.7	9.9	0.60	29	19
9.3	57.30	4.1	2.49	0.79	14.4	4.2	10.0	0.70	19	20
11.3	57.55	3.7	2.88	0.81	16.7	6.1	10.6	0.63	27	21
9.3	57.35	3.2	2.48	0.77	14.2	3.2	11.0	0.78	14	22
					184.1	94.7	89.4		415	175

1. Blood sugar concentrations were measured by methods now known to give high values owing to the detection of compounds other than glucose in the assay. The values are still useful as a relative guide.

2. The listed totals differ somewhat from the sum of the figures given, owing to the rounding off of the latter figures.

3. This is an experiment reported by Edwards, Margaria, and Dill (Am. J. Physiol, *108*: 203 (1934)), and their data have been recalculated to determine high-energy phosphate production. Protein metabolism accounted for only 0.02 of the total energy production and has been neglected.

First let us describe the circumstances. The man was previously well fed on a mixed diet, which means that we can expect his glycogen reserves were reasonably typical. He was somewhat on the lean side, and therefore did not have gross rolls of fat. He worked quite hard, running at 3.1 meters per second for 25 minutes out of each hour and at 2.6 meters per second for another 25 minutes. However, this is not the maximum possible effort for six hours of work, because the comparable world record for continuous running is 5.4 meters per second.

Now, let us look at the data. They are simple, but they have a wealth of information concealed within them that is applicable to the general question of fuel economy.

1. Despite his leanness, the man used his stored triglycerides and glycogen for 98 per cent of his energy production, with only 2 per cent supplied by protein. This is the usual circumstance in the well-fed.

2. His body weight (column 2) fell by 2.3 kilograms, even though he was supplied with water. This is a guide to the magnitude of the effort.

3. His blood sugar concentration (column 3) fell by nearly one half, and the drop was precipitate during the last hour.

4. His oxygen consumption (column 4) went up and down with the varying rates of running, as might be expected, but increased during periods of equivalent effort as time went on, despite the fall in body weight. This would indicate that more oxygen was needed to maintain the same supply of ATP.

5. His respiratory quotient (column 5) declined throughout, indicating a shift from carbohydrate oxidation to fat oxidation. The RQ was generally higher during the periods of greater exertion, indicating increased mobilization of glycogen.

6. The total yield of high-energy phosphate (column 6) calculated from the gas exchange was remarkably constant throughout for periods of equivalent effort, even though the nature of the fuel was changing. This reinforces confidence in the validity of such calculations for assessing energy metabolism.

7. During the first period, 90 per cent of the ATP was supplied by oxidation of glycogen, but the fraction declined steadily, and the oxidation of fat increased until it was supplying nearly 80 per cent of the total ATP during the final period (columns 7, 8, and 9). This shift accounts for the increased oxygen consumption, because the ATP yield per mole of O_2 is less for fats than it is for carbohydrates. Using the values of 6.2 and 5.6 moles of ATP produced per mole of O_2 consumed in oxidizing glycogen and stored triglycerides, respectively, the calculated yield per mole of O_2 consumed for the first and final periods of running at 9.3 km hr^{-1} are 6.12 and 5.73 moles $\sim$P/mole O_2. To maintain constant ATP supply, the O_2 consumption of 2.34 moles during the first period at that rate would have had to rise to 2.50 moles during the final period. The observed consumption during the final period was 2.48 moles. This is remarkable agreement. (There would be a slight drop in the requirement for ATP owing to the decline in body weight, but the variation with body weight during running is small—not at all proportional to the weight.)

8. The final two columns show the calculated weights of the glycogen and fat consumed. In discussing the storage of glycogen (Chapter 15), we noted that the total in the skeletal muscles and liver of a male adult would be in the neighborhood of 450 grams. The total consumption shown in the table is over 90 per cent of this value. This agrees with the conclusion that might be drawn from the sharp fall in blood sugar concentration. The man had almost exhausted his carbohydrate reserves, and any further effort would have had to be sustained almost completely by the oxidation of triglycerides.

(The weight of the glycogen and fat consumed, together with the loss of the water associated with glycogen, accounts for approximately 60 per cent of the observed weight loss. However, it ought not to be assumed that loss in body weight can be predicted accurately from such data for a few hours of exertion. Changes in water balance have a large effect, and are not directly dependent on the loss of fuels.)

The conclusions we can come to with this experiment are reinforced by much more evidence than we have shown. Glycogen reserves are preferentially used for physical activity, but this activity is sustained with equal efficiency by either glycogen or triglycerides. That is, the metabolism of carbohydrate produces the same fraction of the predicted high-energy phosphate as does the metabolism of triglycerides. *There is no basis for believing that either fat or carbohydrate is an inherently inefficient fuel. There are no "empty calories".* This in itself cuts the ground from under most dietary fads, which reverse the usual ambition and try to get nothing for something.

There is an important qualification that must be added. The availability of a glycogen reserve enables more intense short-term activity. We pointed this out in Chapter 14, showing that lactate production reflects a more rapid formation of pyruvate and associated production of ATP than can be accommodated by oxidative phosphorylation. In addition, the production of ATP is greater per mole of O_2 consumed during the oxidation of carbohydrates. It follows that mitochondria in which electron transport is proceeding as rapidly as possible will generate more ATP from the oxidation of pyruvate, which is derived from glycogen, than they will from the oxidation of fatty acids. The availability of glycogen therefore enables a greater maximum effort up to the time the glycogen is exhausted. This accounts in part, but not completely, for the inability of individuals with glycogen storage diseases of the muscle to do heavy work. Normal humans can get along reasonably well on a diet containing little or no carbohydrate. Their RQ falls to levels consistent with almost total dependence on fatty acids as fuels. They cannot be expected to win short races in competition with people on a mixed diet, but they can carry on normal physical activities without difficulty.

Utilization During Starvation

Starvation differs from heavy exertion primarily in the protracted demand on the stored fuels. The glycogen supply is largely depleted within a few days during total starvation, with nearly all of the glycogen disappearing from the liver, and the content in the muscle falling to the level that can be sustained by glucose derived from the metabolism of amino acids in the liver.

After the loss of the original glycogen stores, the starved person becomes totally dependent upon the metabolism of triglycerides from the adipose tissue and amino acids derived from his tissue proteins. The length of time he can survive is a function of the amount of triglycerides he carries, and the ability to exist on stored fat is demonstrated nicely by the use of starvation in the treatment of the obese. Patients have been deprived of all food for periods as long as eight months, taking only water and vitamin supplements. One such patient lost 33.7 kilograms of body weight. This can be a risky procedure, but in the main, most of the obese people who are subjected to such a regimen get along well. Within 30 days, the daily loss of nitrogen falls to less than 0.3 mole, so there is only rarely any difficulty from failure to maintain protein synthesis. (However, the adjustment of the kidney to the decreased demands on excretion does lead to an increase in the uric acid concentration, occasionally causing gout.)

Many of the starved have blood glucose concentrations as low as 2 mM without ill effect. This leads us to a consideration of a major change in the metabolic economy occurring with starvation. The production of acetoacetate and 3-hydroxy-butyrate by the liver sharply increases, and these compounds become a major fuel for the brain.

Glucose is the major fuel for the brain in a well-fed individual on a mixed diet. Since the RQ of the brain remains near 1.0 under all circumstances, it was thought until very recently that the brain always depended upon an available glucose supply, believed to be derived by gluconeogenesis from amino acids during starvation. However, George Cahill did some simple calculations with long-available data and showed that the total supply of glycogen at the beginning of a fast and the amount of degradation of amino acids during a fast could account for only a small fraction of the total metabolism of the brain, which is responsible for 20 per cent of the total oxygen uptake at rest. His laboratory then showed, by placing catheters in the neck vessels of volunteers and measuring the arterial-venous differences in metabolite concentrations, that the brains of starved humans use the 3-

oxybutyrates as fuels. The RQ values for the oxidation of acetoacetate and 3-hydroxybutyrate are 1.00 and 0.89 respectively, which partially explains why the utilization of these compounds was not observed earlier.

We saw earlier that acetoacetate may be formed from several compounds. It is a direct product of the metabolism of tyrosine and phenylalanine. It is also formed from 3-hydroxy-3-methylglutaryl coenzyme A, which in turn is created by condensation of acetyl coenzyme A and acetoacetyl coenzyme A, in addition to being a product of the metabolism of leucine. We previously emphasized the utilization of these routes in disposing of the carbons of amino acids. However, acetyl coenzyme A and acetoacetyl coenzyme A are also formed by the oxidation of fatty acids, and acetoacetate and 3-hydroxybutyrate are potential end products of this process.

Under ordinary circumstances, the output of the 3-oxybutyrates by the liver is relatively small. The heart can effectively utilize these compounds at low concentrations, as will the skeletal muscles at somewhat higher concentrations. Together, these tissues will remove the compounds from the blood fast enough to keep the concentrations low. Current estimates indicate that this process accounts for only a few per cent of the total energy yield on mixed diets.

Starvation causes a sustained massive mobilization of fatty acids from the adipose tissue. At the same time, owing to the declining supply of glucose, an increased fraction of any citric acid cycle intermediates available in the liver is diverted to the formation of glucose. This combination of circumstances results in an increased oxidation of fatty acids by the liver with a resultant increased elaboration of the 3-oxybutyrates. The exact mechanism of these events is still in doubt. It seems likely that there are two contributing factors. One would be a decline in the concentration of oxaloacetate, owing to its removal for gluconeogenesis, which would slow the oxidation of acetyl coenzyme A. Unfortunately, there is at present no way of distinguishing between oxaloacetate in the cytosol and in the mitochondria, and existing measurements of the total concentration are contradictory. The second factor might be an increased oxidation of fatty acids to supply ATP within the liver for protein synthesis and other normal demands, thereby compensating for the decreased utilization of lactate and amino acids.

Estimates indicate that the metabolism of the 3-oxybutyrates may account for as much as one-quarter of the total energy production during starvation. The formation of these compounds provides a mechanism for feeding the brain and conserving the meager supply of glucose generated from amino acids.

Whether obese or not, total deprivation of food must eventually result in the exhaustion of the triglyceride supply. As this occurs, the only remaining source of energy is in the metabolism of the amino acids derived from tissue proteins. This is not adequate to sustain anything near normal activity, which would require the loss of something like 400 grams of protein (2 kg of tissue) per day, and death occurs rapidly after the triglyceride supply is gone.

In less severe starvation, with some food available, a person with ordinary amounts of adipose tissue gradually becomes more apathetic as the triglycerides disappear, thereby diminishing his energy requirements. He may lose so much tissue that most of the external anatomical features of his skeleton are plainly molded by his skin, and still be alive.

Utilization of Ingested Fuels

Most animals fit in ecological niches by confining themselves to foods of restricted types, and their survival hinges upon the maintenance of an appetite for tissues from particular classes of organisms, plant or animal. We don't think of cows eating meat or cats eating grass as a significant part of their dietary intake. This is in part due to biochemical differences. Cats don't have the intestinal flora necessary for degradation of cellulose in grass, and probably don't have the enzymatic capacity to handle propionate in the concentrations absorbed by the intestines of cows. Cows probably can't handle amino acids in the concentrations absorbed by cats. Much of the difference, however, is a result of adaptation and habits to fit the ecology, with the qualitative nature of the biochemistry not being so altered as might appear. Cows can utilize meat and cats can utilize grains if they are presented in forms that the animals will eat.

The problem of the relative value of energy sources in food comes into its own with the omnivores such as man, who may be gulping down the still-warm hindquarters of a wild pig (1.5 per cent carbohydrate, 25 per cent fat, and 16 per cent protein) brought down by a spear to provide the week's ration, while aspiring to a life of self-contemplation in the land of milk (4.9 per cent carbohydrate, 3.9 per cent fat, and 3.5 per cent protein) and honey (79.5 per cent carbohydrate, 0.3 per cent fat, 0 per cent protein).

Since man is capable, both by anatomy and by taste, of eating a wide variety of foods, we might expect the proportion of carbohydrate, fat, and protein available from the diet for utilization as fuels to vary quite widely. How much variation can man tolerate? Let us first examine variation in the content of protein.

We know that there is a minimum limit on the amount of protein in the diet, which is fixed by the necessity of precursors for proteins and other nitrogenous compounds, not by the utilization of amino acids for energy production. Is there a maximum limit? Several lines of evidence suggest that there is. We should expect it, because we have seen that the processing of amino acids begins in the liver, and involves the consumption of oxygen. The liver is like other organs in having a finite capacity for electron transport.

Good measurements of the oxygen consumption of human liver are not available, but by extrapolation from other animals and from *in vitro* measurements with human specimens, we find that the liver of a 70 kg man ought to consume between two and five moles of oxygen per day. Since the metabolism of amino acids according to the stoichiometry we previously cast involves the consumption of one mole of oxygen per mole of amino acid nitrogen handled in the liver, we can conclude that the total capacity of the liver is the metabolism of two to five moles of amino acid nitrogen per day if *all* of its oxidative capability is devoted to amino acid metabolism. Total metabolism of amino acids containing this quantity of nitrogen is sufficient to supply from 40 per cent to all of the basal requirements.

Since the metabolism of the liver cannot be completely devoted to amino acids, we might expect that the actual ability of the body to handle amino acids as fuels is limited to somewhere near one-half of the total energy requirements. (Of course, any degeneration of liver tissue would reduce this value even further.)

Human experience reinforces the idea of a limited capacity to handle protein. William Clark and Meriwether Lewis, two boys from Albemarle County, Virginia, led a group of men from the Mississippi River to the Pacific Ocean and back in 1803–1805. They lacked scientific credentials, but proved themselves to be reasonably accurate and astute observers. They were dependent upon animals as the sole source of food during much of the journey, and fat dogs became not only a staple, but the preferred diet. However, they were able to obtain only lean deer at some stages during the early spring. They found that, although the meat was available in sufficient quantity, they lost weight and developed gastrointestinal distress along with other symptoms, justifying the conclusion they made: fat is a necessary component of a meat diet. (The word protein hadn't been invented.)

The conclusion is supported in a somewhat better study in more recent times in which an Arctic explorer, Vilhjamur Stefannson, and friends ate an animal tissue diet for more than a year to counter skepticism that such diet could be consistent with health. The results of the study are published in the papers by McClellan *et al.* cited at the end of the chapter. In McClellan's words, "At our request, he began eating lean meat only, although he had previously noted, in the North, that very lean meat sometimes produced digestive disturbances. On the 3rd day nausea and diarrhea developed. When fat meat was added to the diet, a full recovery was made in 2 days. This disturbance was followed by a period of persistent constipation lasting 10 days. The subject had a craving for calf brain, of which he ate freely. [Calf brain has 9 grams of lipid and 11 grams of protein per 100 grams.] On March 12, poor appetite, nausea, and abdominal discomfort were present and a second but milder attack of diarrhea occurred which responded quickly to a proper proportionment of lean and fat meat."

As is always true of these kinds of observations, there is room for conjecture. Was the discomfort with lean meat a result of mental bias or of biochemical aberration caused by the diet? In light of other observations, and the general proclivity of those in pure hunting cultures to gulp fat, it seems likely that protein can not be used as the primary source of energy by humans.

Experiments with rats, also omnivores, give similar results. Rats can be induced to eat a diet made up mostly of proteins only with difficulty, and even then consume little, whereas they will avidly eat the same ingredients if the proportions are altered so as to increase the amount of carbohydrate or fat.

The discussion is somewhat academic as it pertains to most people in the world. We shall see in the later discussion of nutrition that the principal problem is supplying enough, rather than too much, protein. Beyond this, the amount of protein consumed by choice is amazingly constant with humans from a variety of cultures, accounting for 10 to 15 per cent of the total energy yield. This is not always true, because many humans still live almost entirely on meat. The Eskimos are prime examples. However, the protein content in such cases rarely reaches one-third of the potential energy yield, with the remainder being mainly fat.

The hunter's diet, then, represents another of the potential extremes, in which there is little carbohydrate and fat represents the bulk of the fuel. The "proper proportionment" of protein and fat referred to by McClellan in the quotation above was such that protein represented approximately one-quarter of the energy yield, and fat represented three-quarters. What happens in these circumstances? The evidence at hand suggests that adult humans can get along reasonably well for extended periods on such diets. Beyond the McClellan experiments, little has been done in careful observation, and there are no reliable data on relative morbidity and mortality of people living on meat compared to others in similar circumstances. Northern American Indians relied heavily on pemmican, which is ground dried meat in melted buffalo fat, as the primary food on long journeys, and the production of pemmican later became a quite substantial enterprise because the trappers of the Canadian trading companies relied on it exclusively during the winter. (There was a little-known "Pemmican War" between two of these companies for control of the supply during the first half of the nineteenth century. Interestingly enough, settlers on Hudson Bay supplied with conventional food from England developed serious signs of nutrional inadequacy, whereas no indication of this sort of difficulty was reported by the trappers living on pemmican.)

However, the meat-fat diet does cause the appearance of an asymptomatic ketosis, with the excretion of some 50 millimoles of 3-oxybutyrates per day in the urine. (There are great individual and daily variations in that amount.) This is akin to the ketosis seen in the starved obese, except that the obese frequently adapt so as to have little ketosis. The primary difference is that the liver of the individual on a meat-fat diet is presented with a several-fold greater load of amino acids for metabolism than is the liver of the starved individual. Since the metabolism of amino acids involves the later steps in the citric acid cycle, it may well be that this reduces capacity for the oxidation of acetyl coenzyme A arising from the fatty acids, which are also present in high concentrations, so that even more of the acetyl groups are diverted to acetoacetate formation.

The excretion of 3-oxybutyrates represents the loss of potential fuel. There have been efforts at taking advantage of this for control of obesity, with the idea that giving limited amounts of a diet low in carbohydrates but containing ample protein, with fat as the major fuel, would result in greater loss of weight, owing to the ketosis, than would a diet of equal potential energy derived from carbohydrates. Having some food would be regarded by many patients as a less rigorous treatment than total starvation. When this was tested with obese sailors at a Naval Hospital, the expected ketosis and greater loss of weight occurred. The ketosis is greater with limited intake of a high-fat diet than it is with total starvation.

The majority of the world's population lives on diets in which cereals are the primary nutrient, and this has probably been true for a long time. The maintenance of the ever-normal granary was a desideratum in most early civilizations; cities and grain grew together. Most cereals are rich in starch and low in fat, so oxidation of carbohydrate generates more than half of the total energy for most humans. Even children, for whom the dietary requirements are most demanding, will thrive on diets in which the content of carbohydrate is 13 times greater than the content of fat. If we define a normal human diet as the diet of most adults alive today, then the normal partition of potential energy is approximately two-thirds from carbohydrate, one-quarter from fat, and one-tenth from protein. For comparison, over 40 per cent of the energy yield from the average United States adult diet is derived from fats, and approximately 14 per cent from protein.

Economy of Fat Storage

Our previous assessments of the relative energy yield of different fuels have been made on the basis of prompt utilization. This is valid with those who have fasted and with those working on stored fuels. It is not valid if a significant fraction of the ingested protein or carbohydrate is diverted to the synthesis of triglycerides.

So much attention has been paid to the effectiveness of foods in staving off loss of weight in those with limited supplies that the question of ability of foods to contribute to gain in weight has been neglected, except with some farm animals. The problem is more complicated because a particular fuel may be preferentially oxidized so as to enable preferential storage of another, and diversion from one route to another will be quite dependent upon the daily activities of the individual and their timing. For example, glucose may be utilized to support the oxidative metabolism while fatty acids are stored as triglycerides during light activity after a meal, with the result that the amount of fat deposited will represent the total excess of potential energy over the energy demands. However, if the amount of glucose available exceeds the energy demands, part of it will also be converted to fatty acids, and we have seen that the potential energy yield of fatty acids derived from glucose is less than the yield from direct oxidation of the glucose. In these circumstances, the amount of fat deposited will represent less than the calculated excess of potential energy in the absorbed foods.

Similarly, amino acids in excess may be converted to triglycerides in amounts that will yield less high-energy phosphate in later oxidation than could be obtained by direct oxidation of the original carbon skeletons. When glucose is in excess, it is possible for carbon skeletons of amino acids to be diverted from gluconeogenesis, as well as from acetoacetate formation, to the formation of fatty acids in the liver.

In short, we need more information before we shall be able to make a really sound assessment of the amount of fat likely to be deposited from various mixtures of ingested foods in individuals of differing habits. We can come closer to making a guess at the effect of extreme diets—those containing high proportions of a single type of nutrient. If we assume the most direct stoichiometry and do the necessary arithmetic, we arrive at the following approximations of the maximum *conversion* of ingested fuels to stored triglycerides:

> 1 gram of excess dietary fat can be stored as nearly 1 gram of fat
> 1 gram of excess dietary starch can be stored as 0.35 gram of fat.

The storage from starch in this way would represent only 77 per cent of the amount calculated from the accepted caloric values, but it is important to note again that this would only be expected when the stored fat is being obtained exclusively from dietary carbohydrate.

As we have seen, if the diet is low in carbohydrate, most of the ingested amino acids will be used for the formation of glucose and acetoacetate, so a calculation of potential yield of fat from ingested protein is meaningless under these circumstances. If the diet is rich in carbohydrate, so that amino acids may be diverted to fat synthesis in the liver, then the maximum yield per 0.16 grams of N (1 gram of "protein") is 0.29 gram of fat.

If the relative efficiency of fat formation were the only consideration, it would appear that the man to whom control of body weight is one of life's major goals would do well to cultivate an appetite for high-protein cereals and cut his fat intake to the minimum, but even this regimen will not permit him to eat an unrestricted amount and maintain constant weight.

CHANGES IN ENERGY DEMAND

Measurements of oxygen consumption show in a general way the partition of energy demand among the various tissues and the changes in demand with work. The following relative values were calculated from a tabulation by Dr. Richard Havel (quoted by permission):

TABLE 22-3.   RELATIVE OXYGEN CONSUMPTION.

(whole body at rest = 1.00; actual value near 0.17 mmoles min^{-1} kg^{-1})

	AT REST	LIGHT WORK	HEAVY WORK
Skeletal muscles	0.30	2.05	6.95
Abdominal organs	0.25	0.24	0.24
Kidneys	0.07	0.06	0.07
Brain	0.20	0.20	0.20
Skin	0.02	0.06	0.08
Heart	0.11	0.23	0.40
Other	0.05	0.06	0.06
SUM	1.00	3.00	8.00

These data reinforce a point already made—the oxygen consumption of working muscles far outstrips all other demands. The demands of the viscera account for about one-third of the total oxygen consumption at rest. The brain utilizes one-fifth. These demands remain relatively constant during work, but the utilization by the skeletal muscle and heart increases dramatically.

Essentially, we are seeing in these data the requirements for fuel represented by the basal metabolism and the additional fuel required for daily activities. The additional amounts required have been measured through calorimetry for a wide range of human activities, and such measurements provide the basis for estimates of the food consumption required for maintenance of body weight. For example, estimates of the relative energy expenditure for several activities are (basal = 1.00):

sitting, reading	1.7	sweeping	1.9
dressing	3.1	typing	1.5
driving car°	2.4	sawing boards	4.9
walking, 3 mph	4.4	shoveling	6.4
		climbing†	21

° manual transmission
† vertical ladder with 50 lb load at 12 meters per min

Such estimates as these are obviously crude. Individuals differ greatly in the amount of waste motion and in the intensity with which they approach a task.

Another factor that influences the energy requirement of a given person is the environmental temperature. This is really self-evident. Since the body temperature must be maintained within narrow limits, and exposure to cold increases the heat loss, there must be a corresponding increase in metabolism. Part of the increase comes from brown adipose tissue, part from shivering, and part from some increase in metabolism of muscle through mechanisms that have not been defined. Acute exposure to cold, such as in immersion in ice-water, will increase metabolic rate as much as two-fold.

Further reading

Lusk, G.: *The Science of Nutrition*, 4th ed. W. B. Saunders Co. (1928).

Swift, R. W., and K. H. Fisher: *Energy Metabolism.* In Beaton, G. H., and E. W. McHenry, eds.: *Nutrition, A Comprehensive Treatise*, Vol. 1 (p. 181). Academic Press (1964).

Kleiber, M.: *Body Size and Metabolic Rate.* Physiol. Rev., *27:* 511 (1947).

Food and Agricultural Organization of the United Nations: *Calorie Requirements, FAO Nutritional Study # 15.* Rome (1957).

Burton, B. T., ed.: *Heinz Handbook of Nutrition.* Blakiston (1959).

McClellan, W. S., et al.: *Clinical Calorimetry:*
  XLV. *Prolonged Meat Diets with a Study of Kidney Function and Ketosis.* J. Biol. Chem., *87:* 651 (1930).
  XLVI. *Prolonged Meat Diets with a Study of the Metabolism of Nitrogen, Calcium and Phosphorus.* J. Biol. Chem., *87:* 669 (1930).
  XLVII. *Prolonged Meat Diets with a Study of Respiratory Metabolism.* J. Biol. Chem., *93:* 419 (1931).

Kinsell, L. W., et al.: *Calories Do Count.* Metabolism, *13:* 195 (1964).

Keys, A., et al.: *Human Starvation*, Vols. 1 and 2. University of Minnesota Press (1950).

Owen. O. E., et al.: *Brain Metabolism During Fasting,* J. Clin. Invest., *46:* 1589 (1967).

Benoit, F. F., R. L. Martin, and R. H. Watten: *Changes in Body Composition During Weight Reduction. Balance Studies Comparing Effects of Fasting and a Ketogenic Diet.* Ann. Internal Med., *63:* 604 (1965).

Altman, R. L., and D. S. Dittmer: *Metabolism.* Feder. Amer. Soc. Exptl. Biol. (1968). This handbook contains many useful tables.

REGULATION OF THE METABOLIC ECONOMY

CHAPTER 23

ARGUMENT

The rate of enzymatic reactions is changed in several ways so as to regulate the metabolic economy. The usual kinetic behavior of enzymes causes reactions to accelerate or slow as the concentration of substrates rises and falls, thereby dampening fluctuations in the concentration of intermediates within a sequence of reactions. This kind of stoichiometric regulation is illustrated in the metabolic response of skeletal muscles to a sequence of work and rest. The drastic changes in oxygen consumption and the shift from oxidation of fats to oxidation of carbohydrates and back again that occurs during the sequence are mainly the results of simple kinetic responses of the enzymes to varying substrate concentrations. This is especially true for the reactions of the Embden-Meyerhof pathway and of mitochondrial oxidations.

Modulations of enzyme activity are superimposed on ordinary kinetic adjustments, making it possible to alter the kinetic behavior of key enzymes in response to changes in the concentration of metabolites that act as signals for the state of the metabolic economy. Such modulations are used to initiate or halt irreversible sequences of reactions and to prevent wasteful recycling of compounds by the simultaneous operation of two opposing reactions. They are used to mobilize the major fuels for the skeletal muscles, sometimes in response to hormone secretion. They are even more important in tissues with more general metabolic functions, such as the liver, in which modulations are used to divert compounds that may react in several ways into the reactions satisfying current needs, whereas ordinary kinetic adjustments would permit all of the reactions to proceed. The enzymes affected in this way

534

include those catalyzing initial steps in the synthesis of a number of compounds and those that determine the formation or utilization of the major fuels.

The kinetic behavior of enzymes toward substrate concentration and the modulating alterations of this behavior determine the effectiveness of individual enzyme molecules as catalysts, but the actual rate of reaction also depends upon the number of molecules present. Enzymes are constantly being degraded and synthesized, and the total quantity of an enzyme can be adjusted by changing the rate of synthesis, thereby enabling a metabolic adaptation to new circumstances. Regulation in this way is slower than the other mechanisms and it is mainly used to adjust to the supply of nutrients to a cell, which may be changing because of altered metabolism in other tissues or because of changes in feeding habits. Enzyme concentrations are changed so as to supply those catalysts necessary for handling a transient abundance of some fuels only when needed. Similarly, enzymes required for the synthesis of compounds frequently appearing in the diet are formed only when it becomes necessary to augment the dietary supply of the compounds.

The secretion of compounds, the hormones, that influence the metabolism of several tissues assists coordination of the metabolic activity of the many organs. The hormones act as a signal that mobilizes the metabolic resources to respond to some alteration in the state of the organism. Most, if not all, hormones evoke their effects by combination with the plasma membrane of particular cells; each hormone has its target tissues with which it will combine. In many cases the hormone either activates or inhibits adenyl cyclase, thereby changing the metabolism of the cell through changes in cyclic AMP concentration. In other cases it is postulated that the hormone causes release of other compounds acting as messengers to alter the synthesis of particular kinds of mRNA. Some hormones may cause direct effects by penetrating the cell and combining with internal constituents.

Varied mechanisms are used to govern the release of hormones. Adrenaline and noradrenaline, which are synthesized from tyrosine, are released by sympathetic stimulus and cause the mobilization of glycogen and fat through the effects discussed in earlier chapters. The pancreas secretes insulin in response to high blood glucose concentrations and secretes glucagon in response to low concentrations. Insulin promotes the uptake of glucose and the storage of glycogen and fat. Glucagon promotes the mobilization of glycogen and fat.

The release of other hormones is mediated by the anterior pituitary, which secretes peptide hormones affecting particular endocrine glands; for example thyrotropin, which stimulates secretion by the thyroid, and corticotropin, which stimulates secretion by the adrenal cortex. The thyroid gland forms an iodinated amino acid, thyroxine, that increases the metabolic rate and is necessary for normal growth. The adrenal cortex secretes steroids with glucocorticoid activity that promote gluconeogenesis from amino acids and steroids with mineralcorticoid activity that maintain salt balance.

The levels of these hormones are partially regulated by negative feedback; the anterior pituitary secretes more thyrotropin if the blood concentration of thyroxine falls and more corticotropin if the concentration of glucocorticoids falls, thereby stimulating the appropriate gland to release more circulating hormone. The anterior pituitary is also stimulated by releasing factors formed in the hypothalamus and passing through the pituitary stalk. The signal represented by the releasing factors is in turn given in response to stimuli from other parts of the brain and to circulating adrenaline. These mechanisms are used to cause the release of more thyrotropin when the body temperature falls and more corticotropin at times of injury or other stress.

One of the great challenges to biochemistry today is understanding the mechanisms by which order is maintained among the multitude of possible metabolic reactions. We have mentioned many of these mechanisms in connection with particular segments of metabolism. Let us now draw this information together and take a more general view.

Our problem is simplified, as it is with many biological questions, because we already know some of the answers. We are not trying to construct a theoretical model of how life might be, but.to explain how it actually is, and we can discard any rationalizations that do not fit what is known about real organisms. Simple facts about the economy of the entire organism frequently provide important clues to the nature of the regulation, and they always provide constraints within which we must reason.

One of the most important guides is the fact that ingested foods are nearly completely utilized in defined ways. They may be oxidized and the products excreted, or they may be stored as fats, carbohydrates, or proteins. We know there isn't significant accumulation or excretion of intermediates in these processes. Nitrogen is excreted at the same oxidation level at which it is ingested, but nearly all of the carbon, hydrogen, and sulfur is excreted in a completely oxidized form. This is a great aid in reasoning, because it means that an observed change in the rate of formation of some intermediate must be counterbalanced by a corresponding change in the rate of utilization of the intermediate. All must balance so that the only major changes in the composition of the organism are in the content of fats, carbohydrates, and proteins.

STOICHIOMETRIC CONTROL

The most widely used adjustment in rates of reactions is through simple stoichiometry. The principle is quite elementary. When more substrates accumulate, the reactions removing them will proceed more rapidly if the responsible enzymes are not already saturated with substrate, which they frequently are not. Contrariwise, if the supply of substrates diminishes, the reactions utilizing them will slow and tend to maintain the substrate concentration. This simple principle is sometimes obscured by more complicated mechanisms of control with less general application.

Control Through Nucleotide Concentrations

The nucleotides, such as the various coenzymes and the adenine nucleotides, are particularly important in stoichiometric regulation. We have repeatedly used the principles, but let us re-emphasize them. The important point is that the quantity of the nucleotides is small compared to the quantities of substrates passing through the metabolic reactions. Reactions utilizing particular forms of the nucleotides, for example NADH rather than NAD, or ATP rather than ADP, would quickly exhaust the supply if there were no other reactions converting them back to their original state.

The most important locus of regulation is the concentration of ADP, because changes in its concentration are responsible for the adjustment of oxygen consumption to fit the utilization of energy. Oxygen is not consumed for mitochondrial electron transport unless ADP is present, and the concentration of ADP is frequently the limiting factor. The ADP appears when ATP is being utilized for muscular contraction, synthetic processes, and so on. As soon as ADP appears, it stimulates oxidation, and also the Embden-Meyerhof pathway, so as to remove the ADP and convert it back to ATP. This is the principle of the *constant high-energy phosphate balance.*

The principle does not imply that the relative concentrations of ADP and ATP remain absolutely stable. If they did, there would be no change in the rate of oxygen consumption. It only means that there can be large changes in the total amount of oxygen consumed over a period of time with only small changes in the total quantity of ADP and ATP. During the time, most of the consumption of ATP will be balanced by production of ATP.

Let us summarize the various major processes we have seen that must be balanced (not all of these occur in a single type of cell).

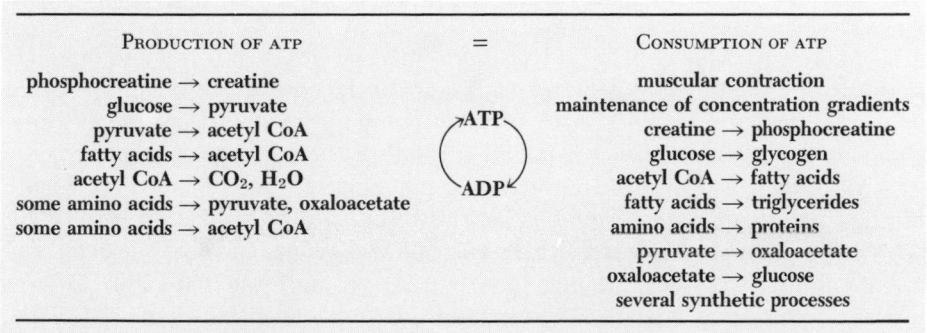

PRODUCTION OF ATP	=	CONSUMPTION OF ATP
phosphocreatine → creatine		muscular contraction
glucose → pyruvate		maintenance of concentration gradients
pyruvate → acetyl CoA	ATP	creatine → phosphocreatine
fatty acids → acetyl CoA		glucose → glycogen
acetyl CoA → CO_2, H_2O	ADP	acetyl CoA → fatty acids
some amino acids → pyruvate, oxaloacetate		fatty acids → triglycerides
some amino acids → acetyl CoA		amino acids → proteins
		pyruvate → oxaloacetate
		oxaloacetate → glucose
		several synthetic processes

ATP is constantly being utilized for one or more of the processes in the right-hand column. The proportion of substrates being oxidized by the processes in the left-hand column adjusts so as to replace the ATP. If there is excess substrate after the requirements for muscular contraction, nerve transmission, and so on have been met, then part of the excess will be stored. The storage itself requires high-energy phosphate, so some of the excess, or an equivalent amount of other substrates, must be oxidized to generate the necessary ATP. We used this principle in discussing the efficiency of the Cori cycle and of glycogen storage. We also saw that storage of excess glucose as fat is a self-contained process in which the reactions producing ATP almost balance the reactions utilizing ATP.

The same kind of principle applies to the *nicotinamide nucleotides.* Many reactions involve the reduction of NAD to NADH, and each reduction must be accompanied by a corresponding oxidation of NADH back to NAD, or metabolism will halt. The converse applies most of the time with NADP and NADPH. If the utilization of NADPH increases, as in increased synthesis of fatty acids, the concentration of NADP will rise and accelerate the oxidations, such as those of the pentose phosphate pathway, that require NADP, thereby balancing the rate of utilization and production of NADPH.

Another potential control is the availability of *coenzyme A*, and the full implications are yet to be assessed. Much of the coenzyme A appears to be combined in the form of acetyl coenzyme A, and there may well be competition for the remainder among the processes forming acyl coenzyme A compounds, including the various synthetases and the 2-ketocarboxylate dehydrogenases.

Stoichiometric Control in Skeletal Muscles

Much of the metabolic economy of the skeletal muscles is concerned with the ability to generate high-energy phosphate at a rapid rate for use in muscular contraction. The muscles, like the neurons, are terminal organs for metabolism—built to receive, not to give—and their functions constitute the purpose that the rest of the organs serve in a quantitative metabolic sense. Because of their specialized metabolism, we can dissect out more clearly the contributions of stoichiometric control to the total metabolic regulation of these tissues, and set aside for the moment the contributions of the modulation of phosphofructokinase, the effects of adrenaline, and similar regulatory mechanisms. Let us also neglect the differences between the red and the white fibers. The character of the control is similar in both types, although the quantitative effects differ.

Consider a muscle in a well-nourished man at rest. We can show by direct measurement that the muscle slowly absorbs glucose from the blood and produces pyruvate and lactate equivalent to half or more of the glucose absorbed. The major part of the energy production is due to a slow consumption of oxygen, and measurements of the RQ show that it is mainly a result of the oxidation of fats.

All of these effects are consistent with the expected stoichiometry. The store of glycogen and phosphocreatine is high at rest. There is utilization of ATP for protein synthesis, for maintenance of salt concentrations, and for sufficient contractile activity to preserve muscle tone, but ADP is not formed at anywhere near the rate seen during repeated contractions; the low concentration of ADP is the primary circumstance affecting metabolism. The [ATP]/[ADP] ratio may exceed 20:1 with full relaxation; ratios of 10:1 have been demonstrated experimentally. The effect is a competition for ADP among the enzymes of the Embden-Meyerhof pathway and of oxidative phosphorylation.

Why should the oxidation of fats be favored in the presence of an abundant supply of glucose? The reason is that flavoproteins are more oxidized than the nicotinamide nucleotides when electron transport is limited by the ADP supply, and the oxidation of substrates by flavoproteins is favored over the oxidations by NAD. We can see this from an abbreviated set of reactions:

1. $NADH + CoQ + (ADP + P_i) \longrightarrow NAD + H_2CoQ + ATP$
2. $H_2CoQ + cyt\ c + (ADP + P_i) \longrightarrow CoQ + reduced\ cyt\ c + ATP$
3. $reduced\ cyt\ c + O_2 + (ADP + P_i) \longrightarrow cyt\ c + H_2O + ATP$

Electrons from NADH must pass through three steps that require ADP, and each of these is slowed by the limiting supply, whereas electrons from the flavoprotein substrate dehydrogenases go to coenzyme Q and pass through only two steps requiring ADP. Looked at another way, the flavoprotein enzymes are slowed only by the unfavorable $[H_2CoQ]/[CoQ]$ ratio created by ADP deficiency; NADH dehydrogenase likewise is slowed by this effect, but it is further slowed by the low ADP concentration. The $[NADH]/[NAD]$ ratio is increased to an additional extent.

All of the electrons from the mitochondrial oxidation of pyruvate to acetyl coenzyme A pass through NADH, whereas half of the electrons from the oxidation of fatty acids to acetyl coenzyme A are transferred directly to flavoproteins. The less-reduced state of the flavoproteins will favor fatty acid oxidation. (The citric acid cycle is also slowed, but this is an equal impediment for oxidation of acetyl coenzyme A from either source.)

Now let us turn to the conversion of glucose to pyruvate and lactate in the cytosol. Since there is an ample supply of ATP, there is no stoichiometric obstacle to the formation of glucose-6-phosphate and its conversion to glycogen or to fructose-1,6-diphosphate. The main controls over these conversions come from modulating effects that we have already considered (Chapters 13 and 15) and shall recapitulate shortly. However, further conversion of fructose diphosphate to pyruvate requires ADP, of which there is little, and involves the reduction of NAD to NADH. The requirement for ADP limits the conversion to a slow rate, but further oxidation of even this small amount of pyruvate by mitochondria is severely limited, as we discussed in the preceding paragraph.

The NADH produced in the cytosol may be reoxidized in two ways. One is by reducing pyruvate to lactate. The other is by the glycerol phosphate shuttle, but the shuttle involves mitochondrial electron transport, already slowed by the low supply of ADP. The result is that much of the pyruvate indeed must be reduced to lactate, which diffuses into the blood along with the remaining pyruvate, even though the rate of pyruvate formation is low.

Skeletal Muscle at Work

When a skeletal muscle begins repeated contractions, the major metabolic event is a sharp rise in oxygen consumption, and measurements of the respiratory quotient show that the oxygen is being utilized mainly for the oxidation of carbohydrate. Combustion of glucosyl residues is favored so long as the supply of glycogen lasts; the oxidation of fatty acids increases as the glycogen disappears. There is also an increased output of pyruvate and lactate with muscular contraction. The amount depends upon the vigor and frequency of contractions, becoming especially large when the work approaches the maximum possible.

The immediate chemical consequence of muscular contraction is the conversion of ATP to ADP, and the events we have described hinge upon the rising concentration of ADP. We can calculate that the wave of increasing concentration generated by the contracting fibrils ought to reach creatine kinase in milliseconds and shortly thereafter reach the enzymes of the Embden-Meyerhof pathway. Prompt changes can be demonstrated experimentally. The additional time required for significant changes in the ADP concentration of the mitochondria is not known, but a high estimate is of the order of tenths of a second.

The height of the initial wave of ADP is damped by creatine kinase, which maintains the equilibrium:

$$ADP^{3-} + (phosphocreatine)^{2-} + H^+ \longleftrightarrow ATP^{4-} + creatine$$

Reasonable concentrations for the components in resting muscle are (millimoles per kilogram):

$$ATP = 6; \quad ADP = 0.3; \quad phosphocreatine = 20; \quad creatine = 10$$

Suppose, for example, that muscular contraction uses all of the original six millimoles of ATP in a burst. Part will immediately be remade by the creatine kinase reaction, and the new concentrations will be:

$$ATP = 5.7; \quad ADP = 0.6; \quad phosphocreatine = 14.3; \quad creatine = 15.7$$

The ADP concentration will have risen by only 0.3 millimole for each 6 millimoles of high-energy phosphate utilized. (These calculations are made on the basis of $K'_{eq} = 10$ for the creatine kinase reaction. The value changes with pH and Mg^{2+} concentration and is not known precisely for intracellular conditions. The uncertainty does not cloud the qualitative effect, which has been demonstrated by measurement.)

Although the rise in ADP concentration may be small initially, any increase accelerates the conversion of triose phosphates to pyruvate in the cytosol and the electron flow in mitochondria. We saw that limiting ADP concentration has the greatest effect on the NAD-coupled dehydrogenases, owing to the cascade effect in electron transfer. Similarly, an *increase* in ADP concentration accelerates the NAD-coupled dehydrogenases more than the flavoproteins, which means that the oxidation of pyruvate is selectively increased over the oxidation of fatty acids. Hence, exercise accelerates the entire pathway for the complete oxidation of triose phosphates, which also are available in greater amounts from the glycogen store, owing to the accompanying modulating effects we discussed in Chapter 15. The consequences of the rising ADP concentration therefore account for both the rising oxygen consumption and the shift of the RQ toward 1.00.

Continuation of work causes a further drop in the phosphocreatine concentration and a rise in the ADP concentration until a steady state is reached at which the new ADP level is high enough to stimulate the Embden-Meyerhof pathway and oxidative phosphorylation so that the rate of ATP formation just balances the rate of consumption. If the work is light, sufficient stimulation is obtained with low concentrations of ADP, and enough phosphocreatine will remain for later emergency effort, if required.

Pyruvate and lactate are nearly at equilibrium in the soluble cytoplasm. The accelerated formation of pyruvate caused by exercise tends to cause a rise in the concentration of both pyruvate and lactate, so that the loss of both compounds into the blood by diffusion tends to increase. The actual balance depends upon the magnitude of the work and the state of the individual muscle. If there are relatively few mitochondria, a high concentration of pyruvate is required before the rate of its removal by mitochondrial oxidation balances the rate of formation less the losses by diffusion. There is an additional physiological factor in that the blood flow through the muscle must increase in order to supply oxygen at a greater rate. Dilation of the vessels requires time, and the resultant lag seems to be greater in sedentary individuals, who also have less mitochondrial capacity for oxidative phosphorylation. These people therefore frequently have a short burst of lactate and pyruvate production at the beginning of effort, whereas the stimulation of oxidative metabolism in the trained individual is sometimes greater than the stimulation of pyruvate formation, so that his output of pyruvate and lactate initially declines at the beginning of a modest effort.

As the rate of work increases in all individuals, the rate of lactate and pyruvate production also increases, owing to the rising concentration of pyruvate. However, the accelerated removal of electrons by the glycerol phosphate shuttle keeps pace with the accelerated formation of pyruvate, so the [lactate]/[pyruvate] ratio remains relatively constant up to moderate work loads. With a further increase in work load toward the maximum, the ADP concentration will reach a saturating level for the enzymes of oxidative phosphorylation. Oxygen consumption will have reached its limit because the mitochondria are functioning at a maximum rate. (The rate of oxygen supply through the blood is probably not limiting for active individuals performing sustained effort.) A person working at a rate near this limit who makes even greater effort cannot counterbalance the resultant additional production of ADP by additional oxidative phosphorylation, and the rate of production of pyruvate sharply increases. Most of the additional pyruvate is reduced to lactate because the mitochondria cannot handle additional electron flow through the glycerol phosphate shuttle, and the [NADH]/[NAD] ratio in the cytosol rises. As we saw in Chapter 14, stimulated glycolysis cannot persist because the yield of ATP is low and the supply of glycogen is limited.

As the glycogen becomes depleted, the supply of pyruvate diminishes, and the flow of electrons through NADH will also fall. This will tend to cause a rise in the amount of oxidized coenzyme Q, which in turn will favor an increased rate of fatty acid oxidation. If the individual adjusts his rate of work to the lower level that can be sustained by the lesser yield of high-energy phosphate per mole of oxygen consumed in fatty acid oxidation, he can continue for extended periods. Otherwise, he will become exhausted.

Recovery from Work

Let us picture the circumstances in a muscle after the abrupt cessation of vigorous contractions. The ADP concentration is high, the phosphocreatine concentration is low, and the glycogen is partially depleted. The rapid conversion of ATP to ADP by muscular contraction has suddenly stopped at a time when oxidative phosphorylation is proceeding at a high rate, owing to the elevated concentration of ADP. However, oxidative phosphorylation does not suddenly slow to the resting rate. The oxygen consumption will remain high until the [ATP]/[ADP] ratio has been restored to the resting level. This will not occur until the stores of phosphocreatine and glycogen have been replenished.

During recovery, the creatine kinase reaction buffers a rise in ATP concentration in the same way that it buffers a fall during contraction. Suppose the work has been severe enough to utilize half of the total phosphocreatine in the body.

We can calculate that more than 50 additional millimoles of oxygen must be consumed in the muscles after the completion of work solely to rebuild the resting level of high-energy phosphate. (For comparison, this would require doubling of the resting rate of oxygen consumption for five minutes.)

As the ADP concentration falls, the glycogen store begins to be rebuilt, owing to the modulating changes in activity of glycogen synthase that we noted in Chapter 15. The formation of glycogen requires additional high-energy phosphate, which will further delay the decline to the resting level of the ADP concentration and the associated oxygen consumption. (For comparison of this effect, assume that a man utilizes only glycogen as a fuel and works for 40 minutes at a rate such that his oxygen consumption is six times his resting consumption. In order to rebuild the glycogen that has been consumed, he would have to keep his oxygen consumption elevated six-fold for an additional two minutes after the work. This estimate neglects any less-productive conversion of glycogen to lactate and pyruvate.)

During the actual course of recovery from work, the oxygen consumption remains very high during the first moments and gradually declines over an extended period as the ADP concentration falls. The extra consumption over the resting level during recovery is sometimes referred to as an *oxygen debt* resulting from the work. It is an unfortunate expression, implying to many an impaired utilization of oxygen by working muscles, which in fact utilize greatly increased amounts of oxygen quite efficiently. It partly came about from efforts to make a direct cause and effect correlation between lactate production (an "anaerobic" process) and the extra consumption. As we see, the extra oxygen consumption during recovery represents, as extra oxygen consumption usually does, an increased demand for high-energy phosphate, used in this case for making phosphocreatine and glycogen. It is not an abstract debit on some biological checking account.

The reconversion of lactate to glucose by the liver and kidney also requires high-energy phosphate, but it is not clear whether this requires an immediate increase in the oxygen consumption of these organs or a diversion of their metabolic processes from other syntheses. Attempts at correlation of the rate of decline of the extra oxygen consumption during recovery with the rate of disappearance of lactate have not been successful.

Stoichiometric Regulation in Other Tissues

The kind of stoichiometric regulation we have outlined here also applies to other tissues. All tissues have a limited supply of nucleotides, and the interconversions of the various forms must be balanced. The slow citric acid cycle in adipose tissue synthesizing fats from glucose can be accounted for by the inherent completeness of the scheme outlined in Chapter 16, both in respect to the nicotinamide nucleotides and high-energy phosphate balance. However, when long-chain fatty acids are synthesized from acetate or other short-chain acylates, the requirement for high-energy phosphate and electrons must be met from another source; hence the citric acid cycle accelerates.

Interpretation of stoichiometric regulation in tissues with more complex metabolism, such as the liver, is more difficult, and frequently cannot be done with present information. More attention is given to modulating effects in such cases.

REGULATION BY MODULATION OF ENZYME ACTIVITY

We have already seen many examples of metabolic regulation through the alteration of enzyme activity. This is the kind of regulation used to balance the many possible sequences of reactions. The "correct" changes in rate are made so that intermediates don't accumulate and there is efficient management of the major stores of fat and carbohydrate. Enzyme modulation is also used to insure the maintenance of the correct content of other cellular constituents.

Modulations Affecting the Major Fuel Economy

Let us consider the modulations of enzymes involved in the metabolism of the major fuels. We shall be concerned with effects rather than mechanisms, which may

TABLE 23-1. MODULATED ENZYMES OF FUEL METABOLISM.

ENZYME	ACTIVATED BY	INHIBITED BY
Acetyl CoA carboxylase	Citrate	Acyl coenzyme A
Citrate cleavage enzyme		ADP
Citrate synthase		Acyl coenzyme A, ATP
CoA—carnitine acyl transferase		Acyl coenzyme A
Fructosediphosphatase		Fructosediphosphate, AMP
Fumarase		ATP (but not MgATP)
Glucose-6-phosphate dehydrogenase		Acyl coenzyme A (?)
Glutamate dehydrogenase	Amino acids, ADP	GTP, NADH
Glyceraldehyde-3-phosphate dehydrogenase		1,3-Diphosphoglycerate
Glycogen synthase	Glucose-6-phosphate (D-form only)	UDP
Hexokinases		Glucose-6-phosphate
NAD—isocitrate dehydrogenase	ADP, citrate	
Phosphofructokinase	Fructosediphosphate, AMP	Citrate, ATP
Phosphorylase b	AMP	
Pyruvate carboxylase	Acetyl coenzyme A	ADP
Pyruvate dehydrogenase		Acetyl coenzyme A, NADH
Pyruvate kinase (liver)	Fructosediphosphate	ATP, alanine

include simple competitive inhibition, association of enzyme complexes, and allosteric changes. Some of the important modulations are listed in Tables 23-1 and 23-2, by the enzyme affected and by the modulating metabolite, respectively.

The way in which these modulating effects serve to divert substrates for the most efficient utilization is obvious in most cases. Consider the purine nucleotides. An accumulation of ATP and GTP inhibits the sources of substrates for the citric acid cycle by blocking the phosphofructokinase and glutamate dehydrogenase reactions. It also inhibits the cycle itself through the inhibition of citrate synthase and fumarase. The inhibition of fumarase is especially interesting because only free ATP, and not the magnesium chelate, is effective. This means that ATP will not influence the reaction until its concentration rises near the value of the concentration of magnesium ions, which serves to sharpen the region of controlling concentrations.

TABLE 23-2. COMPOUNDS MODULATING FUEL METABOLISM.

MODULATING COMPOUND	ACTIVATES	INHIBITS
Acetyl coenzyme A	Pyruvate carboxylase	Pyruvate dehydrogenase
Acyl coenzyme A		Acetyl CoA carboxylase, citrate synthase, glucose-6-phosphate dehydrogenase, CoA—carnitine acyl transferase
ADP	NAD—isocitrate dehydrogenase, glutamate dehydrogenase	Citrate cleavage enzyme, pyruvate carboxylase
AMP	Phosphofructokinase, phosphorylase b	Fructosediphosphatase
ATP		Citrate synthase, fumarase, phosphofructokinase, pyruvate kinase (liver)
Alanine		Pyruvate kinase (liver)
Amino acids (*e.g.*, Leu)	Glutamate dehydrogenase	
Citrate	Acetyl CoA carboxylase, NAD—isocitrate dehydrogenase	Phosphofructokinase
1,3-Diphosphoglycerate		Glyceraldehyde-3-phosphate dehydrogenase
Fructose-1,6-diphosphate	Pyruvate kinase (liver), phosphofructokinase	Fructosediphosphatase
Glucose-6-phosphate	Glycogen synthase D	Hexokinases
GTP		Glutamate dehydrogenase
NADH		Glutamate dehydrogenase, pyruvate dehydrogenase
UDP		Glycogen synthase

If the generation of ATP lags behind utilization, the resultant rise in the concentrations of ADP and AMP activates isocitrate dehydrogenase, glutamate dehydrogenase, phosphofructokinase, and phosphorylase b, thereby promoting the mobilization of substrates for the increased oxidative phosphorylation that will also be stimulated by stoichiometric regulation. At the same time, these nucleotides inhibit gluconeogenesis and fatty acid synthesis through the effects on the citrate cleavage enzyme, pyruvate carboxylase, and fructosediphosphatase; these inhibitions prevent the removal of substrates needed for oxidative phosphorylation and the concomitant extra demand for electrons and high-energy phosphate in the syntheses themselves.

Many of the other effects serve to prevent accumulation of intermediates. For example, an accumulation of acetyl coenzyme A activates pyruvate carboxylase, which is used in fatty acid synthesis and as a source of oxaloacetate to prime the citric acid cycle. Pyruvate dehydrogenase, which would supply even more acetyl coenzyme A, is inhibited by the accumulation. The same sort of control can be seen in the effects of glucose-6-phosphate and citrate on the respective enzymes by which the compounds are formed and removed.

The effect of alanine on pyruvate kinase in the liver is a more subtle example. The effect is quite specific for alanine. Now, alanine is in equilibrium with pyruvate through the glutamate—alanine transamination. Endocrine effects, which we shall discuss shortly, cause a mobilization of amino acids when there is a demand for gluconeogenesis. Efficient gluconeogenesis requires that phospho-*enol*-pyruvate be diverted back up the Embden-Meyerhof pathway and not be converted to pyruvate by the pyruvate kinase reaction, which would be wasteful recycling in the case of the carbon skeleton of alanine. The inhibition of pyruvate kinase by alanine prevents this useless diversion back to the starting compound.

A modulation that may turn out to be one of the most important is not shown in the table because it has only recently been discovered and the modulating agents have not been pinpointed. Lester Reed's laboratory found that the pyruvate dehydrogenase complex of liver and kidney mitochondria also contains a protein kinase and phosphatase. The kinase catalyzes phosphorylation of the pyruvate decarboxylase component of the complex by ATP, and the phosphatase catalyzes the hydrolysis of the phosphate groups from the protein. The phosphorylated enzyme is completely inactive.

This observation is important for several reasons. The mechanism is similar to those used in regulating phosphorylase and glycogen synthase, and we now see that such mechanisms may be used for regulating processes in parts of the cell other than the glycogen particles.

We have seen that pyruvate dehydrogenase catalyzes a pivotal reaction in the metabolic economy, and the newly-discovered regulatory effect may be an important part of the causes for a shift from fatty acid synthesis to gluconeogenesis. (It helps explain why isolated mitochondria deprived of other metabolites will form acetoacetate from pyruvate, whereas the intact organ or mitochondria supplemented with other substrates does not.) The information missing at present is the nature of the compound or compounds that trigger the action of the kinase and phosphatase. Reed's group speculates that it hinges on the concentration of ATP itself through a stoichiometric control, but it is likely that the effective agents are yet to be demonstrated.

Modulation of the Synthesis of Other Components

Many of the specific routes used for the synthesis of cellular components are controlled through modulation. The modulation frequently, but not always, involves an inhibition of the first enzyme peculiar to the route. Here are some illustrative examples:

NADP Formation.   NAD kinase is inhibited by NADPH and NADH. The limiting steps in the turnover of NADP generally involve the utilization of NADPH for reductive syntheses, rather than the formation of NADP. Hence, it is reasonable to shut off the formation of additional NADP when the accumulation of reduced nucleotides is adequate for these syntheses.

One-carbon Pool and Serine Metabolism.   The use of serine as a source of one-carbon units is controlled through an inhibition of serine transhydroxymethylase by 5-methyl-H_4folate or 5-formyl-H_4folate, which is a by-product in the one-carbon pool. The formation of serine itself is controlled through an inhibition of phosphoserine phosphatase by serine.

Porphyrin Synthesis.   The initial enzyme of porphyrin synthesis, 5-amino levulinate synthase, is inhibited by protohemin IX, which will accumulate if the total amount of porphyrins exceeds the amount of apoproteins, such as globin, with which they combine.

Purine Synthesis.   The initial enzyme, phosphoribosyl pyrophosphate amido-transferase, is inhibited by any one of the purine nucleotides, and the inhibition is synergistic with guanine and adenine nucleotides. The effect is to make the formation of new purines especially sensitive to a rise in concentration of a nucleotide pool that is balanced in its content of adenine and guanine.

Pyrimidine Synthesis.   The carbamyl phosphate synthetase of the soluble cytoplasm, which generates the precursor of pyrimidines, is inhibited by CTP. CTP is formed from UTP, so the presence of an adequate amount assures the presence of both pyrimidine nucleotides.

Deoxyribonucleotide Formation.   The ribonucleotide reductases of mammalian cells appear to be under complex control to insure a balanced supply of all four precursors of DNA. As we noted before, the reductases have not been characterized, so the exact enzymatic locus of each effect is unknown. The modulations demonstrated with crude preparations are shown in the following:

REDUCTION	INHIBITED BY	ACTIVATED BY
CDP → dCDP	dTTP, dUTP, dGTP, dATP	ATP
UDP → dUDP	dTTP, dUTP, dGTP, dATP	ATP
GDP → dGDP	dGTP, dATP	dTTP, ATP
ADP → dADP	dATP	dGTP, dTTP

These effects would appear to show that the primary control is through the regulation of pyrimidine nucleotide reduction by the balance between the total deoxyribonucleotide pool and available ATP. The dTTP will stimulate purine nucleotide reduction unless there is already a surfeit or the ATP supply is falling.

The inhibition of ADP reduction by dATP and its stimulation by dGTP will tend to keep the supply of purine deoxyribonucleotides balanced.

The deoxycytidine phosphates are not modulators of any of the reductions, and it may be that the concentration of these compounds is adjusted in specific ways. For example, dCMP, dCDP, and dCTP inhibit deoxycytidine kinase, which is an enzyme used for recovering the nucleoside for re-use. Deoxycytidine mono-phosphate deaminase is inhibited by dTTP and activated by dCTP. The dUMP created by the action of this enzyme may be used for dTMP formation or degraded.

METABOLIC ADAPTATIONS

When cells are presented with increased loads of particular kinds of compounds for extended periods, they frequently adapt to the new circumstances by changing the content of enzymes so as to metabolize the load or to cut off further formation of the compounds present in excess. The adaptation is an alteration of the actual number of enzyme molecules present, and ought not to be confused with modulating effects that change the activity of a fixed amount of enzyme.

Metabolic adaptations are supplementary controls over metabolism, usually effective within a few hours, that are especially important in adjusting to dietary changes. Stoichiometric regulation and the modulating effects we discussed in the preceding sections provide prompt moment-to-moment adjustments, but they are only effective when the capacity for catalysis can cope with the largest load imposed. When the capacity is exceeded, intermediates will accumulate.

Consider the cells in the liver of an omnivore. They must be capable of making fats from glucose or glucose from amino acids; they must rearrange amino acids and dispose of the excess nitrogen; they must handle lactate from the muscles; and while doing all of this they must turn out a variety of minor constituents and maintain the plasma proteins. It would be wasteful, perhaps impossible, to constantly maintain a full complement of all of the required enzymes at the levels necessary to handle loads that might occur only occasionally. The problem has been solved by providing mechanisms for changing the content of some of the enzymes, especially those involved in the metabolism of compounds that can safely accumulate or otherwise be stored temporarily, and those forming compounds that frequently appear in the diet so that synthesis becomes unnecessary.

We shall primarily be concerned with the effects of these changes, but let us briefly consider the possible mechanisms. A change in the amount of an enzyme could be caused by a change in the rate of its synthesis or in the rate of its destruction. Most of the observed adaptations are due to an *induction* or *repression* of synthesis, with a constant rate of destruction.

We can imagine several ways in which the rate of formation of a particular protein can be altered, involving any of the steps in protein synthesis. There might be a change in the rate of peptide formation on a polyribosome, in the ability of ribosomes to complex with mRNA, in the number of ribosomes available, or in the rate of mRNA formation. Many inductions and repressions of enzyme synthesis in bacteria clearly are a result of alterations in the rate of mRNA synthesis. Indeed, the terms induction and repression properly are used only in this sense. Some argue that similar effects are the only mechanisms by which the content of an enzyme is altered in mammalian cells, but there is substantial evidence that this is not always true, and control sometimes involves the effectiveness of pre-existing mRNA for peptide synthesis.

However, many of the mammalian adaptations probably do resemble those of bacteria, in which genetic analysis has shown the presence of specific regulator genes that cause the eventual synthesis of regulators, presumably proteins. The regulator genes are quite distinct from the operons responsible for the coding of peptides in the enzyme whose level is to be adjusted. In most cases, the protein regulator is an inhibitor of the controlling site on the operon, preventing the formation of mRNA molecules.

A compound that induces enzyme formation, which may be the initial substrate or a by-product that reflects the substrate concentration, prevents combination of the inhibiting regulator with the operon, thereby enabling enzyme synthesis to proceed. In this mechanism, induction is the prevention of an inhibition.

With those enzymes whose formation is repressed, for example by an already available supply of the product of the enzymatic reaction, the metabolite itself may act as an inhibitor of operon action.

Finally, there are less common instances of positive regulators, which stimulate the transcription of an operon.

With this background, let us turn to a consideration of the actual changes observed in response to dietary variations.

Adaptations to Starvation

When an animal doesn't eat, many of the enzymes responsible for handling transient excesses of foodstuffs become excess baggage, and we might expect them to disappear so that their constituent amino acids can be used for more pressing purposes. Starvation sharply diminishes the need for digestion, for conversion of glucose to fatty acids, for storage of fatty acids in adipose tissue, and for adjustments in fatty acid composition. At the same time, there is an increased mobilization of fatty acids from adipose tissue, part of which is handled by the liver with an accompanying increase in formation of acetoacetate and 3-hydroxybutyrate, and a corresponding shift of the metabolism of the brain toward greater dependence on these compounds. Even though the total utilization of the amino acids may fall with starvation, it is important that the liver divert as much as possible of the carbon skeletons toward glucose formation when gluconeogenesis provides the only source of carbohydrate for emergency effort or for the construction of cellular components.

The study of dietary adaptations is particularly satisfying because the observed changes nearly always are what we think they ought to be. Table 23-3 shows some examples from starved rats (everything we know about the omnivore, man, leads us to believe he has similar changes).

It is especially impressive to note the selective cut-off in synthesis of enzymes peculiar to fat storage, even including the liver desaturatase that introduces more double bonds into fatty acids, and the lipoprotein lipase of adipose tissue that serves to clear transported fats from the blood for storage. The pyruvate carboxylase of liver does not decrease, even though it may sometimes be involved in fatty acid synthesis from glucose; indeed it increases in some cases because it is an essential

TABLE 23-3.   ENZYME ADAPTATIONS TO STARVATION.[*]

INCREASED	DECREASED
Enzymes secreted by the pancreas	
None	All hydrolytic enzymes
Enzymes of fatty acid synthesis (adipose tissue and liver)	
None	Acetyl CoA carboxylase Fatty acid synthase Acyl CoA desaturase (liver) Lipoprotein lipase (adipose tissue) NADP—malate dehydrogenase Citrate cleavage enzyme
Enzymes of fatty acid utilization	
CoA—carnitine acyl transferase (liver) β-hydroxybutyrate dehydrogenase (brain)	None
Enzymes of glucose utilization	
None	Glucokinase (liver)
Enzymes of gluconeogenesis from amino acids	
Serine dehydratase Glutamate—alanine transaminase (liver cytosol) Pyruvate carboxylase Phospho-*enol*-pyruvate carboxykinase Glucose-6-phosphatase	None
Miscellaneous	
AMP kinase (liver)	

[*] The levels of only a few of the hundreds of enzymes have been measured, so the absence of any from the list does not mean its level is unchanged. This is also true of the later tabulations.

component of the route of gluconeogenesis, and the enzymes peculiar to this pathway are generally increased. The rise is more striking with phosphopyruvate carboxykinase, which is required for gluconeogenesis but not for fatty acid synthesis.

Glucose-6-phosphatase, responsible for releasing glucose from the liver, increases during starvation, but glucokinase, which functions when the blood glucose concentration is high, declines. This is as it ought to be, because a starved animal does not have a high concentration of glucose. The hexokinases do not decline during starvation, and this reinforces the idea that these enzymes serve to insure a minimal supply of glucose for the liver at all times, whereas the glucokinase is used when storage of glucose is appropriate.

To show that adaptations are not always easily explained, Table 23-3 includes a demonstrated rise in the liver AMP kinase level during starvation. This ought to give us pause. Why is it necessary to increase the rate of attainment of equilibrium between ATP, ADP and AMP during starvation? Could it be that the rate of ATP utilization is greater in the liver at such times, or is it a response to a greater load of purines delivered from other tissues?

When we think on it, we see that most animals go through a period of food deprivation every day while they sleep. The levels of many of the enzymes in the preceding tabulation have been shown to rise and fall every day in response to the diurnal variation in habits. It is a little unsettling to know that our metabolic machinery is being taken apart and rebuilt to new specifications while we sleep, only to have the alterations rescinded when we eat, but if it weren't so we should really be stuck in a rut.

Adaptations to Dietary Glucose and Fat

There are only two degrees of freedom for variation in the proportion of major fuels in a diet. A diet can't be simultaneously rich in carbohydrates, fats, and proteins. If the carbohydrate content is high, the fat content frequently will be low, and the enzymes necessary for handling carbohydrates will increase, while those peculiar to routes utilizing *dietary* fat and to routes of gluconeogenesis from amino acids decrease. The reverse effects occur if there are little glucose and abundant fat.

The responses shown in Table 23-4 are very much along the lines that would be predicted, with especially dramatic responses in the enzymes necessary for synthesizing fatty acids from glucose and for synthesizing glucose from amino acids. As a minor point, the selective response of the pancreas to the two types of diet, so that only the appropriate hydrolytic enzyme is elaborated in increased amounts, indicates that the decline in synthesis of all hydrolytic enzymes during total starvation mentioned earlier is a real adaptation and not merely the result of some general debilitation of the secretory cells.

TABLE 23-4.   ENZYME ADAPTATIONS TO GLUCOSE OR FAT DIETS.

INCREASED WITH HIGH GLUCOSE, DECREASED WITH HIGH FAT	INCREASED WITH HIGH FAT, DECREASED WITH HIGH GLUCOSE
Amylase (pancreas)	Lipase (pancreas)
Glucokinase (liver)	Glucose-6-phosphatase (liver)
Glucose-6-phosphate dehydrogenase (liver and adipose tissue)	Fructosediphosphatase (liver)
6-Phosphogluconate dehydrogenase (liver and adipose tissue)	CoA—carnitine acyl transferase (liver)
Acetyl CoA carboxylase (liver and adipose tissue)	Serine dehydratase (liver)
Fatty acid synthase (liver and adipose tissue)	Glutamate—tyrosine transaminase (liver)
Citrate cleavage enzyme (liver and adipose tissue)	Glutamate—ornithine transaminase (liver)
NADP—malate dehydrogenase (liver and adipose tissue)	

Adaptations to Diets Rich in Protein

Eating diets in which proteins supply a major fraction of the oxidizable substrates mainly causes increases in the activity of enzymes involved in nitrogen metabolism. Many of the changes are exaggerated if the diet also is low in carbohydrates. All of the following changes occur in the liver unless otherwise noted.

In some cases a specific amino acid is required for the change, rather than a general increase in nitrogen intake, and these are also noted parenthetically:

INCREASED:

peptidases (pancreas)
ornithine transcarbamylase (Arg)
argininosuccinate synthetase
argininosuccinase
arginase
serine dehydratase (repressed by glucose)
glutamate—tyrosine transaminase (separate inductions by Tyr and by mixed
    amino acids)
glutamate—alanine transaminase (cytosol)
glutamate—aspartate transaminase (cytosol)
cystathionase (Met)
glutaminase (kidney)
tryptophan pyrrolase (Trp)
fructosediphosphatase
glutamate—ornithine transaminase (kidney)

DECREASED:

glutamate—ornithine transaminase (liver: Arg)
methionine adenosyl transferase (Cys)

The only surprises in the list are the two enzymes decreased by particular amino acids. The transamination of ornithine in the liver evidently is used primarily as a means of making ornithine when the arginine supply is low; the major degradation of excess ornithine derived from dietary arginine occurs in the kidney. Consequently, opposing adaptations in the two organs are appropriate to the same stimulus.

As we have seen, methionine can be used as a source of cysteine when the diet is deficient in that amino acid. It therefore seems appropriate for cysteine to inhibit the initial enzyme required for the conversion of methionine to cysteine. However, the catch is that activation of methionine to form S-adenosyl methionine is necessary for transmethylation reactions, and it is presently not at all obvious how these reactions are stimulated so as to consume methyl groups when there is a requirement for cysteine.

Adaptations to Birth

The biggest change most mammals experience occurs when they are born. A fetus is supplied with necessary nutrients through the placental circulation, which also disposes of waste products. All of a sudden the fetus becomes an infant, and must cope with metabolic problems on its own. We are just beginning to explore the metabolic adaptations that result.

In general, it now appears the fetus is well equipped for the oxidation of glucose and fats, but not for the metabolism of amino acids or for the storage of fuels. For example, the fetal liver has the non-specific hexokinases, but not the specific gluco-kinase used for storing excess glucose. It has higher levels of phosphoglucoisomerase, phosphofructokinase, and phosphofructoaldolase than does the liver of an adult, but it almost totally lacks many of the enzymes of nitrogen metabolism.

Near term, frequently within a few hours after birth, the enzymes necessary for complete metabolism of ingested foods make their appearance. A striking example is the appearance of *mitochondrial* carbamyl phosphate synthetase and ornithine transcarbamylase required for the initial steps of urea synthesis. Before birth, the fetal liver contains greater than adult levels of the *cytosol* carbamyl phosphate synthetase, which uses glutamine as a nitrogen donor, and aspartate transcarbamylase. These are enzymes required for pyrimidine synthesis, and the fetus is undergoing rapid growth with a concomitant high demand for precursors of nucleic acids, whereas it has no need for the similar enzymes necessary to dispose of urea. The placental circulation removes accumulated nitrogen to the maternal tissues for disposal.

The enzymes appearing near or shortly after term include the following additional examples: glucose-6-phosphatase, fructosediphosphatase, pancreatic amylase, glutamate—tyrosine transaminase, tryptophan pyrrolase, and serine dehydratase. There undoubtedly are many more that will be discovered when more complete measurements have been made.

Other Adaptations

Many changes in enzyme concentration result from the action of hormones, some of which we shall discuss in the following sections. There are also adaptations in the metabolism of carbohydrates other than glucose that we shall mention later. Since the study of metabolic regulation is a young field, it may well be that there are hundreds of examples yet undiscovered in the mammals alone. Two examples may serve to illustrate the range of effects to be sought.

As we have seen, excess absorbed iron is stored in the form of ferritin. Apoferritin must be synthesized for this purpose, but it serves no other function. Hence, the formation of the peptides is tightly regulated through an induction of synthesis by ferric ions. No excess iron to be stored; no synthesis of apoferritin.

Another example is the formation of creatine for use as an energy store in skeletal muscles. Creatine is also provided through eating the muscles of higher animals, reducing or eliminating the need for synthesis of the compound. Creatine represses the formation of the initial enzyme peculiar to its synthesis, glycine transamidinase.

In general, we may expect to find inductions or repressions of protein synthesis when the requirement for the protein hinges on the presence or absence of some dietary component.

REGULATION BY HORMONES

The action of hormones adds an additional level of metabolic regulation, affecting the interrelationships of tissues. We shall defer discussion of those affecting mineral metabolism until the following chapters; we are concerned here with those hormones affecting the general metabolic routes. These include *adrenaline, glucagon,* and *insulin,* which we have already mentioned, *thyroxine, glucocorticoids, prostaglandins,* and *corticotropin.* Casual mention will be made of the metabolic effects of still others. (The subject of endocrinology is a science of its own, and we must pick and choose those aspects illuminating our general understanding of the metabolic economy.)

Let us begin by a general discussion of the mechanisms by which hormones act and then go on to a consideration of the circumstances in which particular hormones are secreted and the results of the secretion.

Mechanisms of Hormone Action

Hormones affect particular tissues more than others. (An example is the male steroid hormone, testosterone, which causes hair to disappear from the scalp and grow on the torso. Let it be hastily said that there are other genetic factors governing the response, and that a man may be capable of performing like a Solomon without these indicators.) On the other hand, the metabolic responses caused by different hormones on their respective target tissues may have great similarities.

It now appears that there are no more than three kinds of mechanisms by which hormones act, and specificity comes from the localization of receptors for the different kinds of hormones in different tissues. These mechanisms are: (a) the stimulation of synthesis of particular kinds of mRNA; (b) the stimulation of adenyl cyclase, the response being due to an increasing concentration of cyclic AMP; and (c) the alteration of transport through the cell membrane.

The first two effects are nothing more than specialized examples of adaptation by induction or repression of enzyme synthesis and modulation of enzyme activity. The especial nature of the response lies in the use of a compound as regulator or modulator that is synthesized by another tissue especially for this purpose. Even the alterations in transport may turn out to be similar, with the affected process being the formation of specific transport proteins or alteration of their characteristics through combination of the membrane with a hormone.

It is all well and good to describe the mechanisms, which are easy to comprehend, but it is often difficult to sort out the possible permutations of response to a single hormone. It is common to find that a hormone changes the rate of metabolite transport, the concentration of cyclic AMP, and the rate of mRNA formation in a single cell. Are these all direct results of the presence of the hormone, or is one of the effects the precipitating event that leads to the others? We find ardent endocrinological unitarians arguing the primacy of a single mechanism and opportunists who use any explanation fitting the immediate data. Let us stand aside from the battle for personal priority, and simply accept that the three changes do occur after target cells are exposed to hormones, that there is a cause and effect relationship between the changes and the presence of the hormone, and that there presently is no evidence for any other direct effect of any hormone.

We ought to be aware, however, that the consequences of hormone action do cause changes in metabolite concentrations. Enzymes can't distinguish changes due to endocrine activity from those caused by the diet or other causes. Therefore, secondary effects of hormones include the whole spectrum of regulatory changes: modulation of enzyme activity, altered stoichiometry, and metabolic adaptations. It is especially difficult to sort out relatively direct effects of a hormone on mRNA formation from adaptive responses also affecting mRNA formation. Timing is sometimes helpful. If administration of a hormone causes a change within a few minutes, the change is not likely to be a metabolic adaptation. If the change only appears a day later, it probably is a metabolic adaptation to the new metabolite levels. In intermediate cases, one can't always be sure, and we shall take a pragmatic view that it is the functional result that matters, not the way in which it is achieved.

Let us turn now to the specific hormones.

Adrenaline and Noradrenaline

Adrenaline and noradrenaline (p. 440) are synthesized and stored in chromaffin cells, from which they are released into the blood upon stimulation by the sympathetic nervous system. The chromaffin cells (named for a characteristic brown color appearing upon treatment of tissue sections with chromic acid) are found in the adrenal medulla, but also are scattered along the aorta and in other locations.

The gross effects of the release of these amines include an increase in the rate and depth of contraction of the heart, a suppression of contraction of many smooth muscles, a rise in the concentration of blood glucose, and a mobilization of fatty acids from adipose tissue. The analysis of these effects from the biochemical point of view is complicated by two factors. One is that noradrenaline is also a neurotransmitter. This sometimes makes it difficult to say whether a response demonstrated experimentally occurs physiologically as a result of direct neural stimulation of the tissue. (Experimental discrimination of these effects is not always easy because the sympathetic nerves not only can synthesize noradrenaline, but are also capable of taking up noradrenaline and adrenaline from the blood, storing them, and later releasing both compounds directly into the tissue upon stimulation.) Although there is a difference in the response to adrenaline and to noradrenaline within a given tissue and among tissues, the circulating hormone in many cases simply mimics the effect of stimulation of sympathetic nerves to a particular tissue. The relative contributions of externally supplied adrenaline and internally supplied noradrenaline to some physiological responses have not been sorted out.

The second complicating factor is that the receptors for the action of adrenaline and noradrenaline are known to be of two types, *alpha* and *beta receptors*. This was shown through the study of pharmacological agents that blocked some of the actions of catecholamines but not others. For example, the increase in the force (inotropic effect) and rate (chronotropic effect) of heart contraction caused by adrenaline is due to combination with beta receptors. Specific beta-adrenergic blocking agents have been developed that will prevent these effects without disturbing responses by alpha receptors.

Many tissues contain both alpha and beta receptors, frequently with opposing actions, and this causes many problems in interpretation of the physiological action of catecholamines. For example, the beta cells of the pancreatic islets (not to be confused with beta receptors) have alpha receptors that inhibit the release of insulin. They also have beta receptors that stimulate the release of insulin. What governs the balance between these opposing effects? Alpha receptors predominate, so the usual effect of release of adrenaline into the blood is decreased release of insulin from the pancreas. One can rationalize that an emergency alarm ought to shut off diversionary use of ATP for taking up glucose, and partially accomplishes this by diminishing the supply of insulin. However, one can also rationalize that circulating glucose ought to be available for use in further effort. Otherwise, what would be the point of mobilizing the glucose supply, which adrenaline does? This appears to make an equally effective case for the value of increasing the release of insulin, which the beta receptors do. This case is also supported by the action of beta receptors in skeletal muscle that increase the permeability to glucose, enabling more rapid utilization. We simply don't know the circumstances and mechanisms for shifting from one response to the other. The important thing to have in mind is that anyone delving into the literature on adrenergic responses will find quite contradictory statements, frequently not mentioning opposing effects; both statements may be correct, but under different conditions.

Fortunately, all is not chaos—biochemistry is bringing some order to our understanding of adrenergic mechanisms. It now appears likely that all *beta receptors*, certainly those causing metabolic responses, are sites at which adrenaline or noradrenaline *activates adenyl cyclase,* causing increased formation of cyclic AMP. Beta adrenergic effects therefore include modulations of activity caused by cyclic AMP, but only in those tissues in which adenyl cyclase is sensitive to adrenaline or noradrenaline.

Caffeine or theophylline inhibits the phosphodiesterase catalyzing the hydrolysis of cyclic AMP. It follows that these agents will cause an increase in concentration of cyclic AMP with a given rate of formation by adenyl cyclase, because they slow its removal. Therefore, they potentiate beta adrenergic effects—the cyclic AMP formed in response to beta adrenergic stimulation will persist longer in the presence of these agents even though they in themselves do not cause the formation of cyclic AMP.

It is less certain, but there is some evidence that *alpha receptors* are sites at which the formation of cyclic AMP is *inhibited,* perhaps also by action on adenyl cyclase. There are various speculations on the mechanisms by which catecholamines may activate and inhibit the same enzyme. Most hinge on interaction with different proteins in the cell membrane, with resultant changes in configuration causing opposing effects on the cyclase, but there is no substantial evidence for this interpretation.

We can come to a tentative conclusion: All of the metabolic effects of catecholamines can be accounted for by changes in the concentration of cyclic AMP in the affected cells. These include the familiar activation of kinases acting on phosphorylase and glycogen synthase in muscles, and the activation of lipase in adipose tissue. The activations are stimulated by beta receptors and prevented by alpha receptors.

The catecholamines may produce similar changes in the phosphorylase and glycogen synthase kinases of liver, but the required concentration is so high that the levels of circulating hormone probably reach an effective level only under strong stimulation. However, the liver is supplied with sympathetic nerves, and the concentration of noradrenaline released upon nerve stimulation probably is sufficient to cause activation of the kinases.

The increases in blood concentration of free fatty acids and glucose following exciting stimuli may be caused both by release of noradrenaline from nerve endings in the tissue and by increased amounts of circulating adrenaline and noradrenaline from the adrenal medulla and other chromaffin cells. The fatty acid concentration rises because of activation of adipose tissue lipase. The glucose concentration rises because of activation of phosphorylase and inhibition of glycogen synthase in the liver, coupled with suppression of insulin release from the pancreas in some cases. There also may be a contribution to elevated blood glucose by modulating regulation in skeletal muscle: Owing to increased phosphorolysis of glycogen, the concentration of glucose-6-phosphate rises, and this may inhibit uptake of glucose by the hexokinase reaction until the rate of glucose-6-phosphate utilization exceeds the rate of its formation.

Glucagon

Glucagon is a small peptide secreted by the alpha cells of the pancreatic islets. The structure is known for the peptide from hogs, which has 29 amino acid residues. We can't deduce anything about its mechanism of action from the structure. In fact, it is amazingly similar to another peptide hormone of entirely different function, secretin, which is produced by the duodenal mucosa and serves to stimulate secretion of pancreatic juice.

We have already seen that glucagon *stimulates adenyl cyclase* in the liver and adipose tissue, but not in skeletal muscles. The amount required is low—10 micrograms per kilogram of body weight is ample to counteract most cases of overdosage with insulin in adults. (Commercial insulin contains small amounts of glucagon, and the effect of glucagon may become briefly predominant when large doses of these preparations are given.) The physiological secretion is controlled primarily by the concentration of glucose in the blood. *Elevated glucose suppresses the release of glucagon* through unknown mechanisms.

Glucagon mimics adrenaline in causing *greater mobilization of fatty acids* and *glucose,* but is without effect on adenyl cyclase of the heart and skeletal muscles, which immediately tells us that the receptors for glucagon action are not identical with beta adrenergic receptors.

Glucagon also *stimulates gluconeogenesis* from lactate or amino acids in the liver. An effect appears in 90 seconds with as little as 2×10^{-10} M, but there is some debate over the possibility that the stimulation is a secondary response, through stoichiometric and modulating effects, to stimulated lipolysis.

Furthermore, glucagon stimulates an adaptive increase in at least one enzyme associated with increased gluconeogenesis from amino acids, glutamate—tyrosine transaminase. Here we have an illustration of the difficulty in sorting out primary and secondary effects. There is no question that the effects of glucagon on induction of the transaminase hinge on cyclic AMP, because glucagon will not cause an increase in the enzyme in certain hepatomas (liver tumors) that lack adenyl cyclase and have no demonstrable cyclic AMP. (However, the transaminase is inducible by cortisol in these tumors, which shows that the induction does not always require cyclic AMP.)

This leaves open the question of deciding whether a response to increased concentrations of cyclic AMP represents a direct effect on protein synthesis or an indirect effect caused by accumulation of substrates due to the modulating action of cyclic AMP on glycogen and fat metabolism. We don't know. However, a stimulation by physiological concentrations of cyclic AMP has been demonstrated for kinases causing the phosphorylation of histones, which gives a little support for the possibility of a direct effect of cyclic AMP on enzyme induction, since the histones are involved in control of mRNA formation.

Insulin

Insulin is formed in the beta cells of pancreatic islets as a single peptide, *proinsulin.* Three disulfide bonds are formed, and a segment of the chain is then removed by hydrolysis, leaving two peptide fragments connected by disulfide bridges (Fig. 23-1). The resultant dipeptide structure, with 51 amino acid residues, is insulin. (Since it is the original proinsulin structure that enables easy formation of the proper disulfide bridges, it is difficult to get the final insulin chains to recombine properly if these bridges are broken.)

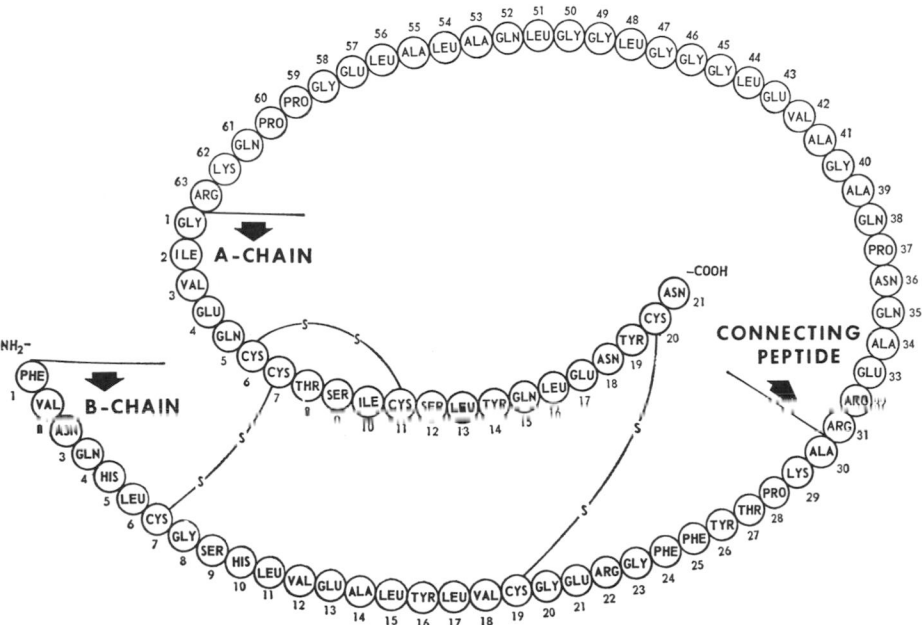

Figure 23-1   The primary structure of proinsulin from pigs. The molecule is synthesized as a single peptide chain. Three disulfide bridges are formed within the molecule from cysteinyl residues. The chain is then cleaved at two points, but the terminal segments are held together by the disulfide bridges and constitute the active insulin molecule, which now has two separate peptide chains linked by the bridges. (Reproduced from Chance, Ellis, and Bromer, Science, *161:* 166. Copyright 1968 by the American Association for the Advancement of Science.)

A rise in the concentration of blood glucose is the primary signal for the secretion of insulin, and the response is prompt. There are measurable changes in the blood insulin concentration within two minutes after experimental elevation of the glucose concentration.

The major target tissues for the action of insulin are the muscles and the adipose tissue, with lesser effects in the liver. Most of the changes caused by insulin are attributable to effects on the plasma membrane *affecting transport* through the membrane. The molecular nature of these effects is not known.

The best-known result of insulin action is the dramatic *increase in the rate of transport of glucose into skeletal muscle and adipose tissue.* We can see that this is not caused indirectly by an effect on Na^+ transport because glucose is taken up by these tissues in the absence of Na^+, unlike the transport in intestinal mucosa that we mentioned earlier. Recent studies with membranes from fat cells have shown that K^+ is required for stimulation of glucose transport, and it is also known that insulin stimulates transport of that ion in muscle. We ought not to leap to the conclusion that movement of K^+ and glucose are directly connected, however, because it is also known that K^+ inhibits glucose uptake in muscles. Decades of work have gone into a search for the mechanism of the effect of insulin on glucose metabolism, and it appears we shall have to wait even longer for an answer.

Insulin also *promotes the uptake of amino acids by skeletal muscles* and *increases protein synthesis*. These are separate effects. It is somewhat disconcerting to find that amino acid transport in muscles, unlike glucose transport, is dependent on concomitant transport of Na^+. Since one naturally seeks simplicity, it is tempting to presume that insulin causes a single kind of change in the architecture of the plasma membrane affecting disparate processes, and that the absence or presence of Na^+ dependence is peculiar to the individual transport mechanisms and is not a part of the general effect of insulin. How does one explain the effect on protein synthesis by this unitarian hypothesis, especially since there is evidence that ribosomes from insulin-deficient muscles are less effective in protein synthesis? Some try to get around this by citing evidence that insulin has no direct effect on the ribosomes themselves, but promotes the formation of some other intermediate modulator that causes changes in the ribosomes.

Insulin at low concentrations *inhibits the hydrolysis of triglycerides in adipose tissue*. It also *causes a fall in the concentration of cyclic AMP*. Is the inhibition of lipolysis a result of the change in cyclic AMP concentration? Probably, since cyclic AMP is known to activate the lipase of adipose tissue, and there is no strong evidence suggesting otherwise. However, there are no demonstrable effects of insulin on cyclic AMP formation with broken cell preparations, and once again one has to postulate that the primary action of insulin is on the plasma membrane, and the changes in cyclic AMP concentration are one of several effects.

The effects of insulin on *liver* metabolism are more diffuse. Glucose transport in the liver is not insulin-dependent. However, insulin does promote the uptake of glucose and the formation of glycogen. These effects appear to be indirect results of a decline in cyclic AMP concentration, which also results in the liver from the action of insulin. The modulating effects of increased glucose utilization then act to shut off gluconeogenesis.

The best working hypothesis one can offer at this time is that insulin combines with the plasma membrane of many types of cells and causes some sort of change in its structure or enables the release of some secondary messenger. The physiological consequences of the membrane interaction vary with the tissue. The transport of glucose in muscle and adipose tissue requires the interaction, but transport in intestinal mucosa, brain, and liver is self-sufficient. In some tissues, the membrane interaction modifies the action of adenyl cyclase so that less cyclic AMP is produced in response to stimuli. This effect is important in adipose tissue and the liver, but perhaps not in muscles.

The interaction of insulin and membrane also affects the transport of K^+ and the Na^+-dependent transport of amino acids in skeletal muscle, and perhaps in other tissues.

Finally, the stimulation of protein synthesis in muscles by insulin may be due to the release of some compound from the membrane, which in turn causes the formation of a positive regulator of ribosome formation.

DIABETES.   Any impairment of the supply of insulin to target tissues causes the condition known as *diabetes mellitus*. (The qualifying adjective is commonly dropped.) Juvenile diabetes is a severe form, with almost complete failure of the pancreas to form insulin appearing near puberty. It is a serious condition, despite the availability of insulin for treatment, with a frequently stormy course resulting from the difficulty in balancing tissue metabolism under all conditions of stress found in normal day-to-day life.

A milder form appears in later life, and adults in the early stages of the disease frequently have functional beta cells in the pancreatic islets. Indeed, the cells are sometimes loaded with insulin, but something has gone wrong with the release of the hormone. (Some sulfonylphenylureas promote the release and synthesis of insulin by these functional cells, and can be used to avoid the necessity for injections of insulin, or at least reduce it, in many patients. These oral antidiabetic drugs are only useful if the patient has functional beta cells.) Without treatment, which may involve nothing more than dietary control for many years, the beta cells gradually atrophy, and the diabetes grows progressively worse.

Genetic analysis of diabetes is difficult. Some believe it is due to an autosomal recessive mutation at a single locus, meaning that only homozygotes develop the condition. However, not all homozygotes show the effect, and it is not clear why the condition should be provoked in some at puberty and yet not have developed at age 50 in nearly three-quarters of those who ought to be homozygous. The mutant gene (or genes) for diabetes is frequent. A conservative estimate would be that it is present in one-fourth of the population, with one in 20 being homozygous.

Untreated juvenile diabetes is lethal, yet it appears in one in 2500 children; adult diabetes is a serious condition even today, and was a formidable one before the discovery of insulin. The normal death rate for all people age 40 is near three per 1000 in the United States today. The death rate for diabetics of the same age is 15 per 1000—what one would find at age 60 in the whole population. At the turn of the century, the death rate for 40-year-old diabetics was 170 per 1000, giving them little hope of seeing age 50.

We see that diabetes is undoubtedly a prime example of balanced polymorphism, with the retention of a potentially lethal gene because of some other advantage it conveys. No obvious difference has been discovered in the metabolic characteristics of the heterozygous carriers, and we must look at the homozygotes for potential advantages. George Cahill suggests that the diabetic may have an advantage during famine. Without effective insulin action, there is less competition by skeletal muscle for the limited supply of glucose that can be made from body protein, so it will be available longer for the brain and for the production of other cellular components. In addition, early diabetes is associated with obesity and greater growth, which aid survival during famine.

The major gross physiological changes in diabetes can be ascribed to the failure to utilize glucose and the increased mobilization of fatty acids. The combination of decreased uptake of glucose in muscles and adipose tissue together with accelerated gluconeogenesis in the liver, owing to the absence of the effects of insulin on that tissue, leads to an elevated concentration of glucose in the blood. This *hyperglycemia* may become so great that the kidney cannot reabsorb all of the glucose from the glomerular filtrate, causing *glycosuria*, and testing the urine for glucose is a common screening device for detecting unsuspected diabetics. However, measurement of blood glucose concentration is the most effective single diagnostic test. If the blood concentration exceeds 0.006 M after an overnight fast, or 0.008 M one hour after eating a high-carbohydrate meal, there is a probable, but not infallible, diagnosis of diabetes. Normal values are 0.004 to 0.005 M under these conditions.

The untreated diabetic may mobilize so much fat that his blood plasma becomes milky; sufficiently so to be apparent in the retinal blood vessels seen through an ophthalmoscope. He then has *hyperlipemia*. Values as great as 0.2 gram of fat per milliliter of blood have been reported in extreme cases.

There is an accompanying *ketosis*, perhaps due in part to the massive load of fatty acids presented to the liver and in part to excessive loading of the citric acid cycle by the superfluous gluconeogenesis caused by lack of insulin in the liver. This is a serious aspect of the disease, because an accumulation of the oxybutyrates represents net formation of H^+. For example:

$$(\text{palmitoyl residue}) + 7\ O_2 \longrightarrow 4\ (\text{acetoacetate})^- + 4\ H^+ + 3\ H_2O$$

Acute diabetes may cause the excretion of up to one mole of oxybutyrates per day, and the accompanying production of H^+ may exceed the capacity of the body to handle it, so that the pH of the body fluids falls. This *acidosis* frequently becomes severe enough to cause unconsciousness—the diabetic coma.

The excretion of glucose and the oxybutyrates in the urine, especially the former, prevents the normal concentration of the urine by absorption of water from the glomerular filtrate. There are frequent urination, or *polyuria*, and excessive thirst, or *polydipsia*, to compensate for the loss. If the acidosis proceeds toward development of apathy, the fluid intake may be insufficient, and there is *dehydration*.

We shall return to a further consideration of the effects of the acidosis in Chapter 25, when we discuss ion balance.

Experimental Diabetes.   It is possible to create insulin deficiencies or mimic the effects in experimental animals by several means. The entire pancreas can be removed, but this has the disadvantages of interfering with digestion, owing to the loss of hydrolytic enzymes secreted by the organ, and of simultaneously creating a deficit of glucagon.

It has been known for some time that *alloxan* causes destruction of the beta cells in the islets of Langerhans, and the alloxan-diabetic rat is a common tool for investigation of the effects of insulin. Unfortunately, alloxan also damages the kidneys and liver in doses large enough to be effective in a high percentage of the treated animals, so there is always a question of the real origin of particular effects on some reported experiments. A combination of dehydroascorbic acid and alloxan has been reported to be more selective (Prahl, J. W., and W. J. Steenrod: *Diabetes,* *14*: 289 [1965]).

In recent years, many investigators have used a serum containing antibodies to insulin as a means of selectively removing the hormone without causing destruction of any tissues. Others have used mannoheptulose, a 7-carbon sugar that cannot be metabolized, but that effectively replaces glucose as a signal to the pancreas not to produce insulin. The effects with these more discriminating methods have in the main confirmed the results obtained earlier by treatment with alloxan. This is reassuring, because the mass of earlier data is enormous, and it would be a pity to throw it all away.

The various procedures cause experimental animals to develop the same hyperglycemia, ketosis, polyuria, and polydipsia as are seen in human patients. It is possible to measure metabolic adaptations in these animals and to extrapolate the results to the human condition with some confidence. Diabetes causes a change in the metabolic economy of the liver, muscles and adipose tissue similar to the results of a high-fat diet, and it isn't surprising to find appropriate inductions and repressions of enzyme synthesis. The enzymes involved in the uptake of glucose and its conversion to fat diminish; those involved in utilizing fat and forming glucose from amino acids increase. Table 23-5 shows some reported changes (all in the liver unless otherwise noted).

TABLE 23-5. ENZYME ADAPTATIONS IN DIABETES.

DIABETES INCREASES (INSULIN DECREASES)	DIABETES DECREASES (INSULIN INCREASES)
Serine dehydratase	Glucokinase
Pyruvate carboxylase	NADP—malate dehydrogenase
Phospho-*enol*-pyruvate carboxylase	Citrate cleavage enzyme
CoA—carnitine acyl transferase	Acetyl CoA carboxylase and other
Glutamate—alanine transaminase	enzymes of fatty acid synthesis
Glucose-6-phosphatase	(liver and adipose tissue*)

* The adipose tissue has not been tested for changes in some of the other enzymes listed.

Thyroxine and Thyrotropin

The thyroid gland elaborates iodinated derivatives of tyrosine, mainly *thyroxine* but also *triiodothyronine* (Fig. 23-2), into the blood stream. These hormones have two important effects. One is to increase the metabolic rate. The other is to promote normal development of the growing organism.

The exact mechanism by which these effects occur is not known. The increase in metabolic rate mimics the effect of uncoupling of oxidation and phosphorylation by agents such as dinitrophenol (p. 204), and thyroxine is known to cause changes in the structure of mitochondria, but we have to admit that we really don't know exactly what is happening.

Normal fetal development cannot occur without a supply of thyroid hormone, and the fetal thyroid gland begins to function early. (Maternal thyroxine does not readily pass through the placenta.) The influence on development has been studied more in amphibia, which will not undergo metamorphosis in the absence of thyroid hormone. We can surmise that physiological amounts of thyroxine promote protein synthesis in some unknown way, but this statement is little more than a paraphrase of the gross physiological effect. Here again, we shall have to wait for the results of intensive studies presently under way.

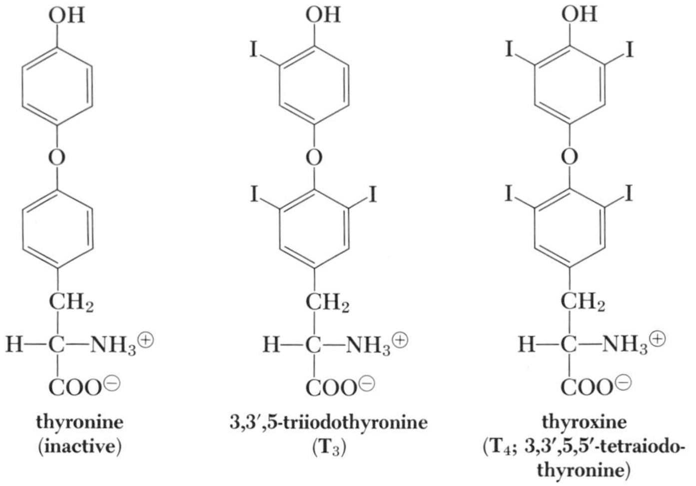

| thyronine (inactive) | 3,3′,5-triiodothyronine (T_3) | thyroxine (T_4; 3,3′,5,5′-tetraiodo-thyronine) |

Figure 23-2   The active thyroid hormone, thyroxine, is in a formal sense the tetraiodo derivative of an amino acid called thyronine. Triiodothyronine is also active, and is formed to a limited extent by the thyroid gland.

Let us turn now to the formation of thyroxine, about which we can be more definite. The thyroid gland creates a particular protein, *thyroglobulin*, stored in cell-lined follicles, and the gland consists of a collection of these sacs. The follicular cells concentrate iodide ion from the blood stream, much more so than any other cells in the body. Indeed, the thyroid gland can be selectively irradiated in a patient with overactive tissue by feeding radioactive iodine (I^{131}), which emits energetic gamma rays as well as electrons. The isotope is also given for diagnostic purposes when the activity of the gland is suspect. (The half-life of the isotope is only 8.05 days, so the danger of tissue damage rapidly declines.)

The stored iodide ion is used to iodinate tyrosyl residues on thyroglobulin. The ion must be oxidized before this can happen, and it is believed that hydrogen peroxide is the oxidizing agent in a reaction catalyzed by a specific peroxidase. (A comparable haloperoxidase has been found in a mold that forms a chlorinated compound, and the thyroid has been shown to have peroxidases capable of iodinating thyroglobulin, but the thyroid enzyme is yet to be isolated.)

When thyroglobulin is iodinated, a substantial part of the tyrosyl residues are converted to thyroxyl residues even though they remain in the protein. This can only happen by migration of an iodinated phenyl group from one tyrosyl residue to the phenolic oxygen of another (Fig. 23-3). Comparison of the effects of iodination of other proteins in the test tube shows that the placement of tyrosyl residues in thyroglobulin especially favors phenyl group migration, and we ought to expect this particular protein to have a structure facilitating the purpose for which it exists.

Thyroglobulin acts as a storage form of thyroxine and is hydrolyzed by a proteolytic enzyme to release the free amino acid for transport by the blood. The control of the protease and of the passage of thyroxine into the blood is not completely understood. Thyroxine is bound to several proteins in blood plasma during transport, but a particular globulin appears to function as the principal carrier.

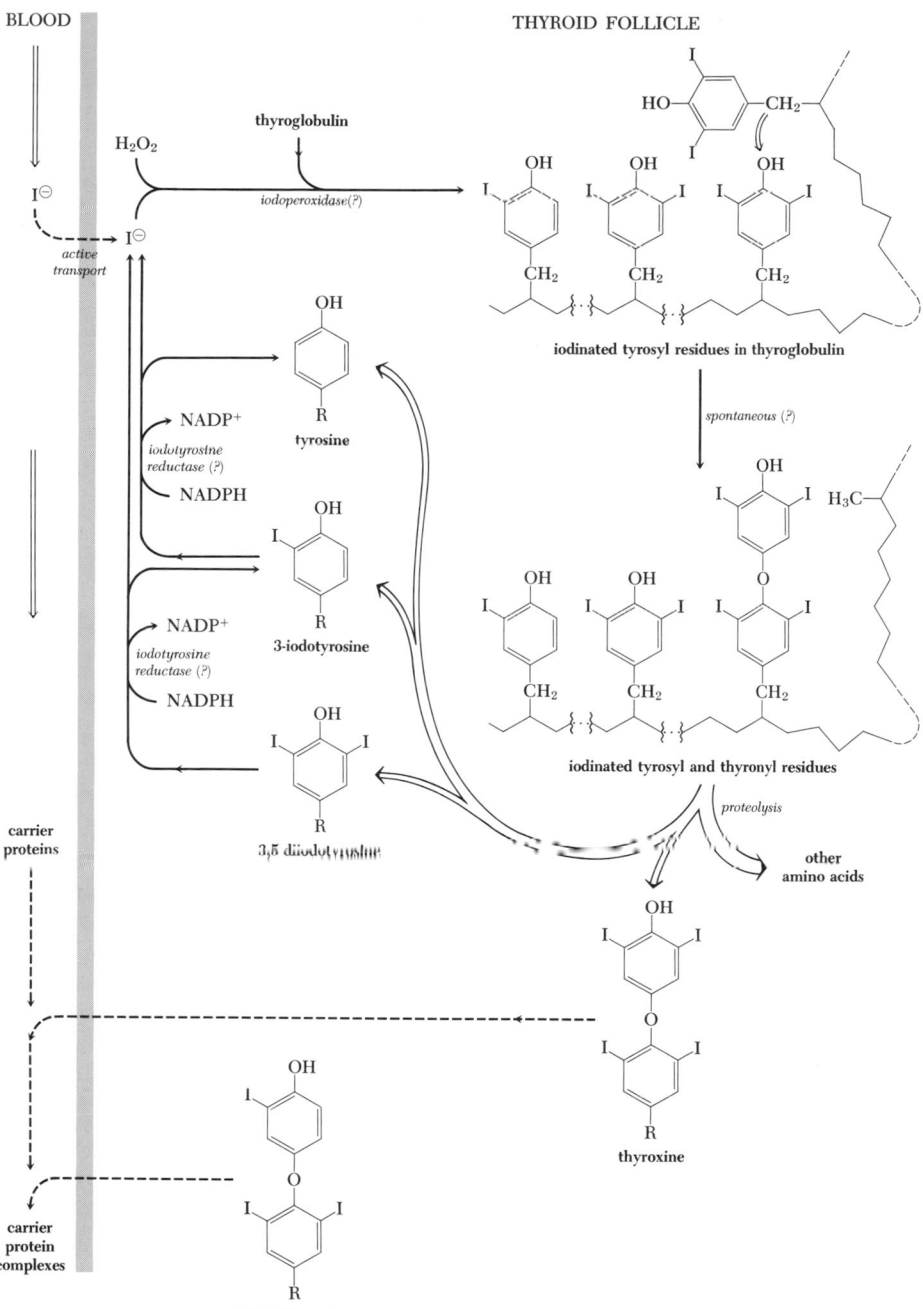

BLOOD

H₂O₂

thyroglobulin

I⊖

iodoperoxidase(?)

active transport

I⊖

NADP⁺

iodotyrosine reductase (?)

NADPH

tyrosine

OH

R

3-iodotyrosine

OH

R

NADP⁺

iodotyrosine reductase (?)

NADPH

3,5 diiodotyrosine

OH

R

carrier proteins

carrier protein complexes

triiodothyronine

OH

R

THYROID FOLLICLE

HO—⟨⟩—CH₂—

iodinated tyrosyl residues in thyroglobulin

spontaneous (?)

H₃C—

iodinated tyrosyl and thyronyl residues

proteolysis

other amino acids

thyroxine

OH

R

Figure 23-3 The formation of thyroxine. The thyroid gland actively concentrates iodide ion from the blood (*upper left*). Iodide is oxidized by hydrogen peroxide and reacts with tyrosyl residues in a peptide, thyroglobulin. The iodination is apparently non-specific, and several residues react, some being substituted with only a single iodine atom, but most with two (*upper right*). Thyroglobulin has a spatial arrangement facilitating the transfer of the diiodophenol group from one tyrosyl residue to another that is also iodinated. The result is a peptide that now contains some thyroxyl residues in addition to mono-iodo- and diiodotyrosyl residues (*center right*). The protein is then hydrolyzed, and the released thyroxine diffuses into the blood, where it is bound by specific carrier proteins. The iodotyrosines that are also released by proteolysis are reduced by NADPH (*center left*) so as to recover the iodide ion. The small amount of triiodothyronine that is formed by the thyroid gland (*lower left*) may be an accidental by-product of the same sort of reduction of thyroxine, or it may be formed by occasional transfer of a monoiodophenol group during the rearrangement of thyroglobulin.

563

Not all of the iodine in thyroglobulin is present as thyroxine. Some tyrosyl residues are iodinated once or twice without later combination with another iodinated phenyl group, and these residues also appear as free amino acids when thyroglobulin is hydrolyzed. Monoiodo- and diiodotyrosine are not effective hormones, and release of these compounds from the gland would represent an unnecessary loss of iodine. Little of the compounds is released; their iodine is removed by a reduction involving NADPH, producing iodide ion that can be stored and used over again. The action of this deiodinating enzyme may also cause the formation of some triiodothyronine from thyroxine. Triiodothyronine is actually more effective than thyroxine as a hormone, but most of the total biological activity is represented by thyroxine.

The primary regulation over the formation of thyroglobulin and the release of thyroxine comes from the activity of the *anterior pituitary gland*, or *adenohypophysis*. Let us digress a moment to consider this important tissue. The pituitary is really two separate organs located in a cavity in the base of the cranium. The posterior gland, or *neurohypophysis*, is part of a specialized development of neurons that elaborates hormones affecting smooth muscle contractions and water balance, which we shall discuss more extensively in Chapter 25. The anterior pituitary is formed from part of the primitive oral cavity, and it elaborates hormones that affect other endocrine glands (especially the thyroid, adrenal cortex, and gonads) as well as the mammary gland and uterus, and a somatotropic hormone causing growth of bones and skeletal muscles. The pituitary serves as a central switchboard that receives many of its signals from the hypothalamus in the form of uncharacterized transmitter substances coming through the stalk of the gland. Other signals reach the pituitary directly in the form of changes in the concentration of blood constituents. The signals are then transmitted to the other endocrine organs in the form of secreted hormones, and the final responses come from the specialized action of the various hormones produced by these organs.

The hormones secreted by the anterior pituitary are moderately large peptides; among them is *thyrotropin*, or *thyroid stimulating hormone (TSH)*. Thyrotropin interacts with the plasma membrane of follicular cells in the thyroid gland. The result is release of thyroxine from already-formed thyroglobulin, an increased turnover of iodine, synthesis of more thyroglobulin, and a general stimulation of the apparatus of protein synthesis in the gland. Thyrotropin also causes an increased formation of cyclic AMP in the thyroid, but we are left with the same problem that we encountered before. Are the changes in the rate of protein synthesis and in cyclic AMP concentration independent primary events, or are they secondary to some alteration of the plasma membrane? We don't know.

There are two ways in which the anterior pituitary is stimulated to release thyrotropin, thereby provoking the formation of more circulating thyroxine. One is through a direct negative feedback control. If the level of circulating thyroxine falls, more thyrotropin is released. When the level rises, less is released. This negative feedback introduces problems in therapy with any hormone produced by a gland under pituitary control. For example, administration of thyroxine depresses the formation of thyrotropin and therefore shuts off production of thyroxine by the patient's own thyroid. The reverse may occur on cessation of therapy. When the external supply of thyroxine is removed, the sudden fall in blood concentration may cause an overshoot of pituitary activity, so that there is undue stimulation of the thyroid.

The other mechanism of thyrotropin release is through neural stimulation of the hypothalamus. Neural control causes thyrotropin to be released in response to lowered body temperature and to less well-defined psychogenic influences.

Many pathological conditions are known in which too much or too little thyroxine is formed. At one extreme are patients with *thyrotoxicosis*, who have excessive secretion of thyroxine; the increased metabolic rate with constant heavy demands on the major fuel supply probably accounts for the excessive degradation of tissue proteins, with severe wastage of the muscles, seen in these patients. At the other extreme are *cretins*, who have a congenital inability to iodinate thyroglobulin and are typically mentally retarded dwarfs. Their thyroid gland has the normal capacity for concentrating iodide ion and for synthesizing thyroglobulin peptides.

Adrenal Glucocorticoids and Corticotropin

An adrenal gland, like the pituitary, is really two tissues, with a medulla that elaborates adrenaline and a cortex that produces a variety of steroid hormones in response to another peptide hormone secreted by the anterior pituitary. These adrenal steroids have two types of activity, a *glucocorticoid* action causing an increased production of glucose by the liver, and a *mineralcorticoid* action regulating the salt balance. It is the glucocorticoid action with which we are now concerned. (We are neglecting the sex hormone activity of some of the adrenal steroids.)

The steroids produced by the adrenal have varying degrees of the two functions—none of the natural compounds exhibits only one activity at the exclusion of the other. However, the effects do differ from one compound to another; some are effective glucocorticoids with much less mineralcorticoid activity and *vice-versa*. The most active of the glucocorticoids in humans is *cortisol*, or *hydrocortisone*.

The adrenal cortex is stimulated to form cortisol by *corticotropin* released from the anterior pituitary. (Corticotropin is also known as *adrenocorticotropic hormone*, or *ACTH*). There is a negative feedback relationship between cortisol concentration and corticotropin output somewhat like the relationship between thyroxine concentration and thyrotropin output. A drop in the concentration of cortisol in the blood causes an increased elaboration of a corticotropin-releasing factor from the brain stem that in turn stimulates the release of corticotropin from the anterior pituitary. There is also another response to conditions of stress in the whole animal that is independent of the circulating cortisol concentration. Burns and other wounds, physical effort, fright, and other exciting stimuli cause an increased secretion of corticotropin-releasing factor with a resultant increased formation of cortisol by the adrenal cortex. Adrenaline promotes the secretion of the factor and some of the stimuli may have their effect through this mechanism.

The gross effects of cortisol are almost the opposite of those of insulin. It causes a *mobilization of amino acids* from the peripheral tissues and *accelerated gluconeogenesis* from amino acids in the liver. It *suppresses peripheral glucose utilization* and *accelerates lipid mobilization*, with a concomitant increase in the production of acetoacetate and 3-hydroxybutyrate by the liver. (Many animals do not develop the lethal ketosis and acidosis of diabetes if both the pancreas and the adrenals, or the pancreas and the pituitary, are removed.)

TABLE 23-6.   EFFECTS OF CORTISOL.

LIVER	HOURS
Increased glycogen	4–6 (*in vivo*)
Increased glucose production	2–6 (*in vitro*)
	2–6 (*in vivo*)
Increased oxybutyrate production	3–24 (*in vivo*)
Increased urea production	4–8 (*in vivo*)
Increased amino acid uptake	1.5–2 (*in vitro*)
	2–4 (*in vivo*)
Increased RNA synthesis	2–4 (*in vivo*)
Increased protein synthesis	8–20 (*in vivo*)
ADIPOSE TISSUE	
Increased fatty acid release	1–2 (*in vitro*)
Decreased glucose utilization	2–4 (*in vitro*)
MUSCLE	
Decreased glucose utilization	2–4 (*in vitro*)
LYMPHATIC TISSUE	
Decreased glucose utilization	2–4 (*in vitro*)
Decreased nucleic acid synthesis	2–4 (*in vitro*)

Claims have been made elsewhere for suppression of glucose uptake by thymus cells in 15 to 20 minutes.

The metabolic effects of cortisol are summarized in Table 23-6 (from James Ashmore in Eisenstein, A. D., *The Adrenal Cortex*, Little Brown [1967], by permission). The table also shows the earliest times at which measurable effects appear; sometimes in the intact animal and sometimes with isolated tissues.

One of the most useful actions of cortisol from a therapeutic standpoint is its suppression of inflammation. How this ties in with the other metabolic effects is not clear, except that it may represent an extension of the general peripheral suppression of nucleic acid and protein synthesis so as to prevent leucocyte infiltration, release of histamine, and collagen formation in the neighborhood of the provoking influence. This is not a mixed blessing, because ordinarily mild infections can suddenly become disasters when normal tissue responses are suppressed, and a little stomach ulcer that has been lightly dismissed may become a grossly hemorrhagic lesion when its continual repair is prevented by treatment with cortisol for other conditions.

All of this fits in a vague way with many observations of the necessity of cortisol for the *synthesis of particular proteins*, and the effects may be more direct than with many hormones because there is no evidence for the participation of cyclic AMP in the response. One of the most frequently studied enzymes is glutamate—tyrosine transaminase, and cortisol will induce its formation in hepatomas containing no adenyl cyclase. Some of the effects of cortisol appear to be a *promotion of substrate induction* of enzyme formation—with both the substrate and the hormone required. Some effects are still different. Cortisol prevents the repression by glucose of serine induction of serine dehydratase.

Many of these phenomena appear to point to effects of cortisol on the *regulators of protein synthesis*—tending in many cases, if not all, to augment the rate of formation of proteins whose synthesis is signalled by other agents. Of course, this doesn't mean that cortisol itself penetrates the nucleus to act directly on sites of mRNA formation—there could be other messengers released by the presence of cortisol.

Formation of Cortisol

The adrenal cortex makes cortisol from cholesterol. Cholesterol may be obtained from the blood or synthesized within the cortex by a set of reactions that we shall discuss in the next chapter because they are a part of the metabolism of many tissues.

Before considering the transformation of cholesterol to cortisol, it will be helpful to review the nomenclature and numbering of sterols, which is summarized in Figure 23-4.

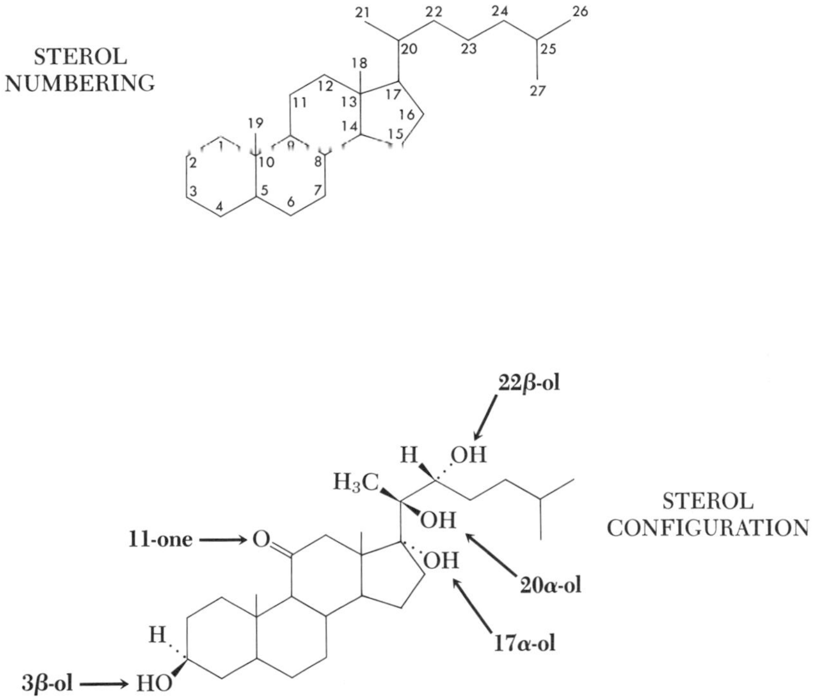

Figure 23-4   The nomenclature of the sterols involves numbering the carbon atoms as shown in the top drawing. Carbonyl and hydroxyl groups are shown by the usual suffixes, -one or -ol. Where stereoisomers can exist, as is the case with hydroxyl groups, substituents in front of the plane of the ring as drawn are designated as having the β-configuration, those behind the plane, as having the α-configuration. The configuration of substituents on the side chain is specified by imagining the chain stretched out vertically with the bonds directed behind the plane of the paper, as in the Fischer convention. Substituents that are then to the right of the chain are designated α, those to the left, β.

The cholesterol molecule contains 27 carbon atoms and a single oxygen atom. Essentially, it is a bulky hydrocarbon except for the lone hydroxyl group. The conversion to cortisol involves the removal of six carbon atoms from the side chain and the introduction of four more oxygen atoms.

We have seen earlier that the oxidation of hydrocarbons at sites away from potential resonating structures frequently requires the direct utilization of molecular oxygen, and that these oxidations are carried out by hydroxylases or other mixed function oxidases through mechanisms involving the activation of oxygen by another reduced cofactor. Examples include the oxidation of prolyl residues, in which the additional electrons come from ascorbate, and the introduction of double bonds in the middle of acyl coenzyme A, in which the additional electrons eventually come from NADPH.

The adrenal cortex contains a battery of such mixed function oxidases, which are used to attack the side chain of cholesterol and later substitute additional oxygen atoms on the ring. The positions attacked and the intracellular location of the responsible enzymes are summarized in Figure 23-5. The intermediate compound, *pregnenolone*, is shown because it is made in the same way as an intermediate for the formation of mineralcorticoids in the adrenal cortex, testosterone in the testis, and estrogens and progesterone in the ovary. The wide divergence in physiological effects of these various steroid hormones comes from the different positions at which oxygen is later substituted and further changes in the side chain of pregnenolone in the various tissues. An important distinguishing feature of cortisol is the 11-hydroxyl group. Oxygen at this position is necessary for glucocorticoid action to be most effective. (The corresponding ketone, *cortisone*, is also active.)

The most effective second donor of electrons for the hydroxylases in experimental systems is NADPH. Since the formation of the steroid hormones is the primary function of the adrenal cortex, we may expect hydroxylations to be a considerably more significant fraction of the metabolic economy than they are in most tissues. This is indeed the case, and this raises the question of the source of the relatively large amounts of NADPH required. There is no problem in the case of the hydroxylations occurring in the endoplasmic reticulum because the adrenal cortex has an active pentose phosphate pathway that can provide NADPH in the cytosol by oxidizing glucose-6-phosphate to 6-phosphogluconate and on to ribulose-5-phosphate.

However, many of the hydroxylases of the adrenal cortex are located in the mitochondria. Most people think that the NADPH is generated in these organelles by NADP—isocitrate dehydrogenase or by an ATP-coupled transhydrogenation from NADH.

None of these observations account for an important fact. The adrenal cortex has the highest concentration of ascorbate of any tissue in the body, and the ascorbate is depleted when corticosteroid secretion is stimulated. (Corticotropin preparations can be assayed biologically by measuring the ascorbate depletion they provoke.) Obviously, ascorbate is donating electrons to something at the time of cortisol formation, and it seems reasonable to conclude that it is somehow acting as a reservoir of electrons for the mixed function oxidases, even though the route of transfer is not known.

The information we do have about electron flow through the hydroxylases tells

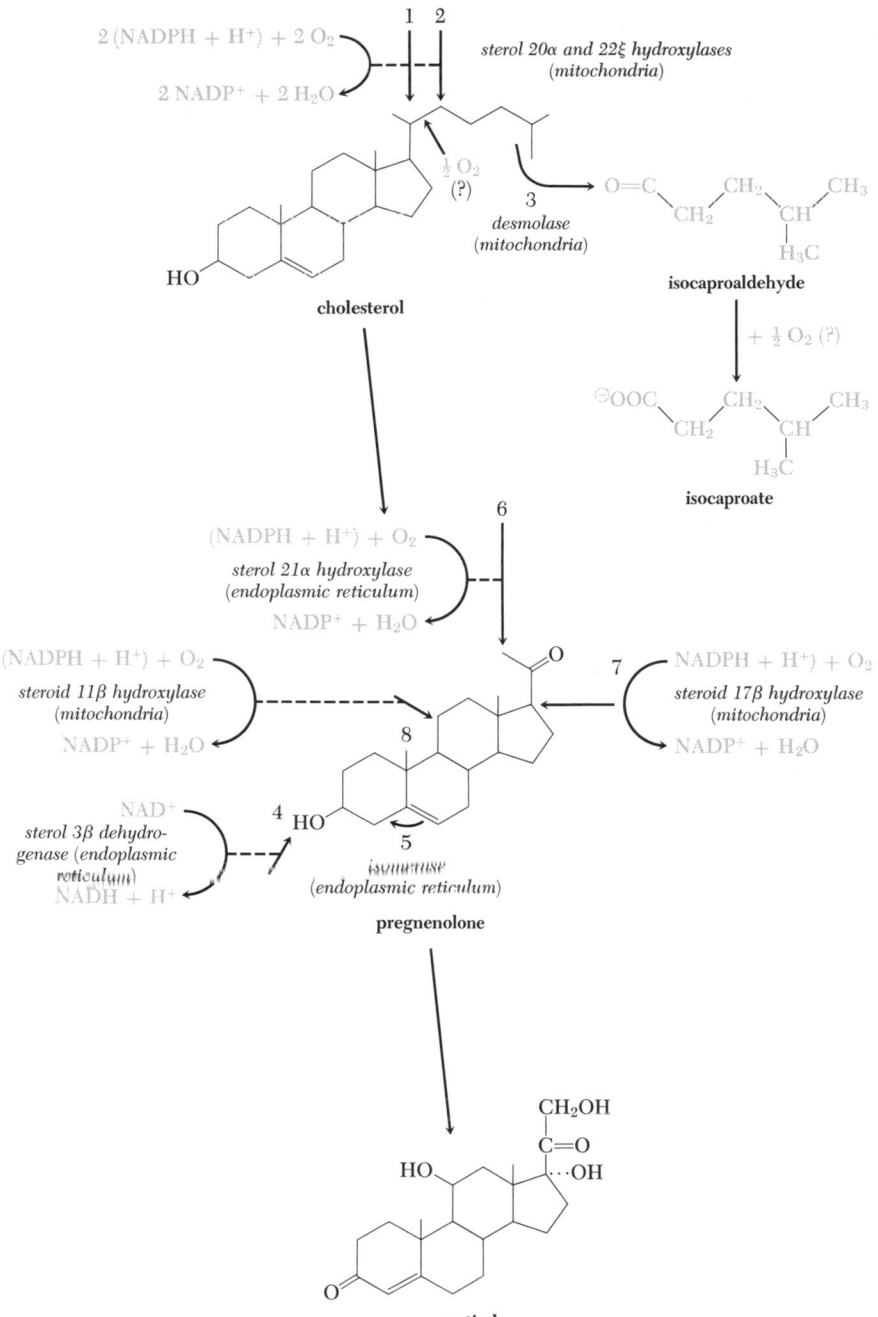

cholesterol

sterol 20α and 22ξ hydroxylases
(mitochondria)

$2\,(NADPH + H^+) + 2\,O_2$

$2\,NADP^+ + 2\,H_2O$

$\frac{1}{2}O_2$
(?)

desmolase
(mitochondria)

isocaproaldehyde

$+ \tfrac{1}{2}O_2$ (?)

isocaproate

sterol 21α hydroxylase
(endoplasmic reticulum)

$(NADPH + H^+) + O_2$

$NADP^+ + H_2O$

steroid 11β hydroxylase
(mitochondria)

$(NADPH + H^+) + O_2$

$NADP^+ + H_2O$

steroid 17β hydroxylase
(mitochondria)

$NADPH + H^+ + O_2$

$NADP^+ + H_2O$

sterol 3β dehydro-
genase *(endoplasmic
reticulum)*

NAD^+

$NADH + H^+$

isomerase
(endoplasmic reticulum)

pregnenolone

cortisol

Figure 23-5 The conversion of cholesterol to cortisol.

1 and 2. Alcohol groups are created on C-20 and C-22 by the action of hydroxylases. The unknown configuration of one of these substituents is indicated by ξ.

3. The C—C bond between the alcohol groups is then oxidatively cleaved to liberate isocapro-aldehyde (which is then oxided to isocaproate for disposition through the regular metabolic pathways) and to form pregnenolone *(center)*.

4. The 3-hydroxyl group of pregnenolone is then oxidized to a ketone, followed by (*5*) a shift of the double bond from C-5 to C-4.

6, 7, and 8. Additional hydroxyl groups are then created by the action of specific hydroxylases. These enzymes, like the ones catalyzing the earlier reactions, are distributed between the mitochondria and endoplasmic reticulum of the adrenal cortical cells.

569

us that cytochrome P450 is involved, and that it occurs in the mitochondria as well as the endoplasmic reticulum of the adrenal cortex. The electron flow also involves a low molecular weight, iron-containing protein, *adrenodoxin*. We have seen many examples of the involvement of non-heme iron proteins in various pathways of electron transfer, and another would not be especially notable were it not for the fact that adrenodoxin closely resembles *ferrodoxin*, which is involved in the light-driven electron transfers of photosynthetic plants. The occurrence of similar proteins in a highly specialized tissue of vertebrates and in the organelles of higher plants titillates many. (This isn't quite fair. The existence of similar functions in organisms of ancient evolutionary divergence always raises the suspicion that the function has a more general significance than is apparent and therefore justifies detailed inquiry.)

Perhaps the most important functional point about the route of cortisol synthesis is that the conversion of cholesterol to pregnenolone is rate-limiting and is stimulated by cyclic AMP. Corticotropin causes an immediate activation of adenyl cyclase in those cells of the adrenal cortex that synthesize cortisol. These two facts appear to account for the mechanism of stimulation of cortisol production by corticotropin. Corticotropin also causes growth of the adrenal cortex and is necessary for maintenance of cellular form in the tissue, but this may be a secondary result of the primary interaction with the plasma membrane that causes stimulation of adenyl cyclase. There is some evidence that the effect on adenyl cyclase is mediated through the rapid synthesis of some protein messenger. As with many of the endocrine effects we have discussed, the intricate interrelationships of mechanism have not been completely untangled, but this ought not to obscure the proven significance of the cyclic AMP concentration in pregnenolone synthesis.

Malfunctions of Glucocorticoid Secretion

Failure of the adrenal cortex causes *Addison's disease*. Some of the symptoms, such as gastrointestinal disorders, probably are due to the failure of mineralcorticoid secretion. One of the most striking symptoms due to failure of glucocorticoid secretion is a bronzy pigmentation of the skin, which results from a stimulation of melanocytes (the pigment-forming cells). Melanocytes are regulated by another pituitary hormone that is similar to corticotropin, and corticotropin itself has some melanocyte-stimulating activity. The Addisonian's drop in cortisol concentration stimulates the pituitary through negative feedback, and the pituitary in turn over-stimulates the melanocytes.

Other symptoms of Addison's disease, including a general feeling of weakness and easy fatigability, probably result from the lowered blood glucose concentration that occurs because of decreased gluconeogenesis. This susceptibility to hypoglycemia that is a consequence of diminished cortisol secretion is more exaggerated in patients with failure of the anterior pituitary, who are also less prone to hyperpigmentation.

Cushing's syndrome is a condition in which effects of excessive glucocorticoid secretion dominate the clinical picture. It may result from some defects, not fully characterized, of regulation of the pituitary, or from the growth of secreting tumors of either the adrenal cortex or the anterior pituitary. There is a general depletion of proteins to supply amino acids for rapid gluconeogenesis with wastage in the muscles, skin, blood vessels, and bone. As a result, there is general weakness, a tendency to bleed on slight injury, and a susceptibility to fractures, especially of the vertebrae.

The excessive output of glucose causes hyperglycemia, even to the extent of glycosuria, and the overload on the pancreas may result in full-blown diabetes. Low tolerance for carbohydrate is common. The hyperglycemia leads to the deposition of fat, with a peculiar distribution causing a characteristic buffalo hump and moon face. Both the excess fat and the diminished supply of protein may contribute to widespread fatty degeneration of tissues. Death frequently results from diabetic coma, vascular accidents, or infections. (Cushing's syndrome ought not to be confused with another type of overactivity of the adrenals, the adrenogenital syndrome, in which there is an accumulation of intermediate steroids with sex hormone activity.)

Prostaglandins

It seems quite reasonable to expect that semen, which is a complex mixture derived from several tissues, might have effects on the female reproductive tract beyond the consequences of the presence of spermatozoa. Many years ago it was indeed found that seminal fluid affected the motility of the uterus. Within the last two decades, the active principles responsible for this effect have been characterized. Swedish laboratories have been the leaders in this effort. The compounds were named prostaglandins because it was assumed they were secreted from the prostate gland. It was later found that they originated from the seminal vesicle in humans, with the concentration in human semen being near 1 millimolar.

The prostaglandins are oxygenated and cyclized 20-carbon fatty acids (Fig. 23-6); the parent compound has the trivial name *prostanoic acid*. They are derived

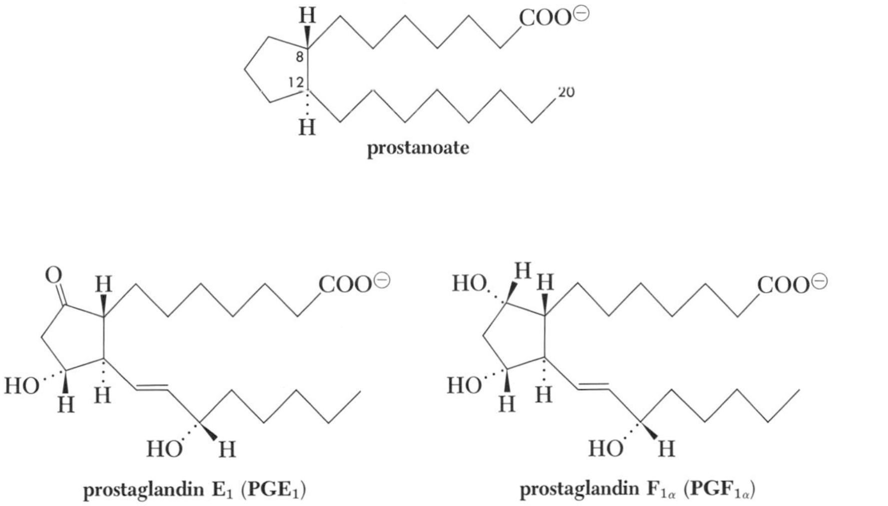

prostanoate

prostaglandin E_1 (PGE$_1$)   prostaglandin $F_{1\alpha}$ (PGF$_{1\alpha}$)

Figure 23-6   The prostaglandins are derivatives of the 20-carbon cyclopentyl acid, prostanoic acid. Several are known, and two of the most common are shown. Prostaglandin E_1 and prostaglandin $F_{1\alpha}$ are interconvertible by a dehydrogenase.

from polyunsaturated fatty acids such as arachidonic acid and perhaps account for part of the physiological function of the essential fatty acids.

Prostaglandins may cause either contraction or relaxation of uterine muscle; the contradictory effects cause difficulties in interpretation reminiscent of the problem of sorting out alpha and beta adrenergic effects. The 9-ketoprostenoates tend to cause relaxation and the 9-hydroxyprostenoates to stimulate contraction. There are additional complications. Uteri from non-pregnant females who have previously borne children are more likely to relax, whereas those from long-sterile women tend to contract with the same agent. There are variations in response during the menstrual cycle with contraction most likely to occur at the time of ovulation. This is all very interesting and no doubt important to promoting fertile matings, but it would be too specialized a topic for our purposes were it not for later discoveries that the prostaglandins are present in a number of tissues, even the lung and brain, and that they have considerably more wide-ranging effects.

The prostaglandins are now known to be additional examples of compounds affecting the response of adenyl cyclase in a variety of tissues. For example, they cause the concentration of cyclic AMP to fall in fat cells, thereby blocking the mobilization of lipid by adrenaline, glucagon, or corticotropin. As little as 4 nanomolar PGE_1 prevents the effects of adrenaline.

On the other hand, prostaglandins cause *increased* formation of cyclic AMP in the lung, spleen, diaphragm, and kidney. They mimic the effects of corticotropin on the adrenal cortex and of thyrotropin on the thyroid gland.

The physiological role of these responses has not been worked out. We have the unusual situation of knowing more about the mechanism of action of a potential hormone than of the circumstances in which it operates. It is possible that the effects are artifacts and the occurrence in many tissues is simply a reflection of the high solubility of the compounds in lipids, with the primary source and function being in the reproductive organs. This possibility appears increasingly remote, and we ought to watch with interest further studies in the next few years for clues to the primary function of prostaglandins.

Further reading

Advances in Enzyme Regulation. Pergamon Press. Annual volumes.

Chance, B., R. W. Estabrook, and J. R. Williamson, eds.: *Control of Energy Metabolism*. Academic Press (1965).

Tager, J. M., S. Papa, E. Quagliariello, and E. C. Slater, eds.: *Regulation of Metabolic Processes in Mitochondria*. Elsevier (1966).

Scrutton, M. C., and M. F. Utter: *The Regulation of Glycolysis and Gluconeogenesis in Animal Tissues*. Ann. Rev. Biochem., *37*: 249 (1968).

Linn, T. C., F. H. Pettit, and L. J. Reed: *Regulation of the Activity of Pyruvate Dehydrogenase Complex from Beef Kidney Mitochondria by Phosphorylation and Dephosphorylation*, Proc. Natl. Acad. Sci., *62*: 234 (1969).

Owen, E. E., et al.: *Liver and Kidney Metabolism During Prolonged Starvation*. J. Clin. Invest., *48*: 574 (1969).

Smith, A. L., H. S. Satherwaite, and L. Sokoloff: *Induction of Brain* D(−) *β-Hydroxybutyrate Dehydrogenase Activity by Fasting*. Science, *163*: 79 (1969).

Drysdale, J. W., E. Olafsdotter, and H. N. Munro: *Effect of Ribonucleic Acid Depletion on Ferritin Induction in Rat Liver*. J. Biol. Chem., *243*: 552 (1968).

Epstein, W., and J. R. Beckwith: *Regulation of Gene Expression.* Ann. Rev. Biochem., *37:* 411 (1968).

Butcher, R. W., and E. W. Sutherland: *Cyclic AMP.* Ann. Rev. Biochem., *37:* 149 (1968).

Moran, N. C., ed.: *New Adrenergic Blocking Drugs: Their Pharmacological, Biochemical and Clinical Actions.* Ann. N. Y. Acad. Sci., *139:* 541 (1967).

Butcher, R. W., C. E. Baird, and E. W. Sutherland: *Effects of Lipolytic and Antilipolytic Substances on Adenosine 3′,5′-Monophosphate Levels in Isolated Fat Cells.* J. Biol. Chem., *243:* 1705 and 1713 (1968).

Rodbell, M.: *Metabolism of Fat Cells, V and VI.* J. Biol. Chem., *242:* 5744 and 5751 (1967).

Pohl, S. L., L. Birnbauner, and M. Rodbell: *Glucagon-sensitive Adenyl Cyclase in Plasma Membrane of Hepatic Parenchymal Cells.* Science, *164:* 566 (1969).

Exton, J. H., and C. R. Park: *Effects of Glucagon, Catecholamines and Adenosine-3′,5′-monophosphate on Gluconeogenesis in Perfused Rat Liver.* J. Biol. Chem., *243:* 4189 (1968).

Birnbaumer, L., and M. Rodbell: *Adenyl Cyclase in Fat Cells. II. Hormone Receptors.* J. Biol. Chem., *244:* 3477 (1969).

Recent Progress in Hormone Research. Pergamon Press. Annual volumes.

Vitamins and Hormones. Academic Press. Annual volumes.

White, Priscilla, ed.: *Diabetes.* Med. Clin. N. America, Vol. 49, no. 4 (1965). An excellent general discussion.

Ostman, J., and R. D. G. Milner, eds.: *Diabetes.* Excerpta Medica Foundation (1969). Proceedings of a congress with inclusive discussions.

McKerns, K. W., ed.: *Functions of the Adrenal Cortex,* Vols. 1 and 2. Appleton-Century-Crofts (1968). Inclusive review.

Chance, R. E., R. M. Ellis, and W. W. Bromer: *Porcine Pro-insulin: Characterization and Amino Acid Sequence.* Science, *161:* 165 (1968).

Taunton, O. D., J. Roth, and I. Pastan: *Studies on the Adrenocorticotropic Hormone-Activated Adenyl Cyclase of a Functional Adrenal Tumor.* J. Biol. Chem., *244:* 247 (1969).

Sih, C. J.: *Enzymatic Mechanism of Steroid Hydroxylation.* Science, *163:* 1297 (1969).

Villee, D. B., J. Rettig, and L. Greenough: *Ribonucleic Acid Control of Steroid Synthesis in Human Adrenals and Testes.* Science, *159:* 1365 (1968).

Von Euler, U. S., and R. Eliasson: *Prostaglandins.* Academic Press (1967).

Horton, E. W.: *Hypotheses on Physiological Roles of Prostaglandins.* Physiol. Rev., *49:* 122 (1969).

STRUCTURAL COMPONENTS

CHAPTER 24

ARGUMENT

The organization of a cell, of cells into tissues, of tissues into organs, and of organs into a body depends upon a structural framework. Within the cells, the framework is made of membranes; the cells are held together by an interstitial cement; the organs are contained by fascia and suspended by ligaments and tendons; and the whole is arranged around a skeleton of bone and cartilage. All of these structural elements are made of chemical compounds, frequently quite different from those involved in the more rapid metabolic interchanges, that must be synthesized both for growth and for replacement. Even the connective tissues of an adult are constantly turning over, and most cellular membranes are rebuilt even more rapidly.

The connective tissues—those structural elements outside of the cells—are made from collections of collagen and elastin fibers embedded in a matrix. The matrix, or ground substance, is made of proteinpolysaccharides in which long chains of sugar derivatives are polymerized on a protein core. The derivatives include N-acetylosamines, in which an acetamido group replaces a hydroxyl group, and uronates, in which the terminal carbon is oxidized to a carboxylate group. Many of the proteinpolysaccharides also contain a number of sulfate groups esterified on hydroxyl groups. The large number of negative charges on the carboxylate and sulfate ester groups keeps the structure open and is responsible for some of the functional properties.

Glycoproteins are related compounds, but with much shorter glycosidic chains attached to the peptides. Many proteins secreted by cells are glycoproteins, and the carbohydrate portion of the molecule is believed to act as a sort of prosthetic group that enables them to pass outside the cell. The sugar derivatives from which glycoproteins are built include deoxysugars and acylneuraminates; the latter are derivatives of a complex 9-carbon acid.

Lipids, rather than carbohydrates, are the characteristic constituents of cell membranes. These include cholesterol, and most of the cholesterol in the body is used as a structural component. Cholesterol is made from

3-hydroxy-3-methylglutaryl coenzyme A, which is converted to isopentenyl groups. Six of these branched-chain 5-carbon residues are combined by a series of reactions to form a 30-carbon sterol, which is then modified to form cholesterol. Much of the membrane lipids is composed of phosphatidyl derivatives of choline, ethanolamine, and serine, but there are also phosphatidyl derivatives of glycerol and of inositol, an isomer of glucose. Still other membrane lipids are derivatives of sphingosine, a long-chain dihydroxy amine; the sphingolipids include combinations of sphingosine with phosphocholine and with carbohydrate residues.

The distinguishing feature of bone is the presence of microcrystals of calcium phosphates distributed throughout a collagen-proteinpolysaccharide matrix. The calcium phosphate deposits are constantly being dissolved and reprecipitated, and the turnover not only enables reshaping of the bones, but it also enables the bones to serve as a reservoir of calcium ions. The concentration of calcium ions in the body fluids is tightly regulated. Calcium passes through cells by combination with a specific transport protein. In order to synthesize the protein, cells require a supply of 25-hydroxycholecalciferol, which is a derivative of vitamin D_3. Even with a supply of the transport protein, the passage of calcium is regulated by two hormones. One is the parathyroid hormone, which is elaborated in response to lowered calcium concentrations and causes transport from the bones. The other is thyrocalcitonin, a hormone elaborated by cells in the thyroid and parathyroid glands in response to a rise in calcium concentration, and thyrocalcitonin causes deposition of calcium phosphate. The two hormones are believed to have contrary effects on the cyclic AMP concentration of affected cells.

Our major emphasis to this point has been on the more rapid metabolic traffic of the body. Let us now turn to the architectural components—the compounds found in membranes, connective tissues, and bone. These are long-lived materials compared to the major fuels or the amino acids, but they are not by any means static. They are constantly being degraded and replaced, and an impairment of the ability to synthesize them can be disastrous at any age.

The leisurely metabolism and the complex compounds found in many structural tissues have until recent years discouraged all but a few devoted experimentalists—a handful compared to the number studying, say, oxidative phosphorylation. The pace has now picked up sharply, partly because of improved methodology with which to tackle the difficult technical problems. General principles are beginning to emerge from the fog of unrelated details.

CONNECTIVE TISSUES: PROTEINPOLYSACCHARIDES

We discussed collagen in Chapter 5, using it to show how the structure of a protein can convey insolubility, strength, and rigidity. Let us begin by a broader consideration of the connective tissues in which collagen occurs.

Connective tissue is a catch-all term: It includes the tendons, cartilage, fascia, basement membranes, and the intercellular matrix surrounding all cells.

Connective tissues contain fibers of collagen and sometimes of elastin embedded in a *ground substance*. The ground substance contains a relatively large proportion of proteins with covalently-bound carbohydrates, which give it a gel-like character. The character of the tissue depends upon the proportion and arrangement of fibers in the ground substance; the intercellular matrix is mostly ground substance with few fibers, whereas a tendon is mostly fibers with minimal ground substance filling the spaces between them.

Let us digress a moment to define a term. We saw in glycogen an example of a *homopolysaccharide*, a polymer composed of one kind of sugar. Polymers can also be made by combining two or more different sugars. These are the *heteropolysaccharides*. The heteropolysaccharides are macromolecules that gain differing functional characteristics from the nature of the kinds of residues that they contain, somewhat analogous to the functional changes in nucleic acids and proteins caused by alterations in the residues. Unlike the nucleic acids and proteins, the heteropolysaccharides contain repeating sequences of residues arranged in the same order, so there are fewer kinds of molecules.

Returning now to the ground substance, it contains *proteinpolysaccharides*. A typical example of these compounds has a protein core about 400 nanometers long to which are attached by covalent bonds some 60 polysaccharide chains, each with a length of 100 nm and containing approximately 200 glycosyl residues. This is a molecule in which the protein component mainly acts as a device for spacing the polysaccharide chains like bristles on a brush, and the functional properties come from the polysaccharides and their spacing. Let us consider these polysaccharides, beginning with the constituents from which they are made.

Constituent Carbohydrates of Proteinpolysaccharides

The heteropolysaccharides are assembled by successive transfer of glycosyl residues from UDP derivatives much in the way that glycogen is assembled by transfer of glucosyl residues from UDP-glucose. The only thing we lack for ready comprehension of the process is the structural detail of the constituents. Even here, we shall see that we have already discussed examples of every type of reaction that is involved in forming the constituents. Essentially, we are only dealing with two new kinds of compounds, *osamines* and *uronates*, which are derivatives of sugars.

Osamines are nothing more than an ordinary sugar with an amino group replacing one of the hydroxyl groups. *Glucosamine* is glucose with an amino group on C-2 rather than a hydroxyl group. More formally, it is 2-deoxy-2-amino-D-glucose. The amino group is acetylated in most of the polysaccharides, so the glucosamine introduced into heteropolysaccharides is obtained by transfer from UDP-N-acetyl-D-glucosamine.

Let us see if we can reason backwards with our present knowledge to see how a compound of this kind is made. We can surmise that the UDP group is attached by the same sort of reaction used for making UDP-glucose, that the acetyl group is obtained by transfer from acetyl coenzyme A, and that the amino group is obtained by transfer from glutamine in a way analogous to the formation of phosphoribosylamine as the first step in purine nucleotide synthesis.

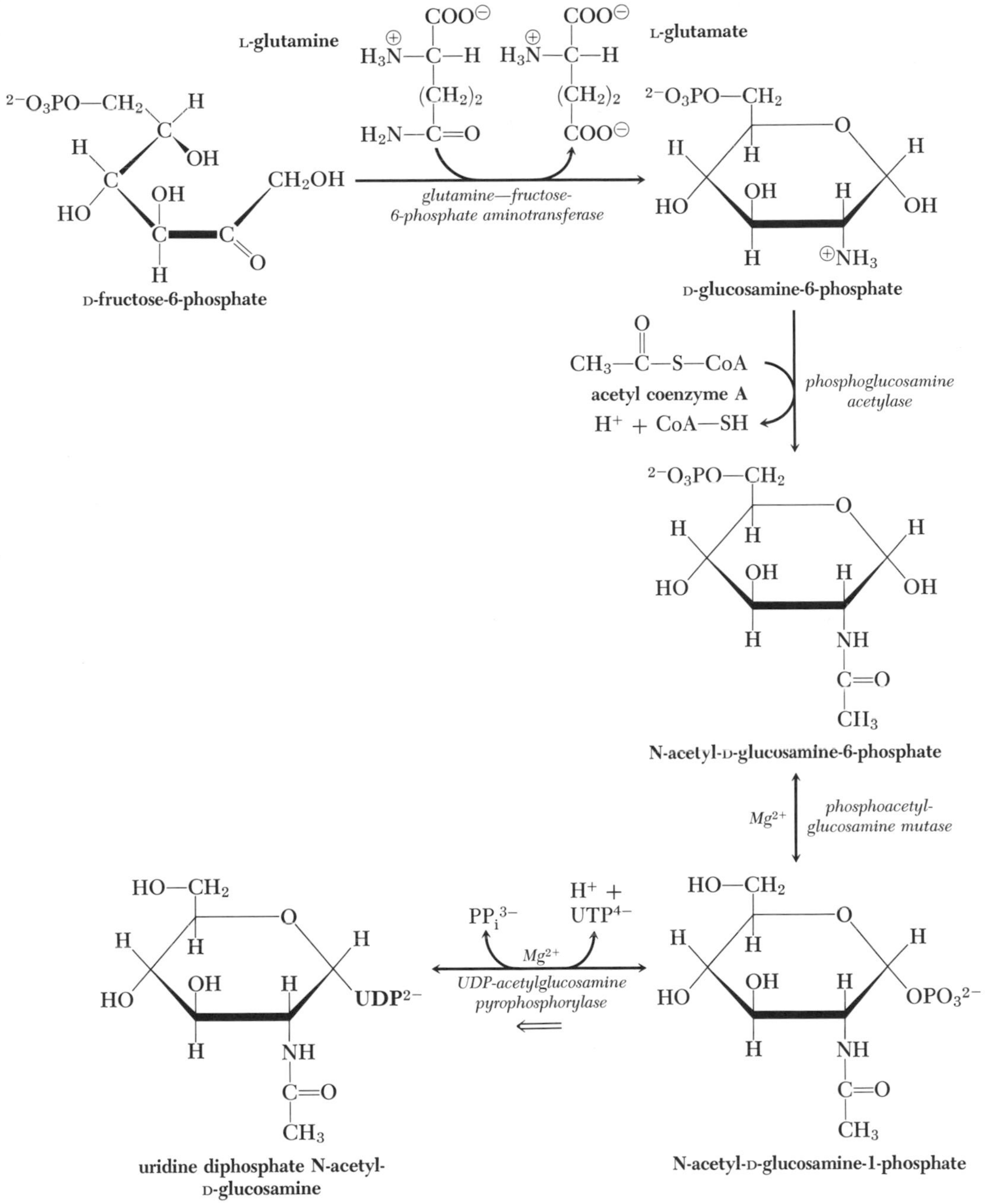

Figure 24-1 The N-acetylglucosamine residues found in heteropolysaccharides are formed from fructose-6-phosphate. The 2-amino group is first formed by transfer of the amide group of glutamine with a simultaneous shift of the carbonyl function to C-1. The amino group is then acetylated, followed by a mutase reaction in which the phosphate group is shifted to C-1. A UMP group is then transferred onto the 1-phosphate to form UDP-N-acetylglucosamine, which is a precursor of heteropolysaccharides in the same way that UDP-glucose is a precursor of glycogen.

When we look at the actual reactions (Fig. 24-1), we see this is almost exactly correct. There is a minor exception in the first step, because the nitrogen of glutamine is transferred onto C-2 of fructose-6-phosphate, rather than the glucose-6-phosphate we might expect, and there is a simultaneous isomerization of the ketose phosphate to the aldose phosphate. The resultant D-glucosamine-6-phosphate is then acetylated, followed by a mutase reaction to transfer the phosphate onto C-1 so that it can react with UTP and form UDP-N-acetyl-D-glucosamine.

Some heteropolysaccharides contain acetylated *galactosamine* residues instead of glucosamine residues. The two compounds differ only in the configuration of C-4, and UDP-N-acetyl-D-glucosamine and UDP-N-acetyl-D-galactosamine are interconvertible through the action of an *epimerase* (Fig. 24-2). The reaction resembles the interconversion of ribulose-5-phosphate and xylulose-5-phosphate in the pentose phosphate pathway. However, the enzyme is like all of the epimerases catalyzing interconversion of UDP-carbohydrate derivatives: it requires NAD. The obvious possibility is that the epimerization involves an oxidation-reduction on the enzyme surface, but all efforts to establish such a mechanism have failed.

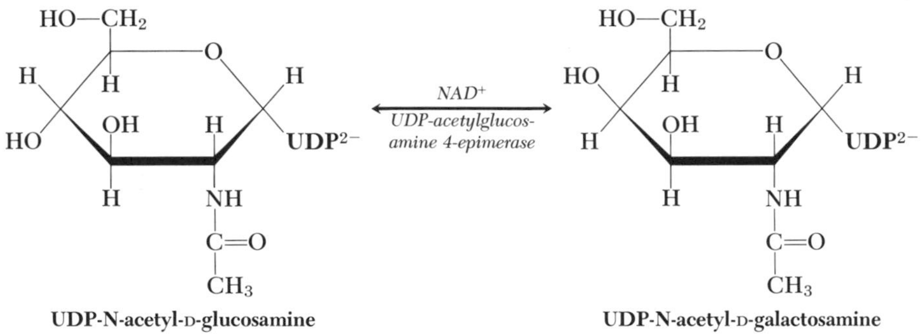

| UDP-N-acetyl-D-glucosamine | UDP-N-acetyl-D-galactosamine |

Figure 24-2   N-Acetyl-D-galactosamine residues are created in the form of the UDP derivative by the action of an epimerase. The enzyme requires NAD, and the isomerization at C-4 probably involves an oxidation at this carbon on the enzyme surface to create a ketone or an enediol, followed by reduction to yield either of the possible configurations.

The formation of the osamines is regulated through an inhibition by UDP-N-acetylglucosamine of the original transfer of an amino group. UDP-N-acetylglucosamine will accumulate if either of the acetylated osamine residues is not being utilized. Here is another example of inhibition of the first step in a synthetic process by an accumulation of the ultimate product.

Uronates are sugars in which the terminal alcohol group is oxidized to a carboxylate group. Two examples occur in mammalian heteropolysaccharides: D-*glucuronate* and L-*iduronate*. UDP-D-glucuronate is obtained by a two-step oxidation of UDP-D-glucose; UDP-L-iduronate is formed by an epimerase (Fig. 24-3). The two-step oxidation catalyzed by UDP-glucose dehydrogenase is somewhat unusual because there is no formation of the intermediate C-6 aldehyde group.

Other carbohydrates occur in heteropolysaccharides. We shall see that residues of galactose and xylose are used at the site of binding of the polysaccharide chain

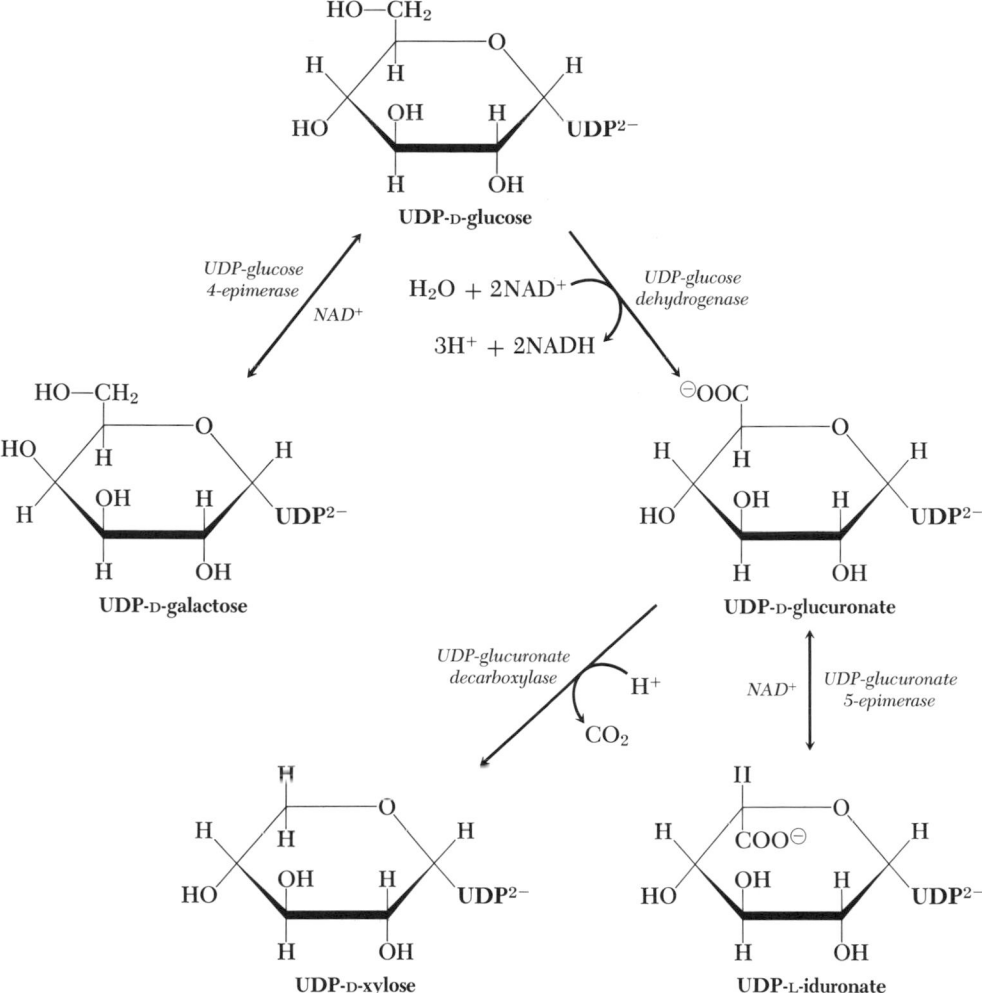

Figure 24-3 UDP-glucose is the source of several precursors of heteropolysaccharides, including the UDP derivatives of D-galactose, D-glucuronate, L-iduronate and D-xylose.

to the peptide core. UDP-D-galactose is formed by a straightforward epimerization of C-4 on UDP-glucose, and UDP-D-xylose is formed by an equally straightforward decarboxylation of UDP-D-glucuronate (Fig. 24-3). UDP-D-xylose is involved in a tricky control mechanism. It inhibits UDP-glucose dehydrogenase, thereby preventing uronate formation. We shall see that this prevents the entire process of heteropolysaccharide formation when there is no core peptide available.

Protein-Chondroitin Sulfates

The most abundant proteinpolysaccharides in the ground substance of many connective tissues, especially cartilage, contain the heteropolysaccharide known as *chondroitin.* The main length of the chondroitin chain is composed of alternating residues of D-glucuronate and N-acetyl-D-galactosamine. Sulfate groups are attached to the galactosamine residues after the protein-chondroitin molecules are synthesized. One kind has the sulfate on C-4, the other on C-6. Chondroitin-4-sulfate was formerly known as chondroitin A, and chondroitin-6-sulfate as chondroitin C. These compounds are especially easy to obtain from cartilage, and the proportion of the two kinds of chondroitins differs from one cartilage to another, as well as among the various types of connective tissue. The reason for the different distributions is not clear, primarily because we don't know the exact properties of the two forms, nor do we know the detailed function of either.

What are the important features of the chondroitin chain? (a) It is a straight chain, unlike the branched structure of glycogen. (b) All of the residues are joined in the β configuration at C-1, unlike the α configuration in glycogen. (c) In the main part of the chain, the glycosidic bonds are alternately 1→3 and 1→4. Figure 24-4 shows how these linkages are usually shown with Haworth formulas; it also shows that this is a case in which pictures are not much help. However, abbreviations are an aid, and here is the way that the chondroitin chain is expressed:

$$\cdots\text{GlcUA}-(1\rightarrow3)-\text{GalNAc}-(1\rightarrow4)-\text{GlcUA}-(1\rightarrow3)-\text{GalNAc}-(1\rightarrow4)\cdots$$

in which GlcUA is an abbreviation for D-glucuronate and GalNAc for N-acetyl-D-galactosamine. (d) Since every other residue over the main length of chain is glucuronate bearing a negative charge and the intermediate galactosamine residues are ionized sulfate esters ($R-O-SO_2-O^-$), chondroitin sulfates carry a large number of negative charges—they are polyanions. (e) Finally, the main chain is attached to a sequence:

$$\cdots[\text{GlcUA}-(1\rightarrow3)-\text{GalNAc}-(1\rightarrow4)]_n-\text{GlcUA}-(1\rightarrow3)-$$
$$\text{Gal}-(1\rightarrow3)-\text{Gal}-(1\rightarrow4)-\text{Xyl}-\text{Ser}$$

in which Gal is the unsubstituted hexose, D-galactose, Xyl is the pentose, D-xylose, and Ser is a seryl group in the core protein, with C-1 of the xylose attached to

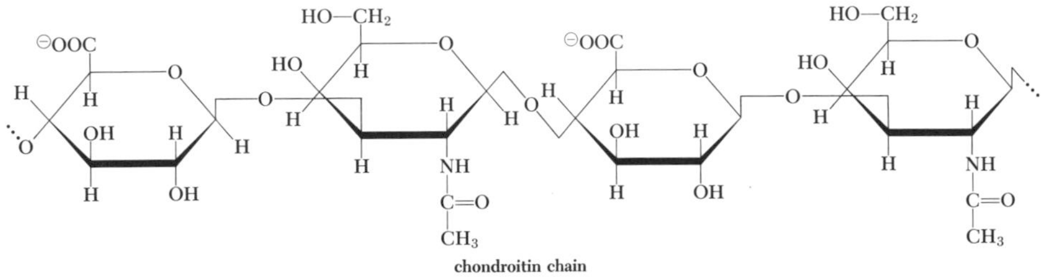

chondroitin chain

Figure 24-4   The chondroitin chain is made of alternating residues of D-glucuronate and N-acetyl-D-galactosamine. The conventional Haworth representation shown here with its impossibly bent bonds does not show the actual conformation.

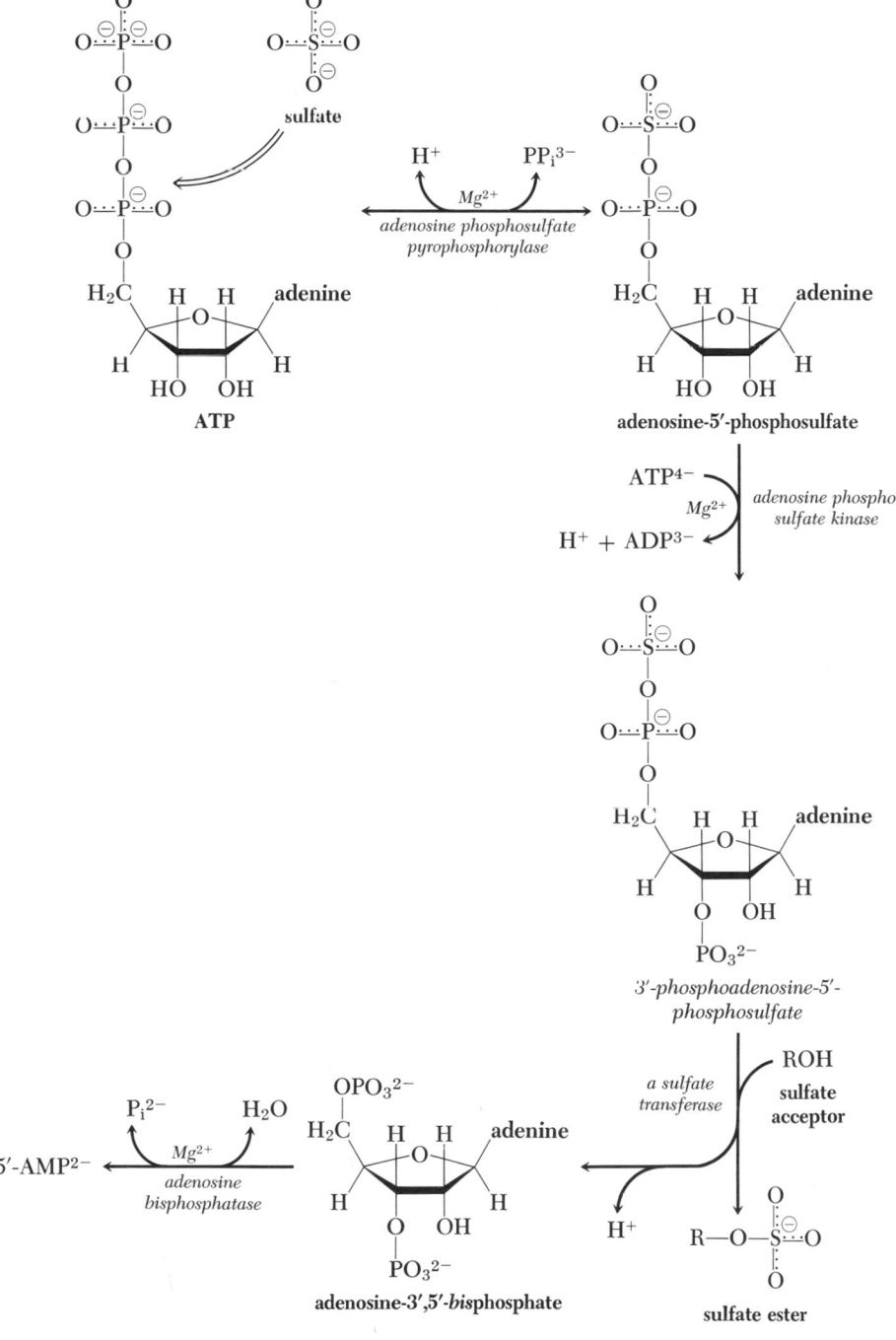

Figure 24-5 The formation of sulfate esters is driven by the transfer of the sulfate group from an energy-rich sulfate-phosphate anhydride, 3'-phosphoadenosine-5'-phosphosulfate. The anhydride bond is created by an exchange of inorganic sulfate for the pyrophosphate group of ATP. The small amount of adenosine-5'-phosphosulfate that is formed is rapidly removed by a further phosphorylation of its 3'-hydroxyl group.

the side chain oxygen of the group. The terminal sequence outside of the bracketed repeating units occurs in every proteinpolysaccharide of known structure.

The formation of the proteinpolysaccharide begins with synthesis of the core peptides by the connective tissue cell (fibroblast, chondroblast, osteoblast, or whatever). Little is known about the regulation of this step or the number of peptide chains involved. A specific enzyme then catalyzes the transfer of the xylosyl residue of UDP-xylose onto a seryl group in the protein.

This transfer, like all of the following, involves an inversion of configuration at C-1, since it is UDP-α-D-xylose that reacts to give a β-D-xylosyl group on the peptide. This is analogous to the inversion that occurs when phosphoribosylamine is made from phosphoribosylpyrophosphate in the first step of purine nucleotide synthesis; it is unlike glycogen formation in which configuration is preserved.

The next step is a transfer of a galactosyl residue from UDP-galactose onto C-4 of the xylosyl residue. This is followed by transfer of another galactosyl residue, now onto C-3 of the galactosyl residue already in place; then a residue of glucuronate is transferred from UDP-glucuronate. Each of these steps is catalyzed by a different enzyme. Finally, the formation of the main chain begins by alternating transfers of N-acetylgalactosamine and glucuronate residues from their respective UDP derivatives. Only two enzymes are presumably required to complete the long chain.

SULFATE ADDITION.   The final step in the formation of the proteinpolysaccharide is the addition of sulfate groups. This is something we have not encountered before. We ought to suspect that sulfate groups are transferred from some high-energy intermediate in the same way that phosphate groups are transferred from ATP to make phosphate esters. This is indeed so; there is a mixed adenosine phosphate-sulfate anhydride, *3'-phosphoadenosine-5'-phosphosulfate*, that acts as a donor of sulfate groups in the formation of proteinchondroitin sulfates (and other sulfate esters). Figure 24-5 shows the synthesis of this compound, the transfer of its sulfate to an acceptor, and the recovery of the by-product adenosine-3'-5'*bis*phosphate as 5'-AMP. Here we see a use for the inorganic sulfate derived from the metabolism of cysteine and methionine as well as from the diet.

Dermatan Sulfates

The structure of dermatan sulfates resembles that of chondroitin sulfates, except that they contain residues of L-*iduronate* in addition to those of D-glucuronate. The proportion of the two uronates varies in preparations from one source to another; the proportion of sulfate groups on C-4 and C-6 also varies. It is not certain whether the preparations contain varying mixtures of more than one compound or compounds with varying substituents along the chain.

The dermatan sulfates (formerly known as chondroitin sulfate B) were originally found in skin, as the name implies. The skin polysaccharides are mainly 4-sulfates. Dermatan was later shown to be present in umbilical cord as a mixture of 4- and 6-sulfates. The reason for the existence of these different heteropolysaccharides is not known; the usual speculation is that it is to create the special arrangement of collagen necessary in the skin.

Hyaluronate

Hyaluronate qualifies as a proteinpolysaccharide because it contains a small amount of protein, but the bulk of the molecule is made of a long heteropolysaccharide chain containing some 5000 glycosyl residues. The chain contains alternating residues of D-glucuronate and N-acetyl-D-glucosamine:

$$\cdots GlcUA—(1{\to}3)—GlcNAc—(1{\to}4)—GlcUA—(1{\to}3)—GlcNAc—(1{\to}4)\cdots$$

No sulfate groups are substituted on the chain.

Hyaluronate therefore is a different sort of molecule than the proteinchondroitin or dermatan sulfates. It is a single long chain with greater conformational mobility—its molecules are able to tangle with themselves or with other molecules. Negative charges appear on alternating residues rather than every residue, so there is somewhat less charge repulsion.

Heparin and Heparan Sulfate

Heparin is a small proteinpolysaccharide that occurs inside mast cells and in the liver and lungs rather than in connective tissue. The polysaccharide is relatively small, with only about 50 residues, and it has several distinctive features. It contains alternating glucuronate and glucosamine residues, but all of the linkages are $1{\to}4$. These are originally joined as glucuronate and N-acetylglucosamine, but the acetyl groups are then removed, and replaced with sulfate groups in an amide bond: $Glc—NH—SO_2—O^-$. Further sulfate groups are added in the regular way onto C-2 of the glucuronate residues, but not at all positions.

A related polysaccharide, heparan sulfate, occurs in association with heparin. The structure is not completely known, but it appears to have the same sort of backbone, except that the acetyl groups still are present on glucosamine, and sulfate is added only to the glucuronate residues. It is possible that it is related to the intermediate acetylated form of heparin.

Heparin is best known because the compound from some species (but not all) inhibits blood clotting (see Chapter 26), and solutions are available commercially for this purpose. However, this is not believed to be its major physiological function. Because of its highly anionic character, it has a strong affinity for cations, and the histamine and serotonin of mast cells are bound to heparin. This complex is believed to be involved in control of the release of these amines.

Control of Proteinpolysaccharide Formation

We know little about the primary regulation of the synthesis of proteinpolysaccharides, which occurs in the synthesis of the protein core. However, the regulation of the synthesis of the major carbohydrate constituents now appears clear. The current picture is as follows: UDP-xylose is the donor of the first carbohydrate residue, so the concentration of this sugar nucleotide rises as the available seryl groups on the core protein become filled and falls as more core protein is made. UDP-xylose inhibits UDP-glucose dehydrogenase, which catalyzes the formation of UDP-glucuronate. The inhibition serves two purposes. UDP-glucuronate is a major precursor of the polysaccharide chains, and the number of chains being built depends upon the availability of seryl sites on which to build them. If the sites are few, UDP-xylose accumulates and shuts off the formation of UDP-glucuronate, which won't be needed. UDP-glucuronate is also the precursor of UDP-xylose, so the inhibition serves as negative feedback to prevent the formation of superfluous amounts of pentose residues.

As the formation of polysaccharide chains accelerates or slows, the concentration of UDP-N-acetylglucosamine and UDP-N-acetylgalactosamine falls or rises. (Since the two are in equilibrium in cells utilizing the galactosamine compound, the concentration of both changes together.) We have already seen that UDP-N-acetylglucosamine inhibits the conversion of fructose-6-phosphate to glucosamine-6-phosphate, so the formation of the original osamine precursor diminishes when the demand decreases, owing to the inhibiting accumulation of UDP-N-acetylglucosamine. The inhibiting concentration will diminish when the formation of polysaccharides accelerates, and more of the precursor will be made.

Function of the Proteinpolysaccharides

We don't know exactly what the proteinpolysaccharides do, and we must fall back on speculations based on their known properties. However, the speculations touch on important functions and are desirable in stimulating further thought. The outstanding properties of the compounds hinge on their polyanionic nature. The repulsion of charges tends to keep the polysaccharide chains extended from the core and separated from each other in chondroitin and dermatan sulfates, but they still can sweep through a part of the surrounding solution; the lesser charge and single-chain character of hyaluronate permit it even greater freedom of movement. The motion of the chains creates a domain that is occupied by the chains. Schubert and Hamerman (see Further Reading) express it beautifully: "Statistically such a domain may have a fairly well defined size and shape. It can be visualized somewhat like the definitely shaped head of a tree, a black oak or a lombardy poplar, with branches extending throughout many cubic yards of space though the wood of its branches may occupy only a few per cent of the volume of the head. Small birds can easily fly through the head of the tree but not through the wood of its branches. Small molecules can easily swim through the domain of a proteinpolysaccharide molecule, larger ones may encounter frequent obstructions, and very large ones could not even enter."

Here we have substances with the ability to sieve out large molecules and let small ones through. They have a large negative charge that will attract cations. The possibility seems good that the ground substance acts as a selective conduit for the transport of materials into the interior of cartilage and tendon, where the cells are widely scattered. The action as a molecular sieve could also serve to exclude unwanted macromolecules as well as cell debris from locations such as the synovial fluid and the chambers of the eye, while still permitting passage of nutrients. The spaces between cells in most tissues is occupied by a polysaccharide-containing material of unknown character, and it may be that the potential transport function is also used at those locations.

Of course, the proteinpolysaccharides also contribute to the physical properties of connective tissue. The chondroitin and dermatan sulfates serve as a matrix for fibers where strength is required, and some believe that the orientation of the fibers in various kinds of connective tissue is determined by the nature of the matrix.

Solutions of hyaluronate have a high viscosity, and it is especially concentrated in the synovial fluid of the joints, where it serves as a lubricant. It is also responsible for the gel-like character of the aqueous and vitreous humors of the eye. Hyaluronate, or some close relative, is a major constituent of the ground substance around cells. Bee venoms and the secretions of some microorganisms contain enzymes that hydrolyze hyaluronate, thereby loosening the cement between cells and permitting the venom or the microorganism to spread beyond the point of penetration.

GLYCOPROTEINS

The glycoproteins are also compounds in which carbohydrate chains are attached to peptides, but they differ from the proteinpolysaccharides in structure, and include compounds of different function. The carbohydrate chains on glycoproteins are *oligosaccharides*, that is, polymers with fewer than 20 constituent residues. These oligosaccharides include residues of carbohydrate derivatives not found in the proteinpolysaccharides.

Some compounds classed as glycoproteins are sulfate esters found in connective tissue along with chondroitin and dermatan sulfates. The others include a whole spectrum of compounds ranging from those studded with so many carbohydrate chains that their function is mainly derived from the carbohydrate portion to those containing only a few chains acting somewhat as prosthetic groups for the function of the protein portion. In other words, glycoprotein is a structural, not a functional, classification. Let us now consider the structural elements on which this classification is based.

Deoxy Sugars

Glycoproteins commonly contain residues of L-*fucose* (6-deoxy-L-galactose), and some have residues of L-*rhamnose* (6-deoxy-L-mannose). The residues are formed by similar reactions (Fig. 24-6), with glucose-6-phosphate as the original source. L-Fucose appears as GDP-L-fucose with intermediate formation of GDP-D-mannose, which is also used as a donor of *mannose* residues in forming some glycoproteins. The mammalian nucleotide carrier of rhamnose is not known for certain. Some microorganisms produce GDP-L-rhamnose and others produce dTDP-L-rhamnose. We might guess that these unusual nucleotides provide a means of distinguishing glycoprotein formation from the synthesis of proteinpolysaccharides, but this is pure speculation.

Acyl Neuraminates: Sialates

Many glycoproteins contain residues of the 9-carbon amino keto acid, *neuraminic acid*, which is synthesized in the form of N-acetylneuraminate-9-phosphate by the condensation of D-mannosamine-6-phosphate and phospho-*enol*-pyruvate (Fig. 24-7), and converted to CMP-N-acetylneuraminate for transfer during glycoprotein formation (Fig. 24-8).

Part of the neuraminates contain an additional 4-O-acetyl group; others have an N-glycolyl rather than an N-acetyl group. All of these acylated neuraminates—the N-acetyl, N-glycolyl, and N-acetyl-4-O-acetyl—are collectively known

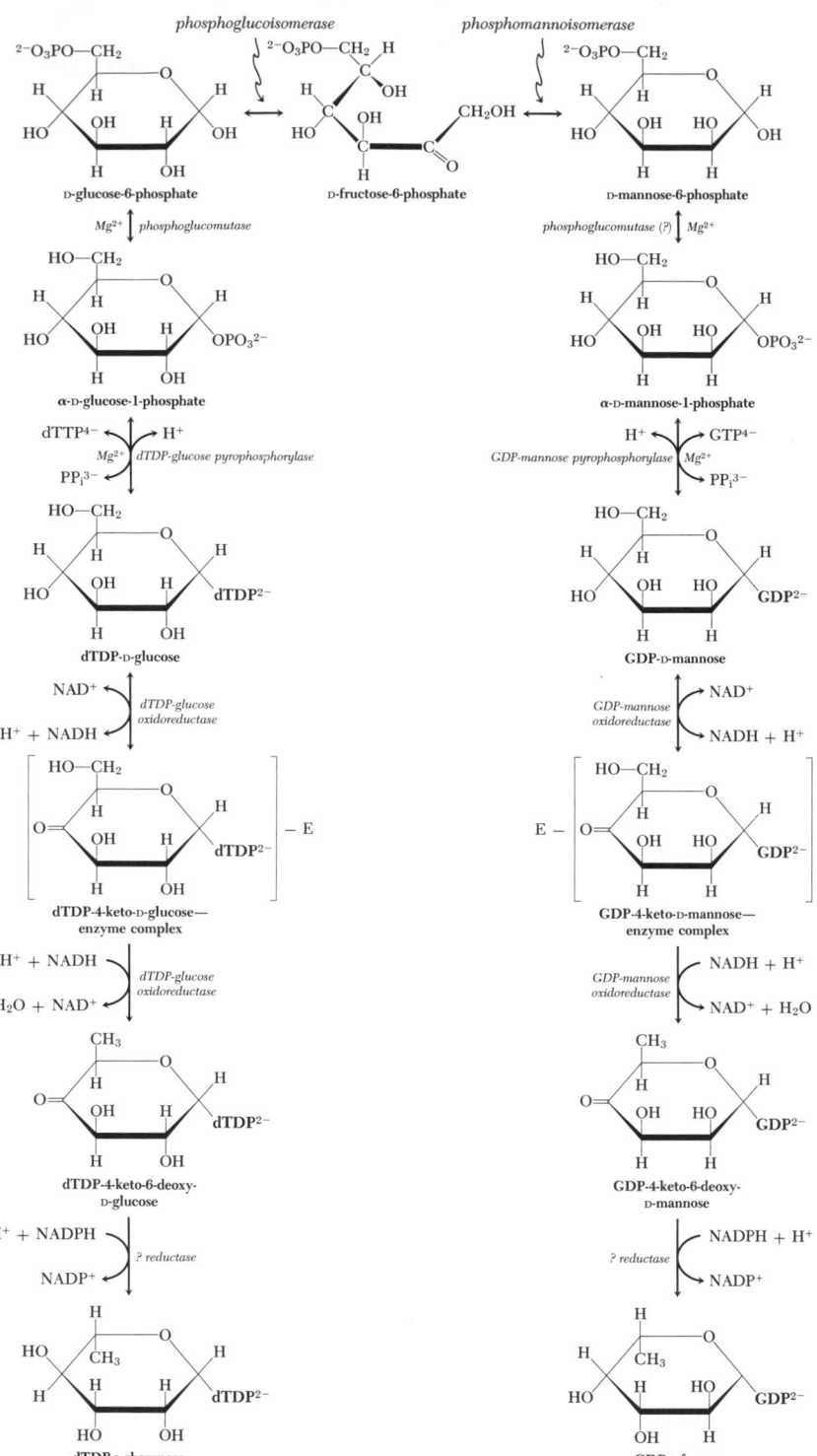

Figure 24-6 Residues of the 6-deoxy sugars, L-rhamnose and L-fucose, are created through a series of oxidations and reductions that result in inversion of the configuration of C-3 and C-4, with a concomitant reduction of C-6 to a methyl group. The starting materials are D-glucose-6-phosphate and D-mannose-6-phosphate that is formed from D-fructose-6-phosphate by a separate isomerase. The residues of the deoxy sugars differ from most in that they are carried on dTDP and GDP for incorporation into polysaccharides.

586

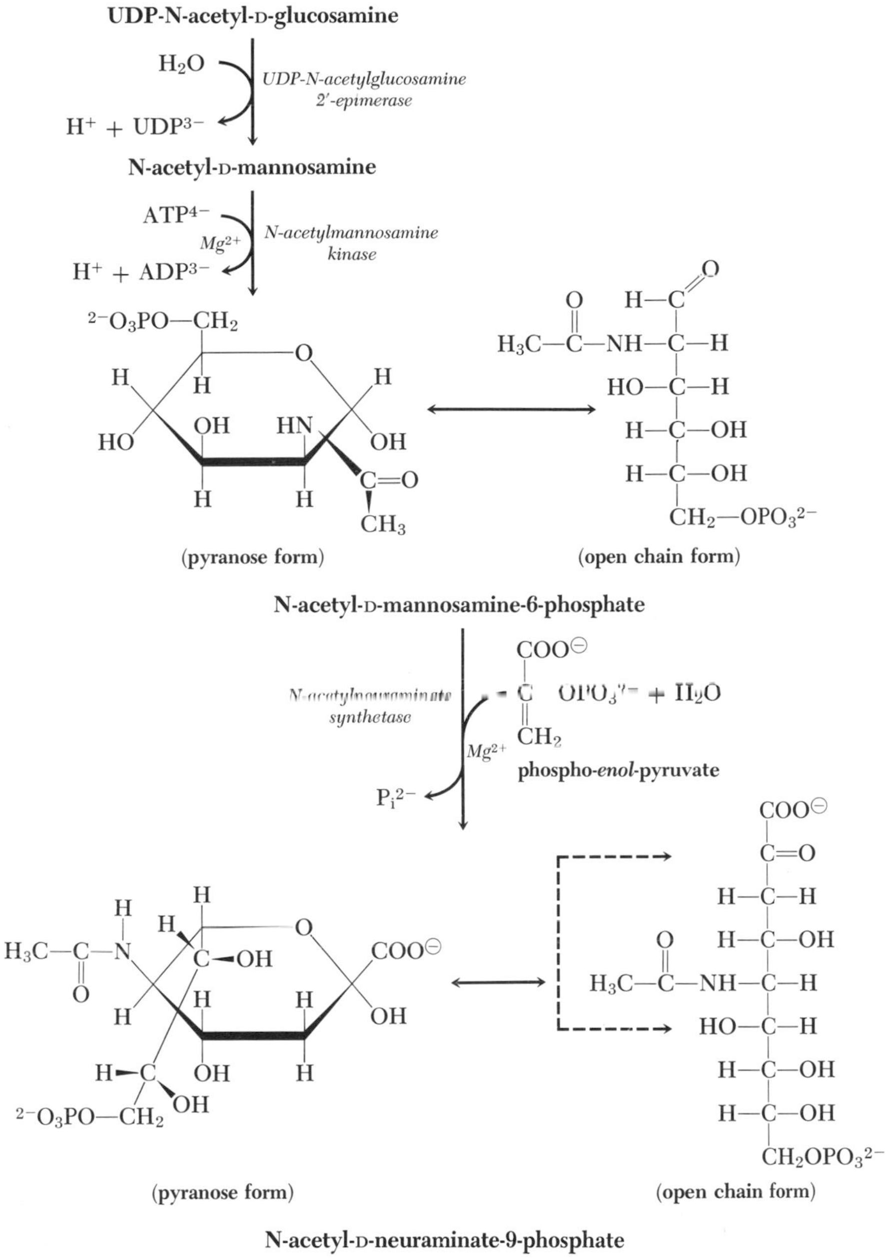

Figure 24-7 Residues of N-acetyl-D-neuraminate, one of the sialates, are created by a condensation of phospho-*enol*-pyruvate with N-acetyl-D-mannosamine-6-phosphate.

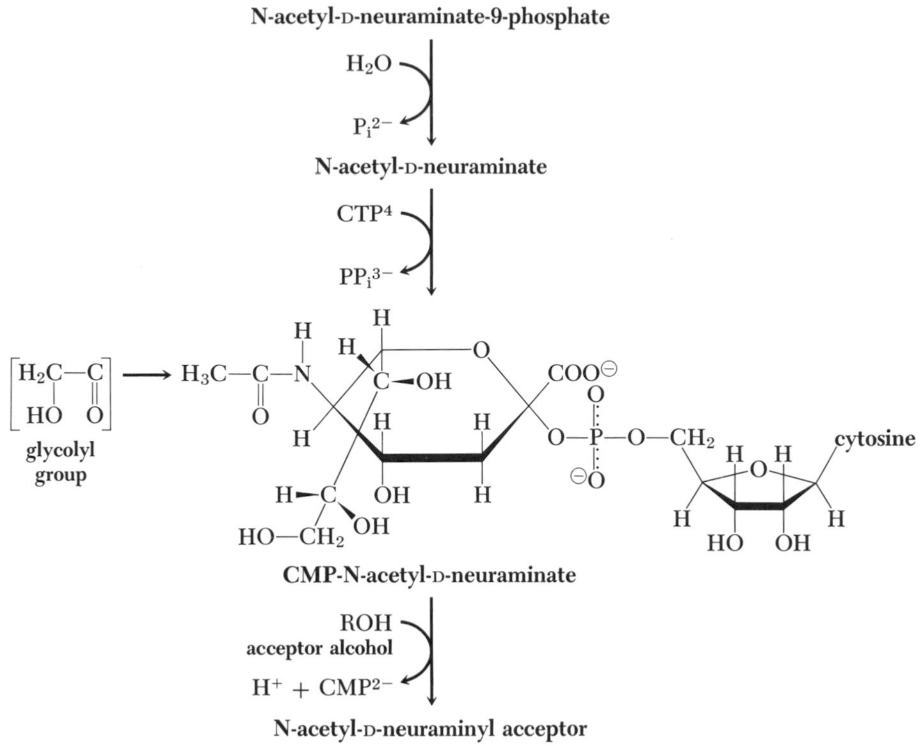

N-acetyl-D-neuraminate-9-phosphate

$$H_2O \searrow$$
$$P_i^{2-} \swarrow$$

N-acetyl-D-neuraminate

$$CTP^4 \searrow$$
$$PP_i^{3-} \swarrow$$

glycolyl
group

CMP-N-acetyl-D-neuraminate

$$ROH \searrow$$
acceptor alcohol
$$H^+ + CMP^{2-} \swarrow$$

N-acetyl-D-neuraminyl acceptor

Figure 24-8   The N-acetylneuraminyl polysaccharides are formed by transfer of the residue from the CMP derivative, which is formed by reaction of CTP with the free compound, rather than the usual reaction of nucleoside triphosphates with sugar phosphates to yield the nucleoside diphosphate sugars. N-Glycolyl-D-neuraminyl groups are also found, and these may be formed by an exchange with the acetyl group of the CMP derivative, although the donor is not known.

as the sialates. Figure 24-8 shows CMP-N-acetylneuraminate as the parent compound, but the actual route of formation of the glycolyl and diacetyl derivatives is not certain.

The Oligosaccharide Chains

A great variety of oligosaccharide chains occur in glycoproteins, ranging from simple disaccharides to complex branched structures. However, there are some characteristics in common. The carbohydrate residue bound to the protein is an acylated osamine, frequently N-acetyl-D-glucosamine. The attachment to the protein may occur by the sort of bond we saw in the proteinpolysaccharides, with the acetylglucosamine residue transferred from the UDP derivative onto the hydroxyl group of a seryl or threonyl residue, but there are also attachments to the amide nitrogen of an asparaginyl residue (Fig. 24-9).

The remainder of the chain is made by sequential addition of the other constituent residues. However, if residues of L-fucose or of the acylneuraminates occur, they are invariably at the end of the chains—capstones, as it were. This may account for the unusual use of the GDP and dTDP, rather than the familiar UDP, forms of these sugars.

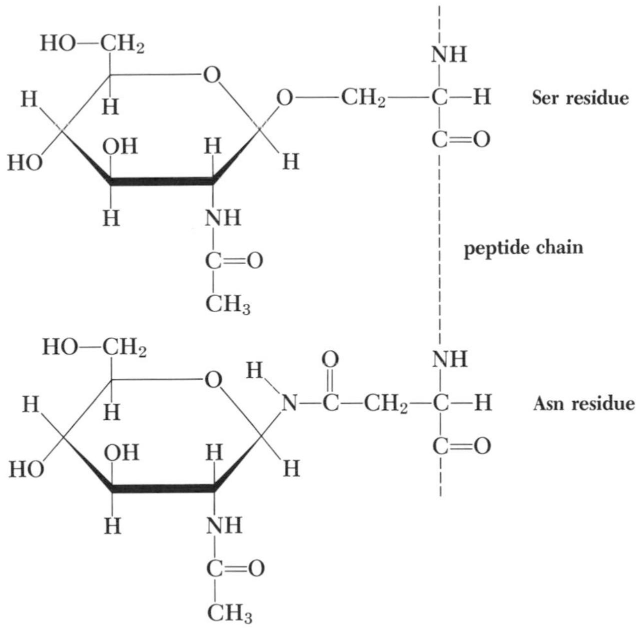

Figure 24-9 The attachment between polysaccharide and peptide in glycoproteins is made by building the polysaccharide chain on N-acetylglucosamine residues that are attached to either the hydroxyl group of a seryl residue in the peptide or the amide nitrogen of an asparaginyl residue.

Function of the Glycoproteins

Collagen might be considered to be a glycoprotein because it has a carbohydrate chain containing glucose and galactose residues bound to the few residues of the unusual amino acid, hydroxylysine, that occur. The function in this case may be to facilitate association with the proteinpolysaccharides of the ground substance, and there may be a similar function for some of the glycoproteins believed to occur on the outer surface of the plasma membranes of cells.

Most of the glycoproteins are secreted from cells, rather than being structural elements, but this is as convenient a point as any to consider them. The presence of glycoproteins in mucous secretions accounts for their viscosity and lubricating quality. (It also accounts for the terms *mucopolysaccharide* and *mucoprotein* originally applied to both the proteinpolysaccharides and the glycoproteins.) These glycoproteins contain large numbers of short-chain oligosaccharides. For example, saliva contains a glycoprotein. The composition of the human protein is not known, but a typical animal salivary glycoprotein contains a single polypeptide chain composed of some 5000 amino acid residues, of which 1300 are seryl and threonyl residues. Approximately 800 disaccharide units are attached to these residues, each disaccharide being α-N-acetyl-D-neuraminyl-(2→6)-β-N-acetylglucosamine. The total carbohydrate accounts for approximately 60 per cent of the molecular weight.

Here we have a very long molecule densely covered with short carbohydrate bristles. The polar character of the carbohydrate attracts water, so there is little hydrophobic interaction between the large surfaces. The many negative charges carried by the terminal N-acetylneuraminyl residues repel each other, serving to keep the molecule extended and overcoming any tendency for them to stack up. The sheath of water and their tendency to keep apart enable them to lubricate.

Many, if not most, of the proteins secreted by cells and retained within the body are also glycoproteins. These proteins usually contain much less carbohydrate than do the glycoproteins of mucus. The predominant speculation, and it seems very reasonable, is that the attached carbohydrate is necessary for secretion. The Golgi apparatus is probably responsible for packaging these proteins for export, and there ought to be some structure enabling the organelle to discriminate between the enzymes that have to remain within the cell and the proteins that are supposed to go.

It isn't necessary for our purposes to go over what is known about the carbohydrate moiety of these proteins—we have seen the basic structures, except for branching, and we shall shortly consider an illustrative example of this. Nor is it necessary to attempt to list all of the known glycoproteins; we can see the diversity of functions that are represented by the following examples of secreted proteins that have carbohydrate "prosthetic groups:"

PROTEIN	FUNCTION	SECRETED BY
immunoglobulin IgG	antibody	lymphocytes, plasma cells
fibrinogen	clot formation	liver
transferrin	iron transport	liver
thyrotropin	hormone	anterior pituitary
ribonuclease B	enzyme	pancreas
ferrooxidase	enzyme	liver

Finally, let us consider a glycoprotein, α_1-acid glycoprotein (orosomucoid), that occurs in the blood plasma to the extent of about 0.75 mg/ml. The function of this protein is not known, despite many efforts to follow its fluctuations in various pathological states and to determine what compounds, if any, are being transported by it. We only mention it to show the current notion of the structure of its carbohydrate side chains as an illustration of how complex they can become (Fig. 24-10). Perhaps this will make us a little more humble as we watch the struggles of those who undertake the determination of glycoprotein structure.

THE STRUCTURAL LIPIDS

As we noted in discussing the organization of mitochondria (p. 179), biological membranes contain a core of lipid. According to the original hypothesis of membrane organization proposed by Danielli, a sheet of lipid made from the hydrocarbon tails of phospholipids as well as other types of lipid is sandwiched between layers of protein. Many now believe the membranes are not continuous sheets, but are made of structural units. However, no one disputes the essential element of the Danielli hypothesis: Much of the core of membranes is composed of lipids, and the non-polar environment created in this way is an important part of the transport processes by which concentration gradients are created and maintained across the membranes.

We shall be concerned here with the lipids that occur in membranes and with their relatives occurring elsewhere. As is the case with the structural polysaccharides, structural lipids and related compounds are also involved in transport func-

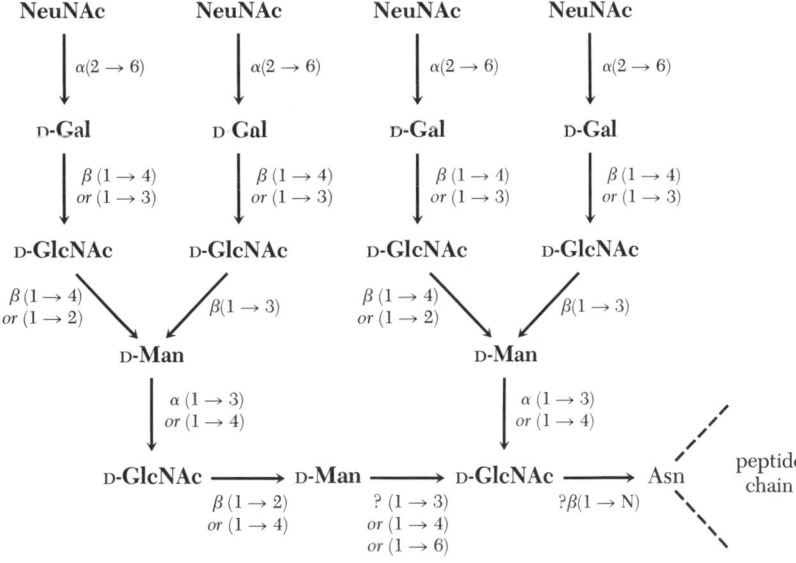

NeuNAc = N-acetyl-D-neuraminate
Gal = galactose
GlcNAc = N-acetylglucosamine
Man = mannose

Figure 24-10 A possible structure for the polysaccharide portion of α_1-acid glycoprotein from human blood, as proposed by Wagh, Bornstein, and Winzler (J. Biol. Chem., *244:* 658 [1969]).

tions, and we have already noted the occurrence of phospholipids in the blood lipoproteins used for moving triglycerides from one tissue to another.

We shall also see that the structural lipids have a variety of complex structures, and we are not yet in a position to say why. Our main concern now is to get the feel of the sort of structures that occur so as to be able to interpret new information as it appears. First, let us examine a tabulation of the lipid content of various membranes (after E. D. Korn in *Feder. Proc., 28:* 7, [1969]).

MEMBRANE	LIPID/PROTEIN	CHOLESTEROL/ POLAR LIPIDS	PRINCIPAL POLAR LIPIDS
	g per g	*mole per mole*	
myelin	4	0.7–1.2	Cer, PE, PC
plasma membrane:			
liver	0.7–1.0	0.3–0.5	PE, PC, PS
intestinal villi	0.22	0.5–1.2	
endoplasmic reticulum	0.8–1.4	0.03–0.08	PC, PE, Sph
(ribosome-free)			
mitochondrial:			
outer	0.8	0.03–0.09	BPG, PC, PE, Plas
inner	0.3	0.02–0.04	
retinal rod	0.7	0.13	PC, PE, PS

BPG = 1,3-*bis*phosphatidylglycerol	PC = phosphatidylcholine
Cer = cerobrosides	PE = phosphatidylethanolamine
Plas = plasmalogens	PS = phosphatidylserine
	Sph = sphingomyelin

We can see from the second column that lipids are indeed major components of all of the examples of membranes; they are predominant in myelin, which is the sheath around some nerves and which appears to be made by wrapping a membrane around and around in a spiral. The membrane lipids contain cholesterol and a variety of other compounds with a greater number of polar groups. From the third column we see that the outer membranes of cells, including myelin and the plasma membranes, contain roughly comparable amounts of cholesterol and the more polar lipids. The membranes of the internal organelles have much lower amounts of cholesterol. The final column shows the various kinds of polar lipids that predominate in the various membranes. We have already discussed the phosphatidyl derivatives of choline, ethanolamine, and serine in Chapter 18, and we shall say more about the others named in the table. Let us now turn to a more detailed consideration of cholesterol.

Cholesterol

Cholesterol is among the least polar of the lipids, with its 27 carbon atoms in a structure containing only one double bond and one hydroxyl group. About 0.2 per cent of the total body weight is cholesterol. The bulk of the total is used for structural purposes; nearly one-quarter is in the myelin and other insulating structures of the central nervous system, and another one-fifth is in the skeletal muscles even though the concentration per weight of muscle is low. The highest concentration is found in the adrenal cortex—some 10 per cent of the total weight —where it is used as a precursor of the steroid hormones, but the weight of the adrenal glands is so small that this only accounts for 1 per cent of the total body content. We have also seen that cholesterol is a constituent of the lipoproteins circulating in the blood. Most of the cholesterol in the blood is in the form of esters of fatty acids, which are formed by transfer of a fatty acid residue of lecithin to the single hydroxyl group of cholesterol within the blood, or by transfer from acyl coenzyme A in the liver. Tissues actively metabolizing cholesterol, such as the adrenal cortex and the liver, appear to hold part of the compound as the esters. Brain also contains substantial amounts of the esters during maturation, but nearly all of the cholesterol in adult brains is in the free form.

SYNTHESIS.   It is likely that all of the tissues of the body are able to synthesize cholesterol for internal use; the liver is more active than the others because it also produces cholesterol for export. The starting material is 3-hydroxy-3-methylglutaryl coenzyme A, which we encountered as an intermediate in the metabolism of leucine and in the formation of acetoacetate by the liver and kidneys. However, most of the compound used to form acetoacetate is made in the mitochondria of the liver and kidneys, whereas that which is used as a precursor of cholesterol is made on the endoplasmic reticulum, where the other enzymes of cholesterol synthesis are located.

The initial step is the reduction of the thiol ester group to an alcohol group, forming *mevalonate* and liberating coenzyme A (Fig. 24-11). This is an unusual double reduction, comparable to the double oxidation of UDP-glucose to UDP-

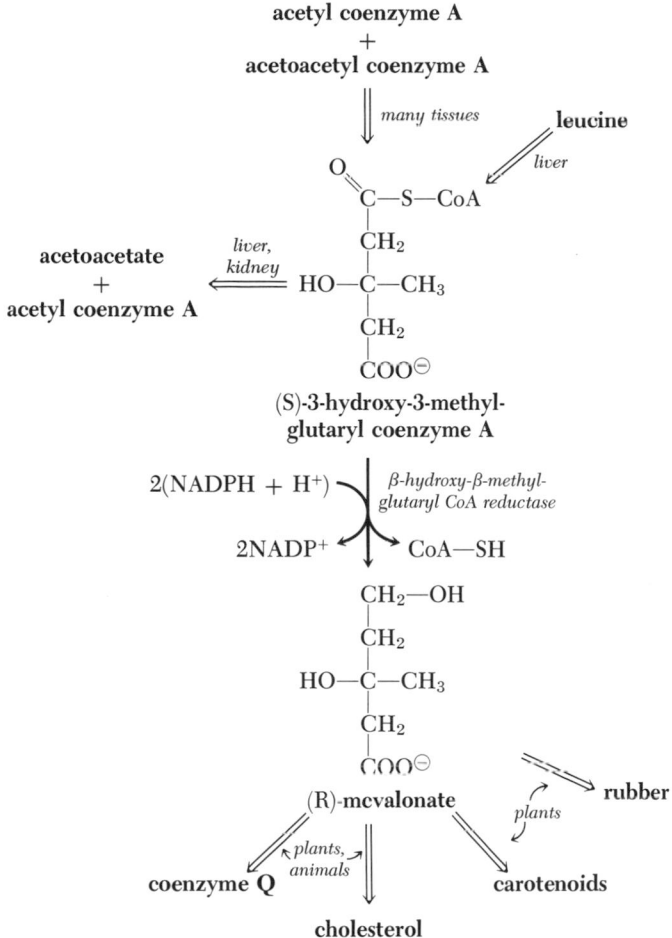

Figure 24-11 (R)-Mevalonate is the precursor of steroids and coenzyme Q in both plants and animals, and of carotenoids and rubber in plants. This 6-carbon acid is formed by reduction of 3-hydroxy 3 methylglutaryl coenzyme A, which is used for this purpose in a variety of tissues as well as acting as a precursor of acetoacetate in the liver and kidney.

glucuronate in that a single enzyme is involved and no aldehyde intermediate has been demonstrated. The simultaneous cleavage of the thiol ester bond makes the reaction irreversible. Feeding cholesterol shuts off cholesterol synthesis, and there is some evidence that the effect is due to slower formation of mevalonate, either through an indirect negative feedback or metabolic adaptation of the reductase, or both. This would seem completely reasonable were it not for the fact that mevalonate is also a precursor of coenzyme Q, and the formation of coenzyme Q is not impaired by cholesterol feeding.

The next steps involve the successive action of three kinases (Fig. 24-12). The first two build a pyrophosphate group (it isn't clear why a pyrophosphate group

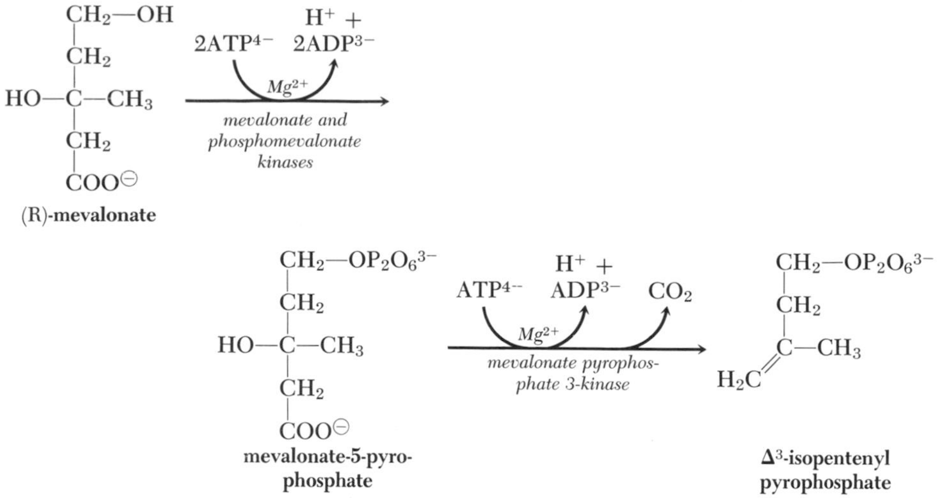

Figure 24-12   The steroids are polymers of prenyl groups. In order to form this 5-carbon unsaturated branched-chain group, mevalonate is phosphorylated three times. The final phosphorylation creates an unstable intermediate on the enzyme that simultaneously loses the additional phosphate group and CO_2.

isn't transferred in one step), and the last phosphorylates the 3-hydroxyl group. However, the final 3-phospho compound isn't released from the enzyme—there is a concerted cleavage of the phosphate group and decarboxylation to form the unsaturated product, Δ^3-isopentenyl pyrophosphate, which has the same carbon skeleton as isoprene. A number of natural products, including rubber and the carotenes in plants and coenzyme Q and the steroids in plants and animals, can formally be regarded as polymers of isoprene (with appropriate oxidations and reductions), and Δ^3-isopentenyl pyrophosphate is in fact the precursor of all of them.

Part of the Δ^3-isopentenyl pyrophosphate is isomerized to the Δ^2 compound to provide a tail upon which to successively add two Δ^3-isopentenyl groups. Each C—C condensation involves the loss of pyrophosphate, and the result is the 15-carbon alcohol, *farnesol*, in the form of a pyrophosphate ester (Fig. 24-13). The next two reactions involve the reductive condensation of two molecules of farnesyl pyrophosphate to form the 30-carbon hydrocarbon, *squalene*, followed by an oxidation that results in a concerted ring closure, thereby forming *lanosterol* (Fig. 24-14). This remarkable series of reactions causes the conversion of six 5-carbon isopentenyl residues to a 30-carbon sterol.

The final stages of cholesterol synthesis involve the removal of three of the side methyl groups, reduction of the side-chain double bond, and isomerization of the remaining double bond (Fig. 24-15).

CHOLESTEROL AND ATHEROSCLEROSIS.   A basement membrane lies beneath the endothelial cells lining the bore of an artery. As people grow older, nodules or plaques tend to develop at scattered locations along their arteries, due to the deposition of lipid in the membrane. The deposition and the secondary changes resulting from disruption of the normal architecture of the vessel cause loss of elasticity and, more importantly, cause an obstruction of blood flow. The obstruction from the plaques themselves may be serious, but there is also a tendency for blood

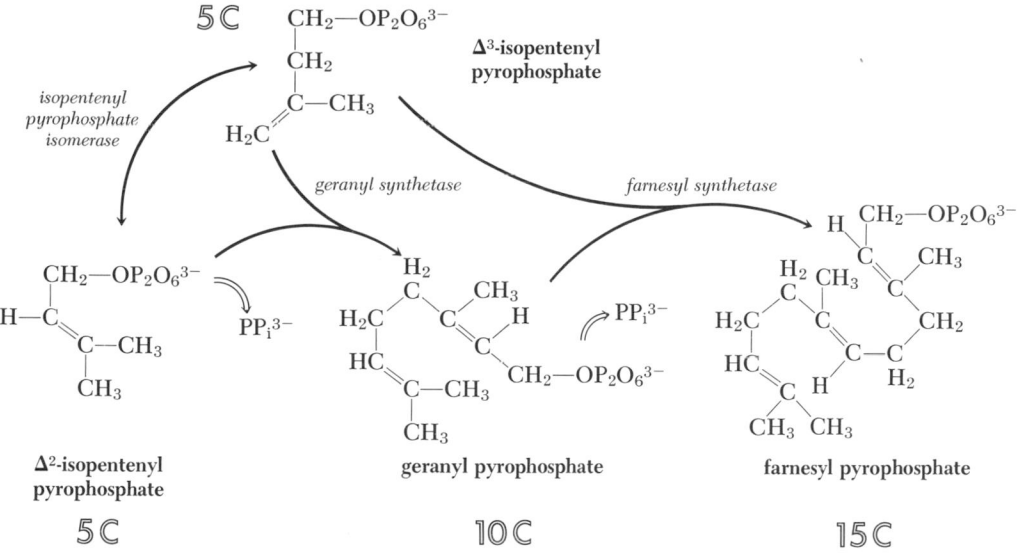

Figure 24-13 Δ^3-Isopentenyl pyrophosphate formed from mevalonate is the primary donor of prenyl groups. One molecule is converted to the Δ^2 isomer to provide a foundation upon which to add two more prenyl groups, creating the 15-carbon farnesyl pyrophosphate. The condensations are driven by the loss of pyrophosphate.

clots, or thrombi, to grow at the points of obstruction. The effects may be gradual, or a clot growing in a large vessel may be dislodged, travel downstream, and suddenly obstruct a smaller but still critical vessel.

Cholesterol and its esters compose a large fraction of the lipid in atheromatous plaques. There is also considerable evidence of a positive correlation between the concentration of circulating cholesterol and triglycerides and the severity of atherosclerosis. Although there is no direct proof that the elevated concentration of cholesterol causes more severe atherosclerosis, many clinicians have advocated dietary measures designed to reduce the concentration as a means of delaying the onset of atherosclerosis, hopefully until the patient is old enough to die of other causes. These measures usually consist of restricting the total fat intake and shifting the nature of the dietary fat from the saturated or short-chain fatty acids found in meat and milk to the polyunsaturated fatty acids that are more abundant in vegetable oils.

Here then is the basis for the public interest in "cholesterol count," safflower oil, and the like. How rational are these measures? Not too bad, really. If A causes B and one removes A, then one prevents B. Suppose A doesn't cause B, but C causes both A and B. If one prevents A, there is at least a chance one is thereby removing the cause of B as well. So long as the measures taken are harmless in themselves, nothing is lost, and something may be gained. In the case at hand, a shift from animal fat to vegetable fat not only decreases the dietary intake of cholesterol, but it also causes less to be synthesized, for reasons that are not clear. We shall see in Chapter 27 that moderate adjustments of the usual American diet in this way are probably not harmful for other reasons.

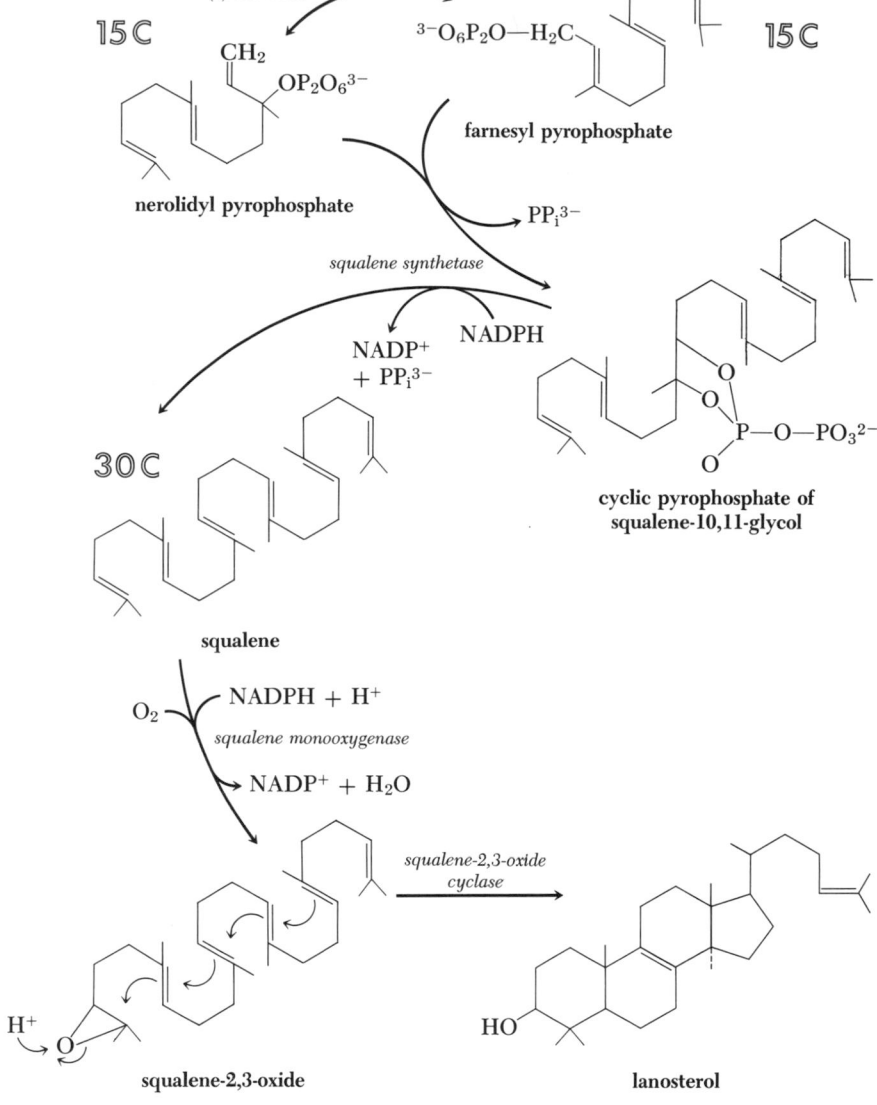

Figure 24-14 One molecule of farnesyl pyrophosphate is converted to nerolidyl pyrophosphate by the shift of a double bond, and nerolidyl pyrophosphate then condenses with a second molecule of farnesyl pyrophosphate. One of the pyrophosphate groups is lost during the condensation, which produces the cyclic pyrophosphate diester of the 30-carbon compound, squalene-10,11-glycol. This compound is then reduced to squalene. Squalene is oxidized to form the 2,3-oxide, and this derivative is attacked at the oxygen ring, which precipitates a concerted shift of electrons causing a condensation that forms a 30-carbon sterol, lanosterol.

MECHANISM OF DEMETHYLATION:

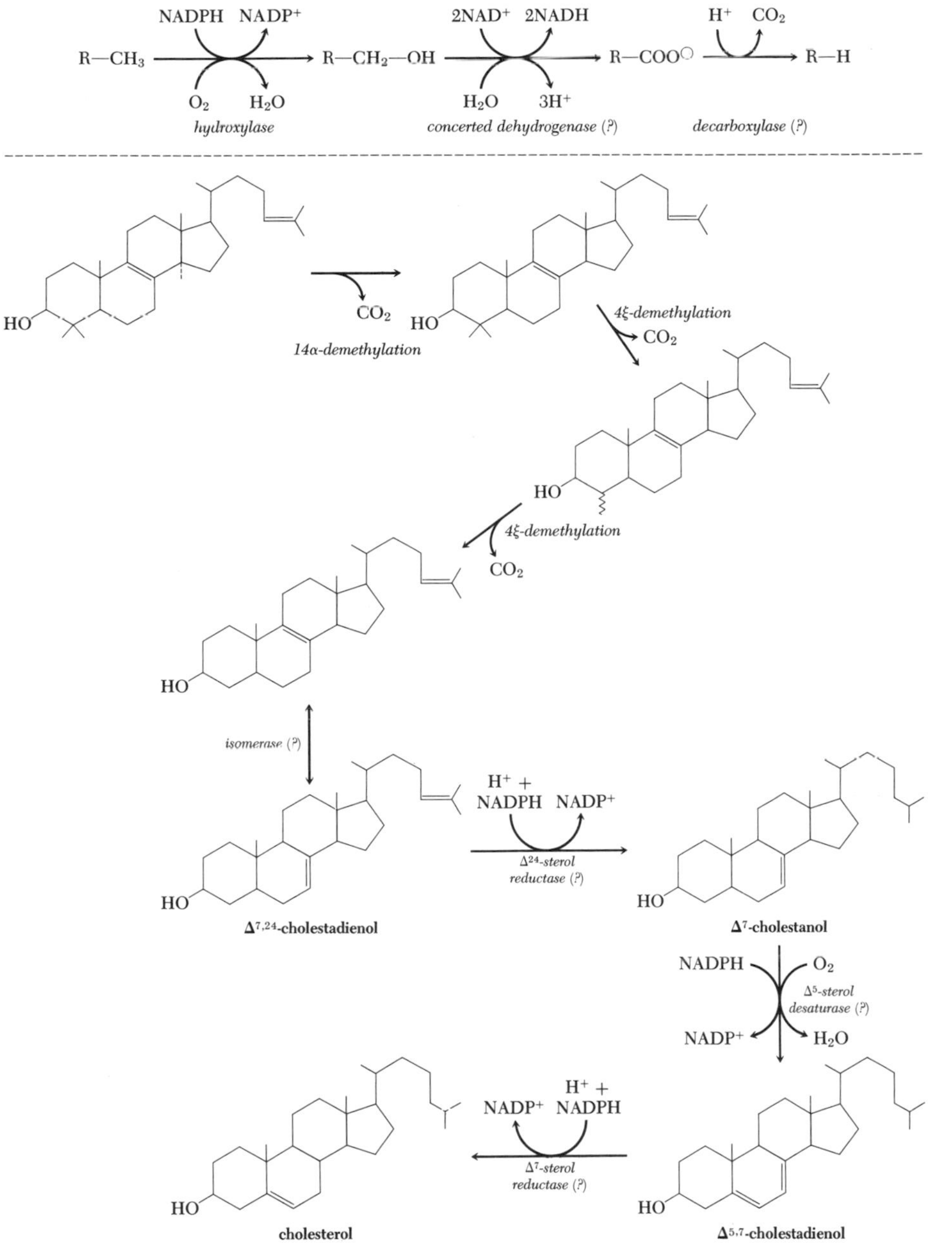

Figure 24-15   In order to convert lanosterol to cholesterol, three methyl groups must be removed. Demethylation is believed to involve the steps shown at the top, in which the initial reaction is a hydroxylation to form an alcohol that can be attacked by dehydrogenases to produce a carboxylic acid, followed by removal of the carboxyl group as CO_2. The later steps involve a reduction of the double bond in the side chain and a shift of the double bond in the ring through an oxidation followed by a reduction. The intermediate in the latter steps, $\Delta^{5,7}$-cholestadienol, is also a precursor of vitamin D.

The Polar Lipids

We have seen that the phospholipids are important elements of membrane structure because they contain the long hydrocarbon side chains of fatty acid residues in combination with regions of electrostatic charge contributed by the phosphate and nitrogenous base constituents. These phosphatidylcholine, ethanolamine, or serine molecules act as links between areas of hydrophobic and hydrophilic nature.

Membrane lipids include other types of molecules capable of performing similar functions. In each case, fatty acid residues are attached to structures contributing polar groups. Most of the lipids are phosphatidates, but carbohydrate residues are the polar groups in some. We don't know the exact purpose of the variation in the nature of the polar substituents, but it presumably enables attachment of different proteins or carbohydrates on the membrane surface. Let us now consider some of the particular kinds of polar groups that occur.

PHOSPHATIDYLGLYCEROLS.   The phosphatidate group contains a phosphate and two fatty acid residues attached to glycerol, but compounds exist in which the phosphate is also attached to still another glycerol residue. The most abundant examples in animals are the *cardiolipins* that occur in mitochondrial membranes; a cardiolipin is a 1,3-*bis*phosphatidylglycerol (Fig. 24-16). The name comes from their discovery in cardiac muscle, which has a high concentration of mitochondria. (We use the plural in talking about these and other phospholipids because there are several variants, differing in the nature of the fatty acid residues but having the same basic structure.) The starting materials for the synthesis of cardiolipins and some other polar lipids are *CDP-diglycerides* that act as donors of phosphatidyl residues. (CDP-diglycerides are formed by a straightforward pyrophosphorylase reaction of CTP and phosphatidates.) In this case, a phosphatidyl unit is transferred to glycerol-3-phosphate, forming 1-phosphatidylglycerol-3-phosphate, the 3-phosphate is hydrolyzed, and a second phosphatidyl unit is transferred to the now-free 3-hydroxyl.

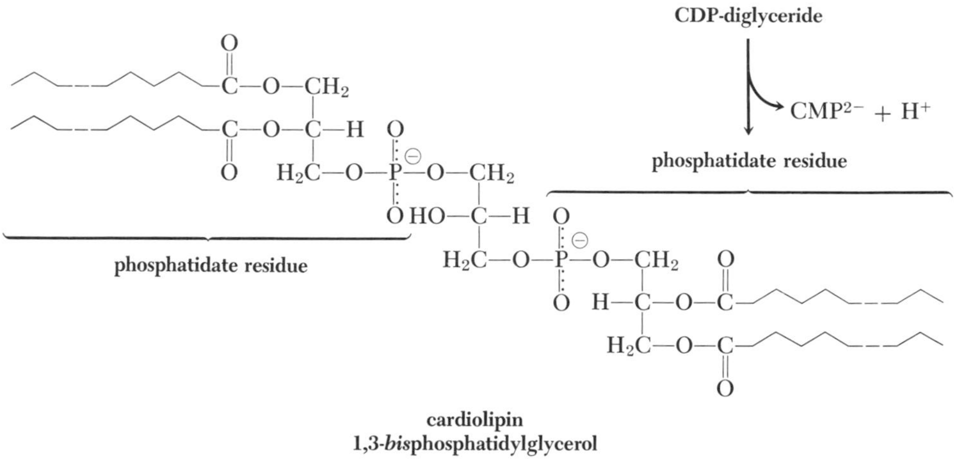

cardiolipin
1,3-*bis*phosphatidylglycerol

Figure 24-16   The cardiolipins are *bis*-phosphatidyl glycerols, formed by transfer of the phosphatidate group from CDP-diglycerides onto the glycerol skeleton. The first acceptor is glycerol-3-phosphate, and the phosphate is removed after the initial phosphatidyl group is transferred to C-1, thereby making it possible to transfer the second phosphatidyl group to C-3.

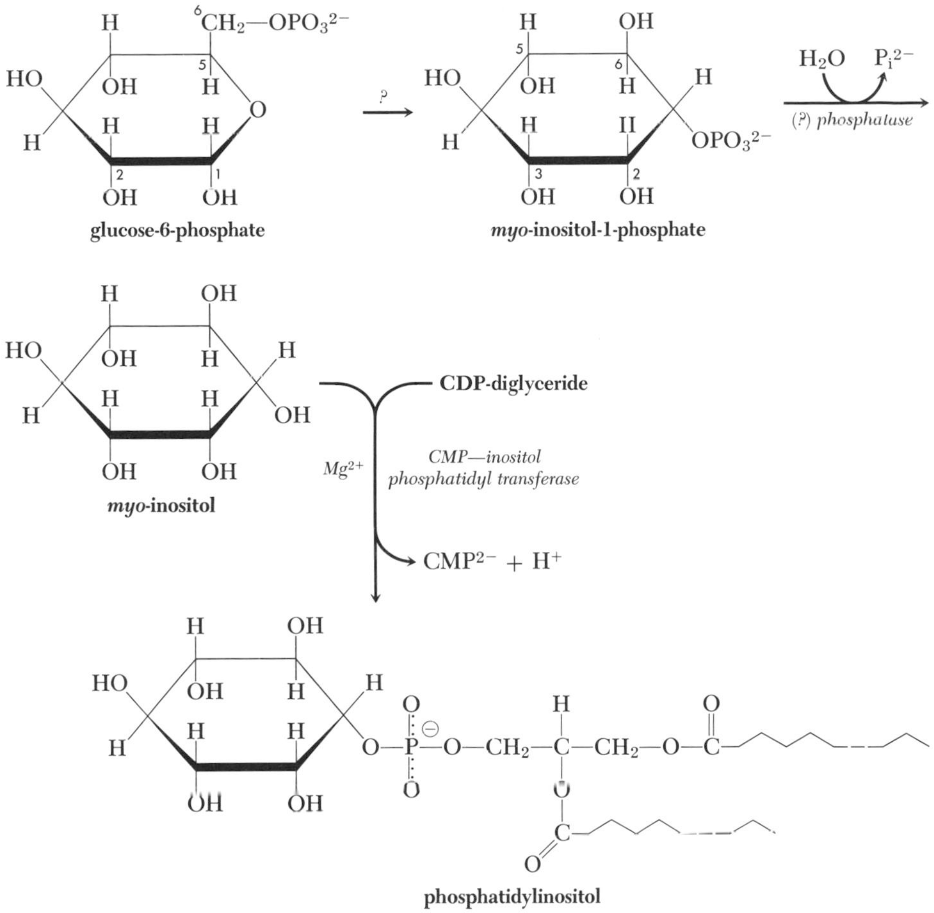

glucose-6-phosphate

myo-inositol-1-phosphate

myo-inositol

CDP-diglyceride

Mg^{2+}

CMP—inositol
phosphatidyl transferase

$CMP^{2-} + H^+$

phosphatidylinositol

Figure 24-17 Inositol residues are created from glucose-6-phosphate by some unknown reaction that appears to involve an intramolecular aldol condensation. No net loss or production of atoms is involved because glucose and inositol are isomers. However, the nomenclature changes, so that what was C-6 of glucose becomes C-1 of inositol. Phosphatidyl groups are transferred from CDP-diglycerides to form phosphatidylinositols.

Phosphatidylinositides are compounds containing residues of *inositol*, a structural isomer of glucose that can be made from glucose-6-phosphate as inositol-1-phosphate. Phosphatidylinositols are made by transfer of a phosphatidyl group from CDP-diglycerides to free inositol. One or more of the free hydroxyl groups are then phosphorylated (Fig. 24-17). The phosphatidylinositides occur in a variety of tissues, and are especially concentrated in the brain. The phosphate groups are turned over rapidly in comparison with those of other lipids, which makes one suspect a functional role, but suspicion is all we have.

Plasmalogens are analogues of phosphatidylcholine and phosphatidylethanolamine in which one of the fatty acid ester groups is converted to a vinyl ether (Fig. 24-18). They are named as *phosphatidal* derivatives of choline or ethanolamine because the structure could be obtained by condensation of the aldehyde analogue of a fatty acid with the hydroxyl group of glycerol. However, the vinyl ether group

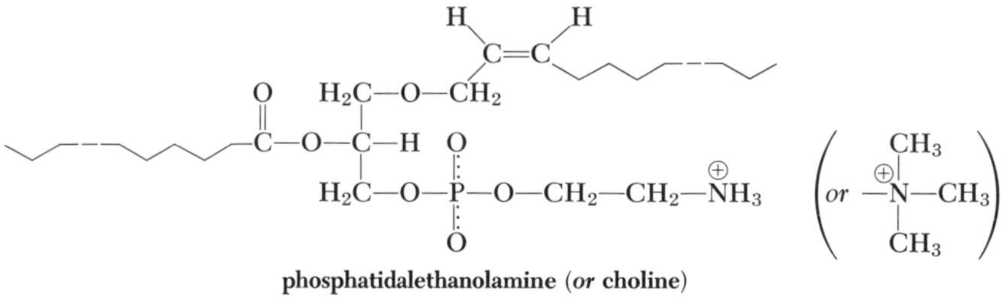

phosphatidalethanolamine (or choline)

Figure 24-18   Phosphatidalethanolamines and cholines are analogues of the corresponding phosphatidyl compounds in which the terminal acyl group has in effect been reduced to an aldehyde (bound as a hemiacetal) followed by dehydration to form a double bond.

is believed to be formed by reduction of a diglyceride or by a reduction concerted with the synthesis of the diglyceride. The resultant plasmalogenic diglyceride reacts with CDP-choline (or ethanolamine) to form phosphatidalcholine (or ethanolamine) in the same way that ordinary diglycerides react in the formation of the phosphatidylcholines.

Sphingolipids are a group of complex lipids containing *sphingosine*, which is a compound with a hydrocarbon tail, two hydroxyl groups, and an amino group (Fig. 24-19). Sphingosine is obtained by a decarboxylating condensation of serine with

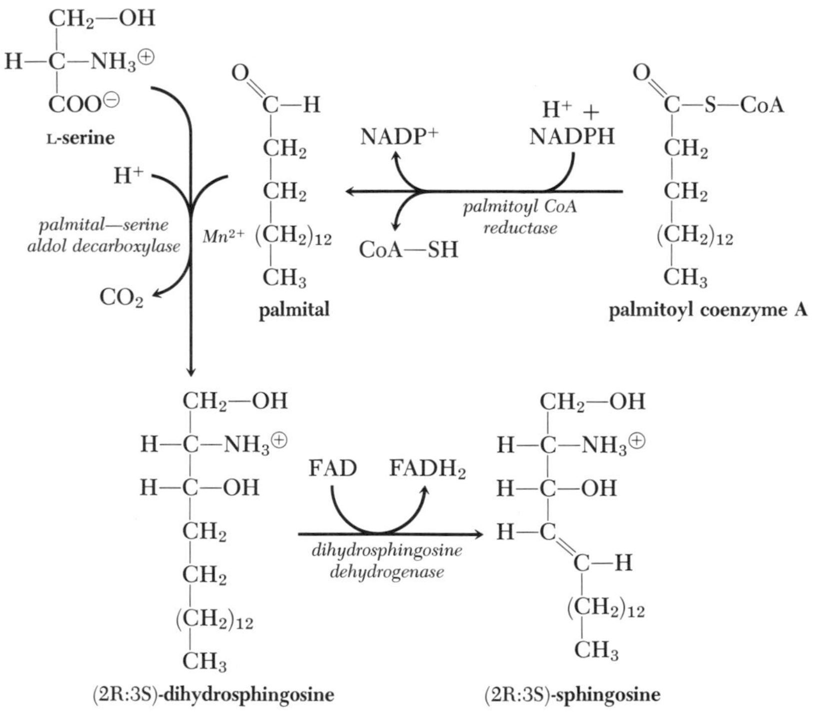

Figure 24-19   The precursors of sphingosine are L-serine and palmital, which is formed by reduction of palmitoyl coenzyme A. These precursors condense with a concomitant loss of the carboxyl group of serine, and the hydrocarbon chain of the resultant dihydrosphingosine is then oxidized to create sphingosine.

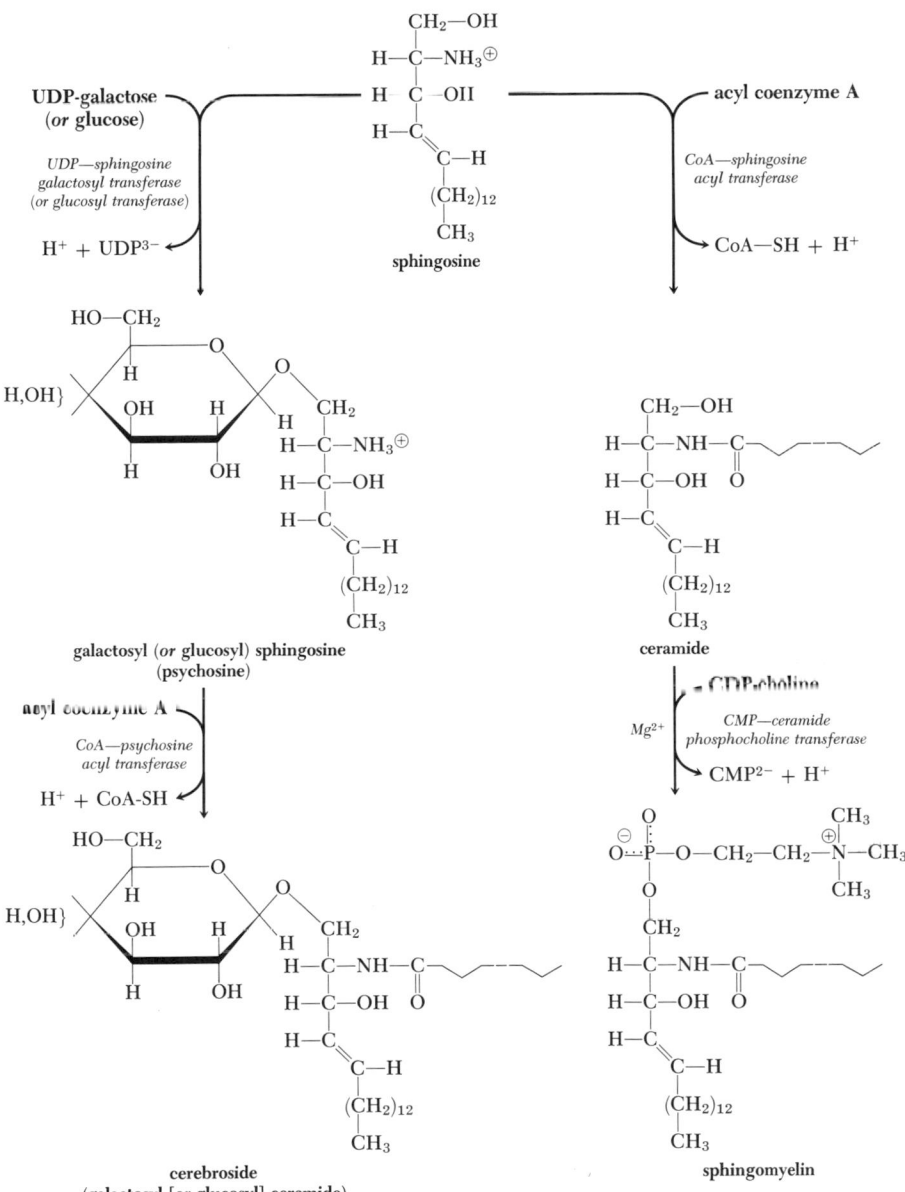

Figure 24-20 Sphingosine is the precursor of a variety of structural lipids. Cerebrosides are formed by first transferring a galactosyl or glucosyl residue to the C-1 hydroxyl group, followed by acylation of the amino group. Sphingomyelins are formed by an initial acylation of the amino group, followed by transfer of a phosphorylcholine group to the C-1 hydroxyl.

palmital (the aldehyde form of palmitate) followed by oxidation of the resultant dihydrosphingosine. The compound invariably occurs in lipids with a fatty acid residue attached to the nitrogen atom, and the N-acylsphingosines are *ceramides*. The various sphingolipids are combinations of different compounds with the 1-hydroxyl group of ceramides. *Sphingomyelins* are formed by the transfer of a phosphorylcholine group from CDP-choline to ceramides. *Cerebrosides* are formed by the transfer of glycosidyl groups from UDP-sugar derivatives to sphingosine, followed by acylation to complete the ceramide residue (Fig. 24-20). Cerebrosides containing galactose account for 4 per cent of the weight of the brain. Other tissues contain cerebrosides with glucosyl groups or even small oligosaccharides of glucose, galactose, and N-acetylgalactosamine.

Gangliosides are another type of sphingolipid that are more abundant in neurons, although they are found in other tissues. They are formed by building branched-chain oligosaccharides on the 1-hydroxyl group of ceramides. The branch is made by the addition of N-acetyl- or N-glycolylneuraminyl residues. A typical example is shown in Fig. 24-21.

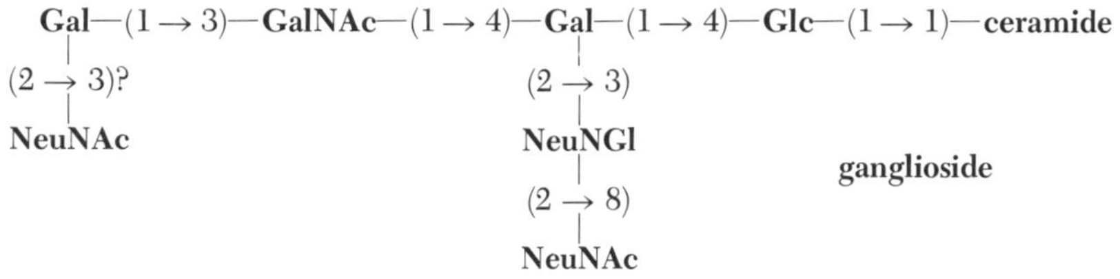

$$Gal—(1 \rightarrow 3)—GalNAc—(1 \rightarrow 4)—Gal—(1 \rightarrow 4)—Glc—(1 \rightarrow 1)—ceramide$$

$$(2 \rightarrow 3)? \qquad\qquad (2 \rightarrow 3)$$

NeuNAc   NeuNGl   **ganglioside**

$$(2 \rightarrow 8)$$

NeuNAc

Gal = D-galactose	**NeuNAc** = N-acetyl-D-neuraminate
GalNAc = N-acetyl-D-galactosamine	**NeuNGl** = N-glycolyl-D-neuraminate
Glc = D-glucose	

Figure 24-21   The arrangement of carbohydrate residues in gangliosides. (After Kaufman, Basu, and Roseman, J. Biol. Chem., *243:* 5804 [1968]).

BONE

The bones are specialized developments of connective tissue. They contain the same arrangement of collagen fibers embedded in proteinpolysaccharides, but with the addition of interspersed deposits of calcium phosphates, and we shall concentrate our attention on the turnover of these inorganic salts. The bones are very much alive. They contain a complex network of tunnels to carry the blood vessels that supply the cells within the solid matrix. The formation of a bone involves initial growth of deposits of calcium salts followed by intensive reworking—removal of deposits here and addition of deposits there—to create the mature architecture. This reworking does not cease when a bone acquires adult form; it only slows.

The turnover of bone tissue enables adjustment of form to the kind of loads an individual routinely imposes on his skeleton and also enables prompt repair of fractures, but it also serves another important purpose as a reservoir of calcium ion. The concentration of calcium in blood plasma is controlled closely; normal values lie between 2.1 and 2.6 millimoles per liter. Part of the ion is chelated with proteins, so the concentration of the free ion is near 1.2 millimoles per liter. It is this value that determines the balance with the tissues, since only the free ion can move into membranes, and it may be even more finely regulated than the total calcium.

The close control appears to be necessary because of the importance of calcium ion in the function of tissues generally. We have mentioned its involvement in muscular contraction, and a fall in calcium concentration to 50 per cent of its normal value will cause tetany. Calcium also appears to be required for normal function of most membranes. The bones of a 70-kg man contain around 1.2 kg of calcium (30 moles), which is an ample supply for maintaining the 0.8 gram (20 millimoles) found in the body fluids.

Deposition of the Mineral Phase

Calcium phosphates have a low solubility, and one might expect the formation of bone to represent precipitation to equilibrium concentrations as expressed by an ordinary solubility product, $[Ca^{2+}][PO_4^{3-}]^x = K_s$, in which x is the ratio of phosphate to calcium in the precipitate. There is evidence that this kind of a relation determines the possibility for bone formation—if the blood calcium concentration is low, the phosphate concentration must be high for deposition to occur, and *vice versa*—but there is more to it than this. The concentration of the ions may be above the saturation level without the occurrence of calcification.

The cells forming bone in ordinary connective tissues (chondrocytes or osteoblasts) somehow create centers of nucleation outside the cells on which ions attach to create crystal lattices having the form of *hydroxyapatite*. The repeating unit in the hydroxyapatite lattice has the composition $Ca_{10}(PO_4)_6(OH)_2$, but the growing crystal in bone also has other ions, especially carbonate, substituted at some of the positions. Some believe that the calcium and phosphate ions are initially concentrated near the center of nucleation in the form of small amorphous precipitates of calcium phosphate, with the ions migrating onto the more stable crystal lattice. (Increasing amounts of amorphous material appear as a bone ages, reaching as much as 40 per cent of the total mineral.)

The crystals do not grow beyond a few unit cells in diameter; typical sizes are five by a few hundred nanometers—too small to be visible in a microscope. The rigidity of bone therefore does not come from a continuous mineral structure, but from the arrangement of minute mineral crystals in a matrix of collagen and proteinpolysaccharides.

The mechanism by which calcium and phosphate ions are concentrated in specific locations continues to be an enigma. Bone contains a high concentration of citrate, which chelates calcium, and of alkaline phosphatase, an enzyme hydrolyzing various phosphate esters. Attempts have been made to construct hypothetical mechanisms invoking these compounds; none to date is convincing.

Calcium Transport and Vitamin D

The movement of calcium ion through cells, whether they be in the intestinal mucosa, bone, or other tissues, requires a specific transport protein. The only purified example of the protein at hand is obtained from chickens, and it has a molecular weight of 28,000 and binds one calcium ion per molecule of protein.

$\Delta^{5,7}$-cholestadienol

cholecalciferol
(vitamin D_3)

The movement of calcium ion through cells also is dependent upon a supply of vitamin D. The physiological form of vitamin D is *cholecalciferol* (vitamin D_3), which is formed in the skin by the action of ultraviolet light on $\Delta^{5,7}$-cholestadienol, which we previously encountered as an intermediate in cholesterol synthesis. The photochemical reaction is a cleavage of the B ring. We see that calling the compound a vitamin is somewhat a misnomer in that a dietary supply is not necessary if sufficient sunlight falls on the skin. (Calciferol, or vitamin D_2, differs in having a double bond and another branched methyl group in the side chain; it is the usual commercial form of the vitamin, prepared from a readily available plant sterol, ergosterol, by radiation.)

There is now strong evidence, although not absolute proof, that the requirement for vitamin D hinges upon the requirement for the transport protein. Cholecalciferol is oxidized by the liver to form 25-hydroxycholecalciferol. 25-Hydroxycholecalciferol (or some related polar metabolite) has been shown to enter intestinal mucosal cells and associate with the nuclear chromatin. The promotion of calcium transport by vitamin D is known to be inhibited by actinomycin, an inhibitor of mRNA formation. 25-Hydroxycholecalciferol has also been shown to be effective in promoting calcium mobilization from bone cells in tissue culture, which cholecalciferol and calciferol are not. Taking all of these facts together, it appears that the biological action of vitamin D hinges upon its conversion to 25-hydroxycholecalciferol, which in turn has its effect by acting as a promoter of mRNA transcription for synthesis of the calcium transport system.

There is also some evidence for the involvement of vitamin D in phosphate uptake by the small intestine, but the mechanism for this effect is not known.

VITAMIN D DEFICIENCY.   Infants and young children who are not exposed to ultraviolet light and who also lack a substitute source of vitamin D from the diet develop *rickets*, a condition characterized by deficient calcification of the bones. Now, cholecalciferol is necessary for the *mobilization* of calcium from the bones as well as for absorption from the gut, and one might think that the deficient children would be depositing too much calcium. However, the young have relatively high requirements for calcium because they are forming new bones, and the effects of improper absorption from the gut with resultant calcium deficiency is the predominant effect in rickets. Calcification of the bones is arrested and deformities appear because of the inability of the remaining cartilaginous structures to resist the increasing mechanical stresses of the growing body. (There is an additional mineral imbalance from the action of parathyroid hormone, which we shall mention below.)

Vitamin D deficiency is rare in adults; when it does appear, the effects of negative calcium balance again predominate, with the development of soft, calcium-deficient bones, an *osteomalacia.*

The relationship between lack of sunlight and the incidence of rickets is quite clear. It is said that all of the children in a New York hospital during the winter months had some signs of rickets in the days before dietary supplementation, whereas few had the signs in the summer. The condition was mainly confined to the very young, who had the least exposure to daylight.

Because of the prevalence of the condition, manufacturers were encouraged to fortify common foods such as milk, bread, and margarine. In addition, mothers were urged to give routine daily doses of the vitamin, principally as modified fish liver oils such as oleum percomorpheum (212 micrograms, or 8500 International Units, per gram of oil). Infants require something in the neighborhood of 5 micrograms per day if they synthesize none, and 10 micrograms (400 I.U.) is usually specified to be safe.

The result of the program was a rapid decline in the incidence of rickets. However, cases of rickets still appeared and the pressure for more widespread use of fortified foods continued. The apparent consequences of this effort to supply vitamin D to those remaining deficient by increasing the intake of the whole population have tarnished the luster obtained from the nearly complete victory over rickets.

It has long been known and repeatedly stated in textbooks that large doses of vitamin D are toxic. The few who stopped to think saw that the intake of some children could be approaching the toxic level, since they had a high content of the vitamin in the diet, plus the amount synthesized internally during the summer, plus the large doses easily given from preparations commonly found in most households. Unfortunately, the few who saw were not heeded until frank cases of vitamin D poisoning were finally recognized in a number of children. Most of the severe cases turned out to be children unusually prone to hypercalcemia, but further inquiry suggested the existence of marginal cases in children of normal response. The result has been a quiet de-emphasis on fortification of foods with vitamin D, a drop in the amount of vitamin D in preparations sold without prescription, and a plea to physicians to make some effort to find out how much an individual is already ingesting before prescribing more.

ENDOCRINE REGULATION OF CALCIUM TRANSPORT. The mobilization of calcium from the bones and its deposition are regulated by two hormones. The *parathyroid* glands secrete a peptide hormone in response to a fall in the ionic calcium concentration in the blood. The parathyroid hormone does two things. It causes dissolution of the bone mineral so as to release more calcium and phosphate ions, provided that 25-hydroxycholecalciferol is available to stimulate synthesis of the transport protein. It also slows the reabsorption of phosphate ion from the glomerular filtrate in the kidney, thereby disposing of the phosphate mobilized from the bone and diminishing the tendency to re-precipitate calcium phosphate. There is strong evidence that the effect on the kidney is due to specific receptors in the proximal tubules that cause an activation of adenyl cyclase with a resultant increase in the concentration of 3',5'-cyclic AMP, although it is not clear how this change affects phosphate transport. There is inferential evidence that the same mechanism is responsible for the mobilization of bone minerals. In short, parathyroid hormone may be another example of the variety of hormones that have their effects through localized changes in adenyl cyclase activity.

There is some doubt about the type of cell affected by parathyroid hormone in the bones. The bones contain osteoblasts that actively lay down mineral, osteocytes that are believed to be mature, "resting" forms of the osteoblast series, and osteoclasts that some think are responsible for dissolution of bone mineral. Parathyroid hormone causes a degeneration of osteoblasts and a proliferation of osteoclasts, but these may be secondary adaptations to the changing metabolism. Some think that the hormone acts directly on the osteoclasts and others believe that it activates the osteocytes, which are also believed to have the capability of degrading the bone. (We have concentrated on the turnover of the mineral, but there is an accompanying degradation of the matrix that is shown by increased secretion of hydroxyproline derived from the hydrolyzed collagen.)

The parathyroid gland causes some difficulty for space travelers. When a limb is immobilized, there is extensive loss of mineral; this *osteoporosis* does not occur if the parathyroid gland is removed. However, astronauts still have their parathyroids and mobilize bone minerals extensively, but the duration of travel has not yet been long enough to require corrective measures in flight.

It was only in 1962 that it was discovered that both the thyroid and parathyroid contain cells secreting a separate peptide hormone, *thyrocalcitonin*. (The thyroid is the larger tissue and secretes most of the hormone.) Thyrocalcitonin is released in response to increases in calcium concentration, and it causes deposition of calcium phosphate, with a concomitant increase in the number of osteoblasts and decline in the number of osteoclasts. Thyrocalcitonin controls the upper limit of calcium ion concentration, while parathyroid hormone controls the lower—somewhat analogous to the control of glucose concentration by the opposing effects of glucagon and insulin. Calcium somehow causes an activation of adenyl cyclase in the secreting cells, and the resultant cyclic AMP causes a release of thyrocalcitonin.

The action of thyrocalcitonin on bone is believed by some to be due to an increased phosphodiesterase activity that causes a fall in the concentration of cyclic AMP. This would seem reasonable—parathyroid hormone causes greater synthesis of cyclic AMP and thyrocalcitonin causes greater destruction—if we were positive that the two hormones are acting on the same cells. Some caution is indicated because thyrocalcitonin has an effect in the absence of parathyroid hormone and in the absence of 25-hydroxycholecalciferol. It is possible that they have contrary effects on the same cells, and it is also possible that they both act by the same sort of mechanism, but on different cells.

A patient has recently been discovered with a tumor of the thyroid that is composed of cells that secrete thyrocalcitonin. The quantity being poured out from the metastases of the tumor overrode the compensating increase in parathyroid hormone secretion, causing frequent lowering of the blood calcium concentration to tetanic levels.

Similar regulatory difficulties may occur in rickets. The fall in blood calcium concentration caused both by failure to absorb calcium from the diet and by failure to mobilize it from the bones causes increased parathyroid hormone secretion. This in turn prevents the retention of phosphate by the kidney, and the loss of phosphate causes even less tendency to precipitate calcium phosphate. It will be interesting to see data on the thyrocalcitonin secretion during this sequence of events.

Chronic renal disease also causes problems in calcium regulation. Destruction of the nephrons diminishes the capacity to remove phosphate by glomerular filtration. Each slight increase in phosphate concentration tends to cause increased precipitation of calcium phosphate, and the resultant fall in calcium concentration causes increased secretion of parathyroid hormone. The hormone restores the ion concentrations to normal, but the increased secretion must be maintained to do this, and the amount required increases as kidney failure progresses. The result is a chronic hyperparathyroidism that can lead to extensive loss of bone minerals. A surgeon transplanting a normal kidney into such a patient must be alert for a subsequent hypercalcemia, because the normal kidney will restore phosphate balance at a time when parathyroid hormone is being secreted at excessive levels.

Further reading

Schubert, M., and D. Hamerman: *A Primer on Connective Tissue Biochemistry.* Lea and Febiger (1968).

Masoro, E. J.: *Physiological Chemistry of Lipids in Mammals.* Saunders (1968).

Balazs, E. A., and R. W. Jeanloz, eds.: *The Amino Sugars*, Vols. 1 to 4. Academic Press (1966). The title is misleading; the work includes general discussions of the proteinpolysaccharides and glycoproteins.

Quintarelli, G., ed.: *The Chemical Physiology of Mucopolysaccharides.* Little Brown (1968).

Greenberg, D. E., ed.: *Metabolic Pathways*, 3rd ed., Vol. 2. Academic Press (1968).

McLean, F. C., and M. R. Urist: *Bone*, 3rd ed. University of Chicago Press (1968).

Ginsburg, V., and E. F. Neufeld: *Complex Heteropolysaccharides of Animals.* Ann. Rev. Biochem., *38:* 371 (1969).

Helting, T., and L. Rodén: *Biosynthesis of Chondroitin Sulfate.* I and II. J. Biol. Chem., *244:* 2790 and 2799 (1969).

Fratantoni, J. C., C. W. Hall, and E. F. Neufeld: *The Defect in Hurler's and Hunter's Syndrome: Faulty Degradation of Mucopolysaccharide.* Proc. Natl. Acad. Sci., *60:* 699 (1968).

Spiro, R. G.: *Characterization and Quantitative Determination of the Hydroxylysine-linked Carbohydrate Units of Several Collagens.* J. Biol. Chem., *244:* 602 (1969).

Ito, S.: *Structure and Function of the Glycocalyx.* Federation Proc., *28:* 12 (1969).

Popják, G., et al.: *Biosynthesis and Structure of a New Intermediate Between Farnesyl Pyrophosphate and Squalene.* J. Biol. Chem., *244:* 1897 (1969).

Swindell, A. C., and J. L. Gaylor: *Investigation of the Component Reactions of Oxidative Sterol Demethylation.* J. Biol. Chem., *243:* 5546 (1968).

Korn, E. D.: *Cell Membranes: Structure and Synthesis.* Ann. Rev. Biochem., *38:* 263 (1969).

Wasserman, R. H., R. A. Corradino, and A. N. Taylor: *Vitamin D-dependent Calcium-binding Protein.* J. Biol. Chem., *243:* 3978 and 3987 (1968).

Kowarski, S., and D. Schachter: *Effects of Vitamin D on Phosphate Transport and Incorporation into Mucosal Constituents of Rat Intestinal Mucosa.* J. Biol. Chem., *244:* 211 (1969).

Blunt, J. W., Y. Tanaka, and H. F. DeLuca: *The Biological Activity of 25-Hydroxycholecalciferol, A Metabolite of Vitamin D_3.* Proc. Natl. Acad. Sci., *61:* 1503 (1968).

The Prophylactic Requirement and the Toxicity of Vitamin D. Pediatrics, *31:* 512 (1963). Report of a committee of the American Academy of Pediatrics.

Behrens, O. K., and E. L. Grinnan: *Polypeptide Hormones.* Ann. Rev. Biochem., *38:* 83 (1969). Review of thyrocalcitonin, parathyroid hormone, and insulin.

G. V. Foster: *Calcitonin (Thyrocalcitonin).* New Eng. J. Med., *279:* 349 (1968).

Bricker, N. S., et al.: *Calcium, Phosphorus and Bone in Renal Disease and Transplantation.* Arch. Int. Med., *123:* 543 (1969).

Melvin, K. E. W., and A. H. Tashjian, Jr.: *The Syndrome of Excessive Thyrocalcitonin Produced by Medullary Carcinoma of the Thyroid.* Proc. Natl. Acad. Sci., *59:* 1216 (1968).

RESPIRATORY EXCHANGE AND H+ BALANCE

CHAPTER 25

ARGUMENT

The maintenance of the ionic environment that is necessary for stable structure and biological function of the proteins and other cellular constituents mainly hinges on control of the concentrations of H^+, Na^+, K^+ and water, which concomitantly control the movement of anions such as Cl^- and HCO_3^-. The concentration of H^+ tends to change as a result of the production and excretion of CO_2 and as a result of the overall metabolism of ingested food. The transport of CO_2 from the tissues to the lungs imposes a transient load that could cause large variations in pH, but these variations are minimized by the Bohr effect, in which hemoglobin takes up H^+ as it loses O_2, and by the buffering action of the charged groups on the blood proteins. However, the baseline concentration of CO_2 in the blood can be altered by changing the rate of ventilation in the lungs, and this is used as a means of correcting the H^+ balance of the entire body when it is altered by other factors. These other factors are termed "metabolic changes," and the most important among them is the result of metabolism of foods. The metabolism of sodium salts of carboxylic acids results in the removal of H^+, while the metabolism of ammonium and sulfhydryl compounds results in the formation of H^+.

CO_2 is in equilibrium with carbonic acid and its dissociated products, H^+ and HCO_3^-. Since the relationship between the acid and its products is fixed by the thermodynamics of the compounds, the measurement of the concentration of any two of the components fixes the third, and these measurements with samples of blood provide a means of assessing the H^+ balance of the whole body. One or more of the measurements, frequently in association with measurements of the Na^+ and Cl^- concentrations, are widely used in medicine.

608

The kidney plays a major role in restoring deviations from normal electrolyte balance. It can exchange H$^+$ derived from carbonic acid in the cells for Na$^+$ in the glomerular filtrate derived from the blood, thereby forming bicarbonate in the body and excreting H$^+$ to lower the total body content. If the H$^+$ content is low, the kidney can substitute K$^+$ for H$^+$ in the exchange, allowing bicarbonate to spill out in the urine and permitting the production of more H$^+$ from carbonic acid within the body. At times of continued excess of H$^+$, the kidney also substitutes NH$_4^+$ for Na$^+$, and derives the NH$_4^+$ from the hydrolysis of glutamine or the metabolism of amino acids. Since this removes nitrogen that would otherwise form urea, the normal production of H$^+$ by the conversion of dietary ammonium compounds, such as the amino acids, is diminished.

The same sort of exchange mechanism is used for the regulation of Na$^+$ concentration. Aldosterone, a steroid hormone that is produced by the adrenal cortex in response to a fall in Na$^+$ concentration, causes the synthesis of more of a specific transport protein in the distal tubule of the kidney, thereby promoting increased reabsorption of Na$^+$ with a concomitant passive reabsorption of Cl$^-$ and HCO$_3^-$.

The active absorption of salts by these mechanisms raises the concentration in the cells above the isotonic level, and water from the lumen of the nephron passively diffuses into the cells, thereby concentrating the urine. The final stages of water absorption involve the existence of leaks in the plasma membrane that are created in the presence of cyclic AMP. The concentration of cyclic AMP is in turn dependent upon an activation of adenyl cyclase by a small peptide hormone, vasopressin, that is elaborated from the posterior pituitary in response to a rise in the osmotic pressure or a fall in the total volume of the blood.

In the previous chapter, we dealt with the chemical processes used to maintain the structural architecture of the body. It is equally important for the existence of the cells that the ionic environment be maintained. We are concerned here with the ions that have general effects on macromolecules, and not so much with those that play a direct part in enzymatic reactions. We shall be especially concerned with the concentration of H$^+$, which determines the charge on proteins and on substrates. A shift in the pH can alter the entire metabolic balance through the effects on the rates of enzymatic reactions; it can also alter every function that depends upon the character of macromolecules, including nerve transmission, muscular contraction, and so on. It is not surprising that the pH is controlled rather finely, and that a shift in the H$^+$ balance reflected by a rise in the arterial blood pH from its normal 7.40 to 7.80 causes tetany, or that a fall in the pH to 7.00 causes coma. Either change of this magnitude results in death if it is not corrected.

We shall see that the regulation of H$^+$ balance is intimately connected with the regulation of salt and water balance. The internal ions of cells mainly include the metabolic anions (phosphate and carboxylate compounds), with the bulk of the cations represented by K$^+$, but the concentration of these ions is regulated by exchange with the blood plasma in which Cl$^-$ and some HCO$_3^-$ are the anions and Na$^+$ is the principal cation. The ionic composition of tissues is regulated in a major way by adjustment of the concentration of Na^+ in the blood, which tends to fix the concentration of K^+ within the cells. However, the success of this regulation depends upon a simultaneous regulation of the pH; let us begin our discussion with a consideration of this fundamental adjustment.

CARBON DIOXIDE TRANSPORT

We have seen that the oxidation of foods to CO_2 is the all-important mechanism for producing ATP, and the quantitative balance of the process in the whole body is expressed by the respiratory quotient, with from 0.7 to 1.2 moles of CO_2 formed for each mole of O_2 consumed. The CO_2 that is formed must be transported from the tissues through the blood to the lungs for excretion, and the quantity that is involved is potentially the largest stress on the maintenance of H^+ balance because, as we all know, the exposure of an aqueous solution to CO_2 results in the production of H^+:

$$CO_2(gas) \longleftrightarrow CO_2(solution) \xleftrightarrow{\pm H_2O} H_2CO_3 \longleftrightarrow H^+ + HCO_3^-$$

If blood behaved like pure water, the concentration of H^+ would bounce up and down as the blood passed through the peripheral tissues and back through the lungs, and the amount of variation would be determined by the changing partial pressures of CO_2 in an equilibrated gas phase. The blood doesn't behave like pure water, and the pH doesn't swing wildly during passage from the arteries to the veins. Let us discuss the mechanisms by which this control is achieved.

The CO_2-Bicarbonate Equilibrium

As with all acidic dissociations, the equilibrium relationship between the concentrations of carbonic acid, H^+, and HCO_3^- is defined by an equilibrium constant:

$$K = \frac{[H^+][HCO_3^-]}{H_2CO_3}$$

However, there are some practical problems we have to consider. One of these is that it is difficult to determine just how much carbonic acid is present in a solution compared to the amount of dissolved CO_2 that isn't hydrated. It is therefore customary to lump together the concentrations of dissolved gas and of carbonic acid and treat them as a single entity, much in the same way as we treat ammonia and ammonium hydroxide as a single entity, and this sum is what we shall mean when we speak of [dissolved CO_2].

The second practical problem is that we are dealing with blood, which is a complex tissue, and it is not possible to sort out all of the possible equilibria involving CO_2, such as the combination with amino groups of proteins that occurs to a small extent spontaneously:

$$R-NH_2 + CO_2 \rightleftharpoons R-NH-COO^- + H^+$$

These unknown reactions prevent us from applying equilibrium constants determined with more pure compounds to the relationship between dissolved CO_2, H^+, and HCO_3^- in blood. The solution is to be pragmatic and not worry about the side reactions, and this is what is done. An apparent equilibrium constant is determined by measuring the actual relationship between the three components of the carbonic acid equilibrium in whole blood, and it is designated K'' to indicate its empirical nature:

$$K'' = \frac{[H^+][HCO_3^-]}{[\text{dissolved } CO_2]} = 10^{-6.10} \text{ (blood at } 38°)$$

in which all of the various forms of CO_2 are lumped together as an entity. The concentration of these forms can be calculated from the partial pressure of the gas, given in mm of Hg, in equilibrium with the blood:

$$[\text{dissolved } CO_2] = 3 \times 10^{-5} \, p_{CO_2}$$

We are primarily interested in the way in which the concentration of H^+ changes with changes in the concentration of dissolved CO_2, so let us rearrange the ionization equation by dividing by $[H^+]$ so as to obtain:

$$\frac{K''}{[H^+]} = \frac{[HCO_3^-]}{[\text{dissolved } CO_2]}$$

Now, let us apply the equation by examining normal arterial blood. The pH is 7.40, so $[H^+] = 10^{-7.40}$. Since we know that $K'' = 10^{-6.10}$, we can now determine the ratio of concentrations of bicarbonate ions and of dissolved CO_2:

$$\frac{10^{-6.10}}{10^{-7.40}} = 10^{1.3} = 20 = \frac{[HCO_3^-]}{[\text{dissolved } CO_2]}$$

We find that there is 20 times as much bicarbonate as dissolved CO_2 in arterial blood.

The arterial blood is equilibrated with CO_2 at the partial pressure existing in the alveoli of the lungs, which is normally near 40 mm Hg. Therefore,

$$[\text{dissolved } CO_2] = 3 \times 10^{-5} \, p_{CO_2} = 3 \times 10^{-5} \times 40 = 1.2 \text{ mM}$$

The concentration of bicarbonate ions is 20 times greater than this, or 24 mM.

As the arterial blood passes through the tissues, it is exposed to the large quantity of CO_2 being produced. We see from the ionization equation that the pH can only be maintained near 7.40 if something is done to convert part of the CO_2 to HCO_3^-, so as to maintain the 1:20 ratio of concentrations. Let us go on to consider how this is done.

Since it is common to express $[H^+]$ in terms of its negative logarithm, the pH, the ionization equation is frequently rewritten by taking the negative logarithm of both sides and rearranging to give the Henderson-Hasselbalch equation:

$$pH = pK'' - \log\frac{[\text{dissolved } CO_2]}{[HCO_3^-]}$$

This equation is a quite useful form for purposes of calculation in terms of pH, but for some reason it obscures the essential simplicity of ionization equilibria for many. (The equation applies to the ionization of all kinds of acids if the concentrations of the appropriate acidic and basic forms are used.) This makes one wonder if pH has perhaps outlived its usefulness. People in the days of Sorensen, who invented the pH scale, used all sorts of stratagems to avoid fractional and negative numbers. We are now growing quite accustomed to the prefixes micro-, nano-, pico-, and femto-, devised for the same purpose, and somehow 80 nanomolar H^+ seems no more clumsy than pH 7.10. Some clinical scientists active in the problem of acid-base regulation have indeed begun to express their results in $[H^+]$ rather than pH.

The Bohr Effect

We described the Bohr effect in Chapter 4 (p. 63), noting that oxyhemoglobin is a stronger acid than hemoglobin, so that the loss of one mole of O_2 from oxyhemoglobin results in the combination of 0.7 mole of H^+ to create more undissociated (that is, non-ionized) hemoglobin. We mostly emphasized in the earlier discussion the way in which the Bohr effect promotes oxygen transport, with the extra H^+ produced by a working muscle causing increased release of O_2. Now let us emphasize the other side of the coin. The H^+ taken up by hemoglobin when O_2 dissociates must come from someplace and it is mainly supplied by converting the carbon dioxide arriving in the blood from the tissues to bicarbonate ions. In short, the Bohr effect carries a major part of the load in maintaining the [dissolved CO_2]/$[HCO_3^-]$ ratio at $1:20$. Let us analyze the effect and some of its consequences in more detail.

Suppose that a tissue is mostly oxidizing fats, so that the RQ is 0.735, which means that 0.735 mole of CO_2 is produced for each mole of O_2 consumed. However, the 0.7 mole of H^+ that is simultaneously taken up by hemoglobin because it has lost one mole of O_2 is enough to cause the conversion of 0.7 mole of the CO_2 to HCO_3^-, leaving 0.035 mole of unchanged CO_2. Now, the ratio of dissolved CO_2 and HCO_3^- is $1:20$ at pH 7.40. The unchanged CO_2 in our example is exactly $1/20$ of the bicarbonate formed, so this ratio is unchanged and the pH remains exactly at 7.40. This action of the Bohr effect to permit the blood to take up CO_2 without a change in pH is known as the *isohydric carriage* of CO_2.

Let us look at it another way. An oxyhemoglobin molecule has roughly 7.9 negative charges at pH 7.25, which is the pH of the interior of the erythrocyte, whereas a hemoglobin molecule has 7.2 negative charges. We can write a typical reaction for the dissociation of oxygen as follows:

$$Hb(O_2)_4^{7.9-} + 0.7\ H^+ \longleftrightarrow H_{0.7}Hb(O_2)_3^{7.2-} + O_2$$

At the same time, we have 0.7 mole of carbonic acid being ionized:

$$0.7\ H_2CO_3 \longleftrightarrow 0.7\ H^+ + HCO_3^-$$

Adding the two processes, we have:

$$Hb(O_2)_4^{7.9-} + 0.7\ H_2CO_3 \longleftrightarrow H_{0.7}Hb(O_2)_3^{7.2-} + 0.7\ HCO_3^- + O_2$$

and there is no change in $[H^+]$ in this stoichiometry.

The above equation describes the overall stoichiometry, but the way in which it is achieved is a little more complicated. CO_2 spontaneously reacts with water to form carbonic acid, but the process is accelerated by an enzyme, *carbonic anhydrase*, that is contained in the erythrocytes. (Carbonic anhydrase contains zinc, and this is one of the reasons for the requirement of trace amounts of zinc in the diet.) The result is that most of the carbonic acid is formed within the erythrocytes where the Bohr effect is also converting most of the acid to bicarbonate. The bicarbonate concentration of the erythrocyte therefore rises rapidly and displaces the ionic equilibrium between the cells and the plasma, leading to a series of changes known as the *chloride shift*. Let us consider the changes step by step.

1. To begin, the erythrocyte has a higher concentration of protein than does the plasma, and therefore must have a lower concentration of salts in order to maintain osmotic equilibrium. Otherwise, water would diffuse one way or the other. The free-energy imbalance represented by the differing protein concentrations is offset by equal free energy imbalances among the ions, so

$$\frac{[HCO_3^-]_{plasma}}{[HCO_3^-]_{RBC}} = \frac{[Cl^-]_{plasma}}{[Cl^-]_{RBC}} = >1$$

2. The formation of bicarbonate within the erythrocytes by the combined action of carbonic anhydrase and the Bohr effect upsets the equilibrium in two ways: It raises the total solute concentration within the erythrocyte above the position of osmotic equilibrium, and it raises the concentration of bicarbonate relative to the concentration of chloride ion.

3. Bicarbonate ion diffuses out of the erythrocyte into the plasma and chloride diffuses from the plasma into the erythrocyte so as to restore the same ratio of concentrations for *both* anions between the erythrocyte and the plasma. (If bicarbonate alone diffused, an electric charge would quickly accumulate because the counterbalancing K⁺ cannot freely diffuse out of the erythrocyte. Hence, there must be an anion for anion exchange.)

4. The bicarbonate—chloride exchange restores ionic balance, but it still leaves an excess of total solute within the erythrocyte. The osmotic pressure that is created causes water to move into the erythrocytes to dilute the solution, and the erythrocytes swell.

In sum, the erythrocytes of venous blood differ from those of arterial blood in that they are larger and contain more bicarbonate and chloride ions. A practical consequence of this effect is the necessity for preventing blood samples from equilibrating with the atmosphere if really accurate values for cell volume and plasma chloride concentration are desired.

The entire sequence of events we have described is reversed in the lungs. The exposure of hemoglobin to the higher oxygen tension causes O_2 to be taken up with a release of H⁺ as the more strongly acidic oxyhemoglobin is formed. The H⁺ is utilized to convert bicarbonate to carbonic acid, and the resultant CO_2 diffuses into the gas space of the alveoli. The fall in bicarbonate concentration within the erythrocyte causes a concomitant fall in chloride concentration, and water diffuses from the erythrocyte into the plasma to restore osmotic balance.

Buffer Action

The Bohr effect in itself would prevent changes in the pH from arterial to venous blood if the RQ were always 0.735 and there were no other sources of H⁺ from the tissues. Of course, the RQ is nearly always greater than 0.735, so that CO_2 is produced in amounts greater than can be accommodated by isohydric carriage, and there are additional sources of H⁺ in the metabolic processes, such as the production of lactate. The maintenance of nearly constant pH therefore requires some additional process that will convert CO_2 to bicarbonate, and this is accomplished through the buffer action of the blood.

Buffer action is a terse way of defining the following circumstance: If a solution has both the acidic and the basic forms of a compound, the effects on the concentration of H^+ will be minimized when H^+ is added to the solution because part of the additional H^+ will be taken up to maintain the ionization equilibrium of the compound. (The effects will also be minimized when H^+ is removed, and for the same reason.) An example will make it more clear.

Let us assume that an imidazole side chain of one of the histidyl residues in hemoglobin has a $K' = 10^{-7.25}$, which is also the $[H^+]$ within the erythrocyte:

$$[H^+] = \frac{K'\ [\text{H-imidazole}^+]}{[\text{imidazole}]}\ ; \qquad 10^{-7.25} = \frac{10^{-7.25}[\text{H-imidazole}^+]}{[\text{imidazole}]}$$

We see that the concentrations of the acidic and basic forms of the imidazole group are equal in this case. (We are also seeing that K' may be regarded as the H^+ concentration at which the concentrations of acidic and basic forms are equal.) Let us say that each form is at 0.001 M in whole blood, which is about right for a girl. Now let us say that we add 10^{-4} M H^+ to the blood. Does the $[H^+]$ of the blood rise to 10^{-4} M? No it does not, because the imidazole group will react with the H^+ to form more imidazolium groups, and the new situation will be:

$$[H^+] = \frac{10^{-7.25}[0.001 + 0.0001]}{[0.001 - 0.0001]} = \frac{10^{-7.25}[0.0011]}{[0.0009]} = 10^{-7.16}$$

The $[H^+]$ changes by only a little over 10^{-8} M because the imidazole group reacts with the remainder of the H^+ to restore its own ionization equilibrium.

Hemoglobin actually contains a number of different acidic and basic groups, but the sum of their effects is equivalent to 20 of our hypothetical imidazole groups, and it requires 2.3 millimoles of H^+ to react with the hemoglobin in a liter of blood before the pH can change by 0.1 unit. (The histidyl residues do account for many of these buffering groups, but the individual ionization constants of the residues vary because of the differing locations in the protein.) The plasma proteins also have acidic and basic groups, and they add an additional buffering capacity for 0.4 millimoles of H^+ per liter of blood per 0.1 pH unit change.

The result of all of this is that the pH of arterial blood falls by only a few hundredths of a unit as it becomes venous blood in a resting individual, even though the p_{CO_2} rises from 40 to 46 mm Hg.

RESPIRATORY CHANGES IN H^+

We have seen in the previous discussion how changes in the pH are minimized between the arterial and venous blood, mainly by effects within the blood itself. In this discussion, we were dealing with a fixed baseline for the concentration of dissolved CO_2 and bicarbonate in the arterial blood. This baseline may be altered; that is, the pH of both arterial and venous blood may be shifted even though the difference in pH between the two parts of the peripheral circulation may be small.

These are the kinds of changes with which we shall now be concerned—changes in which the level of $[H^+]$ is shifted in the entire body. An important distinction of such changes is that they involve all of the tissues, in addition to the blood, because the concentration of H^+ in the tissues and in the blood is equilibrated; the internal pH of cells is not the same as the pH of blood plasma, but a rise or fall in one causes a rise and fall in the other. As a result, the buffers of the entire body come into play in minimizing changes in $[H^+]$. Measurements show that the tissues contribute twice as much buffering power as does the blood.

Changes in p_{CO_2}

We have already noted that the pH of the blood, and therefore of the tissues equilibrated with the blood, depends upon the ratio of $[\text{dissolved } CO_2]/[HCO_3^-]$. If we change the concentration of either component without also adjusting the concentration of the other, the pH will change. We shall consider changes in the bicarbonate concentration in the next section. Let us now consider changes in the concentration of dissolved CO_2.

We noted that $[\text{dissolved } CO_2]$ is determined by the partial pressure of CO_2 in the gas phase equilibrated with the blood, and this in turn is determined by conditions within the alveoli of the lung. The alveolar air spaces do not completely collapse with each expiration; some gas rich in CO_2 remains to be mixed with the inspired air. The depth and rate of respiration is usually adjusted at rest so that the average p_{CO_2} in the alveolar gas is 40 mm Hg, and this is the normal baseline value for arterial blood.

We see that the partial pressure of CO_2 can be altered by the simple expedient of changing the rate of gas flow through the lungs. If one breathes more rapidly or more deeply, there will be greater dilution of the residual CO_2 in the alveoli, and the partial pressure will fall with a resultant fall in $[\text{dissolved } CO_2]$ in the blood. Similarly, if one breathes less rapidly or less deeply, the contrary changes occur with a rise in the partial pressure of CO_2 in the alveoli and a rise in the $[\text{dissolved } CO_2]$ in the arterial blood.

Now, these changes in the $[\text{dissolved } CO_2]$ in the blood caused by changes in the pulmonary ventilation must cause changes in the $[H^+]$, but how extensive? The necessary quantitative information is shown in Figure 25-1, in which the pH, p_{CO_2}, and concentration of total CO_2, that is, $[\text{dissolved } CO_2] + [HCO_3^-]$, is given. (Total CO_2 is shown rather than $[HCO_3^-]$ alone because it is easily measured by acidifying blood and discharging all of the CO_2. We have already seen that there is 20 times as much bicarbonate as dissolved CO_2 at pH 7.40, so total CO_2 is a rough guide to the bicarbonate concentration.) Each point in Figure 25-1 therefore defines the relationship among the three quantities that will occur in whole blood.

The truncated diamond in the right center of Figure 25-1 defines the range of values found in most individuals. The shaded band extending away from this diamond shows the values that will be obtained when pulmonary ventilation is altered in a normal individual so as to change the p_{CO_2}, and therefore the [dissolved CO_2] in his blood. Now, this band applies to blood within the individual, and shows the effects of the tissue, as well as the blood, buffers in diminishing the changes in pH caused by changes in p_{CO_2}.

Suppose that an individual hyperventilates so as to blow off half of his dissolved CO_2. If his $[HCO_3^-]$ remained constant, it would now be 40 times the concentration

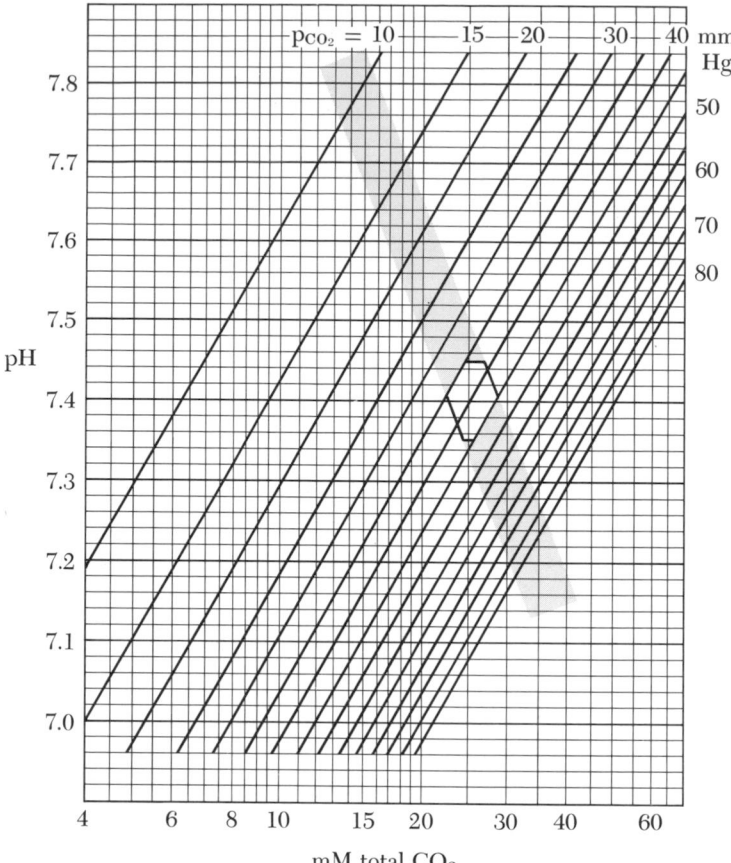

Figure 25-1 The relationship between pH, p_{CO_2}, and the concentration of total CO_2 in the plasma of circulating arterial blood in a human. The concentration of total CO_2 includes the concentrations of dissolved CO_2 and other forms, such as carbamino compounds, but it is mostly the concentration of bicarbonate ion. The diamond-shaped area centered on pH 7.40, 40 mm Hg partial pressure of CO_2, and 25.2 mM total CO_2 includes the usual range of values found in resting adults. The shaded band shows the changes that occur if the p_{CO_2} is altered in a short period of time by changes in the rate of gas exchange in the lungs; that is, the band is the "buffer line" of normal blood against changes in pH resulting from changes in carbonic acid concentration.

of dissolved CO_2, and the H^+ concentration would also have to fall to half of its original concentration (a pH rise of 0.3 unit) to satisfy the equilibrium of carbonic acid. However, a slight fall in H^+ concentration causes all of the body buffers to shift from their acidic to their basic forms:

$$\text{buffer acid} \longrightarrow \text{buffer base} + H^+$$

and the H^+ that is released will cause a concomitant conversion of bicarbonate to carbonic acid:

$$HCO_3^- + H^+ \longrightarrow H_2CO_3$$

The buffering action against a change in [dissolved CO_2] is therefore seen in two ways: less change in pH than would be expected from the carbonic acid equilibrium alone, and a concomitant change in the blood bicarbonate concentration. Figure 25-1 illustrates both of these effects. When hyperventilation causes a fall in the p_{CO_2} from 40 to 20 mm Hg (corresponding to a drop in the [dissolved CO_2] from 1.2 to 0.6 mM), the pH rises from 7.40 to only 7.59, rather than 7.70, because the buffer action has caused the total CO_2 concentration to fall from 25 mM to 19 mM, and 0.9 of this fall is due to a drop in the bicarbonate concentration.

Changes in the rate of ventilation are used as a means of *respiratory control* over the H⁺ concentration of the body. The respiratory center of the brain stem responds to changes in either p_{CO_2} or [H⁺] so as to cause more ventilation in response to a rise in either. The rate of respiration also increases in response to a fall in the oxygen concentration, but the oxygen receptors in the carotid arteries and aorta that are responsible for these effects are only sensitive to large changes.

The effects of respiratory control are dramatically demonstrated by two types of patients with inadequate pulmonary ventilation, which diminishes the oxygen tension and increases the carbon dioxide tension of the blood. The normal response to such a circumstance is to increase the respiratory rate so as to maintain normal p_{CO_2}, and if this is accomplished normal supplies of oxygen are also assured. Patients reacting in this way are the "pink puffers," who are working hard merely to breathe but have a normal skin color. (Pink puffers frequently have severe emphysema, a condition in which the alveoli are distended into large inelastic sacs with much less surface for exchanging gases. They have normal erythrocyte counts because they maintain nearly normal oxygen tension, but they are underweight because they are so busy breathing that they don't take time to eat enough to supply the additional energy required for the constant effort to get more air.)

Some patients, the "blue bloaters," adjust to the slow development of chronic bronchitis and its restricted air flow by diminishing the sensitivity of the respiratory center to the elevated p_{CO_2}. It is as if the center were permitting the partial pressure to rise rather than committing so much of the fuel supply to respiration. The result is that the respiratory rate is controlled primarily by the oxygen-sensitive receptors, which only respond enough to maintain the oxygen tension at near half its normal level. The rate of respiration may appear normal, but it is inadequate for effective gas exchange with the restricted air flow. The blue bloaters therefore have an elevated p_{CO_2} in their blood. Attempting to relieve the oxygen deficiency (which attracts immediate attention because of their blue skin color) by having them breathe pure oxygen can be disastrous, because the elevation in oxygen concentration may suppress the rate of breathing even more—enough to cause a critical rise in the p_{CO_2}. Cautious elevation of atmospheric oxygen is indicated, with frequent monitoring of the blood gas content. (The oxygen deficiency of the blue bloaters causes a secondary polycythemia, a proliferation of erythrocytes, but they usually have nearly normal weights because they aren't working so hard at breathing. The pink puffers and blue bloaters represent extreme examples; most patients with defective ventilation have intermediate responses, but the principles are the same.)

METABOLIC CHANGES IN H⁺

Dietary Balance

We have seen that respiratory exchange does not involve a net change in the hydrogen ion content of the body. However, CO_2 production is the end result of the metabolism of ingested foods. What comes in must go out if the composition of the body is not to change, and this principle applies to H^+. It is an easy principle to apply because we need only to look at the overall balance without worrying about the intermediate processes.

Consider, for example, the carboxylic acids. Suppose we dilute some acetic acid in water, call it vinegar, and eat it on a salad. It will be completely oxidized to CO_2 and H_2O:

$$CH_3-COOH + 2\ O_2 \longrightarrow 2\ CO_2 + 2\ H_2O$$

The CO_2 will be blown off in the lungs, the water will be lost in the urine or sweat, and there is no production or consumption of H^+ even though the ingested compound is an acid.

Now suppose the acetic acid is neutralized before we eat it, so that we are ingesting acetate ions (along with Na^+ or K^+). The overall stoichiometry then becomes:

$$CH_3COO^- + H^+ + 2\ O_2 \longrightarrow 2\ CO_2 + 2\ H_2O$$

The necessary H^+ must be supplied within the body, where it will be replaced by the Na^+ or K^+ eaten with the acetate. There is no escaping it, eating compounds with carboxylate groups tends to lower $[H^+]$ within the body. This accounts for the "alkalinizing" effect of fruit juices, in which the pH is never low enough to have all of the polycarboxylic acids in the non-ionized form.

By the same sort of reasoning we can show that compounds with ammonium groups lead to the production of H^+, whereas those with amino groups do not:

$$R-NH_3^+ \longrightarrow urea + H^+$$
$$R-NH_2 \longrightarrow urea$$

Since most of such compounds have the ammonium group at the pH of ingested food, the production of H^+ is the predominant process. We see that the effect of ingesting proteins will depend upon the relative number of carboxylate and ammonium groups that they contain, and the net effect is usually near a stand-off.

However, proteins also contain sulfur, and the complete metabolism of cysteine or methionine produces H^+:

$$R-SH \longrightarrow SO_4^{2-} + 2\ H^+$$

This represents a substantial load of H^+, easily amounting to 70 millimoles per day.

Incomplete Metabolism

Hydrogen ions may also be produced when compounds are not completely metabolized. We have seen two significant examples. Glucosyl residues of glycogen may be converted to lactate by working muscles:

$$(glucosyl\ residue) + H_2O \longrightarrow 2\ (lactate)^- + 2\ H^+$$

The K' for lactic acid is $10^{-3.74}$, so it will be nearly quantitatively ionized at pH 7.4. The concentration in blood may rise to 5 millimolar or more during heavy work, causing the arterial pH to fall to 7.3 or below. We have seen that the appearance of lactate is transient during exercise. However, we shall also see that lactate is sometimes formed under other circumstances in which the level remains high.

The formation of acetoacetate and 3-hydroxybutyrate also represents the production of H^+:

$$(\text{palmitoyl residue}) \longrightarrow 4\,(\text{acetoacetate})^- + 4\,H^+$$
$$or\ \ 4\,(\text{hydroxybutyrate})^- + 4\,H^+$$

The K' values are $10^{-3.58}$ for acetoacetic acid and $10^{-4.39}$ for 3-hydroxybutyric acid, so the H^+ production will also be nearly quantitative for these compounds. A diabetic continually produces these compounds and the associated H^+. The total load may approach one mole per day.

MAINTENANCE OF H+ STOICHIOMETRY

If the nature of the food or the internal metabolism is causing the net formation or removal of H^+, the excreted products must be changed so as to correct what would otherwise eventually become a lethal change in the internal pH. The important way in which this is done is through changing the composition of the urine.

The cells of the kidney tubules are capable of exchanging H^+ for Na^+. (Either H^+ or K^+ may be exchanged for Na^+ in the distal tubule.) The cells also contain an active carbonic anhydrase. These are the principal tools used for routine regulation of the H^+ stoichiometry. The ultrafiltrate of blood formed by the glomerulus passes into the tubule. This filtrate contains the same proportion of ions as that found in the blood plasma. The tubular cells can inject H^+ into the stream of filtrate in this way: Na^+ is removed from the filtrate in exchange for H^+. The H^+ is derived from carbonic acid within the cells, which now becomes sodium bicarbonate by the Na^+ exchange. The H^+ converts the bicarbonate in the filtrate to carbonic acid, and the resultant CO_2 diffuses into the tubular cells. The fall in the bicarbonate concentration of the filtrate represents a net excretion of H^+.

The capacity for concentrating H^+ in the urine is not unlimited—the pH of the urine cannot fall below 4.5. However, this doesn't mean that the excreted H^+ can be no more than $10^{-4.5}$ moles per liter of urine. For one thing, a pure solution of carbonic acid has a pH of 4.5 at a partial pressure of 40 mm of Hg, which means that all of the bicarbonate can be converted to carbonic acid at this pH and still maintain equilibrium with the CO_2 in arterial blood; this would require 25 millimoles of H^+ for each liter of plasma "cleared" of bicarbonate. (Of course, bicarbonate is replaced in the plasma by the tubular cells, so the kidney really isn't clearing it from the blood in the way that it does with some compounds that aren't reabsorbed.) For another thing, the glomerular filtrate contains other acids and bases, especially phosphate. The second dissociation constant of phosphoric acid is near $10^{-6.8}$; therefore the ratio $[HPO_4^{2-}]/[H_2PO_4^-]$ is $4:1$ at pH 7.40 in blood. Phosphate is almost completely $H_2PO_4^-$ at pH 4.5, and 0.8 mole of H^+ must be added for each mole of phosphate excreted before the pH can fall to 4.5. This can account for another 16 to 20 millimoles of H^+ excretion at the levels of phosphate found in normal urines, and the mobilization of bone that occurs in severe acidosis adds even more phosphate to the urine.

If the pH of the blood is too high, the kidney's adjustment to this condition is quite simple: The exchange of H^+ for Na^+ in the filtrate is shut off and K^+ is substituted instead; all of the free CO_2 is recovered from the filtrate and the remaining bicarbonate is allowed to spill out in the urine. (There is also some evidence that the kidney may actively remove H^+ from the filtrate, thereby converting carbonic acid to bicarbonate.)

However, the body is more likely to be confronted with too much H^+ than it is with too little, and the kidney has additional mechanisms for coping with acidosis. We have seen that the conversion of dietary ammonium compounds, such as the amino acids, to urea results in the simultaneous production of one H^+ for each nitrogen atom converted. A prolonged acidosis results in decreased urea synthesis in the liver, which diminishes one of the principal sources of H^+ but leaves the problem of disposing of the excess nitrogen. This is solved by the kidney, which can form NH_4^+ by direct metabolism of amino acids or by the hydrolysis of the glutamine formed during metabolism of the amino acids in other tissues. The NH_4^+ is then exchanged for Na^+ in the glomerular filtrate, thereby getting rid of the nitrogen. The kidney is not directly removing H^+ by adding NH_4^+ to the urine; it is making it possible for the liver to diminish a process normally producing H^+. (One occasionally still sees a formerly common error in which the hydrolysis of glutamine by the kidney was supposed to be removing H^+ according to this formulation:

$$H^+ + glutamine + H_2O \longrightarrow glutamic\ acid + NH_4^+$$

The error evidently arose from the practice of writing structural formulas with uncharged carboxylic and amino groups for amino acids.)

It is still not certain whether the primary event in the compensatory mechanism is an adaptive decrease of the hepatic enzymes of urea synthesis in direct response to the lowered pH, or a change in the metabolism of the kidney, causing increased excretion of NH_4^+, followed by a secondary adaptation in the liver to the lowered nitrogen load. In any case, the result is that the excretion of NH_4^+ rises from the 2 to 5 per cent of the total nitrogen in the urine of individuals with normal pH to as much as 50 per cent of the nitrogen in the urine of individuals with severe acidosis (arterial pH approaching 7.0).

The acidosis accompanying ketosis is also diminished by the appearance of 3-hydroxybutyrate in the urine. Since the dissociation constant of the acid is $10^{-4.39}$, approximately 45 per cent of the compound is present as the undissociated acid at pH 4.5, which means that an additional 0.45 mole of H^+ must be added to the urine for each mole of 3-hydroxybutyrate excreted if the pH is brought down to 4.5. (The dissociation constant of acetoacetic acid is too low for it to be a significant factor in this way. However, solutions of acetoacetate spontaneously decarboxylate in acidic solutions:

$$(acetoacetate)^- + H^+ \longrightarrow acetone + CO_2$$

and it really isn't known how much H^+ is taken up in this way. There is no reason from present information to suspect that it is a major fraction of the total.)

ACIDOSIS AND ALKALOSIS

An acidosis is any condition causing an excessive formation of H^+; an alkalosis is any condition causing an excessive loss of H^+. There are two extreme types of these conditions, labelled as *respiratory* or *metabolic*, and Figure 25-2 has simplified

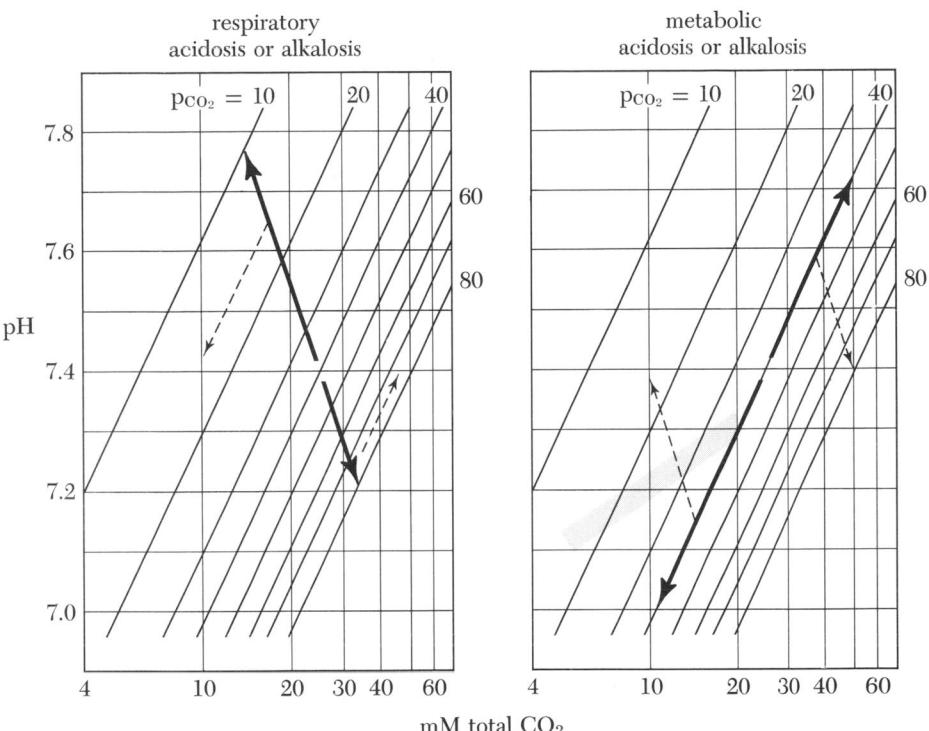

Figure 25-2 Changes in the composition of circulating arterial blood plasma during acidosis or alkalosis. The figure at the left shows the results of changes in pulmonary ventilation that affect the p_{CO_2}. The dark arrow is the center of the buffer band shown in Figure 25-1, and the primary effects of a respiratory acidosis or alkalosis are to shift the pH of the blood along this line to a value corresponding to the new p_{CO_2}. However, compensation will begin to occur, primarily by changes in the composition of the urine, to compensate for the changed pH by altering the ratios of acid to base in the body buffers. This will be reflected by a change in the bicarbonate concentration of the plasma, and therefore of the concentration of total CO_2. Since the p_{CO_2} is held at the value fixed by the changed rate of pulmonary ventilation, the compensatory shift will occur along the line designating the partial pressure, as shown by dashed arrows.

The figure to the right shows the results of changes in the ratios of acid to base in body buffers caused by the addition or removal of H⁺. These metabolic changes, if occurring alone, would change the pH along the line designating a normal p_{CO_2} of 40 mm Hg, as shown by the dark arrow. However, respiratory compensation by altering the rate of pulmonary ventilation usually begins promptly. The primary metabolic acidosis or alkalosis has, in effect, created a new buffer line paralleling the normal line, and the respiratory compensation will therefore proceed along these new buffer lines, as shown by the dashed arrows. There also is renal compensation, especially for metabolic acidosis, and the values found in arterial plasma, even during the early stages of severe metabolic acidosis, usually lie in the shaded band shown on the curve, which reflects the sum of the initial acidosis and the compensating effects of increased pulmonary ventilation and early changes in kidney function.

versions of Figure 25-1 for the two types in which the corresponding changes in blood bicarbonate concentrations are shown. It will be helpful to refer to these figures frequently during our further discussion.

The respiratory conditions are those in which the primary change is in the pulmonary ventilation, causing abnormal changes in p_{CO_2}. If the change occurs so rapidly that there is no time for compensation by the kidney to come into play, then the change in pH will be limited only by the buffer responses of the body, which we had earlier noted in the shaded zone of Figure 25-1. In other words, the pH of the blood will shift to a new value near the intercept of the heavy line shown in Figure 25-2 (the center of the buffer zone) and the line designating the abnormal p_{CO_2}. Measured values that fall near this line indicate the existence of respiratory acidosis if the pH is low and respiratory alkalosis if it is high.

A respiratory acidosis can readily be produced by holding a bag over one's face and rebreathing expired air, and a respiratory alkalosis by forced rapid breathing. These artificial measures cause rapid changes and are the basis for determination of the buffer line drawn on the figures.

A respiratory acidosis also occurs, as we noted in the case of blue bloaters, in patients with obstructed airways or damaged lungs, but these cases usually develop slowly enough that the kidney can at least partially compensate for the elevated $[H^+]$ by excreting the ion and recovering bicarbonate. The resultant rise in bicarbonate reflects a rise in the buffering bases in the whole body; the quantitative effect is a shift of the buffer lines on our graph, as indicated by a dashed line. If the compensation is completely effective, normal pH will be restored, but with elevated p_{CO_2} and HCO_3^-.

Respiratory alkalosis is a common result of anxiety states because the anxious person tends to breathe more rapidly. Two things happen to correct the fall in $[H^+]$: One is the adjustment by the kidney allowing bicarbonate to be excreted that we already mentioned. The other is an increased formation of lactate and pyruvate by the tissues. Here we have a metabolic process ordinarily used for rapid generation of ATP with the concomitant formation of H^+ representing a threat to normal pH now being employed primarily for its production of H^+ to relieve a deficiency of the ion. We see that there must be an additional control over the fraction of the high-energy phosphate generated by the conversion of glycogen to pyruvate beyond those we mentioned in Chapter 23. (The mechanism is unknown, but an effect of pH on the phosphofructokinase reaction has been suggested.) In an experimental test, forced breathing for 15 minutes at 4.6 times the normal rate of ventilation caused the arterial blood lactate concentration to rise from 0.77 to 1.24 millimolar. (Even so, the pH rose from 7.39 to 7.53.) When the forced breathing was done with an atmosphere containing 5 per cent CO_2, neither the pH nor the lactate concentration changed, showing that the increased exertion of breathing was not responsible for the effect.

Metabolic Acidosis and Alkalosis

Any addition or subtraction of H^+ due to causes other than changes in p_{CO_2} is termed a "metabolic" effect. The causes include such diverse things as ingestion of acids, alkalies, or materials that are metabolized so as to change the $[H^+]$; alterations of metabolism so as to accumulate intermediate products with concomitant alteration of [H⁺]; and disturbances in renal function so that undue amounts of H⁺ are retained or escape in the urine. Those changes classed as metabolic cause a shift in all of the buffer systems toward the acidic or the basic forms and are reflected by a change in the bicarbonate concentration of the blood even at constant p_{CO_2}. Metabolic changes without respiratory changes shift the pH along the heavy solid lines shown in the right graph of Figure 25-2. However, metabolic changes nearly always occur slowly enough that some compensation accompanies the change, and the real shift in pH is less than would be predicted by the metabolic effect alone.

Metabolic acidosis is easily produced experimentally by feeding NH_4Cl; the conversion of NH_4^+ to urea produces H⁺. We have also seen that it is a transient result of exercise owing to the formation of lactate. It also occurs with the elevated formation of oxybutyrates present in normal individuals who are fasting or eating a high-fat diet and in diabetics. It commonly occurs in patients with kidney failure, and is the primary problem in rare individuals who produce excessive lactate because of some unknown congenital defect.

Metabolic acidosis usually results in increased respiratory ventilation to lower the p_{CO_2}, and the effect of this compensation is indicated by a dashed line in Figure 25-2. This line really traces the new carbonic acid buffer line created by the change in total buffer concentration that results from the acidosis. The kidney also compensates for the acidosis in the way we indicated earlier, by exchanging internal H⁺ for Na⁺ in the glomerular filtrate and by increasing the output of NH_4^+ if the acidosis persists for more than a few days. The combined effect of the compensations is such that the actual pH found in the blood of patients with acute metabolic acidosis caused by a variety of conditions usually lie in the shaded band shown on Figure 25-2.

Metabolic alkalosis may be the result of ingestion of sodium bicarbonate—a common practice—or of vomiting. It is also a side-effect of congenital defects in urea synthesis because the normal production of H⁺ by this mechanism is missing. The amount of respiratory compensation for metabolic alkalosis is quite variable, and this has to do with associated changes in K⁺ concentration. For example, vomiting often directly expels H⁺ because the gastric juice is strongly acidic, but alkalosis will also occur if the vomitus is neutral, and the reason for this is that K⁺ and Cl⁻ are also lost in the vomitus. Now, most of the K⁺ in the body is contained within the cells; less than 2 per cent of the total four moles is found in the blood, and most of the loss in the vomitus must be replaced by a movement of K⁺ from the cells to the blood. The transport across the plasma membrane involves an exchange of K⁺ inside for Na⁺ outside similar to that seen in the kidney, and an increase in K⁺ concentration in one direction causes a simultaneous increase in H⁺ concentration in the other direction. (It is not clear if the movement of K⁺ competes with the movement of H⁺ in the same direction with both being exchanged for Na⁺, or if Na⁺ and H⁺ are moved together in exchange for K⁺ in this case.)

The result of all this is that the simultaneous loss of K⁺ and Cl⁻ by vomiting causes the intracellular pH to fall while the extracellular pH is rising. Since the respiratory control center is composed of cells that respond to internal changes, the effect on respiration may be the sum of two contrary changes—an elevated extracellular pH created by other tissues that tends to raise the intracellular pH, and a lowered extracellular [K⁺] that tends to lower the intracellular pH. The result is that there may be little or no respiratory compensation to the alkalosis.

A simple lack of K^+ due to dietary deprivation rarely causes marked alkalosis in itself. A simultaneous loss of a non-buffering anion, such as Cl^-, is necessary. For example, loss of bicarbonate would tend to lower the pH and counteract the effect of a loss of K^+.

Quantitative Assessment of Acidosis and Alkalosis

The determination of the state of H^+ balance is one of the more important guides to the management of a variety of clinical conditions. We have seen that the theoretical principles involved are not very complicated, and the major problem is one of being able to make measurements in a simple enough way for routine application and prompt application in the medical emergencies that the conditions sometimes represent. To begin with, we see that any two of the variables pH, p_{CO_2}, and total CO_2 define the other in blood. Larger hospitals have equipment available only in recent years for direct measurement of pH and p_{CO_2}, as well as p_{O_2}, on small samples (arterial blood is the best sample because it reflects the base line conditions of the body). These measurements require skilled operators, and many clinicians must still rely on more routine methods. The simple determination of the total CO_2 in plasma or serum from venous blood can tell a great deal. It can be determined directly ("bicarbonate" or "total CO_2") or after equilibration of the sample with air containing CO_2 at 40 mm Hg ("alkali reserve," that is, the buffering capacity of the blood at constant CO_2 tension). A low value is usually the result of a metabolic acidosis, although it may be due to a respiratory alkalosis, and the magnitude of the loss is a good guide to the severity of the imbalance in diabetics. A high value may be the result of metabolic alkalosis or respiratory acidosis.

If simple measurements of the blood bicarbonate do not give the answer in association with other clinical signs and symptoms, other data are needed. Determinations of the Cl^- and Na^+ concentrations are in the reach of most laboratories. The normal electrolyte concentrations in venous plasma range around the following:

$$Na^+ = 140 \text{ mM}; \quad K^+ = 4 \text{ mM}; \quad Cl^- = 103 \text{ mM}; \quad HCO_3^- = 27 \text{ mM}$$

We see that the cations total 144 mM and the anions 130 mM. Evidently there are at least 14 mM of anions that are not accounted for; this is not surprising, since the blood contains phosphates, lactate, amino acids, and so on. Now, suppose that another sample has 20 mM of unaccounted anions. What do these represent? They almost certainly are due to an accumulation of lactate and pyruvate, or of the 3-oxybutyrates, and the simultaneous observation of a low total CO_2 would almost certainly indicate a metabolic acidosis. (In practice, the unaccounted anions are calculated without measurement of K^+ because there is so little of that ion.)

If all of this fails to illuminate the situation, then the physician needs a measurement of pH or p_{CO_2} or both on a blood sample carefully collected to avoid loss of gases. Let us look at a few examples:

NORMAL VALUES FOR COMPARISON

Venous plasma:
   total CO_2 = 26–28 mM Na^+ = 136–145 mM Cl^- = 100–106 mM
Arterial plasma:
   p_{CO_2} = 35–45 mm Hg   pH = 7.35–7.45   total CO_2 = 23–27 mM by calculation

Case 1.   Arterial pH = 7.58; p_{CO_2} = 28 mm Hg; total CO_2 = 25.5 mM
Analysis: The high pH indicates an alkalosis and the p_{CO_2} is remarkably low. Tracing these values on the normal buffer line (shaded area of Figure 25-1) shows the bicarbonate concentration to be slightly higher than the normal range at the measured p_{CO_2}. Conclusion: Respiratory alkalosis with possible slight metabolic alkalosis. Actual finding: Psychogenic hyperventilation. The pH occasionally rose so high in this 18-year-old boy that tetany resulted.

Case 2.   Venous total CO_2 = 8 mM; Na$^+$ = 119 mM; Cl$^-$ = 87 mM
Analysis: The low total CO_2 suggests metabolic acidosis, and the high unaccounted anions (119 − 87 − 8 = 24) makes diabetes a likely possibility. Further data: Arterial pH = 7.14; p_{CO_2} = 19 mm Hg. Further analysis: These values fall within the typical range for primary metabolic acidosis. (This was a diabetic previously controlled by dietary means who had incurred a severe middle ear infection.)

Case 3.   Arterial pH = 7.46; p_{CO_2} = 56 mm Hg; total CO_2 = 40 mM
Analysis: The pH is slightly elevated, suggesting a mild alkalosis. Is it mild because the primary cause is mild or because of compensation? The high p_{CO_2} would cause a mild acidosis if it were the primary event. The high bicarbonate concentration would represent a moderate alkalosis if it were the primary event. Since there is a mild alkalosis, it is the respiratory change that is compensatory, assuming there is no physical evidence of impaired ventilation. Conclusion: Compensated metabolic alkalosis. (This patient had a virus infection causing persistent vomiting.)

Case 4.   Venous total CO_2 = 4 mM; Cl$^-$ = 120 mM; Na$^+$ = 136 mM. Patient unconscious and breathing rapidly.
Analysis: The very low total CO_2 in an unconscious patient without physical injury suggests a metabolic acidosis, and diabetic coma is a likely diagnosis, but this diagnosis is not supported by the low unaccounted anions (136 − 120 − 4 = 12). Further data: Arterial pH = 6.97; p_{CO_2} = 15 mm Hg; total CO_2 = 3.3 mM. Further analysis: There is a very severe metabolic acidosis, with less than usual compensation despite the hyperventilation. Actual finding: Renal tubular acidosis. This is an unusual condition in which H$^+$ is not exchanged for Na$^+$ by the kidney, and it accounts for the deficient compensation. This case illustrates the necessity of measurements; the admitting diagnosis was psychogenic respiratory alkalosis (!), made on the basis of a psychiatric history.

REGULATION OF SALT AND WATER BALANCE

The maintenance of H$^+$ concentration is inextricably interwoven with the maintenance of the concentrations of Na$^+$ and K$^+$ and of the body water content. Two hormones not previously considered are involved in this regulatory process, and these will be our primary concern. (More detailed examination of the subject can be found in some of the references cited at the end of the chapter.)

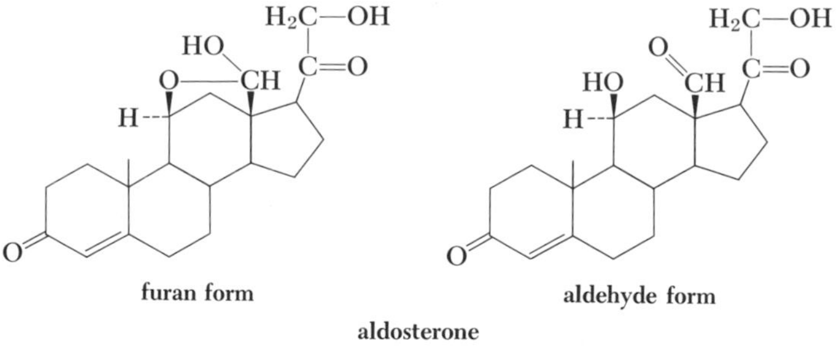

furan form   aldehyde form

aldosterone

Figure 25-3   Aldosterone, the adrenal steroid controlling sodium ion absorption by the kidney tubules, has C-19 oxidized to an aldehyde group. It exists in both the aldehyde form and a furan form in which an inner hemiacetal is formed from the aldehyde group and the 11-hydroxyl group.

Salt Balance

The pair of kidneys in a man filter the equivalent of 125 ml of blood each minute, or 360 liters per day. Most of the constituents of the filtrate are reabsorbed, and these include the electrolytes of the blood. Most of the salts are reabsorbed in the proximal tubule and the loop of Henle, but fine control is exerted through a different mechanism for reabsorption in the distal tubule. All of the evidence suggests that a specific transport protein is needed for the reabsorption of Na^+; Cl^- and HCO_3^- passively follow the Na^+. We have already noted that part of the reabsorbed Na^+ is exchanged for K^+ or H^+ in the distal tubule. The synthesis of the specific transport protein appears to require the presence of a steroid, *aldosterone* (Fig. 25-3), secreted by the adrenal cortex. This steroid is a mineralocorticoid hormone, not a glucocorticoid, known for some time to promote the retention of Na^+. Aldosterone has recently been demonstrated to combine with proteins in the nuclear fraction of kidney cells, and the combination is prevented by spironolactone, a steroid derivative known to inhibit the action of aldosterone in the whole animal. This evidence makes it likely that aldosterone is one of the hormones that has a direct effect on the synthesis of a specific protein—in this case the transport protein for Na^+.

A fall in the $[Na^+]/[K^+]$ ratio of blood circulating through the adrenal cortex directly causes the tissue to secrete more aldosterone, and there also is evidence for a secondary regulation by the anterior pituitary through the formation of a hormone differing from corticotropin, the regulator of glucocorticoid formation. However, there is still a third stimulus causing secretion of aldosterone that many now believe is usually the predominant control. The kidney glomeruli contain a structure, the vascular pole, that secretes a proteolytic enzyme, *renin*, when the blood volume or Na^+ concentration falls. Renin acts on a particular globulin in the blood plasma to clip off a peptide containing 10 amino acid residues. This decapeptide, known as *angiotensin I*, has the sequence: Asp-Arg-Val-Tyr-Ile-His-Pro-Phe-His-Leu. As the venous blood carrying the decapeptide passes through the lungs, it is exposed to at least one more enzyme that causes a hydrolysis liberating the C-terminal His-Leu fragment. The remaining octapeptide, known as *angiotensin II*, is the physiologically active compound, and it has the properties of causing an increase in the arterial blood pressure and of stimulating the secretion of aldosterone by the adrenal cortex.

It is apparent that aldosterone will also affect the water and H^+ balance. Patients with adrenal cortical insufficiency (Addison's disease) do not retain sufficient salt, and the resultant fall in osmotic pressure affects the secretion of vasopressin, a hormone regulating water balance that we shall shortly describe. With sodium not being reabsorbed, less K^+ and H^+ are excreted. The rising concentration of K^+ causes this ion to be exchanged for H^+ from the tissues, so there is an accompanying metabolic acidosis (at least from the standpoint of the blood). Dehydration, high blood $[K^+]$, and some degree of acidosis are characteristic of the condition.

Tumors may arise from the aldosterone-secreting cells of the adrenal cortex, causing aldosteronism. Those with the condition, and also those who are treated with large amounts of the compound for other conditions, transport too much Na^+ back into the blood, causing water retention and excessive loss of K^+ and H^+. We have noted before that conditions causing the loss of both K^+ and Cl^-, which is the case here, result in an apparent metabolic alkalosis regardless of the overall balance because of the movement of K^+ from the cells in an apparent exchange for H^+. (Diarrhea will produce the same result for the same reasons.)

Complications like these make the therapy of many conditions a real challenge. The tissue-blood exchange of K^+ makes it very difficult to assess the real condition of the total K^+ supply. If the supply is low, trying to correct any coexisting imbalances in salt concentration or an acidosis by giving solutions containing only sodium salts may aggravate the K^+ deficiency through an exchange of the added Na^+ for K^+ in the tissues.

Water Balance

The absorption of sodium salts from the glomerular filtrate raises the concentration of salts in the tissues. The nephron is so constructed that the now dilute filtrate passes again through the region of high salt concentration *via* the collecting duct, and the difference in osmotic pressure causes most of the water to pass out of the lumen into the tissues. In other words, the water is passively diffusing—following the salts out of the filtrate. As in the case of Na^+ absorption, most of the water is removed before the distal tubule, but the fine control is exerted in the collecting duct. The water is reabsorbed in this segment by passive diffusion, but the existence of "channels" through which the water can diffuse depends upon the simultaneous presence of *cyclic AMP*, functioning in some unknown way. The adenyl cyclase of the collecting duct is activated by a small peptide hormone, *vasopressin*, or *antidiuretic hormone*, elaborated from the posterior pituitary gland. (This action is not to be confused with the activation of adenyl cyclase in the proximal tubule by parathyroid hormone.)

Vasopressin contains but eight residues: Cys-Tyr-Phe-Gln-Asn-Cys-Pro-Arg-Gly-NH_2 terminating in an amide group as indicated. (This is the arginine vasopressin of most mammals. Pigs have lysine instead.) This small peptide is made by cells of the hypothalamus, perhaps as part of a larger molecule, and migrates down the axons of these cells that terminate in the posterior pituitary, where it is stored in granules. If there is a rise in the osmotic pressure of the blood or a fall in the total blood volume, the hypothalamus sends signals down these same axons to cause the release of the peptide. Either of the precipitating events is an indication that conservation of fluid is needed, and the vasopressin causes more water to leak back into the hyperosmotic cells of the loop of Henle in the kidney. (Vasopressin also causes constriction of peripheral arterioles—hence the name.) The action of vasopressin on adenyl cyclase is counteracted by prostaglandins, but the physiological role of these hormones is, as we indicated in Chapter 23, unknown.

Damage to the hypothalamus causes diabetes insipidus, a condition in which the failure to form vasopressin results in the loss of several liters of water per day in the urine. The urine in diabetes insipidus does not contain glucose; the word diabetes is derived from a Greek word meaning "to pass through" and has reference to the large volume of urine that also occurs in diabetes mellitus (sweet copious urine) as a result of the high concentration of solutes in the urine.

Further reading

Christensen, H. N.: *Body Fluids and the Acid-Base Balance.* Saunders (1964). A detailed analysis in learning program format.

Andersen, O. S.: *The pH-log pCO$_2$ Blood Acid-Base Nomogram Revised.* Scand. J. Clin. Lab. Invest., *14:* 598 (1962). Also see p. 587.

Arbus, G. S., et al.: *Characterization and Clinical Application of the "Significance Band" for Acute Respiratory Alkalosis.* New Eng. J. Med., *280:* 117 (1969). Also see p. 162.

Albert, M. S., et al.: *Quantitative Displacement of Acid-Base Equilibrium in Metabolic Acidosis.* Ann. Internal Med., *66:* 312 (1967).

Elkinton, J. R.: *Clinical Disorders of Acid-Base Regulation: A survey of Seventeen Years' Diagnostic Experience.* Med. Clin. N. America, *50:* 1325 (1966).

Huckabee, W. E.: *Relationships of Pyruvate and Lactate During Anaerobic Metabolism.* J. Clin. Invest., *37:* 244 (1958).

Halperin, M. L., et al.: *Factors that Control the Effect of pH on Glycolysis in Leukocytes.* J. Biol. Chem., *244:* 384 (1969).

Schwartz, I. L., and W. B. Schwartz, eds.: *Symposium on Antidiuretic Hormones.* Am. J. Med., *42:* 651 ff. (1967).

Herman, T. S., G. M. Fimognari, and I. S. Edelman: *Studies on Renal Aldosterone-binding Proteins.* J. Biol. Chem., *243:* 3849 (1968).

Moore, F. D., et al.: *Body Sodium and Potassium V.* Metabolism, *4:* 379 (1955).

Eisenstein, A. B., ed.: *The Adrenal Cortex.* Little Brown (1967).

Earley, L. E., and T. M. Daugharty: *Sodium Metabolism.* New Eng. J. Med., *281:* 72 (1969).

Haber, E.: *Recent Developments in Pathophysiologic Studies of the Renin-Angiotensin System.* New Eng. J. Med., *280:* 148 (1969).

SPECIALIZED BIOCHEMICAL MECHANISMS

CHAPTER 26

ARGUMENT

The metabolism of some compounds may not always rank among the major metabolic processes and yet may prove to be of critical importance to the whole organism at some time in the life cycle or to the maintenance of function in specialized tissues. Closer examination sometimes shows that the specialized processes may conceal a more general function that is neglected because we don't understand it. This is particularly true of the metabolism of fructose. Glucose is converted to fructose through a reduction to the alcohol, sorbitol, followed by re-oxidation at a different point. This route is best known through its occurrence in accessory reproductive glands in the male that provide fructose as an energy source for spermatozoa, but it also occurs in the lens of the eye and in nerves, where the function is not known. Fructose also may be a major fuel in some diets, and a separate metabolic route for its conversion to intermediates of carbohydrate metabolism involves a specific kinase and a modified phosphofructoaldolase that occur in the liver. Some people lack the aldolase and are intolerant of fructose because of an accumulation of fructose-1-phosphate that leads to a low blood glucose concentration and an acidosis.

The formation of lactose by the mammary glands is a means of reserving carbohydrate fuel for the nursing infant, analogous to the use of fructose as a fuel for spermatozoa. The galactosyl residue in lactose is formed by an epimerization of UDP-glucose and then transferred to free glucose to form the disaccharide. The enzyme catalyzing the transfer is found in many tissues, but its normal substrate is not glucose; the mammary glands form another protein, α-lactalbumin, that combines with the transferase in a way that changes its specificity, and this is the only example presently known of metabolic regulation in this way. The infant begins the metabolism of the ingested lactose by hydrolyzing it in the gut to the constituent monosac-

629

charides. Some adults lack the lactase necessary for this hydrolysis and are intolerant of milk. Once galactose is absorbed, it is converted to galactose-1-phosphate by a specific hepatic kinase, and the sugar phosphate is exchanged for the glucose-1-phosphate portion of UDP-glucose, which now becomes UDP-galactose. Since UDP-galactose is in equilibrium with UDP-glucose, the net result is the conversion of the absorbed galactose to glucose-1-phosphate. The transferase catalyzing the hexose phosphate exchange is missing in a number of infants, and the resultant accumulation of galactose-1-phosphate causes symptoms resembling those of fructose intolerance.

The retina of the eye fulfills its function through the presence of retinal, which is the aldehyde form of vitamin A alcohol, or retinol. An isomer of retinal combines with a protein, opsin, that is associated with phosphatidylethanolamine to form a light-sensitive pigment. The absorption of a quantum of light by the pigment causes a shift of a Schiff's base bond that is believed to be the source of the neural stimulation, and also causes a reversion of the retinal isomer to a more stable configuration that dissociates from the opsin complex. This function of vitamin A is well known, but it does not explain why animals die from lack of the vitamin and can be prevented from dying if the acid form of the vitamin, retinoate, is given.

The clotting of blood begins with an aggregation of platelets in the blood. These cells have an affinity for the collagen fibers that are exposed upon injury of a blood vessel, and agglomerate to make a loose plug that is strengthened by the polymerization of a protein, fibrin, into a strong lattice. Fibrin is formed by a partial hydrolysis of fibrinogen that removes the negatively charged regions that ordinarily prevent agglomeration, followed by the formation of covalent bonds through the combination of lysyl and glutaminyl residues. The partial hydrolysis is catalyzed by thrombin, a constituent of a glycoprotein prothrombin complex that is activated upon damage to the vessel or the surrounding tissue by mechanisms not yet completely elucidated. The prothrombin complex is formed in the liver after the assembly of the peptides on mRNA. This assembly requires menaquinone, a form of vitamin K, for reasons that are not clear. Although the only known symptoms of vitamin K deficiency are a failure of blood clotting, the ubiquitous occurrence of the vitamin and its known ability to undergo oxidation-reduction reactions like those of the structurally related coenzyme Q make it likely that it has some presently unsuspected general function.

Our concern in this chapter will be with some biochemical reactions that are used for specialized purposes in particular tissues. There is no particular interrelationship between the examples beyond their common importance in maintaining the species. However, a close study of the special case has in most instances exposed an ignorance of more general phenomena. We shall talk about the metabolism of fructose and lactose, the sugars that seem to serve as a means of protecting reproduction by being distinctive fuels for spermatozoa and infants, respectively. Each of these sugars is metabolized by the adult. Genetic defects are known in the metabolism of each that have quite similar effects even though the routes are different. We shall talk about vision, which depends upon the special structure contributed by vitamin A in the diet, and we shall talk about the role of vitamin K in blood clotting.

THE METABOLISM OF FRUCTOSE

Fructose in Semen

The semen is a complex mixture of fluids, contributed mainly by the seminal vesicles and prostate gland, that is designed to assure the delivery of spermatozoa and their maintenance after delivery. Spermatozoa have mitochondria and are capable of obtaining the ATP necessary for motility by oxidizing a variety of substrates, including fatty acids, but the principal source of energy for the spermatozoa of humans and many other mammals appears to be the metabolism of D-fructose. The fructose in human semen is formed by the seminal vesicle at a final concentration ranging around 12 mM, but it is formed by the prostate or still other accessory glands in some species.

The seminal fructose is created directly from glucose supplied by the blood through the intermediate formation of the corresponding alcohol, D-*sorbitol*. Sorbitol is formed from glucose by reducing C-1 with NADPH, and the sorbitol is then oxidized on C-2 by NAD to form fructose (Fig. 26-1). We see that this route bypasses the ordinary pathways of glucose metabolism, in which phosphorylated derivatives are intermediates. The formation of fructose is favored in two ways. One is through the kinetic characteristics of the enzymes involved. The "aldose reductase" catalyzing the first reaction is really a polyol dehydrogenase, but with a favorable K_M for the reduction of glucose and a correspondingly unfavorable K_M for the oxidation of sorbitol. The NAD—sorbitol 2-dehydrogenase has the opposite characteristics, with the K_M for sorbitol being relatively low even though the free energy change for the reaction is nearly the same as that of the first reaction. Here we have another case in which two enzymes catalyzing thermodynamically similar reactions are built with kinetic characteristics favoring opposite reactions. The "aldose reductase" is analogous to the lactate dehydrogenase of muscle, built to reduce pyruvate to lactate, and the NAD—sorbitol 2-dehydrogenase is analogous to the lactate de-

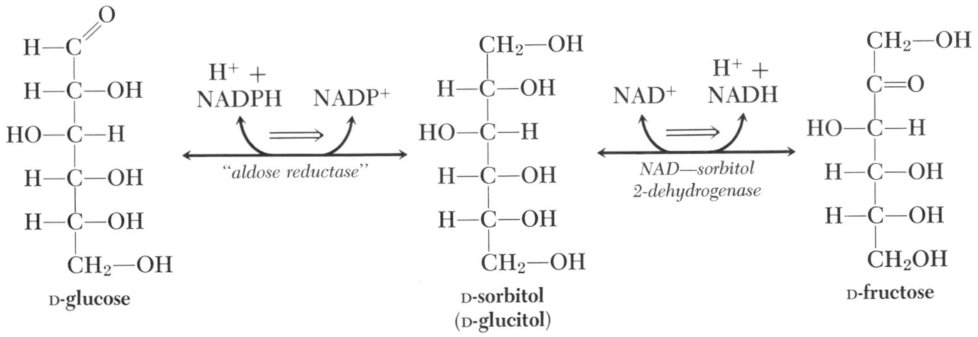

Figure 26-1   D-Glucose is converted to D-fructose in the seminal vesicle by reducing C-1 to an alcohol group, followed by oxidation of C-2 in the intermediate sorbitol. The enzyme catalyzing the initial reaction is an NADP—polyol 1-dehydrogenase commonly named "aldose reductase."

hydrogenase of liver, built to oxidize lactate to pyruvate. (This also illustrates why "aldose reductase" is bad nomenclature for what ought to be called an NADP—polyol 1-dehydrogenase. We don't call the lactate dehydrogenase of muscle pyruvate reductase, even if that might indicate the physiological direction of the reaction.)

The other factor favoring the formation of fructose is the state of oxidation of the two nicotinamide nucleotides. The ratio of [NADPH]/[NADP] is high and the ratio of [NADH]/[NAD] is low in the cytosol fraction where these enzymes occur. In other words, the simple pair of dehydrogenase reactions provides a means for concentrating fructose in the seminal plasma at the expense of blood glucose without any energy expenditure beyond that required to maintain favorable concentrations of the corresponding nicotinamide nucleotides.

The rationalization for the presence of fructose rather than glucose is that it provides a reservoir of fuel for the spermatozoa that is not readily utilized by the cells of the seminal vesicles or the vagina without preventing the cells from utilizing glucose in the ordinary way. The spermatozoa have an active hexokinase that converts fructose to fructose-6-phosphate, which is then metabolized by the usual Embden-Meyerhof pathway. The end products are lactate when the oxygen supply is low, and CO_2 and H_2O when adequate oxygen is available, thereby enabling survival of transient exposure to anaerobic conditions with utilization of the more efficient complete oxidation when possible.

The Sorbitol Pathway in Other Tissues

Some herbivorous animals, notably sheep, use the sorbitol pathway as a device for supplying the fetus with carbohydrate. The placenta of these animals forms sorbitol and then fructose from the maternal glucose, with both of these compounds appearing in the cord blood. Man and the other primates do not use this device, and fetal man utilizes glucose in the same way as do his parents. It is not clear why sheep need more protection.

The sorbitol pathway does occur in other human tissues for reasons that are not clear. The lens of the eye and some nerves make sorbitol and fructose. In the case of the nerves, the "aldose reductase" is believed to occur in the Schwann cells surrounding the nerves, while the sorbitol dehydrogenase is in the axon. The point in mentioning these currently inexplicable locations is that they are believed to be responsible for the damage to the tissues that sometimes accompanies high blood concentrations of hexoses. Cataracts form in the lens during diabetes and when excessive amounts of galactose, whose metabolism we shall shortly consider, are eaten. The cause appears to be due to reduction of the sugars, with excessive accumulation of sorbitol in diabetes and of dulcitol, the hexitol analogue of galactose, in galactose toxicity. The elevated concentration of the hexitols is believed to cause osmotic swelling of the lens. It has also been suggested that the damage to peripheral nerves in diabetes (diabetic neuropathy) is caused by a similar excessive formation of sorbitol.

Metabolism of Fructose in Adults

Fructose is sometimes a major constituent of the diet of adults, since half a mole is formed from each mole of ingested sucrose. Honey also contains a high

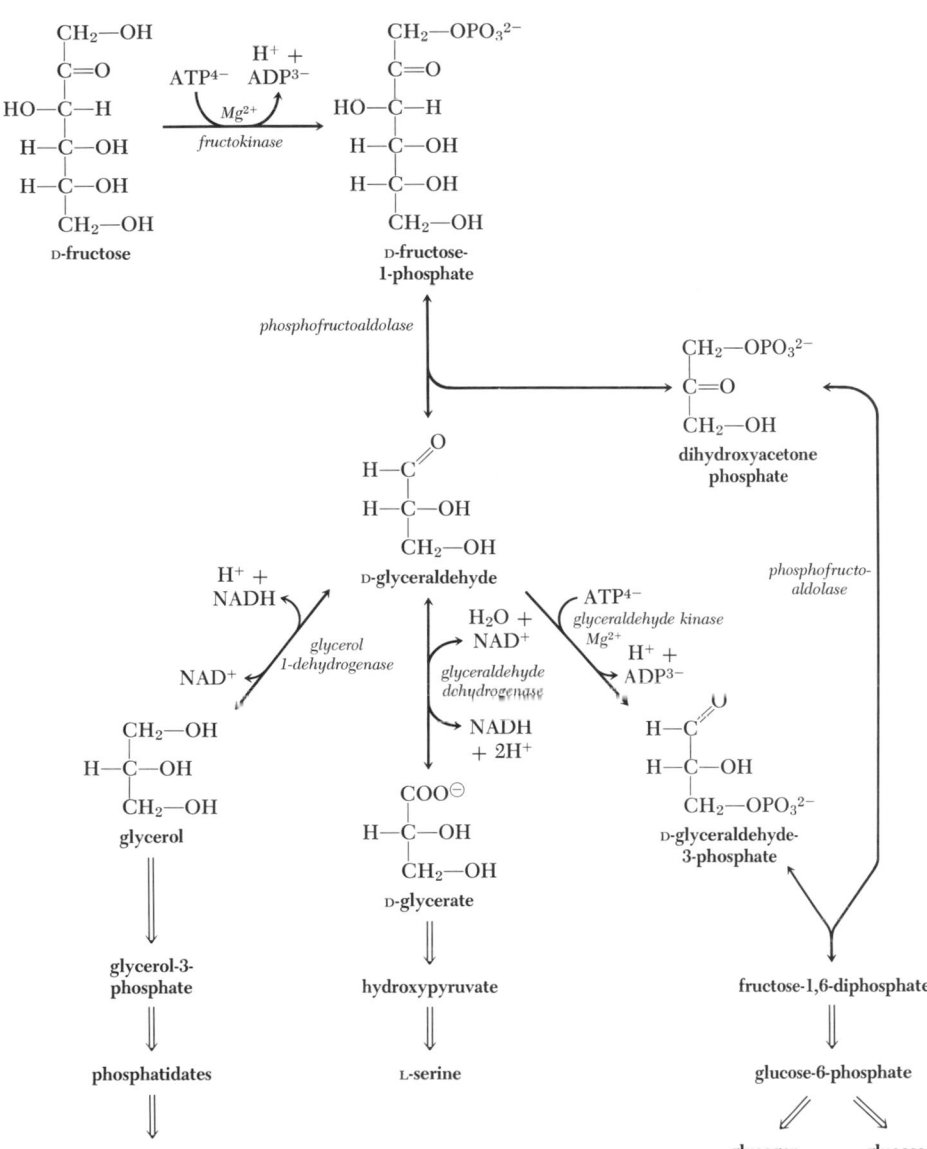

Figure 26-2 The metabolism of fructose in the liver begins with a phosphorylation at C-1. The resultant fructose-1-phosphate is cleaved by a hepatic isozyme of phosphofructoaldolase to form free glyceraldehyde and dihydroxyacetone phosphate. Glyceraldehyde may be metabolized by several routes, including an initial oxidation or reduction, or a phosphorylation.

proportion of fructose. We know that people can eat large quantities of sucrose and thrive, so it isn't surprising to discover that the liver has a high capacity for handling fructose. (The liver can also oxidize sorbitol to fructose at a fast rate, but it isn't able to convert glucose to sorbitol and therefore doesn't manufacture fructose from glucose.) The route for doing this is quite different than the route of fructose metabolism in spermatozoa. The liver has a specific fructokinase that catalyzes the phosphorylation of C-1, forming fructose-1-phosphate.

Fructose-1-phosphate is attacked by an isozyme of phosphofructoaldolase that catalyzes the cleavage of the compound almost as rapidly as it does the cleavage of fructose-1,6-diphosphate (Fig. 26-2). The nature of the reaction is the same except that free glyceraldehyde, rather than glyceraldehyde-3-phosphate, is a product in addition to the usual dihydroxyacetone phosphate. Glyceraldehyde may be metabolized in a variety of ways. Part is phosphorylated by a specific kinase and becomes a precursor of glucose. Part is reduced to glycerol and then phosphorylated to provide glycerol-3-phosphate as a precursor of lipids. Still another part is oxidized first to glycerate and then to hydroxypyruvate, as a precursor of serine.

In addition to all of these routes, there is some evidence for an independent pathway that may involve a fructose nucleotide, in which fructose is converted to glucose without passing through sorbitol or the triose phosphates. In any event, fructose is readily converted to the normal intermediates of carbohydrate metabolism in the liver and can act as a replacement for other sources of glucose. Indeed, dogs treated with phlorhizin spill glucose in their urine almost as rapidly if they are fed fructose as they do when they are fed equivalent amounts of glucose, and 95 per cent of the fructose can be recovered as glucose.

Some humans have genetic defects causing deficiencies in the enzymes necessary for handling fructose. Nothing much happens to those lacking the specific fructokinase beyond the appearance of fructose in the urine. Those who lack the fructose-1-phosphate aldolase isozyme are in serious difficulty. They evidently have the isozymes that are more active toward fructose-1,6-diphosphate, which are necessary for normal gluconeogenesis, so they get along reasonably well until they are exposed to fructose. This frequently happens during infancy when an artificial formula is supplemented with sucrose. According to current thought, the deleterious effects are a result of accumulation of fructose-1-phosphate in the liver and kidneys. The accumulation is believed to suppress gluconeogenesis; if so, this accounts for the observed low blood glucose concentration caused by fructose feeding. Vomiting and loss of appetite are common. The kidney function is seriously disturbed; amino acids appear in the urine, and there is also a renal tubular acidosis, with plasma total CO_2 often falling below 10 mM. The treatment is simple: Rigorously avoid any source of fructose in the diet, including fruits as well as candies and cakes.

THE METABOLISM OF LACTOSE

The infant living on the milk produced by his mother's mammary glands generates approximately 60 per cent of his ATP by oxidizing the fat in the milk and 40 per cent by oxidizing the disaccharide, lactose, which is β-D-galactosyl-(1→4)-D-glucose (Fig. 26-3). Lactose occurs in human milk at a concentration of

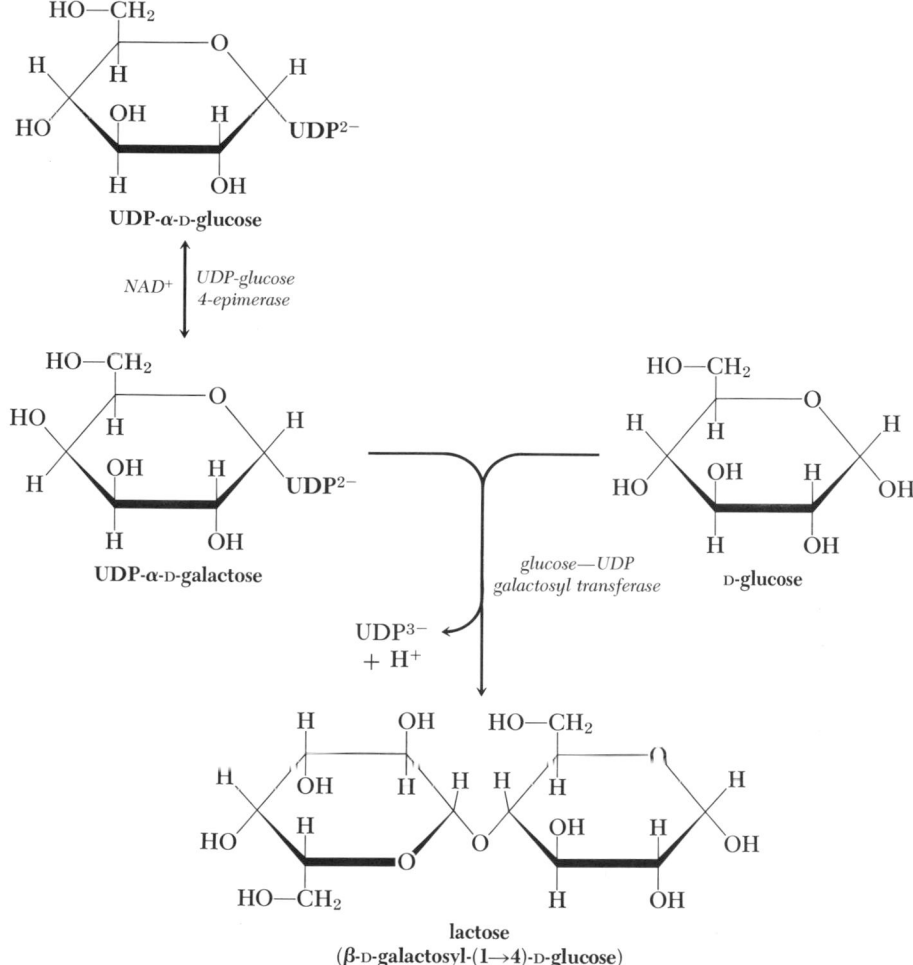

Figure 26-3 Lactose is formed in the mammary gland by transfer of a galactosyl residue to glucose. The galactosyl residue is created by epimerization of UDP-glucose. Note that the galactosyl residue has been turned over from the usual Haworth representation in the formula for lactose so as to have the β-configuration directed down.

200 mM, compared to the 4.5–5.5 mM glucose in the maternal blood from which it is made. Here again we have an example of the concentration of a fuel that is not utilized by the parent tissue as a means of protecting the new generation.

The mammary glands form the galactosyl portion of the lactose molecule by a straightforward epimerization of UDP-glucose, and then form lactose by a transfer of the galactosyl residue to glucose. The enzyme responsible for the transfer is activated in a remarkable way, quite unlike anything we have seen. It may be unique or it may be the first known example of a more general mechanism of control.

Two proteins must be present for the synthesis of lactose. One of these is the galactosyl transferase. However, this enzyme, or something very like it, occurs in

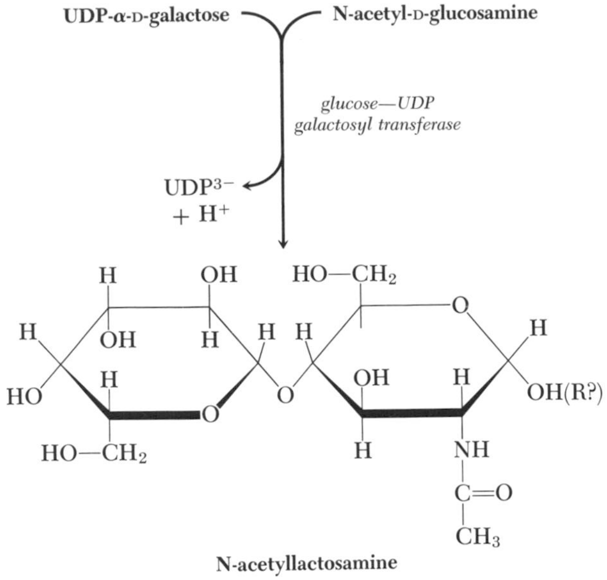

Figure 26-4   Glucose—UDP galactosyl transferase is used to form lactose in the mammary gland as shown in the previous figure, but it also occurs in other tissues, where the normal acceptor for the galactosyl residue is N-acetyl-D-glucosamine rather than glucose. The specificity is different in the mammary gland because the enzyme is modified by the presence of α-lactalbumin. It may be that N-acetyllactosamine is not created as such in other tissues, but as a part of glycoprotein polysaccharides.

a variety of tissues in addition to the mammary glands, and its normal substrate is N-acetyl-D-glucosamine to which the galactosyl residue is transferred with the formation of N-acetyl-D-lactosamine (Fig. 26-4). The enzyme is believed to be involved ordinarily in the formation of glycoproteins rather than lactose. The second protein is an otherwise inert constituent of milk, α-lactalbumin, which comprises about 2 per cent of the total milk proteins. In addition to being one of the secreted constituents, the intracellular lactalbumin acts as a modifier of the galactosyl transferase, combining with the enzyme and *changing its specificity*, so that glucose, rather than N-acetyl-D-glucosamine, becomes its substrate. If one isolates the transferase from liver and adds lactalbumin to it, then the liver transferase also forms lactose. In other words, the mammary glands use a widely distributed enzyme and change it into a very specific enzyme by producing another protein that combines with the enzyme, and the uniqueness of the mammary glands in producing lactose depends not upon the unique occurrence of the enzyme, but upon the unique occurrence of the modifying protein, lactalbumin.

The development of the mammary glands is stimulated during pregnancy by a peptide hormone from the anterior pituitary gland, *prolactin*. Prolactin also causes an increased formation of both the galactosyl transferase and α-lactalbumin, so that one would think the mammary glands would be producing lactose during pregnancy, but they don't to any significant degree, and the reason that they don't is regulation by still another hormone. After ovulation occurs, the ovarian follicle develops a structure known as the corpus luteum, which secretes a steroid hormone, *progester-*

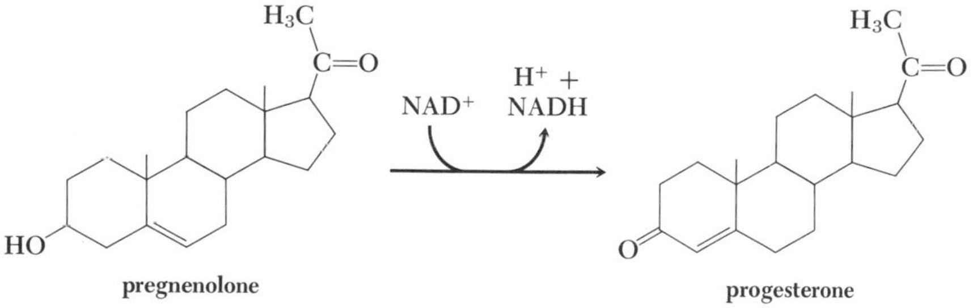

Figure 26-5 The steroid hormone, progesterone, is formed from pregnenolone by the ovarian corpus luteum and by the placenta.

one (Fig. 26-5), under stimulation by prolactin. This structure enlarges during pregnancy and then declines during the last few months, but the placenta then takes over the formation of progesterone. Now, progesterone *inhibits the formation of* α-lactalbumin by the mammary glands even though prolactin is being produced in increasing amounts. Therefore the mammary glands make more enzyme as term approaches, but lack the modifier that permits them to make use of the available substrate, glucose. When the placenta is delivered, the source of progesterone is removed and the increased formation of prolactin can trigger prompt synthesis of α-lactalbumin, permitting lactose synthesis to get under way.

The Utilization of Lactose

The first step in the utilization of lactose is hydrolysis to the constituent galactose and glucose monosaccharides. The brush border of the intestinal mucosa contains *lactase*, a particular kind of a *β-galactosidase* that attacks lactose. Infants have one kind of this enzyme in high concentration, as is appropriate for their specialized diet, while adults develop another to handle the usually lighter load. We should say some adults, because many humans lack the adult form of the enzyme and cannot absorb lactose.

In a study made on adult Americans of various ancestries, 19 out of 20 Orientals, 14 out of 20 Negroes, and two out of 20 Caucasians developed symptoms of lactose intolerance after eating 50 grams of lactose (the amount in a quart of cow milk). The symptoms in Negroes have been shown to be due to a deficiency of intestinal lactase, and it is likely this is also the cause of the symptoms in the others. (The symptoms are abdominal cramps and diarrhea, caused by the osmotic effect of the lactose and its fermentation by intestinal microorganisms.) In short, milk is not a very good food for many adults, and it has been pointed out that shipping dried milk to impoverished lands may cause more griping guts than gratitude.

Be this as it may, children of all kinds are well equipped to handle lactose, and let us now consider what they do with the products of hydrolysis. Glucose is metabolized in the usual way, but galactose must be converted to the normal metabolic intermediates. The conversion begins with a phosphorylation catalyzed by a specific galactokinase to form galactose-1-phosphate (Fig. 26-6). The next step is a transfer of a UMP group from UDP-glucose to galactose-1-phosphate, forming UDP-galactose and leaving glucose-1-phosphate. Finally, UDP-galactose is con-

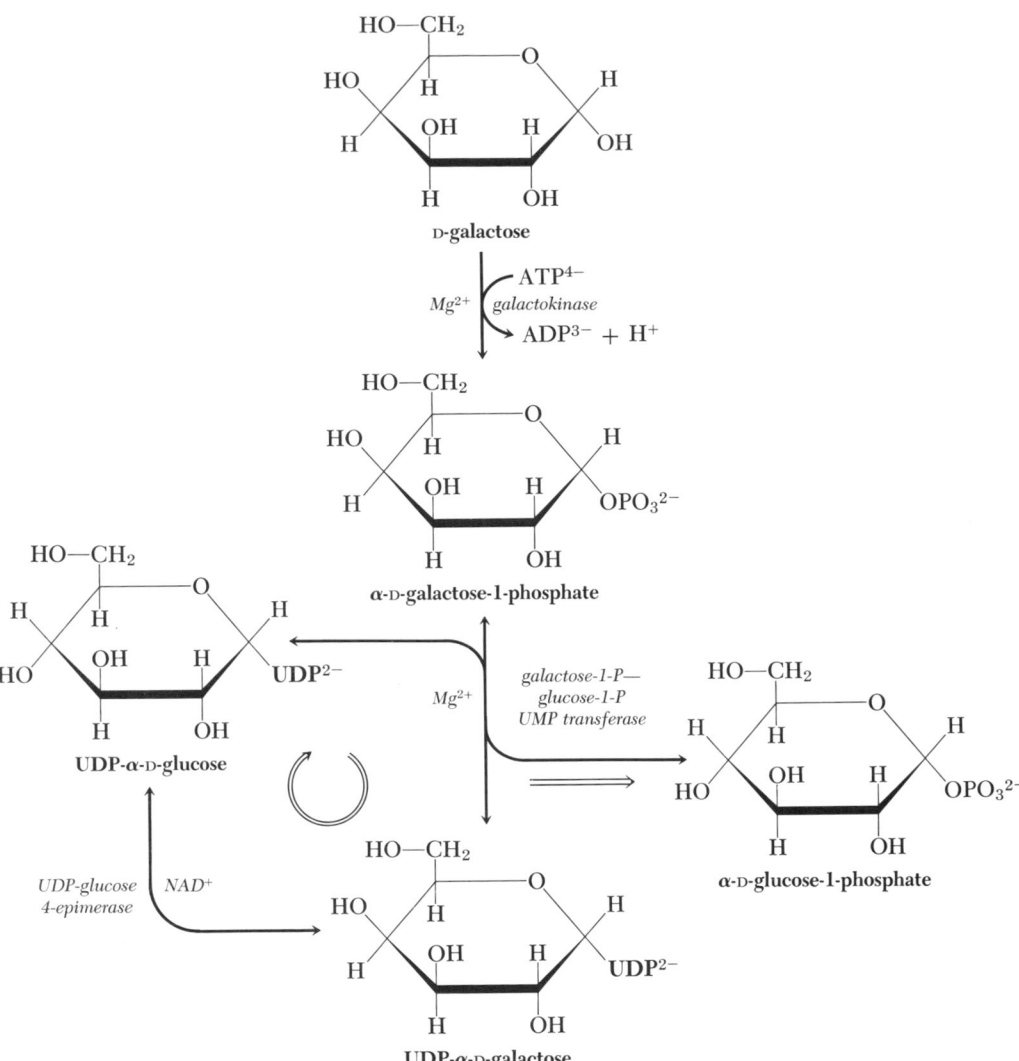

Figure 26-6 The utilization of the D-galactose obtained by hydrolysis of ingested lactose involves an initial phosphorylation at C-1, followed by a transfer of UMP from UDP-glucose to form UDP-galactose. The transfer releases glucose-1-phosphate, which is the net product of the sequence, since the UDP-glucose consumed is replaced by an epimerization of UDP-galactose.

verted to UDP-glucose by the same epimerase reaction that is used in a reverse direction in the mammary gland to make galactosyl residues. The sum of all this is the conversion of galactose-1-phosphate to glucose-1-phosphate:

1. galactose-1-phosphate + UDP-glucose ⟷ UDP-galactose +
                                                   glucose-1-phosphate
2.   UDP-galactose ⟷ UDP-glucose
SUM: galactose-1-phosphate ⟷ glucose-1-phosphate

Some infants have defects of galactose metabolism resembling those affecting fructose metabolism in other infants. If the kinase is missing, galactose is excreted, and the accumulation of galactose leads to formation of cataracts through reduction to the corresponding hexitol, dulcitol. If the transferase is missing, the resultant accumulation of galactose-1-phosphate has serious consequences resembling those caused by the accumulation of fructose-1-phosphate in hereditary fructose intolerance, except that the symptoms usually appear at an earlier age owing to the exposure of the infant to lactose within a few days after birth. The same vomiting, loss of appetite, low blood glucose concentration, and excretion of amino acids in the urine appear, along with early and irreversible mental retardation. This hereditary galactose intolerance is a surprisingly common condition, estimated as occurring in one out of 18,000 infants. Since it is an autosomal recessive condition, the heterozygotes are perhaps one out of 140 in the population. This high incidence leads us again to wonder if there is some unforeseen advantage conveyed by the heterozygous condition, but there is no clue to such an advantage at present.

VISION AND VITAMIN A

The image that we "see" in our brain is constructed by the organization of impulses that are generated by the absorption of quanta of light in the rods and cones near the surface of the retina. In order to translate incoming light into impulses, the eye must contain compounds that will absorb quanta of particular wavelengths, in this case 400 to 700 nanometers This is a useful, indeed a necessary, range of wavelengths for good vision—useful not only because the peak energy of sunlight at the earth's surface lies in the middle of the range, but also because it is the range enabling the greatest distinction between objects. Light of much shorter or longer wavelength is absorbed by a variety of chemical structures, and we distinguish objects because of selective absorption by the unusual chemical compounds they contain. In the biological realm, peptides, carbohydrates, and lipids are colorless. In the surroundings, water and the common minerals, except those containing iron, are colorless. We distinguish rubies and sapphires not by the colorless aluminum oxide crystal lattice, but by contaminants of the lattice, and we detect the jaundiced person through the selective absorption of quanta by the relatively small concentrations of bilirubin.

It is apparent from this that the light-sensitive compound in the eye must also have an unusual structure; it must exist in a number of electronic states of relatively small energy difference so that quanta of light in the visible range can cause excitation, and at the same time we would expect this important compound to differ sufficiently in structure from other components of the retinal cells so that it is not easily removed by the usual metabolic reactions.

We have noted that animals frequently do not synthesize unusual structures when compounds containing the structure are available from the diet, and this is the case with the visual pigment. The necessary highly resonant structure is an isoprenoid hydrocarbon chain containing conjugated double bonds that is contributed by vitamin A in the diet. Let us consider the nature of the vitamin and then go on to consider how it functions.

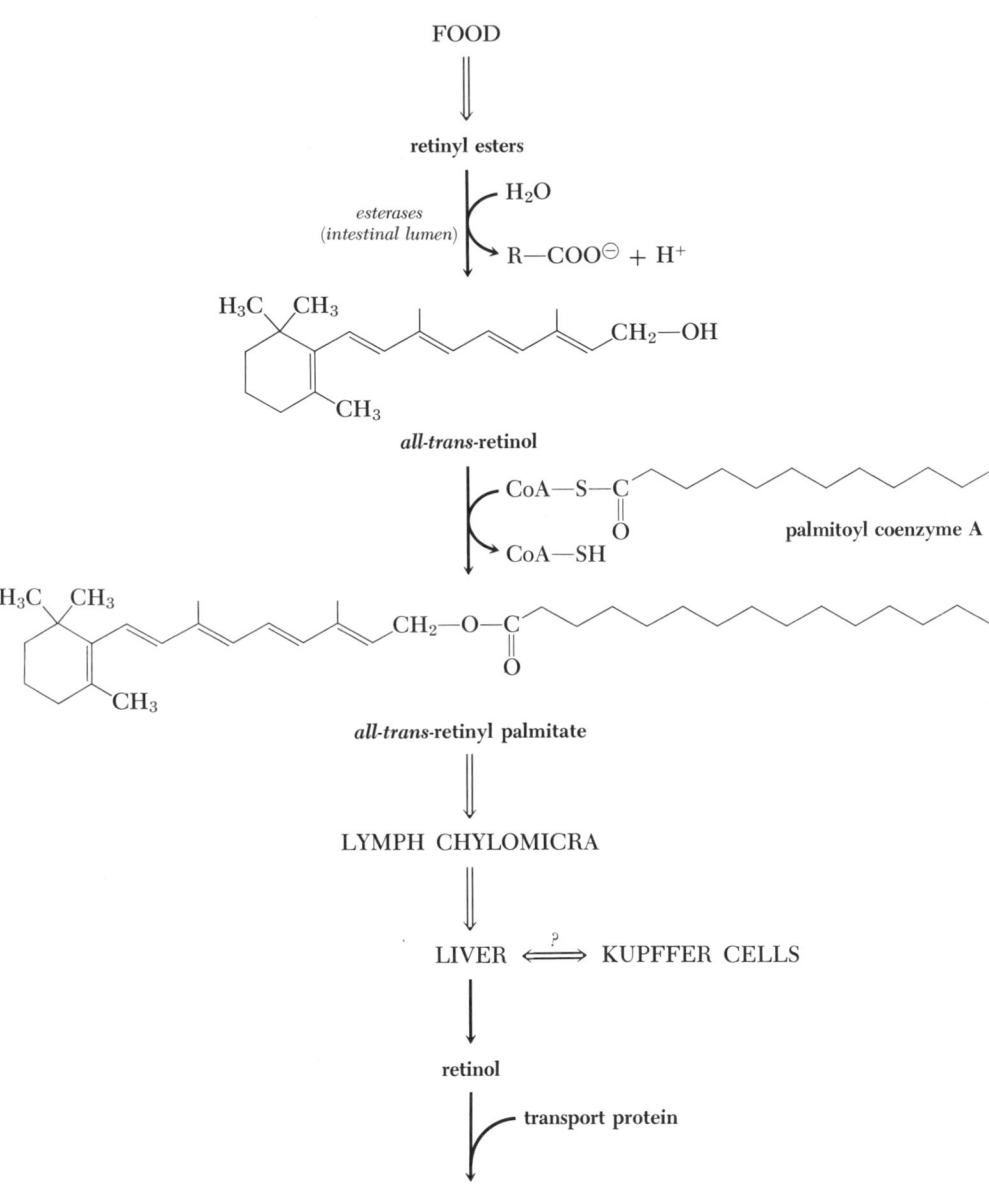

Figure 26-7   Much of the vitamin A ingested in the food is in the form of fatty acid esters, which are hydrolyzed in the gut to liberate free retinol. Retinol is converted specifically to the palmitoyl ester in the intestinal mucosa, and this non-polar compound is transported by solution in the triglyceride droplets of chylomicra. The liver takes up retinyl palmitate, perhaps storing most of it in the Kupffer cells, but also hydrolyzing part of it for release into the blood, where it is bound to a specific transport protein.

The Nature of Vitamin A

The vitamin A of mammals is a 20-carbon alcohol, *retinol* (Fig. 26-7), that occurs in the diet largely as the *all-trans* isomer esterified with fatty acids, with the major component frequently being *all-trans*-retinyl palmitate. The dietary esters are hydrolyzed in the lumen of the small bowel, forming free retinol that passes into the intestinal mucosa, where it is again esterified by reaction with palmitoyl coenzyme A. The palmitoyl ester dissolves in the triglyceride portion of the chylomicra and passes into the lymph. The ester is eventually removed from the blood by the liver and stored, mostly by the Kupffer cells, and a gram of liver may contain as much as 2 micromoles of retinyl esters, although 0.25 micromoles is a more common value. The ester is constantly being broken down and rebuilt, and the parenchymal cells of the liver contain free retinol, which is exported by a mole for mole combination with a particular transport protein. There are only 40 mg of this retinol carrier protein per liter of blood plasma, and the normal plasma concentration of retinol is between 1 and 2.5 micromolar.

The original source of vitamin A is a class of isoprenoid compounds known as *carotenes* that are synthesized by plants. (The synthesis differs from that of cholesterol in that two molecules of the 10-carbon geranyl pyrophosphate are condensed to make 20-carbon geranylgeranyl pyrophosphate, which then condenses with itself in a reaction reminiscent of squalene synthesis to form a 40-carbon carotene precursor.) Many of these carotenes that occur in plants have the vitamin A structure on half of the hydrocarbon chain, but *β-carotene* (Fig. 26-8) is equivalent to two molecules of retinol linked head to head with a concomitant reduction.

Some of these carotenes are absorbed as such and dissolve in the lipid portion of lipoproteins. (The normal plasma concentration is between 2 and 4 micromolar. Some leporine people turn yellow from the much higher concentrations obtained through their proclivity for eating carrots and the like in wholesale quantities. This carotenemia is otherwise harmless, as is the lycopenemia rarely seen in some tomato enthusiasts. The latter become a bright red-orange from the absorbed lycopene, which is not useable by humans.) Most of the absorbed dietary carotene is attacked by an oxygenase in the intestinal mucosa that splits the molecule into two aldehydes; β-carotene is converted in this way to two molecules of *retinal*, the aldehyde derivative of retinol. Other carotenes yield but one molecule of retinal or none at all, depending upon the presence or absence of the retinene structure in one half of the molecule. The retinal that is produced is partly transported as such in the portal blood, but both the liver and the intestinal mucosa have alcohol dehydrogenases that catalyze the reduction of the aldehyde to the alcohol by NADH. In short, some of the carotenes have a biological vitamin A activity when they are eaten. The activity is usually less than that of retinol itself because the low solubility of the compounds makes them poorly absorbed.

The Visual Pigment

The transport of retinol into the retina appears to involve the removal of the compound from its plasma protein carrier and the formation of a fatty acid ester within the retinal cells. The mechanism of concentration is quite effective, because the level of retinol must drop to half its normal range before any loss of the store

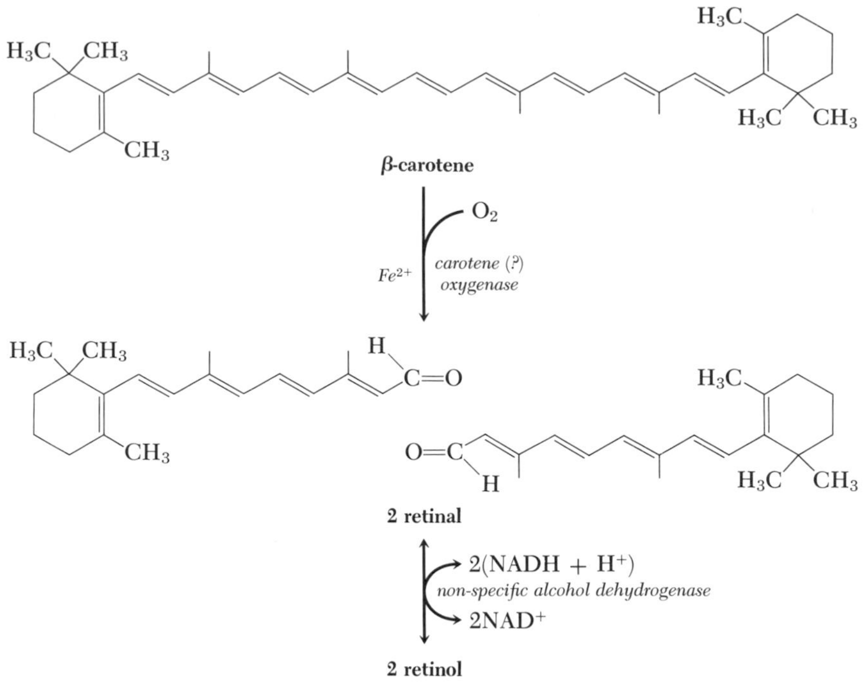

β-carotene

2 retinal

2 retinol

Figure 26-8   The carotene pigments of plants can serve as source of retinol. They are cleaved in the middle by an oxygenase to form two aldehydes. β-Carotene is a symmetrical molecule in which both halves have the retinene configuration, so the cleavage of this carotene results in the formation of two molecules of retinal, which is reduced to retinol. Other carotenes have the retinene structure in only one half of the molecule, or in neither.

Figure 26-9   Vision results from the absorption of a quantum of light by a complex of an isomer of retinal with a lipoprotein in the retina (*bottom right*). The lipoprotein appears to be constructed so that 11-*cis*-retinal will form a number of hydrophobic bonds with the peptide portion, known as opsin, and in a way facilitating the formation of an aldimine bond with the amino group of phosphatidyl-ethanolamine in the lipid portion. The bonding is such that the conjugated double bonds of retinal now have maximum light absorption in the middle of the visible spectrum.

The excitation created by absorption of a quantum of light apparently disrupts the hydrophobic bonding so that the 11-*cis* configuration reverts to the more stable *trans* form. Associated with this change—perhaps as a result of it—is a shift of the aldimine linkage from phosphatidylethanolamine to a lysyl residue on the opsin peptides. These transformations are believed to cause ionic movements that result in neural stimulation.

The resultant all-*trans*-retinal does not have a configuration permitting tight bonding to the opsin-phosphatidylethanolamine complex, and readily dissociates.

The original source of retinal is retinol arriving from the blood (*top*). Much is stored within the retina as fatty acid esters, but the free compound is partly converted to 11-*cis*-retinal by a dehydrogena-tion followed by an isomerization. The equilibrium position of both of these reactions is unfavorable, but the bonding between 11-*cis*-retinal and the opsin-phosphatidylethanolamine complex is so strong that only a small concentration of the retinal isomer is necessary to permit regeneration of the light-sensitive pigment.

See illustration on opposite page.

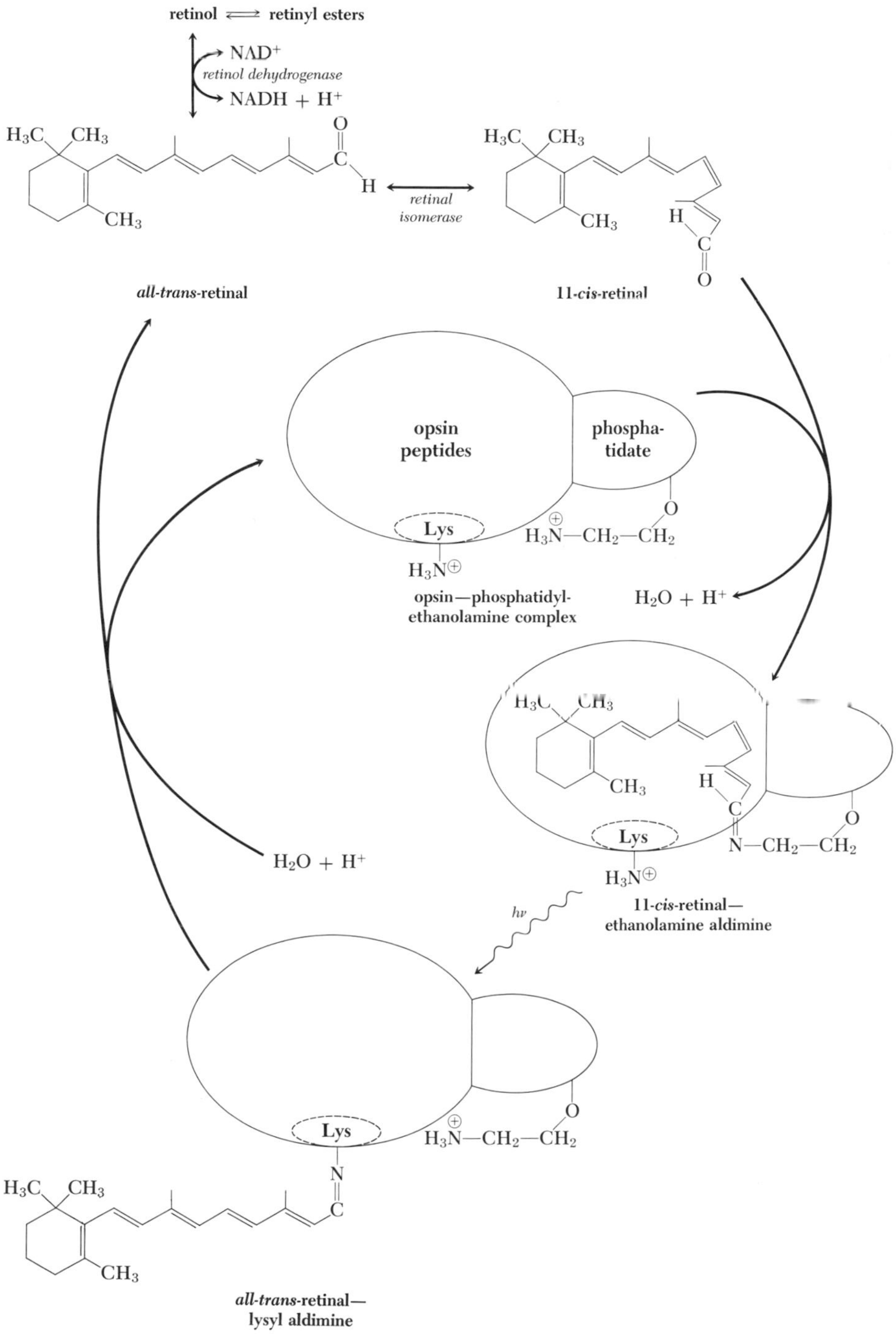

retinol ⇌ retinyl esters

NAD$^+$

retinol dehydrogenase

NADH + H$^+$

all-trans-retinal

retinal isomerase

11-*cis*-retinal

opsin peptides

phospha-tidate

Lys

H$_3$N$^\oplus$—CH$_2$—CH$_2$

H$_3$N$^\oplus$

opsin—phosphatidyl-ethanolamine complex

H$_2$O + H$^+$

Lys

H$_3$N$^\oplus$

N—CH$_2$—CH$_2$

11-*cis*-retinal—ethanolamine aldimine

H$_2$O + H$^+$

hv

Lys

H$_3$N$^\oplus$—CH$_2$—CH$_2$

N

all-trans-retinal—lysyl aldimine

Legend on opposite page

in the eye becomes apparent. The retinyl esters of the eye are subject to hydrolysis, and the retinol liberated within the cells is oxidized by NAD in a reaction catalyzed by a specific dehydrogenase.

The equilibrium position of retinol dehydrogenase does not favor retinal production. However, the retinal that is formed is rapidly removed to form complexes with proteins, the *opsins*, of the rods and cones, and oxidation of retinol continues until no more of these complexes, which are the light-sensitive pigments, can be formed—until the opsins are saturated.

One additional step is necessary before the visual pigment is made. The opsins are constructed so that they combine with the *11-cis* isomer of retinal, and there is an isomerase in the retina catalyzing the equilibration of the *all trans* isomer with this form. The position of equilibrium here also favors the original isomer, but the strong affinity of the *11-cis* form for the opsins pulls the whole sequence toward pigment formation (Fig. 26-9).

The opsins are embedded in a lipid environment that may serve the same function as that which we proposed for the electron transfer systems of mitochondrial membranes: the protection of intermediate energetic states from decomposition by more polar compounds. It now appears that the combination of *11-cis*-retinal with an opsin involves the formation of a Schiff's base with the amino group of phosphatidyl ethanolamine that is also associated with the opsin, together with a space-filling hydrophobic interaction of the remainder of the retinal chain with the opsin, which is a strongly hydrophobic protein.

The interaction of the conjugated hydrocarbon chain with the opsins creates the particular absorption spectrum of the visual pigments, and therefore the spectral sensitivity of the eye. The more sensitive rods have but one opsin, and the retinal-rod opsin complex is known as *rhodopsin*, or visual purple, because of its characteristic color. The cones in the central field appear to have three different opsins creating the three different spectra that are necessary for trichromatic vision (except for a very small area in the very center that lacks the blue-sensitive pigment).

The effect of an absorbed quantum of light that results in a neural impulse is an excitation that results in two events: a transfer of the Schiff's base linkage from phosphatidylethanolamine to an adjacent lysyl residue of the opsin peptide, and a shift of the configuration of the retinal chain from the *11-cis* form back to the more stable *all trans* form. The resultant shift of charge position is believed to be the cause of a cation transfer that initiates the neural impulse.

In any event, the final result is a change in the conformation of the opsin back to its free form that is not able to effectively combine with *all trans* retinal, which is released and mixes with the small pool in equilibrium with retinol and with *11-cis*-retinal (see Figure 26-9), which will react with the now liberated opsin to re-make the complex.

Other Functions of Vitamin A

Animals deprived of vitamin A not only go blind; they die. Before they die, they develop abnormal deposition of keratin in the mucous membranes, a failure of bone remodelling leading to thick, solid (cancellous) long bones like the bones in the skull, lesions of the nerves, an increased pressure of the cerebrospinal fluid causing hydrocephalus, testicular degeneration in the males, and abortion or malformed offspring in the females. Obviously, vitamin A is involved in some critical function in most tissues. We don't know what it is, but we do know that it probably doesn't involve the retinol-retinal interconversion, because the corresponding acid, *retinoate*, will prevent malfunction in all of the tissues except the reproductive tract and the eye. Since it won't save vision, it is apparent that retinoate is not readily reduced to retinal. We may be dealing here with a considerably more fundamental and primitive function of the carotenes, with their use in the development of vision being a fortuitous result of the accompanying absorption spectrum.

Supply of Vitamin A

It is difficult to make all but the very young deficient in vitamin A for two reasons. One is the common occurrence of the vitamin and the carotene precursors in plants, in most fish tissues, in eggs, and in mammalian and chicken liver. Most weaned humans eat one or more of these food classes. The other reason is the large capacity for storing the retinyl esters in the liver. The estimated requirement for humans past the weaning age is on the order of 6 micrograms per kilogram body weight per day (or 20 International Units; 1 I.U. $= 0.3\,\mu g$ retinol). This is on the order of 1.5 micromoles per day for an adult male. Now, the usual adult male will have around 0.25 micromole in each gram of liver, and some store as much as 2 micromoles per gram of liver. Given a liver ranging around 2.3 per cent of the total body weight, we see that our usual male carries around a nine-month supply, and some may have a four-year supply of retinol.

However, there are conditions in which the supply can be depleted, and most of these hinge on improper absorption. We shall have more to say about some of these conditions in the next chapter when we discuss severe malnutrition. The other conditions are any pathological changes that cause deficient absorption of lipids. If a quantity of fat is not digested and passes out in the stools, it will carry with it much of the dietary retinyl esters and carotenes (along with dietary vitamin D and vitamin K). This is no problem with transient diarrheas, but potential vitamin A deficiency is something to consider in any chronic disease of the pancreas, intestine, or liver. A failure of rod vision (night-blindness) is an early symptom, but it is also a result of other conditions, and vitamin A deficiency must be established by careful measurement, not by subjective report. The really characteristic sign is *xerophthalmia*, a dry keratinization of the cornea and conjunctiva of the eye; this is a late symptom and an indication for prompt therapy.

Retinol, like cholecalciferol, is toxic in excess. Acute poisoning has long been known from the fatal result of eating polar bear livers, which contain as much as 30 micromoles of retinyl esters per gram—a 20-year supply for a human in each pound. Severe headache, vomiting, and prostration result within a few hours. Doses of 300 micromoles in infants produced a transient hydrocephalus.

The more usual cases are those in which the intake has been relatively high for a long period of time, and chronic toxicity is frequently manifested by fatigue, loss of appetite, an enlarged liver, diffuse pains in the muscles, coarsening and loss of hair, scaly skin eruptions, and attenuation of the long bones that sometimes results in fractures. Diagnoses of pyschoneurosis or even schizophrenia seem to be made quite frequently. Most of the severe cases have been ingesting on the order of 500 micromoles of retinol per day (500,000 I. U.) for a few years—sometimes from overzealous treatment for other conditions, and sometimes from self-dosage with easily available commercial preparations (or from maternal over-dosage, in the case of infants). The minimum quantity necessary for the appearance of chronic toxicity has been suggested as 50 micromoles per day for 18 months (about 10 times the current recommended daily allowance).

BLOOD CLOTTING

The formation of clots to plug defects in blood vessels is one of the more important defense mechanisms of the body. We readily appreciate the potential loss of blood from gaping wounds because we can see it pouring out, but the constant minor injuries to the small vessels are only appreciated when we see the multiple little hemorrhages that occur in people with a defective blood clotting mechanism. These injuries can result from the simple stresses of motion and the accompanying contacts with physical objects, from ordinary chewing, from the normal motion and loss of cells of the gastrointestinal tract, and so on.

Minimizing blood loss is accomplished by three events, and it is important not to lose sight of these three in pursuing the subject in more detail. One is through a vasoconstriction of the injured vessel that immediately reduces the flow through the break; this physiological reaction doesn't completely stop the loss and we shall not discuss it further. The second event is a clumping of platelets from the blood at the site of injury to plug the opening temporarily. The third event is the aggregation of a protein, fibrin, into a stable three-dimensional lattice that is strong enough to seal the damaged vessel while the injury is being repaired. It is the chemistry of the latter two events with which we shall be concerned.

The Platelet Plug

The platelets, or thrombocytes, of the blood are disk-shaped cells about 1×3 microns in size that lack nuclei but contain mitochondria and other cytoplasmic organelles. The blood normally contains about one platelet for each 20 erythrocytes, and they are believed to be derived by a pinching off of pieces of the megakaryocyte, a large cell found in the bone marrow. The plasma membrane of the platelets is coated with what is believed to be a sulfated polysaccharide. In any event, the platelets have the remarkable property of adhering tightly to collagen fibers. So long as the endothelial lining of the vessel is intact, platelets and collagen don't meet, but if the lining is disrupted by overt injury or by a pathological change, such as atherosclerosis, the collagen fibers of the vessel itself or the neighboring tissue are exposed and platelets stick to them, perhaps by the same sort of forces that develop between collagen fibers and the proteinpolysaccharides of the ground substance.

The platelets undergo a remarkable morphological change when they become stuck. The granules of the cytoplasm disappear and intracellular components, notably ADP, appear outside the cell. The shape of the cell changes from a disk to a spiny sphere. These changes also make the cell adhere to still more platelets arriving in the blood, which in turn undergo the same sort of changes. The ADP released is a critical component for the development of stickiness, but how it acts is not known. The growth of the platelet plug will continue until blood flow is stopped, thereby preventing the arrival of more platelets, or until the formation of the fibrin lattice (to be described next) engulfs the plug.

The agglomeration of platelets serves two functions. One is the formation of a barrier, although a weak barrier, to the flow of blood, and the other is the exposure of the lipoprotein component of the platelet plasma membrane, which acts as an initiating agent for clot formation. In other words, the platelets are also necessary for the formation of the fibrin lattice.

The Formation of Fibrin

The blood plasma contains 1.5 to 3.0 mg per ml of a protein, *fibrinogen*, that is the precursor of fibrin. Fibrinogen is a molecule with a large surface, being about 70 nanometers long but only 3.8 nanometers in diameter, but it doesn't agglomerate in the way many long protein molecules do because it contains regions at each end that have a number of glutamyl and aspartyl residues. The resultant concentrated negative charge causes repulsion of approaching molecules that exceeds the other forces tending to cause interaction. The molecule is made of three pairs of unlike peptide chains (M.W. = 340,000).

When clotting begins, fibrinogen is modified so that it can associate, and this is done by simply hydrolyzing short segments ($\sim$2000 M.W.) from each of two of the three kinds of peptide chains. Since the chains occur in pairs, this involves the removal of four small peptides. Now, the peptides that are removed are the segments containing the concentration of carboxylate side chains, so the remaining portion of the fibrinogen molecule, the fibrin monomer, is no longer prevented from associating by charge repulsion. The hydrolysis of the peptides is catalyzed by an enzyme, *thrombin*, that is discussed in the next section. (The fibrin monomer still contains three peptide chains, but two of these have been shortened by hydrolysis.)

The result is an aggregation of fibrin monomers held together by hydrogen bonds, hydrophobic interaction, and similar forces. This array is still too weak to serve as a semi-permanent plug until final repairs are made, and it is strengthened by the formation of covalent bonds between the monomers (Fig. 26-10). These bonds are formed by the action of a *transamidase* that catalyzes the substitution of the amino group of a lysyl residue on one fibrin monomer for the amide nitrogen of a glutaminyl residue on another monomer; the result is the formation of a 6-(5-glutamyl)-lysyl cross-link between the monomers. (The cross-link is also designated as ϵ-(γ-glutamyl)-lysyl.) The cross-links give rigidity to the clot, which now consists of a lattice in which varying numbers of platelets, erythrocytes, and leukocytes are trapped.

The final step is a shrinkage of the clot due to the action of a contractile protein, thrombosthenin, in the platelets. The contraction of thrombosthenin, like that of myosin in the muscles, occurs at the expense of ATP hydrolysis. The physiological function of contraction is not known.

The Formation of Thrombin

The scheme we have outlined would cause the formation of fibrin clots whenever the substrate fibrinogen and the enzymes, thrombin and fibrin transamidase,

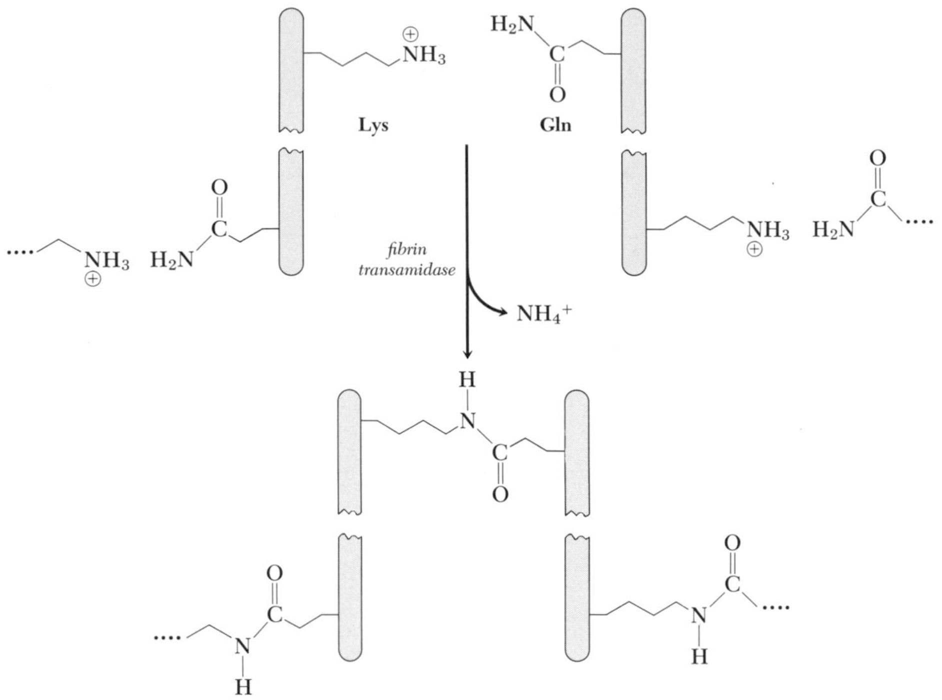

Figure 26-10 The formation of a fibrin polymer involves reaction between lysyl and glutaminyl residues on separate monomer molecules, with a substitution of the amino group on the lysyl side chain for the ammonia moiety of the carboxamide group on the glutaminyl side chain. The result is the formation of amide bonds between monomers and the liberation of ammonia.

are present in the same solution. This would be a disastrous event if it occurred in the normal parts of the circulation; since fibrinogen is always available in the blood, it appears that the enzymes must not be present under normal conditions, and this is indeed the case. Both thrombin and the transamidase exist in the form of *proenzymes* that must be modified by partial hydrolysis of the peptide chains before they become active, much in the way that fibrinogen is modified before it can aggregate. It turns out that the protransamidase is hydrolyzed by the action of thrombin, so the control of clotting hinges upon the conversion of prothrombin to thrombin.

   Prothrombin is a glycoprotein that occurs in a complex with three other glyco-proteins. There is still some dispute as to whether the prothrombin complex is to be regarded as a molecular entity made of several kinds of peptides that can be artificially separated, or as an association of four types of molecules. In any event, one of the other glycoproteins, the Stuart factor or Factor X, is also a proenzyme, and it is activated to become an enzyme that hydrolyzes prothrombin to convert it to thrombin.

   So far so good: The active Stuart factor attacks prothrombin to form thrombin; thrombin attacks fibrinogen to form fibrin monomer; thrombin also attacks fibrin

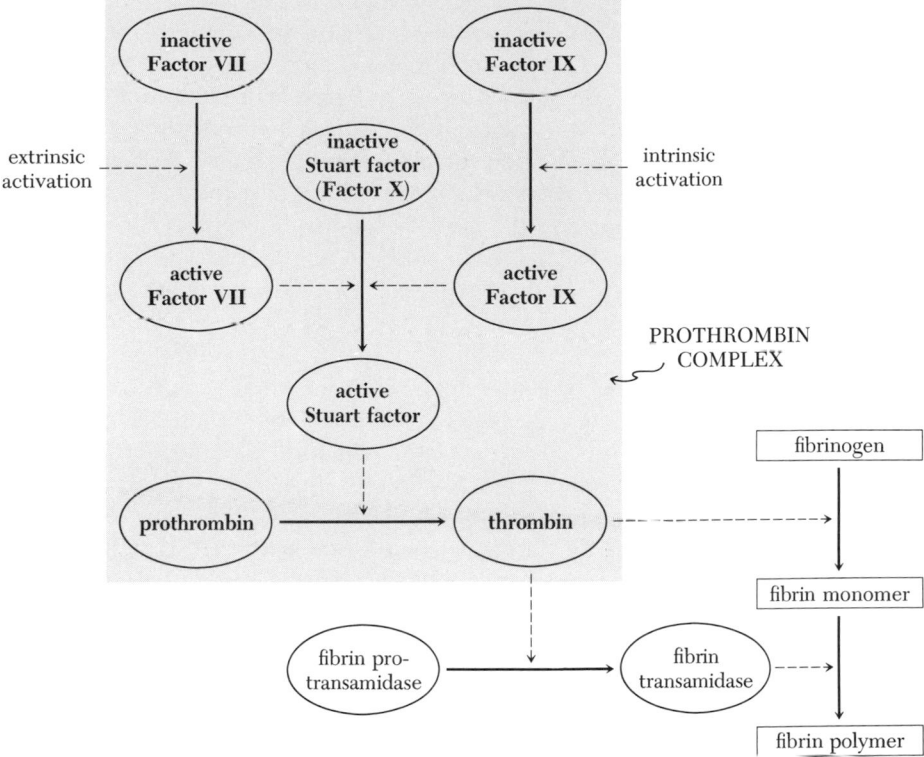

Figure 26-11 The formation of fibrin polymer (*lower right*) to make a strong clot involves the activation of enzymes in a glycoprotein complex, the prothrombin complex (*shaded area*). The complex includes proenzymes (Factors VII and IX) that are converted to active forms by tissue damage (extrinsic activation) or by alterations of the blood itself (intrinsic activation). When either of these factors is activated, it catalyzes the conversion of an inactive Stuart factor (Factor X) to its active form. The active Stuart factor is a peptidase that converts prothrombin to thrombin. Thrombin is another peptidase with two functions: It hydrolyzes fragments from fibrinogen to form a fibrin monomer that can aggregate, and it activates the fibrin transamidase that creates amide bonds between the fibrin monomers, as outlined in Figure 26-10.

The figure does not show the nature of the extrinsic and intrinsic activations by which these events are precipitated within the prothrombin complex. The extrinsic activation is believed to hinge on the combination of lipoproteins from damaged tissue with a component of the blood platelets. The intrinsic activation hinges on changes in the platelets upon contact with exposed collagen. Some believe that the intrinsic pathway involves successive activation of still other proenzymes (Factors XII→XI→IX) before the prothrombin complex is activated. Conflicting views on this point are cited at the end of the chapter.

protransamidase to form fibrin transamidase; fibrin transamidase forms covalent bonds between the fibrin monomers. This leaves the question of how active Stuart factor is created in the prothrombin complex. Here there is a divergence. Both of the remaining glycoproteins in the complex appear to be proenzyme forms of enzymes that catalyze the conversion of inactive Stuart factor to the active form. One of these glycoproteins is responsive to a chain of events initiated by exposure of the blood to surrounding tissues (the extrinsic clotting mechanism), and the other is responsive to events occurring within the vessel (the intrinsic clotting mechanism).

The situation at this point in our discussion is summarized in Figure 26-11.

We see that the prothrombin complex acts as a means of converting fibrinogen to fibrin in response to signals from two sources, and the remaining question is the nature of the signals. Here we are in a considerably more foggy area. Most would agree that activation of the extrinsic-sensitive glycoprotein of the complex requires the presence of lipoproteins derived from the damaged surrounding tissues and something derived from the platelets clumping at the point of damage (perhaps nothing more than exposed plasma membrane as a reactive surface). Most would also agree that activation of the intrinsic-sensitive glycoprotein requires the presence of lipoproteins on the plasma membrane of platelets and one or more additional activating proteins from the plasma. In other words, there is broad agreement that the formation of thrombin as a result of extrinsic or intrinsic initiation requires the presence of lipoproteins and one or more additional activating proteins in each case. Beyond this, argument is rife. Several experts believe that each of the factors found to influence the rate of coagulation is a proenzyme, with the reactions arranged in a cascade sequence in which each active factor converts the proenzyme form of the following factor to its active form, with the sequence culminating in the activation of Stuart factor. Those taking this view believe that there is still another factor, the Hageman factor or Factor XII, that is the true initiating substance in both the extrinsic and intrinsic mechanisms.

The presumption of the existence of this multiplicity of factors hinges in most cases upon the discovery of patients who bleed too long and too often and the subsequent demonstration that their blood will clot properly in a test tube if a component of normal blood is added. This would seem to be a reasonable approach if it were not for the Hageman factor. Individuals who lack this protein component are known, and clotting is not initiated in their blood under laboratory conditions unless the factor is added from normal blood. However, the deficient individuals have no demonstrable defect in their physiological clotting response! In short, they get along all right without the compound that is supposed to be the key to turning on the whole clotting mechanism. This makes one wonder if the key fits the right door.

Some people argue that the whole cascade scheme is an intellectual exercise rather than a physiological fact, and that the proposed activating enzymes are in many instances nothing more than modifiers of the process by which the Stuart factor is activated or of the action of the Stuart factor itself. We shall have to wait and see how the discussion is resolved.

The important point is that the entire mechanism, no matter how simple or elaborate it may turn out to be, is normally turned off because the necessary enzymes are originally synthesized as overly long peptides that lack enzymatic activity, and the mechanism is turned on by a series of removals of extraneous peptides that requires the presence of clumped platelets. The end result of a deficiency of any of the factors or of normal platelets is excessive bleeding except in the case of Hageman factor.

Vitamin K and Clotting

The prothrombin complex and fibrinogen are synthesized by the liver. For some reason, the peptides that later become the various glycoproteins of the prothrombin complex cannot be synthesized on the appropriate mRNA molecules unless the liver

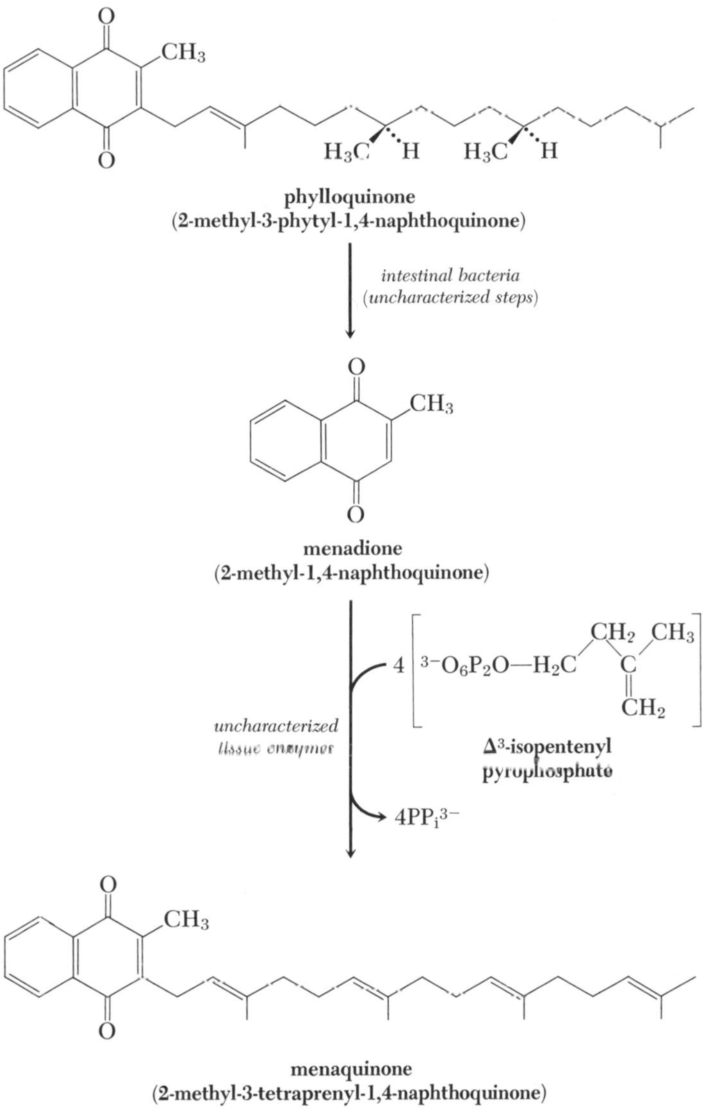
CH3
H3C H H3C H
phylloquinone
(2-methyl-3-phytyl-1,4-naphthoquinone)
intestinal bacteria
(uncharacterized steps)
CH3
menadione
(2-methyl-1,4-naphthoquinone)
4
3−O6P2O—H2C
CH2 CH3
C
CH2
uncharacterized
tissue enzymes
Δ3-isopentenyl
pyrophosphate
4PPi3−
CH3
menaquinone
(2-methyl-3-tetraprenyl-1,4-naphthoquinone)

Figure 26-12 Phylloquinones from plants are absorbed to some extent as such, and they have vitamin K activity. However, the side chain is removed from most of the ingested compounds by intestinal bacteria; the resultant menadione is absorbed and a new side chain is constructed to create menaquinone, the principal form of vitamin K found in animals.

also contains *menaquinone*, but there is no such requirement for fibrinogen synthesis. Menaquinone is a form of a class of naphthoquinones usually known as vitamin K (Fig. 26-12). The substituted naphthoquinone ring cannot be synthesized by animals, and they depend upon plants in the diet or synthesis by intestinal microorganisms for a supply. The form found in plants is *phylloquinone*. Part of the dietary phylloquinone is absorbed directly and apparently can substitute for the usual menaquinone in animals. Another part of the phylloquinone is attacked by intestinal microorganisms that strip off the isoprenoid side chain, leaving *menadione* (*2-methyl-1,4-naphthoquinone*), which is also absorbed. The cells of the body that utilize menaquinone have the capacity to add four isoprene groups to convert menadione to the active menaquinone. (The process appears to be the same as that used for creation of the side chain on coenzyme Q.)

Because of all the different sources of vitamin K, a deficiency in humans appears only when the diet lacks green vegetables and the growth of intestinal microorganisms is simultaneously suppressed by administration of antibiotics, or when the absorption of lipid is impaired by one of the many conditions we mentioned in connection with absorption of vitamin A. None of the forms of vitamin K is stored, and an effective tissue concentration is on the order of only 10 picomoles per gram of tissue, so a deficiency of the vitamin is frequently the earliest to appear during lipid malabsorption. It is especially important to test the clotting time or to measure the prothrombin content of the blood in patients scheduled for surgery who have impairment of digestion or who have been treated with antibiotics.

The only known symptoms of vitamin K deficiency are due to an impaired synthesis of the prothrombin complex, and yet the menaquinones occur in all kinds of animals and a number of microorganisms. They are present in a wide variety of tissues that have nothing to do with blood clotting in the higher animals. Why? No one knows. Most of the speculation has hinged on a possible participation in electron transfer reactions analogous to the function of coenzyme Q, which has a similar structure, and oxidation-reduction reactions of the naphthoquinones have been demonstrated, but it has not been possible to pinpoint electron transfers for which they are required to the exclusion of other known electron carriers. It may be that we only see the impairment of blood clotting because tissue components of more general function retain their naphthoquinones more tightly and the deficient animal bleeds to death before other functional impairments appear. We don't know if this is so, and the answer is only likely to come from studies of cultured cells or perfused organs other than the liver.

Prevention of Clotting

Being able to prevent clotting is a very practical need. Much of the development of modern medicine has hinged on data obtained from analysis of blood, and most of the methods depend upon keeping the blood in a liquid state that can be measured accurately. Beyond this prosaic but all-important application are methods for preventing the formation of clots within a patient so as to avoid obstruction of the blood supply.

Calcium ion is required for the formation of active Stuart factor and for the conversion of prothrombin to thrombin. The most commonly employed methods for preventing clotting in blood withdrawn for analysis or transfusion hinge upon the removal of calcium ions from the sample. A simple way to do this is to add citrate, which not only forms a chelate with calcium, but also interferes with the aggregation of platelets, perhaps by competing with the anionic surface of the platelet for attachment to an initiating surface. A newer method for removing calcium is to add the magnesium chelate of EDTA; since EDTA binds calcium more strongly than magnesium, the effect is an exchange of magnesium for the calcium in the blood.

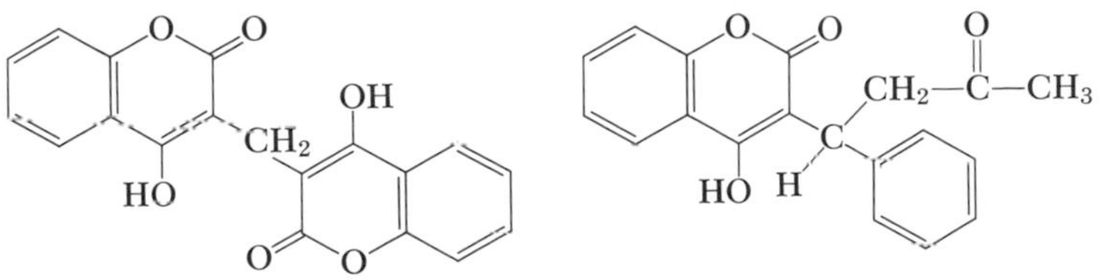

bishydroxycoumarin
(Dicoumarol)

Warfarin

Figure 26-13 Dicoumarol prevents the utilization of vitamin K. After it was discovered in spoiled sweet clover, analogues were synthesized and tested for similar activity. One of these, Warfarin, is an effective rat poison.

It ought to be apparent from our discussion in Chapter 24 that it wouldn't do to attempt to prevent internal clotting in a patient by lowering his blood calcium concentration. The formation of active thrombin must be prevented by other means. One that is commonly employed is *heparin*. The multiple anions represented by its sulfate groups apparently interfere at several points in the chain of events by which prothrombin is converted to thrombin, and it also interferes with the aggregation of platelets. The value of heparin is that its effect is immediate, and it is used to prevent clots from forming on the ends of catheters inserted in the blood vessels, during heart valve replacements, and for quick prevention of further clot growth in someone suffering a heart attack (myocardial infarction). Heparin is also frequently used as an anti-coagulant for blood samples in place of citrate or magnesium EDTA. Of course, it wouldn't do to prevent clotting in patients indefinitely, particularly in surgical cases, but the effects of heparin can be reversed by giving protamine, the highly basic small protein associated with nucleic acids in fish sperm. The tight combination of the polyanion and the polycation removes heparin from effective contact with the components of the blood clotting mechanism.

Cattle eating spoiled sweet clover develop a hemorrhagic disease that was found to be due to the formation of *bis*-hydroxycoumarin in the clover. This compound was found to interfere with the formation of the prothrombin complex by the liver by competing with the utilization of vitamin K. It has recently been suggested that the competition is for transport into the cell. In any event, the original inhibitor and other coumarin analogues of vitamin K are now used to prevent prothrombin formation, and therefore to diminish susceptibility to clot formation (Fig. 26-13). Since the response depends upon the disappearance of the already existing prothrombin, it takes two or so days for a useful effect to appear, and heparin is frequently used in the interval. The advantage of these vitamin K antagonists is that their effect is more prolonged than that of heparin. The effects can be reversed by administering vitamin K, but recovery also hinges upon the somewhat slow synthesis of the prothrombin complex. One of the coumarin derivatives (Warfarin) is an effective rat poison because it is difficult for the rats to associate the delayed death of their companions from internal bleeding with the original cause.

Another agent that can prevent normal clotting is ordinary aspirin, which interferes with platelet aggregation. It isn't used intentionally for this purpose, but loss of blood sometimes occurs in chronic users.

Further reading

Dickens, F., P. J. Randle, and W. T. Whelan, eds.: *Carbohydrate Metabolism and Its Disorders*, Vols. 1 and 2. Academic Press (1968).

Stanbury, J. B., J. B. Wyngaarden, and D. S. Frederickson, eds.: *Metabolic Basis of Inherited Disease*, 2nd ed. McGraw-Hill (1966). Includes discussion of basic mechanisms as well as pathological changes of many processes mentioned in this chapter.

Mann, T.: *The Biochemistry of Semen and the Male Reproductive Tract*. Wiley (1964).

Levin, B., et al.: *Fructosemia*. Am. J. Med., *45:* 826 (1968).

Heinz, F., W. Lamprecht, and J. Kirsch: *Enzymes of Fructose Metabolism in Human Liver*. J. Clin. Invest., *47:* 1826 (1968).

Lerman, S.: *Metabolic Pathways in Experimental Sugar and Radiation Cataract*. Physiol. Rev., *45:* 98 (1965).

Gabbay, K., and J. B. O'Sullivan: *The Sorbitol Pathway. Enzyme Localization and Content in Normal and Diabetic Nerve and Cord*. Diabetes, *17:* 239 (1968).

Turkington, R. W., and R. L. Hill: *Lactose Synthetase: Progesterone Inhibition of the Induction of α-Lactalbumin*. Science, *163:* 1458 (1969).

Huang, S., and T. M. Bayless: *Milk and Lactose Intolerance in Healthy Orientals*. Science, *160:* 83 (1968).

Abraham, H. D., and R. R. Howell: *Human Hepatic Uridine Diphosphate Galactose Pyrophosphorylases*. J. Biol. Chem., *244:* 545 (1969). An alternate pathway for utilizing galactose.

Wald, G.: *Molecular Basis of Visual Excitation*. Science, *162:* 230 (1968).

Bonting, J. S.: *The Mechanism of the Visual Process*. Current Topics Bioenerg., *3:* 351 (1969).

Olson, J. A.: *Some Aspects of Vitamin A Metabolism*. Vitamins and Hormones, *26:* 1 (1968).

Sebrell, W. H., Jr., and R. S. Harris, eds.: *The Vitamins*, 2nd ed., Vol. 1. Academic Press (1967).

Fidge, N. H., and D. S. Goodman: *Enzymatic Reduction of Retinal to Retinol in Rat Intestine*. J. Biol. Chem., *243:* 4372 (1968).

Marcus, A. J.: *Platelet Function*. New Eng. J. Med., *280:* 1213, 1278, and 1330 (1969).

Esnouf, M. P., and R. G. MacFarlane: *Enzymology and the Blood Clotting Mechanism*. Adv. Enzymology, *30:* 255 (1968).

Seegers, W. H.: *Blood Clotting Mechanisms: Three Basic Reactions*. Ann. Rev. Physiol., *31:* 269 (1969).

Tishkoff, G. H., L. C. Williams, and D. M. Brown: *Preparation of a Highly Purified Prothrombin Complex*. J. Biol. Chem., *243:* 4151 (1968).

Pisano, J. J., J. S. Finlayson, and M. P. Paxton: *Cross-link in Fibrin Polymerized by Factor XIII*. Science, *160:* 892 (1968).

Hill, R. B., et al.: *Vitamin K and Biosynthesis of Protein and Prothrombin*. J. Biol. Chem., *243:* 3930 (1968).

Taggat, W. V., and J. T. Matschiner: *Metabolism of Menadione-6,7-^{3}H in the Rat*. Biochemistry, *8:* 1141 (1969).

Vitamins and Hormones. 24: 293ff. (1966). Symposium on vitamin K.

NUTRITIONAL REQUIREMENTS

CHAPTER 27

ARGUMENT

Nutrition is the branch of biochemistry that provides the information neces-sary for the adjustment of the food supply to achieve particular goals. These goals may be the greatest economic gain in raising domestic animals, or greatest growth in the case of humans, but the pressures of the expanding population and limited food supply have led many to define human goals in terms of a minimal intake that will prevent functional impairment. The greatest difficulty in feeding man is supplying sufficient protein. The desired state in an adult is nitrogen equilibrium, with the losses of nitrogen balancing the intake at a total tissue mass enabling a functional life. A child must be in positive nitrogen balance, with intake exceeding losses, if he is to grow. Continued negative nitrogen balance, which is seen when there is little protein in the diet and during illness or injury, eventually must be fatal.

The maintenance of nitrogen balance requires the consumption of suffi-cient total nitrogen to meet all needs, and also requires sufficient quantities of each essential amino acid. The other amino acids, except tyrosine and cysteine, can be made from such simple precursors as ammonia and glucose. The amount of a protein food that must be eaten to supply the total require-ment depends upon its content of essential amino acids, and proteins are rated in various ways for their ability to meet the requirement.

A starving person depends mainly on his fat stores for the energy supply until these are exhausted and then begins to utilize the protein of skeletal muscle. Children with inadequate total food intake become emaciated and are said to have marasmus. Some children in tropical areas have a greater energy supply in the form of carbohydrate but lack a sufficient supply of protein; these children develop kwashiorkor, a condition that causes perma-nent damage and is characterized by edema, diarrhea, and a scaly or ulcerating skin. Diets of this sort lead to secondary deficiencies of a variety of vitamins and minerals, owing to ineffective absorption and utilization.

Most of the vitamin and mineral deficiencies seen in more affluent societies are also due to failures in intestinal absorption that may result from

655

various diseases as well as from bizarre dietary habits. However, instances of primary dietary deficiencies of certain vitamins and minerals are sometimes seen, but these also frequently result from a diet that is limited in both quantity and the nature of the foods. Thiamine deficiency occurs in those living mainly on highly refined foods and in alcoholics. Folate deficiency is sometimes seen with a restricted food intake associated with a style of cooking that depends upon extended heating of foods, which destroys the vitamin. Ascorbate deficiency is an especial problem for the newly weaned infant who has yet to eat the more varied adult diet, but it also is seen in recluse males who choose a limited number of foods. There sometimes is insufficient calcium in the diet for full growth, although the metabolism of calcium tends to adjust to maintain balance with a lesser supply. The iron supply is inadequate for maintenance of normal hemoglobin turnover in many individuals, who become anemic.

The problem of folate and iron deficiencies has received especial attention in pregnant women, but assessment of the incidence of the deficiencies is complicated by normal changes in circulating metabolites that occur with pregnancy and it is difficult to draw the line between the physiological and pathological changes. However, the pregnant woman must have sufficient supply of all nutrients with which to build the placenta and fetus that will later be lost to her.

Recommended dietary allowances are issued in the United States, and the occurrence of dietary deficiencies ought to be extremely rare in those consuming the amount of nutrients stated. The recommendations can be met by a variety of foods, and there is no particular food that is required by all.

Nutrition is the study of the effect on the organism of changes in the diet. It is fitting that we close our discussion of biochemistry with this aspect of the subject because the search for food is one of the things life is all about, and proper analysis of what happens when we consume the results of the search requires the broadest sort of background. We are all aware of the growing problems of feeding the human population, and we are also aware of the many admonitions to the affluent against overeating. The very importance of nutrition creates some ancillary problems that we may as well face head-on.

Many biological scientists and clinicians avoid discussing or even thinking about nutrition, despite its obvious importance as a practical matter and as an ultimate test of abstract rationalization. Nutrition has acquired the taint of a pseudo-science in the eyes of many, who therefore deny themselves one of the more useful bodies of information in understanding the living organism. To avoid falling in the same pit, let us take the time to discuss how this came about.

Part of the difficulty comes from confusing the science of nutrition with the art of dietetics, a confusion that is augmented by the little maxims about good food that are first delivered as the revealed word to nearly everyone in the primary schools and repeated at frequent intervals thereafter without any examination of their basis. To put it bluntly, cooks, menu-planners, and hygienists frequently know little about nutrition.

The next part of the problem paradoxically comes from the eminent practicality of much nutritional knowledge. It provides a basis for value judgments on the character of the food supply. The provision of the supply and its delivery is a major part of the economy of all nations; changes in any aspect can bring financial ruin to some people and great wealth to others, and can make or break a political career. Even a President of the United States has been known to publicly consume a certain food and proclaim its necessity for good health when a decline in the market threatened the income of the producers. The result of all this is that the hard facts of the science are sometimes misstated, ignored, or presented out of context to promote a particular product or justify a particular action, and for some reason these misuses are rarely challenged. Our task here is not to assess economic or political decisions, but to uncover some of the scientific bases on which they can be made.

The task is made more difficult by our own tendency to confuse objectives with proven facts. We talk about malnutrition and frequently forget that it means bad nutrition, with its implication of failure to reach some desired goal. The science of nutrition does not define goals; it only gives information on effects. When we talk about a *minimum requirement* for a nutrient we are stating an estimate based on scientific information to achieve some purpose that is not defined by the science, and there will inevitably be discrepancies in the definition of the requirement because of discrepancies in the objectives. An adviser in a rich nation may strive to keep all of his people saturated with the nutrient as defined by laboratory measurements while one in a poor nation might be content with avoiding obvious functional impairment in 99 per cent of his people. The major defined objective for children frequently is maximum growth—somewhat as if people were Angus cattle. This is an easily measured criterion and there is no question that marked retardation of growth caused by nutritional deprivation results in partial loss of some of the most valued human traits: intelligence, initiative, and so on. However, there is also some evidence from experimental animals that maximum intelligence and maximum longevity are achieved at something less than the maximum possible growth rate. Here is a case in which the science of nutrition has yet to give us all the necessary information we need to make a considered judgment in light of our personal aims.

Perhaps the most difficult misconception for an American layman to shake is the idea that there is some essential value in a given type of food—that without fruit, or leafy vegetables, or meat, or milk, and so on, a diet is inadequate, if not immoral. *There is no nutritional basis for an absolute requirement of any particular food.* Humans have been around for some time, and we have evidence from archeological explorations, historical records, and current observations to show that cultures can thrive and their adherents perpetuate their kind on a wide variety of dietary intakes—in some cases almost completely deficient in one or more of the "basic foods" touted in some quarters.

Having now shaken our preconceptions, let us go on to consider some of the truly usable information. Nearly all of the various components of foods have been encountered in the earlier chapters, and we have seen how they are utilized within the body. Our remaining problem is one of organizing the information and gaining a more quantitative conception. Let us begin by considering the dietary supply of amino acids.

FOOD PROTEINS

Digestion

The large protein molecules found in food are hydrolyzed to the constituent amino acids for absorption by the intestinal mucosa, and they are lost in the feces if this does not occur. Some of the enzymes that accomplish this are *endopeptidases*, or *proteases*, that attack peptide bonds in the central portion of the chain. Others are *exopeptidases*, sometimes simply called peptidases without a clarifying prefix, that attack terminal peptide bonds; they may be *aminopeptidases* or *carboxypeptidases* that specifically attack only the N- or C- terminal bonds.

These enzymes constitute a powerful threat to the tissue that makes them, and this is especially true of the endopeptidases. Consequently, they are formed as proenzymes with a redundant segment of peptide chain that must be removed before they become active, analogous to the formation of prothrombin and its conversion to active thrombin. For example, the gastric mucosa forms *pepsinogen*, which is the proenzyme form of an endopeptidase, *pepsin*, with a specificity favoring hydrolysis of peptides of the aromatic amino acids and other bulky non-polar residues. The conversion of pepsinogen to pepsin is initiated by the high concentrations of H^+ in the gastric lumen, but pepsin itself also catalyzes the conversion of additional pepsinogen to pepsin. This autocatalytic process does not begin in the mucosal cells because the concentration of H^+ is too low within the cells for effective conversion of pepsinogen, and even if some pepsin is formed, it won't catalyze hydrolysis at neutral pH.

The contents of the stomach, containing peptides exposed by the denaturing effects of the high acidity as well as partially hydrolyzed peptides, are discharged into the small intestine, where they are neutralized by the bile and pancreatic secretions. The peptides are attacked in the lumen of the small bowel by a battery of enzymes, mainly derived from the pancreas, and there is an elaborate scheme of activations of the enzymes to protect the pancreas from digestion. The key event is the formation of the powerful endopeptidase, *trypsin*, from the proenzyme form secreted by the pancreas. The intestinal mucosa secretes a small amount of another proteolytic enzyme, *enterokinase*, that catalyzes the conversion of trypsinogen to trypsin. Once trypsin is formed, it also will catalyze its own production from trypsinogen. In order to prevent accidents, a small protein that acts as a *trypsin inhibitor* is secreted in low concentrations by the pancreas, and it combines with any molecules of the active enzyme formed within the pancreas. Once the secretion reaches the small intestine, enterokinase will catalyze the formation of enough trypsin to remove all of the inhibitor, and the excess will then begin catalyzing the formation of additional active enzyme. The pancreas also secretes a number of additional endo- and exo-peptidases as proenzymes. These are converted to their active forms by the action of trypsin, so the pancreas is safe from them so long as trypsin remains inactive.

This battery of enzymes converts ingested peptides to the free amino acids and small peptides. The small peptides can enter the intestinal mucosa, where they are attacked by another battery of exopeptidases contained within the mucosal cells. The free amino acids obtained in this way, along with those absorbed from the lumen, are attached to specific transport proteins and concentrated within the cells for diffusion into the portal blood.

Now, the amino acids being absorbed are really derived from two sources. One is the ingested food and the other is the digestive secretions. The pancreas, for example, puts out the equivalent of one-quarter of its total content of protein every day. Estimates of the total quantity of protein put into the tract by secretion and by loss of mucosal cells range from a conservative 0.35 moles of protein nitrogen per day to an enthusiastic 3.5 moles per day. A range of 0.6 to 1.2 moles per day seems more likely. However, even this is comparable to the dietary intake of nitrogen by many people. The mixing of the dietary amino acids with the secreted amino acids tends to dampen any fluctuations in composition of the proteins in the diet.

Quantitative Expression

It is always difficult to describe the composition of foods in useful and yet simple terms. This is especially true with the proteins. The information we are after is the content of amino acids, but dealing with quantitative values for some 20 amino acids would be clumsy even if the necessary analyses were available, which they usually aren't. The pioneers in the field solved the problem by developing analyses for the total nitrogen content of food. Since the proteins and free amino acids contain over 95 per cent of the total nitrogen in most foods, this is a useful guide to the relative content of amino acid residues in various foods. Unfortunately, the early workers also decided to translate the nitrogen analysis into weights of protein even though there is too much variation in amino acid composition to permit this to be done accurately. (The nitrogen content of natural peptides may range from 14 to 19 per cent of the weight, and the association of prosthetic groups, carbohydrates, and lipids with many proteins makes the extrapolation to weight even more dubious.) After comparing the samples available to them, they decided that an "average" protein contains 16 per cent nitrogen. To this day, tables of composition of foods and of tissues contain figures for the percentage of protein by weight that are really the percentage of nitrogen multiplied by 6.25 (100/16). These numbers are sometimes instructive in making rough assessments of the proportion of the mass occupied by protein, but they create extra work for most purposes, since it is the nitrogen content itself that must be used for assessing quantitative metabolism.

Since we want to be able to make direct comparisons between the metabolic processes and the content of amino acid residues in tissues and foods, it is more convenient to speak of moles of amino acids in individual cases, and of moles of protein nitrogen when discussing total balance. (For comparison with conventional weights of "protein," the corresponding values will sometimes be given parenthetically in the later discussion.) The following table summarizes the relationships between the common units:

MOLES N	GRAMS N	GRAMS "PROTEIN"
0.071	1	6.25
1	14	87.5
1.14	16	100

$$(\text{moles N}) \times 87.5 = (\text{grams "protein"})$$
$$(\text{grams "protein"}) \times 11.4 = (\text{millimoles N})$$
$$(\text{grams "protein"}) \times 0.16 = (\text{grams N})$$

To get an appreciation of the scale, one mole of protein N is a more than ample daily intake for most adults, according to anyone's scale of requirements.

NITROGEN BALANCE

If the total number of cells in the body is increasing, or if proteins are being synthesized at a faster rate than they are being destroyed, the total nitrogen content of the body, which is mostly in protein, increases. It follows that the intake of nitrogen into the body is greater than the losses when this is the case, and the individual is in *positive nitrogen balance*. This is the characteristic state of the young, growing animal, and of the convalescent adult who is repairing injured tissues.

Nitrogen equilibrium is the characteristic state of the adult, with the losses just balancing the intake so that the body composition remains relatively constant. Obviously, there will be moment to moment and day to day variations. Hard work causes a loss of proteins not only to meet energy requirements, but also from cellular damage, so there will be a transient negative nitrogen balance made up later by a transient positive nitrogen balance.

Negative nitrogen balance is characteristic of food deprivation, of illness, and, in a less perceptible way, of aging.

The route of nitrogen intake is simple—through the mouth—but the routes of loss are more complex. The major part of nitrogen loss occurs in the urine, mostly in the form of *urea*, which is typically around 85 per cent of the total urinary nitrogen. *Creatinine* accounts for another 5 per cent of the nitrogen in urine. The remaining 10 per cent is distributed among a variety of compounds including *ammonia*, some free *amino acids*, and *uric acid*.

Significant amounts of nitrogen are lost in the feces and from the skin. Simple loss of epidermis is not negligible—and is aggravated by the scrubbing our standards of cleanliness demand. More importantly, sweat contains free amino acids and urea, and the amount of sweat produced per day is highly variable. The fecal content of nitrogen is also variable—depending upon the ease of digestibility of ingested food and the motility of the bowel, among other things. A typical partition of the nitrogen losses is as follows:

	PERCENTAGE OF THE TOTAL NITROGEN	
	with minimal sweating	*with heavy sweating*
Urine: as urea	70	54
other	12	12
Feces:	12	9
Skin: sloughed epidermis	3	3
sweat	3	22

Heavy sweating due to elevated temperature or exercise is generally associated with greater total nitrogen losses, and if nitrogen balance is to be maintained, the intake must be correspondingly greater. When balance is maintained, the amount of nitrogen excreted in the urine and feces tends to remain the same, and the extra intake is used mainly to replace the losses in sweat.

Maintenance of Protein Balance

Sufficient amino acids must be supplied to enable maintenance of body composition. Amino acids are withdrawn from the body pools for synthesis of proteins, nucleic acids, and lesser components of cells and removed by degradation through the oxidative pathways, which also result in energy production. The oxidative pathway can be minimized by supplying sufficient fat and carbohydrate to meet the requirements for high-energy phosphate, but even in these conditions, there is a constant leakage of nitrogen through the metabolic interchanges necessary to maintain balanced composition. The replacement of this supply therefore represents an absolute minimum requirement for amino acids in the diet.

There are two aspects to the minimum requirement. There must be sufficient *total nitrogen* with which to make all of the required components used in maintaining cells. There also must be a sufficient supply of all of the components that cannot be synthesized within the cells, in other words, an adequate amount of each of the *essential amino acids*. These two factors affect each other, and are also influenced by the nature of the remainder of the diet, so it must be emphasized that there is no rigorous definition of the nitrogen requirements applicable at all times to a given human—there is no number that can be memorized with its basis safely forgotten. It must also be emphasized that definition of nitrogen requirement is a matter of intense and practical human concern, because a shift of a few grams one way or another defines how many people may die if one crop or another is planted in an area of limited food supply.

The Formation of Non-essential Amino Acids

The non-essential amino acids, by definition, are those that can be formed within the body. We have already seen the metabolic routes by which most are made. The key reaction is the formation of glutamate from ammonia and α-ketoglutarate by reduction with NADH or NADPH (p. 355), so that nitrogen supplied as ammonia may be transferred to a variety of carbon skeletons by transamination. We formerly emphasized the use of these enzymatic reactions for degradation, but since they are reversible, it follows that an amino acid in short supply will be synthesized as long as the precursors are available. We may summarize these and other routes available for formation of amino acids:

Alanine—formed by transamination of pyruvate (from glucose) with glutamate (p. 355).

Arginine—formed by the reactions of urea synthesis, with the necessary or-nithine made from glutamate, *via* glutamate semialdehyde, by transamination with glutamate (pp. 360 and 395).

Aspartate—formed by transamination of oxaloacetate (from glucose) with glutamate (p. 374).

Cysteine—the carbon skeleton and nitrogen are provided by serine, and the sulfur is derived from methionine (p. 433).

Glutamate—formed by reductive amination of α-ketoglutarate (from glucose).

Glycine—formed by removal of the hydroxymethyl group of serine (p. 413).

Proline—formed from glutamate *via* glutamate semialdehyde (p. 394).

Serine—formed by transamination of hydroxypyruvate or phosphohydroxy-pyruvate (from glucose) with alanine or glutamate (p. 417).

Tyrosine—formed from phenylalanine by oxidation (p. 385).

In theory, if a diet contained enough of the essential amino acids to balance their utilization, along with enough extra phenylalanine from which to make tyrosine and enough extra methionine from which to make cysteine, the balance of the amino acid requirements could be met by simply feeding ammonium salts along with glucose to provide the carbon skeleton. The ammonium salts could be replaced by urea if the intestinal flora could break down this compound at a rapid enough rate.

In fact, this is nearly correct. Experimental animals grow quite nicely on a diet containing essential amino acids and ammonium salts as the sources of nitrogen, although not quite at the rate obtained when the additional nitrogen is supplied by one or more of the non-essential amino acids. Humans can maintain themselves on diets containing nothing but the essential amino acids, provided that there is sufficient excess to supply nitrogen with which to make the others.

However, natural diets are not deficient in all of the amino acids except the essential ones. The ability to interconvert amino acids is used primarily to adjust intracellular concentrations for synthetic reactions and maintain balanced formation and degradation.

The Essential Amino Acids

An essential amino acid, by definition, is one that must be supplied in the diet. The first critical experiments on human requirements were performed by W. C. Rose at the University of Illinois with volunteer graduate students maintained on artificial diets. One amino acid at a time was omitted, and if negative nitrogen balance resulted, the compound was restored until the requirement, as indicated by a slightly positive balance, was satisfied. In most cases, deprivation of one or more of these essential amino acids has immediate deleterious effects. However, this is not necessarily true. Histidine is often omitted from the list of amino acids essential for the human, because the young adult males in Rose's experiments could maintain nitrogen balance without the presence of the amino acid. This implies that humans either can make the amino acid or do not degrade the compound. There is strong evidence that neither of these possibilities is correct, and in the long run, histidine must be supplied in the diet. Rose's results are contrary to observations with every other mammalian species tested, and it seems highly unlikely that young Midwestern men suddenly remade the genes that have been gone so long. The probability is that during the limited length of time in which the experiments were conducted, the histidine supply represented by anserine and carnosine in the skeletal muscles was sufficient to sustain proper protein synthesis. To recapitulate from our previous discussion of amino acid metabolism, the essential amino acids for vertebrates are *leucine, isoleucine, valine, methionine, threonine, lysine, histidine, phenylalanine,* and *tryptophan.*

QUANTITATIVE ASSESSMENT OF PROTEIN NUTRITION

We have noted that the quantities of total nitrogen and of the individual essential amino acids that must be ingested in order to fulfill some defined expectation are interdependent. In order to simplify comparison of dietary proteins in a practical way, various expedients have been tried. The oldest, and in many ways the most useful, is to assign a *biological value* to the protein, which is arrived at in this way: A person eating a diet free of protein will still excrete some nitrogen in his urine and feces. If he is now fed a small amount of a protein containing a proportion of essential amino acids exactly equivalent to the proportion necessary for rebuilding the tissues that were depleted during the period of nitrogen starvation, there will be no increase in his nitrogen excretion, because all of the ingested amino acids can be used to make protein. However, if the protein is relatively lacking in one of the essential amino acids, all of the remaining essential amino acids cannot be utilized, and part will be degraded and will appear in the urine as extra nitrogen above the basal starving level. The *fraction of the nitrogen in the protein retained* under these conditions is defined as the biological value of the protein. If only half of the nitrogen is retained, the biological value of the protein is 0.5, and it would theoretically require eating twice as much of that protein to get the same result in terms of synthesis of tissue proteins.

The biological value is not completely valid. For example, part of the nitrogen in a protein that completely lacks one essential amino acid will still be retained by the body under the test circumstances, and such a protein would be assigned a biological value even though it is not capable of sustaining life, no matter how much is eaten. For this reason, the idea of a *chemical score* based on known amino acid composition has been introduced. The chemical score is the relative quantity

TABLE 27-1.   RATIO OF CONTENT OF PARTICULAR AMINO ACIDS IN FOOD PROTEINS TO CONTENT IN CHICKEN EGG PROTEINS.

	ILE	LEU	VAL	THR	MET	MET + CYS	TRP	LYS	PHE	PHE + TYR	HIS	BIOLOGICAL VALUE	CHEMICAL SCORE
Egg, chicken	1.0	1.0	1.0	1.0	1.0	1.0	1.0	1.0	1.0	1.0	1.0	0.94	1.0
Milk, human	1.1	1.4	1.0	1.0	0.6	1.1	1.6	1.0	0.9	1.0	0.9	0.95	0.9
Milk, cow	1.1	1.3	1.0	0.9	0.9	0.7	1.3	1.2	1.0	0.9	1.1	0.90	0.7
Muscle, beef	0.8	0.9	0.7	0.9	0.8	0.9	0.9	1.4	0.7	0.7	1.6	0.76	0.7
Soybean meal	1.0	0.9	0.8	0.8	0.5	0.6	1.3	1.1	0.9	1.0	1.4	0.75	0.6
Rice, whole	0.8	0.9	0.9	0.8	1.1	0.9	1.2	0.5	0.9	1.2	0.8	0.75	0.5
Wheat, whole	0.6	0.8	0.6	0.7	0.8	0.8	1.1	0.4	0.9	0.8	1.0	0.67	0.4
Potatoes	0.6	1.1	0.8	1.3	0.8	0.6	1.9	1.4	1.0	0.8	1.1	0.67	0.6
Oats, whole	0.8	0.9	0.8	0.7	0.6	0.6	1.2	0.6	1.0	1.0	1.1	0.66	0.6
Corn, whole	1.0	1.7	0.8	0.7	1.0	1.1	0.5	0.4	0.9	1.0	1.2	0.60	0.4

of the most limiting amino acid compared to the quantity in an ideal protein. This obviously requires some protein for comparison, which is usually taken to be the mixture of proteins in whole chicken eggs. Eggs are readily available and their proteins are almost completely absorbed and retained by humans under the test circumstances.

An example will illustrate how the chemical score is determined. Chicken eggs have 53 millimoles of valyl residues per mole of protein N, whereas beefsteak has but 39. Therefore, the chemical score of beefsteak proteins with respect to valine is 39/53 or 0.73. This process is repeated for each of the essential amino acids, and the lowest ratio is taken as the score of the entire protein.

Table 27-1 shows the content of particular amino acids in a variety of foods compared to the content in the proteins of chicken eggs, that is, the chemical score for each amino acid. The table also shows the chemical score of the whole protein based on the limiting amino acid, and the biological value of the proteins as demonstrated by feeding experiments.

Close inspection of Table 27-1 will show some apparent discrepancies. Human milk proteins contain only sixty per cent as much methionine as do egg proteins, and yet the biological value of human milk is higher than that of eggs, whereas cow milk has more methionine but a lower biological value. The answer probably lies in the total methionine and cysteine content of the proteins. Human milk is much richer in cysteine, and the load on methionine is probably spared more by this milk than by cow milk. For this reason, the sum of methionine and cysteine has been used for final assessment of chemical scores. Similar logic can be applied as a justification for using the sum of the phenylalanine and tyrosine content. The logic presumes that the individual contents of methionine and phenylalanine exceeds the requirement for these amino acids as such in addition to acting as sources of cysteine and tyrosine. We have no data on the requirement for the latter amino acids on low protein intakes and can't test the logic directly. However, the low chemical score of cow milk compared to its biological value indicates that the total sulfur content of the reference egg proteins is greater than necessary, thereby making the sulfur content of other proteins appear unduly inadequate. There are similar discrepancies in regard to lysine content. Despite all of this, the kind of comparison made here is a good first approximation, rough as it may be, in assessing how much of particular proteins people must eat to remain functional.

Defining a Total Protein Requirement

We have noted the difficulty in agreeing on the aims that are to be accomplished by feeding a "required" amount of any nutrient, and this also applies to any statement on the daily requirement for protein in the diet. However, one of the major problems in feeding the human race is providing enough protein, and this has led in recent years to a greater concentration on a definition of the minimum amount of protein capable of preserving nitrogen balance in an adult. Of course, this also varies. A person already living on meager rations requires less protein to maintain balance than does a blooming specimen who has always been well fed, and if the ration of either is cut down, he will tend to lose tissue until he can again maintain balance at the lower level.

It is especially difficult to define the required intake for growing children. Hegsted approached this question in a purely pragmatic way by defining the way things actually are, rather than the way they ought to be, with a sample of American children presumably in reasonable health and growing the way a random sample of the population in Boston does. He used the expected basal loss of nitrogen and the increment in the total nitrogen content of the body because of growth to estimate the required dietary supply. A plot of the estimates is given in Figure 27-1; two curves are shown, one for girls in the tenth percentile of weight (only 10 per cent of the girls are smaller) and one for boys in the ninetieth percentile (only 10 per cent of the boys are larger), so these two curves encompass a range including the requirements for all but 10 per cent of the children.

The figure also shows the calculated minima of daily consumption of a completely utilizable protein (biological value = 1.00) necessary to sustain the rates of growth, both in total moles of protein nitrogen per individual and in moles per kilogram of body weight. The relatively high requirement of protein during the early years is especially striking; small girls need half as much protein when they are 3-kg infants as they do when they are 47-kg adults. It is also apparent that girls and boys have nearly the same protein requirement per unit body weight despite their different sizes.

Similar calculations have been made for adults. The upper limit of estimates of requirement for maintenance of nitrogen balance in an adult 70-kg male is 0.37 moles of protein N (33 grams of "protein") per day, if the protein has a biological value of 1.0, and this is based on high estimates of fecal loss, skin loss, and so on. A more conservative estimate, although not the lowest, is 0.30 moles of protein N (26 grams of "protein").

With real mixed diets there is more uncertainty due to variable efficiency of digestion, destruction of amino acid residues during preparation of the food, and so on, than there is in the theoretical calculations. Even so, actual experience with humans shows that the theoretical results are in the correct range. The best estimates available for the minimum requirement of protein necessary for nitrogen balance range from 4 to 5.3 millimoles of *retainable* protein nitrogen per kg body weight in an adult (0.35 to 0.46 grams of "protein" per kg). In the remainder of the discussion, we shall be conservative and take the high range, assuming the protein requirement for a 70-kg man to be met by 270 millimoles of mixed amino acids containing 376 millimoles of N with an average residue formula weight of 110. (The daily requirement is therefore met by 30 grams of peptide, or 33 grams of

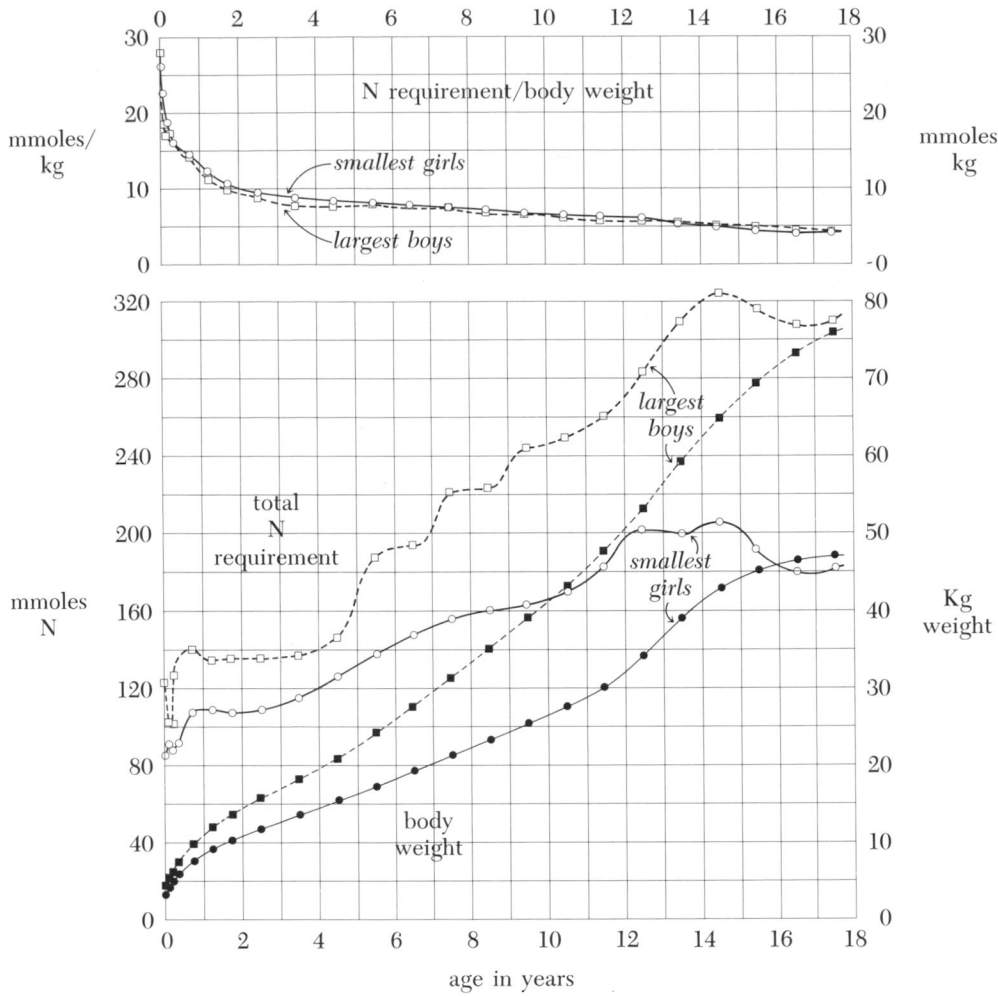

Figure 27-1   Theoretical estimates of the protein requirement of children as a function of age. The amount that must be absorbed in order to balance the utilization for tissue formation and the losses in the urine and feces is shown for the smallest girls (tenth percentile of weight) and the largest boys (ninetieth percentile), both in terms of the total amount and the amount per kg of body weight. The body weights are also plotted. (Plotted from the data of Hegsted, J. Am. Diet. Assn., 33: 225 [1957].)

"protein." The figure of 110 is close to the average formula weight of the residues in most meats. It varies sharply from this in individual proteins of the meats and in the mixtures found in other foods, depending upon the relative abundance of small and large residues.)

Essential Amino Acid Requirements

We pointed out that the requirement for an essential amino acid varies with the total protein content of the diet. Less will suffice when the nitrogen intake is high. The values obtained by Rose and plotted in Figure 27-2 are therefore probably too low for people existing on low daily intakes, because his subjects were fed a total of 7 to 8 millimoles of protein N per kg of body weight. However, his

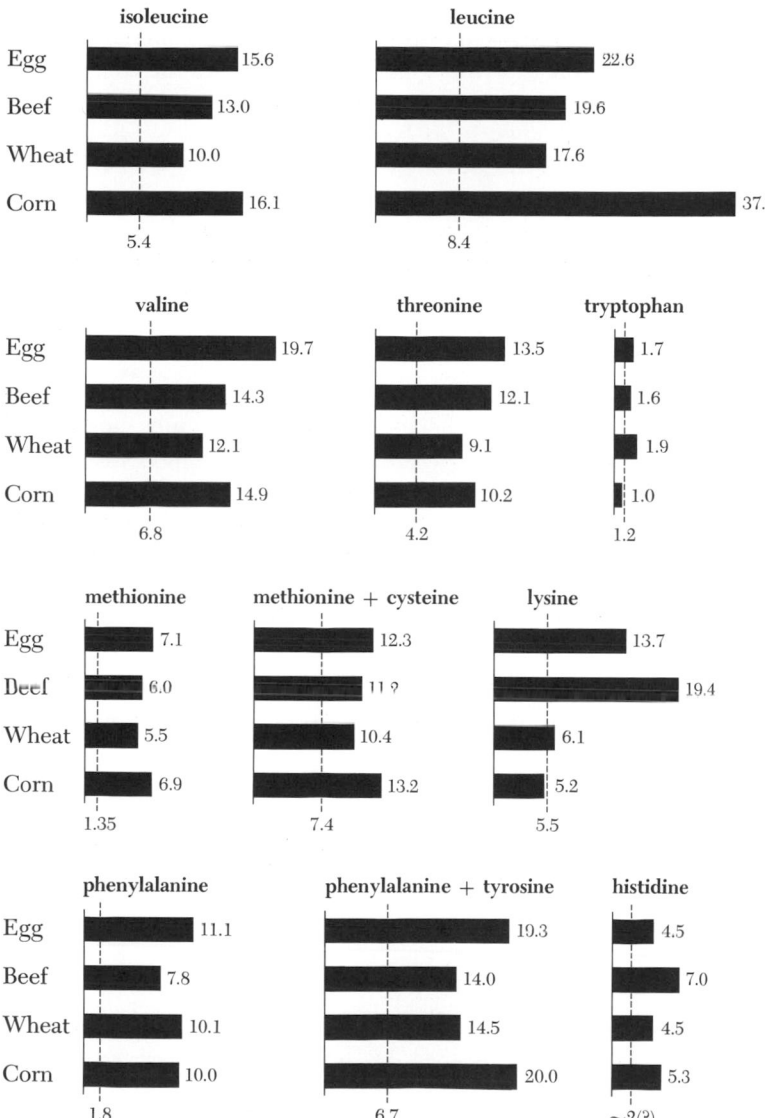

Figure 27-2 The essential amino acid content of various proteins. The bars represent the millimoles of amino acid in the quantity of protein containing 0.37 mole of total nitrogen. This amount of nitrogen, consumed daily, is· sufficient for most adults if it also includes an adequate supply of every essential amino acid. The daily requirement of each essential amino acid, as estimated by Rose's laboratory (except for histidine), is shown by a dashed line. These values were estimated at high nitrogen intakes and are probably low for the requirement if the total intake is much less than 0.4 mole of nitrogen. No long-term test of the requirement for histidine has been made, and the value shown is a highly speculative inference made on the basis of tissue composition and total turnover.

data are the best available and provide a basis for comparison of the various amino acids. The bars in Figure 27-2 show the amount of each essential amino acid represented by 0.37 moles of total protein N in various food proteins. It can be seen that all of the proteins contain quantities of the essential amino acids in this "minimal" intake that exceed the individual minimal levels, with the exception of whole corn, which is deficient in tryptophan and lysine. However, wheat is also precariously close to the minimal level in its lysine content.

It ought to be emphasized that the data for cereals in Figure 27-2 and in Tables 27-1 and 27-2 are for the whole grains, and make no allowance for losses during milling and cooking. It also ought to be emphasized again that there is a *prima facie* case for a requirement of most of the essential amino acids at the levels shown in egg proteins when the total protein intake is at a minimal level, because these proteins are the only ones of those shown that have a biological value approaching 1.0.

There is one more factor of major practical importance. It is all well and good to say that so much protein of given biological value will supply the essential amino acid requirements, but it also must be possible to eat enough of the food to obtain that amount of protein. If the content of protein in the food is low, it may not be physically possible to stuff in enough to meet the requirements. Table 27-2 compares various common raw foods, showing how much must be consumed to match 0.37 moles of protein N of biological value and chemical score of 1.0. It is necessary to note carefully that the weights for the cereals are for the dry grains, not the cooked food.

It is apparent that it is difficult to maintain nitrogen balance with some of these foods because of their low content of one or more of the essential amino acids. Given the various tastes of humans, it is not impossible to imagine wading through mounds of cooked rice or wheat bread each day, but even the appetites of the undernourished might falter before 12 pounds of potatoes or 3 liters of corn mash. People forced to use cereals as the sole source of protein frequently will not eat enough to maintain nitrogen balance, even if the whole grain is consumed. Condi-

TABLE 27-2.   PROTEIN EQUIVALENCE OF FOODS.*

	MILLIMOLES N PER GRAM	GRAMS OF FOOD EQUIVALENT TO REFERENCE PROTEIN IN:	
		BIOLOGICAL VALUE	CHEMICAL SCORE
Egg, chicken, whole	1.47	270	250
Milk, human	0.17	2300	2400
Milk, cow	0.40	1000	1300
Muscle, beef	2.06	240	250
Soybean meal	5.1	100	120
Rice, whole	0.86	570	830
Wheat, whole	1.43	390	590
Potatoes	0.23	2400	2500
Oats	1.26	440	490
Corn	1.14	540	650

* Amounts of entire food containing protein equivalent to 0.37 mole of N in a protein with a biological value of 1.0 and a chemical score of 1.0. Data for raw foods.

tions must be difficult indeed before a cereal is the only available source of nitrogen, but conditions *are* difficult for many humans today.

One of the truly impressive applications of science to improvement of the human condition is being made today in an effort to correct the problem illustrated in Table 27-2. Recognizing that attempts to upset the cereal economy would probably lead to even more rapid disaster, attention has been focused on improving the cereals themselves. This has already led to the introduction of a high-yield strain of rice, but the result with the greatest promise for the future is the discovery of genes that can be introduced into corn to increase the quantity of tryptophan and lysine in the seed proteins. Application of plant genetics in this way to all of the major cereals may revolutionize protein nutrition.

Experience in Renal Failure

An individual whose kidneys have failed tends to accumulate the end products of metabolism of all ingested compounds. The accumulation of urea and other products of nitrogen metabolism is an especially serious consequence. The accumulation can be relieved by dialysis, now sometimes performed by simply flushing large volumes of fluid through the peritoneal cavity. Considerable attention has been given to elimination of the need for dialysis in patients retaining a small fraction of normal kidney function, or at least prolongation of the interval between dialyses, by restricting the protein intake to the minimal level consistent with survival.

Superficially, this appears to be an excellent opportunity for assessing the protein requirements of humans, but the circumstances are not quite that simple. On the one hand, there is more opportunity for re-utilization of nitrogen because the increasing concentration of urea causes more to appear in the intestines, where it is broken down to ammonia by the bacteria and reabsorbed. On the other hand, large quantities of nitrogen are removed by the dialysis treatment. The loss is sometimes counteracted by giving blood transfusions that add nitrogen to the system. Good quantitative data assessing all of these factors have not been obtained.

The usual approach to management of patients with renal failure is to provide an ample supply of essential amino acids, frequently in the form of two eggs, and then to sharply restrict further protein intake. Some prescribed regimens allow an intake of only 0.2 mole of protein nitrogen (18 grams of "protein") per day, but cooperation by the patients appears to be better with diets permitting consumption of 0.46 moles of N (40 grams of "protein") per day. Specialists in the treatment of renal failure have now become aware of the necessity for better definition of the character of the nitrogen intake in terms of amino acid composition as well as total nitrogen balance, and it seems likely that the results of their experience will tell us a great deal about the normal nitrogen economy within a few years.

Amino Acid Imbalance

Consideration of the usefulness of a diet for supplying amino acids is complicated by the fact that an excessive proportion of some amino acids is deleterious. It has been known for some time that all of the essential amino acids must be present within the meal if the diet contains rapidly absorbed foods and if there is an impending deficiency of an essential amino acid. It is not possible under these circumstances to make up a deficiency later and we should expect this to be true from our knowledge of the sequence of events in protein synthesis in which complete peptides are laid down in a short period of time without provision for storage of incomplete fragments.

It is also known that animals fed a diet deficient in one essential amino acid, and therefore having a relative excess of the others, eat less than they will if a diet has no protein. What was not so readily appreciated was that force feeding of the deficient diet put the animals in even worse condition because an accumulation of some of the amino acids beyond the capacity to utilize them is not tolerated —some of the amino acids are frankly toxic in excess. *Tyrosine, methionine, tryptophan,* and *histidine* are especial offenders in this way. (Tyrosine, of course, isn't an essential amino acid because it can be formed from phenylalanine, but it is like the essential amino acids in having an irreversible metabolism without the rapid turnover of most of the non-essential amino acids.) For example, 250-gram rats that were fed a diet upon which they gained 0.88 grams a day, lost 4.6 grams per day when they ate the same amount of diet to which 4.8 per cent of DL-methionine had been added. It appears that refusal to eat a diet with an unbalanced amino acid content is a protective mechanism. The exact mechanism of toxicity has not been worked out. The more adequate the diet, the less toxic the excess amino acids become.

Even if a diet contains excess essential amino acids, an addition of some amino acids beyond a balanced supply is deleterious because of an impairment of absorption of related amino acids from the small intestine, rather than because of an inherent toxicity. Leucine, isoleucine, and valine are related in this way, with an excess of one inhibiting the absorption of the others because of the utilization of a common transport mechanism. Lysine and arginine are also antagonistic. No clear-cut examples of competition of this sort are known with humans on natural diets, but a study of the data given earlier will show that people on corn diets are likely candidates because of the high content of leucine in corn proteins.

DEPRIVATION OF THE MAJOR NUTRIENTS

Lack of sufficient total food is so common that it may be considered a part of the normal human condition, now and at all past times for which we have evidence. More people die from starvation than are killed in wars, and even more people experience starvation and then recover. Despite its prevalence, starvation has received relatively little study until recent decades. The information we have has been obtained in three ways: from experimental study of volunteers of approximately median body composition and with a history of sufficient food intake before the experiments, from the use of starvation in the treatment of grossly obese humans, which we mentioned earlier, and finally from direct observation of individuals experiencing natural famines, who in many cases have had restricted food intakes over most of their lives.

The precipitating factors for the changes in body chemistry during starvation are simple. The rate of oxidation is geared to the rate of utilization of high-energy phosphate, and lack of ingested food does not shut off the utilization. Fats, carbo-hydrates, and amino acids are still metabolized, and metabolism in excess of the dietary supply comes at the expense of constituents of the body itself.

We have seen that the glycogen stores of the body cannot amount to much more than three to four moles of glucosyl residues—a quantity barely able to supply the total metabolic demand for a single day with a modest work load—so glycogen is not an important fuel in sustained starvation. We have also seen that the fats of adipose tissue represent a more impressive reservoir of oxidizable substrates; the 7.7 kilograms in a relatively trim 25-year-old male is enough in itself to produce a month's supply of high-energy phosphate.

The remaining supply of oxidizable substrates is represented by the proteins of the body, but this supply cannot be tapped at will because of the necessity of retaining the architecture and catalytic function of the cells. A substantial loss of protein from the brain, for example, would be lethal. It is in this respect that the presence of inherent controls during starvation become most apparent. One need only look at photographs of humans still living after being subjected to long periods of starvation and note the plainly visible outlines of the long bones to realize that the skeletal muscles have lost a considerable fraction of their protein.

There is a rapid loss of protein up to a maximum of about 3 per cent of the total store of 80 moles of N during the initial days of acute starvation. The loss mainly comes from the liver in experimental animals, and the amount depends upon the previous protein intake, with those having ingested high protein diets losing more upon sudden deprivation. The proteins lost quickly without measurable impairment of function are designated "labile" proteins. After the initial high rate of loss, nitrogen excretion tapers to a level ranging around 0.14 moles per day (12 grams of "protein") at the close of a month of starvation. The tissues losing protein most rapidly during this period are those with high rates of protein synthesis at times of normal nutrition: the liver, pancreas, and intestinal mucosa. The skeletal muscles lose protein more slowly per unit weight of tissue, but represent a major supply during the initial weeks because of their large mass.

People adjust to starvation. Part of the adjustment comes from adaptations of the biochemical economy of the type we noted in Chapter 23. Part comes from a slowing of processes utilizing high-energy phosphate at rest, as is reflected by a drop in the basal metabolic rate. Another part comes from changes in habits and temperament. Starving people don't like to move, and become self-centered and withdrawn. The often-invoked specter of starving hordes ravaging their well-fed neighbors probably never occurs; the starved aren't good candidates for group efforts. (A fear of starvation is something else again.) Charles Dickens was quite correct in suggesting that rambunctious boys might be more tractable if they were fed less meat and more gruel.

Perhaps the most surprising thing about acute starvation is the relative lack of specific deficiencies in many instances. Emaciation may be severe without symptoms ascribable to the lack of a particular coenzyme or amino acid, although this is not a hard and fast rule. It is this efficiency of adaptation to starvation that permits its use in the treatment of the grossly obese.

The duration of starvation that can be withstood obviously depends upon the quantity of stored fat. An extremely lean person faces a quick death with total starvation because he will begin to lose as much as 6 per cent of his protein per day when the fat is exhausted. Young children are also in a precarious position because of a more limited fat store and a greater metabolic rate.

Kwashiorkor: Protein Deficiency

Simple starvation is due to a generalized restriction of food supply. In some areas of the world, especially in or near the tropics, the total quantity of food may not be so obviously limiting, but the material available may consist mostly of starchy plant substances without an adequate content of protein. The diet is usually one of necessity rather than choice, and in an undernourished family it is frequently the children who are most affected, not only because lack of food tends to destroy any feeling of selflessness by the adults, but also because the very young have not built up a reserve of protein and fat upon which to live.

A long-standing inadequacy of protein intake in children leads to a condition known as kwashiorkor, from the Bantu word meaning displaced child; the symptoms appear in infants after they are no longer suckled by their mother (owing to the appearance of still another baby). Lack of protein is a central part of the condition —if the intake of carbohydrates and fats is also deficient, the child is simply being starved and wastes away. Such infant starvation has been termed *marasmus*, and there obviously is a spectrum of conditions ranging from total starvation with "pure" marasmus to a "pure" kwashiorkor in which the total supply of ingested fuels would be ample for maintenance of high-energy phosphate under ordinary conditons, but in which there is not sufficient protein to maintain cellular constituents. The latter results when the mother has access to quantities of starchy vegetables or sugar but not to foods containing enough protein, and the starchy part of the original food is frequently fed to the infant as a thin gruel. This circumstance is common throughout Latin America and Africa.

The deficiency of quantity of protein can also be aggravated by lack of quality. The deficiency of tryptophan and lysine in corn and the deficiency of lysine in wheat are practical examples of imbalances that make the already limiting quantity of protein even less adequate.

Unlike children who are simply starved, those fed a protein-deficient diet may live for a considerable period, perhaps surviving into adulthood even though irreversibly impaired by the consequences of inadequate cellular development. Since kwashiorkor essentially represents a failure to synthesize normal amounts of protein, its consequences could theoretically appear in every metabolic process of the body, and in many ways do. However, some of the more striking phenomena can be given speculative interpretations in terms of isolated segments of the metabolic economy. The metabolic load in kwashiorkor is quite different than that in marasmus. Carbohydrates are still being supplied and the metabolic machinery for handling these compounds will tend to remain intact. At the same time, it is not possible to maintain all of the proteins in the carcass because of the lack of dietary amino acids. The result is an uneven depletion affecting some processes more than others, rather than the relatively smooth, general decline seen in marasmus.

A striking finding in kwashiorkor is the deceptively plump appearance of the youngsters; they are called "sugar babies" in the Caribbean area because they are fed on sugar and starches, but the name also evokes an image of round cheeks and bellies. The plumpness is not an expression of overfeeding and storage of fat, but is due to edema. The youngsters have a general accumulation of water to such a degree that their weight actually falls when they are put on restorative diets.

What can cause edema? One suspects congestive failure of the heart, but that is usually not the case in kwashiorkor; a mild failure is more likely to occur during initial recovery than it is during the active condition. Another possibility is a fall in the protein concentration of the blood, especially of serum albumin, causing a drop in the osmotic pressure. The synthesis of serum albumin is indeed impaired in kwashiorkor, sometimes falling from its normal level of 40 mg per ml to less than 10 mg per ml, and this kind of fall will invariably cause loss of fluid from the blood into the tissues. However, edema may appear without a precipitous fall in the albumin concentration (this is usually the case in adult starvation), and additional factors must be sought. Another possibility would be disturbances in the electrolyte and water balance, and these events also occur, both at the cellular level and in the kidney. There is an especially marked loss of potassium from the cells. It may well be that some of the proteins involved in ion transport are being destroyed and utilized for the formation of other proteins elsewhere, or to put it more accurately, there may be less of the transport proteins formed from the diminishing amino acid pool than there is of some other proteins.

Diarrhea is almost invariably a result of kwashiorkor, and contributes to the potassium loss. Runny bowels are so frequent an accompaniment of disease in general that we give little thought to cause. Many nutritional deficiencies are accompanied by diarrhea. This indicates a failure of intestinal function. The intestinal mucosa has the highest known rate of cell turnover in the whole body, and therefore is especially vulnerable to any failure in the supply of nutrients needed for constructing cells or to an interference in protein synthesis.

The stools in kwashiorkor may also be fatty. Fatty stools may result from the loss of emulsifying agents secreted in the bile, which are steroid derivatives formed by the liver, or from a failure in secretion of lipase by the pancreas. Since the pancreas requires a high rate of protein synthesis for activity, one might expect a decrease in function when there is protein deficiency, and this is the case. The pancreas, along with the intestinal mucosa, atrophies in kwashiorkor.

The skin, which is constantly being shed and replaced in a normal individual, develops gray and scaly or ulcerating patches. This is probably due to a mixture of protein deficiency and nicotinamide deficiency (see the following sections). Similarly, the growing hair becomes fine, dry, brittle, and abnormally light in color. Most of the people in areas where kwashiorkor occurs have naturally dark hair, so that the reddish or even blonde sparse hair of kwashiorkor is striking. In some cases the hair will be banded with light color along its length; the bands indicate times of inadequate protein supply in the way that tree rings indicate the passage of the seasons.

All of the above symptoms are readily interpreted in terms of inadequate protein synthesis, but there are other results that require more cautious analysis. Among these is the accumulation of fat and hemosiderin in the liver. These two changes in liver composition occur together in a variety of conditions, but apparently are separate results. To begin with, an accumulation of iron can only occur when the iron supply in the diet is relatively high. There also must be a failure of intestinal mucosa regulation of iron absorption, and most of the conditions that result in hemosiderosis, including kwashiorkor, lead to defective protein synthesis. An accumulation of fat in the liver, on the other hand, is a result of failure of mobilization from the liver. This may be caused by a failure to form the phospholipids or the protein component of the lipoproteins necessary for transport. A failure of phospholipid formation is known to occur when the supply of methyl groups from the diet or from the one-carbon pool is depleted. However, a low supply of methyl groups also implies a low supply of methionine that is also needed for protein synthesis. Comparison with the effects of experimental depletions makes one suspect that failure of phospholipid formation is the primary cause for fat accumulation, and that failure of protein synthesis in the intestinal mucosa is the primary cause for hemosiderin accumulation, but we can't be positive.

Finally, kwashiorkor is frequently accompanied by the symptoms of deficiency of one or more other nutrients, including many of the minerals and vitamins (except vitamin D). The primary cause appears to be the general failure of protein synthesis, so that the mechanisms for absorption, transport, and utilization are defective. The dietary supply of the nutrients is indeed a factor, but one child may have symptoms of folate deficiency, another of iron deficiency, and still another of cobamide deficiency, all in the same area, and it is likely that the particular deficiency depends as much upon the particular proteins that are not being made as it does on the diet.

DEFICIENCIES OF VITAMINS

When is the supply of a vitamin inadequate? Here we have one of the more pronounced divergences of approach toward definition of requirement of a nutrient. Let us dispose of one peremptorily. Some are inclined to say that an individual is deficient in a vitamin if he does not consume a given amount in his daily diet. It is this approach that gives rise to occasional statements that a third or so of the population of the United States is suffering from nutritional deficiency, a statement that any lay observer has difficulty in accepting and that therefore brings discredit on the entire science.

The remaining approaches fall into two general classes. Some take a laboratory approach: They measure the concentration of a metabolite in the blood or the daily excretion of a metabolite related to the vitamin or its function and define the requirement of the vitamin for an individual as the quantity necessary to maintain a stated level of the metabolite. Others take a functional approach and define the requirement for an individual as the quantity that will prevent the appearance of functional impairments. Those using the laboratory approach tend to set higher requirements; the approach has the advantage of assuring a greater reserve for adverse circumstances because the criteria hinge on the maintenance of saturating concentrations in most cases. It has the disadvantage of making no allowance for individual variation in the efficiency of utilization of a supply—experience indicates that some people are built to supply all of their metabolic needs with lower concentrations of a vitamin or its metabolites. The functional approach has the advantage of a pragmatic assessment of each person's needs that inherently allows for the individual variations, but the disadvantage of setting a minimum requirement at a level that may teeter on the edge of frank pathological changes.

It is apparent that the choice of definition of requirement will depend upon economic circumstances and personal philosophy and temperament. Shall we note each sparrow's fall or concentrate on feeding the multitude? Let us for the moment thread our way between these choices, neither crying doom over rare occurrences of deficiency symptoms nor advocating the elimination of all margin of safety.

We also ought to keep in mind the lesson to be learned from kwashiorkor: The appearance of symptoms of deficiency of a minor nutrient may really reflect an imbalance in the consumption of the major nutrients. The existence of so many dietary fads and crash programs for reduction of body weight may well contribute to the appearance of deficiency symptoms of a vitamin when the actual intake of the compound is adequate for another individual on a more balanced diet. This also may be a factor in some pregnant women. There is a high correlation between large rates of weight gain and the development of preeclampsia during pregnancy. (Preeclampsia is the appearance of premonitory signs, such as the appearance of protein in the urine and an elevated diastolic blood pressure, of a severe toxemia of pregnancy with a high death rate.) Consequently, obstetricians frequently try to keep pregnant women from eating all they may like. Too much emphasis on quantity and too little on composition by either party may cause trouble to the mother. (The fetus is surprisingly resistant to the effects of mild maternal deprivation.)

The point here is the one that we made in an earlier chapter: The ability to make the compounds that we call vitamins has been lost during evolution because the structures are freely available in foods, and all species arising since that time would of necessity have eating habits insuring an adequate dietary supply of the compounds or the growth of an intestinal flora that would synthesize them for the host. We therefore don't expect to find widespread occurrence of a vitamin deficiency unless the choice of foods has been limited, or the foods themselves have been changed, or something has happened to the gastrointestinal physiology. Unusually severe changes in any of these factors can lead to the appearance of symptoms of a deficiency of any of the vitamins, and frequently of several of them. It ought to be evident that the proper solution to the resultant malfunction is not piecemeal correction of the individual deficiencies, but a general change in the diet or restoration of normal gastrointestinal function. (This does not preclude temporary administration of massive doses of the deficient substances to speed restoration of normal function, although such measures sometimes lead to the development of still other deficiencies not previously exposed.)

However, there are circumstances in which the deficiency of a particular vitamin is the major, if not the predominant, cause of pathological changes. We have considered most of these, along with the function of the vitamins, throughout the book. Before reviewing and extending this information, let us summarize what we have learned about the occurrence and function of the vitamins.

Function of the Vitamins

Thiamine pyrophosphate is required for the oxidative decarboxylation of pyruvate, α-ketoglutarate, and the 2-ketocarboxylates derived from amino acids, including 2-ketobutyrate derived from threonine and methionine, and the keto analogues of leucine, isoleucine, and valine. It is also a coenzyme for the transketolase functioning in the pentose phosphate pathway. Thiamine is widespread among foods, but there is little synthesis by intestinal microorganisms, and symptoms readily appear after dietary deprivation.

Flavin adenine dinucleotide and *riboflavin monophosphate* are required for mitochondrial electron transport and for the oxidations in the endoplasmic reticulum and other organelles that are not coupled to phosphorylation. Riboflavin not only occurs in most foods but is also synthesized by the intestinal flora. As a result, a primary deficiency of riboflavin is very rare, and it is not possible to set a minimum dietary requirement. Human volunteers developed nondescript symptoms after several months on a total intake of 1.5 micromoles per day.

Pyridoxal phosphate is the coenzyme for transaminases, serine and threonine dehydratases, amino acid decarboxylases, serine transhydroxymethylase, the hydroxymethyl transfer during glycine oxidation, kynureninase, 5-aminolevulinate synthase, and glycogen phosphorylase. Here again is a ubiquitous compound rarely deficient in the human diet and probably also made by the intestinal flora. Indeed, the only clear-cut primary dietary deficiencies of pyridoxine have been seen in experimental subjects and in a group of infants who were fed a commercial formula that the manufacturer had neglected to test for pyridoxine content. The infants developed convulsions and the experimental adult subjects became irritable; some suggest that these indications of disturbances of neural function result from an early impairment of the glutamate decarboxylase reaction that forms 4-aminobutyrate, but they could as easily be explained by a neural deficiency of almost any of the other reactions requiring pyridoxal phosphate. Kynureninase is depleted early in a deficiency, and urinary excretion of xanthurenate, a by-product of tryptophan metabolism, can be used as an index. Since this enzyme is also involved in the conversion of tryptophan to nicotinamide, a secondary deficiency of this compound may result.

We have noted some genetic defects that apparently result in a lowered affinity of particular enzymes for pyridoxal phosphate, thereby resulting in an increased requirement for the precursors to maintain a saturating concentration. These include defects in cystathioninase (p. 433), a possible defect in 5-aminolevulinate synthase that causes anemia and iron over-loading, a defect in some brain enzyme that causes convulsions in infants, a defect in kynureninase that causes the same results with normal diets that are noted above for deficient diets, and uncharacterized apparent defects in tryptophan metabolism that also result in anemia. Of course, none of these conditions would be of concern to those interested in nutrition of the population at large, in which pyridoxine deficiency is so difficult to provoke that the human requirements are unknown.

Biotin is a part of the enzymes catalyzing the carboxylation of acetyl coenzyme A to form malonyl coenzyme A, methylcrotonyl coenzyme A to form methylglutaconyl coenzyme A (in leucine metabolism), propionyl coenzyme A to form methylmalonyl coenzyme A, and pyruvate to form oxaloacetate. Since it is widely distributed in foods and synthesized by intestinal microorganisms, the only people in danger of a deficiency are those who eat raw eggs in quantity. Egg whites contain a small protein, *avidin*, that has the property of forming a tight complex with biotin, and enough escapes digestion to remove the vitamin. No other cases of a primary deficiency have been demonstrated.

Fatty acid synthase and coenzyme A contain *pantothenate,* and it is therefore widely distributed in foods. It is also synthesized by the intestinal flora, and even an experimental deficiency in humans has not been produced. (Symptoms are created by administration of analogues, but inhibitors may have effects of their own and are rarely likely to deplete the various tissues in the same proportions as those that occur in a natural deficiency.)

Nicotinamide nucleotides are so widespread that one might think a nicotinate deficiency would never occur. If we mentally review the position occupied by these nucleotides in the metabolic economy, we see that they are more like internal substrates than any of the other coenzymes. The cellular concentrations are high because they must be free to move from one enzyme to another, which implies a lower affinity of apoenzyme for coenzyme than is seen with any of the other coenzymes. There is a correspondingly high turnover of the coenzymes to keep the concentrations adjusted. All of this no doubt accounts for the retention of the ability to convert tryptophan to nicotinamide nucleotides over the course of evolution. Strictly speaking, nicotinate isn't a vitamin because it can be synthesized, but as a practical matter a number of humans can't maintain a supply of tryptophan adequate for this purpose and the balance must be supplied as the preformed compound from the diet or a deficiency will result.

Tetrahydrofolate is required for a wide variety of one-carbon transfers, including the interconversion of serine and glycine, the oxidation of glycine, the recovery of homocysteine as methionine, the degradation of histidine, the formation of the purine nucleotides, and the formation of deoxythymidine phosphates. *Dioptorin is* a coenzyme for the hydroxylation of phenylalanine to tyrosine and for other aromatic hydroxylations. Folate compounds have a wide distribution and are synthesized by intestinal microorganisms. However, the plant and microbial compounds frequently contain additional glutamyl residues; that is, they are pteroylpolyglutamates. The extra glutamyl residues are removed by hydrolytic enzymes found in a variety of, if not all, animal tissues. Despite the wide availability and the intestinal synthesis of folate, folate deficiency does occur in humans, partly as a result of poor absorption resulting from a generally inadequate diet, partly because the compound is destroyed by cooking food for extended periods, and partly for reasons that are still not clear.

Cobamide coenzyme is required for the conversion of methylmalonyl coenzyme A to succinyl coenzyme A and for the transfer of methyl groups from 5-methyltetrahydrofolate to homocysteine. We have noted that neither animals nor the higher plants synthesize the cobalamin ring, and the ultimate source for the animals is from those microorganisms that do. People who are strict vegetarians will get no vitamin B_{12} if they also wash their food well to get rid of the soil microorganisms, and are known to become frankly deficient. Those who do eat flesh have no problem unless they lack the intrinsic factor necessary for cobalamin absorption (p. 368), have more general disturbances of intestinal absorption, or are infested with fish tapeworms, which have a large requirement for the compound and rob the host's supply. The cobalamins are truly micronutrients. The daily requirement is not really known, but estimates range around one nanomole per day. The blood concentration is only 0.2 nanomolar, and the total body content is estimated at 2 micromoles. Even so, people who have had total gastric resection, so that they lack intrinsic factor and cannot absorb the cobalamins, take as long as 17 years to exhaust their internal supply to the point where symptoms of pernicious anemia appear. There evidently is some reserve included in that 2 micromoles of total compound, and it evidently is in the liver, which has the highest concentration.

Ascorbate, even more than the nicotinamide nucleotides, ought to be regarded as a substrate. All of the animals that have been studied are able to make the compound except the primates and guinea pigs; the route involves either glucuronyl or galacturonyl residues (Fig. 27-3). We have noted that ascorbate is involved in the hydroxylation of prolyl residues during collagen formation (p. 93) and may be involved in steroid hydroxylation reactions (p. 568), but the general function of the compound is still an enigma. It is not even known how the oxidized compound, dehydroascorbate, is again reduced in animal systems. It is clear that it is, because dehydroascorbate is nutritionally equivalent to ascorbate. We must qualify that somewhat, because dehydroascorbate spontaneously hydrolyzes to form 2,3-diketo-L-gulonate (Fig. 27-4), and this creates a problem in terms of food supply

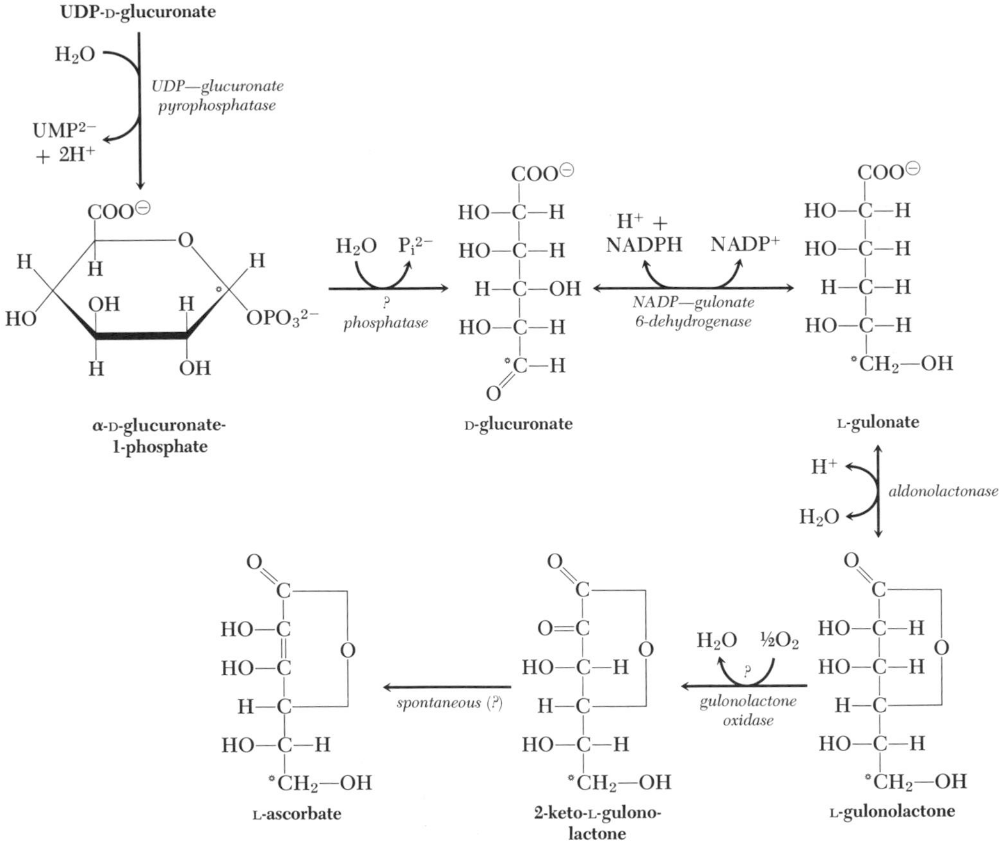

Figure 27-3   Most animals can synthesize ascorbate from glucose by a route involving glucuronate as an intermediate. Primates have lost the final enzyme of the route, the gulonolactone oxidase, and depend upon dietary sources of ascorbate.

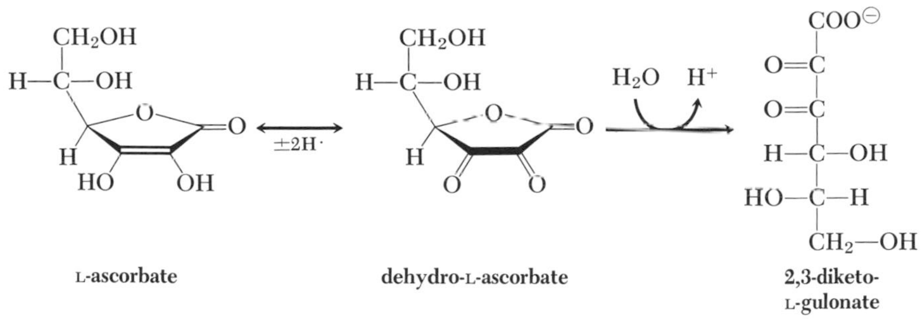

CH₂OH
H—C—OH
H
HO   OH
L-ascorbate

±2H·

CH₂OH
H—C—OH
H
O   O
dehydro-L-ascorbate

H₂O   H⁺

COO⁻
O=C
O=C
H—C—OH
HO—C—H
CH₂—OH
2,3-diketo-
L-gulonate

Figure 27-4 The physiological action of ascorbate involves its oxidation to dehydroascorbate, which can be re-converted to ascorbate by reductions that are not fully characterized. However, part of the dehydroascorbate is lost through a spontaneous hydrolysis of the lactone structure, and the resultant 2,3-diketogulonate is excreted. Ascorbate in food can also be lost in this way after spontaneous oxidation of the compound to dehydroascorbate.

because the isomerization is irreversible and the latter compound will not substitute for ascorbate in the diet. Even traces of contamination by ions of the transition metals catalyze the oxidation of ascorbate by atmospheric oxygen, and the resultant dehydroascorbate slowly isomerizes and loses biological activity. The process is accelerated at higher temperatures, so that much of the ascorbate in foods is lost by cooking, but it will also disappear from processed foods in the refrigerator in time if they are exposed to the air. (Moral: Don't buy the large economy-size cans of orange juice for the baby.) Fruits and vegetables are good sources of ascorbate; meat eaters can get an adequate supply if they eat the liver, kidneys, or brains of animals.

We discussed the metabolic role of *retinol*, its formation from *carotenes*, and the function of *menaquinone* at some length in the preceding chapter. *Cholecalciferol* was discussed in Chapter 24. Let us turn briefly to the *tocopherols*, or vitamin E, which we have not had occasion to mention because they have no known function. The compounds are various derivatives of a reduced triprenylchromanol (Fig. 27-5) that are widely distributed in plant and animal tissues. Experimental animals fed diets lacking the compounds typically become sterile with eventual degeneration (dystrophy) of the skeletal muscles. Many of the symptoms are aggravated by feeding diets rich in polyunsaturated fatty acids.

The polyunsaturated fatty acids are especially susceptible to spontaneous attack by molecular oxygen through an autocatalytic mechanism involving free radicals

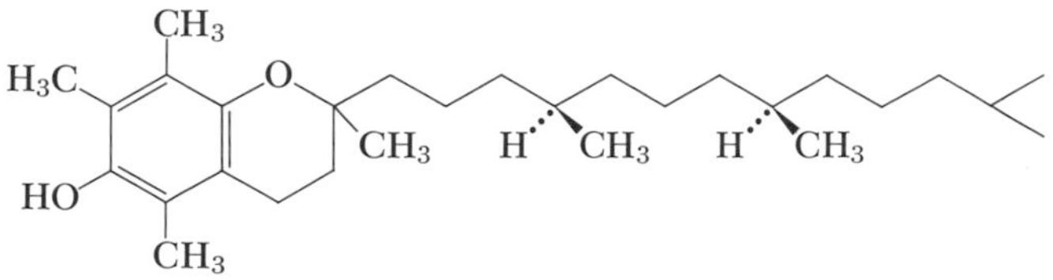

CH₃
H₃C
O
CH₃   H CH₃   H CH₃
HO
CH₃
(2ξ:4′R:8′R)-α-tocopherol

Figure 27-5 The structure of one of the tocopherols, or vitamin E. This compound is believed to serve as an anti-oxidant that prevents free radical chain mechanisms.

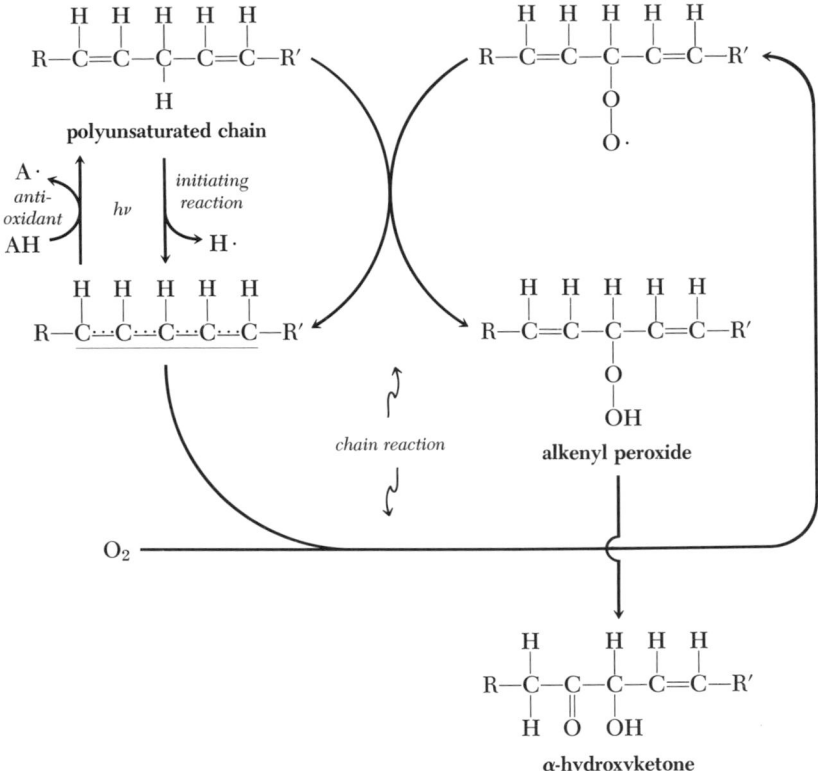

Figure 27-6   Polyunsaturated fatty acids are subject to continued oxidation by a chain reaction after the initial formation of a free radical, which may occur by absorption of radiation. The free radical is a conjugated structure that readily reacts with molecular oxygen to form a free radical peroxide, which in turn is stabilized by the generation of still another molecule of the original free radical. The peroxides that are formed can decompose in a variety of ways, and the intermediate formation of a hydroxyketone, as shown, is only one of the possibilities.

The sequence can be interrupted by the addition of an anti-oxidant that will react with the free radical derived from the polyunsaturated fatty acid and convert it to a more stable form. Many believe the tocopherols serve this function in tissues.

that may be initiated through exposure to light or by the presence of ions of the transition metals. The exact sequence of events is not known, but the general principle is shown in Figure 27-6. The tocopherols have the ability to interrupt this chain of events by removing the intermediate free radical, and many believe that the effects of tocopherol deficiency are the results of accumulation of the fatty acid peroxides in the tissue, which then react to destroy other cellular components. This view is consistent with known information, but there are still some enigmas. Rats on a low-protein diet lacking tocopherols develop a necrosis of the liver, and the effect of tocopherol in preventing the condition is augmented by small amounts of selenium. (Selenium is highly toxic in larger quantities.) When the story is told, it may turn out that selenium will have to be added to our list of elements essential to life, or it may be that its effects are an accident of the particular experimental conditions.

Foods containing unsaturated fatty acids become rancid because of auto-oxidation. Anti-oxidants are frequently added to prepared foods such as potato chips and cooking oils to prevent rancidity by reacting with free radicals. Those commonly employed for this purpose include butylated hydroxyanisole (2 or 3-*tert*-butyl-4-methoxyphenol, commonly abbreviated BHA) and butylated hydroxytoluene (2,6-di-*tert*-butyl-*p*-cresol, or BHT). The possible long-term effects of the present high rate of consumption of these compounds by a large proportion of the population ought to receive more critical review than it has.

Dietary Deficiencies of Vitamins

Let us consider those cases in which vitamin deficiencies occur as a result of a direct dietary lack, and put aside those occurring because of general protein-calorie malnutrition, genetic defects, and disturbances of gastrointestinal function. In short, we are concerned here with deficits that occur in an otherwise normal general population.

We noted that *thiamine deficiency* may appear as beri-beri or Wernicke's encephalopathy (p. 258). In our earlier discussion we emphasized the failure in the metabolism of pyruvate that results from lack of thiamine pyrophosphate. It so happens that a decline in the activity of transketolase is an earlier event in the deficiency. Since transketolase occurs in erythrocytes, which are readily available for analysis, the assay of the enzyme is a tool for assessing the state of the thiamine supply. However, the failure in the metabolism of the 2-ketocarboxylates is a more likely cause of the symptoms of beri-beri.

The amount of thiamine required for normal function varies with the amount of carbohydrate in the diet and ranges around 0.9 micromoles (300 μg) per 1000 kcal equivalent of ingested food for most people. Chronic deficiency is mainly characterized by subjective symptoms and is not well defined. Supplementation with extra thiamine probably does little harm because the injected lethal dose in experimental animals is greater than 300 micromoles per kilogram of body weight. Thiamine deficiency is most likely to occur in those who eat highly refined foods or who get a substantial portion of their caloric intake from alcohol. Prolonged cooking also destroys the compound. A frank deficiency is rare in those who eat at will.

Nicotinate deficiency was formerly common in the South where corn provided a major part of the diet for many. Mexicans apparently are less likely to get pellagra, despite their dependence on corn, because they treat the seeds with lime. This doesn't improve the tryptophan content, but it does make the nicotinate in the grain more accessible, perhaps by destroying some inhibitory substance (p. 456) of unknown character. Pellagra still occurs, and the mild form usually begins with a burning sensation in the tongue, which then becomes inflamed and swollen. Dermatitis follows, with a loss of skin at points of friction, for example, on the elbows. The gastrointestinal tract and nervous system are disturbed in ill-defined ways that later develop into diarrhea and dementia. Preventive treatment obviously consists of assuring a supply of tryptophan in the diet, and the deficiency of this amino acid will not be corrected by dietary supplementation with nicotinate.

Folate deficiency is difficult to define. The consequence that usually attracts attention is an anemia of a macrocytic type; that is, the total amount of hemoglobin is abnormally low and it is contained in abnormally large cells. There also is an appearance of large, nucleated red cells, or megaloblasts in the bone marrow, as the anemia progresses. Another consequence that is revealed by laboratory examination is an increased excretion of formiminoglutamate in the urine, especially after ingestion of a test dose of histidine. This is all well and good and there would be no problem in declaring an individual deficient were it not for the fact that a large fraction of pregnant women have a fall in the concentration of hemoglobin, especially noticeable during the last three months of pregnancy. There is frequently a concomitant fall in the concentration of folate compounds in the blood. Are we to conclude from this that most pregnant women become deficient in folate? Some people do, and say that the incidence is as great as two out of three even in well-fed populations. Let us examine the basis for this kind of statement. If one studies a population of pregnant women and measures the concentration of folate in the blood or the excretion of folate metabolites in the urine, most of the women will have lower values than their non-pregnant peers. If one defines the non-pregnant levels as indicating normal status, then most of the pregnant women are indeed deficient, and this impression is reinforced in many instances by the restoration of "normal" values by vigorous supplementation with folate.

If one rejects this basis and attempts to find impending functional impairment by examination of marrow biopsies, changes in the developing erythrocytes suggestive of folate deficiency will be found in some 2 to 17 per cent of the specimens from pregnant women approaching term in Western countries. Now, if one goes even further and says that there is no folate deficiency until the characteristic macrocytic anemia appears in the circulating blood, the incidence is cut even further and will be substantially higher in the poor than in the well-to-do.

What is one to make of all this? It appears that some degree of folate depletion is a normal concomitant of pregnancy, despite the widespread occurrence of folate in foods and its synthesis by microorganisms. Males usually have enough in the body so that it requires a few months for deficiency symptoms to appear after total deprivation under experimental conditions, but the pregnant female must supply enough coenzyme for a substantial synthesis of nucleic acids needed to form the placenta and infant as well as a considerably larger one-carbon pool. There is no doubt that those on marginal food intakes, particularly if they are devoted to cooking up stews that steam so long that the folate is destroyed, are susceptible to a frank folate deficiency that requires treatment when it appears, but what about the others who do not have the characteristic hematological changes?

Here we ought to ask some difficult questions, most of them presently unanswerable. The most important is: Is the declining folate store in the pregnant female a normal event, normal in the sense that it occurs in most pregnant women now and during the past millennia in which modern man evolved? If the answer is yes, we ought to be cautious about tinkering even though we see no obvious disadvantage in raising the folate levels to the non-pregnant levels (that is, no disadvantage beyond the possible elimination of evolutionary discard *in utero* of metabolically incompetent mutations). If the answer is no, then we need an immediate and searching examination of our present dietary practices to discover what has changed to cause the decline in folate levels.

Any incipient deficit of folate can be aggravated by a deficit in cobamide coenzyme, owing to the increased accumulation of 5-methyltetrahydrofolate in some cases, and perhaps to an adaptive decline of folate metabolism in others. (The effects of cobalamin deficiency on erythrocyte formation and on histidine metabolism are indistinguishable from those of folate deficiency.)

Ascorbate deficiency (scurvy) is the price we sometimes pay for abandoning the free-living fruit-picking life of our primate ancestors. Most of the effects are those of defective collagen synthesis. Infants are afflicted with painful tenderness in their limbs, so that they draw up their legs and lie quietly. Bone formation is defective, and the blood vessels are weakened, so that hemorrhage is common. Adults also tend to bleed readily, especially from the gums, and wounds remain open. Scurvy has been known for millennia as an affliction of organized traveling groups of men: soldiers on the march, sailors at sea, and exploring parties who have depended upon easily shipped supplies of non-perishable foods. Western civilization didn't make the connection between the disease and diet until the eighteenth century, when a series of studies by the British Navy brought scurvy under control on its vessels.

Scurvy is now relatively rare in more developed countries, but large hospitals will still see a few cases a year, mainly in infants, after they are weaned and no longer have the maternal supply, and in older men living alone.

Scorbutic symptoms are prevented by an intake of approximately 0.1 millimole (18 mg) of ascorbic acid per day. Older statements of much higher requirements were based on the amount necessary to saturate the body so that most of an ingested dose was excreted, or on the amount necessary to maintain maximum concentration in the blood. These definitions of requirement were also responsible for the notion, still persisting in some quarters, that a daily intake of ascorbic acid is essential because the compound cannot be stored. The fact is that there is a substantial reservoir available, but the concentration in the blood falls rapidly. When someone got around to doing an actual experiment with humans, it was discovered that the blood concentration falls to unmeasurable levels about six weeks after deprivation, whereas the first signs of scurvy don't appear for another 13 weeks. (In retrospect, a study of the annals of many exploring expeditions would have led to the same general conclusion.) It is now known that measuring the ascorbate concentration in the platelets of the blood, rather than the whole blood, provides a reliable index of the all-important tissue stores.

The minimum requirement can be met by approximately 100 grams of whole potato (50 grams freshly dug or 300 grams after six months' storage), 60 grams of liver from a variety of animals, and from 10 grams to 400 grams of various fruits, tubers, and leafy vegetables.

Scurvy also causes a megaloblastic anemia typical of folate deficiency, and it appears that there is a double action: a primary ascorbate deficiency and a secondary deficiency of folate. Since the function of ascorbate and one of the functions of folate is to act as a second substrate for mixed function oxidases, there is at least an intriguing possibility that part of the folate is being removed to replace ascorbate in its function or that ascorbate is sometimes involved in regenerating tetrahydrobiopterin. (The argument for a primary effect of ascorbate is based on the conversion of the megaloblastic anemia to a normoblastic type, but without complete remission, when folate alone is given. A deficiency of ascorbate may also cause an iron deficiency anemia through disturbances in iron absorption along with increased losses from the small hemorrhages that result.)

MINERAL NUTRIENTS

We have previously encountered the various inorganic ions that are required in the diet. Some of the ions occur so commonly in foods that they are not of practical concern. At one extreme are sodium, potassium, and chloride ions, which are needed in quantity, but which also occur in quantity and are efficiently retained as needed in normal people. (However, we have seen that management of salt concentration is an important part of the treatment of many pathological conditions.) At the other extreme are elements such as molybdenum, chromium, zinc, manganese, and copper, which occur in minute quantities in the diet, but which are needed in even more minute amounts. We have noted that deficiencies of these trace elements are a problem to stock growers and farmers in some areas of the world, but human deficiencies are rare. Magnesium is an element that is required in intermediate amounts, but without known deficiencies of supply in normal humans. Let us now consider those elements that occasionally do become deficient in the diet.

Calcium

A human is born with some 0.7 mole of calcium within his body, and gains another 24 to 29 moles as he grows. His increment of calcium is therefore near 3.8 millimoles per day on the average, and this increment must be supplied by the diet. The nursing infant obtains his calcium supply from his mother, and human milk is 8 millimolar to calcium. After weaning, the supply must come from other foods. If there is too little, the general effect is not so much one of defective bone formation as it is of lessened growth.

It is difficult to estimate a calcium requirement beyond the amount required for growth. There is a very efficient adaptation of absorption to the internal need and the customary dietary intake, and the amount absorbed also depends upon the character of the diet. The result of all this is that an adult, who has no net deposition of calcium to confuse the picture, will "require" 10 millimoles of calcium per day to remain in balance if he has been ingesting 10 millimoles per day, but will only require 8 millimoles per day if that is all he has been eating. The fraction of the intake that is absorbed may range from a few percent to as high as 65 per cent.

It now appears that an intake of 10 to 12 millimoles (0.4 to 0.5 gram) of calcium per day is sufficient to maintain the deposits in the bones of adults, with sufficient margin for losses and incomplete absorption. Pregnant and lactating women must provide the calcium for the growing child and need approximately double the usual adult intake to avoid mobilizing calcium from their own bones in order to supply the infant.

Where is the calcium to come from? Lake and river water ranges from 0.1 to 2 mM in calcium, so only the very hard waters supply a significant fraction. Most of the major foods, including the cereals and meats, have such a low calcium content that it would require eating from 2 to 5 kilograms before 10 millimoles of calcium were ingested. Cow's milk is an obvious source because it contains 30 millimoles per liter. Fish eaters are in a better position because many fish and shellfish have 1.5 or more millimoles per 100-gram portions, and chicken eggs have almost as much. Soybeans are an excellent source (6 millimoles per 100 grams), and peas and beans range around 1 millimole per 100 grams.

All of this indicates that calcium deficiency is most likely to be a problem of people who are restricted to a limited range of foods, and experience suggests that it is indeed the poor who are most likely to be deficient.

The other major constituent of bone, *phosphate*, is rarely deficient as such in the diet, and disturbances in its supply are usually the result of disturbances in calcium turnover.

Iron

The body content of iron also increases from birth to adulthood. Table 27-3 shows the values at various ages, and the portion of the content that is storage

TABLE 27-3.   BODY CONTENT OF IRON.

AGE years	BODY CONTENT millimoles				PHYSIOLOGICAL REQUIREMENT millimoles/year	
	STORAGE		TOTAL			
	male	female	male	female	male	female
birth	1.0	1.0	5.1	5.3	2.6	2.0
4	2.7	2.7	13.9	13.3	2.5	2.6
19	13.0	8.3	62	40	5.0	7.1
60–70	12.3	9.0	60	44	1.2	1.4

iron, available for use in synthesis of the hemoproteins and iron-containing oxidases. It also shows the estimated physiological requirement, that is, the amount necessary for replacement of losses and the production of new tissue. Of course, we must not apply these data too rigidly, because of the variation in iron storage with consumption that we noted in Chapter 21. Even so, they constitute a guide and could be used as an index of the requirement for iron in the diet if it were not for the fact that the efficiency of absorption of iron is even more variable than that of calcium. The usual thinking today is to assume that something like 8 atoms of iron must be ingested in order to absorb one. Referring to Table 27-3, then, this would make the requirement in the food of our 19-year-old female 57 millimoles per year, or 0.16 millimoles (9 milligrams) per day.

In our discussion of iron metabolism we emphasized the iron overload that can occur through indiscriminate ingestion of iron tonics, cheap wines, and so on. However, it is a mistake to assume that there is never a case for increasing the iron intake. Iron deficiency occurs in the United States, and is primarily manifested by an anemia in which the red blood cells are small and pale (microcytic hypochromic anemia), and the discovery of such a condition is a *prima facie* case for undetected bleeding or for insufficient ingestion of iron. Those who are eating too little iron usually are eating a relatively restricted selection of refined foods. They may be poor or they may be teen-age girls worrying about their waistlines and living on soft drinks and potato chips.

We are also faced here with the same sort of problem in defining a requirement as that which we saw with folate. Some would like to emphasize the maintenance of specified levels of circulating iron or the saturation of the blood transferrin as an index. Others take the functional approach of asserting that the iron supply is adequate in the absence of the characteristic anemia. These difficulties again become acute in assessing the status of pregnant women, who inevitably deplete their iron stores to supply the fetus. Pregnant women also have an undisputed lowering of the concentration of hemoglobin due to an increased volume of circulating blood. This hemodilution causes a compensatory increase of some 15 to 20 per cent in the total number of erythrocytes, even though the number of erythrocytes in a given volume of blood is still below the non-pregnant level. The formation of the new erythrocytes requires an additional 5 millimoles of iron.

The result of all this is that a large fraction of the pregnant women have a decline in the blood concentration of iron even after allowing for hemodilution. Some become frankly anemic with pronounced hematological changes, and we may expect that these are the ones who already had low iron stores before the pregnancy. The number in this condition increases with the number of pregnancies, because each infant formed causes a further drain on the maternal stores. If we put those who obviously need iron supplementation aside and consider the others, we have the same question posed with folate, but in sharper form because the iron deficiency anemia is much more common. Is the depletion of iron, along with the depletion of folate, a normal physiological event, or a pathological deficiency that demands treatment whenever any indicator of the condition appears? There is no glib answer. Administration of iron causes increased formation of red blood cells, even in normal, definitely non-pregnant, males, so the response is not necessarily a reliable guide to an existing impairment of function. Some pregnant women are quite refractory to administered iron even though they appear to be deficient, and if their responses are pushed so that the blood measurements approach"normal" levels, they sometimes promptly fall back as soon as the extra iron load is removed. We need open eyes, an open mind, and more information.

RECOMMENDED DAILY ALLOWANCES

Let us close our discussion with the presentation of the recommended daily intake of food prepared by the Food and Nutrition Board of the National Research Council. A comparison of the recommendations with the minimum requirements we have discussed will show that they are, in the main, conservative. They are not a guide for stretching a limited food supply over the greatest possible number of people, but a guide for the almost complete elimination of any symptoms of dietary deficiency in an affluent society. The executive secretary of the board, LeRoy Voris, has put it nicely:

Deviations of individual intakes from the recommended nutrient intakes of the recommended dietary allowances are significant only in terms of the individual's total health status. Food consumption survey data cannot be used alone as a measure of nutritional adequacy. In determining nutritional status, the current and past nutrient intake must be taken into consideration, as well as an evaluation of clinical signs and symptoms, growth and development, and biochemical data on tissue and excretory levels of nutrients. Since the RDA are designed to be adequate for the maintenance of good nutrition for practically all of the population of the United States, they allow a margin of safety for individual variations. Individuals whose diets do not meet the RDA are not necessarily suffering from malnutrition and diets should not be judged as "poor" on an arbitrary figure based on comparison with the RDA, which are revised about every five years in order to include new research findings.

The original presentation by the Board states the allowances in terms of weights and I.U.; these have been recalculated in terms of molar dimensions so as to be more easily compared with other information. Conversion factors are given.

Notes: In converting the Board allowances to a molar basis, the integer increments with age used by the Board have sometimes been translated to the closest integer molar increment. The Board's estimate for the protein allowance is on the basis of proteins equivalent to those of human milk in supporting growth. The fat-soluble vitamin allowances have been calculated in terms of the equivalent moles of retinol, calciferol, and (d)-α-tocopherol. The nicotinate allowance is calculated on the basis of 1 micromole of nicotinate being equivalent to 36 micromoles of tryptophan. The folate allowance is determined on the basis of a microbiological assay of the foods, and less than one-quarter of the stated amount given as pure folate will be effective.

TABLE 27-4.   RECOMMENDED

	AGE	WEIGHT kg	HEIGHT cm	ENERGY YIELD kcal	PROTEIN N mmoles	RETINOL EQUIV. μmoles	CALCIFEROL EQUIV. nmoles	TOCOPHEROL EQUIV. μmoles	ASCORBATE μmoles
Infants	0–2 mo.	4	55	120/kg	26/kg	1.6	25	8	200
	2–6 mo.	7	63	110/kg	23/kg	1.6	25	8	200
	6–12 mo.	9	72	100/kg	21/kg	1.6	25	8	200
Children	1–2 yr.	12	81	1100	290	2.1	25	16	230
	2–3	14	91	1250	290	2.1	25	16	230
	3–4	16	100	1400	340	2.6	25	16	230
	4–6	19	110	1600	340	2.6	25	16	230
	6–8	23	121	2000	400	3.7	25	23	230
	8–10	28	131	2200	460	3.7	25	23	230
Men	10–12	35	140	2500	510	4.7	25	31	230
	12–14	43	151	2700	570	5.2	25	31	260
	14–18	59	170	3000	680	5.2	25	39	310
	18–22	67	175	2800	680	5.2	25	47	340
	22–35	70	175	2800	740	5.2		47	340
	35–55	70	173	2600	740	5.2		47	340
	55–75	70	171	2400	740	5.2		47	340
Women	10–12	35	142	2250	570	4.7	25	31	230
	12–14	44	154	2300	570	5.2	25	31	260
	14–16	52	157	2400	630	5.2	25	39	280
	16–18	54	160	2300	630	5.2	25	39	310
	18–22	58	163	2000	630	5.2	25	39	310
	22–35	58	163	2000	630	5.2		39	310
	35–55	58	160	1850	630	5.2		39	310
	55–75	58	157	1700	630	5.2		39	310
	Pregnancy			+200	740	6.3	25	47	340
	Lactation			+1000	860	8.4	25	47	340

The stated protein nitrogen allowance may be converted to grams of "protein" by multiplying by 0.0875. The fat-soluble vitamin allowances may be converted to I.U. by multiplying by the following conversion factors:

$$\text{I.U. vitamin A} = \mu\text{moles retinol} \times 950$$
$$\text{I.U. vitamin D} = \text{nmoles calciferol} \times 16$$

Daily Allowances.

Folate µmoles	Nicotinate Equiv. µmoles	Ribo-flavin µmoles	Thiamine µmoles	Pyri-doxine µmoles	Cobalamin nmoles	Calcium mmoles	Phosphate mmoles	Iodine µmoles	Iron mmoles	Magnesium mmoles
0.1	40	1.1	0.7	1	0.7	10	6.5	0.2	0.09	40
0.1	60	1.3	1.3	2	1.1	12.5	13	0.3	0.18	60
0.2	65	1.6	1.7	2.5	1.5	15	16	0.35	0.27	70
0.2	65	1.6	2.0	3	1.5	17.5	23	0.45	0.27	100
0.5	65	1.9	2.0	3.5	1.8	20	26	0.5	0.27	150
0.5	75	2.1	2.3	4	2.2	20	26	0.55	0.18	200
0.5	90	2.4	2.7	5.5	3.0	20	26	0.65	0.18	200
0.5	105	2.9	3.3	6	3.0	22.5	29	0.8	0.18	250
0.7	120	3.2	3.7	7	3.7	25	32	0.85	0.18	250
0.9	140	3.5	4.3	8	3.7	30	39	1.0	0.18	300
0.9	145	3.7	4.7	9.5	3.7	35	45	1.05	0.32	350
0.9	165	4.0	5.0	10.5	3.7	35	45	1.2	0.32	400
0.9	150	4.3	4.7	12	3.7	20	26	1.1	0.18	400
0.9	150	4.5	4.7	12	3.7	20	26	1.1	0.18	350
0.9	140	4.5	4.3	12	3.7	20	26	1.0	0.18	350
0.9	115	4.5	4.0	12	4.4	20	26	0.85	0.18	350
0.9	120	3.5	3.7	8	3.7	30	39	0.85	0.32	300
0.9	120	3.7	4.0	9.5	3.7	32.5	42	0.9	0.32	350
0.9	130	3.7	4.0	10.5	3.7	32.5	42	0.95	0.32	350
0.9	120	4.0	4.0	12	3.7	32.5	42	0.9	0.32	350
0.9	105	4.0	3.3	12	3.7	20	26	0.8	0.32	350
0.9	105	4.0	3.3	12	3.7	20	26	0.8	0.32	300
0.9	105	4.0	3.3	12	3.7	20	26	0.7	0.32	300
0.9	100	4.0	3.3	12	4.4	20	26	0.65	0.18	300
1.8	120	4.8	+0.3	15	5.9	+10	+13	1.0	0.32	450
1.2	160	5.3	+1.7	15	4.4	+12.5	+16	1.2	0.32	450

I.U. vitamin E $=$ µmoles tocopherol $\times$ 1.6

The molar allowances for the other vitamins and the minerals can be converted to a weight basis by multiplying by the molecular weights: ascorbate $=$ 176, folate $=$ 441, riboflavin $=$ 376, thiamine $=$ 301, pyridoxine $=$ 170, cobalamin $=$ 1355, calcium $=$ 40, phosphate $=$ 31 (as P), iodine $=$ 127, iron $=$ 55.6, and magnesium $=$ 24.

Further reading

Beaton, G. H., and E. W. McHenry, eds.: *Nutrition—A Comprehensive Treatise*, Vols. 1 to 3. Academic Press (1964).

Munro, H. N., and J. B. Allison, eds.: *Mammalian Protein Metabolism*, Vols. 1 and 2. Academic Press (1964).

McCance, R. A., and E. M. Widdowson, eds.: *Calorie Deficiencies and Protein Deficiencies*. Little Brown (1968).

Sebrell, W. H., Jr., and R. S. Harris: *The Vitamins: Chemistry, Physiology, Pathology, Methods*, 2nd ed. Academic Press (1967– Several projected volumes now slowly appearing).

Vitamins and Hormones. Academic Press. Annual volumes.

Nutritional Reviews. Useful short summaries of current developments.

Nutrition Today. Quarterly with non-technical descriptive reviews. The issue for Summer, 1969 has two instructive articles on iron deficiency.

Hein, R. E.: *Heinz Nutritional Data*, 5th ed. H. T. Heinz Co. (1964). Very useful.

Keys, A., et al.: *The Biology of Human Starvation*, Vols. 1 and 2. University of Minnesota Press (1950).

Autret, M., and M. Behar: *Sindrome Policarencial Infantil (Kwashiorkor) and Its Prevention in Central America*. FAO Nutrition Studies, No. 13. FAO (1954).

Thomson, T. J., J. Runcie, and V. Miller: *Treatment of Obesity by Total Fasting for up to 249 Days*. Lancet, 2: 992 (1966).

Drenick, E. J., C. B. Joren, and M. E. Swendserd: *Occurrence of Acute Wernicke's Encephalopathy During Prolonged Starvation for the Treatment of Obesity*. New Eng. J. Med., 274: 937 (1966).

Robinson, F. A.: *The Vitamin Co-Factors of Enzyme Systems*. Pergamon (1966).

Rosenberg, L. E.: *Inherited Aminoacidopathies Demonstrating Vitamin Dependency*. New Eng. J. Med., 281: 145 (1969).

Symposium: Nutritional Aspects of Uremia. Amer. J. Clin. Nutr., Nos. 5 and 6 (May-June, 1968).

Hegsted, D. M.: *Theoretical Estimates of the Protein Requirements of Children*. J. Am. Diet. Assn., 33: 225 (1957).

Hegsted, D. M.: *Present Knowledge of Calcium, Phosphorus and Magnesium*. Nutr. Rev., 26: 65 (1968).

Symposium on Prospects of the World Food Supply. Proc. Natl. Acad. Sci., 56: 305ff. (1966).

Chandler, R. E.: *Improving the Rice Plant and Its Culture*. Nature, 221: 1007 (1969).

Nelson, O. E., E. T. Mertz, and L. S. Bates: *Second Mutant Gene Affecting the Amino Acid Pattern of Maize Endosperm Proteins*. Science, 150: 1469 (1965).

PART VI

Appendices

SOME CHEMICAL PROPERTIES

CHAPTER 28

The purpose of this appendix is to summarize some of the modern techniques that are available for studying amino acids, sugars, and fatty acids, and also to clarify the stereochemistry where applicable. It does not give a comprehensive review of the structure and chemical reactions of all natural compounds. The references at the end of the appendix give more complete accounts of the organic and analytical chemistry of natural compounds, including the many colorimetric analyses available for particular kinds of compounds. Modern textbooks of organic chemistry also treat some aspects of the chemistry of natural products.

AMINO ACIDS AND PEPTIDES

Structural Formulas (See page 694)

Ionization

Consider the simplest amino acid, glycine. It can equilibrate with H^+ in two ways:

$$^+H_3N-CH_2-COOH \xrightleftharpoons[K_1 = 10^{-2.35}]{H^+} \ ^+H_3N-CH_2-COO^- \xrightleftharpoons[K_2 = 10^{-9.78}]{H^+} H_2N-CH_2-COO^-$$

The form shown in the middle is a *zwitterion*, meaning hermaphrodite ion, that has no net charge because it has equal numbers of positive ammonium groups and negative carboxylate groups. This form can be crystallized as such and is what one has in a bottle of glycine. Suppose one adds HCl to a solution of the zwitterion form. Nothing will happen to the ammonium group; it already has H^+ affixed to

693

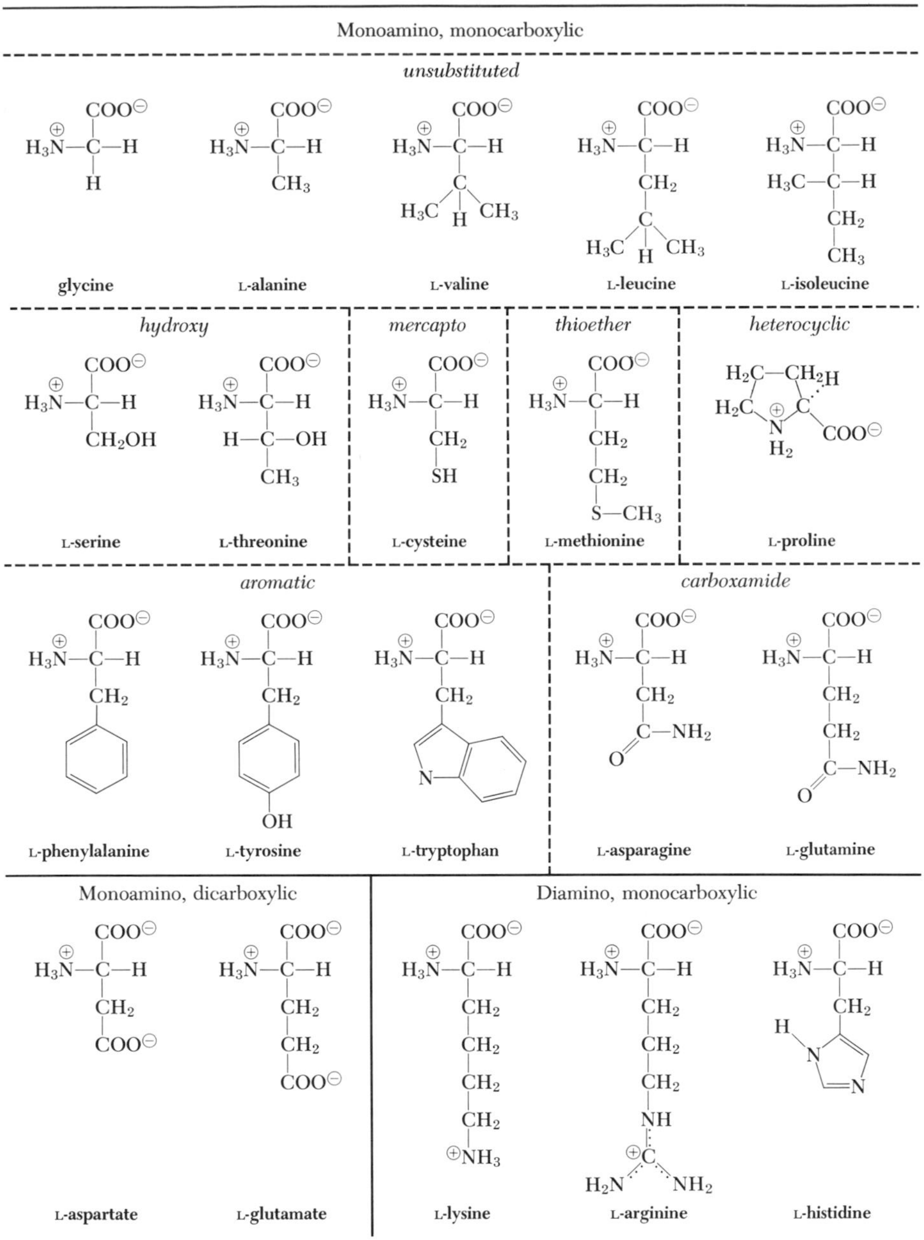

Monoamino, monocarboxylic

unsubstituted

| glycine | L-alanine | L-valine | L-leucine | L-isoleucine |

hydroxy   *mercapto*   *thioether*   *heterocyclic*

| L-serine | L-threonine | L-cysteine | L-methionine | L-proline |

aromatic   *carboxamide*

| L-phenylalanine | L-tyrosine | L-tryptophan | L-asparagine | L-glutamine |

Monoamino, dicarboxylic   Diamino, monocarboxylic

| L-aspartate | L-glutamate | L-lysine | L-arginine | L-histidine |

the nitrogen. However, the increasing concentration of H^+ will cause the carboxylate group to pick up H^+ to form the undissociated carboxylic group to an extent that is defined by the dissociation constant:

$$\frac{[R—COO^-][H^+]}{[R—COOH]} = K_1 = 10^{-2.35}$$

When $[H^+]$ equals K_1, the pH is 2.35 and half the glycine will be in the zwitterion form (designated Z in the figure below), and half will be in the acidic form (designated A). Since the acidic form has a net positive charge, it cannot be isolated as such, but must be associated with some anion; in our example, it could be precipitated as glycine hydrochloride if the pH is low enough to convert most of the amino acid to that form.

If H^+ is removed from a solution of the zwitterion by adding NaOH, for example, H^+ will dissociate from the ammonium group to an extent defined by a second dissociation constant:

$$\frac{[R—NH_2][H^+]}{[R—NH_3^+]} = K_2 = 10^{-9.78}$$

As in the first dissociation, the concentration of the basic form (designated B in the figure below) will equal the concentration of the zwitterion when $[H^+]$ equals K_2, which will be at pH 9.78. The basic form still has the negative charge of the carboxylate group, and therefore can only be isolated as the salt of a cation, in this case as sodium glycinate.

The fraction of the zwitterion present at a given pH is shown in this graph:

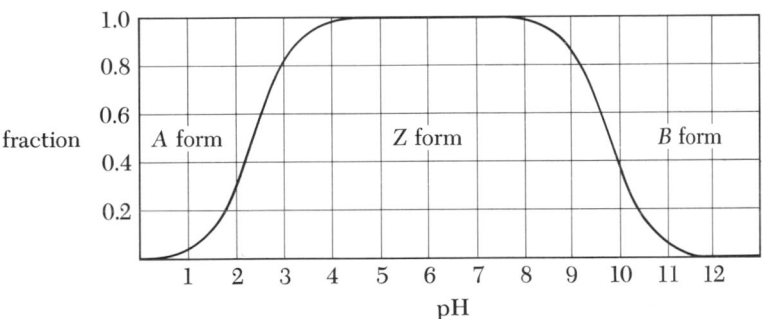

We see that glycine exists mostly as a zwitterion over a broad range centered near pH 6. The curves at the end are almost the shape of titration curves showing the change in pH of a solution of glycine as one adds successive increments of a strong acid or strong base (reading from the top down); the titration curve isn't quite the same because it takes a little extra reagent to titrate water to the extreme pH values at which glycine is present mostly in the acidic or basic forms.

According to the equations given above for the two dissociations, there must be at least a small loss of H^+ from the ammonium group and a small uptake of H^+ by the carboxylate group at any pH, leaving some molecules with net negative and positive charges, respectively. We see from the graph that there must be some pH at which these charges are just balanced, so that most of the molecules are zwitterions with no net charge and the minute balance is evenly divided between molecules with net negative and positive charges. This pH is the *isoelectric point* of glycine, at which there will be no net movement of the molecules through an electric field, and it is exactly at the midpoint between pK_1 and pK_2 (pH 6.06 in this instance). This can be shown by setting $[R—NH_2]$ equal to $[R—COOH]$ in the two equations, solving for $[H^+]$, and taking the negative logarithm of the resultant $[H^+] = \sqrt{K_1 K_2}$.

The description we have given of the ionization of glycine applies to all of the monoamino, monocarboxylic acids, and most of them have ionization constants close to those of glycine. Now let us consider the amino acids that have additional ionizable groups on their side chains, using aspartate as an example:

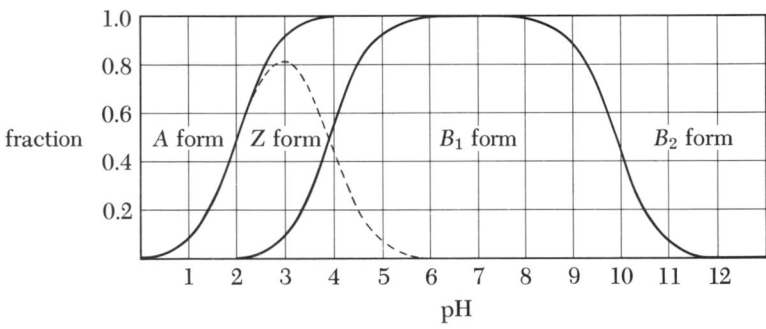

Here we have three distinct equilibria. The ammonium group doesn't behave much differently than the ammonium group of glycine, but the side chain carboxylic group is a weaker acid than the group on glycine and the 1-carboxylic group is a stronger acid. The way in which the proportion of these forms shifts with pH is shown in the following graph:

The bell-shaped solid curve in the center shows the fraction existing as the B_1 form, in which all of the groups are charged, and the dashed bell-shaped curve to the left shows the fraction existing as the Z form, in which one of the carboxylic groups is not ionized and the remaining negative charge of the other carboxylate group neutralizes the positive charge of the ammonium group. The other solid curve to the left shows the fraction of the total that is not in the A form (that is, it shows the sum of the Z and B_1 forms). Here we see that aspartate has a net charge of -1 over a broad range near the physiological pH values.

What is the isoelectric point of aspartic acid? It will be the pH at which there is an average of one negative charge to balance the positively charged ammonium group. This balance will occur when the number of molecules with no negative charges (form A) equals the number with two negative charges (form B_1). When we solve the equilibrium equations for this condition, we find again that the isoelectric pH is exactly halfway between pK_1 and pK_2, which is pH 2.94 in this case, the pH at which the curve for the Z form reaches its maximum in the graph. In other words, if we want to crystallize aspartic acid, we ought to bring the solution to pH 2.94 so that there will be no charges to attract other ions. If we add one mole of NaOH to aspartic acid, the pH will rise to 6.90, and if we now precipitate the compound, we will be isolating monosodium aspartate. (The pH of a solution of the monosodium salt is halfway between pK_2 and pK_3 for all practical purposes. The first dissociation constant can be neglected because it is 100-fold greater than the second, which means that near neutral pH the amount of compound with no undissociated carboxylic groups will be only $1/100$ of the amount with one undissociated carboxylic group. This approximation can't be made for multibasic acids in which the dissociation constants are more nearly equal.)

The same kind of reasoning we have applied here can also be used in treating the diamino, monocarboxylic acids such as lysine and arginine. The zwitterion of these compounds will exist at high pH values because there are two acidic forms and only one basic form. (The pK values for lysine are 2.16, 9.18, and 10.79.) The isoelectric point of these compounds lies midway between pK_2 and pK_3.

Stereoisomers

Let us briefly review what organic chemistry tells us about stereoisomers as it applies to the amino acids. All of the amino acids that are introduced into peptide chains by protein synthesis have one carbon (C-2) that is bonded to four different groups, except glycine, which has two H atoms. There are two possible arrangements of groups around such an asymmetric center:

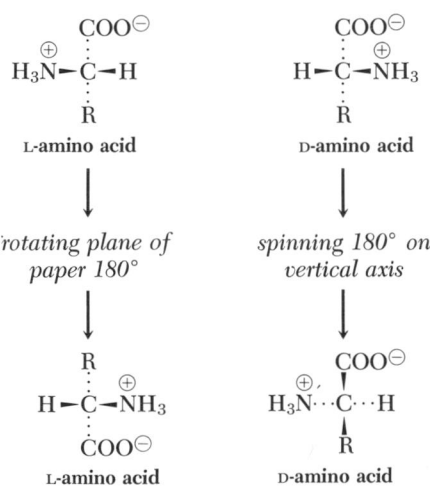

Amino acids having one of the arrangements are said to have the L-configuration; those with the other have the D-configuration. The two configurations are mirror images of each other, and like all mirror images of asymmetric objects cannot be superimposed no matter how they are turned. We have noted that the natural amino acids used for protein synthesis have the L-configuration.

D and L isomers have the same chemical and physical properties in many cases. Many chemical reactions involve reagents with symmetrical molecules that can approach a molecule of either the D or the L isomer at an orientation that will make contact with the same reacting groups. However, this is not true of an asymmetrical reacting compound or of attachment to a surface. It is not possible to set down models of D and L isomers on a sheet of paper so that the contacts are identical, and this is the reason that many enzymes only catalyze the reaction of one stereoisomer.

The isomers also differ in their effect on polarized light passing through the compound. Each isomer has a spatial sense analogous to the sense of a screw; a right-hand screw is still a right-hand screw if one views it from the opposite end. The spatial sense represents a different order of arrangement of electron fields in the compounds, and the result is that the electron field of a D-isomer oscillates to shift the electric vector of the light quantum in one direction and the L-isomer causes it to shift an equal amount in the opposite direction. We see the result as a change in the plane of polarization of linearly polarized light, and the measurement of this *optical activity* can be used to measure the relative amounts of D and L isomers in a solution, and is also used now to make deductions about the structure of molecules. (For example, the helix found in proteins also has a spatial sense and causes changes in the optical activity of a peptide compared to that of a random arrangement of the same primary structure.)

It is common to designate a stereoisomer in structural formulas by a convention in which all vertical bonds are directed behind the plane of the paper and all horizontal bonds are directed in front of the plane of the paper. (We have followed the convention in this book, but a word of caution is necessary. Many people do this so long as they want to draw a carbon skeleton in a vertical position, but they simply turn the representation sideways if they want the carbon skeleton to be horizontal; of course, this now makes the vertical bonds in front of the paper and the horizontal bonds behind. The practice can cause great confusion with branched-chain molecules, and has been avoided here.) There are 12 ways in which a molecule of L-alanine, for example, can be written with the vertical bonds behind the plane of the paper. Some of them are shown in the following:

$$\underset{\overset{|}{CH_3}}{\overset{COO^{\ominus}}{\overset{|}{\underset{H_3\overset{\oplus}{N}-C-H}{}}}} \qquad \underset{\overset{|}{COO^{\ominus}}}{\overset{CH_3}{\overset{|}{\underset{H-C-\overset{\oplus}{N}H_3}{}}}} \qquad \underset{\overset{|}{H}}{\overset{CH_3}{\overset{|}{\underset{H_3\overset{\oplus}{N}-C-COO^{\ominus}}{}}}} \qquad \underset{\overset{|}{CH_3}}{\overset{\overset{\oplus}{N}H_3}{\overset{|}{\underset{H-C-COO^{\ominus}}{}}}} \qquad \underset{\overset{|}{\overset{\oplus}{N}H_3}}{\overset{CH_3}{\overset{|}{\underset{{}^{\ominus}OOC-C-H}{}}}}$$

A useful rule of thumb is that interchanging any two substituents on each asymmetric carbon in a conventional structural formula gives a representation of the other stereoisomer; making any two such interchanges gives another representation of the same stereoisomer.

Some amino acids have more than one asymmetric carbon atom. Consider L-threonine, which has four possible stereoisomers:

L-threonine   D-threonine   L-allothreonine   D-allothreonine

Each of the four is a different compound, and only one, L-threonine, occurs naturally in proteins. There are two mirror-image pairs represented, and the two pairs are given different names. Why is this so? D- and L-threonine behave in the same way toward symmetrical chemical reagents, and D- and L-allothreonine behave the same, but L-threonine and L-allothreonine behave differently and have different melting points, solubilities, and so on.

Imagine a reacting molecule approaching the region of the two asymmetric carbon atoms in L-threonine as we have drawn it. The $-NH_3^+$ and $-OH$ groups are on opposite sides of the C—C bond. The same reagent, if it is symmetrical, can rotate and approach D-threonine in the same region and encounter the same groups at exactly the same spacing. This isn't true of either isomer of allothreonine. If one of the carbons is rotated from the position we have drawn so as to place the $-NH_3^+$ and $-OH$ groups on opposite sides of the C—C bond, either the $-COO^-$ or the $-CH_3$ group will also be rotated into the position formerly occupied by $-H$, and this will change the reactivity toward a reagent. Since most organic chemists deal with symmetrical reagents, they note this different reactivity of the two pairs of the four different compounds and give one pair one name and the other pair another name. This practice also points up the difference in optical activity, because it is only the mirror images that cause equal and opposite rotation of plane polarized light. L-Threonine and L-allothreonine cause different amounts of rotation.

We designate all of the amino acids having the same arrangement of groups around C-2 found in L-alanine as L-amino acids regardless of the arrangement about other asymmetric centers. Some L-amino acids cause a rotation of plane polarized light to the left, others to the right. The direction of rotation is shown when desired by lower case italic letters (d) or (l) in parentheses to designate dextro- or levorotatory, respectively. (It can also be shown by (+) for dextro- and (−) for levo-.) The older literature is confusing in this respect because there was a time of transition when the lower case letters were used for configurational family as well as for actual rotation.

Because of the difficulty sometimes created in designating configurational family with many types of compounds, a new nomenclature has been invented. Briefly, each of the four constituent groups about an asymmetric carbon is arranged in order of increasing atomic number of the nearest constituent atom or in order of increasing valence electron density. (N ranks higher than C, O higher than N; ethylene carbons rank higher than saturated carbons, a $-CH_2-COO^-$ group ranks higher than a $-CH_2-CH_3$ group, and so on.) One looks at the asymmetric center in such a way as to peer directly down on the substituent of lowest rank order, which is frequently

—H with the amino acids. When this is done, the remaining three substituents will be arranged as spokes on a wheel, and one goes around the wheel from the lowest rank order to the highest. If this is a clockwise direction, the configuration is *rectus* or (R), if it is counterclockwise, the configuration is *sinister* or (S). The process is repeated for each asymmetric center. It sounds complicated, and it does take some practice in visualization of the structures, but it has the advantage of creating an unambiguous designation of the absolute configuration, no matter how many asymmetric centers there are. Under this system, L-threonine is (2S:3R)-threonine, or more systematically, (2S:3R)-2-amino-3-hydroxybutyrate (Fig. 28-1). The system has a disadvantage in that the actual configuration of an asymmetric center may be unchanged through a series of reactions, but the order of atomic number of the substituent groups may be changed by a reaction involving one of the groups so that an (R) compound yields an (S) product, or vice-versa. (For example, simple hydrolysis of the phosphate group from an (R)-phosphatidate forms an (S)-digly-ceride.)

Qualitative Detection of Amino Acids

A widely used reaction for the detection of amino acids involves an oxidative decarboxylation upon heating with *ninhydrin* (triketohydrindene hydrate), which is followed by a condensation of the hydrindamine formed with a second molecule of ninhydrin:

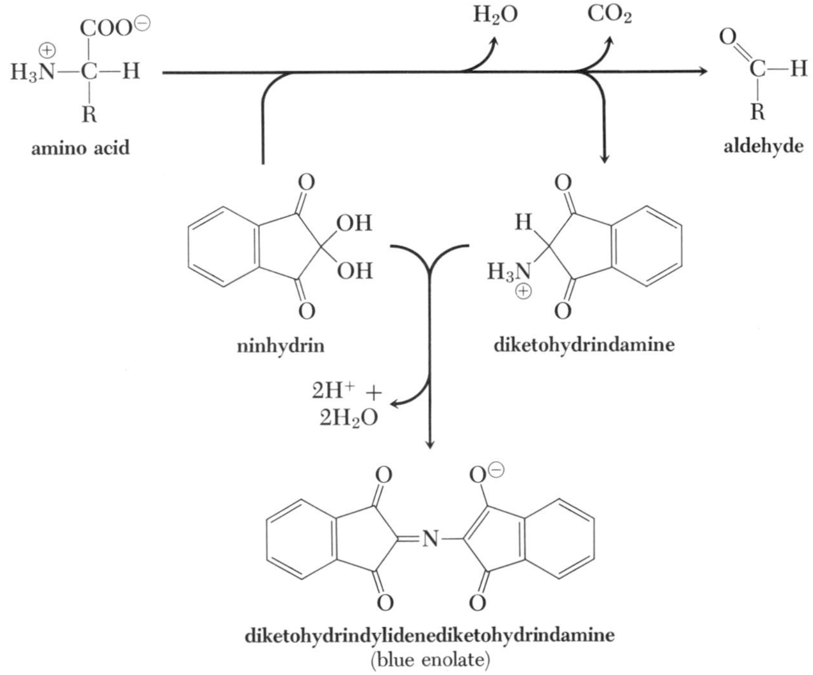

ninhydrin

diketohydrindamine

$2H^+ + 2H_2O$

diketohydrindylidenediketohydrindamine
(blue enolate)

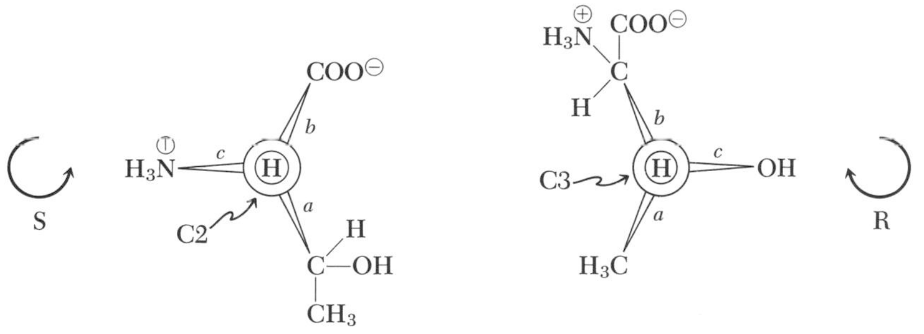

Figure 28-1 Configuration in the (R) and (S) system can be determined by looking down on an asymmetric atom arranged so that the substituent of lowest atomic number is facing the viewer. In the example on the left, one is looking past the substituent H atom at C-2 of L-threonine. In order to go from the lowest to the highest substituent among the remaining three, one must go in a left-hand circle (COO outranks CH—OH(CH₃) because it has two O atoms attached to the carbon, and these higher atomic number atoms outrank the single O and C in the other group; the directly attached N of the amino group has a higher atomic number than C, and therefore outranks both of the other groups). The left-hand circle is designated as (S) configuration.

The example on the right shows the configuration about C-3 of L-threonine. One is again looking down on H as the substituent of lowest rank order in terms of atomic number. In this case, the methyl group is the lowest ranking of the remaining three substituents, and the hydroxyl group is the highest. Traveling in a right-hand circle carries one from lowest to highest ranking substituent, so this carbon has an (R) configuration.

The condensation product has a bright blue color that can be detected when nanomolar quantities of amino acids on paper are sprayed with ninhydrin and heated in an oven. This technique is used by police for developing fingerprints on porous materials. The reaction is used in two ways in the laboratory. A quantitative estimate can be made of the total amino acid content of a sample from the amount of color developed when the sample is treated with ninhydrin in solution. (More exact determinations can be made of individual amino acids separated from the others, because the color yield differs from one to another, but a correction can be made if the nature of the amino acid is known.) The second use is for the detection of amino acids that have been separated by paper chromatography or similar methods. In these methods, a small sample is spotted on the corner of a sheet of filter paper or on a plate coated with a thin layer of some adsorbent such as silica gel. An edge of the sheet or plate near the spot is then dipped in a solvent, typically a phenol-water mixture. As the solvent rises across the sample spot because of capillary action, it dissolves different amounts of the various amino acids in the sample. Those that go readily into the solvent will be carried rapidly across the fresh adsorbent being passed by the moving solvent. There is a succession of solution of the compound by fresh solvent arising from below and adsorption on the fresh adsorbent being reached by the compound. The result is a series of spots of amino acids spaced up the sheet in the direction of the solvent front (Fig. 28-2). Each amino acid will move a specific fraction of the total distance traversed by the solvent front. (This fraction is designated as the R_F for the amino acid in that solvent system.) Not all of the amino acids are separated in one dimension, and more complete separation can be achieved by drying the sheet and then turning it so that the band of spots is down. It is then dipped into a different solvent system that will separate those that moved together in the first system. This is frequently a butanol-water-acetic

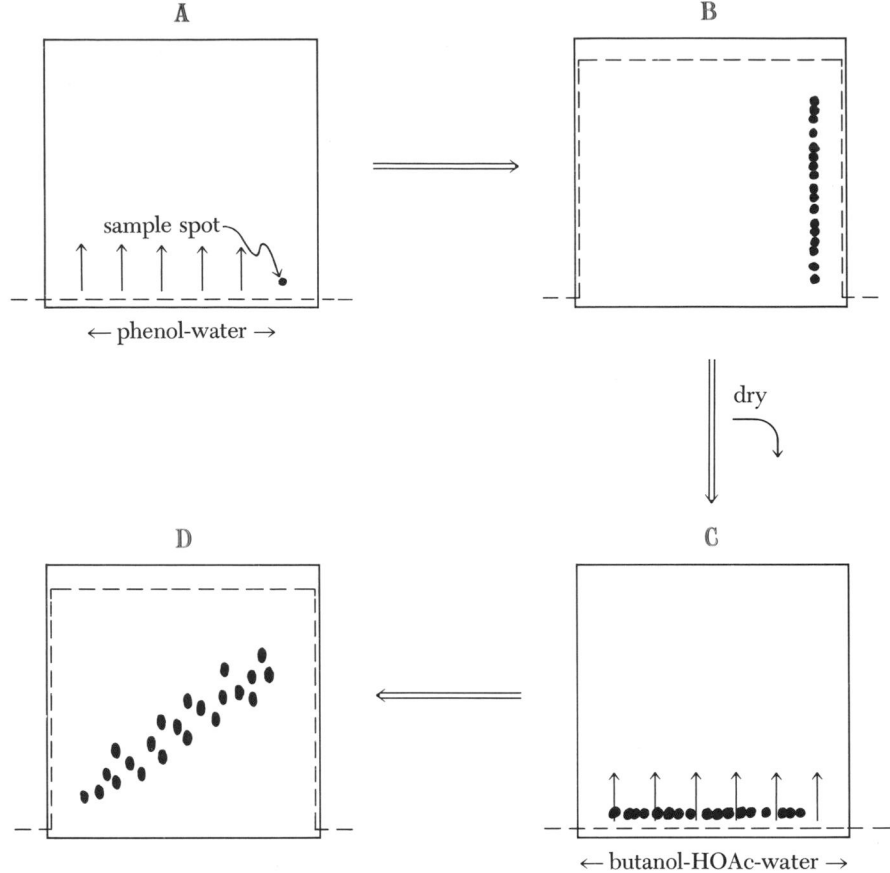

Figure 28-2 Amino acids can be separated by chromatography on paper or on thin layers of silica gel or other adsorbents.

A. A sample containing a mixture of amino acids is applied as a spot to one corner of the adsorbent and the sheet is lowered into a solvent, typically a phenol-water mixture. The solvent moves up the adsorbent by capillary action, as shown by arrows.

B. After the solvent has migrated up the paper, as shown by the dashed lines, the amino acids will be partially separated as a series of spots.

C. The paper (or other layered adsorbent) is then dried, turned 90°, and immersed in a different solvent that migrates across the band of spots.

D. Amino acids not separated in the first solvent move at different rates in the new solvent, so that the result is a roughly diagonal array of spots, each containing but one amino acid in an ideal case. The chromatogram is then dried, and usually is sprayed with ninhydrin and heated to make the location of the amino acids visible.

acid mixture. The second solvent carries the amino acids at right angles to their original direction of motion so that the final result is a sheet with a roughly diagonal scatter of spots, which is then dried, sprayed with a ninhydrin solution, and heated to make the presence of particular amino acids in the original sample known through the appearance of blue-colored spots. (The color varies somewhat, and proline gives a yellow spot.)

Quantitative Analysis of Amino Acids

Only a few of the amino acids present in a mixture such as might be obtained by hydrolysis of a protein can be determined directly. Tyrosine and tryptophan have characteristic absorption spectra in the ultraviolet with maximum absorption near 280 nanometers, so measuring the amount of light absorbed at this wavelength can provide an estimate of the amount of these amino acids, but even in this case the absorption of each interferes with the determination of the other.

Current practice for the determination of the amount of each amino acid in a mixture involves their quantitative separation by ion exchange chromatography. This method involves the use of insoluble resins substituted with ionizable groups; the resins are typically polymers of styrene made with variable amounts of cross-links between the linear polystyrene chains through the incorporation of divinylbenzene in the monomer mixture. Typical ionizable groups that are incorporated include substituted amine groups for anion exchange and carboxylic or sulfonic groups for cation exchange. The sulfonic cation exchange resins are most widely used for chromatography of amino acids.

The separation of compounds on the ion exchange resins hinges upon selective displacement by other ions. Those that are bound most weakly will be displaced more readily, and if the resin is placed in a cylindrical column that is washed from the top with a solution containing other cations, the weakly bound cations will move down the column and appear in the effluent at the bottom most rapidly.

Let us consider the case in which amino acids are fractionated on a sulfonic resin. There are three variables that can be manipulated to determine the affinity of the resin for an amino acid. One is the temperature at which the column is operated, which will change the equilibrium of association of a cation with the resin and will also change the dissociation constants of the amino acids, thereby altering the proportion of acidic, neutral, and basic forms at a given pH. Another is the pH of the solution. For example, only lysine, arginine, and histidine would have a net positive charge at pH 7 that could interact with the negative charge of the sulfonate groups on the resin. In order to bind all of the amino acids, the pH of the solution must be very low—below the isoelectric point of even aspartic and glutamic acids. The final variable is the concentration of other cations in the medium. For example, if a buffer at a given pH is made from sodium salts, and sodium citrate is commonly employed, the pH will determine what fraction of the amino acid is in cationic form, but the concentration of sodium ions in the buffer will determine the extent to which the amino acid cations will be displaced by competitive binding of the sodium ions to the sulfonate groups on the resin. The general principles are as follows:

1. The amino acids with two potential positive charges, such as lysine, arginine, and histidine, will tend to be bound more tightly than the others, and those with one potential positive charge and two potential negative charges, such as glutamic and aspartic acids, will be bound least tightly. If the pH of a buffer passing through an ion exchange column is raised, the amino acids with the lowest isoelectric points will be displaced most rapidly and will appear earlier in the stream of effluent from the bottom of the column.

2. Raising the concentration of an eluting cation, such as Na^+, in the solution passing through the column will cause displacement of all of the other cations on

the column, but this displacement will be greatest for those least tightly bound. The effect of the increased concentration is to sharpen the bands of eluted compounds, as well as to make them appear more rapidly; if enough of even a dilute solution of buffer is passed through a column all of the other cations will eventually appear, but the tightly bound ones tend to appear in very wide bands under these conditions.

3. Elevating the temperature of the eluting buffer causes greater displacement of the amino acids. This is partially a result of the effect on H^+ equilibria, but it also changes the relative affinity of the resin for various amino acids, owing to the fact that part of the binding is determined by interaction of the side chains of the amino acids with the hydrophobic matrix of the resin, and this binding is decreased by elevated temperature.

All of these variables have been used to develop procedures for the complete separation of amino acids on ion exchange columns. The effluent from the column is collected in fractions, which are analyzed for amino acid content by the ninhydrin method. Commercial amino acid analyzers are now available that include the necessary columns and means for recording the quantity of each amino acid as it appears in the effluent. This has greatly facilitated determinations of the primary structure of proteins, to be discussed shortly, as well as the amino acid composition of food proteins, and the quantitative analysis of the free amino acids in biological samples.

Quantitative Determination of Protein Content

The classic method for determining the amount of protein in a sample is through the measurement of the total nitrogen by the *Kjeldahl* method. This involves digesting the sample in boiling sulfuric acid, which converts most of the nitrogen to ammonium ion. The amount of ammonium ion is then determined by colorimetric analysis, or by making the sample alkaline and then distilling the ammonia into standard acid, followed by titration of the remaining acid. The Kjeldahl method has the advantage of being applicable to insoluble materials, such as most foodstuffs, and the disadvantage of not distinguishing proteins from other sources of nitrogen, such as nucleic acids, urea, and the like.

A quick estimate of both the protein and the nucleic acid content of soluble samples can be made by measuring the light absorption at 260 and 280 nanometers. The double measurement is necessary because both types of compound absorb light over that range; however, the nucleosides have maximum absorption near 260 nanometers, in contrast to the absorption of the tyrosyl and tryptophanyl residues of the proteins with a maximum at 280 nanometers. By solving the simultaneous equations for the relative absorption of each compound at the two wavelengths, factors can be obtained for determining the amount of each. This method is good only as a relative guide to the content of protein in mixed samples, because it depends upon the amino acid composition of the protein.

Another widely used method for determining the quantity of any protein that can be dissolved in strong NaOH is the *biuret method*. Polypeptides and cupric ions react in strong alkali to develop a violet color that is more dependent upon the number of peptide groups present than on the nature of the side chains, and measurement of the colored material by its light absorption is therefore a general method for measuring concentration of proteins of varying kinds.

Determination of the Primary Structure of Proteins

The number of residues of each of the amino acids in a protein can be estimated by hydrolyzing a sample and performing an ion exchange analysis. (Of course, this presumes that a purified sample is available and that the molecular weight is approximately known, but it wouldn't pay to begin a determination of the primary structure if these conditions are not met.) The hydrolysis is usually performed by heating a sample in hydrochloric acid with rigorous exclusion of oxygen, but even these conditions result in the destruction of tryptophan and partial destruction of serine and threonine. Tryptophan is separately estimated after alkaline hydrolysis of another sample, and the destruction of serine and threonine is corrected for by following the kinetics of loss.

Knowing the number of amino acid residues says nothing about the order in which they are arranged. Before attacking the determination of amino acid sequence, it is first necessary to discover how many different kinds of peptide chains are present in the protein molecule, because the sequence on each must be determined separately. This can sometimes be done by chromatographic or electrophoretic separation of the dissociated chains, but important clues can be discovered by *end-group analysis*, that is, by the determination of the kinds of amino acid residues that are at the ends of the peptide chains. A protein with a single kind of peptide chain will only have one kind of N-terminal residue and one kind of C-terminal residue. The discovery of the existence of more than one kind in either case establishes the existence of more than one kind of peptide. Quantitative measurements will then establish the proportion of the two kinds.

The nature of the N-terminal residues is easily established because there are a number of reagents that will react with free amino groups in aqueous solution. The one that is most widely used today is *dansyl chloride* (1-dimethylaminonaphthalene-5-sulfonyl chloride):

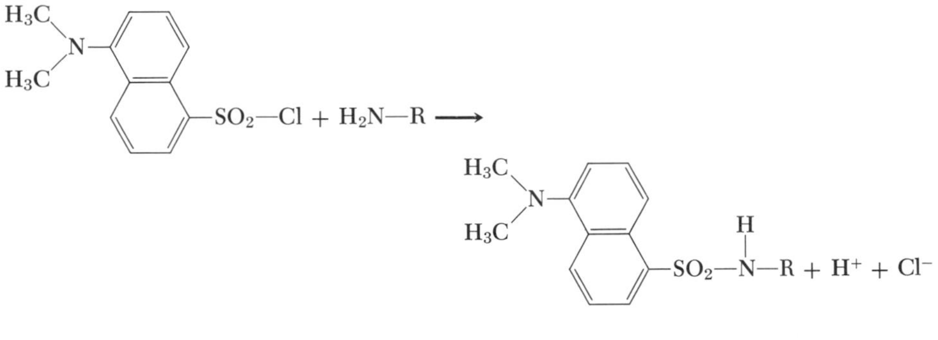

This reagent has especial value because the strong fluorescence of the dansyl group makes it possible to detect as little as 100 picomoles of derivative on chromatograms, and useful results can be obtained with 1 nanomole of the original protein. The procedure is to treat the protein with dansyl chloride in mildly alkaline solution, and the acid chloride reacts to add dansyl groups to the terminal amino group of the peptide chain, the side chain amino groups of lysyl residues, the phenolic hydroxyl group of tyrosyl residues, the mercapto group of cysteinyl residues, and the imidazole group of histidyl residues. The solution is then acidified and heated to hydrolyze the peptide bonds. The hot acid also causes the release of the dansyl groups from the cysteinyl and imidazole residues. The mixture is then subjected to paper chromatography and the position of the fluorescent spots is compared with those of known derivatives of the various amino acids. (If either a lysyl or a tyrosyl residue is at the N-terminal position, the isolated derivative will have two dansyl groups, one on the 2-amino position and the other on the side chain. These are easily separated from the derivatives of the same residues in the middle of the chain, which only acquire dansyl groups on the side chain substituents.)

The C-terminal residues can be identified by heating the protein with hydrazine, which reacts to form the hydrazides of carboxylic acid residues that are combined in amide linkage:

$$H-\left[\begin{array}{ccc} H & R & O \\ | & | & \| \\ N-CH-C \end{array}\right]_n \begin{array}{ccc} H & R' \\ | & | \\ N-CH-COO^{\ominus} \end{array} + nH_2N-NH_2 \longrightarrow$$

$$n\left[\begin{array}{ccc} R & O & H \\ | & \| & | \\ H_2N-CH-C-N-NH_2 \end{array}\right] + \begin{array}{c} R' \\ | \\ H_2N-CH-COO^{\ominus} \end{array}$$

The C-terminal residues do not react because they have free carboxylate groups, and they are easily separated and identified. If one finds that a protein has only one kind of N-terminal residue and one kind of C-terminal residue, the chances become much higher that there is only one kind of peptide chain in the molecule.

Many proteins do have more than one kind of peptide chain, and before anything further can be done, the individual kinds of chains must be isolated in pure form. Any covalent bonds between chains must first be broken. These are usually disulfide bridges, and one way of breaking them is to treat the protein with performic acid, which oxidizes all of the cysteinyl and cystinyl residues to residues of the corresponding sulfonic acid, cysteic acid. It also oxidizes methioninyl residues to the corresponding sulfone and plays hob with the tryptophanyl residues, breaking the indole ring to form a variety of products. A more gentle method is to reduce the disulfide bridges by adding another mercaptan, such as mercaptoethanol. The liberated sulfhydryl groups are frequently protected by reacting the protein with iodoacetate to create the carboxymethyl derivatives (p. 250).

The protein is then dissociated into its individual chains, frequently by adding guanidine or urea (p. 147), which are then separated by ion exchange chromatography or by molecular sieving. Molecular sieving frequently uses *dextrans*, which are 1→6 polymers of glucose with frequent branching that are formed by micro-organisms; dextrans hydrate to form a gel-like structure with domains that are resistant to penetration by large molecules. (Dextrans occur with various kinds of structures, so it is possible to obtain some that exclude molecules of rather small size, whereas others will only exclude very large molecules.) A column packed with beads of dextran, available under the trade name *Sephadex*, can be used to separate molecules of varying size, ranging near the size that can partially penetrate the beads. Large molecules will not enter the beads and will be washed through early. Very small molecules will not appear in the effluent until they have equilibrated with the water within the beads. Those of intermediate size will appear at varying positions, depending upon the amount of water in the beads that is available to them.

Once the peptides are purified, determination of the primary structure can begin. The terminal residues can be identified by end-group analysis, but there may be 200 or more residues between the ends. It is desirable to break this long chain into identifiable pieces, and hydrolysis by trypsin is a valuable tool for this purpose because it only catalyzes the hydrolysis of lysyl and arginyl peptide bonds. The result is a reproducible mixture of peptides, each of which terminates in a lysine or arginine residue except for the peptide obtained from the original C-terminal end:

$$\overset{\oplus}{H_3N}—(R)_a—Lys—(R)_b—Arg—(R)_c—Lys—(R)_d—COO^{\ominus}$$

$$\Big\downarrow \textit{trypsin}$$

$$\overset{\oplus}{H_3N}—(R)_a—Lys—COO^{\ominus} + \overset{\oplus}{H_3N}—(R)_b—Arg—COO^{\ominus} + \overset{\oplus}{H_3N}—(R)_c—Lys—COO^{\ominus}$$

$$+ \overset{\oplus}{H_3N}—(R)_d—COO^{\ominus}$$

It is often desirable at this point to determine how many peptide fragments one is dealing with by preparing a peptide map. This is done by spotting a sample on the corner of a sheet of filter paper and developing a one-dimensional chromat-ogram as we outlined in amino acid analysis. However, instead of proceeding to develop a chromatogram in the second dimension, the sample is then subjected to electrophoresis in the second dimension, which gives greater resolution of most mixtures of peptides. The number of spots on the resultant "fingerprint" is charac-teristic of the original peptide. (The same technique is used for identifying mutations in a peptide. The rough location of the change in a hemoglobin variant is often determined in this way.)

In any event, it is now necessary to separate the different tryptic peptides, which may be done by ion exchange chromatography and by molecular sieving. Each peptide must then be analyzed further, and an end-group analysis of each is often made at this point, as well as analysis for amino acid composition. Some of the tryptic peptides will frequently be small enough to proceed directly with sequence determination. The most powerful tool for this purpose is the *Edman degradation* using phenyl isothiocyanate, which combines with amino groups in alkaline solution to form a thiocarbamyl derivative. Derivatives of the terminal amino group cleave upon exposure to acid to form the phenylthiohydantoin of the N-terminal amino acid, liberating the remainder of the peptide chain, which now has the next amino acid exposed in the N-terminal position:

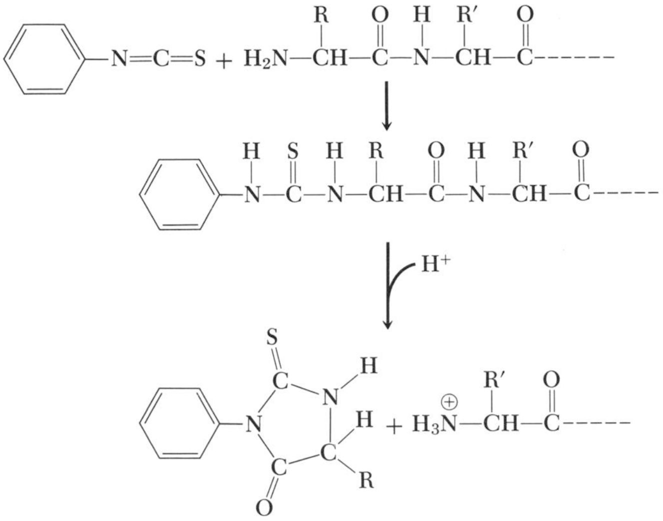

The liberated phenylthiohydantoin can be identified by comparison with known derivatives. So far this appears like ordinary end-group analysis, but the important difference is that the remainder of the peptide chain is still intact, and can be isolated from the reaction mixture. The shortened chain can then be exposed to fresh phenyl isothiocyanate in an alkaline solution, followed by exposure to acid to release the phenyl thiohydantoin of what was originally the second amino acid from the terminal amino group. The degradation can be repeated as many as five times before the yield is too low, so that one obtains a sequence for the same number of residues in the N-terminus of each of the tryptic peptides.

Some information can be obtained about the C-terminal sequence by treating each peptide with carboxypeptidase, which catalyzes the hydrolysis of the residue with a free carboxylate group. Of course, the enzyme begins an attack on the next residue as soon as one is liberated, but by following the time course of appearance of particular amino acids it is sometimes possible to deduce the order in which the residues occur.

The tryptic peptides are frequently too large to permit complete analysis by these methods, and must be further broken into smaller peptides that are purified for determination of structure. Other endopeptidases are sometimes used. For example, chymotrypsin, which is also obtained from the pancreas, preferentially cleaves tyrosyl and phenylalanyl peptide bonds under some conditions. Another technique is to treat the peptide with cyanogen bromide, which causes a rupture of methionyl peptides:

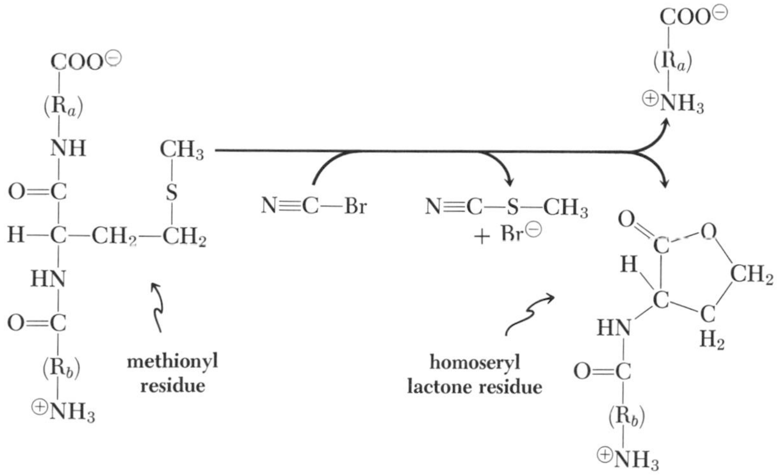

Even after the sequence of all of the tryptic peptides is known, the story is not complete, because the order of arrangement of these fragments of the original chain is not known except for those that contain the original terminal residues identified by end-group analysis. In order to define a final structure, it is necessary to attack the original peptide in a different way, so as to create fragments that overlap the positions at which trypsin cleaves the chain. Short cuts are frequently possible. Suppose, for example, that a peptide chain is known to have only two Met residues, and that one tryptic peptide is found to end in the sequence -Met-Ala-Lys and another tryptic peptide is found to begin with the sequence Gly-Trp-Met. The question is, were these two peptides originally linked -Met-Ala-Lys-Gly-Trp-Met- before hydrolysis of the lysyl peptide bond by trypsin, or were they attached to other peptide fragments? If a peptide fragment created by another procedure is found to have both methionyl residues, as well as Ala, Lys, Gly, and Trp, and perhaps two or three other residues, it becomes fairly certain that the two peptides are linked in the original chain. (The situation is rarely that simple, but the logical principles are the same.)

SUGARS

The sugars, by definition, are polyhydroxy aldehydes or ketones. The simplest aldose and ketose are glyceraldehyde and dihydroxyacetone. In a formal sense, all of the other sugars are considered to be derivatives of either D- or L-glyceraldehyde, made by adding the elements of formaldehyde to the aldehyde carbon. For example, adding one formaldehyde unit would create the tetroses:

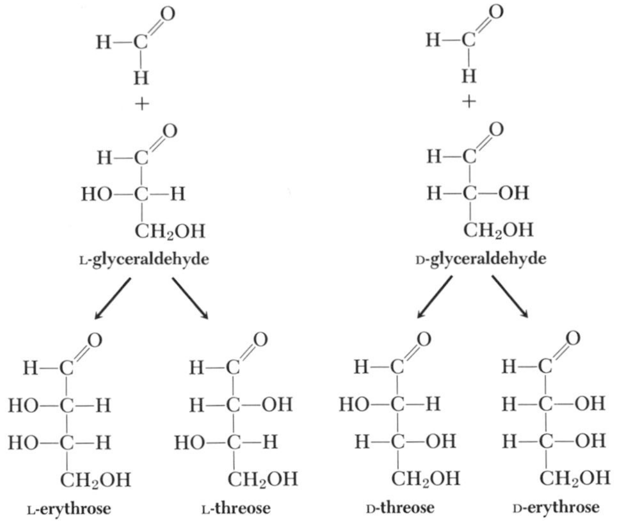

[Structural formulas: L-glyceraldehyde and D-glyceraldehyde, each branching to two aldotetroses: L-erythrose, L-threose (from L-glyceraldehyde) and D-threose, D-erythrose (from D-glyceraldehyde).]

Looked at in this way, the configurational family (D- or L-) is determined by the configuration of the asymmetric center most distant from the aldehyde group. We also see that the addition of the equivalent of a formaldehyde unit to the structure creates two new mirror-image pairs, each given a different name. Thus, there is only one pair of aldotrioses, two of aldotetroses, four of aldopentoses, and eight of aldohexoses.

The ketoses could have the carbonyl group on any of the central carbon atoms, but most of the natural ketoses have it on C-2. Some of the ketoses, such as fructose, have trivial names of their own, but many are named by combining the prefix of the name of the corresponding aldose with the suffix -*ulose*. Some of the common aldoses and ketoses are illustrated in the following:

pentoses

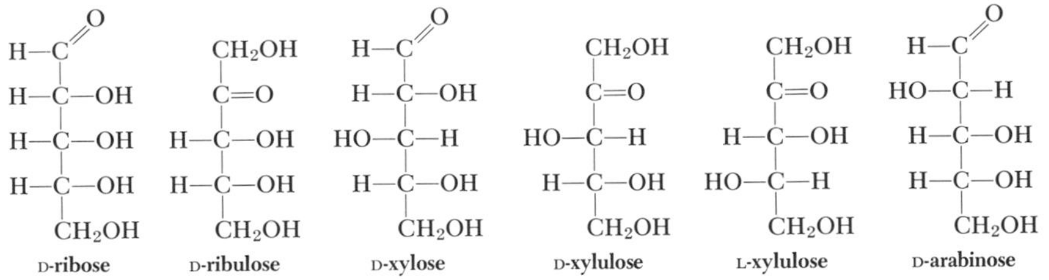

[Structural formulas: D-ribose, D-ribulose, D-xylose, D-xylulose, L-xylulose, D-arabinose]

hexoses

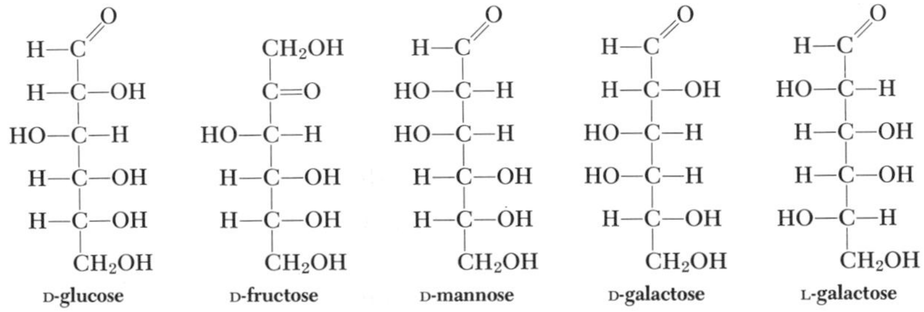

[Structural formulas: D-glucose, D-fructose, D-mannose, D-galactose, L-galactose]

We have already seen that sugars tend to form internal hemiacetals, creating the *furanose* or *pyranose* forms. This cyclization creates another asymmetric center, and the isomers formed in this way are known as the α or β *anomers* of the sugar. Which name applies depends upon the configurational family. If the configuration of the anomeric carbon is the same as that of the final asymmetric center (the center that determines D- or L- designation), it is the α-anomer; if not, it is the β-anomer. In other words, mirror images will have the *same* anomeric designation. The anomers are really different compounds in the same way that glucose and galactose are different compounds, and the only reason they aren't given different trivial names is that they are so freely interconvertible in solution.

The furanose and pyranose forms of the sugars are usually depicted as pentagons or hexagons lying on edge, as is done in this book. These Haworth formulations are not very accurate representations of the true conformation of the molecule because the ring atoms do not lie in a plane, and the substituent hydroxyl groups are not above or below the plane, but are *equatorial* or *axial* to the center of the molecule:

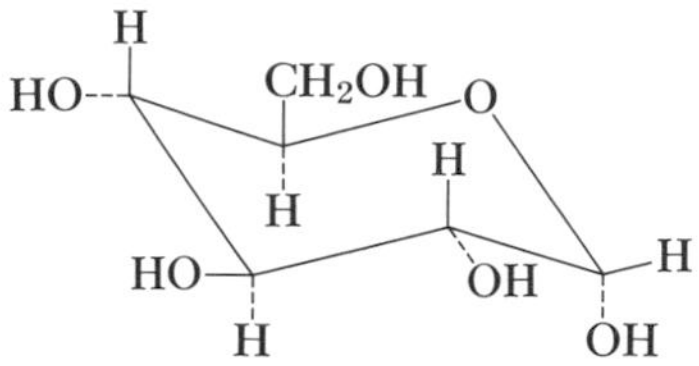

α-D-glucose

The use of these conformational representations will undoubtedly become common when the actual conformation of the residues in natural compounds is discovered.

Reducing Sugars

Hydroxy aldehydes and ketones are more easily oxidized than ordinary aldehydes and ketones, which is the same as saying that they are more powerful reducing agents. This property has long been used in making both qualitative and quantitative analyses. Even those sugars existing mainly in furanose or pyranose forms have sufficient of the open-chain modification present to have the property. Reagents for determination of reducing sugars contain cupric ion associated with a chelating agent, usually tartrate, or they contain ferricyanide as the oxidizing agent. The amount of the reduced agent, either cuprous ion or ferrocyanide ion, that is formed upon heating with the sample is determined by various means. One involves iodimetric titration. Another involves the use of phosphomolybdate or phosphotungstate, which deserves a little more discussion because of the more general application of the principle. The complex formed between phosphate and molybdate (or tungstate) ions is a moderately strong oxidizing agent that produces a highly colored complex, molybdenum blue, upon reduction. In the example at hand, the actual measurement is the absorption of light by molybdenum blue; the amount of molybdenum blue formed depends upon the amount of cuprous ion, and the amount of cuprous ion is a measure of the amount of reducing sugar. This kind of reaction is the basis for a wide variety of colorimetric methods, and is especially common in the determination of phosphate; this involves mixing the sample with molybdate and an excess of some reducing agent so that the amount of color formed is dependent upon the limiting amount of phosphate.

The determination of reducing sugar was formerly widely employed for the estimation of glucose in the blood and urine. However, the method does not distinguish between the various reducing sugars, nor between sugars and other compounds that will reduce cupric ions or ferricyanide, and it has been replaced by more convenient methods utilizing a glucose oxidase from microorganisms that catalyzes the specific oxidation of glucose to gluconate with the production of hydrogen peroxide. The peroxide is then determined by the action of a peroxidase on an organic amine to produce a colored compound.

The glucose oxidase method is inherently less precise, so the older methods are still valuable under some circumstances, and they are also of great value in structural analysis of oligo- and polysaccharides. Compounds in which another group is attached to the anomeric hydroxyl group of a sugar residue cannot form an open-chain isomer, and therefore are not susceptible to oxidation in this way. Glycogen, for example, has very little reducing power because there is only one unbound anomeric carbon in the entire molecule. All of the other glucosyl residues are linked through C-1. Lactose, on the other hand, has roughly half as much reducing power per glycosyl residue as does free glucose or galactose, because its glucose residue can form the open chain:

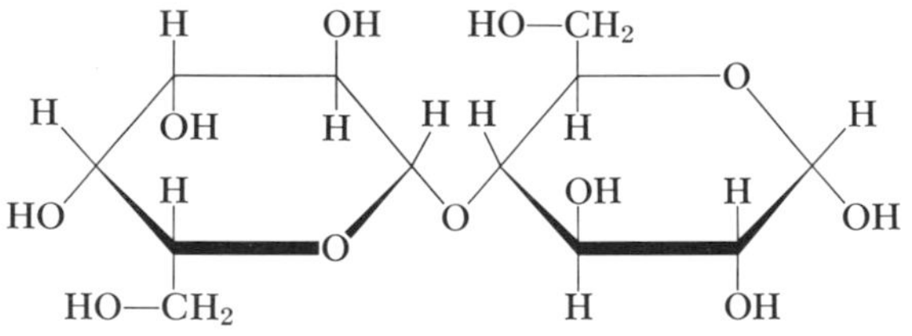

α-pyranose form of D-lactose

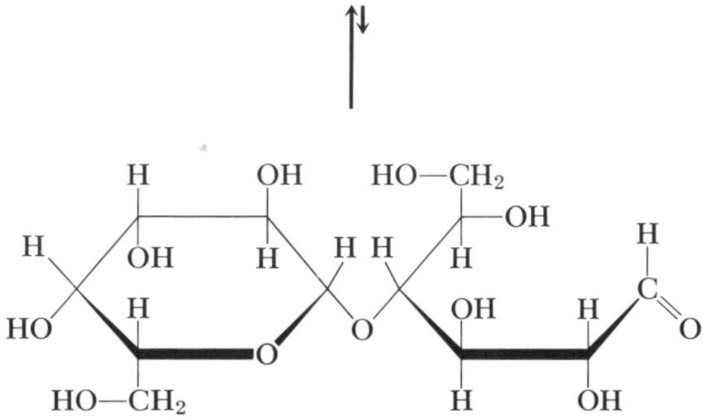

open chain form of D-lactose

This kind of information on an unknown oligosaccharide can give important clues to structure. The fact that sucrose is not a reducing sugar was an important clue for deducing that its unusual structure involves an anomeric carbon to anomeric carbon bond:

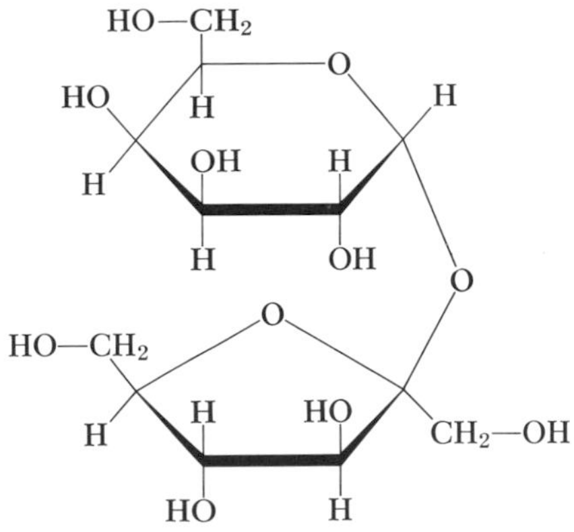

sucrose
(α-D-glucopyranosyl-(1 → 2)-β-D-fructofuranoside)

Analysis of Total Polysaccharide

There are no good methods for analyzing the total content of carbohydrate polymers in a biological sample. Heating a soluble sample in a strongly acidic solution of *anthrone* produces a color with many glycosyl compounds, as well as the free sugars; even the pentosyl residue of nucleoside derivatives will react, but quantitative application requires some knowledge of the kind of carbohydrate involved. Glycogen and starch are relatively stable in strong alkali, and digestion of tissues under these drastic conditions was formerly employed as a preliminary to the determination of these homopolysaccharides. Estimation of the heteropolysaccharides usually hinges upon colorimetric determination of the uronate or osamine content.

The structural analysis of heteropolysaccharides is a complicated business beyond the scope of our discussion. Those who tackle this kind of problem must be prepared to use partial acid hydrolysis and attempt to identify the various fragments that are produced, any of a variety of hydrolytic enzymes that will remove specific groups, specific tests for particular sugars, and a number of organic reactions that will degrade the sugars so as to enable identification of the linkages between the residues. One of the more general methods of degradation is discussed in the following section.

Periodate Oxidation

Periodic acid oxidizes compounds that have one oxygen atom on each of adjacent carbon atoms to cause a C—C cleavage. For example, glycerol is attacked by periodic acid in this way:

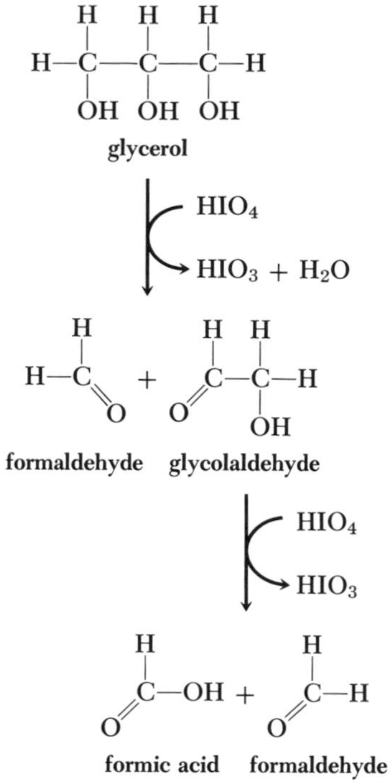

glycerol

HIO$_4$

HIO$_3$ + H$_2$O

formaldehyde   glycolaldehyde

HIO$_4$

HIO$_3$

formic acid   formaldehyde

Reactions of this kind are valuable in determining the structure of carbohy-drates. (Periodic acid will also attack compounds in which one or both of the requisite oxygen atoms is replaced by an amino group, as in the osamines, but not if the amino group is combined in amide linkage, as in the N-acetylosamines.)

Here are two examples of the kind of information that can be obtained:
(1) Furanosides and pyranosides can be distinguished by the number of moles of periodate consumed in their oxidation:

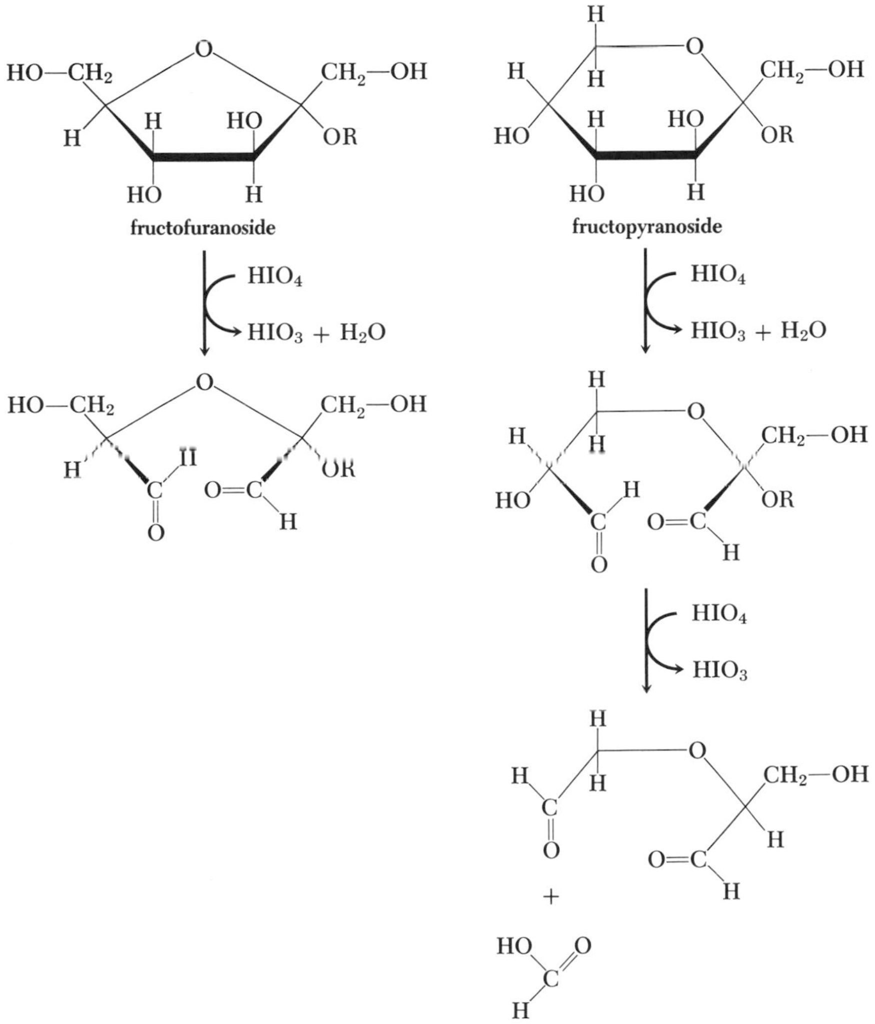

(2) The number of end-groups in some homopolysaccharides can be determined by noting the number of moles of formic acid produced. Consider the oxidation of one terminal chain in glycogen:

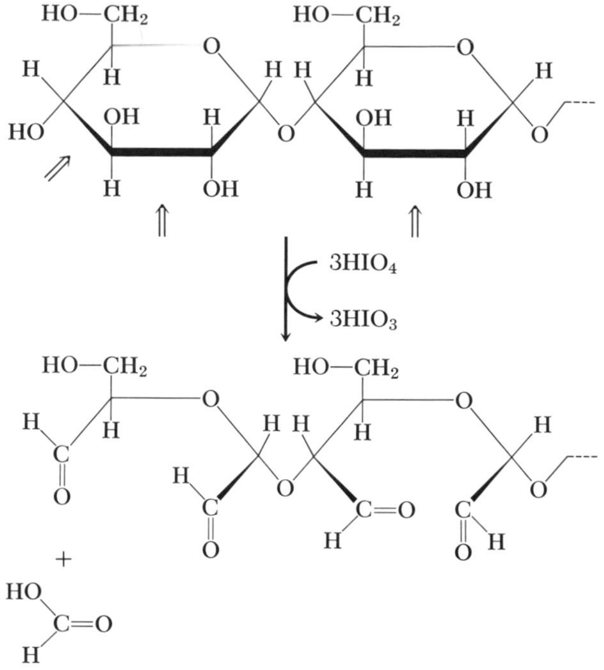

We see that the presence of bonds on the 1 and 4 carbons of the internal residue prevents further oxidation to release formic acid.

Lead tetraacetate also acts in the same way as does periodate, with the additional property of being soluble in organic solvents, which makes it more useful for some preparative purposes. It also attacks *cis*-glycols much more rapidly than the *trans* isomers, so it can sometimes be used to indicate the relative configuration of adjacent hydroxyl groups in sugars having a ring structure. However, lead tetraacetate is less useful than periodate in applications requiring quantitative analysis.

Oxidized and Reduced Derivatives of Sugars

The oxidation of C-1 of an aldose forms an aldonic acid; this is readily accomplished by the action of bromine in acid solution. Oxidation at C-6 forms an alduronic acid; although this can be done chemically in one step, the yields are very small, and other routes are employed for realistic synthesis. The aldonic and alduronic acids are named after the parent aldose:

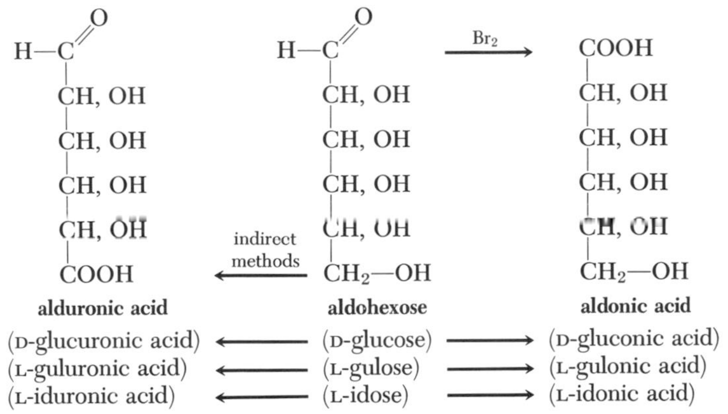

$$
\begin{array}{ccc}
\text{alduronic acid} & \text{aldohexose} & \text{aldonic acid} \\
\end{array}
$$

(D-glucuronic acid)	(D-glucose)	(D-gluconic acid)
(L-guluronic acid)	(L-gulose)	(L-gulonic acid)
(L-iduronic acid)	(L-idose)	(L-idonic acid)

The reduction of the carbonyl group of sugars by agents such as sodium borohydride forms an alcohol group:

This leads into a sticky problem of nomenclature. Since the product of monosaccharide reduction is a polyhydric alcohol without distinguishing groups at either end, there is no inherent reason for looking at it from a particular end, and D- and L- nomenclature becomes ambiguous. Consider the alcohol formed by reducing glucose:

$$
\begin{array}{cccc}
\text{H—C}{\overset{\displaystyle O}{\Vert}} & & \text{CH}_2\text{—OH} & \text{CH}_2\text{—OH} \\
\text{H—C—OH} & & \text{H—C—OH} & \text{HO—C—H} \\
\text{HO—C—H} & \xrightarrow{\text{NaBH}_4} & \text{HO—C—H} & \text{HO—C—H} \\
\text{H—C—OH} & & \text{H—C—OH} & \text{H—C—OH} \\
\text{H—C—OH} & & \text{H—C—OH} & \text{HO—C—H} \\
\text{CH}_2\text{—OH} & & \text{CH}_2\text{—OH} & \text{CH}_2\text{—OH}
\end{array}
$$

D-glucose   "D-glucitol"

We see that merely turning the molecule end-for-end changes it in a conventional sense from a derivative of D-glucose to a derivative of the sugar, L-gulose. Indeed, the reduction of either of these aldoses forms the same compound. What shall we call the hexitol? D-Glucitol or L-gulitol? It is sometimes called D-glucitol, but it was originally named sorbitol because of its occurrence in the mountain ash, *Sorbus aucuparia*, along with a corresponding ketose, L-sorbose:

$$
\begin{array}{cc}
\text{CH}_2\text{—OH} & \text{CH}_2\text{—OH} \\
\text{C}=\text{O} & \text{HO—C—H} \\
\text{OH—C—H} & \text{OH—C—H} \\
\text{H—C—HO} & \text{H—C—HO} \\
\text{HO—C—H} & \text{HO—C—H} \\
\text{CH}_2\text{—OH} & \text{CH}_2\text{—OH}
\end{array}
$$

L-sorbose   "D-sorbitol"

You would think from this it might be known as L-sorbitol, but no, it is frequently designated even today as D-sorbitol, so all rules are out the window. (The odd name undoubtedly persists because L-gulose and L-sorbose are not common natural sugars and we are therefore not frequently reminded that they, as well as D-glucose and D-fructose, are relatives of sorbitol in the sense that they can all be reduced to the same compound.)

The alcohol obtained by the reduction of D-galactose illustrates an interesting point:

$$
\begin{array}{cccc}
\text{H—C}\overset{*}{\overset{\displaystyle O}{\Vert}} & \text{*CH}_2\text{—OH} & \text{CH}_2\text{—OH} & \text{H—C}\overset{\displaystyle O}{\Vert} \\
\text{H—C—OH} & \text{H—C—OH} & \text{HO—C—H} & \text{HO—C—H} \\
\text{HO—C—H} & \text{HO—C—H} & \text{H—C—OH} & \text{H—C—OH} \\
\text{HO—C—H} & \text{HO—C—H} & \text{H—C—OH} & \text{H—C—OH} \\
\text{H—C—OH} & \text{H—C—OH} & \text{HO—C—H} & \text{HO—C—H} \\
\text{CH}_2\text{—OH} & \text{CH}_2\text{—OH} & \text{*CH}_2\text{—OH} & \text{*CH}_2\text{—OH}
\end{array}
$$

D-galactose   galactitol   L-galactose
                   (dulcitol)

This galactitol, also known as dulcitol, has no optical activity because it is symmetrical. We can see this by turning it end-for-end, as is done in the figure. The mirror image is the compound itself. It follows from this that galactitol is the alcohol corresponding to both D- and L-galactose.

The reduction of the aldehyde group of an alduronic acid also causes difficulties in nomenclature. The rules for naming these compounds require one to view the aldehyde carbon of the alduronic acids as C-1 and the carboxylic carbon of the aldonic acids as C-1. Therefore, when one reduces an alduronic acid to an aldonic acid, what was C-1 becomes C-6, and vice versa. This makes the simple reaction appear more formidable than it actually is because it usually necessitates a change in the type name and sometimes in configurational designation from reactant to product. We already encountered the reduction of D-glucuronate to L-gulonate, a shift of names that confuses the fact that the actual configuration has not been changed on a single asymmetric carbon atom:

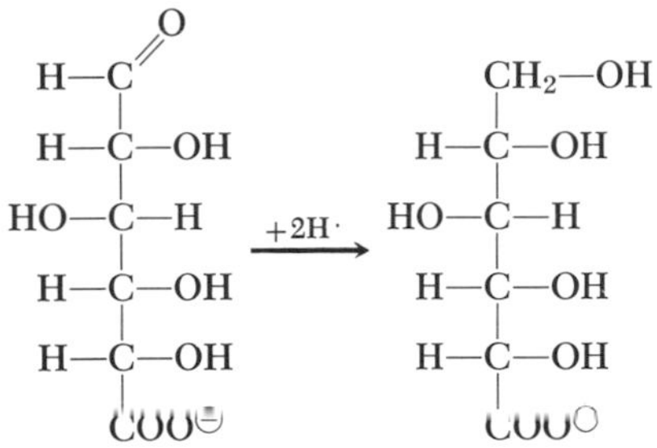

$$
\begin{array}{ccc}
& \text{H—C}{\overset{\displaystyle O}{\diagup}} & \text{CH}_2\text{—OH} \\
& \text{H—C—OH} & \text{H—C—OH} \\
& \text{HO—C—H} \quad +2\text{H}\cdot & \text{HO—C—H} \\
& \text{H—C—OH} \longrightarrow & \text{H—C—OH} \\
& \text{H—C—OH} & \text{H—C—OH} \\
& \text{COO}^{\ominus} & \text{COO}^{\ominus} \\
& \text{D-glucuronate} & \text{L-gulonate}
\end{array}
$$

LIPIDS

The salient property of the lipids that is exploited for their isolation and determination is their greater solubility in non-polar solvents. Hydrocarbon solvents, such as n-hexane, extract triglycerides from aqueous suspensions, and also extract free fatty acids if the pH is low enough to suppress their ionization. Simple titration of the extract will give an estimate of free fatty acid concentration in the latter case, and can also be used to determine the fatty acids liberated from triglycerides by hydrolysis.

The phospholipids require more polar solvents, such as a chloroform-methanol mixture, but even these leave the carbohydrates and proteins behind. The initial step in the determination of the distribution of lipids in a tissue therefore is an extraction with this sort of solvent, usually with great care to exclude oxygen so as to avoid destruction of the unsaturated compounds. Various techniques are available for further fractionation, and ion exchange chromatography is a useful first step that separates triglycerides and other relatively uncharged lipids from the phospholipids. This procedure employs a derivative of cellulose, diethylamino-ethylcellulose (DEAE-cellulose), as an anion exchanger because it can be used with non-polar solvents that attack the hydrocarbon polymer in the more common types of resin ion exchange materials. (DEAE-cellulose is also a very useful material for fractionating proteins in aqueous solution.) The various groups of lipids separated by ion-exchange chromatography can then be further fractionated on silicic acid columns. For example, the neutral fraction can be divided into triglycerides, sterols, and sterol esters in this way.

The fatty acid constituents of the various types of lipids can be liberated by hydrolysis and the nature of the fatty acids determined by a method that has revolutionized lipid chemistry: gas-liquid chromatography. This is a method that hinges upon the conversion of the components of a sample to a gas, and it can be applied to any compound that can be converted to stable volatile derivatives. There are even methods for converting amino acids and sugars to volatile derivatives, but gas-liquid chromatography is especially applicable to the fatty acids because of their inherent low reactivity and the ease with which they are converted to volatile derivatives. The mixture to be analyzed is converted to the methyl esters by reaction with diazomethane and is then introduced into a heated column that contains a layer of some non-volatile solvent, such as various silicones, hydrocarbon greases, or organic polymers.

The principle is to hold the non-volatile solvent in position, and pass a current of inert gas, such as helium, over it to sweep along the gaseous sample. Those compounds in the sample having the lowest vapor pressure at the operating temperature and the greatest solubility in the stationary solvent will tend to remain behind.

The stationary phase is created in the column by suspending the non-volatile solvent on materials like crushed fire-brick or diatomaceous earth, or by coating the walls of a long piece of capillary tubing. The gas emerging from the column is passed through an ionization detector that creates a signal when ionizable molecules appear in the gas stream. The application of gas chromatography has made available for the first time routine and reliable assays of the amounts of the individual fatty acids in natural lipids, and it is this technique that has made it possible to follow changes in the composition of human adipose tissue by analysis of the small samples obtained by plunging a needle through the skin of very much alive subjects.

Further reading

Cahn, R. S., C. K. Ingold, and V. Prelog: *The Specification of Asymmetric Configuration in Organic Chemistry.* Experientia, *12:* 81 (1956). The definition of (R) and (S) nomenclature.

Colowick, S., and N. O. Kaplan, eds.: *Methods in Enzymology.* Academic Press. The many volumes of this work, the first of which appeared in 1955 and the last of which is yet to come, contain excellent discussion of the applications as well as the procedural details on ion exchange chromatography, paper chromatography, thin-layer chromatography, gas-liquid chromatography, general and specific analyses, and the procedures used in determining structure of proteins and complex polysaccharides.

Davidson, E. A.: *Carbohydrate Chemistry.* Holt, Rinehart, Winston (1967). Discusses newer conformational determination and chemistry.

Pigman, W. W., and R. M. Goepp, Jr.: *Chemistry of the Carbohydrates.* Academic Press (1948). More complete discussion of the general reactions then available.

Rouser, G., et al.: *Lipid Composition of Beef Brain, Beef Liver, and the Sea Anemone: Two Approaches to Quantitative Fractionation of Complex Lipid Mixtures.* J. Am. Oil Chem. Soc., *40:*425 (1963).

THE
NOMENCLATURE
OF
ENZYMES

CHAPTER 29

Until this decade, the creation of names for enzymes was mainly the responsibility of the discoverer, subject to modification by later investigators and the taste of the various editors of the scientific journals. Decision between alternative names for the same enzyme was a matter for the marketplace; those names that weren't used disappeared. The only agreement was that all enzyme names ought to have the suffix *-ase*, but this might be preceded by the name of a substrate, the name of a product, or the type of reaction.

During the first years of this past decade, an effort was made by a commission of the International Union of Biochemistry to create a more systematic nomenclature. (By-products of this effort included agreement on the International Unit for enzyme activity and the designation of kinetic parameters, as well as the introduction of NAD and NADP as replacements for the older DPN and TPN.) The proposed nomenclature, like the systematic nomenclature for organic compounds, has had a mixed reception in terms of actual usage. Most of the older trivial names still persist, and some of the newer systematic names are so clumsy that it doesn't seem likely that they will ever be a part of the working language of the subject. The names used in this book are in the main the trivial names, because these are the designations most common in the current and older literature. Even so, some of the systematic names appear to be convenient for every-day use, and the readers of this book ought to be familiar with the principles of the nomenclature, if for no other reason than because they have the power through their own choice of usage in the coming years to decide which of the proposals have merit.

The IUB nomenclature uses six type designations as the final part of an enzyme name:

1. **Oxidoreductases.** These are enzymes catalyzing all of the reactions in which one compound is oxidized and another compound is reduced. This includes all of the enzymes now named as:

> *dehydrogenases*
> *reductases*
> *oxidases*
> *peroxidases*
> *hydroxylases*
> *oxygenases*

2. **Transferases.** These are enzymes catalyzing reactions not involving oxidation and reduction in which a group containing C, N, P, or S is transferred from one substrate to another. This includes all of the enzymes now named as:

> *transferases*
> *trans————ases* (such as transaminase, transketolase, transaldolase, trans-methylase)

3. **Hydrolases.** These are enzymes catalyzing hydrolytic cleavages or their reversal. (They do not include enzymes catalyzing the addition or removal of water from one compound, such as fumarase or enolase.) This includes all of the enzymes now named as:

> *esterases*
> *amidases*
> *peptidases*
> *phosphatases*
> *glycosidases*

4. **Lyases.** These are enzymes catalyzing the removal of a group from one substrate without a hydrolysis, usually to leave a double bond. These include all of the enzymes now named as:

> *decarboxylases*
> *aldolases*
> *synthases* (but not properly named synthetases)
> *cleavage enzymes* (such as citrate or 3-hydroxy-3-methylglutaryl CoA cleavage enzymes)
> *hydrases* or *hydratases* or *dehydratases*
> *deaminases*

5. **Isomerases.** These are enzymes catalyzing intramolecular rearrangements not involving a net change in the concentration of compounds other than the substrate. They include all enzymes now named as:

> *isomerases*
> *racemases*
> *epimerases*
> *mutases*

6. **Ligases.** These are enzymes catalyzing all of the reactions involving the formation of bonds between two substrate molecules that are coupled to the cleav-

age of a pyrophosphate bond in ATP or another energy donor. They include all of the enzymes now named as *synthetases*, except in cases where this name has been misapplied. The name synthetase was originally defined to include those enzymes now being defined as ligases, but it was later applied to enzymes catalyzing the formation of compounds by other mechanisms. Thus, one sometimes sees *glycogen synthetase* or *δ-aminolevulinate synthetase* even though the reaction catalyzed by these enzymes does not involve the cleavage of a high-energy phosphate bond. The term *synthase* was later coined to cover these enzymes, which are lyases in the IUB nomenclature.

This is the general outline of types. Within each type, further subdivisions are made and given group numbers. The groups are divided into numbered subgroups, and the individual enzymes within a subgroup are given another number. Thus, each enzyme is designated by four numbers. Some examples will make it clear, and will also show advantages and disadvantages of the systematic nomenclature.

1. Oxidoreductases
   1. Acting on the CH—OH group of donors
      1. With NAD or NADP as acceptor
         27. L-Lactate:NAD oxidoreductase

According to this, our familiar lactate dehydrogenase is now EC 1.1.1.27. (The EC stands for Enzyme Commission.)

Here are some more examples with the current trivial name shown in italics:

1. Oxidoreductases
   4. Acting on the CH—NH₂ group of donors
      1. With NAD or NADP as acceptor
         2. L-Glutamate:NAD(P) oxidoreductase (deaminating)
            EC 1.4.1.2 (*glutamate dehydrogenase*)

2. Transferases
   4. Glycosyltransferases
      1. Hexosyltransferases
         18. α-1,4-Glucan:α-1,4-glucan 6-glycosyltransferase
            EC 2.4.1.18 (*branching enzyme*)

   7. Transferring phosphorus-containing groups
      1. Phosphotransferases with an alcohol group as acceptor
         2. ATP:D-glucose 6-phosphotransferase
            EC 2.7.1.2 (*glucokinase*)

3. Hydrolases
   1. Acting on ester bonds
      3. Phosphoric monoester hydrolases
         11. D-Fructose-1,6-diphosphate 1-phosphohydrolase
            EC 3.1.3.11 (*fructosediphosphatase*)

   4. Acting on peptide bonds
      4. Peptide peptidohydrolases
         4. No specific name given
            EC 3.4.4.4 (*trypsin*)

4. Lyases
   1. Carbon-carbon lyases
      1. Carboxy-lyases
         22. L-Histidine carboxy-lyase
             EC 4.1.1.22 (*histidine decarboxylase*)

   2. Carbon-oxygen lyases
      1. Hydro-lyases
         13. L-Serine hydro-lyase (deaminating)
             EC 4.2.1.13 (*Serine dehydratase* or *serine deaminase*)

5. Isomerases
   1. Racemases and epimerases
      3. Acting on carbohydrates and derivatives
         1. D-Ribulose-5-phosphate 3-epimerase
            EC 5.1.3.1 (*ribulose phosphate 3-epimerase*)

   3. Intramolecular oxidoreductases
      1. Interconverting aldoses and ketoses
         9. D-Glucose-6-phosphate ketol-isomerase
            EC 5.3.1.9 (*phosphoglucoisomerase*)

6. Ligases
   1. Forming C—O bonds
      1. Amino acid-RNA ligases
         1. L-Tyrosine:tRNA ligase (AMP)
            EC 6.1.1.1 (*tyrosylRNA synthetase*)

   2. Forming C—S bonds
      1. Acid-thiol ligases
         2. Acid:CoA ligase (AMP)
            EC 6.2.1.2 (*Acyl CoA synthetase*)

Further reading

Webb, E. C.: *Nomenclature of Enzymes and Coenzymes.* In Florkin, M., and E. H. Stotz, eds.: *Comprehensive Biochemistry*, Vol. 13. Elsevier (1964).

INDEX

Page numbers in **boldface** indicate figures illustrating structures of listed compounds, reactions catalyzed by listed enzymes, or other listed phenomena. Isomeric designations are neglected in the primary alphabetization. Thus, *cis*-Aconitate, S-Adenosyl-L-methionine, L-Alanine, and *p*-Aminobenzoate are listed as beginning with A.

Albumin(s) (*Continued*)
  serum, edema and, 673
    fatty acid complexes, 341
    formation, 461
Alcaptonuria, 389
Alcohol dehydrogenase(s), **272**
  equilibrium, 251
  mechanism, 185, **187**
Alcoholism, hemosiderosis in, 506
  hepatic cirrhosis in, 362
  thiamine deficiency in, 258
Aldehydes, condensation, **211**
  oxidation, 443, **444, 446, 447, 450**
    free energy change, 214
    to acyl phosphates, **261**
Aldimines. Also see *Schiff's base*
  in transamination, **117,** 118, **119**
  in visual pigment, **643**
Aldol(s), 248
  formation, 211, **223**
Aldol condensation, collagen cross-links from,
  88
Aldolase(s), reaction of, **261.** Also see *Phos-
  phofructoaldolase*
Aldonic acids, **717,** 719
Aldonolactonase, **678**
Aldopyranoses, C-1 oxidation, **347**
Aldose(s), **710**
  group transfer to, **347**
  ketoses from, 261
Aldose reductase, **631**
Aldosterone, **626, 627**
Aldosteronism, 627
Alduronic acids, **717,** 719
Alkaline phosphatase, in bone, 603
Alkalosis, blood analyses in, **621,** 624, 625
  in aldosteronism, 627
  in diarrhea, 627
  metabolic, 622-625
  renal compensation, 619, 620, 622
  respiratory, 620-622, 625
    lactate formation in, 622
  respiratory compensation, 623, 624
  tetany in, 609
Alkyl carboxylates. See *Fatty acid(s)*
Alkylating agents, in cancer therapy, 481
Allenes, 235
Allergy, histamine in, 450
Allopurinol, **486**
Allosteric effects, 138-141. Also see *Enzyme
  modifiers, Metabolic regulation*
  aspartate transcarbamylase, bacterial, 139-
    141, 489
  defined, 139
  phosphofructokinase, **246,** 247, 279
Allothreonine, **699**
Alloxan diabetes, 560
Alpha receptors, 554, 555
Amidase, definition, 112
Amidinio group, **423**
Amine(s), from amino acids, 427, 438
  nerve function and, 439-448

Amine oxidase (mitochondrial). See *Monoamine oxidase*
Amino acid(s), **694.** Also see *Amino acid
  pool, Essential amino acids,* and specific
  compounds
  absorption, 659, 670
  absorption spectra, 702-703
  acetylation, 458, 490
  antagonists of, 404, 405
  as fuels, 353. Also see *Protein(s), as fuels*
  chromatography of, 701-704, **702**
  coding for, 79-81, 80 (table)
  combination with tRNA, 35-37, **37**
  dansyl derivatives, **705,** 706
  deamination, 353-357
  dietary imbalance, 669, 670
  fatty acids from, 354
  function in hemoglobin, 66, 70, 71, 74-78
  glucogenic, 364
  glucose from, 354
  hydrazides, 706
  in beef proteins, 396
  ionization, 693, **695, 696,** 697, 703
  isoelectric point, 696, 697
  D-isomers, in antibiotics, 403
    metabolic fate, **403,** 404
  ketogenic, 372
  mobilization by cortisol, 565
  ninhydrin reaction, **700,** 701
  non-essential, formation, 662
  oxidative metabolism, 351-407, 418-421,
    432, 433
    ketosis and, 530
    location, 364
    regulation, 356
    stoichiometry, 395-403, **400,** 528
  3-oxybutyrates from, 354
  peptide bond between, 8
  peptide chains from, 18-21, **19**
  stereoisomerism, **697-699, 700, 701**
  transport, insulin effects, 558
Amino acid decarboxylase(s), 427, 428, **438.**
    Also see specific enzymes
  pyridoxal phosphate in, 427
D-Amino acid oxidase, **403,** 404
  mechanism, 404
L-Amino acid oxidase, 420
  as experimental tool, 404
Amino acid pool, 352, 353, 418
Aminoaciduria, in fructose intolerance, 634
  in galactose intolerance, 639
Aminoacyl tRNA, 19, 21, 34-37
  formation of, 36, **37**
  free energy of synthesis, 167
Aminoacyl-tRNA synthetase, **37,** 48
L-2-Aminoadipaldehyde, 392, **393**
L-2-Aminoadipate, 392, **393,** 394
p-Aminobenzoate, in folate, **411,** 412
4-Aminobutyrate, concentration in brain, 447
  from glutamate, **446**
  metabolic fate, **447**

2-Amino-3-carboxymuconaldehyde, in nicotinamide synthesis, 452, **453**
  in tryptophan metabolism, **390, 391**
Aminocarboxymuconaldehyde   decarboxylase, **391**
5-Aminoimidazole-4-carboxamide,   excretion
    in folate deficiency, 469
β-Aminoisobutyrate, **488**
5-Aminolevulinate, **496**
5-Aminolevulinate synthase, **496**
  in porphyrias, 500
  regulation, **496, 499**
2-Aminomuconaldehyde, **391**
Aminomuconaldehyde dehydrogenase, **391**
2-Aminomuconate, **391**
Aminomuconate reductase, **391**
Aminopeptidases, 658
β-Aminopropionitrile, 91, **92**
Aminopterin, **480**
Ammonia. Also see *Ammonium ions*
  carbamyl phosphate from, **360, 361**
  diffusion, 359
  disposal, evolution, 359
  formation in intestines, 363
  from alanine, **355**
  from AMP, 483, 484
  from cytosine, **488**
  from glutamate, **354, 355**
  from glutamine, 362
  from glycine, 418, 419
  from guanine, 483, 484
  from histidine, **420, 421**
  from homocysteine, **432, 433**
  from serine, **357**
  from threonine, **357**
  from urate, 484
  from uroporphyrinogen synthesis, **496, 497**
  glutamine from, 361, **362**, 363
  sites of formation, 361, 362
  toxicity, 357, 359, 361-363
  transport as glutamine, 361, 362
  urea from, **354, 359, 360, 361**
Ammonium ions. Also see *Ammonia*
  amino acids from, 662
  blood concentration, 359
  dietary, acidosis from, 623
    glutamine and, 362
  excretion in acidosis, 620, 623
  urea from, H⁺ stoichiometry, 618
AMP. See *Adenosine monophosphate*
AMP kinase, 247
  in starvation, 549, 550
AMP pyrophosphorylase, 482
  in brain, 487
  in liver, 487
Amphetamine, 448
Amylo-(1→6)-glucosidase, 296-299
  deficiency disease, 307
  specificity, 296
Amylose chains, 288, **289**, 294
Anaerobes, facultative, 271

Anaerobes (*Continued*)
  fermentations in, 271, 272
  obligatory, 271
Anaerobic process defined, 268
Anemia. Also see *Pernicious anemia, Sickle-cell anemia*
  in salmon, xanthopterin in, 385
  in scurvy, 684
  iron-deficiency, 686
    in scurvy, 684
    protoporphyrin in, 499
  macrocytic, 682, 684
  megaloblastic, 682, 684
  microcytic, 686
  murine caprilactogenic, xanthopterin in, 385
  normoblastic, 684
Aneurism, in lathyrism, 91
Angiotensins, 626, 627
Anomers, 240, 711
Anserine, **457**
  formation, 457
  in skeletal muscles, 457
Anterior pituitary, 564
  corticotropin production, 343, 565
  prolactin secretion, 636
  thyrotropin production, 564, 565
Anthrone, 713
Antibiotic(s), in cancer therapy, 46, 47
  inhibitors, of nucleic acid synthesis, 46, 47
    of protein synthesis, 46, 47
    vitamin K and, 652
Anticodon(s), definition, 32
  in tRNA, 39-41
  wobble hypothesis and, 41
Antidiuretic hormone, 627. See *Vasopressin*
Anti-histaminics, 449-451
Antimycins, inhibition of electron transfer, 203, 204
Anti-oxidants, **680, 681**
Anxiety state, respiratory alkalosis in, 622
Aphonia, in infantile beri-beri, 258
Apoenzyme, definition, 116
Apoferritin, 504
Apoprotein, defined, 10
D-Arabinose, **710**
Arachidate, 226
Arachidonate, **233**
Arachidyl coenzyme A, from stearoyl CoA, 334
Argentaffine cells, 5-hydroxytryptamine formation by, 457
Arginase, **359**, 360, 361
L-Arginine, **11**, 59. Also see *Arginyl residues*
  creatine from, 422, **423**
  formation, 662
  from argininosuccinate, **360, 361**
  glutamate from, **395**
  metabolic fate, **395**
  ornithine from, **359**, 395, 451
  urea from, **359**, 395

Caproyl coenzyme A, 230
Carbamino groups, 610
N-Carbamyl-β-alanine, **488**
N-Carbamyl-β-aminoisobutyrate, **488**
N-Carbamyl-L-aspartate, in pyrimidine synthesis, 474, **475**
Carbamyl phosphate, citrulline from, **360**, 361
  from ammonia, **360**, 361
  from glutamine, 474, **475**
  in pyrimidine synthesis, 474, **475**
Carbamyl phosphate synthetase, and birth, 552
  cytosol, 361, 474, **475**
    inhibition by UTP, 474, **475**
  mitochondrial, **360**, 361
    deficiency disease, 363
Carbohydrate(s). Also see *Sugars*
  as fuels, 530
  fat equivalent, 531
  respiratory quotient, 515
Carbon dioxide. Also see *Carbon dioxide tension*
  blood transport, 610-614
  carbonic acid equilibrium, 610, 619
  H^+ concentration and, 610, 611, 619
  one-carbon relationship, 414
  production, and calorimetry, 514, 515, 516(table)
  purine ring from, **463, 467**
  pyrimidine ring from, 474, **475**
Carbon dioxide tension, blood pH and, **621**
  dissolved CO_2 and, 611
  H^+ concentration and, 611, 615-617, **616**, 619
  respiratory rate and, 615, 617
Carbonic acid, carbon dioxide equilibrium, 610, 619
  dissociation constant, 610, 611
  one-carbon relationship, 414
Carbonic anhydrase, in erythrocytes, 612, 613
  in kidneys, 619
Carbonyl group, one-carbon relationship, 414
Carboxyl group, one carbon relationship, 414
Carboxylation(s), biotin-requiring, **282**
  of enol-phosphate, **282**
Carboxypeptidases, 658
  use in peptide analysis, 708
Carcinoid, nicotinate deficiency in, 456, 457
  tryptophan metabolism in, 456, 457
Cardiolipins, **598**
Carnitine, **228**
  function, 228
Carnosine, **457**
  formation, 457
  in skeletal muscles, 457
Carotene oxygenase, **642**
Carotenemia, 641
Carotenes, absorption, 641
  formation by plants, 641
  retinol from, 641, **642**
Carotenoids, 593. Also see *Carotenes, Retinol*

Carrier, in regulated diffusion, 242
Cartilage, collagen in, 84
Catalase, 404
Catalysis, acid-base, 109-112
Catechol transmethylase, **444**
Catecholamines. Also see *Adrenaline, Noradrenaline*
  3-methoxy-4-hydroxymandelate from, 443, **444**
  nomenclature, 448, 449
CDP. See *Cytidine diphosphate*
CDP-choline, in phosphatidyl choline synthesis, **425**
  sphingomyelin from, **601, 602**
CDP-choline pyrophosphorylase, **425**
CDP-diglycerides, phosphatidyl glycerols from, **598**
  phosphatidylinositides from, **599**
CDP-ethanolamine, in phosphatidyl ethanolamine synthesis, **425**
CDP-ethanolamine pyrophosphorylase, **425**
Cellulose, metabolism in ruminants, 369
Central nervous system. See *Brain.*
Centrifugation, of DNA, 30
  of proteins, 15, 16
Cephalin(s) 426. Also see *Phosphatidylethanolamine(s)*
Ceramides, **601, 602**
Cereals, as fuels, 530
  as nitrogen source, 668, 669, 672
Cerebrosides, formation, **601, 602**
Ceruloplasmin, 505
Cesium chloride, use in centrifugation, 30
Chelation. See *Metals*
Chemical score, of proteins, 663, 664(table), 668
Chloramphenicol, action of, 47
Chloride ions, blood concentration, 624
Chloride shift, 613
p-Chloromercuribenzoate, **148**
Cholecalciferol. Also see *Vitamin D*
  formation, 604
  25-hydroxycholecalciferol from, 604
Cholera, management, 243
$\Delta^{7,24}$-Cholestadienol, **597**
$\Delta^{5,7}$-Cholestadienol, cholecalciferol from, **604**
  cholesterol from, **597**
  formation, **597**
Δ^7-Cholestanol, **597**
Cholesterol, atherosclerosis relationship, 594, 595
  cortisol from, 567-570, **569**
  esters of, **592**
  formation, 592, **593-597**
    regulation, 593
  occurrence, 591, 592
  pregnenolone from, 568, **569**
Choline, **100**
  acetylcholine from, 424, 440
  betaine from, 429, **430**
  dietary, 424, 425
  formation, 424, **426-428, 429**

Dihydroxyacetone phosphate, as electron carrier, 251, 252, **253**
   from D-glyceraldehyde-3-phosphate, **248**, 249
   from fructose, 633, 634
   from fructose disphosphate, 248, 249
   glycerol-3-phosphate from 251, **252**, 253, 338, **339**
   in serine synthesis, **417**
2,5-Dihydroxyphenylacetate. See *Homogentisate*
3,4-Dihydroxyphenylacetate, in tyrosine metabolism, 385, **386**
3,4-Dihydroxy-L-phenylalanine, 3-4 dihydroxyphenylethylamine from, 440, **441**
   from tyrosine, 440, 441
Dihydroxyphenylalanine decarboxylase, **441**, **445**
   pyridoxal phosphate in, 440
3,4-Dihydroxyphenylethylamine, in brain, 440
   metabolism, 440, **441**, 442
   noradrenaline from, 441, 442, 443
Dihydroxyphenylethylamine β-hydroxylase, **441**, 442, 443
   intracellular location, 443
   mechanism, 442
Diiodotyrosine, **563**, 564
Diisopropyl phosphofluoridate, **114**, 115
2,3-Diketo-L-gulonate, 678, **679**
2,3-Dimercaptopropanol, therapy with, **259**
1-Dimethylaminonaphthalene-5-sulfonyl chloride, **705**
Dimethylammonium ion, **104**
Dimethylbenzimidazole group, in cobalamins, **368**
3,3-Dimethylbutyl acetate, 104
N,N-Dimethylglycine, in choline metabolism, 430, **431**
Dimethylglycine dehydrogenase, **431**
6,7-Dimethylisoalloxazine, in riboflavin, **188**
2,4-Dinitrophenol, uncoupling of oxidative phosphorylation, 204
Diphenhydramine, 449, 451
1,3-Diphospho-D-glycerate, **249**
   2,3-diphosphoglycerate from, **254**, 255
   from glyceraldehyde-3-phosphate, **249**, 250, 252
   high-energy phosphate from, **249**, 252, 253
   hydrolysis, free energy change, 249
   3-phosphoglycerate from, **249**, 250, **252**
2,3-Diphospho-D-glycerate, from 1,3-diphosphoglycerate, **254**, 255
   phosphoglycerate 2,3-mutase and, **254**
Diphosphoglycerate mutase, reaction of, **254**
Diphosphopyridine nucleotide, 188. Also see *NAD*
Dismutations, defined, 270
Disulfide bonds. Also see *Cystinyl residues*
   chemical cleavage, 706
   exchange in proteins, **95**, 96
   formation, 94, 95, 96
   in insulin, **557**

D:N ratio, 399
DNA. See *Deoxyribonucleic acid(s)*
DNA ligase, (DNA sealase), 27-30, **28**
DNA polymerase, **26**, 27, 29, 30
DNA sealase. See *DNA ligase*
Dodecatrienoyl coenzyme A, from linolenoyl CoA, 234
   isomerization, **234**
Dogs, running fits in, 405
Dopa, **441**. Also see *3,4-Dihydroxy-L-phenylalanine*
Dopamine, **441**. Also see *3,4-Dihydroxyphenylethylamine*
DPN, 188. See *NAD*
DPNH. See *NADH*
dTDP-D-glucose, **586**
dTDP-glucose pyrophosphorylase, **586**
dTDP-glucose oxidoreductase, **586**
dTDP-4-keto-6-deoxy-D-glucose, **586**
dTDP-L-rhamnose, 585
   formation, **586**
Dulcitol, 632, 639, **718**

Edema, causes, 673
   from histamine action, 449
   in kwashiorkor, 672, 673
   in thiamine deficiency, 258
Edman degradation, **708**
EDTA, 152
Eggs, avidin in, 676
   proteins in, as reference, 664, 668
      essential amino acid content, **667**
Eicosanoate, 226
Elastin, desmosine in, 87
   in ground substance, 576
Electron transfer(s), citric acid cycle, **210**, 211
   endoplasmic reticulum, 334, 335
   from cytosol to mitochondria, 251, 252, **253**
   from NADH to NADPH, 321, 322, 329
   in stoichiometric regulation, 538-541
   ion pump and, 197-199
   mitochondrial, **184-194**, 195
      site, 182
Electron transfer flavoprotein, **230**, 430
Electrophoresis, 707
   of proteins, 73, 74
Embden-Meyerhof pathway, 240, 241, 243-255
   in gluconeogenesis, 274
   in serine synthesis, **417**, 418
   relation to fatty acid synthesis, 321, 322, 329, 331, **332**
   stoichiometric regulation, 538-541
Embryomas, actinomycin with, 47
Emphysema, 617
End-group analysis, 705, 706
Endopeptidases, 658
   definition, 113

Endoplasmic reticulum, acyl desaturases in, 334, 335
  electron transfer in, 334, 335
  glucose-6-phosphatase in, 280
  lipids in, 591(table)
  ribosomes on, **Endpaper**, 20
  sulfite oxidase in, 382
Energy. Also see *Free energy*
  caloric equivalents, 513-523
  in foods, 528-531
  total turnover, 511-533
Energy-rich bonds, in thiol esters, 208, 209
  methyl groups as, 422, 430
  pyrophosphates as, 165-168
  sulfate anhydrides as, 582
Energy supply, 523-531
  for exercise, 523, 526, 539-541
  recommended, 688(table)
Enol phosphates, as high-energy phosphate, 255
  from glycol phosphates, **262**
  high-energy phosphate from, **262**
Enolase, **253**, 255
Enoyl CoA oxidase, **336**
2-*trans*-Enoyl coenzyme A, from acyl coenzyme A, 228, **229**, 230
  hydration, **229**, 230
3-*cis*-Enoyl coenzyme A, isomerization, **234**, 237
Enoyl CoA hydratase, 229, 230, 235, 236
  reaction of, **229, 236**
  stereochemistry, 229, 235
Δ^3 *cis*$\rightarrow$ ^{2} *trans* Enoyl CoA isomerase, **234**, 235, 236
  equilibrium of, 235
Enterokinase, 658
Enthalpy, free energy and, 161
Entropy, free energy and, 161
Enzyme(s), 98-121
  assay of, 127, 128, 130-132
  catalytic subunits, 140
  concentration and equilibria, 136
  mechanism, 107-119
    ping-pong, 129
    sequential, 129
  nomenclature, 100, 101, 721-724
  pH optimum, 110-112
  regulatory subunits, 140
  units, 124
Enzyme adaptations, 547-552
  at birth, 551, 552
  glucagon effect, 556
  insulin effects, 561(table)
  to dietary fat–carbohydrate load, 550(table)
  to dietary protein load, 550, 551
  to starvation, 548-550, 549(table)
Enzyme kinetics, 123-141
  effect of substrate concentration, 126-132
  Haldane relationship, 130, 132
  Lineweaver-Burk plot, 128, 129, 133, 134
  initial velocity, 125

Enzyme kinetics (*Continued*)
  maximum velocity, 125
    determination, 128, 129
  Michaelis constant, 126-131, 137, 139, 140
    determination, 128, 129
  Michaelis-Menten equation, 126, 127
  product concentration and, 130-132
  rate-limiting step, 135
  reversible reactions, 136
  steady state, 125
Enzyme modifiers, 543-547, 544(table)
  acetyl CoA, 276, 277, 545
  acyl CoA, 343, 344
  ADP, 356, 545
  alanine, 545
  AMP, **246**, 247, 279, 300, 471, 545
  ATP, **246**, 247, 279, 465, 544, 545
  citrate, 247, 343
  3′,5′-cyclic AMP, 302-305
  deoxynucleotides, 473
  glucose-6-phosphate, 244, 305, 306, 556
  GTP, 356, 465, 544
  leucine, 356
  NADH, 356
  palmitoyl CoA, 343
  protohemin IX, 496, 499
  UDP-N-acetylglucosamine, 578
  UDP-xylose, 579
  UTP, 474, 475
Enzyme modulation, 139, 543-547, 543 (table). Also see *Allosteric effect(s)*
Enzyme specificity, definition, 99
Enzyme-substrate complex, 102, 125
Epidermis, 84. Also see *Skin*
Epinephrine, **301**. Also see *Adrenaline*
Epimerization, 324, 325, **348**, 578
Equilibria, enzyme concentration and, 136
  rate-limiting step and, 136
Equilibrium constant, dimensions and equilibrium position, 160
  free energy and, 156-160, 162
    numerical values, 158, **159**, 160
  kinetics relationship, 130-132
  of aconitase reaction, 158, 160
  of aminoacyl-tRNA synthesis, 167
  of creatine kinase reaction, 265
  of fructose diphosphate–phosphopyruvate interconversion, 278
  of malate oxidation, 202
  of peptide hydrolysis, 163, 164
  of phosphopyruvate carboxykinase reaction, 277
  of protein synthesis, 168, 169
  of thiol ester hydrolysis, 208
Ergosterol, 604
Erythroblasts, protein synthesis in, 33
Erythrocytes. Also see *Reticulocytes*
  development, 32, 33
  hemoglobin in, 64
Erythropoietic porphyrias, 500
D-Erythrose, isomers of, **710**

Hemoglobin(s) (*Continued*)
  oxygen dissociation curve, **62**
  oxygen transport, 60-64
  paramagnetism, 59
  poryphyrin bonding, 59
  regulation of formation, **499**, 500
  S, 73, 74-76
  Seattle, 78
  solubility, 64
  species variation, 66-69(table)
  structure of, **52**, 60-78
  Sydney, 78
  Yakima, 77, 78
  Zurich, 77
Hemolysis, jaundice and, 504
Hemorrhage, in scurvy, 93, 683
  iron deficiency and, 686
Hemosiderin, 504, 505, 506
Hemosiderosis, 506
  in kwashiorkor, 673, 674
Heparan sulfate, 583
Heparin, amine binding, 583
  blood clotting and, 653
  cells forming, 449
  structure, 583
Hepatic cirrhosis, ammonia toxicity in, 362, 363
  antibiotics in, 363
  hemochromatosis in, 506
Hepatic coma, 362, 363
Hepatic porphyrias, 500
Hepatomas, adenyl cyclase in, 556, 566
Herbivores, copper deficiency in, 193, 194
Heteropolysaccharides, defined, 576
  regulation of formation, 578, 579
Hexadecenoyl coenzyme A, from palmitoyl CoA, **229**
  hydration, **229**
Hexenoyl coenzyme A, **236**
Hexokinase(s), **243**, 244, 275
  brain, Michaelis constants, 137
  in starvation, 549
  inhibition by glucose-6-phosphate, 244
  isozymes, 243
    Michaelis constants, 243
  reaction of, free energy change, 163
Hexosamines. See *Osamines* and specific examples
Hexose(s), **710**
Hexose monophosphate shunt. See *Pentose phosphate pathway*
Hexuronates. See *Uronates* and specific examples
High-energy bonds. Also see *Energy-rich bonds*
  definition, 167, 168
High-energy phosphate, 164-168. Also see *Energy-rich bonds*, and specific nucleoside triphosphates
  caloric equivalents, 520(table), 521
  constant balance, 537
  consumption in exercise, 268-270

High-energy phosphate (*Continued*)
  formation in brown adipose tissue, 346
  from acyl phosphates, **262**
  from creatine phosphate, **265**, 266
  from 1,3-diphosphoglycerate, **249**, **252**, 253
  from enol phosphates, **262**
  from phosphopyruvate, **253**, 255
  from protein metabolism, 353
  in exercise, 524-526
  invertebrate storage, 266
  use in carboxylations, 276
  use in fatty acid synthesis, 319, 329-331
  use in gluconeogenesis, 273, 280, 281
  use in glycogen storage, 308
  use in NADH-NADP electron transfer, 322, 329
  use in protein synthesis, 19, 21, **26**, **28**, 34, **37**
  use in urea synthesis, 361
  yield, amino acid oxidation, 396-399, 403
    citric acid cycle, 222
    fats *vs.* glycogen, 345
    fatty acid oxidation, 235-237, 341
    glucose oxidation, 260
Histamine, antagonists, 449-451
  cells forming, 449
  effects, 449, 450
    and cortisol, 566
  from histidine, **449**
  heparin binding, 583
  metabolic fate, **450**
Histidase, 420, **421**
L-Histidine, **59**. Also see *Histidyl residues*
  as essential amino acid, 422, 663, 664, **667**
  carnosine from, 457
  formiminoglutamate from, 682
  histamine from, 449
  metabolic fate, 420, **421**, 422
  one-carbon groups from, 410
  toxicity, 670
Histidine decarboxylase, **449**
Histidyl residues, acid-base catalysis by, 109-111
  as metal ligands, 148
  buffer action, 614
  in chymotrypsin, 113
  in cytochrome *c*, 194
  in hemoglobin, 13, **14**, 59, 66, 76, 77, 111
  in ribonuclease, 107-110
  in trypsin, 113
Histones, acetylation of, 490
  function, 31, 556, 557
  in metabolic regulation, 490
  methylation, 490
  phosphorylation, 490
L-Homocysteine, excretion as homocystine, 433
  from S-adenosylhomocysteine, 429
  metabolic fate, 432, **433**
  methionine from, 429-432, **430**, **432**

Pentoses, **710**
  hexose phosphates from, 329
Pentose phosphate pathway, 322-329
  in adrenal cortex, 568
  stoichiometry, 327-329
Pentose phosphates, disposal, 324-329
Pentose shunt. See *Pentose phosphate pathway*
Pepsin, 658
Pepsinogen, 658
Peptidases, 113
  in lysosomes, 307
  protein digestion and, 658, 659
Peptide(s). Also see *Protein(s), Protein structure*
  aggregation, 144
  dissociation, 144
  fingerprinting, 707
  free energy of hydrolysis, 163, 164
  free energy of synthesis, 168-170
  nomenclature, 10
  separation, 706, 707
  structure of chain, **8**, 9, 10
Peptide bond, **8**
Peptide synthetase, 34, **44, 48**
Perchloric acid, 151
Performic acid, 706
Periodic acid, in sugar analysis, 713, **714-716**, 717
Pernicious anemia, 368-370
  gastric resection and, 678
Peroxidases, 404, 420, 562, 563
Pesticides, acetylcholinesterase and, 101, 115
  arsenite, 258
  cyanide, 203
  rotenone, 204
pH. Also see *Hydrogen ion concentration*
  arterial blood, 609
  arterial-venous difference, 614
  blood, CO_2 tension and, **621**
  pK relationship, 611
Phage, ΦX174, replication, 27, 29, 30
Phenyl isothiocyanate, **708**
Phenylacetate, from phenylpyruvate, 388
Phenylacetylglutamine, **388**
L-Phenylalanine, **57**. Also see *Phenylalanyl residues*
  as essential amino acid, 384, 663, 664, **667**
  in phenylketonuria, 387, 388
  metabolic fate, 384, **385**
  phenylpyruvate from, 387, **388**
  tyrosine from, 384, **385**
Phenylalanine hydroxylase, 384, **385**
  deficiency disease, 387, 388
  tetrahydrobiopterin in, 384, **385**
Phenylalanyl residues, in hemoglobin, 70, 71
Phenylketonuria, 387, 388, 389
Phenyllactate, from phenylpyruvate, 387, **388**
Phenyllactate dehydrogenase, 387, **388**
Phenylpyruvate, from phenylalanine, 387, **388**
  in phenylketonuria, 387

Phenylpyruvate (*Continued*)
  o-hydroxyphenylacetate from, 387, **388**
  phenylacetate from **388**
  phenyllactate from, 387, **388**
Phenylthiohydantoins, **708**
Pheochromocytoma, catecholamine formation by, 443
Phlorhizin, 399, 634
Phosphatases, **282**
  alkaline, in bone, 603
  in lysosomes, 307
Phosphate esters, dissociation constant, 111
  hydrolysis, **282**
Phosphate ions, absorption, 604
  analysis, 711, 712
  dietary supply, 685
  dihydrogen, dissociation constant, 619
  H^+ excretion and, 619
  recommended allowance, 689(table)
  renal absorption, 605, 607
Phosphatidalcholine, 599, **600**
Phosphatidalethanolamine, 599, **600**
Phosphatidate(s), diglycerides from, 339
  from lysophosphatidates, **339**
Phosphatidate phosphatase, **339**
Phosphatidylcholines, **182**
  formation, 424, **425**, 426, **427**
  from phosphatidylethanolamines, **427**, 429
  hydrolysis, **428**, 429
  in membranes, 591, 592, 598
  mitochondrial, 181-183
Phosphatidylethanolamines, formation, **425, 426**, 427
  hydrolysis, **428**, 429
  in membranes, 591, 592, 598
  in visual pigment, **643, 644**
  phosphatidylcholine from, **427**, 429
Phosphatidylglycerols, **598**
Phosphatidylserine, formation, **426**, 427
  in membranes, 591, 592, 598
  phosphatidylethanolamine from, **426**, 427
Phosphoacetylglucosamine mutase, **577**
3'-Phosphoadenosine-5'-phosphosulfate, as sulfate donor, **581**, 582
  formation, **581**
Phosphoarginine, **266**
Phosphocholine, in phosphatidylcholine synthesis, **425**
Phosphocholine transferase, **425**
Phosphocreatine, **265**
  creatinine from, 423, **424**
  formation, **265**, 266
  high-energy phosphate from, 265, 266
  in skeletal muscles, 540, 541
Phosphodiesterase(s), polynucleotide hydrolysis by, 481, 482
Phospho-*enol*-pyruvate, as glycogen precursor, 293
  from amino acids, 364, **365**
  from oxaloacetate, 273, 274, **277, 278**
  from 2-phosphoglycerate, **253**, 255
  from pyruvate, 273, **274**

Succinyl coenzyme A (*Continued*)
  from methylmalonyl CoA, 366, **367**
  from propionyl CoA, 366, **367**
  glucose from, 364, **365**
  porphyrins from, 495, **496**
  succinate from, 219, **219**, 373, **375**
Succinyl phosphate, in succinyl CoA synthetase reaction, 218, **219**
D-Sucrose structure, **713**
Sugars, acids from, **717**, 719
  alcohols from, **717-719**
  chemistry, 709-719
  configuration, 709, **710, 711**
  periodate oxidation, 713, **714-716**, 717
  reducing, 711-713
Sulfanilamide mechanism of action, **412**
Sulfate esters, formation, **581**, 582
  in protein polysaccharides, 580, 582, 583
Sulfate ions, from amino acids, stoichiometry, 396, 398
  from cysteine, 382, **383**
  sulfate ester from, **581**, 582
Sulfenic acids, formation, 149
Sulfhydryl groups. Also see *Cysteinyl residues*
  arsenite complexes, **259**
  as metal ligands, 148
  exchange with disulfides, 95
  in coenzyme A, 207, 208
  in fatty acid synthase, 316, 317
  mercuric ion complexes, 147, 148, 259, 260
  oxidation, 149
    H$^+$ stoichiometry, 618
    to sulfinates, **406**
  reaction with iodoacetate, 706
  thiohemiacetal formation, 250
Sulfinates, oxidation, **406**
Sulfinic acids, formation, 149
Sulfisoxazole, mechanism of action, **412**
Sulfite ions, in cysteine metabolism, 382
Sulfite oxidase, 382, **383**
  deficiency disease, 382
Sulfonamides, mechanism of action, **412**
  methemoglobin formation with, 77
Sulfonic acids, formation, 149
Sulfonium group, **436**
  free energy of hydrolysis, 422
Sulfonylphenylureas, in diabetes, 559
Sulfur dioxide. See *Sulfite ions*
Sulfur transferase, **384**
Surface area, 516-518
Svedberg units, 15, 16
Sweat, nitrogen in, 660, 661
Synovial fluid, 585
Synthases, nomenclature, 723
Synthetase, definition, 218
  nomenclature, 723, 724

Taurocyamine, **266**
Temperature, kinetics and, 145. Also see *Transition temperature*

Temperature (*Continued*)
  metabolic rate and, 532, 533
  thyroid function and, 565
Terminator codons, 45
Testes, in vitamin A deficiency, 644
  embryoma, 47
Testosterone, 553
Tetany, in alkalosis, 609
  in calcium deficiency, 603
Tetracosahexenoyl coenzyme A, formation, 337
Tetracycline, action, 47
Tetrahydrobiopterin, 384, **385**
  from dihydrobiopterin, **385**
  in phenylalanine hydroxylase, 384, **385**
  in tryptophan hydroxylase, 445
  in tyrosine hydroxylase, 440, 441
Tetrahydrofolate. Also see specific derivatives
  from H$_2$folate, **411**, 477, 479
  function, 409, 410, 411-416, 677
  reactions summarized, **434-437**
Tetrahydropteroylglutamate. See *Tetrahydrofolate*
Tetramethylammonium ion, **104**
Thalassemias, 72, 73, 76
Theophyllin, 555
Thiamine, 214
  as vitamin, 675, 676
  deficiency, 257, 258, 681
    blood α-ketoglutarate concentration, 257
    blood pyruvate concentration, 257
    α-ketoglutarate dehydrogenase in, 257
    pyruvate dehydrogenase in, 257
    signs and symptoms, 258
  recommended allowance, 689(table)
  requirement, 257, 681
  thiamine pyrophosphate from, 256
Thiamine pyrophosphate, 214, **215**
  formation, 256
  functions, 675, 676
  in α-ketoglutarate dehydrogenase, 214, 215
  in pyruvate decarboxylase, 272
  in pyruvate dehydrogenase, 256-258
  in transketolase, 324, 325
  mechanism, **215**
Thiamine pyrophosphokinase, 256
Thin-layer chromatography, 701
Thiocyanate ions, from cysteine, 383, **384**
Thioethanolamine. See *β-Mercaptoethylamine*
Thioethers, formation, **437**
  hydrolysis, **436**
Thioguanosine, from 6-mercaptopurine, 480
Thiohemiacetal, formation, 250
  oxidation to thiol ester, 250
Thiokinases, **223**
Thiol esters, formation, **223**
  from hemiacetal oxidation, 250
  hydrolysis, free energy change, 208
Thiols. See *Sulfhydryl groups*
Thiolysis, of 3-hydroxycarboxylates, **346**
  of 3-ketoacyl CoA, 230, **231**, 237

RANGE OF CONCENTRATIONS

Clinical analyses may be made on whole blood, separated plasma, or the serum remaining after clotting, and these are indicated by (B), (P), or (S). Venous blood is used unless otherwise specified. Standard values differ from one laboratory to another according to the conditions of analysis; those cited here are from an extensive tabulation in the New England Journal of Medicine, Vol. 276, p. 167 (1967). The more useful molar concentrations are calculated where appropriate, but the commonly used clinical units are also shown. The values are for adults.

Acetoacetate + acetone (S)	0.05—0.35 mM	0.3—2.0 mg/100 ml
Alpha amino nitrogen (P)	2.1—3.9 mM	3.0—5.5 mg/100 ml
Ammonia (B)	25—40 μM	40—70 μg/100 ml
Blood volume (B)	8.5—9.0 liters/100 kg body weight	
Calcium (S)	2.1—2.6 mM	8.5—10.5 mg/100 ml
CO_2, total (S)	26—28 mM	26—28 meq/liter
Chloride (S)	100—106 mM	100—106 meq/liter
Cholesterol, total (S)	3.9—7.3 mM	150—280 mg/100 ml
Cholesterol, esterified (S)	60—75% of total	
Cobalamins (S)	0.15—0.6 nM	200—800 pg/ml
Creatinine (S)	0.06—0.13 mM	0.7—1.5 mg/100 ml
Fat, neutral (S)	0—2.4 mM	0—200 mg/100 ml
Fatty acids, total (S)	7.0—15.5 mM	190—420 mg/100 ml
Glucose, fasting (B)	3.9—5.6 mM	70—100 mg/100 ml
Hematocrit: volume occupied by RBC (B)	males: 42—50% females: 40—48%	
Hemoglobin (B)	males: 2.0—2.5 mM females: 1.8—2.2 mM	13—16 g/100 ml 12—14 g/100 ml
Iron (S)	9—27 μM	50—150 μg/100 ml
Lactate (B)	0.7—1.8 mM	6—16 mg/100 ml
Magnesium (S)	1.5—2.5 mM	1.5—2.5 meq/liter
Oxygen saturation, arterial (B)	96—100%	